Under the General Editorship of Paul F. Brandwein

JUNIOR HIGH SCHOOL *Science for Better Living Series*

YOU AND YOUR WORLD · Teacher's Manual · Teaching Tests

YOU AND YOUR INHERITANCE · Teacher's Manual · Teaching Tests

YOU AND SCIENCE · *Experiences in Science* (Workbook) · Teacher's Manual · Teaching Tests, Forms A and B · Harbrace Science Filmstrips

SCIENCE FOR BETTER LIVING: COMPLETE COURSE · Teaching Tests · Film Guide · Workbook

SENIOR HIGH SCHOOL

EXPLORING BIOLOGY: THE SCIENCE OF LIVING THINGS · *Experiences in Biology* (Workbook plus Laboratory Manual) · Teacher's Manual · Teaching Tests, Forms A and B

YOUR BIOLOGY · Teacher's Manual (Including Unit Tests)

YOUR HEALTH AND SAFETY · Teacher's Manual · Teaching Tests

LIFE GOES ON

THE PHYSICAL WORLD: A COURSE IN PHYSICAL SCIENCE · Teacher's Manual · Teaching Tests

EXPLORING CHEMISTRY · *Laboratory Manual in Chemistry* · *Experiences in Chemistry* (Workbook plus Laboratory Manual) · Teacher's Manual · Teaching Tests (In Preparation)

▶ EXPLORING PHYSICS · *Laboratory Manual in Physics* · *Experiences in Physics* (Workbook plus Laboratory Manual) · Teacher's Manual · Teaching Tests, Forms A and B

A SPECIAL BOOK FOR THE STUDENT

HOW TO DO AN EXPERIMENT

BOOKS FOR THE TEACHER

TEACHING HIGH SCHOOL SCIENCE: A BOOK OF METHODS

TEACHING HIGH SCHOOL SCIENCE: A SOURCEBOOK FOR THE BIOLOGICAL SCIENCES

TEACHING HIGH SCHOOL SCIENCE: A SOURCEBOOK FOR THE PHYSICAL SCIENCES (In Preparation)

THE GIFTED STUDENT AS FUTURE SCIENTIST

RICHARD F. BRINCKERHOFF

JUDSON B. CROSS

ARTHUR LAZARUS

UNDER THE GENERAL EDITORSHIP OF

Paul F. Brandwein

HARCOURT, BRACE AND COMPANY

New York Chicago

DRAWINGS BY *Mildred Waltrip*

Sample of zinc oxide smoke bombarded by electrons. Deviation from the usual circular diffraction pattern has been caused by attraction of bombarding electrons toward a U-shaped magnet.

EXPLORING Physics

NEW EDITION

THE AUTHORS

RICHARD F. BRINCKERHOFF *Instructor in Physics, Phillips Exeter Academy, Exeter, New Hampshire*

JUDSON B. CROSS *Instructor in Physics, Phillips Exeter Academy, Exeter, New Hampshire*

ARTHUR LAZARUS *Instructor in Physics, Forest Hills High School, New York City*

GENERAL EDITOR

PAUL F. BRANDWEIN *General Editor and Consultant to Schools, Harcourt, Brace and Company; formerly Chairman, Science Department, Forest Hills High School, New York City*

GRATEFUL ACKNOWLEDGMENT is made to the authors of the First Edition of *Exploring Physics,* Hyman Ruchlis and Harvey B. Lemon, to those teachers who assisted in the preparation of the First Edition, and to Mr. Barton F. Sensenig of the Haverford School, Haverford, Pennsylvania, who read and offered suggestions on the manuscript for this revised edition.

PICTURE CREDITS: *Front cover:* James B. Minnick, Fairchild Aircraft. Smoke tunnel test of "jet wing" air foil. The transit in the foreground is used to measure angles of deflection as the slipstream, shown by smoke particles, passes over the wing. *Back cover:* Radiation Laboratory, University of California (see page 724). *Title page:* Selection by Robert B. Leighton from *Scientists' Choice,* Basic Books, 1958.

Printed in the United States of America

[b · 12 · 59]

CONTENTS

UNIT ONE

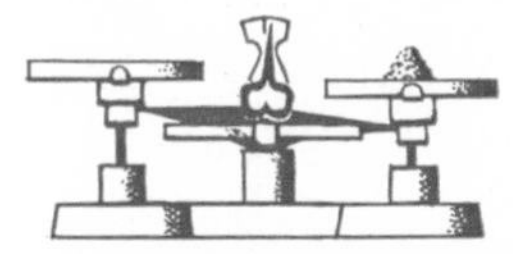

The Science of Physics

UNIT TWO

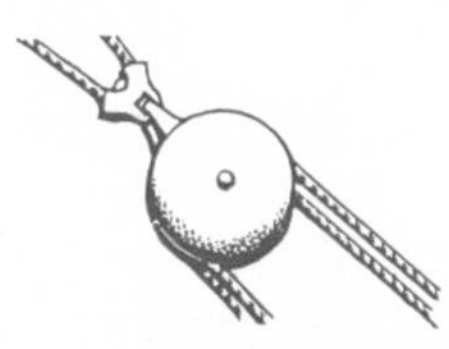

The Machine Age—Force and Motion

UNIT THREE

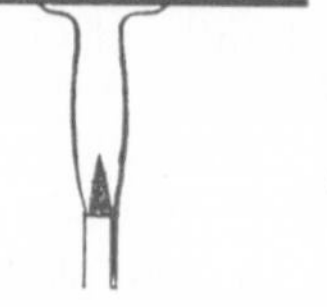

Heat Energy— The Motion of Molecules

UNIT FOUR

Sound Energy— The Motion of Vibrations

UNIT FIVE

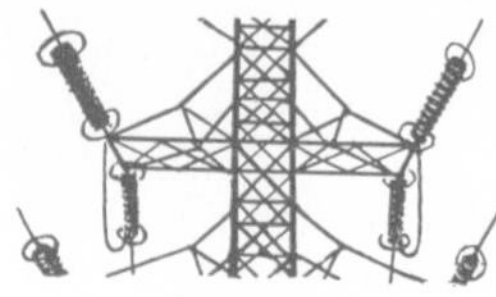

Electrical Energy— The Motion of Electrons

UNIT SIX

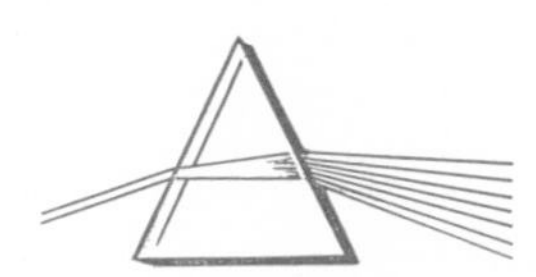

Light Energy—The Motion of Electromagnetic Waves

UNIT ONE

The Science of Physics

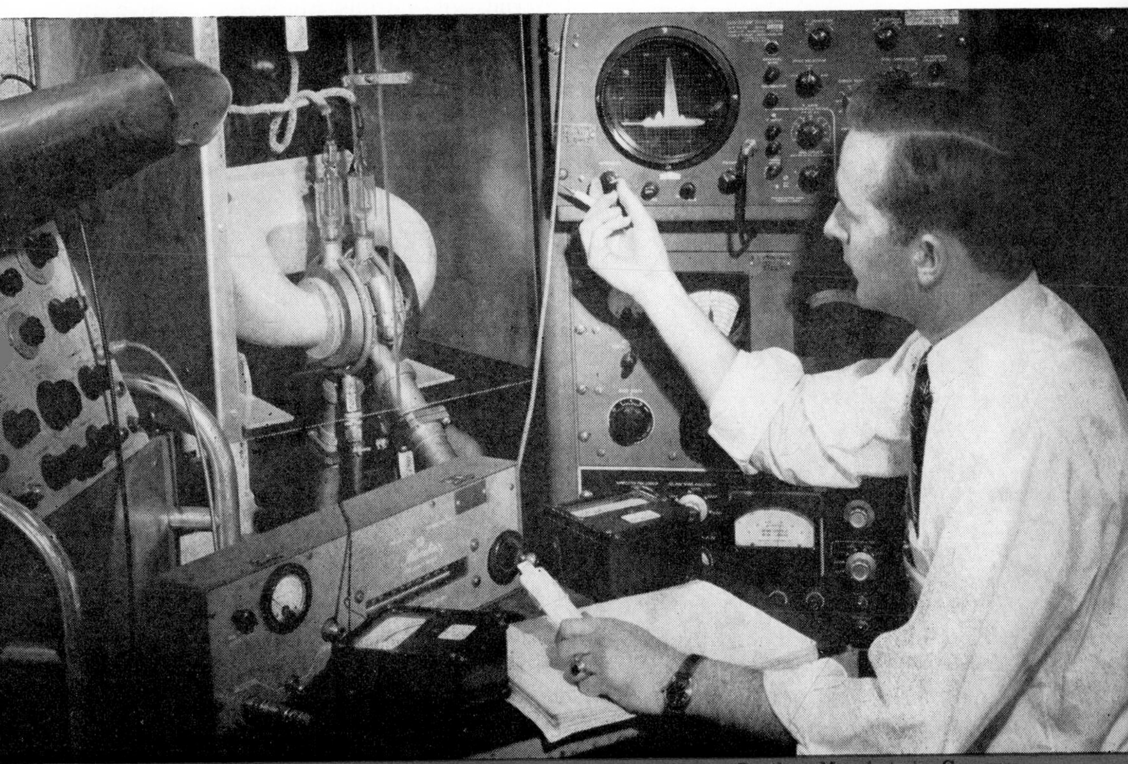

Raytheon Manufacturing Company

Have you ever tried something to see if it would work? Of course you have. So has everyone else, whether merely to test a new recipe, try how a different bicycle works—or test the operation of a new type of electron tube as the technician is doing here.

When the tube (you can see it between the poles of the white magnet) is perfected, it will increase the range of a new kind of radar set.

Many physicists spend most of their time trying out things to see if they will work. Whether they are trying to perfect a useful new instrument such as a radar tube or whether they are trying to verify a new theory that seems to have no practical applications, physicists always turn ultimately to the final test: "Will it work?"

In this unit you will learn about some of the kinds of quantities a physicist must measure as he tests his ideas and theories.

CHAPTER

1

Physics—as the Science of Energy

What would happen if you tried to hit a home run using a baseball bat made of glass and a ball made of putty? The bat would splinter in your hands, and the ball itself would flatten out like a pancake and fall near your feet.

No one would ever make a baseball of putty, because we know the properties of putty; we know what it can do and what it cannot do. It is useful for filling holes in wood but not for making baseballs. Likewise we know what glass can do; it is good for making windows but not for making baseball bats.

The concern of the physicist

The physicist is very much concerned with the properties of the different materials which make up our world. To all these materials he gives the name of **matter.** Wood, coal, and steel, for example, are matter. So are all liquids, such as water and oil, and all gases, like air.

Matter is anything that has mass; that is, it must have weight and occupy space.

By studying the properties of different kinds of matter man learns to use materials of various kinds to improve the world around him—to become master of his environment. Look up from your book. How many different kinds of materials can you see which have been used in the manufacture of objects in the room around you?

But the study of the properties of matter is only half the physicist's job. To describe the world we live in, the physicist needs to do more than list the properties of the materials in it. He must know how matter behaves when forces of various kinds are exerted on it. This involves the other half of the physicist's job, the study of energy. When the physicist uses the word "energy," he is talking about the ability of a force to cause motion. Thus:

The science of physics is the study of matter and energy.

Let us now take a closer look at energy in order to find out what it is and to learn about some of its simpler forms. Then, in the next chapter, we shall look at the other half of the physicist's job, the study of matter.

1. Kinds of energy

Energy has a variety of forms. There are heat energy, electrical energy, radiant energy (light, radio waves), mechanical energy, chemical energy, and still other kinds which you will study.

Mechanical energy is sometimes due to the fact that an object has been elevated or distorted and sometimes due to the motion of an object. For example, the elevated position of a weight or the distortion of a wound-up spring provides mechanical energy that can drive a clock or produce motion in other ways. Or again, moving air (wind) and moving water possess mechanical energy that can be used to turn generators which produce electrical energy. The motion of a pile driver can ram a pole into the ground and do work for us. We shall study mechanical energy in Unit Two.

Heat energy is produced by the vibrating motion of tiny particles of matter called **molecules.** It is the motion of these molecules that provides the heat energy which moves trains and turns the propellers of an airplane. You will study this form of energy in Unit Three.

Sound energy results from the vibrating motion of particles of matter, and will be discussed in Unit Four.

Electrical energy results from the motion of tiny particles of matter called **electrons.** We use this form of energy to run electric motors and to operate lamps and radio sets. You will study electrical energy in Unit Five.

Radiant energy, such as light, is energy which is in motion at a speed of about 186,000 miles per second. Radiant energy enables us to see, to take photographs and X-ray pictures, to receive warmth from the sun, and to receive radio broadcasts. You will study radiant energy fully in Unit Six.

Chemical energy is the energy locked up in combinations of atoms called **molecules.** When molecules fly apart or atoms combine with one another to form new combinations, energy is often released as heat energy and sometimes as light energy. The burning of a fuel or the explosion of dynamite are typical examples of the release of chemical energy.

Atomic energy, which is more accurately called nuclear energy, results from extremely rapid motion of particles released from within atoms. This form of energy will be considered in Unit Seven.

National Park Service

1-1 Unharnessed energy. Yosemite Falls could light a million lamps and turn a thousand motors.

All of these forms of energy can make changes in our environment; all of them can be harnessed to do work. From this fact comes our definition of energy: **Energy is the capacity for doing work.**

2. Energy: potential and kinetic

It may seem as though a rock poised on the edge of a cliff has no energy. But if we push it over the edge, it will fall faster and faster, striking the ground traveling at high speed. Let us assume it crushes something, perhaps an abandoned jalopy parked on the flat ground below. Whether or not the owner wanted the car destroyed, the rock has accomplished work in crushing it. Thus, the rock, while it was on the edge of the cliff, actually contained a form of stored-up energy, waiting to be released. This stored-up energy is called **potential energy.** When energy is due to actual motion, we call it **kinetic energy.** While the rock was motionless on the edge of the cliff, its energy was potential. While it was falling through space, its potential energy was changing into kinetic energy (compare Fig. 1-1).

All forms of energy may be classified as either potential or kinetic. A wire spring has potential energy when it is stretched out; if you fasten the ends down, the spring is motionless. But as soon as the spring is released, its energy becomes kinetic as it jumps back. We can use the potential energy of a coiled spring, as it is slowly released, to operate a clock.

If someone, in the days before history, had found a piece of coal for the first time, he probably would have thought of it as nothing but a piece of black rock. But actually this ordinary-looking black stuff has great potential energy; when burned in a furnace it produces heat energy which can warm our houses, run trains and ships, and operate dynamos to generate electricity. Dynamite also has potential energy and can do useful work (Fig. 1-2). The food you eat is another familiar example of potential energy. The kind of potential energy possessed by coal and dynamite and food is called **chemical energy.**

Lifted weights, compressed air, and coal have different kinds of potential energy. The potential energy of the rock on the edge of the cliff is due to its position high above the base of the cliff. The potential energy of compressed air is due to its pressure. As for coal, it has chemical potential energy. When coal is burned, a chemical action is set up in which molecules in the coal tend to join with the molecules of oxygen in the air. The resulting rapid motion of the newly formed molecules is heat energy.

3. Energy converters

We have already had several examples of how one form of energy can be changed into another. For instance, we know that the mechanical energy of moving water can turn generators which then produce electrical energy. In our study of physics, we shall consider many devices which are used to change energy from one form into another and which we call **energy converters.** Figure 1-4 shows several interesting conversions of energy. Below is a list of some common energy changes.

A gasoline engine	*changes* chemical energy	*into* mechanical energy.
A house heating system	*changes* chemical energy	*into* heat energy.
An electric light	*changes* electrical energy	*into* light energy.
A dynamo	*changes* mechanical energy	*into* electrical energy.
A battery	*changes* chemical energy	*into* electrical energy.
An electric bell	*changes* electrical energy	*into* sound energy.
A telephone	*changes* sound energy	*into* electrical energy *and back into* sound energy.

E. I. du Pont de Nemours and Co.

1-2 Chemical energy changing to mechanical, heat, and sound energy. These dynamite blasts provide a cheap way to move many tons of rock at one time.

4. The Laws of Conservation of Mass and Energy

When you burn coal in a furnace most of the matter or substance of the coal seems to disappear, leaving only a small amount of ash. From this we might conclude that the matter in the coal has been "destroyed." But we would be wrong.

When we take into account the gases resulting from the burning of the coal, we find that the mass of all the materials after the burning takes place seems to be the same as the mass of all the materials before burning.

We can prove this ourselves by burning the coal in a closed container with oxygen or air. After the coal has been burned, the total mass of the box and its contents seems to be the same as before. We have less of the solid matter of the coal, but more gas than before. Because the gas is invisible, early investigators made the mistake of assuming that the matter in the coal had simply "disappeared."

After many experiments such as this, chemists arrived at the **Law of Conservation of Mass.** This laws states that:

In a chemical change, mass is not created or destroyed but is merely changed from one form into another.

A **law,** in science, is the statement of an important fact or rule which has been verified many times until it is accepted by all scientists. Laws have been known to be in error. In fact, even this Law of Conservation of Mass requires some changes as a result of recent discoveries. These changes are discussed later in this chapter. But for ordinary chemical reactions the law as stated above is accurate enough.

Some time after chemists discovered the Law of Conservation of Mass, physicists arrived at a similar law for energy. It is called the **Law of Conservation of Energy,** and it states that:

Energy can be neither created nor destroyed, but is merely changed from one form into another.

In later chapters we shall discover how scientists were able to measure en-

1-3 Could you make wood from wind? Trace the energy conversions in this picture.

ergy and how they used these measurements to arrive at this law. Let us look into an illustration of how the law works.

Suppose that we put some fuel oil into a diesel locomotive and use it to pull heavy freight a distance of several miles. Most of the oil is gone when the train arrives at its destination. What has happened to the potential energy the oil possessed at the start of the trip? Part of the heat energy produced by burning the oil was turned into the mechanical energy of the moving train. Some heat energy went into the air with the exhaust gases. Still more heat energy warmed the air around the engine. The tracks became warmer because of friction. A tiny portion of the energy appeared as sound when the engineer blew the horn. Another small amount may have turned up as light in the headlamp. Energy was needed as the front of the train pushed the air away, thereby causing motion of air all along the track.

The energy originally in the fuel was not destroyed but was simply changed into other forms. Eventually all of this energy is converted into heat and scattered over a large area of the earth.

The sun—our source of energy

To the scientist a living plant is not only a thing of beauty. It is also a factory which captures the energy of the sun and puts it into suitable form for our use. The light energy from the sun enables the leaves of the green plant to take carbon dioxide from the air and water from the soil and convert them into sugar and starch. The new substances thus created have chemical potential energy (Fig. 1-3).

This is the energy that you get for your own use when you eat food. In your body the energy that originally came from the sun's light is changed into body heat as a result of many chemical reactions, and into several kinds of energy of which the mechanical energy of movement is an important one. The energy you are using to move your eyes as you read this page and to think about what you are reading is energy that came from the sun, probably within recent months.

Just consider for a moment what would happen if the sun were suddenly to stop radiating energy. Midnight darkness would descend on the earth, leaving only faint starlight. Within a matter of days the temperature all over the globe would drop to far below freezing. Rain and snow would fall for the last time and soon all rivers, lakes, and oceans would be frozen. Plants and animals other than

Department of the Interior

1-4 SOME KINDS OF ENERGY CONVERTERS. What conversion takes place in the boiler of the steam locomotive? In water falling from the top of a dam? What kind of energy does a tightly coiled spring have? When the spring is released what kind of energy results?

Pennsylvania Railroad

Buick Division, General Motors Corp.

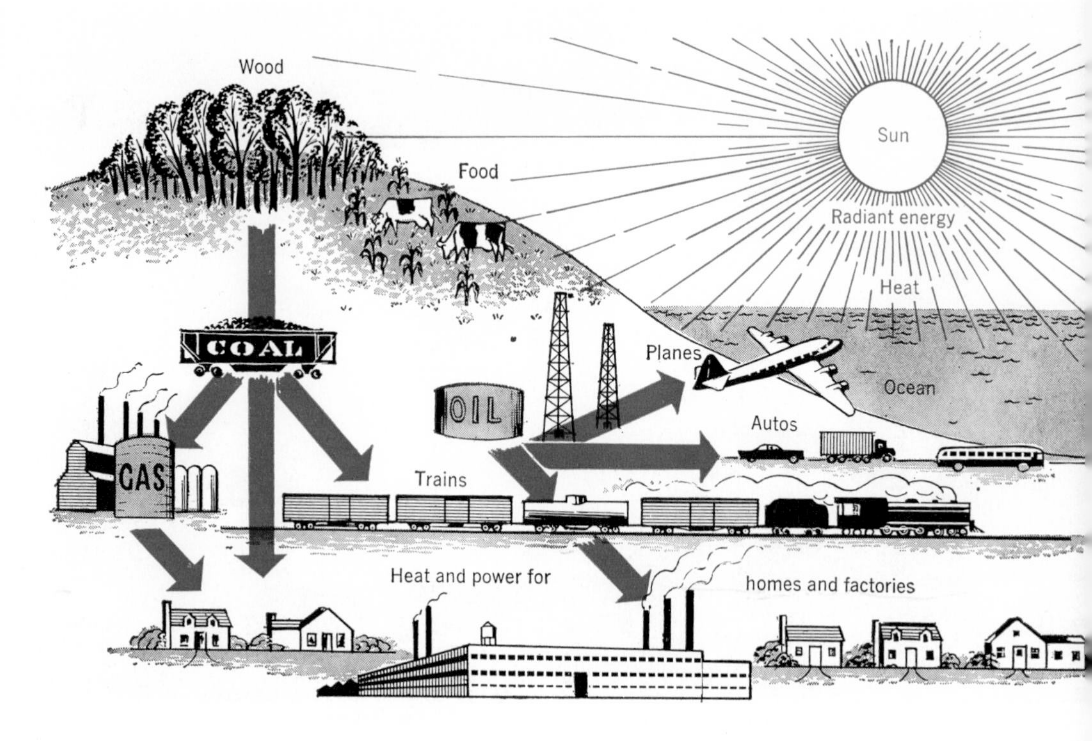

1-5 The sun—basic source of energy for all life and all industry.

man would quickly freeze to death. Man himself might manage to keep alive for a little longer by using all the knowledge at his command. But eventually even he would succumb to the subzero temperature. Finally even the air would become liquid, and it would be impossible to maintain any life whatever.

5. Energy from falling water

The energy of the sun can be harnessed through falling water (Fig. 1-4). The heat of the sun evaporates the water from the oceans, lifting it high in the air so that it has potential energy. This water vapor is then carried inland where some of it falls as rain, which flows downhill into rivers and back into the oceans.

But if we dam up the rivers we create reservoirs. Now we can direct the flow of the water so that it turns a wheel as it falls from the level of the reservoir to the level below the dam. In this way we convert kinetic energy of moving water into the kinetic energy of the moving electric generator. With the electrical energy thus produced we supply energy for homes and factories.

6. Packaged energy

The sun is also the source of potential energy locked up beneath the earth's surface in valuable deposits of coal. These deposits were once trees and plants which covered the earth in far greater abundance than they do today.

Coal today is our most important single source of energy and may well remain so for years to come. In 1956 the annual coal production of the United States was about 530 million tons, or about 3 tons per person. The value of this production each year is over 2 billion dollars.

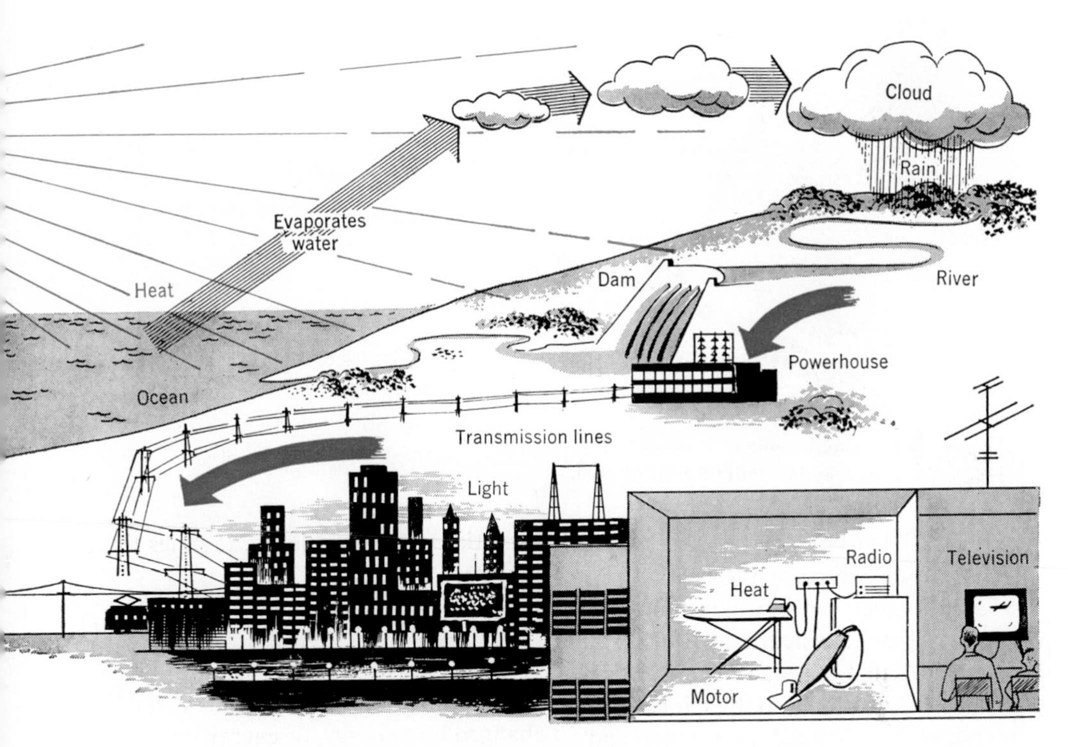

Find six forms of energy in this picture. Would life be possible without the sun's energy?

We find other remains of ancient living things in the form of petroleum. This is one of our most useful sources of energy. We use it to heat our homes and to operate diesel engines. Refined into gasoline, it provides energy for automobiles and ordinary airplanes; refined into kerosene, it is used for heating purposes and for propelling jet planes. In 1956 the total production of petroleum in the United States was about 110 billion gallons a year, or nearly 650 gallons for each person in the country.

The world's coal supply will probably last a long time, but the same cannot be said for oil. At the present rate of increase of consumption, it may last only a few score years more unless we take pains to use the precious supply wisely.

In 1956 about one quarter of the electrical energy in the United States was produced by falling water. The other two thirds came from the burning of fuels such as coal and petroleum.

Figure 1-5 shows some of the energy changes that start with the sun. Can you think of any source of energy that does not start with the sun?

7. Mass-energy relationship

We see how important the sun is to our energy supply. But where does the sun itself get its energy? Scientists have long asked themselves that question. At first it was assumed that the sun got its heat by burning some kind of fuel. But when the size of the sun was finally established, this hypothesis had to be discarded. Calculation showed that by burning fuel the sun could provide energy for only about 6,000 years. And we are fairly certain that the sun is billions of years old. Today we think we have the answer.

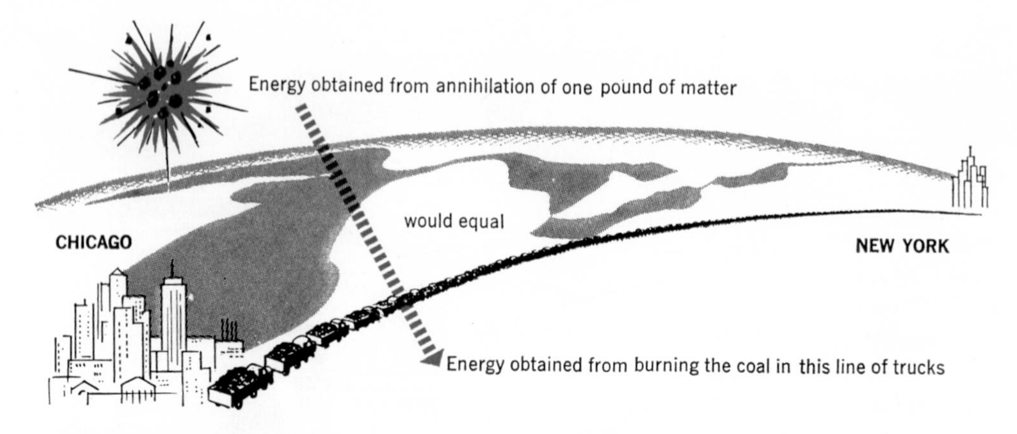

1-6 If we could convert a pound of matter into energy, then it would give as much energy for industry as 1½ million tons of coal.

8. Mass into energy

With the discovery of radioactivity in 1896, and other work which followed (such as Albert Einstein's in 1905), it became clear that mass and energy were equivalent. According to this idea, energy might be changed into mass and mass into energy. In other words, it might be possible to take a pound of mass and make it disappear; in its place would appear a definite amount of energy. Or we could take a certain amount of energy and change it into a definite amount of mass!

And how much energy might we get from a given amount of mass? From Einstein's theory we could predict that the complete destruction of 1 pound of any material would produce energy equal to that produced by burning 3 billion pounds of coal ($1\frac{1}{2}$ million tons) (Fig. 1-6). In the atomic bomb a small amount of mass is destroyed. In its place appears an enormous amount of energy—most of it occurring as heat and light and some of it occurring in other forms (Fig. 1-7).

As a result, the Law of Conservation of Mass and the Law of Conservation of Energy have been combined into a new law. We might call this the **Law of Conservation of Mass-Energy.** It states that:

The total amount of mass plus energy in the universe always remains the same.

According to this law, mass can be changed into energy, or energy into mass.

9. Source of the sun's energy

These discoveries help us to understand how the sun gets its energy. Today we believe that the sun derives its energy from the destruction of its mass. According to this theory, hydrogen gas is changed into helium inside the sun at temperatures of about 20 million degrees C. In this process about $\frac{1}{100}$ of the mass of each hydrogen atom is transformed into energy. Our calculations show that about 4 million tons of the mass of the sun are being changed into energy every second. This seems like an enormous destruction of mass. But it amounts to little in comparison with the total mass of the sun, which is about 2,200,000,000,000,000,000,000,000,000 tons (2.2×10^{27} tons).

10. Significance of atomic energy

Until the Atomic Age the sun remained the original source of all the energy man had at his command. It is only 250 years since he began to use the stored-up energy of coal to operate steam engines and do work for him. It is only a century since he harnessed the energy of petroleum.

But now he is faced with a totally new source of energy—one that allows him to by-pass the sun! By changing mass into energy, man is using the methods of the sun itself. He can get millions of times more energy from a pound of atomic "fuel" than he ever got from a pound of coal or oil.

The possibilities of this new development are almost endless. They are possibilities for a richer life—and also for destruction. We should be warned. Every new form of energy that man has discovered has made possible new machines, new gadgets. But each advance has also made war more terrible (Fig. 1-7). Now we have a source of energy so powerful that we can destroy civilization itself. Just how we use this newest form of energy is the vital concern of every one of us.

Atomic Energy Commission

1-7 **In an atomic bomb explosion a tiny amount of mass is converted into energy. But the total amount of mass plus energy remains unchanged.**

Going Ahead

To strengthen your understanding of physics you need to use the *vocabulary of physics* that you have learned in this chapter. Physicists use these terms constantly to communicate with each other and with the public through newspapers, magazines, and books.

You also need practice in applying the *principles* you have just learned so that you can use them in the rest of this book.

Studying the text of each chapter is therefore only a part of your work in physics. Practice with the terms, principles, applications, and problems is part of physics too. You should do as many of the following exercises as you can. Some are easy, and others are more challenging.

THE VOCABULARY OF PHYSICS

What does each of the following mean?

matter
potential energy
electrical energy
physics
atomic energy
energy converters
sound energy
energy
radiant energy
mechanical energy
heat energy
conservation of energy
kinetic energy
conservation of mass
law, in science
conservation of mass-energy
chemical energy

THE PRINCIPLES WE USE

Explain why

1. Heat energy is a form of kinetic energy, while the energy of a coiled spring is a form of potential energy.
2. The sun is the source of energy for life.
3. The sun is the original source of energy for industry.
4. The sun has been radiating energy for thousands of years without being noticeably changed.

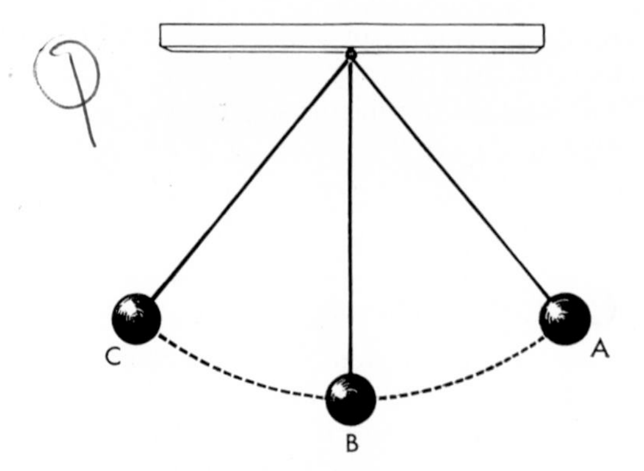

1-8 **In what position does this pendulum have its greatest potential energy? greatest kinetic energy? Why does it swing to equal heights on each side?**

APPLYING PRINCIPLES

1. Name three different sources of energy for industry. Show how each source of energy originated in the sun.
2. How many devices in your home can you name that convert energy from one form to another? What kind of energy is changed to what other kind in each device?
3. Write a paragraph or two discussing the forms of energy you think we shall use in the future.
4. State the energy change that takes place in each of the following devices: furnace, airplane, electric light, steam engine, dynamo.
5. The standard of living of a country is often measured by the amount of electrical energy consumed per person per year. Do you think this is a reliable measure of the standard of living? Why?
6. Find out how aluminum is made, and show how the energy for making it came from the sun.
7. Figure 1-8 shows how a pendulum swings if pulled to one side and released. (a) What form of energy does it have at A, the point to which it is lifted? (b) What form of energy does it have at the bottom of the swing, B? (c) How does the fact that it swings to an equal height, C, on the other side follow from the Law of Conservation of Energy? (d) After swinging for a while, the pendulum slows down and stops. Why does it stop? (e) Where does its energy go?

A PROJECT TO DEMONSTRATE ENERGY CONVERSION

List the various forms of energy *twice* in parallel columns, as follows:

Mechanical	Mechanical
Heat	Heat
Sound	Sound
Electrical	Electrical
Radiant	Radiant
Chemical	Chemical
Atomic	Atomic

Now from "Mechanical" in the left-hand column draw arrows to each entry in the opposite column (except Mechanical). Repeat, drawing arrows from "Heat," etc.

You will be surprised to find that you can almost always invent a simple classroom demonstration of the energy conversion represented by each arrow (except perhaps "Atomic"). For example, Mechanical → Heat is represented by the heat generated by friction as you rub your hand rapidly and firmly back and forth across the table top. Mechanical → Electrical can be shown by a hand-cranked generator, and Mechanical → Light by arranging the hand-cranked generator to operate a flashlight bulb.

Hold a competition to see which group of your classmates can demonstrate the greatest number of conversions.

CHAPTER

2

Physics—as the Science of Matter

You will remember that physics is the science of matter and energy. Now that we have looked at the nature of energy, let us turn to the other half of the physicist's job, the study of matter.

What is matter made of?

Imagine that you are holding in your hand the piece of aluminum pictured in Fig. 2-1. Obviously with very little trouble you could saw the piece of aluminum in half. It would not be very hard to saw one of the halves in half again. Perhaps you could even cut one of the smaller pieces in half once again. How many times do you suppose you could repeat this process? By the time the piece of aluminum had been reduced to the size of a pea you would have a hard time cutting it in half, but at least you can *imagine* continuing to do it. You would eventually need a microscope to see what you were doing. Finally, your pieces would be so small that even a microscope would no longer be of help to you. Are these the smallest pieces of aluminum that can exist? Surely you can *think* of making still smaller pieces. But if you could continue your slicing you would finally reach a piece of aluminum so small that it could not be cut in half again and still be aluminum! This smallest possible piece of aluminum that can exist is called an **atom** of aluminum. As you know, aluminum is only one of a number of different elements.

1. The Atomic Theory

In 1808, an English chemist, John Dalton, proposed the theory that all the matter in the world is made of such atoms, which cannot be cut into smaller bits. This idea is called the **Atomic Theory.** As time went on chemists identified the atoms of more and more different elements, until today the list contains over 100 different kinds. (See table on page 681.) We know now that not only the earth, but all the stars, including the sun, are made of these same kinds of atoms. For example, our knowledge of physics gives us every reason to believe that the structure of the atoms of hydrogen in the sun is identical with that of those on earth. These 100 or so different kinds of atoms are the building blocks of our universe.

We know too that atoms can combine with each other in many different combinations called **molecules.** For instance, one atom of oxygen is often found attached to two atoms of hydrogen to form a molecule of water. Perhaps you have heard water called H_2O (H-two-O). Table salt (sodium chloride) is made of molecules that consist of one atom of sodium combined with one atom of chlo-

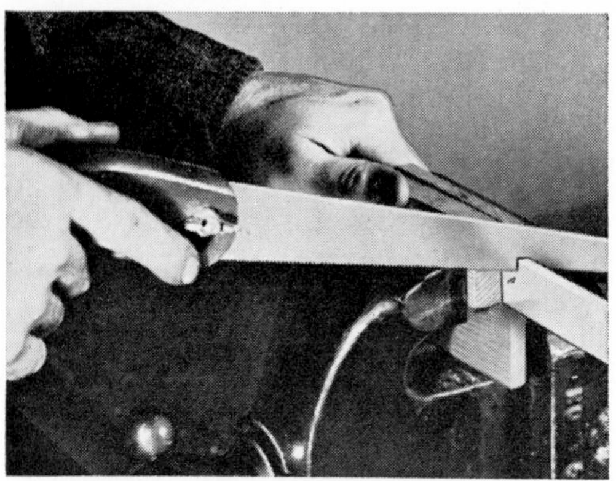

Reynolds Metals Co.

2-1 Pieces of this aluminum cannot be cut in half repeatedly forever. Eventually a piece will be reached that is so small it cannot be divided and still be aluminum. This smallest piece is an atom of aluminum.

rine.* Such a molecule of salt is the smallest piece of salt you can have that is still salt. Obviously if you divide it still further you no longer have salt; you have two atoms, whose properties are very different from salt.

Most of the materials you see in the room around you are made of such combinations of atoms. Your body contains thousands of different combinations.

2. Mathematics of the molecule

Molecules are extraordinarily tiny. It takes about one million molecules in ink side by side to stretch across the period at the end of this sentence.

From their observations, scientists have calculated that there are about 7,000,000,000,000,000,000,000 molecules (the physicist usually writes 7×10^{21}) in a glassful of air. In most solid materials the distance between molecules is about $\frac{1}{100,000,000}$ inch (10^{-8} in). (See Mathematics for Physics, page 666, for a description of this easy way of writing very large and very small numbers.)

Look at the matter another way. Uranium has the heaviest atoms found in nature. A single pound of uranium contains about 1 million billion billions of atoms (10^{24} atoms). If each atom were enlarged to the size of a period on this page (.), this number of atoms would fill a giant cube 15 miles on each edge. Such a cube would hold millions of skyscrapers the size of the Empire State Building.

Do such figures seem to belong to the world of make-believe? Remember that these figures describe the matter of which your own body and everything around you are made. Moreover, it was just such calculations that went into the conquest of atomic energy.

* Actually salt has a crystal structure of atoms interlocked in such a pattern that it is impossible to say which of several chlorine atoms a particular sodium atom is attached to. But it is easier here to think of them as forming simple molecules.

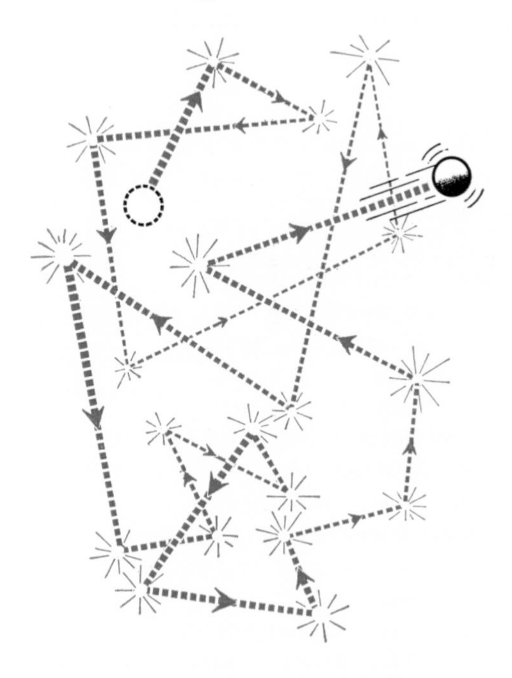

2-2 The Brownian movement is a weird dance observed in microscopic particles suspended in liquids or in air. These zigzag movements provide important evidence for the theory that heat is the motion of molecules.

3. The vibration of particles under a microscope

In 1827 the Scottish botanist Robert Brown was studying some tiny plant cells, called spores, under a microscope. He noticed that when the spores were placed in water, they danced around in an irregular, zigzag manner (Fig. 2-2). He decided that these small dancing particles were the molecules which scientists had been looking for. It was discovered later that the spores themselves were not molecules; their motion resulted from bombardment by still smaller particles, the molecules of water, which were far too small to be seen under the microscope. But Brown's discovery did give some evidence of molecular motion and so added to our understanding of the nature of matter. The peculiar movement of small particles which he noticed is called **Brownian motion.**

All through the nineteenth century, scientists labored to prove or disprove the existence of molecules. By 1900 the weight of evidence seemed overwhelming. Although no one had actually *seen* a molecule, there was every reason to believe that molecules existed—and that their motion was the cause of heat. Today, the modern electron microscope can actually make visible certain kinds of "giant" molecules of which living things are composed. Figure 2-3 shows the molecules of collagen, an important constituent of your skin, magnified 100,000 times. These molecules are thousands of times larger than those of air or water. Their magnification is so great that a hair magnified a like amount would have the dimensions of a freight train.

Massachusetts Institute of Technology

2-3 **Collagen molecules in the skin, magnified 100,000 times.**

4. Wandering molecules

You yourself can perform experiments which point to the existence of molecules. Pour a small amount of liquid ether into an open dish. In a few seconds the liquid is gone, but there is now a distinct odor of ether in all parts of the room (Fig. 2-4). Apparently the ether has escaped into the air of the room in tiny, invisible particles. Obviously, these particles of ether must have been moving. And there must have been space between the similar tiny particles of air which permitted the ether to move about. The particles of the ether that move through the air in this way are molecules, and the particles of air they move among are also molecules.

Solids can evaporate in the same way. Moth balls placed in a closet disappear by evaporation and produce a strong odor in the air. Similarly, when sugar dissolves in water, the molecules of sugar are dispersed among the molecules of water.

These molecules move at tremendous speeds. The molecules in a glass of water are endlessly zigzagging about at an av-

2-4 The odor of ether will be noticed in a far corner of the room almost as soon as it is poured into a dish. This indicates that there is rapid motion of ether particles from the dish into the air. What precautions should be taken in this experiment?

erage speed of about 1,200 miles per hour. Imagine the number of collisions between molecules in this enormous commotion! Each molecule in the air of your next breath is traveling at more than 1,500 miles per hour and makes ten thousand million collisions with its neighbors in each second.

These facts fit in with the theory that all materials are made of molecules; that these molecules can be separated from one another; that there are spaces between them; and that the molecules are in constant motion.

What holds molecules together?

The facts about evaporation and the dissolving of solids in liquids indicate that molecules tend to separate. But solid objects tend to hold together. What force holds them together? Let's look into the question.

Dip a pencil into water. When you pull the pencil out, some water still clings to it (Fig. 2-5A). You know, from earlier studies, that the force of gravity is acting to pull the water down. Therefore, if some of the water still clings to the pencil, there must be a force of attraction between the pencil and the water. In other words, the different molecules which make up pencil and water are attracted to each other. This force of attraction between *different* kinds of molecules is called **adhesion.**

The molecules of water on the pencil stick together in the form of drops. This indicates a force of attraction between molecules of the *same* kind which we call **cohesion.**

5. Cohesion and adhesion

If it were not for the force of cohesion—attraction of like atoms and molecules for each other—all materials in the world would be gases. This cohesive force is not the same in all materials. The cohesion of atoms in steel, for example, is greater than that of the atoms in copper. The greater cohesion of atoms explains, in part, the greater strength of steel.

Adhesive forces also vary with different materials. Figure 2-5B shows what happens when a glass plate is placed on a mercury surface and then pulled away by a spring. The spring has to be stretched a little bit before the glass comes away from the surface; this shows that there is a definite molecular force between the glass and the mercury. When the glass finally does pull away, however, we find that there is no mercury sticking to it. In other words, the mercury does not wet the glass. In this case the force needed to pull the glass away from the mercury was due mainly to adhesion between mercury and the molecules in glass. The adhesive force between glass and mercury was not so large as the cohesive force of the atoms of mercury for each other—and so no mercury was pulled away by the glass.

Now when the same experiment is performed in water (Fig. 2-5B), the spring stretches much less, and this time the glass comes out wet. Since the water clinging to the glass was pulled away from water, we can say that the adhesive attraction of glass for water was greater than the cohesive attraction of water for water. In other words, water wets glass because the cohesion between water molecules is less than the adhesion of water molecules to glass. What evidence would lead you to decide that glass adheres to water less than it does to mercury?

If glass is coated with grease, water does not wet it. In that case the cohesion between water molecules is greater than their adhesion to grease.

6. Joining materials together

You can see that an understanding of adhesion and cohesion is of great practical importance when we have the job of joining two materials together. The great usefulness of glue arises from its ability to exert a high adhesive attraction for other materials. Also, since glue is a liquid, it is able to penetrate cracks and crevices. Once the glue hardens, a strong cohesive force is set up in the hardened glue which acts as a bridge of molecular attraction between the materials it joins.

Figure 2-6 shows an interesting use of cohesion to hold blocks together. No glue joins the blocks. Why is it that they don't fall?

7. The "skin" of a liquid

If you take a razor blade and gently lower it with the broad side downward onto a water surface, the blade will not

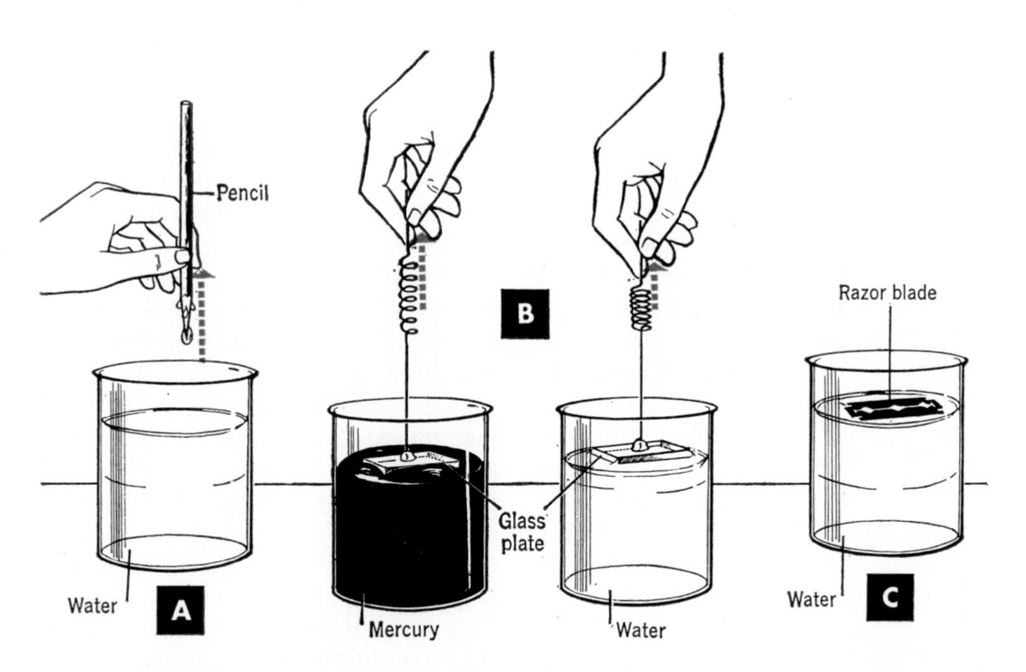

2-5 These simple experiments provide evidence that there are attracting forces between molecules. In A water is attracted to the wood of a pencil. In B we can measure the molecular forces between different liquids and glass. In C we can observe how the cohesive force of water molecules at the surface of the water creates a film that supports a razor blade.

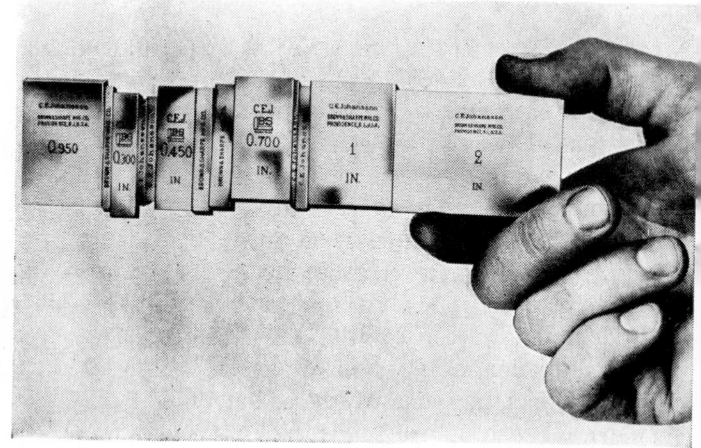

2-6 These accurately polished gauge blocks are used by machinists in making measurements. The surfaces are so flat that when the blocks are placed together, large numbers of the surface molecules are close enough to attract and make the blocks stick together. A force of 220 pounds is needed to pull this combination of blocks apart.

Brown and Sharpe Mfg. Co.

sink (Fig. 2-5C). This is explained by the fact that the blade is resting, as it were, on a delicate "skin" of water.

It will help you to understand what this "skin" is if you look at a high-speed photograph of a single drop of milk (Fig. 2-7). You notice that the drop is perfectly round. The cohesive force on the molecules on the surface of the drop is greater than that on the molecules inside, because a molecule on the surface is being pulled in only one direction—inward—while a molecule on the inside is being pulled in every direction at once. Thus the molecules on the outside tend to be pulled toward the molecules inside the drop. The molecules therefore crowd inward until the drop has the most compact shape possible—and that shape happens to be a sphere. Since the molecules of the surface layer are crowded closer together the cohesive force between them is increased, and the surface of the drop acts like an elastic skin holding the liquid together. The cohesive forces acting on the surface molecules exert what is called **surface tension.**

The top layer of molecules of liquid in a dish acts like an elastic skin for the same reason. In the case of water, this surface tension is sufficient to support light objects such as a razor blade.

If the water wets the object (if the adhesion is greater than the cohesion), then the razor blade will not remain on top of the water. Try this experiment: After the razor blade has rested on the surface of the water, remove it; notice that most of it has not been wetted by the water. Now try placing the blade on the surface of soapy water. This time the blade sinks, and when it is removed from the water you will find that it is wet.

What happens in this case is that the soap (and this is also true of detergents) reduces the cohesion of the water solution. This reduced cohesion reduces the surface tension, which can no longer support the razor blade, and the razor blade sinks. Of course it is this property of reducing cohesion that enables soaps and other detergents to allow objects to get wet. Adhesion becomes greater than the reduced cohesion, and the soapy solution clings to a material (wets it). This property of wetting materials also gives soap its ability to clean. Greasy particles on the body or on clothes can dissolve only in a solution which wets them. If the solution did not wet the greasy particles, it would be because cohesion was greater than adhesion, and then the grease would tend to stick to the surface of the object rather than to dissolve.

8. Capillary action

If you put the corner of a blotter into ink, the ink rises into the blotter and spreads out. The ink is actually rising by itself above its original level. This happens because of the adhesive attraction between molecules of the porous blotter and molecules of the liquid. This process, by which a liquid can be made to rise in a porous material or a narrow tube, is called **capillary action.**

We can study capillary action in greater detail by dipping thin tubes of glass into a container of colored water (Fig. 2-8A). We notice that the water, both inside the tubes and out, climbs up the vertical glass sides, making a concave (inward-curving) surface wherever it touches the glass. If the tube is very narrow, the water may rise several feet. It rises because the adhesion of water to glass is greater than the force of cohesion between molecules of water. Hence the water will climb the side of the glass tube until the upward force of adhesion is equaled by the increasing weight of the column of water.

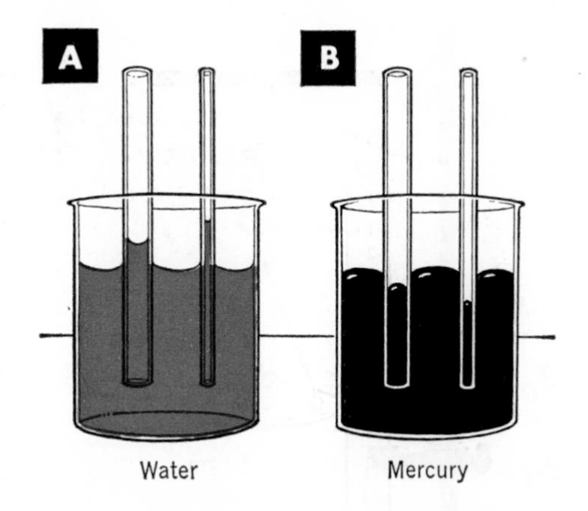

2-8 Water rises in narrow tubes because its molecules are attracted more strongly by glass molecules than by nearby water molecules. On the other hand, the attraction of mercury molecules for each other is much greater than that of the glass molecules, and so the mercury tends to be pulled down the tube toward the main body of the liquid.

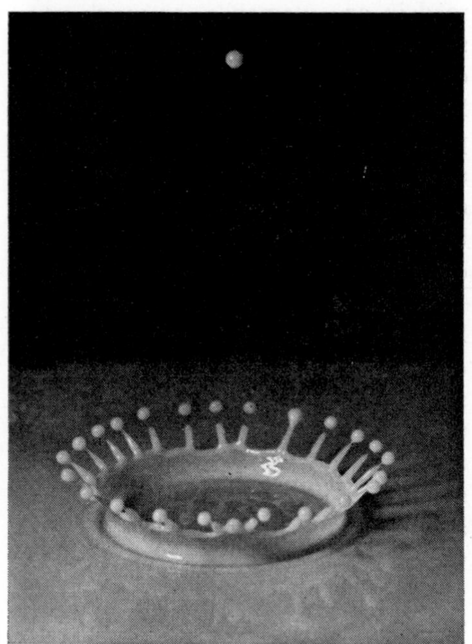

Harold E. Edgerton

2-7 High-speed photography catches this fleeting crown made by an object dropped into a bowl of milk. Note the almost perfect spherical shape of the droplets. Why are they spherical in shape?

Plants make use of capillary action (in part) to draw water from the soil. The water goes up in the plant through thin tubes essentially like those we have been considering. The same principle operates in raising water from the moist earth several feet below ground to the upper layers of soil.

All of the absorbent materials we use in our daily life make use of capillary action. Towels, blotters, absorbent cotton, and wool socks are examples. All are able to soak up liquids because the tiny spaces between the fibers of these materials form tubes in which capillary action can take place.

If we put a narrow tube into a container of mercury, we notice something odd (Fig. 2-8B). The level of the liquid in the tube is actually lower than the level outside. And instead of curving up-

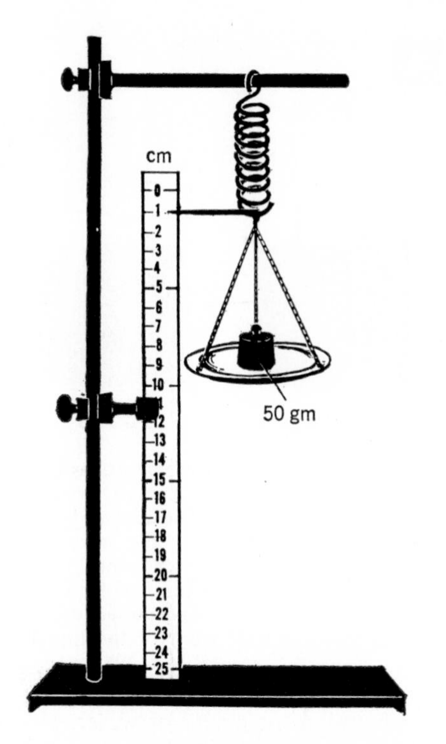

2-9 The length of a spring increases in proportion to the weight suspended from it—until the elastic limit of the spring is exceeded.

ward at the sides of the glass tube, as water does, the liquid mercury curves downward, forming a convex surface. This fact is easily explained if we remember what we have already learned about mercury: the cohesive attraction of its molecules for each other is greater than their adhesive attraction to glass. Hence the sides of the glass tube do not pull the molecules of mercury up; on the contrary, the greater attraction of the mercury molecules below the surface tends to pull the molecules of mercury down the tube.

For the same reason, the surface of the mercury in the tube is convex (bulges outward): the adhesive attraction of the glass is not great enough to counteract the cohesive force from beneath which tends to pull the mercury surface into a spherical shape.

All these examples of adhesion, cohesion, capillary action, and surface tension provide further evidence for the theory that there are forces acting between molecules. It is because of these forces that molecules in a solid or a liquid can be in motion (as they are continually) and yet stick closely together.

9. Hooke's Law

Thus far we have been concerned with the behavior of molecules in gases and liquids, and on the surfaces between liquids and solids. The forces of cohesion in solid objects are considerably stronger than they are in gases and liquids. Indeed, it is this greater force of cohesion that makes a solid "solid."

You observe these cohesive forces in solids every time you bend a stick, wind up a clock spring, bounce over a bumpy road in an automobile, or weigh yourself on a scale. In each situation you are changing the shape of a solid object and hence pushing molecules together or pulling molecules apart. The stretched stick or spring usually returns to its original shape in the end, pulled back by the cohesion of neighboring molecules.

To observe this fact more closely, hang up a spring with a weight suspended from it as in Fig. 2-9. Arrange a ruler so that you can measure the position of the bottom of the spring. In the figure the weight of 50 grams has already stretched the spring 1 centimeter. If you add another 50-gram weight, the spring will stretch another centimeter. Adding another 50 grams will stretch the spring still another centimeter. If you record your observations, your results will approximate those shown in Table 1.

As you study the numbers in Table 1, you will notice that during the early part of the experiment the spring increases in

TABLE 1

F Weight hung on spring (gm)	L Increase in length (cm)
0.0	0.0
50.0	1.0
100.0	2.0
150.0	3.0
200.0	4.0
250.0	5.0
300.0	6.0
350.0	7.5
400.0	9.1

length at a constant rate as the weight is increased by constant amounts. This behavior of the spring is an illustration of a law discovered nearly three hundred years ago by the English scientist, Robert Hooke. It is known as **Hooke's Law** and states that:

The distance an object is stretched is proportional to the force exerted on the object (provided the elastic limit is not exceeded).

In other words, the size of the stretch, L, in column 2 of the table is proportional to the weight, F, in column 1. We can write this fact:

$$F = kL$$

where k is a constant number. You will notice that in the first part of the table F is always 50 times as large as L. Thus, substituting any pair of values for F and L, we get:

$$k = \frac{F}{L} = 50 \text{ gm/cm}$$

Knowing the value of the constant, k, you can find the stretch of the spring for any weight up to about 300 grams. For instance, how much will the spring be stretched when you hang on it 183 grams? Substituting into the formula:

$$F = kL$$

we get:

$$183 \text{ gm} = 50 \text{ gm/cm} \times L$$

$$L = \frac{183 \text{ gm}}{50 \text{ gm/cm}}$$

$$L = 3.66 \text{ cm}$$

Engineers make use of Hooke's Law every time they design a bridge or an automobile, or any other structure whose bending, stretching, or compressing must be kept within safe limits.

SAMPLE PROBLEM

If an automobile spring is compressed 3.0 inches by a load of 1,800 pounds, how much should an engineer expect it to be compressed by a load of 450 pounds?

First find k.

$$k = \frac{F}{L} = \frac{1{,}800 \text{ lb}}{3 \text{ in}} = 600 \text{ lb/in}$$

Substituting this value of k into Hooke's formula

$$F = kL$$

$$450 \text{ lb} = 600 \text{ lb/in} \times L$$

$$L = \frac{450 \text{ lb}}{600 \text{ lb/in}}$$

$$L = 0.75 \text{ in}$$

STUDENT PROBLEM

If a spring is compressed 2 inches by a force of 100 lb, how much will it be compressed by a force of 400 lb? (ANS. 8 in)

. . . .

The ordinary spring balance in your school laboratory or in the grocery store is an application of Hooke's Law (Fig. 2-10).

If you go back now and inspect Table 1 once more, you will see that weights greater than 300 grams stretch the spring more than you might expect. The last 50 grams in the experiment did not stretch the spring 1.0 centimeter; it stretched the spring 1.6 centimeters. Why is this so?

When more than 300 grams are hung on the spring, the molecules of the spring are pulled so far apart from one another that the forces of attraction between them can no longer act as strongly. More weight might stretch the spring permanently. Indeed, if enough weight were hung on the spring it would pull some of the molecules apart altogether. The spring would break. Clearly, for weights of more than 300 grams, Hooke's Law does not apply to this spring. Here, then, 300 grams is called the **elastic limit** of the spring. The elastic limit of an object is the force on it beyond which its stretch is no longer proportional to the force. Once the elastic limit has been exceeded, the spring will no longer return to its original length when all the weight is removed. It will be permanently stretched.

Hooke's Law applies to many other objects besides springs. The law describes the bending of beams, the twisting of rods (such as propeller shafts), the coiling of springs (such as you find in your watch), and the compression of pillars. Indeed, as you look at the buildings and machinery in the world around you, you will quickly find dozens of places where objects are stretched, squeezed, or twisted in proportion to the forces exerted upon them.

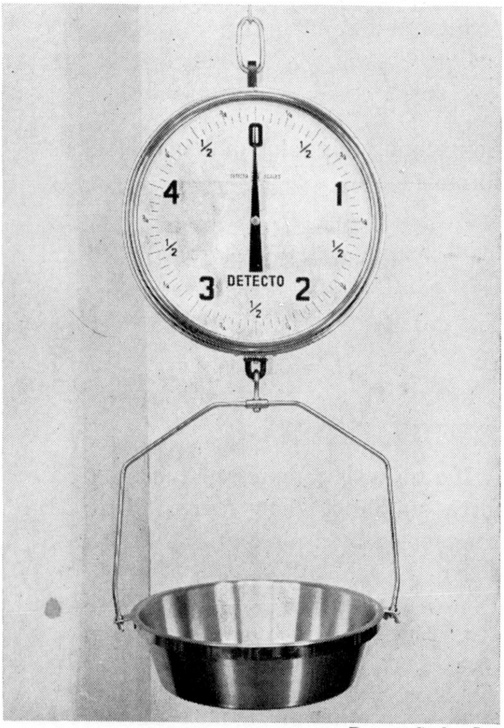

Detecto Scales, Inc.

2-10 You are depending upon Hooke's Law when you accept the reading of the familiar grocery store scales.

Graphs and doing problems

Scientists have found that a collection of data such as that in Table 1 can be understood much more quickly if it is arranged in a **graph.** You will find it useful to draw graphs too.

10. Drawing graphs

The data in Table 1 have been graphed in Fig. 2-11. There are several important points for you to notice.

1. Along the bottom of the graph (the **horizontal axis**) we write down the quantity which we control in the experiment (in this case the weight, W, on the spring). Along the **vertical axis** we write the quantity whose changes we are interested in observing (in this case the increase in length, L).

2. We number the spaces along the horizontal axis in such a way that the numbers from 0 to 400 grams will spread across most of the width of the graph paper. Similarly the lengths, L, from 0.0 to 9.1 centimeters are spaced so as to cover most of the length of the vertical

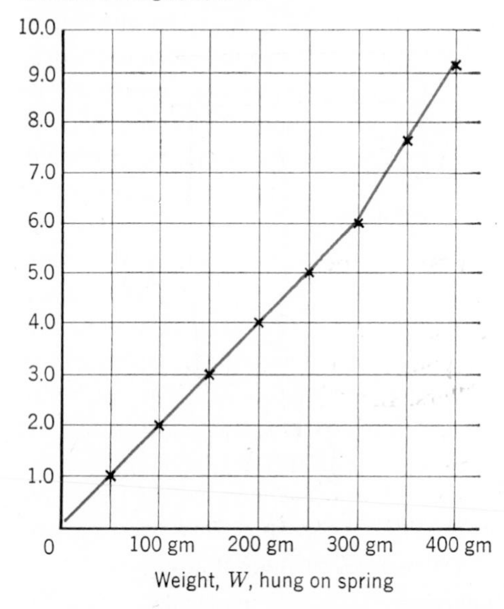

2-11 **A graph of load, W, against increase in length, L, shows that this spring obeys Hooke's Law.**

axis. In this way we make sure that the graph will fill the graph paper and thus be as large and clear as possible.

3. Each pair of numbers in the table is now located on the graph and marked as an "x" or an "o." (Do not use a dot, for it will be covered up when the line is drawn connecting the points.) Sometimes a graph will have two or three lines plotted on it. For instance, you might plot the data taken from two or three different springs. You should then use different kinds of marks for points plotted on different lines.

4. Connect the points you have plotted by drawing the smoothest curve you can.

Now you can see how all the points lie in a straight line (that is, they indicate that W is proportional to L) for weights up to 300 grams. The graph shows clearly where the elastic limit is. One of the most useful properties of graphs is that they show at a glance relationships between two sets of physical measurements.

11. Doing physics problems

One of the best ways of learning physics is to solve problems. You will do your physics problems most successfully if you follow the orderly steps in the box on page 25.

Don't try to get along on memorized formulas alone. As you try to figure out how to solve your problem, think through the situation, trying to imagine the principle of physics that applies. Only *after you have the principle clearly in mind* should you think at all about formulas. Physics is a body of related principles, NOT a collection of formulas.

SAMPLE PROBLEM

Suppose you try to make your own spring balance out of a discarded spring. You have found that each pound hung from the spring stretches it 2.0 inches. A bag of apples stretches the spring 12 inches. How much does the bag of apples weigh?

Step 1. What information are you given? What are you asked to find out?

Step 2. Your diagram may look something like the one on page 24. Notice that it is drawn clearly and labeled as completely as possible.

Step 3. Since the problem is concerned with the connection between the load on a spring and the amount the spring stretches, Hooke's Law applies to this situation. The law says that the amount a spring stretches, L, is proportional to the force stretching it, F. Thus

$$F = kL$$

Step 4. Since we are given the fact that each pound hung from the spring stretches it 2.0 inches, $k = 0.50$ lb/in. Substituting in the formula

$$F = 0.50 \text{ lb/in} \times 12 \text{ in}$$
$$F = 6.0 \text{ lb}$$

which is the answer we require. Notice

page 17 problem 4 John Doe

$$F = kl$$

$$k = 0.5 \text{ lb/in}$$

$$l = 12 \text{ in}$$

$$F = 0.5 \text{ lb/in} \times 12 \text{ in}$$

$$\boxed{F = 6.0 \text{ lb}}$$

2 inches
1 pound
12 inches
F pounds

how the proper unit, lb, is written after the number 6.

Drawing a box around the answer will help it to stand out more clearly.

In writing up this problem to be handed in it is unnecessary to include all the written statements. It can be condensed into the form shown in the figure above.

Going Ahead

With this chapter, as in the last, you again use some new terms and principles. You also need to use mathematics. Most of the problems are quite easy, especially if you have followed the solutions of the sample problems in the chapter. One of them, however, is more challenging. It is separated from the others by a horizontal line. Throughout this book the more challenging problems will be separated from the easier ones in this way.

Remember that mathematics is one of the most important tools of physics. Its use is absolutely necessary in order to calculate such things as the amount of force needed to lift an object, how much heat to apply to melt a given mass of metal, how much electric current is needed to run an electrical appliance, or how bright a light source is needed to protect your eyesight as you work or read.

THE VOCABULARY OF PHYSICS

Define each of the following:

adhesion
Brownian motion
atom
capillary action
elastic limit
horizontal axis
graph
molecule
Atomic Theory
Hooke's Law
cohesion
surface tension
vertical axis

THE PRINCIPLES WE USE

Explain why

1. Water wets wood.
2. Mercury does not wet glass, but does wet copper.
3. The surface of a liquid seems to act like an elastic sheet.
4. A razor blade floats on water.
5. Water rises in narrow tubes.
6. The level of mercury in narrow glass tubes drops.

APPLYING THE PRINCIPLES

1. Explain the part that adhesion plays in gluing two surfaces together.
2. Mention two practical applications of surface tension.
3. Mention two practical applications and two important natural uses for capillary action.
4. State whether you would classify these as matter and give your reason: carbon dioxide, glass, light, vacuum, sound, air.

PROBLEMS TO DO

Engineers designing bridges and other structures often use Hooke's Law to be sure they meet the specifications. Here are some problems of the sort they frequently must solve.

1. If a diving board is pushed down 4 in when a 120-lb boy stands on its end, how much will it be pushed down when a 180-lb boy stands on it?

2. If the elastic limit of the board in Problem 1 were $5\frac{1}{2}$ in, would it be safe for the 180-lb boy to stand on it?

3. If the constant, *k*, of a certain spring is 30 gm/cm, how much will a weight of 150 gm stretch it? What will be its new length if its original length was 20 cm?

4. What is the constant, *k*, of a spring that is stretched 25 cm by a force of 4 gm?

5. A spring is stretched 2 cm when a 1000-gm weight is hung on it. How much will it be stretched by a load of (a) 500 gm, (b) 3 kg, (b) 4.2 kg?

6. If the elastic limit of the spring in Problem 5 is 6.5 cm, which of the given weights, if any, will exceed it?

7. A spring is compressed 4 inches when a 100-lb boy puts his weight on it. How much will it be compressed by a weight of (a) 1 lb, (b) 96 lb, (c) 500 lb?

8. In an experiment to measure the relation between the bending of a beam and the load on it, the following data were obtained:

Load (in pounds)
50 100 150 200 250 300 350

Bending (in inches)
0.13 0.26 0.39 0.54 0.65 0.78 0.91

Plot these values on a graph. Has the beam been loaded beyond its elastic limit?

9. Draw a graph of the following data and find the elastic limit of the spring.

Load on spring (in grams)
40 80 140 226 295 320 390

Stretch (in centimeters)
0.80 1.6 2.8 4.5 5.8 6.6 8.4

STEPS IN DOING PROBLEMS

1. Read the problem carefully twice, trying to form *a picture of the situation* as clearly as you can, but making no attempt to solve it.

2. Draw and label a neat diagram if one applies to the problem situation.

3. After you understand the situation, ask yourself what principle or law applies. Not until you have the principle clearly in mind should you write it down, preferably in the form of a formula.

4. Usually you will be able to find the answer by substituting numbers in the formula you have written. (Harder problems may require two or three formulas.)

5. Solve for the unknown quantity in your formula. A 10-inch slide rule is an excellent short-cut to problem solving and gives answers quite as accurate as the data provided throughout this book.

6. Go over your substitutions, check your arithmetic, and be sure you have written the units such as feet, grams, or seconds after your answer. Be sure your answer is reasonable.

10. When the front wheels of an automobile are clamped so they cannot turn to left or right, it is found that a 10-lb force applied to the steering wheel will turn it through an angle of 7°. How much force would be needed to turn it through an angle of 19°?

---

11. Two springs of equal length are hung side by side. If the constants of the springs are 5 lb/in and 7 lb/in, how far will they stretch when their bottom ends are tied together and a single 20-lb weight is hung from them?

CHAPTER

3

Standards of Measurement

Look at Fig. 3-1. Which line is longer? But if you measure the two lines with a ruler you will find that one is slightly longer than the other.

Every day we have experiences which show us how our senses can fool us. We can tell that an express train is passing us at great speed, but the moon appears to be standing still. Actually, however, the moon is traveling much faster than the express train. It is whizzing through space at a speed greater than that of a rifle bullet. On a cold day a steel post feels colder to the touch than a piece of wood, even though both may be at exactly the same temperature. We can get a sunburn from certain kinds of light which are invisible. Our sense of hearing can be deceived by echoes.

Obviously we cannot rely on our senses alone.

Fundamental units of measurement

Scientists and engineers rely on measurements made with accurate instruments to aid their senses. Figure 3-2 shows the enormous number of such instruments which even an airplane pilot needs to help him in flying. They tell him what is happening inside his plane and in the air outside.

1. Numbers and units

All measurements have two distinct parts. Whenever you write down the measurement of something, you must use a standard unit such as an inch for length or hours for time and a number to tell how many of these units there are in what you are measuring. For example, you may observe that the time a train takes to go past you is 9 seconds, or that the length of a board is 3 feet. Here the units are "seconds" and "feet."

The unit of your measurement is important, and no measurement is complete without it. If you say only that the length of a board is "3," no one knows whether you mean 3 inches, 3 feet, or 3 yards long.

2. The fundamental units

The three most important quantities we measure are length, mass, and time. Exact definitions of these quantities are difficult to give in a simple way. However, the following definitions are good enough for our purposes. **Length** is the

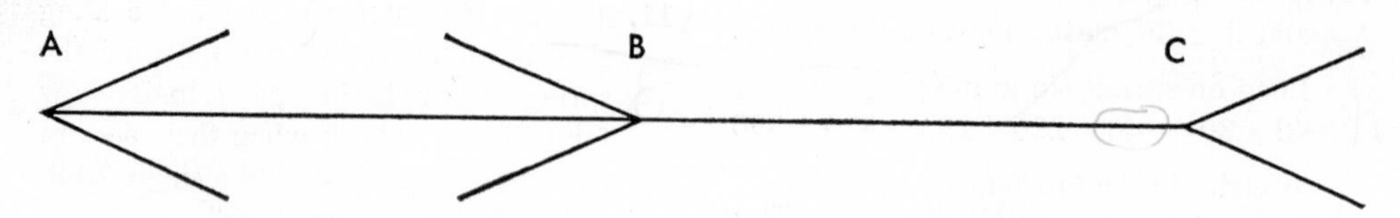

3-1 Which line is longer—AB or BC? Or are they the same length? Guess, and then measure.

distance between two points. **Mass** may be thought of as the "quantity of matter" in an object. Although the mass of an object is closely related to its weight, we shall see later (Chapter 10) that an accurate definition of mass is based on the fact that the greater the mass the harder it is to set into motion. **Time** measures our sense of things happening one after the other. Let us consider these quantities in the order named.

3. Standards of length: the English and metric systems

Primitive man took his standards of length from the human body. For instance, he measured distance—just as you do sometimes—by counting the number of paces needed to walk between two points. For ordinary rough measurements, this system was accurate enough. But as civilization and industry developed, a more accurate system became necessary. Also, the growth of trade between one region and another made it important to arrive at some universally accepted standards of measurement. The chief system of measurement of length and mass in the United States today is the English system. We are all familiar with the units of length in the English system: inch, foot, yard, mile, etc. It is an awkward system because of the selection of such unhandy numbers as 5,280 (the number of feet in a mile). Calculation becomes difficult with such num-

Trans World Airlines, Inc.

3-2 Measuring at 400 miles an hour! To fly a mammoth modern plane, a pilot needs all the measuring instruments shown here and a copilot and a flight engineer to help him read them. Only with the precise information about temperature, pressure, speed, direction of flight, revolutions of engine, and so on, from these instruments, can the plane be flown safely.

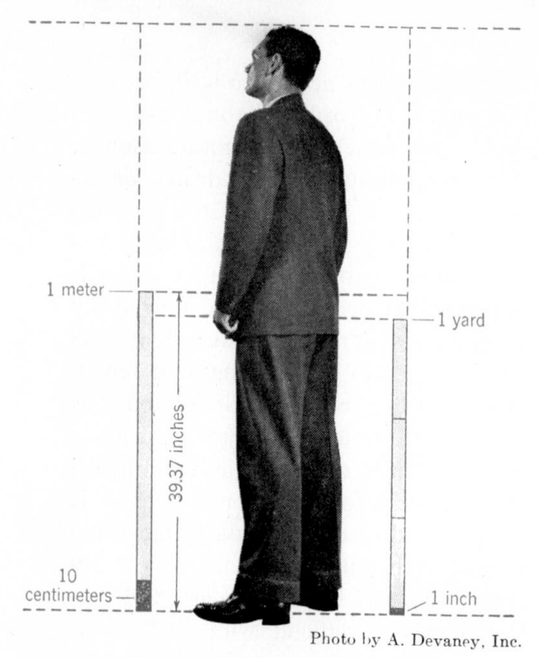

3-3 **Compare the English and metric systems in terms of this six-footer.**

bers, because measurements are not easily converted from one unit into another. The **standard unit of length** in the English system is the **yard.**

Before the French Revolution, the French likewise had an unwieldy system for measuring length. But in 1791 the new French government decided to create a more orderly system of measurement based on the decimal system. As we shall soon see, such a system makes it possible to express one measurement in larger or smaller units simply by moving the decimal point the proper number of places. This system is called the metric system. Because of the ease with which calculations can be made, scientists in all countries have adopted the metric system.

The **standard unit of length** in the metric system is the **meter** (Fig. 3-3). This is a length originally intended to equal one ten-millionth of the distance from the Equator to the North Pole. A standard bar of durable platinum-iridium alloy was marked to what was thought to be this length. Copies of the bar were then made to be used as rulers in everyday measurements. Today we know that the original meter bar is not quite one ten-millionth of the distance between the Pole and the Equator. But this is unimportant so long as the world has one carefully specified length that all can use. This original bar is kept under carefully controlled conditions in a vault of the International Bureau of Weights and Measures at Sèvres, near Paris, and serves as the standard for the entire world. Extremely accurate copies of this standard meter bar are kept in other countries for use in science and industry.

Scientists hope that the more convenient metric system (which, as we shall see, is also used to measure mass) will someday completely replace the cumbersome English system. This would establish a single convenient system as a world-wide standard.

4. Units of length in both systems

It is not hard to see why the metric system is much handier than the English system. Say you want to express 42,638 feet in miles. You have to go through the tedious job of dividing by 5,280. But suppose you want to express 42,638 centimeters in meters. You simply move the decimal point two places to the left to get your answer, 426.38 meters.

The following are the units most commonly used to measure length in the two systems.

THE METRIC SYSTEM OF LENGTH

1 kilometer (km)	= 1,000 meters (m)
1 meter (m)	= length of standard bar
1 decimeter (dm)	= 0.1 m
1 centimeter (cm)	= 0.01 m
1 millimeter (mm)	= 0.001 m

Notice that in the metric system the prefix in front of each unit shows what part of a meter it is. For multiples of a meter, Greek prefixes are used. In the word "kilometer," the prefix "kilo-" comes from the Greek word for 1,000. For fractions of a meter, Latin prefixes are used ("deci-" meaning $\frac{1}{10}$, "centi-" meaning $\frac{1}{100}$, "milli-" meaning $\frac{1}{1,000}$).

THE ENGLISH SYSTEM OF LENGTH

1 mile (mi) = 5,280 ft
1 yard (yd) = length of standard bar
1 foot (ft) = $\frac{1}{3}$ yd
1 inch (in) = $\frac{1}{12}$ ft

5. Converting lengths from one system to the other

If we measure the length of a meter stick in inches, we find that

1 meter = 39.37 inches

Using this fact, we can convert any measurement of length in the metric system into the English system, and vice versa. Say we want to find how many centimeters there are in 1 inch. We know that:

1 meter = 39.37 inches

and also that

1 meter = 100 centimeters

therefore

100 centimeters = 39.37 inches

Dividing both sides by 39.37 we find:

$$\frac{100 \text{ cm}}{39.37} = \frac{39.37 \text{ in}}{39.37}$$

2.54 cm = 1 in

STUDENT PROBLEM

Express 16 cm in inches. (ANS. 6.30 in)

6. Units of mass in both systems

The standard unit of mass in the metric system is the **kilogram**. Since scientists want standards that are unchangeable and indestructible, the kilogram is defined as the mass of a carefully preserved block of platinum kept by the International Bureau of Weights and Measures near Paris. Its mass was intended to be the same as the mass of pure water at 4°C that would just fill a cubical box 10 centimeters on each side.

Often the mass of an object is measured by weighing the object on spring scales. We shall see later in the chapter that the weight measured in this way is not exactly the same as the mass. But for most practical purposes we can consider the value of the mass to be the same as the value of the weight.

METRIC SYSTEM OF MASS

1 kilogram (kg) = mass of standard block
= 1,000 gm
1 gram (gm) = 0.001 kg
1 milligram (mg) = 0.001 gm

ENGLISH SYSTEM OF MASS

1 ton = 2,000 lb
1 pound (lb) = mass of standard block
1 ounce (oz) = $\frac{1}{16}$ lb

By comparing the standard masses of the two systems (the kilogram and the pound), we find by measurement that

1 kilogram = 2.2 pounds
(approximately)

If we want to find how many grams there are in a pound, we go about it in the following way:

1 kg = 2.2 lb

or

1,000 gm = 2.2 lb

Dividing both sides by 2.2, we get:

$$\frac{1,000 \text{ gm}}{2.2} = \frac{2.2 \text{ lb}}{2.2}$$

454 gm = 1 lb

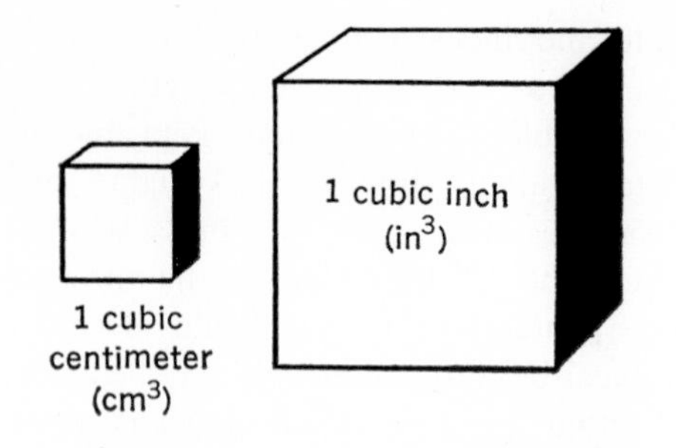

3-4 These are the actual sizes of the cubic centimeter and the cubic inch.

7. Units of time

We have defined time as the measure of our sense of things happening one after the other. The units of time are the same throughout the civilized world. We measure time by the regular swing of a pendulum or by the regular vibrations of the balance wheel of a watch. But the basic unit of time is the day, which is based on the regular rotation of the earth on its axis. Actually, the length of the day changes slightly from century to century, and scientists today are devising new time standards based on vibrations of atomic particles.

Smaller units of time are obtained by dividing the day into 24 hours, the hour into 60 minutes, and the minute into 60 seconds. Today we even have need for time units as small as a billionth of a second.

Derived units of measurement

From the three fundamental units of length, mass, and time, we can derive a great many others. **Velocity** is an example. If a policeman wants to check the velocity of cars on a highway, he measures off distances on the road and uses his watch to time the cars as they cover this measured strip. The velocity, v, of any one car will be equal to this measured distance, s, divided by the time, t, the car takes to cover it:

$$v = \frac{s}{t}$$

The value of the velocity depends upon the units we are using. If a car travels 600 feet in 10 seconds, its velocity is 60 feet per second.

$$v = \frac{s}{t} = \frac{600 \text{ ft}}{10 \text{ sec}} = 60 \text{ ft/sec}$$

If it travels 80 kilometers in 2 hours, its velocity is 40 kilometers per hour. If it travels 500 feet in 5 seconds, its velocity is 100 feet per second (ft/sec). (The symbol / means "divided by" and is often expressed by the word "per." When we say "feet per second" we mean feet divided by seconds.)

8. Area and volume

When we are interested only in measuring length, we use units such as feet, miles, or kilometers. If we are measuring **area,** we may use square feet (sq ft or ft^2) or square centimeters (sq cm or cm^2). But sometimes we are interested in the **volume,** or amount of space an object occupies. Then we may use the cubic foot (cu ft or ft^3), the cubic inch (cu in or in^3), or the cubic centimeter (cc or cm^3). Each of these units represents a cubical box with the length of one side equal to 1 foot, 1 inch, or 1 centimeter as the case may be (Fig. 3-4).

9. Measuring the volume of liquids

Let us take a hollow cubic centimeter (a box which measures 1 centimeter on each inside edge) and fill it with water. We now have 1 cubic centimeter of liquid. Then we pour this water into a tall cylinder and put a 1-cc mark at the level it reaches. Every time we fill the cylinder up to this level, we know that we have exactly 1 cubic centimeter of liquid. We can make marks for each additional cubic centimeter of liquid until we have reached the top of the cylinder. Such a measuring instrument is known as a **graduated cylinder.** This process of marking off correct readings on a measuring instrument is called **calibration** (kăl'ĭ·brā'shŭn). Some typical calibrated cylinders used in the home for measuring volume are the measuring cup and a baby's milk bottle.

Some of the familiar units for measuring volume are the gallon, quart, pint, and fluid ounce (32 fl oz = 1 qt). In engineering work the cubic foot and the cubic yard are also commonly used. Scientists usually use metric units of volume, the most common being the liter (l) and the cubic centimeter. A **liter** (lē'tẽr) is the volume of a cubical box 1 decimeter (10 cm) on each edge. A liter contains 1,000 cubic centimeters ($10 \times 10 \times 10$). A milliliter (ml; 0.001 liter) is thus equal to 1 cubic centimeter. A liter is approximately equal to 1.06 quarts. That is:

$$1 \text{ l} = 1{,}000 \text{ cm}^3$$
$$1 \text{ ml} = 1 \text{ cm}^3$$
$$\mathbf{1 \text{ l} = 1.06 \text{ qt}}$$

10. Measuring the volume of a box

The rectangular box (each side of which is a rectangle) is the simplest shape of container. The volume of such a box is easily obtained by measuring the length, width, and height and multiplying them together. No matter which dimensions you multiply first, the answer is always the same.

SAMPLE PROBLEM

Find the volume of a rectangular box 5 inches long, 4 inches wide, and 3 inches high:

$$\text{Volume} = \text{Length} \times \text{Width} \times \text{Height}$$
$$\text{Volume} = 5 \text{ in} \times 4 \text{ in} \times 3 \text{ in}$$
$$= 5 \times 4 \times 3 \times \text{in} \times \text{in} \times \text{in}$$
$$= 60 \text{ in}^3$$

STUDENT PROBLEM

Find the volume of a box 25 in $\times$ 35 in $\times$ 40 in. (ANS. 35,000 in^3)

11. Measuring the volume of irregular solids

How would you measure the volume of an irregular object such as a potato? Surely you could not do it with a ruler.

Figure 3-5 shows one way to measure the volume of an irregular solid. Some water is poured into a graduated cylinder and its volume measured. Then the irregular object is also placed in the cylinder, raising the level of the water. The new level shows the volume of the water plus that of the object. So we find the volume of the object by subtracting our first reading from this new reading. If the object is too large to fit into the cylinder, we can find the volume of the object by measuring the amount of water it displaces into a catch bucket, as shown in Fig. 3-5.

12. Density

The **density** of a material is the mass (or weight) of 1 unit of volume of the material. For instance, 1 cubic centimeter of mercury weighs 13.6 grams. Hence, the density of mercury is 13.6 grams per cubic centimeter (13.6 gm/cm^3). Using different words we might say that the density is the mass per unit volume.

13. Density of common materials

What is the density of water? We know that a kilogram is defined as the mass of a cubical box of water measuring 10 centimeters on each edge. Such a box contains 1 liter by volume, or 1,000 cubic centimeters. Since 1 kilogram, or 1,000 grams, of water has a volume of 1,000 cubic centimeters, then the mass of 1 cubic centimeter of water must be 1

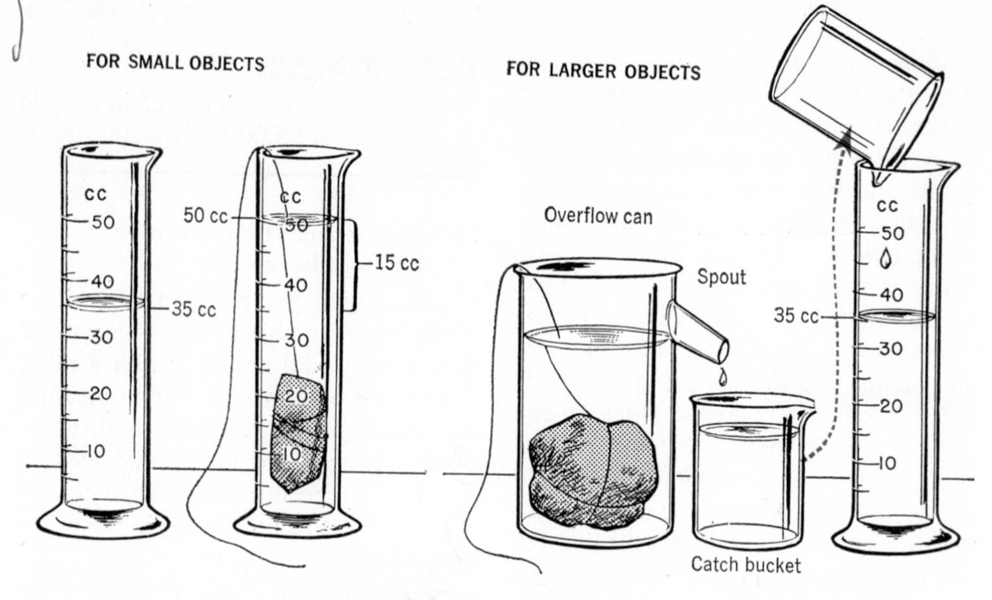

3-5 Two reliable methods of measuring the volume of an irregular object.

gram. Thus, we say that water has a density of 1 gram per cubic centimeter (1 gm/cc or 1 gm/cm^3).

By actual measurement in the English system, we find that the mass of a cubic foot of water is about 62.4 pounds. We can say that the density of water is 62.4 pounds per cubic foot (62.4 lb/ft^3).

Table 2 gives densities of some common materials. A more complete table will be found on page 677.

You will notice that each number in the lb/ft^3 column is 62.4 times as great as the number in the gm/cm^3 column. Hence, you can always find the density of a material in pounds per cubic foot if you know its density in grams per cubic centimeter. All you need to do is to multiply the density in grams per cubic centimeter by 62.4.

14. Solving density problems

Suppose you, as an engineer, are planning to build something which will require 2,000 cubic feet of iron. But you must place your order in pounds. How many pounds of iron will you order from the mill? From Table 2 you know that 1 cubic foot of iron weighs 490 pounds. Hence, 2,000 cubic feet will weigh 2,000 × 490 pounds. Thus:

$$\text{Mass of iron} = \underset{\text{(volume)}}{2{,}000\ \text{ft}^3} \times \underset{\text{(density)}}{490\ \text{lb/ft}^3} = 980{,}000\ \text{lb}$$

To get the mass we have multiplied the volume by the density. This is expressed by the formula:

$$m = VD$$

When solved for D, this formula becomes:

$$D = \frac{m}{V}$$

When solved for V, the formula becomes:

$$V = \frac{m}{D}$$

SAMPLE PROBLEM

A certain rock has a volume of 300 cubic centimeters and a mass of 1,200 grams. What is its density? We know that density is the mass (or weight) of 1 cubic unit of the material. So if 300 cubic centimeters of the rock weigh 1,200 grams, 1 cubic centimeter will weigh $\frac{1}{300}$ as much, or 4 grams. The density is therefore 4 gm/cm^3. The same answer may be obtained from the formula, as follows:

$$D = \frac{m}{V} = \frac{1{,}200 \text{ gm}}{300 \text{ cm}^3} = 4 \text{ gm/cm}^3$$

STUDENT PROBLEM

Find the density of a piece of metal whose mass is 4,200 gm and whose volume is 750 cm^3. (ANS. 5.60 gm/cm^3)

This is the usual method for finding the density of any given material. You can see that usually it is too difficult to get exactly 1 cubic centimeter, or 1 cubic foot, of a material and weigh it. It is much more convenient to take a sample of whatever size is available, find the mass and the volume by measurement, and then calculate the density. In this way you can find the weight of large immovable objects such as brick walls and telephone poles.

15. A word about units

Most of the numbers you will see in your study of physics have units that can be reduced to various combinations of mass, m, length, l, and time, t. For example, you will notice that since density is a mass, m, divided by a volume (which is a length times a length times a length, or l^3), all densities *in whatever system of units* will involve units in the form of m/l^3. Similarly, velocity is always a length, l, divided by a time, t, whether we are talking about miles per hour or kilometers per second. Hence the units of velocity are always in the form l/t.

Physicists call the expression l/t the **dimensions** of velocity; similarly, m/l^3 are the dimensions of density, and the

Westinghouse Electric Corp.

3-6 This tremendous chunk of new plastic foam, used as insulating material, weighs only 8 pounds. Its density is less than 1 pound per cubic foot.

TABLE **2**

DENSITIES OF COMMON MATERIALS

Material	In lb/ft^3	In gm/cm^3
Compressed gases in centers of dense stars	6×10^6	10^5
Gold	1,200	19.3
Mercury	849	13.6
Compressed iron in core of earth	about 750	about 12
Lead	708	11.3
Copper	555	8.89
Iron	490	7.85
Granite	170	2.72
Aluminum	168	2.70
Glass	about 150	about 2.40
Water	**62.4**	**1.00**
Methyl alcohol	50.5	0.810
Gasoline	42.0	0.673
Air at 0°C at sea level	0.0807	0.00129
Air 15 miles up	5.6×10^{-3}	9.0×10^{-5}
Gases in interplanetary space	about 10^{-19}	about 10^{-21}

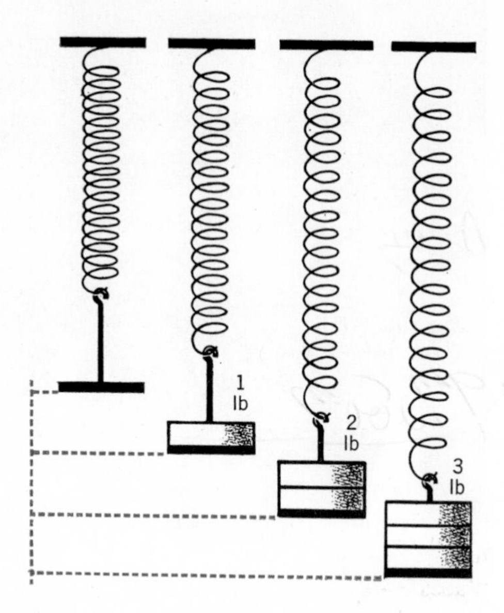

3-7 A spring scale in action. The stretch of a spring can be used to measure forces, according to the amount of stretch.

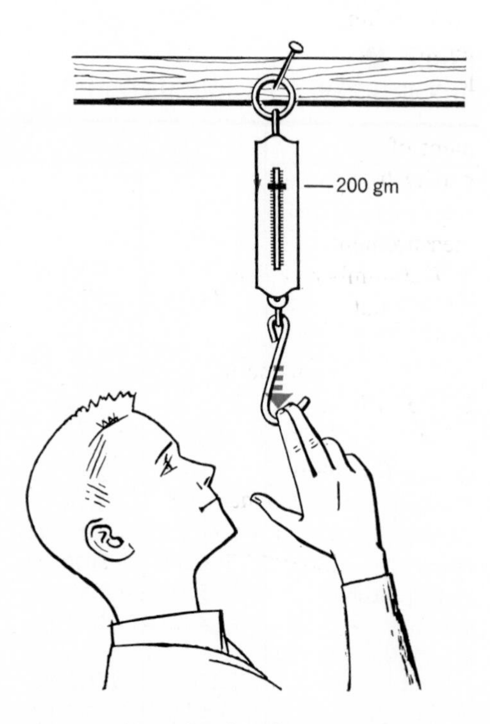

3-8 Does this boy have a mass of 200 grams? No. The spring balance measures the amount of pull (force) the boy is exerting.

dimensions of area and volume are l^2 and l^3.

You will be interested to know that even electrical quantities, which you will study in Unit Five, have dimensions which involve m, l, and t.

Force and mass

16. The spring scale

You ask the grocer for 2 pounds of apples. He puts the apples on a spring scale, and you watch the pointer move over toward the 2-pound mark. Since your apples probably do not weigh exactly 2 pounds, you want to know how much more or how much less you will have to pay. A housewife operating on a small budget will pay a good deal of attention to the spring scale. How does this important little device operate?

In the spring scale, the stretch of the spring increases as more and more weight is added to it, as shown in Fig. 3-7. You will remember from your study of Hooke's Law in the last chapter that the distance the spring stretches is proportional to the force on it.

17. Force versus mass

A spring scale, as shown in Fig. 3-8, does not measure mass, but a force, usually the force of gravity, which is called the **weight** of an object. We shall see in Chapter 11 that the pull of gravity changes slightly at different points on the earth. Therefore, the spring balance will give slightly different readings in different locations.

To avoid this difficulty, scientists who wish to measure the mass of an object accurately use the type of equal-arm balance shown in Fig. 3-9. With this equal-arm balance we use a standard set of masses (commonly called a set of weights) on one pan to balance against

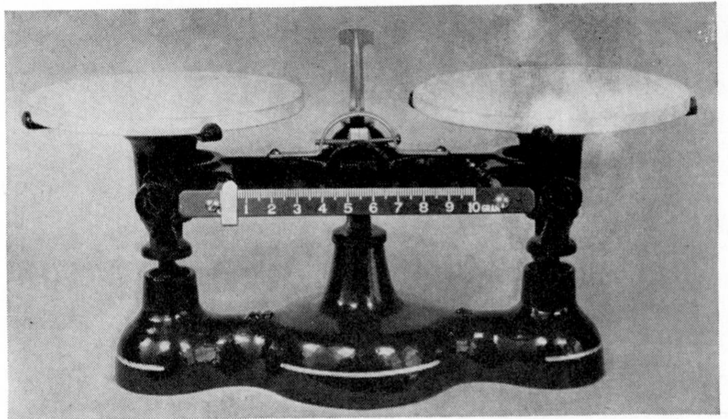

3-9 **The equal-arm balance does not measure the force of gravity on an object—which changes slightly from place to place. Instead, it compares masses. Thus its action is independent of gravity.**

Detecto Scales

the object on the other pan whose mass we are measuring. The true mass of the object is always equal to the mass it balances, no matter where it is located or what the force of gravity may be.

Accuracy of measurement

18. Going beyond our senses

Man owes some of his most remarkable achievements to the fact that he *can* measure, that he is able to go beyond the immediate evidence of his senses (Fig. 3-10). As we shall see later, it was accuracy of measurement that enabled a French scientist, Urbain Leverrier (lĕ-vĕ′ryā′), to predict the position of the planet Neptune before that planet had ever been seen by telescope. Today, accurate measurement is so necessary that modern industry could not be carried on without it. Even the average citizen must understand our system of measurement in order to get along in everyday life.

19. Significant figures

Three boys each measured the length of a block of wood. One boy's measurement was 10.7 millimeters, another boy read 1.07 centimeters, and the third 0.0107 meter. Which boy do you think was the most accurate?

You might vote for the last measurement "because it is to four decimal places." Actually, all three numbers represent the same degree of accuracy. One can be converted to another merely by changing units. In other words, the *number of decimal places of a measurement has nothing to do with its accuracy*. Even 10.7 millimeters and 0.0000107 kilometer are equally accurate statements of the length of the block.

To make a more accurate measurement of the block, each boy would have had to find the number that comes after the 7. That is, he would have to make his measurement to one more digit.

The number 10.7 is accurate to 1 part in 107, but if the measurement is carried out to the next digit and yields 10.70 millimeters, then the measurement is accurate to 1 part in 1,070 (or ten times as accurate). The more digits in a measurement, the more accurate it becomes. Thus, a measurement of 10.70 millimeters is ten times as accurate as a measurement of 0.0000107 kilometer. The number of digits in a measurement is called the number of **significant figures.**

Of course the first zeroes in 0.*0000*107 kilometer do not count as significant figures since they merely serve to locate the position of the decimal point. But both zeroes in 10.70 *are* significant figures

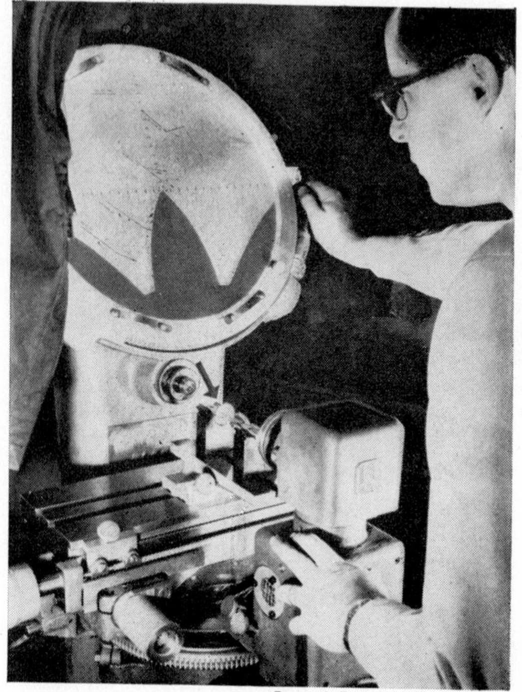

Jones & Lamson Machine Co.

3-10 Modern industry requires precision. Here we see a "comparator" being used to detect flaws in the cogs of a gear wheel (arrow). A shadow many times the original size is cast on the screen so that a slight error can be detected from the magnification and corrected.

since each is the result of an actual measurement.

For example, in 25.20 there are 4 significant figures; in 0.0252 there are 3 significant figures. The first number is ten times as precise as the second.

The rule for determining the significant figures in any number is as follows:

Reading from left to right, the first digit that is not a zero is the first significant figure. The next digit is the second significant figure, whether it is a zero or not; the next digit is the third significant figure whether it is a zero or not, etc.

Go back over the previous paragraphs and determine the number of significant figures in each number, using the above rule.

STUDENT PROBLEM

Find the number of significant figures in (a) 1,270, (b) 100, (c) 100.00, (d) 0.1350, (e) 0.00130. [ANS. (a) 4, (b) 3, (c) 5, (d) 4, (e) 3]

■ ■ ■ ■ ■ ■ ■ ■

You should notice that the measurements you make at the left end of a meter stick (say 1.3 cm) are much less accurate than measurements made to the same number of decimal places at the right-hand end (say 93.6 cm) since the larger number contains more significant figures than the smaller one.

In multiplying or dividing two measurements, an important rule to remember is that your answer can never be more accurate than the least accurate number in your computation. Thus, if the dimensions of one face of the wooden block are 10.7 millimeters by 69.2 millimeters, the area is $10.7 \times 69.2 = 740.44\ \text{mm}^2$. But our data are at best accurate to only 1 part in 692. Hence we can hardly expect our product to be accurate to 1 part in 74,044! Those last two digits would be correct only if the dimensions of the block had been given as 10.700 millimeters and 69.200 millimeters—that is, accurate to 5 significant figures. Since our data are stated to only 3 significant figures, we must "round off" our answer to 740 square millimeters. We really can have no idea what the next digit is until we refine our measurements of length. Likewise $\frac{740.44}{69.2} = 10.7$, not 10.700.

In our textbook *we shall suppose that all numbers are accurate to 3 significant figures unless otherwise stated*. Thus, one side of a sheet of paper 2 centimeters by 3 centimeters may be considered to have an area of $2.00 \times 3.00 = 6.00\ \text{cm}^2$. You should get in the habit of writing your answers to 3 significant figures whenever possible.

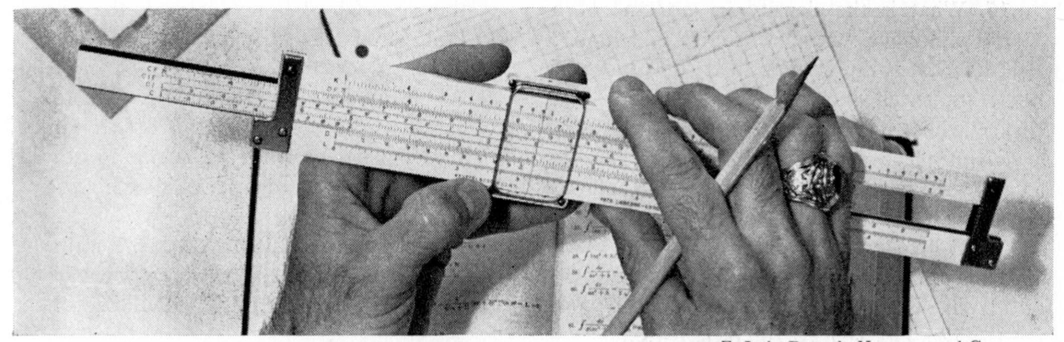

E. I. du Pont de Nemours and Co.

3-11 Answers may be computed by arithmetic or the use of a slide rule. Accuracy to 3 significant figures is readily obtainable when you use a 10-inch slide rule. It is strongly recommended that you learn how to use this handy labor-saving device. The time you spend learning to use it will be amply repaid by the speed of calculation that it makes possible.

Going Ahead

THE VOCABULARY OF PHYSICS

Define each of the following:

standard of length	calibration
density	volume
liter	mass
time	kilogram
standard of mass	dimension
area	significant figures
weight	

THE PRINCIPLES WE USE

Explain why

1. The metric system is more convenient to use than the English system.
2. Mass is measured with an equal-arm balance, while weight is measured with a spring balance.

PROBLEMS OF MEASUREMENT AND DENSITY

1. Change

(a) 5,496.3 cm into m	(g) 5 yd into m
(b) 90.93 kg into gm	(h) 30 ft into cm
(c) 428 cm into mm	(i) 4 in into mm
(d) 36,000 cm into km	(j) 300 lb into kg
(e) 50 m into in	(k) 50 kg into lb
(f) 60 inches into m	(l) 3 oz into mg

2. How many square feet are there in a rectangle 15 ft long and 6 ft wide? How many square inches are there in this area?
3. How many cubic centimeters are there in (a) 5 qt, (b) 1 qt, (c) 1 gal?
4. How many cubic centimeters are there in (a) 8 liters, (b) 5 liters, (c) 11.4 liters?
5. What is the volume of a box 40 inches high, 2 ft long, and 25 inches wide?
6. Find the unknown in each line below:

	Density	*Mass*	*Volume*
(a)	5 gm/cm^3	?	30 cm^3
(b)	?	4,500 lb	15 ft^3
(c)	40 lb/ft^3	1,800 lb	?
(d)	8 gm/cm^3	15 kg	?
(e)	150 lb/ft^3	?	40 ft^3
(f)	? gm/cm^3	3 kg	2 liters

7. A plane travels 2 mi in 10 sec. What is its velocity in mi/sec?
8. How many cubic centimeters are there in 5.00 liters?
9. How many cubic inches are there in 3.00 ft^3?
10. How many liters are there in 7.2 qt?
11. What is the volume of 200 gm of lead?
12. What is the weight of 3 ft^3 of aluminum?

---

13. A block of material has been sold as pure aluminum. It weighs 6,250 gm and measures 40.0 cm long, 10.0 cm wide, and 5.00 cm high. (a) What is its density in gm/cm^3? (b) Is it pure aluminum? How do you know?
14. What is the density in lb/ft^3 of a substance whose density is 15.6 gm/cm^3?

15. What is the density in gm/cm^3 of a substance whose density is 620 lb/ft^3?

16. What is the volume in ft^3 of a piece of material whose weight is 100 gm and whose density is 8.7 gm/cm^3?

17. A piece of glass tubing 100 cm long has an internal cross-sectional area of 2 mm^2. What weight of mercury will it hold?

18. How many significant figures are there in (a) 1,000,000, (b) 0.0001, (c) 123.456, (d) 10,700.0, (e) 432?

---

19. An alloy of copper and aluminum contains 2 parts by volume of aluminum to 1 part by volume of copper. What is its density?

20. An alloy of lead and copper contains 30% lead by weight. What is its density?

21. An alloy is sold as 80% aluminum and 20% copper by weight. A customer measures the density of a sample and finds it to be 4.2 gm/cm^3. Is he getting the right alloy? If not, what is the actual percentage of each metal in the alloy?

22. A barrel contains 100 lb of crushed stone whose density is 180 lb/ft^3. To fill the barrel 2 ft^3 of water are added. What is the *average* density of the contents of the barrel?

23. A piece of glass tubing 100 cm long is found to hold 2 gm of mercury when filled. What is the inside diameter of the tubing?

24. 1 gm of oil (density 0.80 gm/cm^3) when carefully dropped on water spreads out to form a circular patch whose diameter is 1.5 m. How thick is the patch of oil?

25. Using the fact that the density of water is 1 gm/cm^3, prove by calculation that the density of water is 62.4 lb/ft^3.

26. Express a speed of 60 mi/hr in ft/sec. (This is a handy equivalence to remember.)

27. How many significant figures are there in (a) 4.36×10^4, (b) 2.01×10^{-3}?

UNIT TWO

The Machine Age—Force and Motion

U.S. Steel Corp.

In trying to learn more about the nature of matter and energy, physicists find that they must study mechanics, heat, sound, electricity and magnetism, light, and the structure of the atom. Taken together these large areas compose the science of physics.

Of all these branches of physics, mechanics is one of the easiest in which to begin; you can *see* what is going on. In this unit you will study simple machines and discover the laws governing their behavior. You will find in later units that the laws you have learned here have application in the study of heat, sound, light, electricity, and atomic energy as well. One good reason for starting our study with machines is that the physics you learn here is fundamental to much that comes later.

CHAPTER

4

Work and Energy

Suppose you were faced with the problem of lifting—without anybody else's help—a weight of 500 pounds. Figure 4-1 shows four different ways you could do it.

The simple machines shown in Fig. 4-1 are the ones most commonly used. They are: the pulley, the lever, the wheel and axle, and the inclined plane. All complex machines, such as automobiles, are combinations of simple machines. Let us consider the simple ones first.

You see from Fig. 4-1 that in order to lift 500 pounds you have to exert an effort force of only 100 pounds. In other words, these machines multiply your effort force by 5. We call this number which multiplies the effort force the **mechanical advantage**, and abbreviate it as M.A. The mechanical advantage shows us how much our effort force is being multiplied by a machine. A machine which has an M.A. of 3 multiplies the effort force 3 times and enables us to lift 300 pounds by exerting only 100 pounds of effort force.

Pulleys

Let us examine one of the machines shown in Fig. 4-1, the pulley system, and see how it works. Then we shall be able to see how 100 pounds of effort can lift 500 pounds of weight.

If you must lift a piece of furniture weighing 100 pounds to the height of a second-story window, Fig. 4-2A shows a simple way to do it. The boy in Fig. 4-2A loops a rope around a smooth pipe above the window and then pulls downward on the loose end. How much effort does he use? We could find out by attaching a weight to his end of the rope. Suppose we find that it takes 125 pounds of weight to start the 100-pound object moving upward. We then conclude that a person pulling downward with a force of 125 pounds can lift the 100-pound weight.

From this point on let us refer to the weight being lifted as the **resistance** and the force applied to move it as the **effort.** In the instance just described, an *effort* of 125 pounds is used to lift a *resistance* of 100 pounds.

In all machines, no matter how complex, an effort force exerted on the machine by some source of energy is converted by the action of the machine into a force exerted on a resistance.

1. Changing the direction of effort

At first glance it might look as though our arrangement of a rope over a smooth pipe had not helped, since the mover must use 125 pounds of effort to lift a resistance of 100 pounds. But actually he has gained the important advantage of being able to *change his direction of effort.* By using the pipe, he can pull down instead of up. Most of us are acquainted with the fact that it is easier to pull downward with a force of 125 pounds than upward with a force of 100 pounds.

The use of the rope has still another advantage. By pulling on the rope instead of on the object directly, the man can stand on the ground while the object moves upward.

We can improve the arrangement still more if we substitute a wheel for the smooth pipe, as illustrated in Fig. 4-2B. Now we find that less effort is needed for the job, perhaps 105 pounds instead of 125 pounds. This is because the wheel, by rolling with the rope, reduces friction. A wheel used in this manner is what is usually meant by a **pulley.** When a single pulley is attached to a fixed object (wall, ceiling), as in Fig. 4-2B, we call it a single **fixed pulley.**

The more we reduce friction by improving the wheel used in Fig. 4-2B, the more closely the amount of effort required to move the object approaches the weight of the object itself. Thus we conclude that the **ideal effort** required in this instance—that is, the effort if there were no friction—would be 100 pounds. This ideal effort is never quite achieved; some friction always remains no matter how nearly perfect the wheel.

In the next few chapters *we shall disregard friction* and consider only ideal effort. We do this to simplify the discussion because the amount of friction (and hence the extra effort needed to overcome it) will be different depending on the type of pulley and rope used, and the way they are used. But it is important to

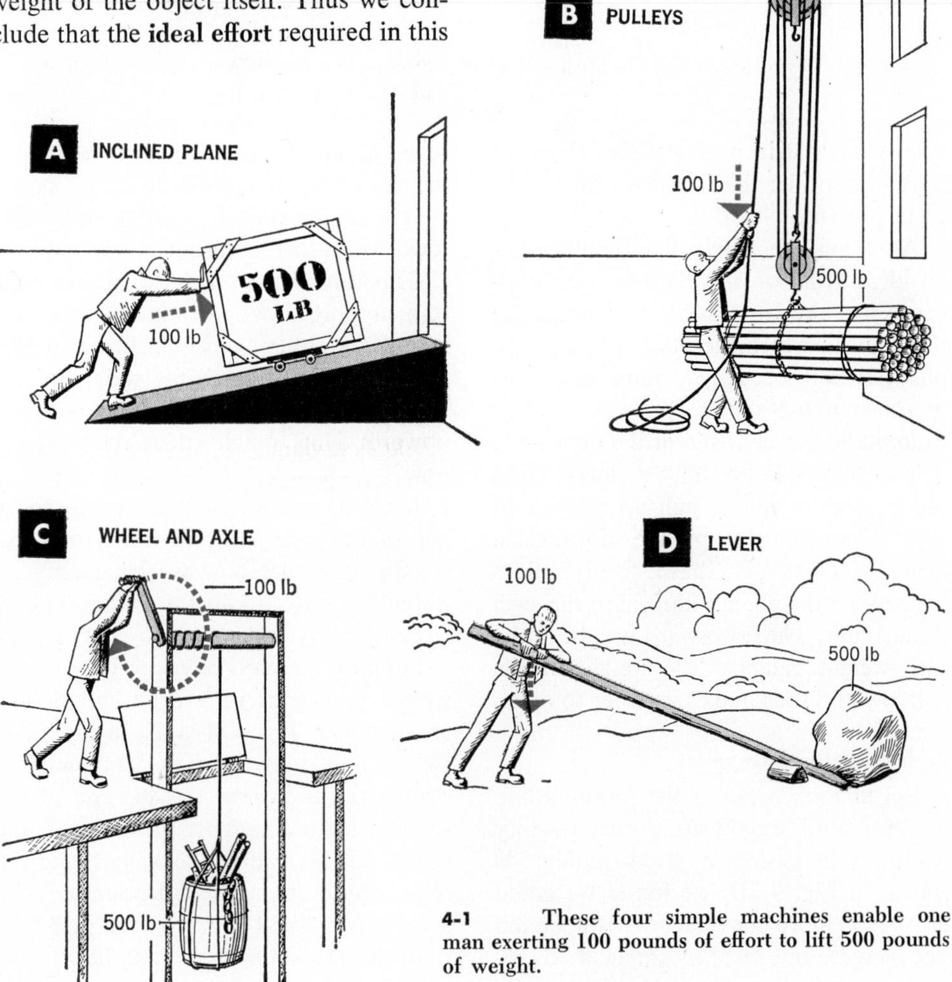

4-1 These four simple machines enable one man exerting 100 pounds of effort to lift 500 pounds of weight.

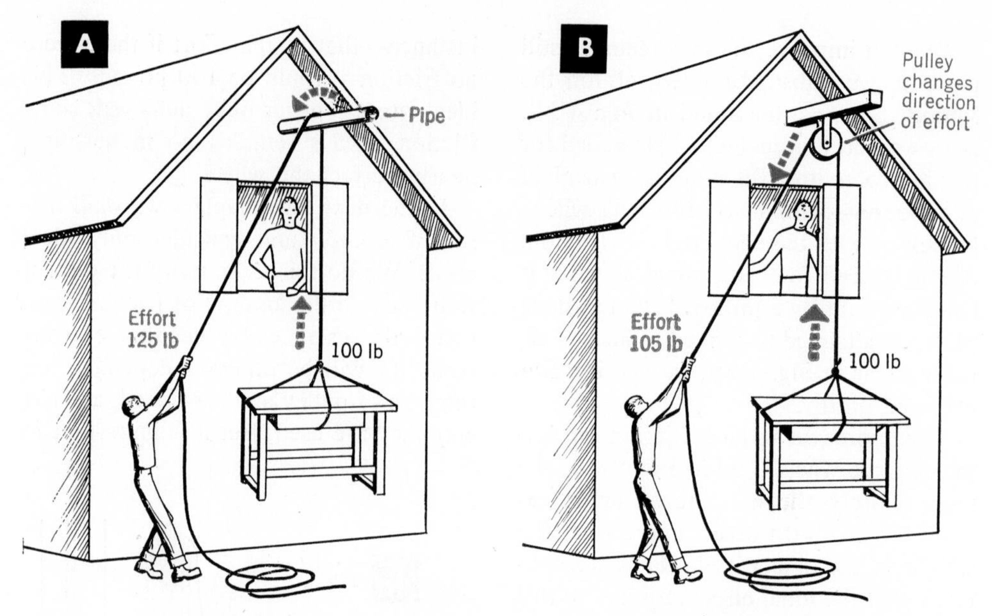

4-2 Two simple ways to lift a weight of 100 pounds without lifting your own body.

remember that in practical situations we have to consider the friction, which is *always* present.

Although the single fixed pulley has an ideal M.A. of only 1, it is useful because it makes it possible to change the direction in which the effort force is applied. It is also used in many machines to transmit motion or force. A familiar example is the dentist's drill (Fig. 4-3).

You may also find that a simple fixed pulley enables you to pull up a shade in your classroom by pulling downward. Most windows also have fixed pulleys over which the chain attached to the sash weight runs. The downward weight of the sash weight balances the weight of the window and thus makes it easier to open.

2. Movable pulleys

Let us look again at the job of lifting a 100-pound weight to a second-story window. By using a fixed pulley, as shown in Fig. 4-2B, we found we could do the job more conveniently. But we did not increase our ideal mechanical advantage (M.A.) because the effort needed to lift the object remained the same as the weight of the object—100 pounds. The ideal M.A. in this instance was 1.

This time, however, let us tie the rope to a support above the window and loop it under a pulley wheel as shown in Fig. 4-4B. Then we attach the weight to the block (the frame of the pulley) and pull upward. How much effort do we need now?

If we tie a spring balance to the loose end of the rope and pull slowly upward, we discover that now we need only 50 pounds of effort to lift the 100-pound resistance. Why is this?

Suppose that two boys are holding the weight by means of a rope and pulley, as shown in Fig. 4-4A. Each boy then carries half the weight, or 50 pounds. If one of the boys now ties his end of the rope to a fixed support on the roof, the roof holds his half of the weight, and the other boy carries only 50 pounds. With a slight additional effort to overcome friction, the remaining boy finds he can lift the 100-pound weight with only a little more than 50 pounds of effort. He is now getting a mechanical advantage of 2, since he has doubled the effect of his effort. As shown in Fig. 4-4C, it is usually easier to pull downward than to pull upward, although the M.A. is the same.

The arrangement shown in Fig. 4-4B is called a **movable pulley.** The ideal M.A. of a single movable pulley is always 2. Remember, we are disregarding friction.

What is the formula for obtaining the M.A.? Since M.A. tells us how many times the resistance is greater than the effort, we find it this way:

$$\text{M.A.} = \frac{\text{Resistance}}{\text{Effort}} = \frac{R}{E}$$

In Fig. 4-4 the resistance is 100 pounds while the effort is 50 pounds. Therefore:

$$\text{M.A.} = \frac{100 \text{ lb}}{50 \text{ lb}} = 2$$

Note that M.A. has no units. They divide out.

STUDENT PROBLEM

An effort of 75 lb is needed to lift a 300-lb desk with a block and tackle. What is the M.A. of the block and tackle? (ANS. M.A. = 4)

■ ■ ■ ■ ■ ■ ■

We can combine fixed and movable pulleys in various ways to increase the M.A. Several pulleys can be built into a single block or frame.

Fig. 4-8 on page 48 shows an interesting combination of pulleys. How do we go about finding the M.A. of such a complex arrangement?

3. Pulley ropes and M.A.

Consider the pulley system in Fig. 4-5. Imagine that all the ropes holding up the weight are attached to a support. Since there are 5 ropes, each must be holding up $\frac{1}{5}$ of the weight. If there were 7 ropes, each would take care of $\frac{1}{7}$ of the total weight. The more ropes we use, the smaller the fraction of the weight that each holds.

In the pulley system shown in Fig. 4-5A, ropes 1, 2, 3, and 4 are held up by the overhead support, while rope 5 is held up by the man. The support holds up $\frac{4}{5}$ of the resistance while the man holds up only $\frac{1}{5}$. Since his effort is only $\frac{1}{5}$ of the total resistance, the M.A. of this pulley system is 5. Note that the number for the M.A. is exactly the same as the number of ropes holding up the weight.

We can obtain the M.A. of most common pulley systems in the same way. Simply count the number of ropes holding up the movable pulley block to which the weight is attached. This number is the M.A.

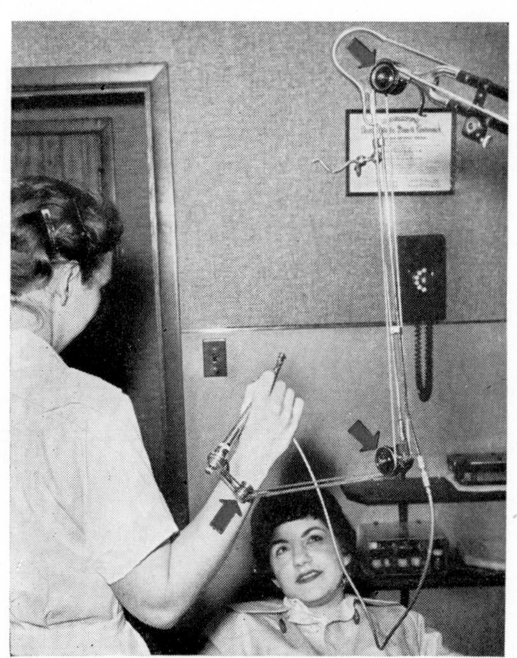

Myles J. Adler

4-3 The dentist uses machines, too. Fixed pulleys (marked by arrows) transmit the motion of the electric motor to the drill.

As we have seen, it is usually easier to pull down on a rope than to pull up. This is why in practical pulley systems the loose rope is usually looped over a fixed pulley and pulled downward, as shown by the rope shown in color in Fig. 4-5A. This rope from the pulley downward does not affect the M.A.

The Principle of Work

If we want to raise a heavy load with little or no effort, we need only to increase the number of pulleys until we have a very large M.A. It looks as though we are getting something for nothing, but there are drawbacks.

Consider, for example, the pulley system of M.A. 5 shown in Fig. 4-5A. A weight of 100 pounds is being lifted with 20 pounds of effort (neglecting friction). The weight is held up by 5 ropes.

4. Mechanical advantage and distance moved

Suppose we want to raise this 100-pound weight 1 foot. We pull on the loose end of the rope so that each of the five ropes attached to the weight is 1 foot shorter, as shown in Fig. 4-5B. But to do this we have had to pull in 5 feet of rope. In other words, the distance the effort moves is 5 times greater than the distance the resistance moves. So what we have gained in effort we have lost in distance. If we had used 7 supporting ropes in order to obtain an M.A. of 7, then we would have had to shorten 7 ropes by pulling on the loose end. Therefore, the effort would have had to move 7 times as far as the resistance. Thus, the more ropes we add the greater the M.A., but the greater the distance the effort will have to move.

This points to another way of measur-

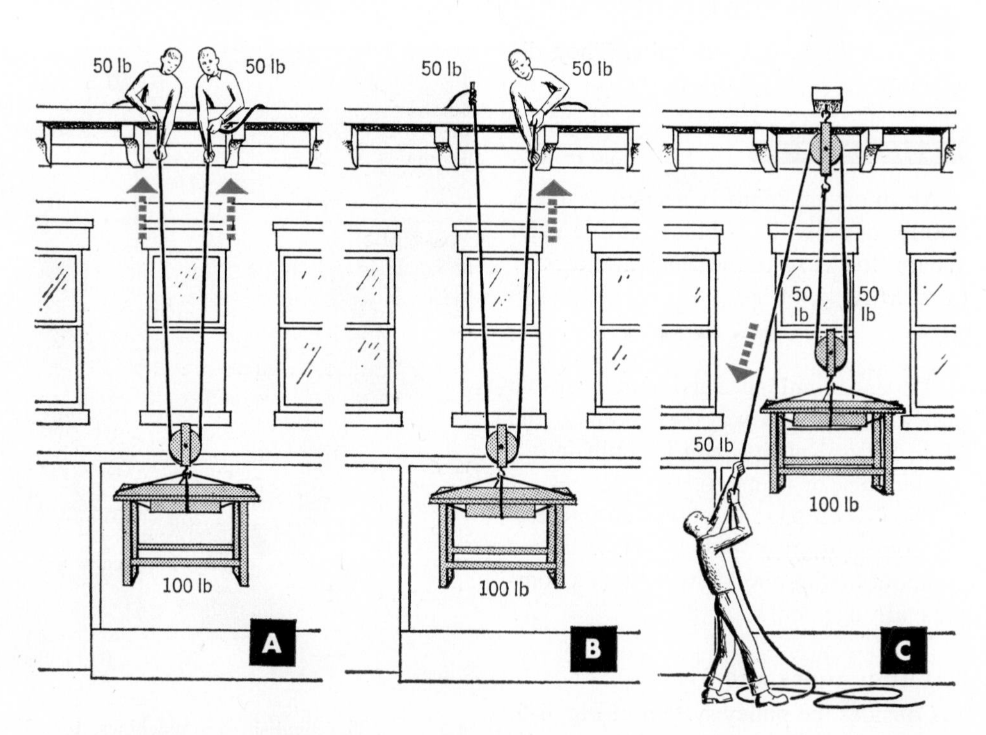

4-4 Three ways of lifting 100 pounds with only 50 pounds of effort. What are the advantages and disadvantages of each?

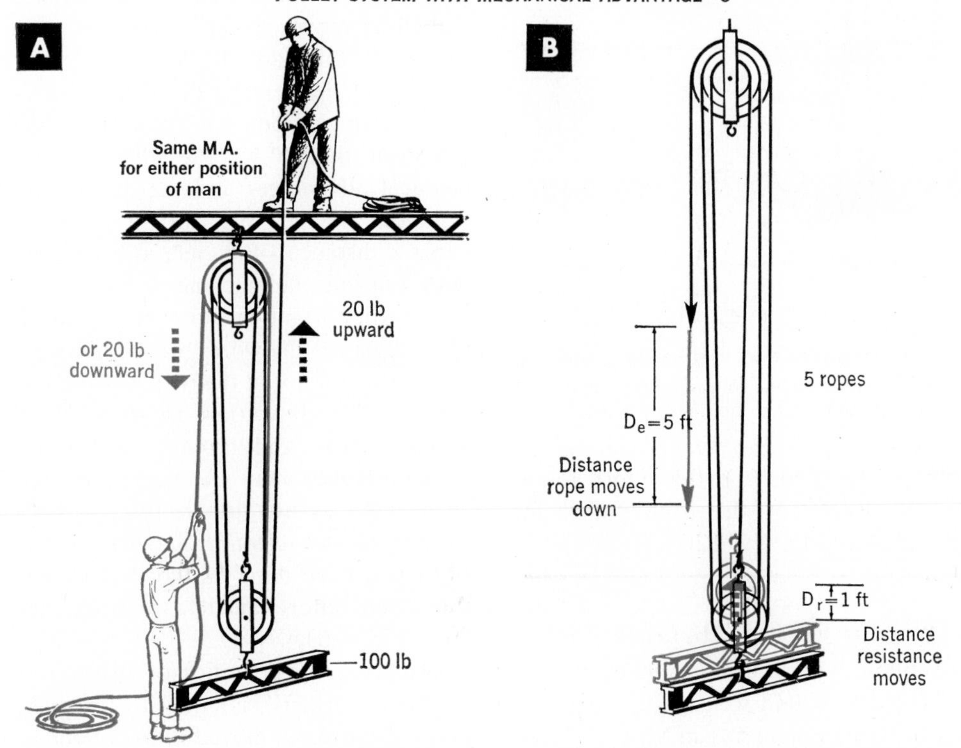

4-5 (A) Twenty pounds lift 100 pounds to give this pulley system a mechanical advantage (M.A.) of 5. Note that 5 ropes hold up the weight. An extra fixed pulley (red lines) changes the upward effort to downward effort for greater convenience—without changing the M.A. (B) To lift the weight 1 foot the effort rope moves 5 feet.

ing the M.A. of a pulley system. If we call the distance that the effort moves (amount of rope pulled in) D_e and the distance the resistance moves D_r, then we find that, in Fig. 4-5B:

$$\text{M.A.} = \frac{\text{Distance moved by effort}}{\text{Distance moved by resistance}}$$

$$\text{M.A.} = \frac{D_e}{D_r} = \frac{5}{1} = 5$$

As we arrange our pulleys to give us a higher and higher M.A., we reduce the effort we must exert—but we increase the distance through which we must exert it. Suppose we have to lift a weight of 400 pounds. With an M.A. of 4, we require an effort of 100 pounds. But we decide this is too exhausting, and we increase the M.A. to 100. To move the 400-pound weight we now have to exert only 4 pounds of effort, which sounds like a pretty good thing. But now to raise the resistance 10 feet, we have to pull in 100 ropes 10 feet each, in other words, a total of 100 × 10 or 1,000 feet. It is not hard to see that it might be a good deal more tiring to pull in 1,000 feet of rope with 4 pounds of effort than to pull the rope a much shorter distance with, say, 50 pounds of effort.

STUDENT PROBLEM

A block and tackle is used to raise a 600-lb piano a vertical distance of 24 ft. If there are 5 ropes attached to the mov-

4-6 Work is done only when a force results in motion. This fellow's force does not result in motion and so he does no work.

able pulley, find (a) the effort force, (b) the length of the rope pulled in. [ANS. (a) 120 lb, (b) 120 ft]

5. Work = Force × Distance

Suppose we are using a pulley system to lift a 300-pound weight to a height of 6 feet. If our system has an M.A. of 2, we will need an effort that is $\frac{1}{2}$ of 300 pounds, or 150 pounds. In this situation we use 2 ropes to hold up the weight and pull in each one 6 feet; all told we pull in 2 × 6, or 12 feet of rope. If we increase the M.A. to 3, the effort required will be $\frac{1}{3}$ of 300, or 100 pounds; in this situation we pull in 3 × 6, or 18 feet of rope. If the M.A. is 6, our effort will be $\frac{1}{6}$ of 300, or 50 pounds; the amount of rope pulled in will be 6 × 6, or 36 feet.

As the effort required to move the weight becomes less, the distance through which the effort moves increases. As shown by the table below, in any given job the effort multiplied by the distance the effort moves always gives the same figure—in this case, 1,800.

What is the meaning of this number—1,800—which does not change for the job even though we use different mechanical advantages and the efforts and distances change? We call this figure—effort × distance—the measure of the **work** put into the machine.

It is important to note here that in physics we use the word "work" in a special sense and not in the way it is used in everyday life. If you are trying to move a heavy desk and do not succeed in budging it, you may *feel* you have been "working" because you are tired, but according to the *scientific meaning* of the word you have done no work at all because your effort failed to produce motion (Fig. 4-6).

The scientist's definition of mechanical work is: work, W, is equal to force, F, times distance, D, moved *in the direction of the force.*

$$W = FD$$

If the 300-pound weight is lifted 6 feet without the aid of any pulley system at all, the work accomplished is found as follows:

$$\begin{aligned} W &= FD \\ &= 300 \text{ lb} \times 6 \text{ ft} \\ &= 1{,}800 \text{ ft-lb} \end{aligned}$$

Since we multiply distance by force (feet by pounds) to find the work done, we call the unit of work a **foot-pound.** One foot-pound of work is done in lifting a

M.A. 1	M.A. 2	M.A. 3	M.A. 6
300-lb effort moved 6 ft	150-lb effort moved 12 ft	100-lb effort moved 18 ft	50-lb effort moved 36 ft
300 × 6 = 1,800	150 × 12 = 1,800	100 × 18 = 1,800	50 × 36 = 1,800

1-pound weight a height of 1 foot (Fig. 4-7).

In the metric system the unit of work is the **gram-centimeter,** which is the work done when a force of 1 gram is exerted through a distance of 1 centimeter.

STUDENT PROBLEM

A 250-lb weight is lifted 20 ft. How much work is done? (ANS. 5,000 ft-lb)

. . . .

We see then that no matter how a job is done, we always accomplish the same amount of work (always remembering that we are neglecting friction and other losses). We can do the job directly or operate a pulley system. We can use any method whatever, but we always come out with the fact that a 300-pound weight has been lifted 6 feet, and that 1,800 foot-pounds of work have been done. This fact is expressed in a statement known as the **Principle of Work.**

6. The Principle of Work

A simplified way to state the facts that we have learned about the pulley system would be as follows:

No matter how the job is done, the work accomplished is the same as the work put in (disregarding friction).

In other words, if we use an effort of 100 pounds and pull in 18 feet of rope, the work we put in is 100 pounds × 18 feet or 1,800 foot-pounds. The work accomplished is the lifting of 300 pounds a height of 6 feet—or 1,800 foot-pounds (300 lb × 6 ft). If we call the work put in at the effort end the **input,** and the work accomplished in moving the resistance the **output,** we can express the Principle of Work in this simple form:

Work output = Work input
(neglecting friction and other losses)

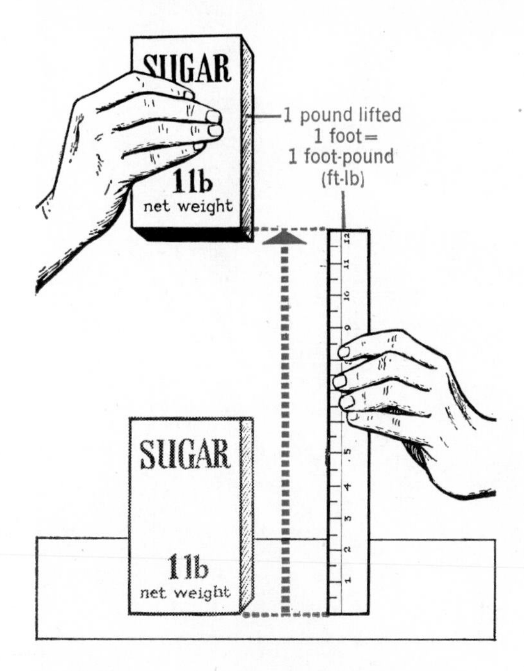

4-7 The foot-pound pictured.

Experiments with many different kinds of machines and under all kinds of conditions have led to the conclusion that the Principle of Work applies in every situation in which machines are used. Although the machine does not change the amount of work done, it does help us by making the job easier to perform (Fig. 4-8), because it can make the effort *force* required less than the resistance *force.* No machine is able to make the *work* input less than the *work* output.

Machines may save force but never work.

We can also arrive at the Principle of Work—the equality of output and input—from our formulas for M.A. For a pulley system:

$$\text{M.A.} = \frac{\text{Resistance}}{\text{Effort}} = \frac{R}{E}$$

We can also say that:

$$\text{M.A.} = \frac{\text{Distance effort moves}}{\text{Distance resistance moves}} = \frac{D_e}{D_r}$$

Plymouth Cordage Co.

4-8 **What is the M.A. of each of these movable pulley blocks? Can you explain how they are being used to bend a steel pipe?**

Since these are both formulas for the same M.A., we can set them equal to each other, thus:

$$\frac{R}{E} = \frac{D_e}{D_r}$$

Multiplying both sides of the equation by ED_r, we get:

$$RD_r = ED_e$$

The left-hand side of the equation, RD_r, is the resistance times the distance the resistance moves. In other words, it is the output, the work which the machine accomplishes. The right-hand side, ED_e, is the effort times the distance the effort moves, that is, the input. So again output is equal to input as stated by the Principle of Work.

We must not forget that there is always some loss due to friction and to other factors, such as the weight of the pulleys and ropes. Hence it follows in any *actual* situation that, even though our machine is the most efficient possible, we will never get out of it quite as much work as we put in.

7. Gaining distance

Up to now we have thought of the pulley system as a means of multiplying our effort. But if we use the machine in reverse we can obtain a gain in resistance distance, D_r, instead of a gain in effort force, E. However, as we would expect from the Principle of Work, we will have to pay for this gain in resistance distance by exerting greater effort force.

Figure 4-9 shows a pulley system in which the usual positions of effort and resistance have been reversed. The effort is now being applied at the pulley block where the resistance usually is, and the resistance is attached to the place, A, where the effort force usually is. In this instance an effort of 40 pounds is needed to lift the 10-pound weight. But when the effort is pulled down 1 foot, it causes all 4 ropes to lengthen 1 foot each. As a result, the 10-pound resistance moves up 4 feet. The weight therefore moves 4 times as far as the effort. Such an arrangement can be used to increase the speed of doing a job, since the resistance moves faster than the effort. In Fig 4-9 we multiply the speed 4 times.

Thus we see that the Principle of Work still applies; we have gained speed but lost force. The work done remains the same.

What is the M.A. in this situation? Using our familiar formula, we find it thus:

$$\text{M.A.} = \frac{R}{E} = \frac{10 \text{ lb}}{40 \text{ lb}} = \frac{1}{4}$$

Our M.A., then, is less than 1. Instead of increasing the amount of weight lifted by our effort, we have decreased it. We have moved the weight 4 feet for a 1-foot motion of the effort, but we have had to pay for this by using 4 times as much effort (Fig. 4-9).

8. Work and energy

The scientist defines **energy** as the **capacity for doing work.** We have already found that we define mechanical work as force × distance. This, then, will be our measure for mechanical energy since the amount of energy possessed by a given source is simply the amount of work it can do.

For example, how much energy is there in a 50-pound weight suspended 4 feet above the ground? We know the weight has potential energy because it can do work by lifting other weights as it falls. If this 50-pound weight is placed on the effort end of a pulley system of M.A. 2, as shown in Fig. 4-11, it can fall 4 feet and lift a 100-pound weight to a height of 2 feet (neglecting friction and the weight of rope and pulleys). In doing this it accomplishes 200 foot-pounds of work (100 lb × 2 ft). So we say that the amount of *potential* energy the 50-pound weight possesses at a height of

4-9 A pulley system in reverse.

Alan J. Bearden

4-10 This lacrosse player's right hand moves with great force through a small distance. What happens to the ball? What is the M.A. of this lever?

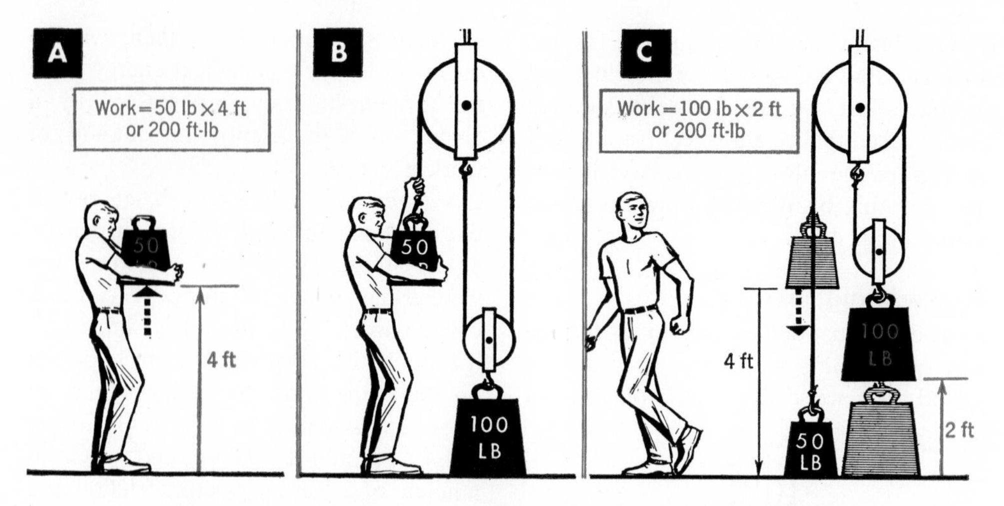

4-11 **The boy gives 200 foot-pounds of potential energy to a 50-pound weight in A when he lifts it 4 feet. In C this weight gives up its potential energy by lifting a 100-pound weight 2 feet, using a pulley system of M.A. 2. Is there any loss of energy?**

4 feet is 200 foot-pounds. This is exactly equal to the amount of work that had to be done originally to lift the weight off the ground to a height of 4 feet (50 lb × 4 ft).

In other words, the potential energy of this 50-pound weight at a height of 4 feet is equal to the amount of work done in lifting it to that height, and is also equal to the amount of work it can do in lifting other weights. Experiments with other weights show that this fact is true for all weights lifted on the earth, no matter to what height they are raised.

Thus we see that potential energy can be measured by the work which must be done to store up the energy. We measure the potential energy of the water behind a dam by measuring the amount of work it would take to get it up to its present height from the base of the dam.

When we use up 200 foot-pounds of energy in raising our 50-pound weight to a height of 4 feet, we have enabled it to do 200 foot-pounds of work. It may use this work or energy in raising another weight, in which case the energy of the first weight has been transferred to the second. But the amount of energy remains the same; it has simply passed from one weight to the other.

For the sake of simplicity we have illustrated the meaning of work by the lifting of weights. The work done was found to be equal to the force exerted (which was *vertical*) times the distance through which it was exerted (which was also *vertical*). But how much work is done when you push a 3,000-pound stalled car 10 feet along a horizontal street? 30,000 foot-pounds? Of course not, for the force you must exert is not the 3,000-pound weight of the car at all, but is much less than that—just enough to overcome friction—say 100 pounds. The work done would be equal to the force exerted, 100 pounds, times the distance through which it was exerted, 10 feet, or only 1,000 foot-pounds. Note that the direction of the 100-pound force and the 10-foot distance are the same.

When calculating work, both the force and the distance must be in the same direction.

Obviously the 30,000 foot-pounds mentioned above cannot be correct, as it was obtained by multiplying the *vertical* force of 3,000 pounds by the *horizontal* distance of 10 feet.

Now what has become of the 1,000 foot-pounds of work used to move the car? It cannot be stored in the stalled car because the car cannot give it back. Actually the work was converted into the heat energy produced when overcoming friction. Thus the energy is not destroyed, but merely inconvenient to get back again.

SAMPLE PROBLEM

It requires 20 pounds to drag a 100-pound desk. How much work is done in (a) raising it to 15 feet, (b) dragging it 15 feet?

(a) $W = FD$
$= 100 \text{ lb} \times 15 \text{ ft}$
$= 1{,}500 \text{ ft-lb}$

(b) $W = FD$
$= 20 \text{ lb} \times 15 \text{ ft}$
$= 300 \text{ ft-lb}$

Going Ahead

THE VOCABULARY OF PHYSICS

Define each of the following:

Principle of Work	resistance
movable pulley	ideal effort
gram-centimeter	effort
work output	work
pulley	fixed pulley
foot-pound	work input
mechanical advantage	

THE PRINCIPLES WE USE

Explain why

1. A pulley system with many ropes gives a large mechanical advantage.
2. A single fixed pulley is often used even though the effort needed is slightly greater than the resistance.
3. It does not always pay to increase the M.A. of a machine too much.
4. The M.A. of a pulley system can usually be obtained by simply counting the number of ropes holding up the weight.
5. You always put in more work than you get out of a machine.
6. When a machine has an ideal M.A. greater than 1, the distance moved by the effort is greater than that moved by the resistance.

APPLYING PRINCIPLES

1. Draw a diagram of a pulley system that has an ideal M.A. of 5.
2. Draw a diagram of a pulley system that will lift 80 lb with 20 lb of effort.
3. Draw a pulley system with an M.A. of 6.
4. Draw a diagram showing a pulley system that could make a 5-lb weight move 3 times as fast as the effort. What is the M.A. of this system? How much effort would be needed?
5. State three different ways of obtaining the ideal M.A. of a given pulley system.
6. Give three examples of the use of fixed pulleys. What is their purpose?

SOLVING PROBLEMS

Assume all M.A.'s are ideal.

1. Find the M.A. in each of the following:

Resistance	*Effort*	*M.A.*
(a) 600 lb	120 lb	?
(b) 950 lb	50 lb	?
(c) 75 lb	375 lb	?
(d) 600 gm	80 gm	?

2. State the Principle of Work. Show how this law applies when a weight of 400 lb is lifted 2 ft with a pulley system of M.A. 8.
3. How much work is done in each of the following jobs: (a) 340 lb lifted 20 ft, (b) 550 lb lifted 40 ft, (c) 600 gm lifted 2 cm, (d) 1,300 lb lifted 150 ft, (e) 4.5 kg lifted 10 m.
4. How much work is done on a 20-lb suitcase in (a) lifting it 2.5 ft, (b) carrying it 25 ft along a station platform?
5. A block and tackle has an M.A. of 3. (a) How heavy a weight can it lift with an effort of 90 lb? (b) How high will it rise if 27 ft of rope are pulled in?

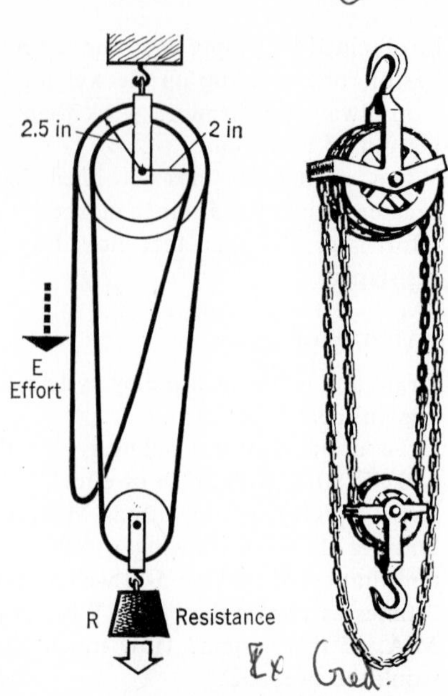

4-12 **You cannot find M.A. of this common pulley system (differential pulley) by counting ropes. As the workman pulls downward and causes the upper pulley to rotate once, how many inches of chain has he pulled in? Both pulleys are fastened together and rotate as a single unit. How many inches has the load been raised? What is the ratio of effort distance to resistance distance?**

6. What is the M.A. of a block and tackle if 60 ft of rope must be pulled in to lift a piano 12 ft?
7. It requires 400 ft-lb of work to raise a box 5 ft in the air. How heavy is the box?
8. A boy weighing 50 kg climbs a 5-m vertical ladder. How much work does he do? What has become of this work?
9. How many foot-pounds of work are done in lifting a 100-lb bag of cement 3 inches off the floor?
10. A block and tackle consisting of two fixed pulleys and one movable pulley is used to raise a 240-lb crate 25 ft. What is (a) the M.A., (b) the effort force, (c) the effort distance?

---

11. How far across a floor will a 300-lb refrigerator be moved if 120 lb are needed to move it and 2,400 ft-lb of work are done in moving it?
12. Assume 20 ft of rope are pulled out to lift a 300-lb weight 5 ft with a block and tackle. (a) Draw a diagram of the block and tackle. (b) What effort was applied to lift the weight?
13. In Problem 12 what is (a) the work input, (b) the work output?
14. Assuming that you could take your pick of any one pulley system, would it be easier to lift 2,500 lb 8 ft or to lift 1,500 lb 15 ft?
15. Calculate the M.A. in Fig. 4-12.
16. Referring to Fig. 4-13, answer the following questions:
    (a) If E equals 100 lb when the house is pulled along on the rollers, what is the resistance overcome?
    (b) What force is exerted on Post A?
    (c) What force is exerted on Post B?
    (d) Through what distance does E act in moving the house 100 ft?
    (e) Is it true that all the work done on the house is used in overcoming friction? Explain.
    (f) Which system of pulleys should have the stronger rope?
    (g) At which point should a steel cable be used in place of a rope?

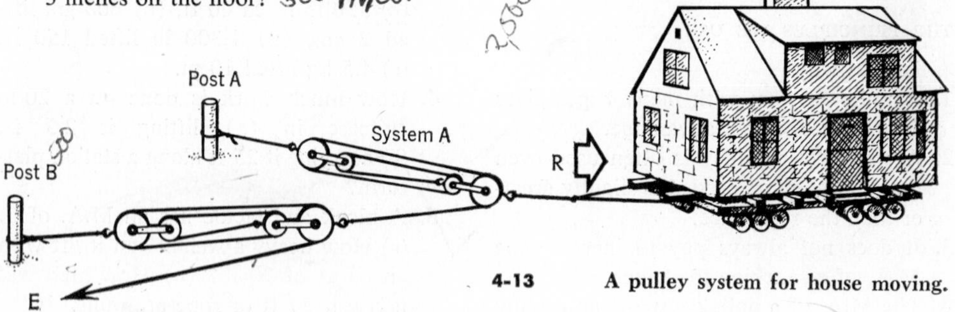

4-13 **A pulley system for house moving.**

CHAPTER

5

Balancing Forces

You balance yourself whenever you ride a bicycle, skate, dive, dance, or tip back in your chair.

This chapter on balance begins with the story of levers and seesaws. We shall then learn that the study of balance is also concerned with objects that are not tipping over, such as bridges, buildings, automobiles, tables, and other objects.

The lever

Have you ever used a crowbar to lift a heavy weight? If you have, you know that a small effort applied to one end of the bar, as shown in Fig. 5-1, can lift a comparatively large weight. In fact, one man can lift a car in this way. How does the bar, which is one kind of **lever,** give us this mechanical advantage?

1. Mechanical advantage of a lever

We can set up a situation in the laboratory to study the nature of a lever. In Fig. 5-2 a 36-inch bar is supported at one end on a platform balance and at the other end by a spring balance. A weight of 24 pounds is hanging from the bar between the two balances. We can lift this 24-pound weight by pulling up on the spring balance. When we do this, the only spot on the bar which is stationary is the place where it rests on the platform balance. The bar may be said to pivot on or around this stationary point. We call this point the **fulcrum.** The end at which the effort is applied is called the **effort end.**

When our 24-pound weight hangs

Philip D. Gendreau

5-1 Do not underestimate the crowbar. It is a lever with a large M.A.

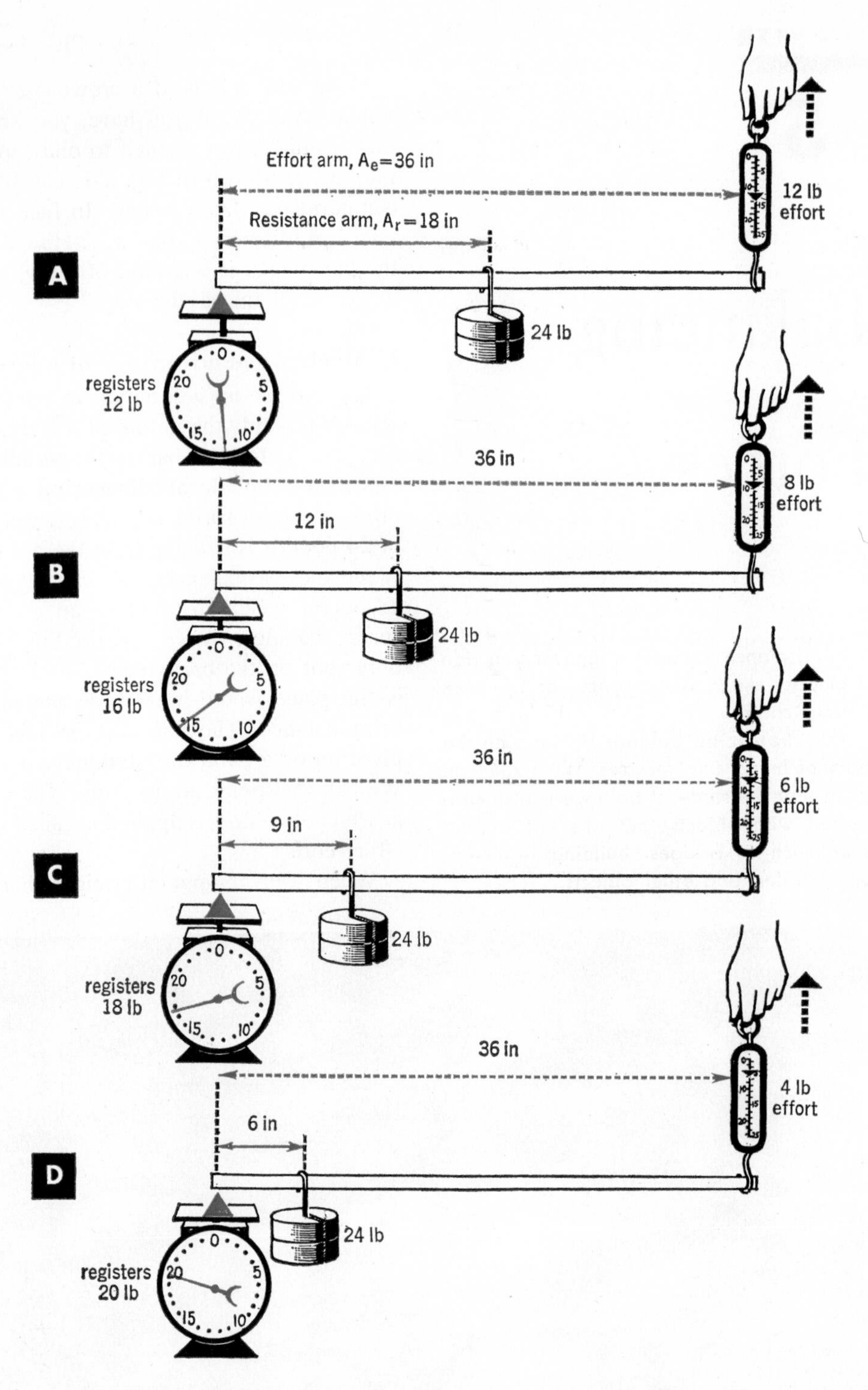

5-2 The closer the weight to the fulcrum (pivot on platform balance), the smaller the effort.

from the center of the bar, 18 inches from either end, the platform balance and the spring balance register 12 pounds each (Fig. 5-2A). This is because the weight is being shared equally by the two supports; each carries half of the 24 pounds. In other words, while we are lifting up on the effort side at the spring balance with 12 pounds of force, the other 12 pounds of weight are being carried by the fulcrum.*

As we move the 24-pound weight away from the center and closer to the fulcrum, the fulcrum carries more and more of the weight, and the spring balance (the effort end) carries less and less (Fig. 5-2B, C, D). As the weight approaches the fulcrum, the reading on the platform balance approaches 24 pounds—the full amount of the weight. At the same time the reading of the spring balance at the effort end approaches zero. We see that the two supports together always carry the full weight. The sum of their readings is always 24 pounds.

We see now why this lever can produce a large M.A. and makes it possible to lift a large weight with a smaller effort. By placing the bar on a fulcrum instead of on a platform balance, we get the fulcrum to carry part of the weight we have hung on the bar. Thus, we can raise the weight by exerting at the effort end a force equal to only a fraction of the weight. If we want to increase our M.A. and lift the weight with even less effort, we simply move the weight closer to the fulcrum, giving the fulcrum a greater share of the weight.

* In this experiment we neglect the weight of the bar itself, since it would be a complicating factor. But if we were solving a practical problem, we would have to consider how much the bar weighs by itself. We would get this by taking a reading of the balance when no weight is hung from the bar; then we would subtract this from each of the readings taken with the weight attached.

2. Calculating the M.A. of a lever

We learned in Chapter 4 that M.A. is equal to the resistance divided by the effort. In the lever example we have just been considering, we found that if we placed the 24-pound weight at the halfway point on the bar, we could lift it with an effort of only 12 pounds (Fig. 5-2A). Hence:

$$\text{M.A.} = \frac{R}{E} = \frac{24 \text{ lb}}{12 \text{ lb}} = 2$$

So our M.A. is 2. In Fig. 5-2B we increase our M.A. by moving the weight closer to the fulcrum. Now it hangs at the 12-inch mark, $\frac{1}{3}$ of the distance from the fulcrum to the effort. The effort we need is only $\frac{1}{4}$ of the weight, or 8 pounds.

$$\text{M.A.} = \frac{24 \text{ lb}}{8 \text{ lb}} = 3$$

When we move the weight up to the 9-inch mark, $\frac{1}{4}$ of the distance between fulcrum and effort, the effort needed is only $\frac{1}{4}$ of the weight, or 6 pounds. Now

$$\text{M.A.} = \frac{24 \text{ lb}}{6 \text{ lb}} = 4$$

So we see that our M.A. for the lever depends upon where on the bar we place the weight. Let us call the distance of the effort from the fulcrum the **effort arm**—abbreviated A_e. Let us then call the distance of the resistance from the fulcrum the **resistance arm**—abbreviated A_r. After many experiments with the weight at different positions on the bar, we always find that:

$$\text{M.A.} = \frac{\text{Effort arm}}{\text{Resistance arm}} = \frac{A_e}{A_r}$$

In Fig. 5-2D the weight is placed at the 6-inch mark. The effort arm, A_e, is 36 inches, and the resistance arm, A_r, is 6 inches. The M.A. is then:

$$\text{M.A.} = \frac{A_e}{A_r} = \frac{36 \text{ in}}{6 \text{ in}} = 6$$

If the M.A. is 6, then we can lift the 24-pound weight with only $\frac{1}{6}$ as much effort, or 4 pounds.

3. Levers and the Principle of Work

Why should the M.A. of a lever be determined by the ratio of the effort arm to the resistance arm? The Principle of Work gives us the answer.

In Fig. 5-2C the distance of the effort from the fulcrum is 36 inches, that of the resistance is 9 inches. In order to lift the resistance 1 inch, it will be necessary for the effort to rise 4 inches. This happens because the effort, which is 4 times as far from the fulcrum as the resistance, must move 4 times as far.

You will see this more clearly if you notice that the moving lever traces out two similar triangles. From what you have learned in geometry about corresponding sides of similar triangles you will see that we can write

$$\frac{A_e}{A_r} = \frac{D_e}{D_r} = \text{M.A.}$$

Hence we have still another way of describing the mechanical advantage of a lever.

Moreover, when you remember that mechanical advantage was defined in the last chapter as

$$\text{M.A.} = \frac{R}{E}$$

you will see that we can set our two expressions for M.A. equal to each other.

$$\frac{R}{E} = \frac{D_e}{D_r}$$

or

$$RD_r = ED_e$$

which says as before that the output work is equal to the input work. Thus the Principle of Work applies to levers just as we saw that it applied to pulley systems. Similarly if we write down the M.A. of other simple machines as a ratio of distances and also as a ratio of forces we always arrive, as here, at a statement of the Principle of Work.

You will note in Fig. 5-2 that if the resistance is moved closer to the fulcrum, then the weight will be raised a shorter distance than before. As with pulley systems, we lose distance but make up for it by an increase of M.A.

4. Other arrangements of levers

Up to this point we have considered the lever as pivoted at one end with the effort at the opposite end—while the weight is suspended somewhere between the two.

But other arrangements are possible. For example, we may have the fulcrum, or pivot, of our lever between the effort and resistance, as in a seesaw or a crowbar.

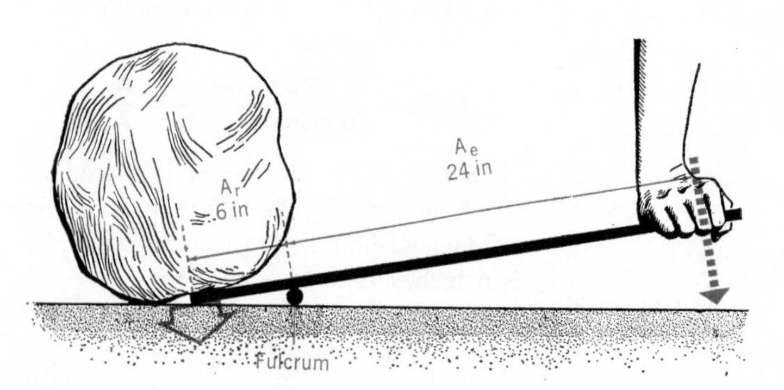

5-3 The M.A. of this lever is $\frac{24}{6} = 4$.

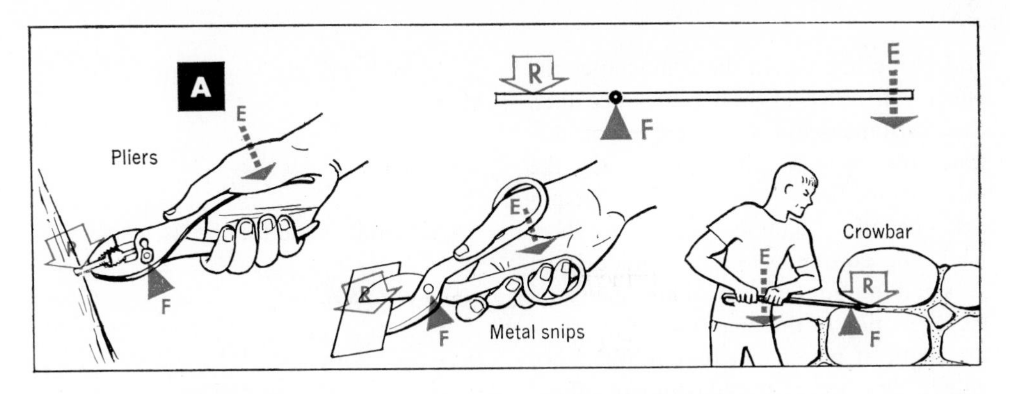

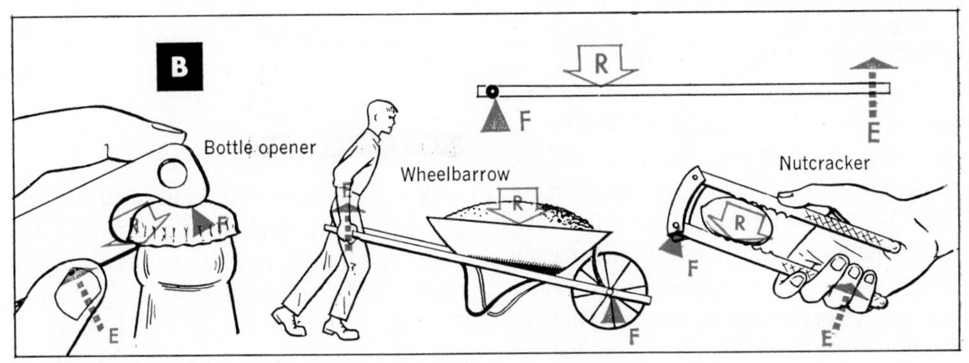

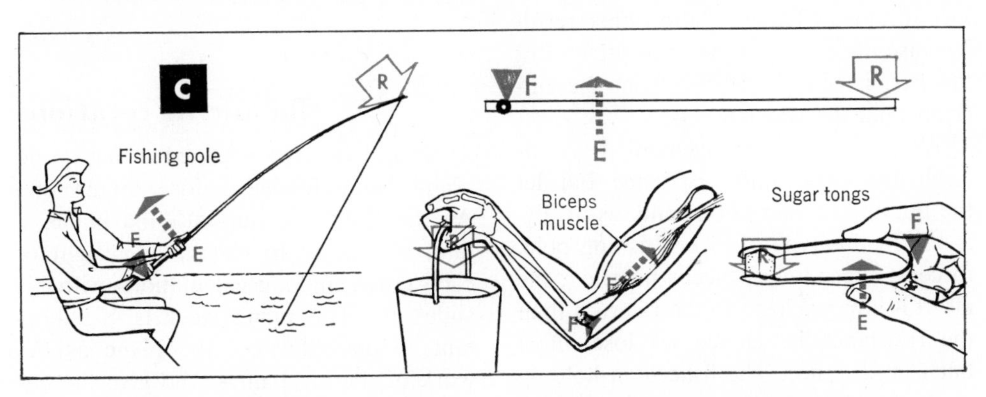

5-4 The three different arrangements of levers have many uses—at home, in our work, in our bodies.

We calculate the M.A. for this arrangement in the same way as for other levers. Thus the M.A. of the lever shown in Fig. 5-3 is:

$$\text{M.A.} = \frac{A_e}{A_r} = \frac{24 \text{ in}}{6 \text{ in}} = 4$$

The effort needed is therefore $\frac{1}{4}$ of the resistance of 60 pounds—or 15 pounds. Note that A_e and A_r are always measured from the pivot or fulcrum.

Different machines which use this type of arrangement are shown in Fig. 5-4A. Metal snips, scissors, pliers, the crowbar, and the can opener are all machines of this kind.

In the pliers and metal-cutting shears shown in Fig. 5-4A, the resistance is placed close to the fulcrum while the effort is placed farther away at the other end of the handle. In these machines the effort will move a greater distance than the resistance, but the sacrifice of distance results in a gain of effort. On the other hand, for scissors used to cut paper, the effort handles are made short while the cutting blades are made long. A small motion of the hands causes a larger motion of the blades—and a greater length of paper can be cut. We have gained distance at the expense of effort.

When the fulcrum and effort are at opposite ends of the bar, as in Fig. 5-4B, we have a second arrangement. In this type of lever effort and resistance move up and down together. Examine all the examples of this lever arrangement in Fig. 5-4B (nutcracker, wheelbarrow, bottle opener). You notice that for each the resistance force overcome is greater than the effort exerted, but that the resistance distance is smaller than the effort distance. These relationships result because the effort, at the opposite end of the bar, is always farther from the fulcrum than the resistance is.

We get a third arrangement if we interchange effort and resistance but let the fulcrum remain at one end, as in Fig. 5-4C. In this type of lever (fishing pole, forceps, sugar tongs, biceps muscle), the effort is always closer to the fulcrum than the resistance is. Hence we lose effort but gain distance. The fisherman holding the rod in Fig. 5-4C can pull in the fish a great distance with a small motion of his hand. But this gain in distance must be paid for by greater effort.

SAMPLE PROBLEM

An 8-foot crowbar is used to lift a 150-pound boulder. The fulcrum is 0.5 foot from the end of the bar where the boulder rests. How much effort is required to balance the boulder? (Assume the crowbar is weightless.)

To find the effort we must know the M.A.

$$\text{M.A.} = \frac{A_e}{A_r} = \frac{(8 - 0.5)\text{ ft}}{0.5\text{ ft}} = 15$$

Using this value for the M.A. we have

$$\text{M.A.} = \frac{R}{E}$$

$$15 = \frac{150\text{ lb}}{E}$$

$$E = \frac{150}{15} = 10\text{ lb}$$

STUDENT PROBLEM

A 6-ft wooden pole is used to lift one side of a crate resting on one end of the pole. A fulcrum is placed 2 ft from the crate, and an effort of 80 lb is applied to the other end of the pole. How much does the crate weigh? (ANS. 160 lb)

Balancing rotations

Have you ever watched a big cargo plane being loaded before taking off? As the freight is put into the plane, a man stands by to check the weight of each crate, making calculations with his slide rule. This man's work is as important to the safety of the plane as the work of the mechanics who give the engine its final checkup.

What this man is doing is figuring out just where every piece of cargo should go, depending on its weight. If too much weight is concentrated at the rear of the plane, it will become tail-heavy. If too much is concentrated at the front, it will be nose-heavy. Either of these conditions could easily bring about a crash. Before the plane takes off, the weight of the whole cargo must be properly distributed so that the plane balances perfectly while in flight.

5. Balancing a stick

In order to learn how one gets the cargo of an airplane to balance, let us begin with a simpler example. Assume that we have a stick which is balanced on a pivot. We place a weight of 600 grams on the right side of the stick, 10 centimeters from the pivot or fulcrum (Fig. 5-5A). As you see, this weight causes the bar to rotate. We say that the bar has a **clockwise rotation** because it turns in the same direction as the hands of a clock. What can we do to prevent this rotation? In other words, how can we restore the stick to a condition of balance?

From our study of the lever we know that we can balance the bar by using 300 grams at 20 centimeters from the fulcrum on the other side (Fig. 5-5B), 200 grams at 30 centimeters (Fig. 5-5C), or 100 grams at 60 centimeters. It is clear that all of these forces have exactly the same turning effect. In each situation, if we multiply the weight by its distance from the fulcrum, we get the same answer. Thus:

$$600 \text{ gm} \times 10 \text{ cm} = 6{,}000 \text{ gm-cm}$$
$$300 \text{ gm} \times 20 \text{ cm} = 6{,}000 \text{ gm-cm}$$
$$200 \text{ gm} \times 30 \text{ cm} = 6{,}000 \text{ gm-cm}$$
$$100 \text{ gm} \times 60 \text{ cm} = 6{,}000 \text{ gm-cm}$$

We can say that the turning effect, therefore, is 6,000 gm-cm.

You might suppose from the units of this answer (gm-cm) that turning effect is the same thing as work, whose units can also be gram-centimeters. This is not true, however, as you will see when you consider what the centimeters measure in each case. In measuring work, the centimeters indicate the distance through which the force actually moves, while in measuring turning effect, the centimeters indicate the distance of the force from the fulcrum. In the English system the units for turning effect are commonly written as pound-feet.

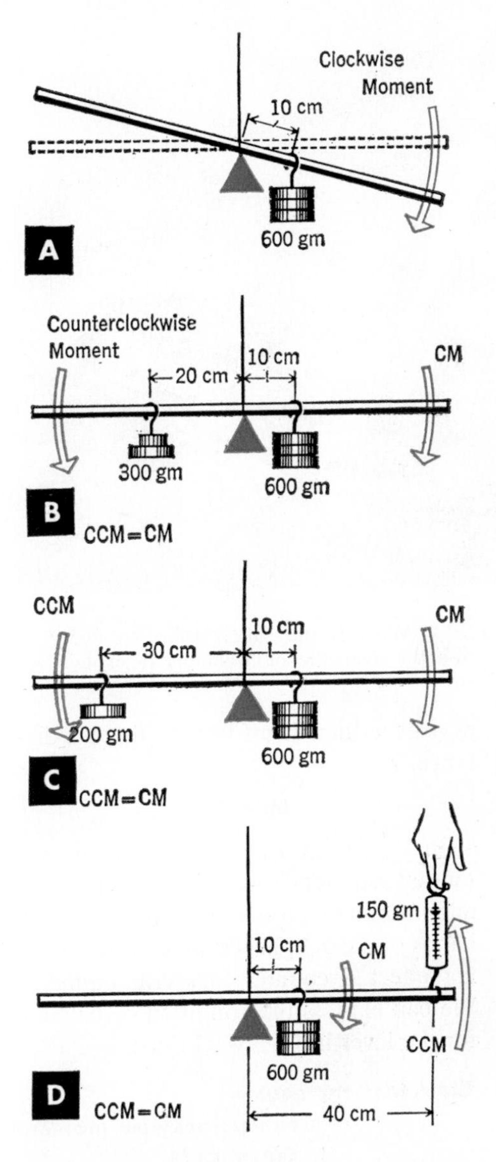

5-5 Three differing ways of balancing a weight are shown here. What do all three ways have in common? In what way does D differ from the arrangements in B and C?

6. Moments

This turning effect, which we obtain by multiplying force times the distance from the fulcrum, is given the special name **moment or torque**. A moment, M,

A. Devaney, Inc.

5-6 What is the sailor trying to do? Can you use the Principle of Moments to explain his action?

may be written as a force, F, times a distance, L, or

$$M = FL$$

In Fig. 5-5C the balancing weight on the left side tends to make the stick rotate in a direction opposite to that of the hands of a clock; hence we call this turning effect a **counterclockwise moment.** We can express the condition of balance for the lever in Fig. 5-5C thus:

$$\text{Clockwise moment} = \text{Counterclockwise moment}$$
$$CM = CCM$$

Moments, or turning effects, may be created by using forces other than hanging weights. An upward force will also produce balance. Thus in Fig. 5-5D an upward force of 150 grams acting 40 centimeters from the fulcrum will balance a weight of 600 grams hanging 10 centimeters from the fulcrum.

$$CCM = CM$$
$$150 \text{ gm} \times 40 \text{ cm} = 600 \text{ gm} \times 10 \text{ cm}$$
$$6{,}000 \text{ gm-cm} = 6{,}000 \text{ gm-cm}$$

Even though the upward force is on the same side as the weight, it counteracts the weight by tending to make the bar rotate in a counterclockwise direction. Hence this upward force causes a moment which tends to make the bar rotate in a direction opposite to that caused by the weight (Fig. 5-5D).

Moments play an important part in many common experiences. For example, the sailor in Fig. 5-6 finds that the force of the wind against his sail causes a turning moment that will tip his boat over. So he prevents this and creates balance by using his weight to create an equal and opposite moment. He increases his moment by getting far out on the end of a long plank.

7. The Principle of Moments

Suppose that we have two weights at different points on the left end of a bar (as shown in Fig. 5-7) and wish to balance these with one weight placed 20 centimeters to the right of the support. How much weight will be needed at this 20-centimeter point on the right side to balance the two weights of 50 grams and 100 grams each on the left side?

The turning effect, or moment, of the 50-gram weight 10 centimeters from the fulcrum is 50 grams × 10 centimeters, or 500 gram-centimeters. The moment of the 100-gram weight at 20 centimeters is 100 grams × 20 centimeters, or 2,000 gram-centimeters. So the total counterclockwise turning effect is 2,000 + 500, or 2,500 gram-centimeters. We find the weight, W, needed on the opposite (clockwise) side in the following manner:

$$CM = CCM$$
$$W \text{ gm} \times 20 \text{ cm} = 2{,}500 \text{ gm-cm}$$
$$W = 125 \text{ gm}$$

We can check the correctness of this prediction by actual experiment.

We see that we can get the total turn-

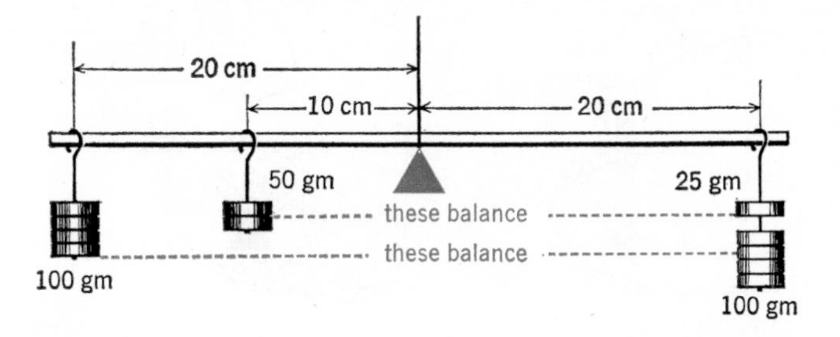

5-7 The Principle of Moments enables us to predict the exact weight and position needed to balance any two other weights on the bar.

ing effect of the two weights on the same side of a bar by adding together their separate turning effects, or moments. Experiments with other weights and distances show that this is always true. These experiments and calculations can be summarized in the **Principle of Moments,** which we may state simply:

A motionless object will not start to rotate if the sum of the clockwise moments acting on it is equal to the sum of the counterclockwise moments.

SAMPLE PROBLEM

Three boys, A, B, and C, are balanced on a seesaw. They weigh 76 pounds, 100 pounds, and 120 pounds, respectively. If A sits 10 feet to the left of the fulcrum and B sits 2 feet to the left of the fulcrum, where must C sit to produce balance?

Here we use the fact that the total counterclockwise moment must equal the total clockwise moment.

$$\text{CCM} = \text{CM}$$

$$76 \text{ lb} \times 10 \text{ ft} + 100 \text{ lb} \times 2 \text{ ft} = 120 \text{ lb} \times L$$

$$L = \frac{960 \text{ lb-ft}}{120 \text{ lb}} = 8 \text{ ft (to the right)}$$

STUDENT PROBLEM

A meter stick, balanced at its center, has a 50-gm weight hanging from the 10-cm mark and a 100-gm weight hanging from the 30-cm mark. Where must a 200-gm weight be placed to balance the stick? (ANS. at the 70-cm mark)

8. Moments around bent levers

Levers in actual machines are often bent in such a way that the effort arm and the resistance arm are not in the same straight line. In using a claw hammer to pull out a nail you use such a lever. The Principle of Moments still applies, however.

Let us suppose that you have to exert a 40-pound pull on the hammer in Fig. 5-8 in order to draw the nail out of the board. How great a pull, *P*, does the hammer exert on the nail? Measuring the length of the handle from your hand down to the fulcrum (where the hammer rests on the board) you find that the effort arm is 10 inches long. The distance from the fulcrum to the nail is 1.6 inches.

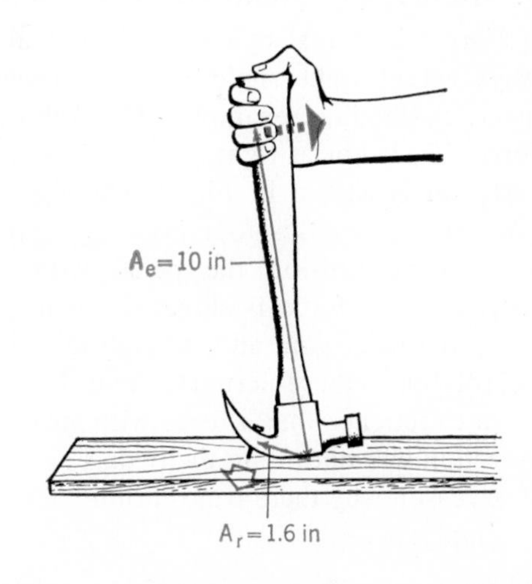

5-8 A hammer is a bent lever. But as long as the effort is applied at right angles to the effort arm A_e and the pull of the claw is at right angles to the nail, the mechanical advantage of this lever is still A_e/A_r.

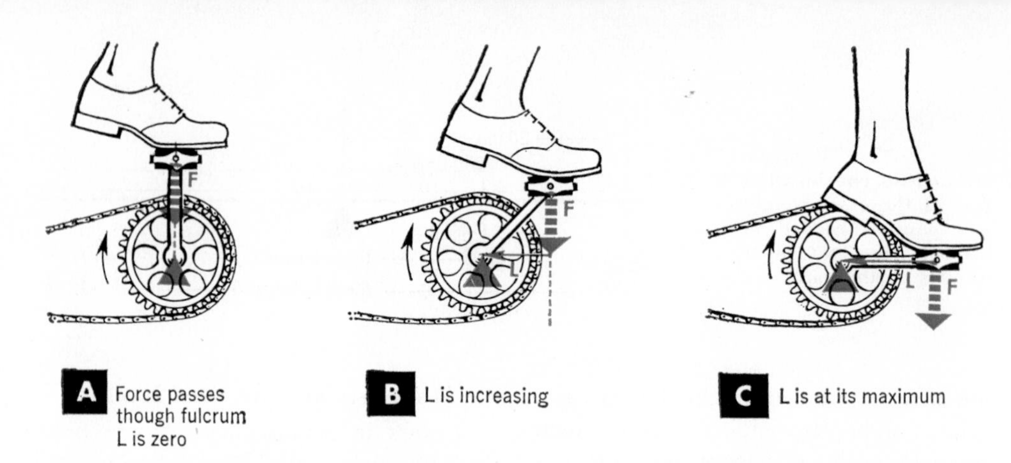

5-9 As the perpendicular distance, *L*, between the fulcrum and the effort force, *F*, grows larger, the moment exerted by the force, *F*, increases. The moment reaches its maximum value when the force, *F*, is exerted at right angles to *L*.

Applying the Principle of Moments:

$$\text{CCM} = \text{CM}$$
$$10 \text{ in} \times 40 \text{ lb} = 1.6 \text{ in} \times \text{P lb}$$
$$\text{P} = \frac{10 \text{ lb} \times 40 \text{ in}}{1.6 \text{ in}}$$
$$= 250 \text{ lb}$$

Even though the two arms of this lever do not lie in a straight line, the Principle of Moments still applies.

9. Moments of oblique forces

Forces in actual machines do not always act at right angles to their lever arms. A familiar example of an effort force that is not at right angles to its effort arm is shown in Fig. 5-9A. There you see the cyclist's foot pressing vertically downward on the pedal with a force *F*. The force is directed squarely along the pedal arm and through its axis of rotation (the fulcrum). You know from experience that a force with such a direction will not rotate the pedals at all. The reason why there is no turning effect is simply:

The turning effect or moment is always the product of the force and the perpendicular distance from the fulcrum to the line of action of the force.

In Fig. 5-9A this perpendicular distance is zero, and so our moment is zero and there is no turning effect.

In Fig. 5-9B the pedals have rotated so that the line of action of the force, *F*, is now at a distance, *L*, which is further from the fulcrum than before. You must know from experience that this arrangement of force and effort arm is effective in producing rotation.

The greatest possible moment is produced when the pedal arm is horizontal, for then, of course, the line of action of the force falls at the greatest possible perpendicular distance from the fulcrum, as in Fig. 5-9C.

The cable holding up the derrick boom in Fig. 5-10 is exerting a moment FL_1 around the point where the horizontal boom rests against the vertical mast. This counterclockwise moment is in balance with the clockwise moment, WL_2, created by the weight, *W*, acting downward at the end of the boom at distance L_2 from the same point. Hence the boom is supported motionless.

Perhaps you can see that the tension, *F*, on the supporting cable, which provides this torque or moment, must grow larger if its point of attachment to the mast is brought lower and L_1 grows shorter. Indeed, if the cable is nearly horizontal the tension in it necessary to keep FL_1 constant becomes too great and the cable will snap.

10. Equilibrium

We have seen that the bar in Fig. 5-7 will not start to rotate as long as the sum of the clockwise moments equals the sum of the counterclockwise moments. But it might be made to move in another way. For example, if the fulcrum, *F*, is being supported by a string, and you cut the string, the bar and weights would fall to the floor. Even while falling, the bar would not rotate because the sum of the clockwise and counterclockwise moments continue to be equal. Try it. Nevertheless, though the bar remains in balance it is no longer motionless.

Obviously the bar and weights fall to the floor because gravity is pulling downward on them. But the force of gravity was acting on them even while they were firmly supported by the string. Why didn't the force of gravity make them fall before you cut the string? The answer is that the tension of the string was exerting an upward force exactly equal to the downward force of gravity. You can easily verify this by tying the supporting string to a spring balance. You will find that the balance is pulling upward on the string with a force just equal to the weight of the bar and weights.

Similar experiments show that:

A motionless object will not start to move as long as the sum of the forces acting on it in one direction is equal to the sum of the forces acting on it in the opposite direction.

For example, when you weigh yourself on scales, the force of gravity pulls downward on you with a force exactly equal to the force exerted upward on you by

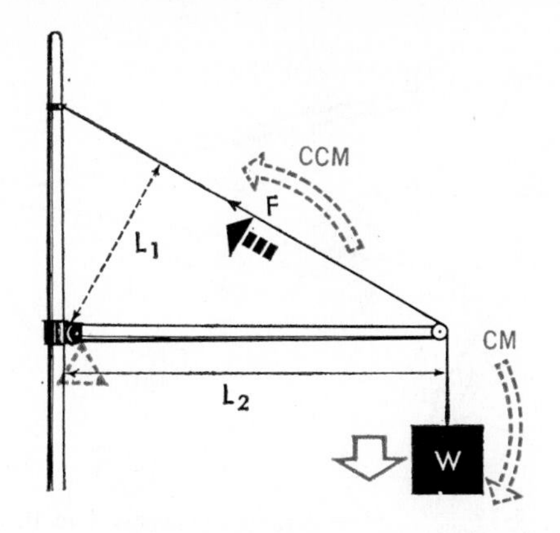

5-10 When the beam is in equilibrium the CW moment around the fulcrum must equal the CCW moment regardless of the directions of the actual forces.

the scales. Similarly a chair pushes upward on you with the same force with which gravity pulls you downward. If the chair should collapse, it would no longer exert its upward force on you. Only one force would then be acting on you, and hence you would begin to move downward.

In the tug of war shown in Fig. 5-11 the rope has many large forces exerted on it. Can you explain why the handkerchief does not move?

When an object is motionless—neither rotating nor moving from one place to another—it is said to be in **equilibrium.** We have just learned two important facts about objects that are in equilibrium. These two facts, taken together, can now be summarized in a more complete statement of the **Principles of Equilibrium,** which were first given in simple form on page 61 and earlier on this page:

When an object is in equilibrium:

1. The sum of the forces acting on it in one direction is equal to the sum of the forces acting on it in the opposite direction, and

2. The sum of the clockwise moments acting about a given fulcrum is equal to the sum of the counterclockwise moments acting about the same fulcrum.

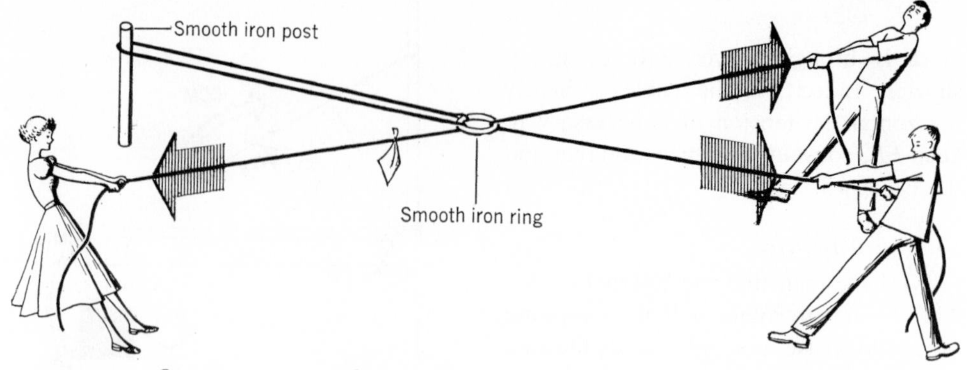

5-11 One person can easily win this kind of tug of war against two friends. Why?

All the motionless objects you see are in equilibrium. Look at a chair across the room and see if you can imagine just how the two principles apply to it. In a later chapter we shall see that moving objects can be in equilibrium too, provided they are moving in a straight line at constant speed.

Look around you, and see if you can find objects that are not in equilibrium —a buzzing fly, someone getting up from a chair, the swaying branches of a tree, an opening door.

SAMPLE PROBLEM

In Fig. 5-12 a 3,000-pound automobile is standing 20 feet from one end of a 100-foot bridge. How much of its weight is being carried by each of the supports at the ends of the bridge? To find out, we apply the two principles of equilibrium.

We begin by drawing an exact and careful diagram, showing on it all the forces and all the distances that we know or wish to know. We use letters for the unknown forces and distances. In such a diagram there is no need to draw a picture of the car; a block will do. But it is important to show *all* the forces and *all* the lengths, and it is important to draw them in neatly. A messy drawing frequently leads to mistakes.

Now we observe that the bridge and car are in equilibrium. Therefore we can be sure that the sum of all the clockwise moments equals the sum of all the counterclockwise moments. Any point can be chosen as a fulcrum. If we choose the left-hand end of the bridge as a fulcrum, we can write down the equation:

$$\begin{aligned} \text{CCM} &= \text{CM} \\ \text{B lb} \times 100 \text{ ft} &= 3{,}000 \text{ lb} \times 20 \text{ ft} \\ \text{B} &= \frac{3{,}000 \text{ lb} \times 20 \text{ ft}}{100 \text{ ft}} \\ &= 600 \text{ lb} \end{aligned}$$

Notice that because we chose end *A* as the fulcrum, force *A* has no moment. Its distance from the fulcrum is 0 feet, and hence the moment is

$$\text{A lb} \times 0 \text{ ft} = 0 \text{ ft-lb}$$

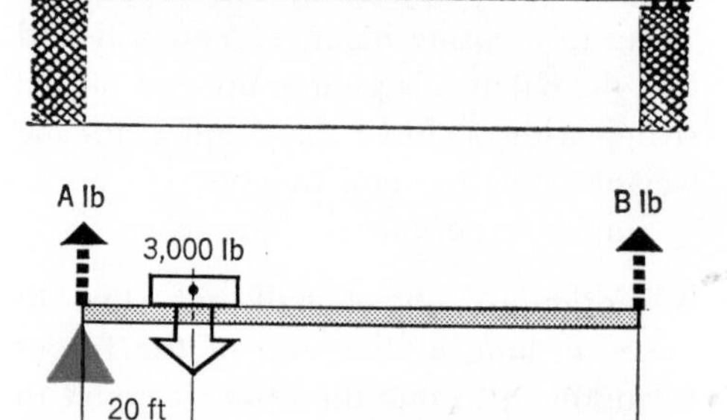

5-12 The principles of equilibrium enable you to find the load on each support quickly and easily, and to check your answers. Engineers use simple principles like these in the design of the most complex structures.

To find the force at A, we choose end B as the fulcrum. Now:

$$\text{CCM} = \text{CM}$$
$$3{,}000 \text{ lb} \times 80 \text{ ft} = A \text{ lb} \times 100 \text{ ft}$$
$$A = \frac{3{,}000 \text{ lb} \times 80 \text{ ft}}{100 \text{ ft}}$$
$$= 2{,}400 \text{ lb}$$

We can check our two answers by using the other principle of equilibrium. Since the bridge and car are in equilibrium we can be sure that the sum of all the upward forces equals the sum of all the downward forces. Hence we can write

$$\text{Sum of forces up} = \text{Sum of forces down}$$
$$A + B = \text{Weight of car}$$

Substituting we find:

$$2{,}400 \text{ lb} + 600 \text{ lb} = 3{,}000 \text{ lb}$$

This use of our two answers has led us to a statement which we can see is true. Hence we can have confidence in our answers.

STUDENT PROBLEM

A 5,000-lb truck is standing 30 ft from the north end of a 75-ft bridge. How much of its weight is carried by each of the supports at the ends of the bridge? (ANS. 2,000 lb, south end; 3,000 lb, north end)

Center of gravity

As you worked through the preceding sample problem of the car parked on the bridge, you may have been disturbed because we ignored the weight of the bridge. To make our problem more realistic we need to know how to take its weight into account.

11. The balancing point

It will help us to understand how to take the weight of the bridge into account if we do a simple experiment. Cut out an irregular cardboard shape like that shown in Fig. 5-13. Then puncture the cardboard at a number of different points: A, B, C, and D. Suspend the cardboard together with a plumb line (a weight on a string) from each of these points, as shown in the diagram. Be sure that the cardboard is free to swing back and forth. The plumb line gives us the vertical line from each point. Now draw these vertical lines on the cardboard from each of the points A, B, C, and D. The lines all meet at a point which we call the **center of gravity.** If you support the card by a pin through the center of gravity, you will find that the cardboard is perfectly balanced in any position in which it is placed. It acts as if all its weight were concentrated at this point (Fig. 5-14).

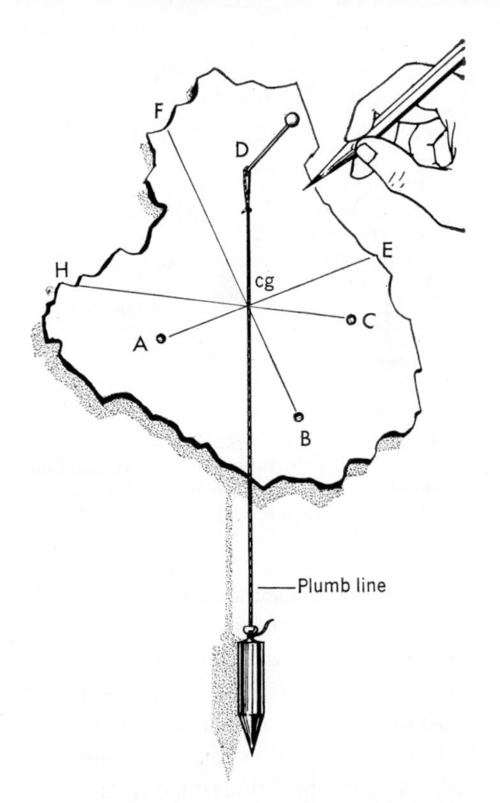

5-13 **Plumb lines from any point on the cardboard intersect at a point known as the center of gravity (cg). If you suspend the cardboard at the center of gravity, it will be in perfect balance and will not rotate at all.**

The center of gravity is of great im-

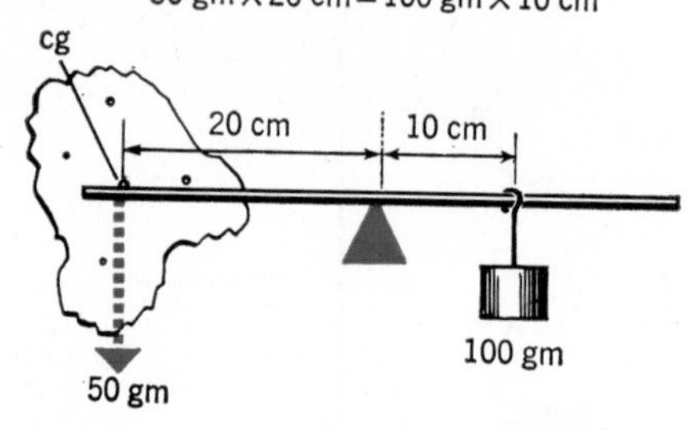

5-14 When an irregular object is being dealt with, all calculations using the principle of moments may be made by considering the weight as concentrated at the center of gravity.

portance to engineers designing automobiles, buildings, bridges, ships, planes, and other structures, for the weight of any object acts as if all of it were concentrated at its center of gravity.

For objects with a regular shape such as a box, or the bridge in our problem, the center of gravity is at the center of the object. For example, you already know that a ruler will balance on your finger if you support it at its center.

M. L. Dalen

5-15 A football player poised to resist a block. Why is he hard to tip over?

12. A center-of-gravity problem

Let us suppose that the bridge span in our problem on page 65 weighs 12,000 pounds. With the 3,000-pound car parked as before, what now will be the load supported at each end?

We draw once more a careful diagram, showing forces as before by means of arrows. This time we add a force of 12,000 pounds acting downward at the center of gravity of the bridge, to represent its weight (Fig. 5-16). Treating the left-hand end of the bridge as a fulcrum, we apply the Principle of Moments:

$$\text{CCM} = \text{CM}$$
$$B \text{ lb} \times 100 \text{ ft} = 3{,}000 \text{ lb} \times 20 \text{ ft} + 12{,}000 \text{ lb} \times 50 \text{ ft}$$
$$100\ B = 60{,}000 + 600{,}000$$
$$100\ B = 660{,}000$$
$$B = 6{,}600 \text{ lb}$$

Around the right-hand end as a fulcrum the moments are

$$\text{CCM} = \text{CM}$$
$$3{,}000 \text{ lb} \times 80 \text{ ft} + 12{,}000 \text{ lb} \times 50 \text{ ft} = A \text{ lb} \times 100 \text{ ft}$$
$$240{,}000 + 600{,}000 = 100\ A$$
$$840{,}000 = 100\ A$$
$$A = 8{,}400 \text{ lb}$$

Again we can check by applying the first principle of equilibrium:

$$\text{Sum of forces up} = \text{Sum of forces down}$$
$$A + B = \text{Weight of car} + \text{Weight of bridge}$$
$$8{,}400 \text{ lb} + 6{,}600 \text{ lb} = 3{,}000 \text{ lb} + 12{,}000 \text{ lb}$$

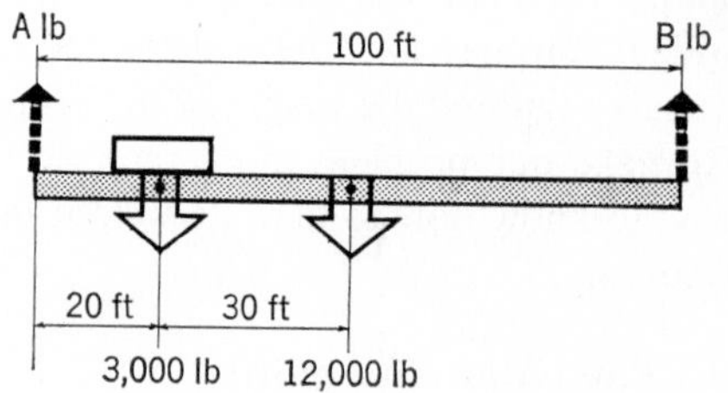

5-16 Here are shown the four forces acting on the bridge span when a car is parked near one end.

STUDENT PROBLEM

A 5,000-lb truck is standing 30 ft from the north end of a 75-ft bridge weighing 30 tons. How much weight is carried by each of the supports at the ends of the bridge? (ANS. 32,000 lb, south end; 33,000 lb, north end)

---

13. A more difficult center-of-gravity problem

Let us apply our principles to one more problem.

A 100-pound diving board 14 feet long, of uniform thickness, is supported by two beams 4 feet apart at one end. A 150-pound boy is about to dive off the end of the board farthest from the supports. A second boy weighing 120 pounds awaits his turn to dive, standing on the board midway between the two supports. How much force does each support exert on the diving board?

Draw a diagram like Fig. 5-17, showing all the forces and distances we know or wish to know. This will give you a clear understanding of the problem.

Since the diving board is motionless, it is in equilibrium, and we can apply both principles of equilibrium. Choosing the left-hand end of the board as a fulcrum:

$$\text{CCM} = \text{CM}$$
$$4 \text{ ft} \times F_2 \text{ lb} = 2 \text{ ft} \times 120 \text{ lb} + 7 \text{ ft} \times 100 \text{ lb} + 14 \text{ ft} \times 150 \text{ lb}$$
$$4F_2 = 240 + 700 + 2{,}100$$
$$4F_2 = 3{,}040$$
$$F_2 = 760 \text{ lb}$$

Choosing the other support as a fulcrum:

$$\text{CCM} = \text{CM}$$
$$4 \text{ ft} \times F_1 \text{ lb} + 2 \text{ ft} \times 120 \text{ lb} = 3 \text{ ft} \times 100 \text{ lb} + 10 \text{ ft} \times 150 \text{ lb}$$
$$4F_1 + 240 = 300 + 1{,}500$$
$$4F_1 = 1{,}560$$
$$F_1 = 390 \text{ lb}$$

Checking our answers by using the other principle of equilibrium:

$$\text{Forces up} = \text{Forces down}$$
$$F_2 = F_1 + 120 \text{ lb} + 100 \text{ lb} + 150 \text{ lb}$$
$$760 \text{ lb} = 390 \text{ lb} + 120 \text{ lb} + 100 \text{ lb} + 150 \text{ lb}$$
$$760 \text{ lb} = 760 \text{ lb}$$

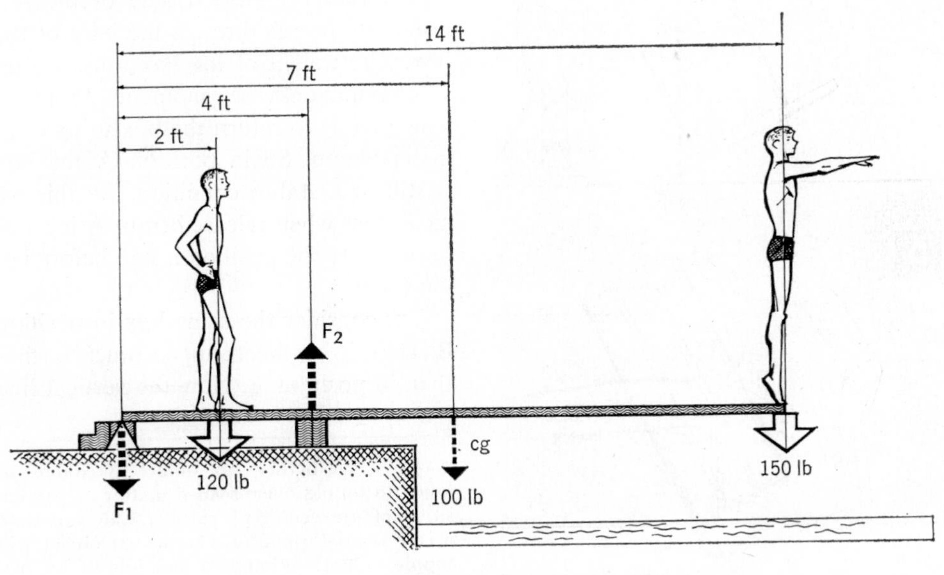

5-17 Forces on a diving board.

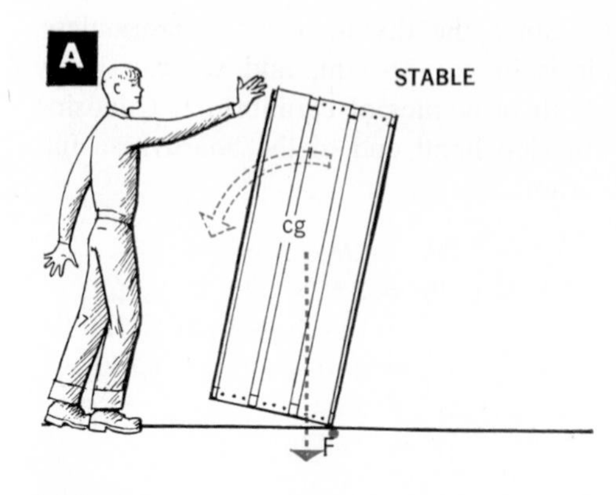

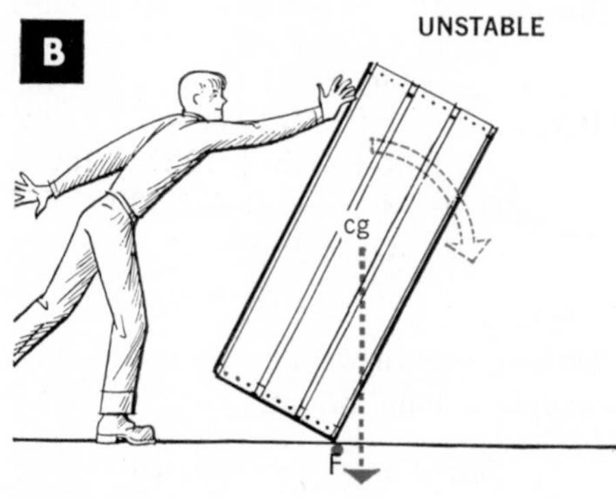

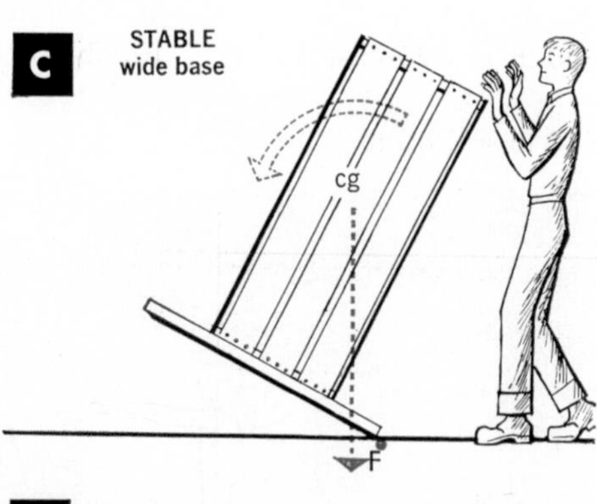

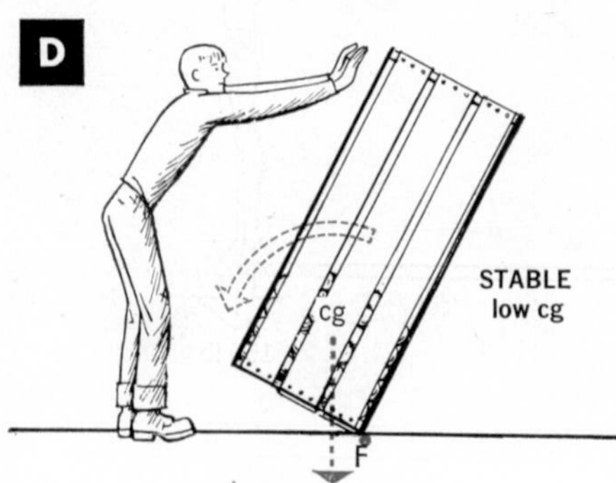

STUDENT PROBLEM

What would be the forces on the two diving-board supports in the sample problem just solved if the boy farthest from the supports weighed 200 lb? (ANS. $F_1 = 515$ lb; $F_2 = 935$ lb)

Stability

If you have ever stood up in a canoe, you know how easy it is to tip it over. But if you are sitting on the bottom of the canoe, the canoe is relatively steady. Why should your position make such a difference? The answer, again, involves center of gravity.

14. Why does an object topple over?

Examine the positions of the object shown in Fig. 5-18. In A the box has been tipped slightly. When released from this position, it tends to return to its original position with its base on the floor. In doing so it pivots about the edge at F. Hence we can consider F as a fulcrum. When we drop a vertical line from *cg,* the center of gravity, we find that it falls on the left side of the fulcrum and passes through the base of the box. The weight of the box thus sets up a counterclockwise moment about F which tends to return the box to its original position. So in position A the box is still in a **stable** condition. By this we mean that when released from A it tends to return to the position it had before being tilted.

But consider the same box in position B. Here it has been tipped much farther than in position A. Now the vertical line

5-18 Whether an object is stable or tends to topple over is all a matter of the position of the center of gravity. Can you state a rule for determining whether an object will topple or not? What two methods of increasing stability are shown here?

from the center of gravity falls on the other side of F (that is, outside the base). The turning moment here is clockwise and tends to topple the box. Thus we see that if we push the box to a point where the vertical line from the center of gravity falls outside the base, the box is **unstable.** When we release the box, it will not return to its original position but will topple over.

15. How to increase stability

In position C (Fig. 5-18) an extra board has been attached to the box to enlarge its base. This shifts pivot F to a point much farther out than before; the box will have to be pushed through a much greater angle before it topples. Hence we see that the larger base gives greatly increased stability.

We can also increase the stability of the box if we give it a lower center of gravity, as shown in position D. If the weight is concentrated at a lower point in the box, the box will remain stable in a position in which it was formerly unstable.

So we see that we can increase the stability of an object in two ways: by widening the base, and by lowering the center of gravity (Fig. 5-18C and D). A canoe tends to topple when you stand up in it because you are raising the center of gravity. On the other hand, the relative instability of a tall stool is due to the fact that it has a narrow base and a high center of gravity.

Both methods of increasing stability are used by designers of cars. Modern cars have wide bases; their weight is also concentrated at a low point so as to keep the center of gravity low. As a result they are far more stable and less likely to tip over on a turn than their old-fashioned ancestors.

In this chapter and in previous ones we have seen how the study of forces is of great importance in operating mechanical devices. But so far we have applied this knowledge only to solid objects. When we consider how forces are applied to fluids (liquids and gases), we shall discover a whole new set of facts which have the greatest practical use.

Bob Smallman, courtesy *Friends* Magazine

5-19 This acrobat may not know it, but he is obeying the laws of physics. Can you explain how?

Going Ahead

THE VOCABULARY OF PHYSICS

Define each of the following:

lever	unstable
effort arm	stable
fulcrum	moment
resistance arm	equilibrium
center of gravity	torque

THE PRINCIPLES WE USE

1. Describe two methods of finding the center of gravity of an irregular piece of cardboard.
2. State two ways of increasing the stability of an object. Give a practical application of each method.
3. State a rule by which one could predict whether or not an object would topple.
4. Mention five examples of the use of a lever.
5. An acrobat is balancing his body on one hand, as shown in Fig. 5-19. Why is his right elbow near the center of his body? Why are the fingers of his right hand spread slightly? Do you think he could balance himself without stretching out his left arm? Why?

PROBLEMS ON FORCES

In this problem section, as in many of the following chapters, the problems are of three levels of difficulty. The first group contains fairly easy problems; the second group is more challenging; and the third group makes the utmost demand on your ability. The three groups are divided by a horizontal line.

1. Find the unknown quantity for each of the machines shown in Fig. 5-20.

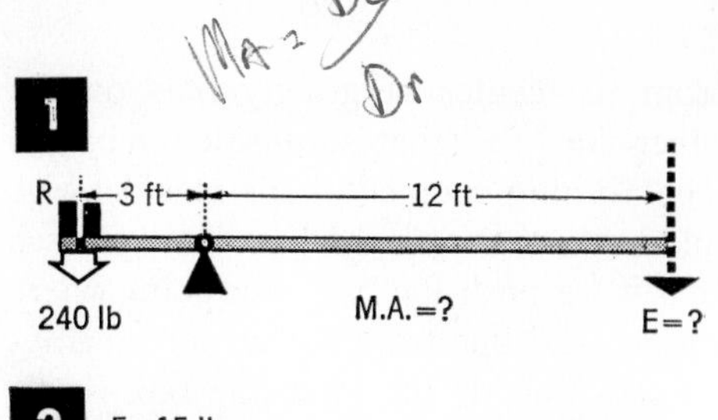

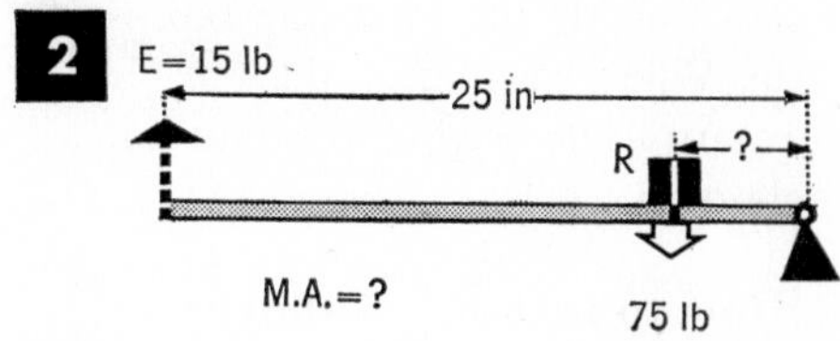

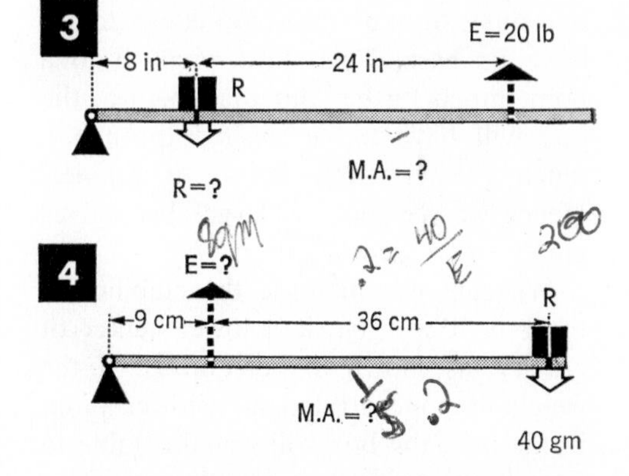

5-20 **Problems to solve. Do not mark the book in any way.**

2. Calculate the missing information in the table at the foot of the page. (Assume the lever to be weightless.)
3. Two boys find they can balance each other on a plank if the heavier boy, weighing 150 lb, sits 4 ft from the fulcrum and the lighter boy sits 6 ft from the fulcrum. How much does the lighter boy weigh?

Machine	*Effort Arm*	*Resistance Arm*	*M.A.*	*Effort Force*	*Resistance Force*	*Effort Distance*	*Resistance Distance*
(a) Pliers	6 in	0.5 in	?	10 lb	?	4 in	?
(b) Scissors	2 in	6 in	?	3 lb	?	1 in	?
(c) Nutcracker	15 cm	3 cm	?	500 gm	?	?	2 cm
(d) Tweezers	3 cm	15 cm	?	?	500 gm	4 cm	?
(e) Wheelbarrow	8 ft	?	10	?	150 lb	?	2 ft

4. If you place a 100-lb load in a wheelbarrow 18 in to the rear of the axle, how much force must you apply to lift the handles of the wheelbarrow if they are 3 ft to the rear of the axle?
5. To weigh a large boulder you insert under it one end of a thick, wide plank 2 m long. When you stand on the other end of the plank you find you can just barely balance the boulder when the fulcrum is 20 cm from the boulder. If you weigh 50 kg, what is the weight of the boulder?
6. A force of 25 kg is exerted on the end of a lever whose total length is 2.0 m and whose M.A. is 4. The fulcrum is at the other end of the lever. (a) How long is the resistance arm? (b) When the effort moves through 4 cm, how far does the resistance move?
7. A force of 80 lb is being exerted on the end of an 8-ft lever whose M.A. is 3. (a) How long is the effort arm? (b) When the effort moves through 6 in, how far does the resistance move?
8. A girl is using a screwdriver to pry up the lid of a paint can. She has placed the tip of the screwdriver under the edge of the lid, and using the edge of the can as a fulcrum 2 mm from the tip, she pushes down on the handle of the screwdriver with a force of 600 gm. The screwdriver is 18.4 cm long. (a) What is the upward force she brings to bear on the lid of the can? (b) What is the M.A. of the screwdriver in this case? (c) When she moves her hand down 1 cm, how far does she raise the lid?

---

9. A 130-lb boy is standing 3 ft from one end of an 8-ft plank whose ends are supported by sawhorses. How much force is exerted on each sawhorse? (Assume the plank to be weightless.)
10. A meter stick weighing 220 gm is lying on the floor. (a) How much force must you exert on one end of it to raise the end slightly off the floor? (b) What is the M.A. of this lever arrangement?
11. A tapered pole 3 yd long balances at a point 1 yd from the thick end. Where is its center of gravity?
12. When a 6-kg tapered pole 2 m long is supported at its midpoint, a weight of 3 kg must be hung on the thin end in order to balance it. Where is the center of gravity of the pole?
13. An 8-m tapered beam weighing 200 kg is lying flat on the ground. Its center of gravity is 2 m from one end. How much force must be applied vertically upward (a) to the heavy end in order to lift it just off the ground, (b) to the light end in order to lift it just off the ground?
14. A force is exerted at the end of a 1.5-m lever whose M.A. is 5. How long is the effort arm? Give two possible answers, illustrating each with a sketch.
15. A 3,000-lb car rests on a 60-ft bridge. The car's center of gravity is 12 ft from one end. Find the upward force exerted on each end of the bridge by its supports. Assume the bridge itself to be weightless.
16. Solve Problem 15 again, assuming that the bridge is uniform and weighs 10 tons.
17. A 150-ft bridge weighing 800 tons has its center of gravity 60 ft from the north end. When a loaded railroad car rests on the bridge, the north abutment supports 500 tons and the south abutment supports 400 tons. (a) How much does the railroad car weigh? (b) Where is it standing?
18. A plank 4 m long weighs 15 kg. A boy finds that the plank will balance if it is supported by a fulcrum 40 cm from the end he is standing on. How much does the boy weigh?
19. Find the value of X in the figure just below.

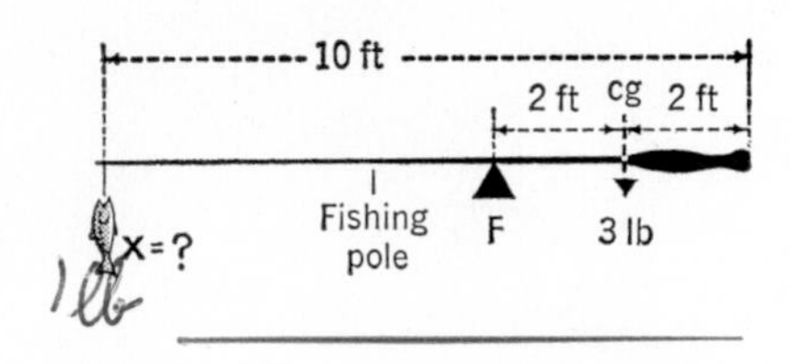

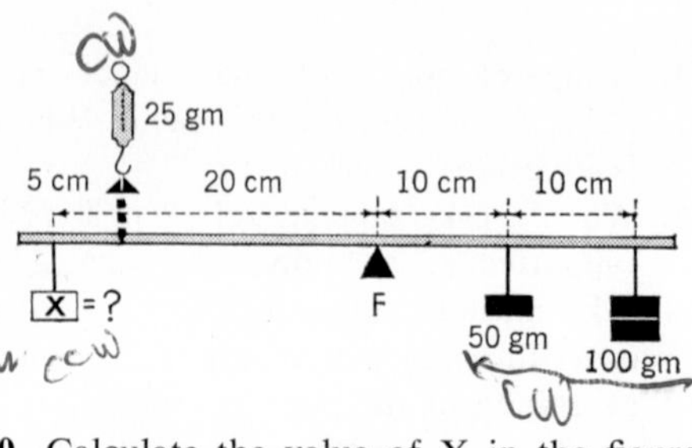

20. Calculate the value of X in the figure above.

21. To raise one end of a heavy crate a boy used as a lever a uniform 14-ft board weighing 25 lb. With the tip of the board under the crate and a wooden block as a fulcrum 1 ft from the crate, the boy found that he could just raise the crate when he put all his 130-lb weight on the end of the board. How great was his lifting force?

22. A telescoping steel pole is made of three sections of tubing, each of a different diameter. The first section is 3 ft long and weighs 12 lb; the second section is 4 ft long and weighs 16 lb; the third section is 5 ft long and weighs 20 lb. Each section is uniform. Where will the pole balance if suspended horizontally?

23. A 5-kg pole 4 m long has its center of gravity 1.5 m from one end. A 10-kg weight is hung on the pole 1.0 m from the heavy end and another 10-kg weight 3 m from the heavy end. If the pole is supported from its midpoint, where must a third 10-kg weight be placed in order to balance it?

24. A meter stick balanced at its center has a 500-gm weight hanging from the 10-cm mark and a 300-gm weight hanging from the 20-cm mark. How much weight must be hung from the 80-cm mark in order to balance it?

25. A meter stick supported at its center has a 500-gm weight hanging from the 10-cm mark and a 300-gm weight hanging from the 80-cm mark. Where must a 400-gm weight be hung in order to balance it?

26. Will the answer to Problem 25 be changed if you assume that the meter stick weighs 250 gm? If not, explain why; if so, compute the new answer.

27. By pulling horizontally on a claw hammer's vertical handle you pull a nail vertically out of a board. Your hand is 10 in above the board and exerts a 20-lb pull. The nail is 1 in from the point where the hammer rests on the board. What is the force exerted on the nail?

28. What is the minimum force needed to raise one edge of a 50-lb cube? Where and in what direction must it be applied?

29. How much work is done in lifting one end of a 150-gm horizontal meter stick until it is vertical?

30. A horizontal beam weighs 50 lb and a tie rope supporting one end makes an angle of 45° with the beam. What is the tension in the tie rope?

31. A tapering telephone pole balances at a point 24 ft from the thicker end. When a 200-lb man stands on the thicker end, the pole balances 20 ft from the thicker end. How heavy is the pole?

32. A girl catches a 2-lb fish with an 8-ft tapered pole weighing 3 lb whose center of gravity is 1.5 ft from the end. She is holding the pole horizontally with her right hand at the thick end and her left hand 8 inches from that end. What force is exerted by each hand?

CHAPTER

6

Applying the Principle of Work

We have already learned that the study of machines includes such simple devices as pulleys and levers. In this chapter we shall examine several more simple machines that we find about us at home, in school, on the street, and even in our own bodies.

Wheel and axle

The lever has one disadvantage as compared to the pulley system: with a lever the resistance (weight to be lifted) cannot be moved very far. Often the weight cannot be lifted more than a few inches before the effort end touches the ground and is stopped.

But suppose we construct what we might call a "spinning lever." If we turn a lever completely around its pivot, or fulcrum, we find that both the resistance arm and the effort arm make complete circles. Suppose we build solid wheels the same size as the two circles, with the radius of one equal to A_e and the radius of the other equal to A_r (Fig. 6-1). Then we attach them rigidly together, so that A_e and A_r rotate around the same center or pivot.

Now instead of applying our effort and resistance directly to the lever, we apply them by means of ropes that are tied to the rims of the wheels and wound around them. We use separate ropes for the effort and for the resistance. In a machine of this type, a **wheel and axle**, the ropes always keep the same distance from the pivot as they wind up on the wheel and the axle (Fig. 6-1). Thus the effort arm, A_e, is always fixed in size and, when the arrangement is like that in Fig. 6-1, is equal to the radius of the large wheel. The resistance arm, A_r, is also fixed in size, and is equal to the radius of the smaller wheel. Moreover, the forces are always applied at right angles to the two radii.

1. M.A. of the wheel and axle

If we think of the wheel and axle as a kind of lever, we can find its M.A. thus:

$$\text{M.A.} = \frac{A_e}{A_r}$$

or

$$\text{M.A.} = \frac{\text{Radius of effort wheel}}{\text{Radius of resistance wheel}}$$

The diameter of each wheel is equal to twice the radius. Hence the ratio of the diameters will be the same as the ratio of the radii. Thus:

$$\text{M.A.} = \frac{\text{Diameter of effort wheel}}{\text{Diameter of resistance wheel}} = \frac{D}{d}$$

If you examine Fig. 6-1, you see that as the effort rope is pulled, it unwinds from the wheel with the large diameter. At the same time the smaller axle winds up the resistance rope and lifts the weight. Since the effort wheel is larger, more rope will be pulled in on the effort end than will be wound up by the smaller axle to which the resistance is fastened. Hence we see again as in all machines that a gain in force is accompanied by a loss of distance.

In many practical wheel-and-axle combinations, no ropes are used. The effort can be applied directly to the wheel either by friction (as in the case of the doorknob), or by means of a crank handle (as in the case of the windlass shown in Fig. 6-1). The handles of faucets and the knobs on radio sets are other examples of this particular arrangement of the wheel and axle. Can you name still other examples?

The wheel and axle is also used in combination with other machines. Examples: the brace and bit for drilling holes, the meat grinder, one kind of auto jack, the steering wheel of a car, the wrench, the screwdriver, the corkscrew, and the vise.

2. The wheel and axle in reverse

We can also use this machine in reverse. The effort is then applied to the axle to cause the larger wheel which supports the resistance to turn. The effort applied is now greater than the resistance, but the small motion of the effort results in a larger motion of the resistance. The rear wheel of a bicycle is a good example of a wheel and axle used to increase speed at the expense of force (Fig. 6-2).

SAMPLE PROBLEM

A wheel and axle has an effort-wheel diameter of 3 feet and a resistance-wheel diameter of 6 inches. (a) What effort force is needed to lift a weight of 300 pounds? (b) If the weight is lifted 20 feet, how far will the effort have to move?

(a) Using the formula for the M.A. of a wheel and axle:

$$\text{M.A.} = \frac{D}{d} = \frac{3}{0.5} = 6$$

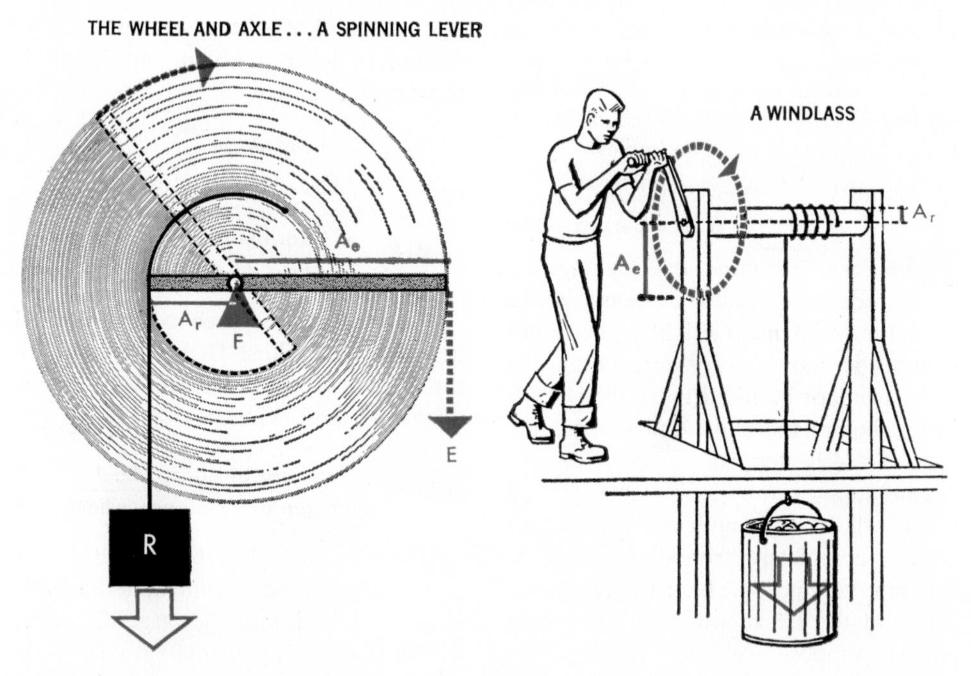

6-1 One use of the wheel and axle, and a diagram of the forces around the fulcrum.

6-2 The rear wheel of a bicycle is a wheel and axle used to gain distance. The force applied by the chain on the small rear sprocket wheel causes the large wheel to turn and move the bicycle. What kind of M.A. has it?

Therefore

$$\text{M.A.} = \frac{R}{E}$$

$$6 = \frac{300}{E}$$

$$E = \frac{300}{6} = 50 \text{ lb}$$

(b) Since for any machine the ideal M.A. equals D_e/D_r:

$$\text{M.A.} = \frac{D_e}{D_r}$$

$$6 = \frac{D_e}{20}$$

$$D_e = 6 \times 20 = 120 \text{ ft}$$

STUDENT PROBLEM

A wheel and axle has a large diameter (wheel) of 48 in and a small diameter (axle) of 8 in. (a) What effort is needed to lift a 50-lb load? (b) If the effort is exerted through 20 ft, how far will the load be lifted? [ANS. (a) 8.33 lb, (b) 3.33 ft]

The inclined plane

Would it be easier to raise a weight by carrying it up a vertical ladder or by hauling it up a gradual incline? Let us analyze the situation by means of an experiment.

3. A sloping board

Figure 6-3 shows a ramp made from a long board resting on a box. This is an **inclined plane.** We want to get a weight to the top of the box; to reduce friction we put it in a rolling cart. Let us assume that the total resistance (weight of load plus cart) is 100 pounds. We call the resistance 100 pounds because this is the force we would have to exert if we had no inclined plane. The length of the board is 8 feet, and the height of the box is 2 feet. If we pull the weight by means of a rope and measure the effort with a spring balance, we find that it takes only about 25 pounds of effort to get the 100-pound resistance to the top of

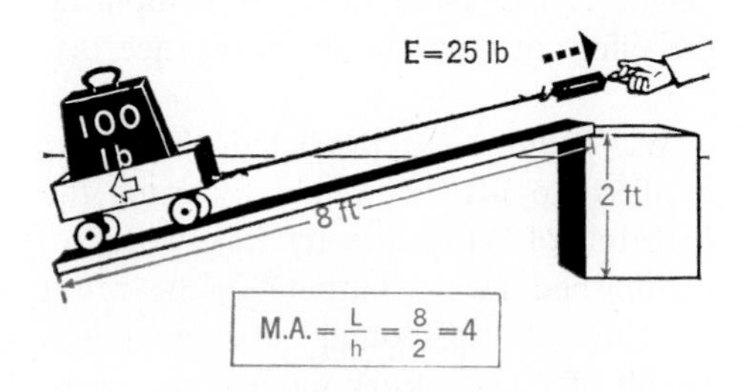

6-3 The inclined plane offers mechanical advantage.

the incline. Thus the M.A. of this arrangement would be:

$$\text{M.A.} = \frac{R}{E} = \frac{100 \text{ lb}}{25 \text{ lb}} = 4$$

Notice that our M.A., 4, is equal to the number of times the length of the board, 8 feet, is greater than the height, 2 feet. This suggests that we might find the M.A. of an inclined plane by merely comparing the length of the incline with the height. Or, in this case:

$$\text{M.A.} = \frac{L}{h} = \frac{8 \text{ ft}}{2 \text{ ft}} = 4$$

Experiments with boards of different lengths and heights show that we can always find the M.A. in this manner.

4. The Principle of Work

Without the aid of the inclined plane, we would have to lift the 100-pound weight straight up to a height of 2 feet to get it to the top of the box. Instead of this, we pull the weight up the incline a distance of 8 feet. Our effort distance, D_e, is 8 feet, while the resistance distance, D_r, is only 2 feet. Our M.A., therefore, is:

$$\frac{L}{h} = \frac{D_e}{D_r} = \frac{8 \text{ ft}}{2 \text{ ft}} = 4$$

So the length of the incline is really the effort distance while the height is the resistance distance. In other words, comparing the length of the incline to the height is the same thing as comparing the effort distance to the resistance distance.

We could have predicted the effort required to lift the weight by using the Principle of Work directly. The work accomplished (work output) is the result of lifting a 100-pound resistance to a height of 2 feet. Work input is the result of pulling the effort a distance of 8 feet along the incline. The Principle of Work tells us that work output is equal to work input. Hence:

$$\text{Work output} = \text{Work input}$$
(neglecting friction)
$$RD_r = ED_e$$
$$100 \text{ lb} \times 2 \text{ ft} = E \text{ lb} \times 8 \text{ ft}$$
$$8E = 200$$
$$E = 25 \text{ lb}$$

The effort is 25 pounds, or $\frac{1}{4}$ of the 100-pound resistance.

You can see why it is easier to roll bricks up an incline in a wheelbarrow than to lift them straight up. The effort needed will be much less. On the other hand, while there is a saving in effort, there is an increase in the distance (the length of the ramp) that the bricks must be moved.

5. Increasing the M.A. for inclined planes

Consider what would happen to the M.A. in our experiment above if we made the length of the board 20 feet instead of 8 feet, while keeping the height of 2 feet the same. In that situation:

$$\text{M.A.} = \frac{L}{h} = \frac{20 \text{ ft}}{2 \text{ ft}} = 10$$

With an M.A. of 10, the effort required to lift the weight would be only $\frac{1}{10}$ of 100 pounds—or 10 pounds. Suppose we make the incline even more gradual by lengthening it to 200 feet, at the same time keeping our height of 2 feet. In this situation:

$$\text{M.A.} = \frac{L}{h} = \frac{200}{2} = 100$$

With an M.A. of 100, we now need only 1 pound of effort to lift our 100-pound weight. But this effort must be exerted along a length of 200 feet to raise the weight to the height of 2 feet.

Thus we see that the steeper the slope of the inclined plane, the less the M.A.

Ewing Galloway

6-4 **Road builders use the principle of the inclined plane.**

The more gradual the slope, the greater the M.A.

6. Uses of the inclined plane

The inclined plane is most commonly used for roads, ramps, and stairs. It takes far less effort to ascend a gradual incline to a high place than to climb straight up. Roads in steep mountain country are usually built to wind back and forth so as to reduce the grade, or steepness (Fig. 6-4). Few roads have a grade of more than 10 per cent. When we say that a road has a 10 per cent grade, we mean that it rises 10 feet for every 100 feet of road length. For such a road the M.A. would be 10. In other words, the tires of the car must push with a force of only $\frac{1}{10}$ of the weight of the car to get to the top of the hill.

Railroad grades are even less steep, seldom more than 3 per cent. With this grade the road rises only 3 feet in 100 feet. The M.A. is then about 33.

Inclined planes are also useful in rolling heavy weights onto platforms or into trucks. If you have watched the construction of buildings, you know how useful long, gradual inclines are to workmen who have to get wheelbarrows full of bricks and other materials up to certain heights.

SAMPLE PROBLEM

A trunk weighing 80 pounds is slid up an inclined plane 12 feet long into a truck 3 feet high. How much effort is required?

We must first find the M.A. of the plane from the formula:

$$\text{M.A.} = \frac{L}{h} = \frac{12}{3} = 4$$

Then, since

$$\text{M.A.} = \frac{R}{E}$$

$$4 = \frac{80}{E}$$

$$E = \frac{80}{4} = 20 \text{ lb}$$

Or even more briefly, combining these two steps,

$$\frac{L}{h} = \frac{R}{E}$$

$$\frac{12}{3} = \frac{80}{E}$$

$$E = 20 \text{ lb}$$

STUDENT PROBLEM

A road up a hill rises 10 ft for every 100 ft of length. How much force is needed to move a 4,000-lb automobile up the hill? (ANS. 400 lb)

---

7. The wedge, a moving incline

Just as a wheel and axle is a moving lever, a wedge is a moving inclined plane. When you chop down a tree, you use an ax. When you cut bread, you use a knife. Strange as it may seem, both these tools are forms of the inclined plane.

In Fig. 6-5 a wedge 6 inches long and 1 inch high is being pushed under a beam which presses down on it with a force of 600 pounds. When the wedge is pushed under the beam a distance of 6 inches (the effort distance), the beam rises on the wedge to a height of 1 inch (the resistance distance). Hence the M.A. is

$$\frac{D_e}{D_r} = \frac{6}{1} = 6$$

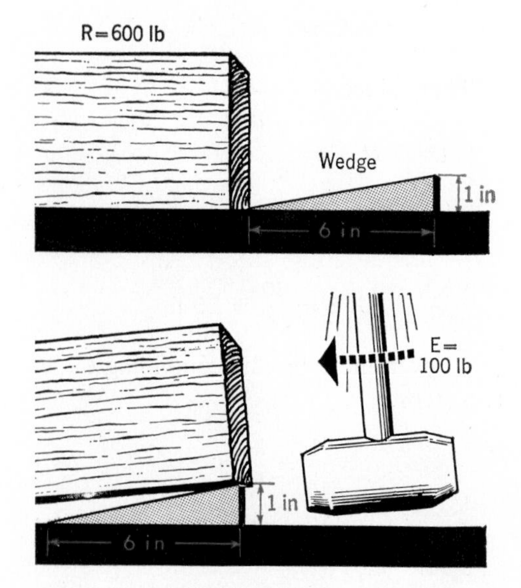

6-5 **A wedge is a moving inclined plane.**

Ideally, a force of 100 pounds is sufficient to raise the 600-pound beam. The effort is multiplied 6 times as a result of the ideal M.A. of 6. In actual practice a job of this kind involves a great deal of friction, and therefore the actual effort required is considerably more than 100 pounds.

The **wedge,** then, is an inclined plane moving under a stationary resistance. In the case of the ramp, by contrast, the inclined plane is stationary while the resistance moves along it.

Cutting tools are among the most important applications of the wedge. Axes, chisels, knives, and planes all have blades that taper to a sharp edge. The two sides of the material to be cut are pushed apart by the sharp edge in exactly the same way that a beam is lifted from the base on which it rests. The sharper the edge, the more gradual is the wedge and the greater the mechanical advantage. (The edge cannot be too sharp, however, or it will break.)

Figure 6-6 shows two interesting uses of inclined planes for clearing roads of snow.

Simple machines in combination

Up to this point we have considered machines as separate units. But if two machines taken separately give us a certain mechanical advantage, we would expect to get a still greater mechanical advantage by combining them. This is exactly what happens.

Suppose you have the problem of lifting a 1,000-pound safe to a height of 5 feet. You have an inclined plane 20 feet long. If you depend on this alone to help you with the job, your M.A. will be:

$$\text{M.A.} = \frac{L}{h} = \frac{20 \text{ ft}}{5 \text{ ft}} = 4$$

Standard Oil Co. (N.J.)

6-6 Two uses of the inclined plane. The rotating screw at the left, which is a spiral incline, pushes snow away. At the upper right a wedge (another inclined plane) forces a path through the snow.

In other words, the job will require an effort of $\frac{1}{4}$ of 1,000 pounds, or 250 pounds. But since this is much too great an effort for one man to exert conveniently, you combine your inclined plane with a pulley system, as shown in Fig. 6-7. Now you have 5 ropes to pull the weight up the incline. Hence the M.A. of your pulley system is 5, and the effort required to pull the safe up the incline is now only $\frac{1}{5}$ of 250 pounds, or 50 pounds.

8. M.A. of combined machines

If we can raise our 1,000-pound safe with only 50 pounds of effort, what is the M.A. of our combined incline and pulley system? We find it thus:

$$\text{M.A.} = \frac{\text{R}}{\text{E}} = \frac{1{,}000 \text{ lb}}{50 \text{ lb}} = 20$$

Notice that this new M.A., 20, is what we would get by multiplying the individual M.A.'s of the incline and the pulley system (4×5). Using the inclined plane alone, we need an effort $\frac{1}{4}$ as great as the resistance. Given the additional help of a pulley system with an M.A. of 5, the effort we now need is $\frac{1}{5}$ of this $\frac{1}{4}$—or $\frac{1}{20}$ of the resistance. We always find the combined M.A. of two machines by multiplying their separate M.A.'s.

As we already know from the Principle of Work, we pay for a saving in effort by an increase in distance. In Fig. 6-7 an effort of 50 pounds must be exerted through a distance of 20 feet to raise a 1,000-pound weight to a height of 5 feet. This means that each of the 5 ropes of the pulley system must be shortened 20 feet. At the effort end we must pull in $5 \times 20 = 100$ feet of rope to lift the weight 5 feet. Although we have increased our M.A. to 20, the effort has to move 20 times as far. Thus work input is equal to work output.

9. M.A. of the screw

Have you ever noticed how a mechanic tightens a bolt on a car? He takes great care not to turn the wrench too hard. If he did, he would break off the head of the bolt. It is amazing that it should be so easy to break a piece of iron a quarter of an inch thick. The fact that he can shows that the machine he is using has a very high M.A.

Our mechanic is using a **screw**, a combination of two machines: the inclined plane and the wheel and axle. When he fits his wrench on the head of the bolt and turns it, he is using the wrench as a large wheel to turn the bolt, which is the axle. The hand which holds the wrench moves through a circle of large diameter while the bolt itself moves in a much smaller circle. In addition, the threads on the bolt act as a kind of spiral staircase (inclined plane) which slowly advances the bolt into the nut (Fig. 6-9).

Figure 6-8 shows why we can think of the spiral threads on the bolt as an inclined plane. As the bolt is made to rotate one complete turn, it advances into the nut a distance equal to the space between the threads. This distance between threads is called the **pitch** (Fig. 6-8).

To find the M.A. of this combination of machines, we must know the length of the wrench and the pitch of the screw. For example, imagine that Fig. 6-9 shows a wrench 8 inches long turning a bolt which has 16 threads to the inch. In this case the pitch of the screw—the distance between threads—is $\frac{1}{16}$ of an inch.

When the mechanic makes one complete turn of the wrench, his effort distance is equal to the circumference of the circle his hand describes. If the wrench is 8 inches long, the radius of this circle will be 8 inches. The circumference is about 50.2 inches ($c = 2\pi r = 2 \times 3.14 \times 8 = 50.2$ in). The effort distance, D_e, is thus about 50 inches.

The resistance distance, D_r, is the distance the bolt advances during one turn of the wrench. In other words, it is equal to the pitch of the screw—in this example, $\frac{1}{16}$ inch. We can find the M.A. by comparing effort distance and resistance distance as follows:

$$\text{M.A.} = \frac{\text{Effort distance } (D_e)}{\text{Resistance distance } (D_r)} = \frac{\text{Circumference of wrench circle } (2\pi r)}{\text{Pitch of screw } (p)}$$

$$\text{M.A.} = \frac{D_e}{D_r} = \frac{2\pi r}{p}$$

$$\text{M.A.} = \frac{50 \text{ in}}{\frac{1}{16} \text{ in}} = 50 \times 16 = 800$$

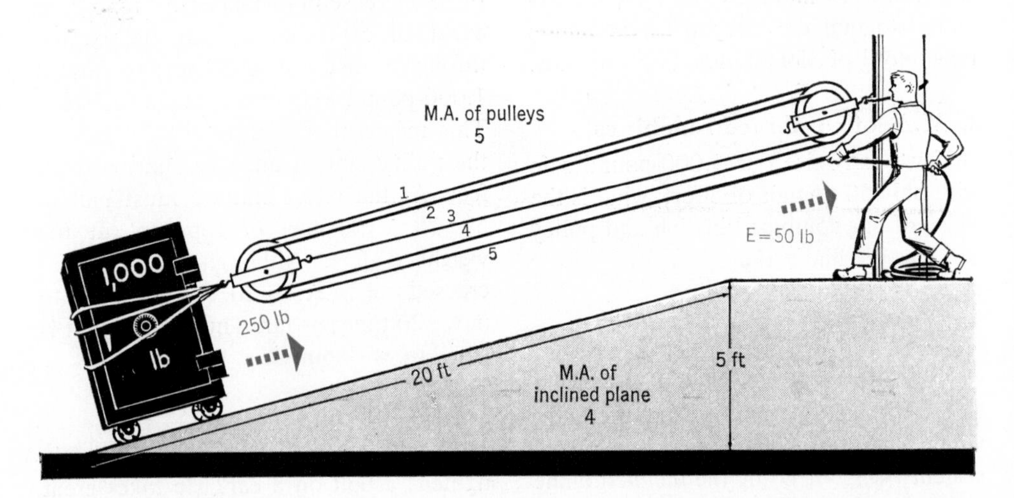

6-7 **This combination of machines gives an M.A. of 20. Can you figure out why?**

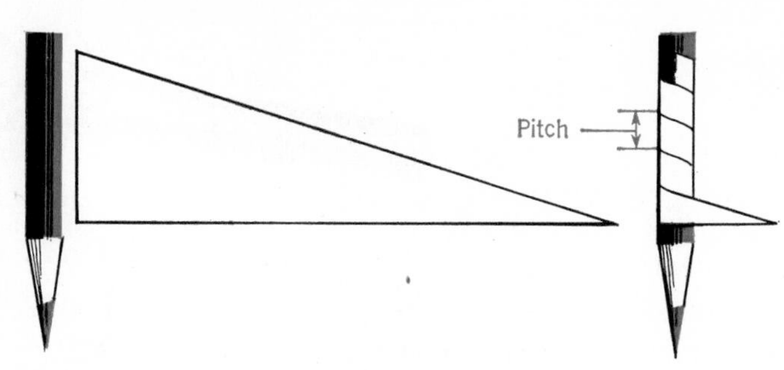

6-8 **The screw is a spiral inclined plane.**

With this particular wrench and bolt we have the enormous M.A. of 800! If our mechanic exerts 50 pounds of effort, his torque creates a tightening force approximately 800×50 pounds, or 40,000 pounds. Such a force as this may easily exceed the force of molecular cohesion in the bolt. Small wonder he has to go easy even though the bolt is made of solid iron!

The screw is one of the most useful machines we have. Its most important use is to fasten parts of a mechanism together. An automobile contains hundreds of bolts and nuts which hold its various parts together with enormous force. But great as this holding force is, the car can be easily taken apart with wrenches. Sometimes a screwdriver is used to turn the bolt. The handle of the screwdriver then turns in a circle just as the handle of the wrench does.

A clamp or vise makes use of the principle of the screw. So does a meat grinder. The same principle operates in corkscrews, in faucets, in some automobile jacks, and in worm gears (Fig. 6-10).

SAMPLE PROBLEM

A mechanic uses a 1-foot wrench to turn a bolt having 20 threads to the inch. With what force does the bolt hold if an effort of 40 pounds is applied to the wrench?

We find the mechanical advantage of the screw by the formula, being careful to change 1 foot to 12 inches so that the units agree.

$$\text{M.A.} = \frac{2\pi r}{p} = \frac{2 \times 3.14 \times 12}{\frac{1}{20}} = 1{,}510$$

Thus, to an accuracy of three significant figures, we have

$$\text{M.A.} = \frac{R}{E}$$

$$1{,}510 = \frac{R}{40}$$

$$R = 1{,}510 \times 40 = 60{,}400 \text{ lb}$$

Or even more briefly, combining these two steps

$$\text{M.A.} = \frac{2\pi r}{p} = \frac{R}{E}$$

$$\frac{2 \times 3.14 \times 12}{\frac{1}{20}} = \frac{R}{40}$$

$$R = 60{,}400 \text{ lb}$$

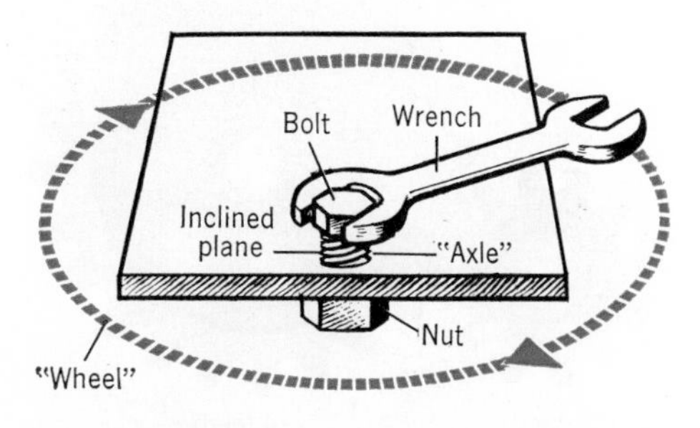

6-9 **Here is a combination of wheel and axle and spiral inclined plane with an M.A. well over 100!**

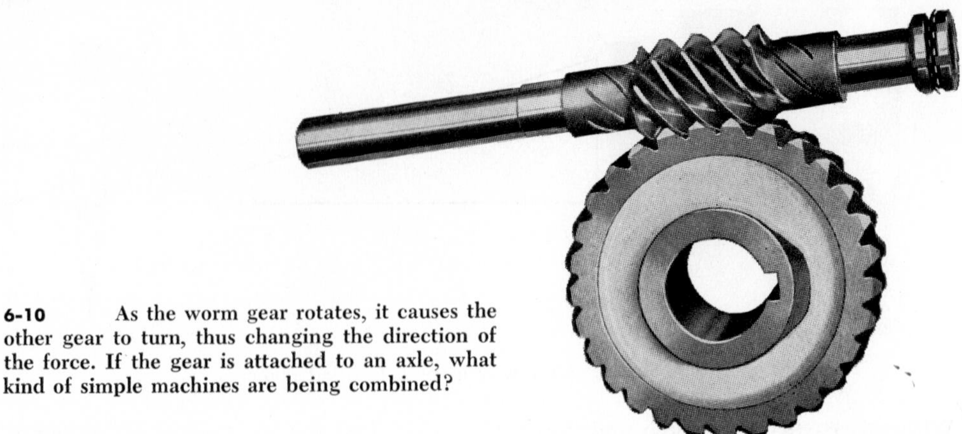

6-10 As the worm gear rotates, it causes the other gear to turn, thus changing the direction of the force. If the gear is attached to an axle, what kind of simple machines are being combined?

Cleveland Worm and Gear Co.

STUDENT PROBLEM

How much force can be exerted by a bolt if it is tightened by a force of 20 lb applied to a wrench 6 in long, and the screw has 10 threads to the inch? (ANS. 7,540 lb)

Gears

Most complex mechanisms contain gears. Automobile and airplane engines are familiar examples. But gears are also necessary to such comparatively simple devices as the hand drill and the egg beater.

Figure 6-11 shows us how one arrangement of gears operates. Here a large gear of 24 teeth meshes with a smaller one of 12 teeth. Every time the large gear makes one complete revolution the smaller gear must make two. Hence the smaller operates at twice the speed of the larger. Gears can be combined to produce any desired M.A. Two gears of different diameters fixed to the same shaft may be considered to be another form of the wheel and axle (Fig. 6-12).

Clocks are the most common example of the usefulness of gears. Inspection of the inside of a watch or a clock shows how gears are used in a timepiece. A train of gears keeps the second hand rotating 60 times as fast as the minute hand, and the minute hand 12 times as fast as the hour hand.

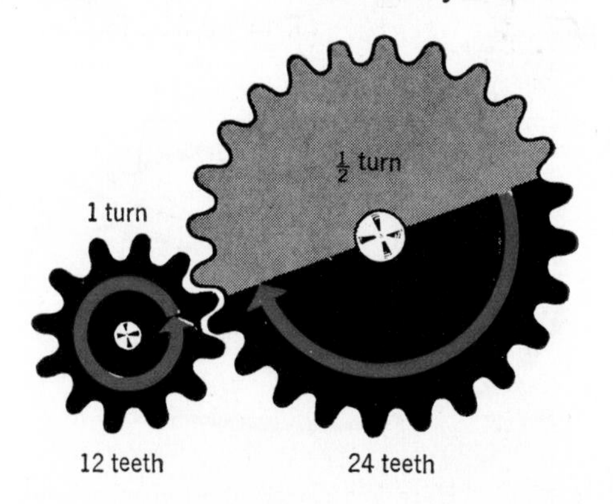

6-11 **The small gear turns twice as rapidly as the large one.**

10. Gears in the automobile

The **transmission** of an automobile illustrates another important use of gears. For the car to operate at highest efficiency it is important that the engine rotate at a certain speed. If it rotates too fast, it tends to wear out and also wastes gasoline. If it rotates too slowly, gasoline is still being wasted; also the car may stall. But the wheels of the car are constantly changing speed. How can the engine, which is connected with them, be kept rotating at a medium speed while the speed of the wheels changes so greatly? The job is done by the gears in the transmission.

When the car is in high ("third") gear (Fig. 6-12C), the motor is connected directly to the shaft that goes to the rear wheels. Thus, in high gear, both engine and car are arranged to be running at normal cruising speed. But if the car has to be slowed down beyond a certain point, the engine must still be kept running at sufficient speed so that it will not stall. Hence it is necessary to shift to "second" (Fig. 6-12B). When the car is in second gear, the engine can still turn at normal cruising speed, but the gears have the effect of reducing the speed of the wheels. Since the gears cause a reduction in speed, there will be an increase in force applied. Speed is lost but force is gained. This is why second gear is used to climb a steep hill.

If the hill is exceedingly steep, it may be necessary to shift to low ("first") (Fig. 6-12A). In low gear the motion of the wheels is slowed down still further as compared with the speed of the mo-

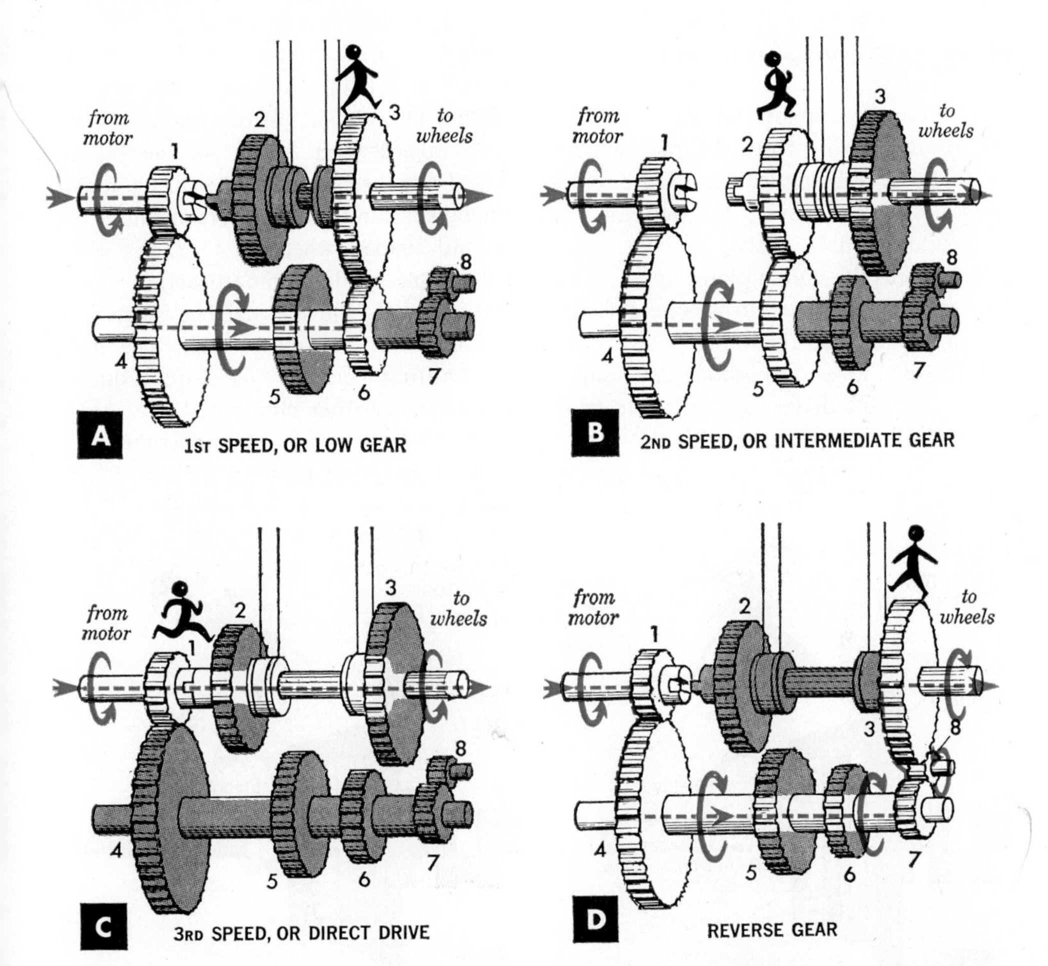

6-12 (A) In first (low) speed, small gear 1 turns large gear 4 to lose speed. There is a double loss of speed as small gear 6 turns large gear 3. (B) In second speed there is only one reduction in speed as small gear 1 turns large gear 4. Gears 2 and 5 are equal in size. (C) In third (high) speed there is no reduction of speed in the gears since the shaft from the motor is connected directly to the shaft from the wheels. (D) In reverse, an extra gear 8 causes an opposite motion of the shaft to the wheels.

tor, resulting in a still greater increase of force. Because a great deal of force is necessary to get the car started from rest, the driver starts in low. Then he shifts to second, which provides less force than low, but enough to enable him to pick up speed. Only when the car is approaching cruising speed does he shift into high.

In addition to the transmission gears, a car usually has a clutch and differential gears. These parts transmit the motion of the engine at the front of the car to the wheels in the rear, which makes the car move.

The **clutch** enables the driver to disengage the engine from the rear wheels while he is shifting gears. Figure 6-13 shows how the driver, by stepping on the clutch pedal, pushes a bent lever which forces the clutch plates apart. Now the rotating engine turns one of the clutch plates, but the other plate, connected to the transmission, no longer turns. Thus the driver can shift into first, second, third, or reverse with the motor no longer connected to the gears and is therefore unable to damage the gears. When he has finished shifting gears, he slowly releases the clutch pedal. The spring then forces the two plates together again. These clutch plates have special high-friction surfaces so that they grip each other strongly. Now, as in Fig. 6-13B, the engine is once again connected, through the clutch, to the transmission and the drive shaft. Many modern cars are equipped with mechanisms ("hydramatic," "gyromatic," etc.) which automatically change gears as the car changes speed. Others have a "fluid drive" clutch in addition to the regular clutch. This consists of two bladed wheels with oil between them. As the bladed wheel attached to the motor turns, it sets the oil spinning and this in turn causes the second bladed wheel to turn. As a result of the gradual smooth action of the spinning oil, the car can be started in second gear or even in high. The need for shifting gears is therefore reduced, but not altogether eliminated.

After the motion of the engine has been transmitted through the clutch and

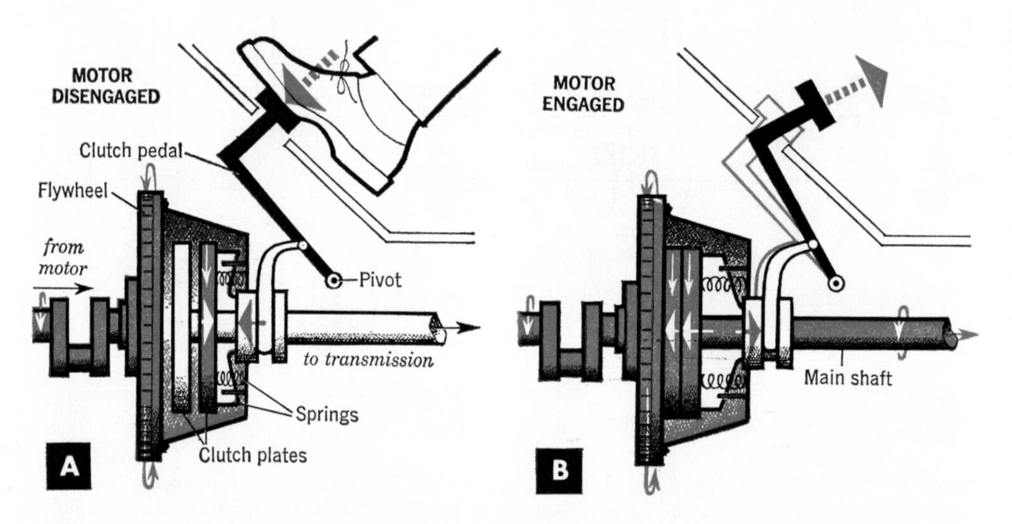

6-13 (A) Stepping on the clutch causes the plates to separate, and the motor can now turn without moving the car. It is then safe to shift gears. (B) After the gears have been shifted, the foot is removed and a spring presses the plates together. Now the motor causes the wheels to turn.

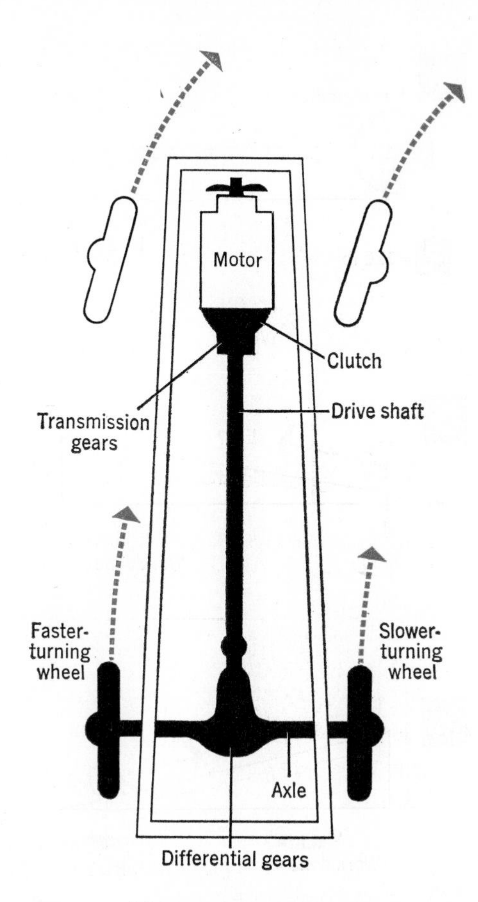

6-14 The rotation of the motor is transmitted to the wheels through the clutch, gears, shaft, differential gears, and axles.

gears to the shaft that goes to the rear wheels, it must be transmitted at right angles to the axle in Fig. 6-14. At the same time the two rear wheels must be able to rotate in such a way that the outside wheel, on a turn, can cover a greater distance than the inner wheel. The **differential** is the complex set of gears at the rear of the car which enables the motion of the engine to reach the rear wheels and which enables each to rotate at a different speed.

From this discussion we can see what an important part gears and other simple machines play in the operation of an automobile.

Going Ahead

THE VOCABULARY OF PHYSICS

Define each of the following:

inclined plane	wheel and axle
clutch	differential
wedge	pitch
screw	transmission

APPLYING THE PRINCIPLES

1. Mention five examples of the use of the wheel and axle.
2. Mention three examples of the use of the inclined plane.
3. Mention three examples of the use of the wedge.
4. Mention five common examples of the use of gears.
5. Can you explain why one turn of the pedals of a bicycle causes the bicycle to move a great distance?
6. Why are transmission gears used in a car? When are first, second, and third gears used?
7. When a car is in first gear, the rear wheels turn very slowly. Explain why this gear is best for climbing a very steep hill.
8. What happens when a driver steps on the clutch pedal and then releases it? Why is it necessary to step on the clutch before shifting gears?
9. What is the purpose of the differential gears in a car?

SOLVING PROBLEMS

1. Find the unknown quantity for each of the machines shown in Fig. 6-15 (on the next page).
2. Draw a diagram of a combination of two machines that will be able to lift 1,200 lb with 30 lb of effort.
3. An effort of 40 lb is applied to a 10-in wrench to tighten a bolt having 12 threads to the inch. How much tightening force is exerted?
4. What is the M.A. of a clamp having a screw with 8 threads to the inch and a handle with a radius of 2 in?

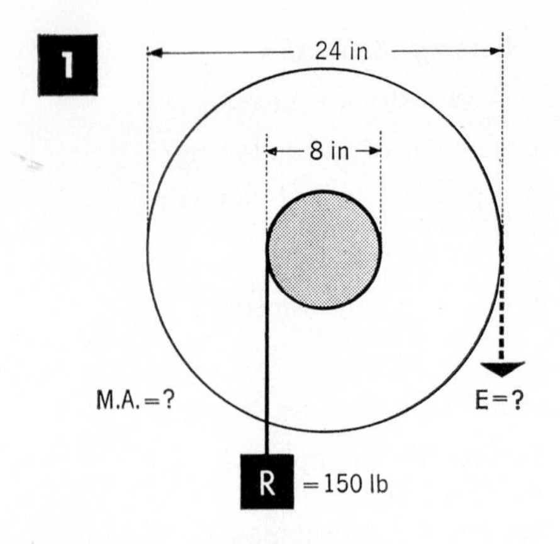

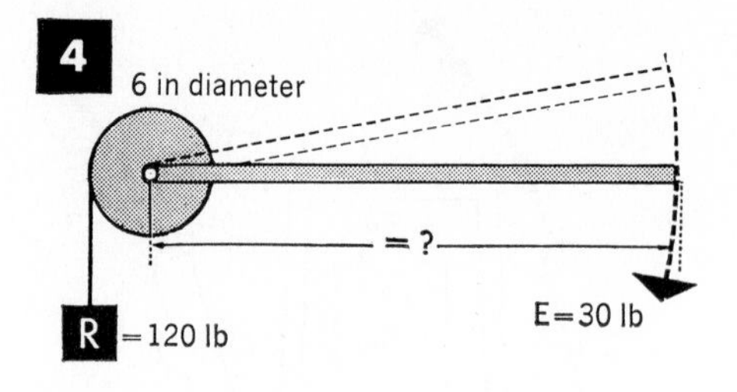

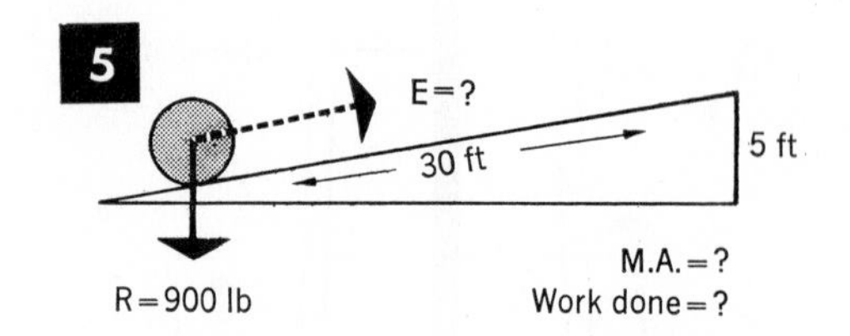

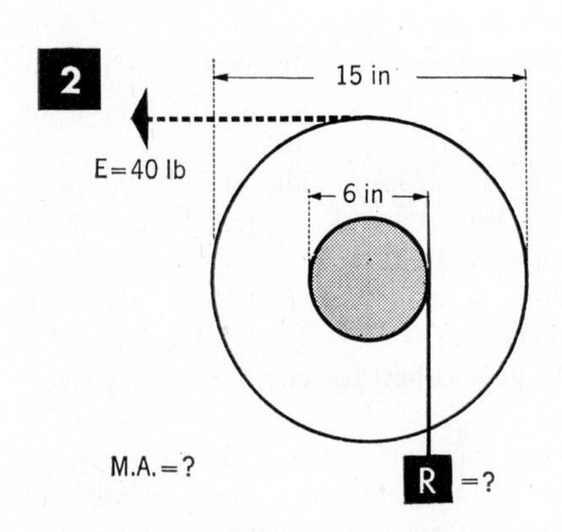

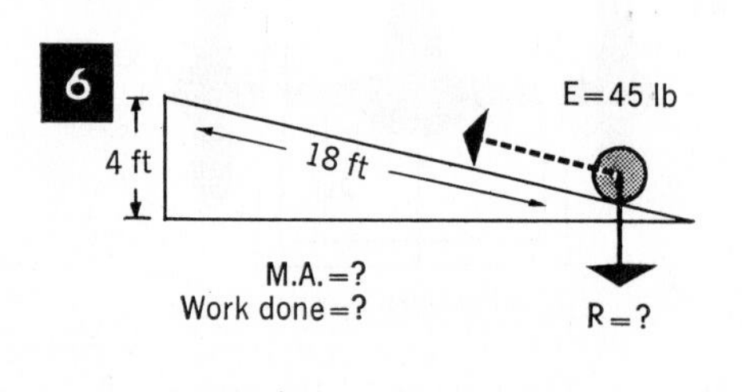

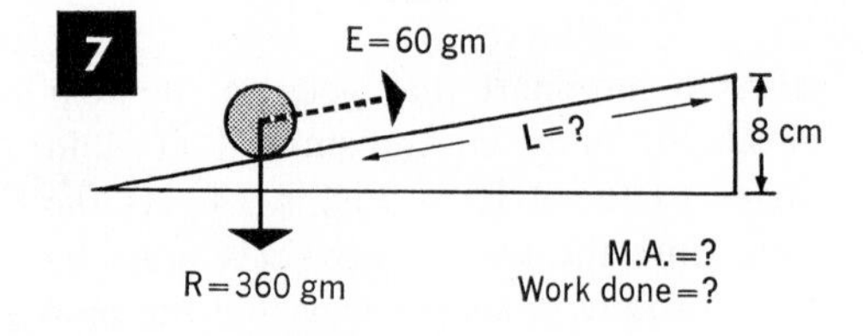

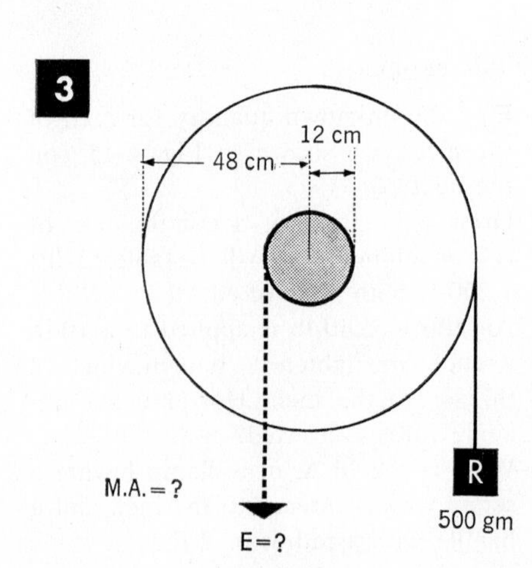

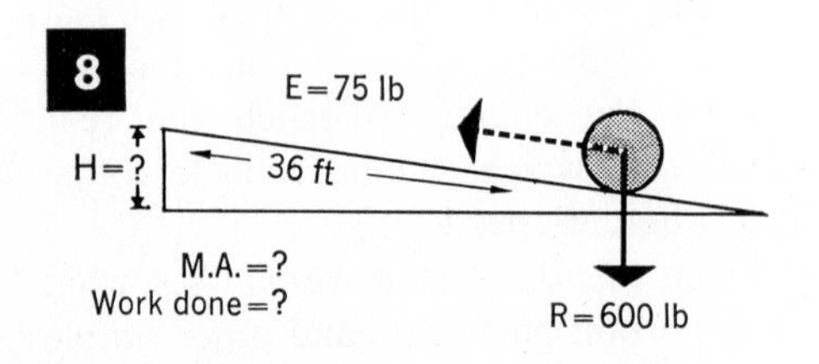

6-15 **Problems to solve. Do not mark the book in any way.**

5. Calculate the missing information for the following wheel-and-axle machines:

	Effort arm	*Resistance arm*	*M.A.*	*Resistance force*	*Effort force*
(a)	20 in	4 in	?	15 lb	?
(b)	?	2 ft	7	?	28 lb
(c)	40 cm	?	8	?	10 gm

6. Find the resistance distance in Problems 5(a) and 5(b) above if the effort distance in each case is 24 in.

7. Calculate the missing information for the following inclined planes:

	Length	*Height*	*M.A.*	*Effort*	*Resistance*	*Work done*
(a)	36 ft	3 ft	?	12 lb	?	?
(b)	15 ft	3 ft	?	?	25 lb	?
(c)	1 m	5 cm	?	?	220 gm	?
(d)	25 in	?	$2\frac{1}{2}$	60 lb	?	?

---

8. How much squeezing force is exerted by a vise having a 10-in handle and 10 threads to the inch when an effort of 25 lb is applied to the handle?

9. A jackscrew has 10 threads per inch. (a) What weight can be raised by an effort force of 50 lb applied to an effort bar 8 in long? (b) How high will the weight be lifted for each inch through which the effort moves?

---

10. A gear having 20 teeth is meshed with a second gear having 60 teeth. How many times must the first gear revolve in order to rotate the second gear twice?

11. A 20-ft chain weighing 80 lb hangs straight down from the resistance wheel of a wheel and axle. How much work will be done in winding up the chain? The diameter of the resistance wheel is 2 ft. (HINT. By how many feet is the center of gravity of the chain raised?)

12. A 200-lb barrel lying on its side is to be rolled up a 12-ft inclined plank into a truck 3 ft high. With what force must you push on the highest part of the barrel in a direction parallel to the plane in order to roll the barrel up the plank?

13. The front sprocket of a certain bicycle has 60 teeth, and the rear sprocket has 15 teeth. The diameter of the rear wheel is 30 in. How far will the bicycle move for one turn of the pedals?

14. A wheel and axle (diameters 2 ft and 0.3 ft) is used as a windlass to haul boats up an inclined plane 300 ft long and 5 ft high. What is the M.A. of the system?

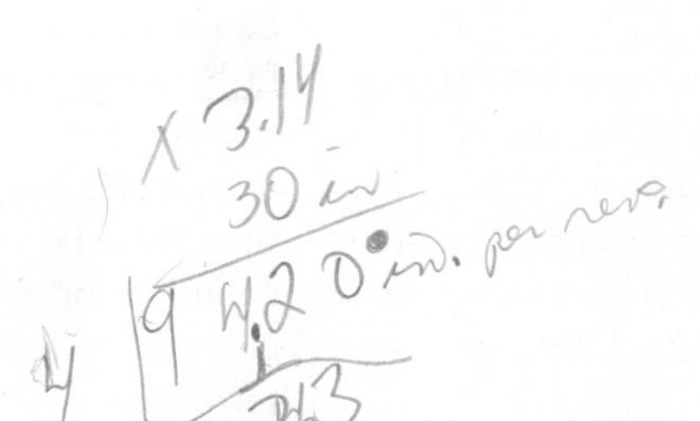

CHAPTER

7

Friction and Efficiency

We have said many times that in calculating the effort needed to do a job we were neglecting friction. In other words, we have been speaking of the ideal effort. In actual practice friction is always present, so that the actual effort necessary to do a job is always greater than the ideal effort.

Friction

1. What is friction?

There is still some disagreement among scientists as to the explanation of friction. But for our purposes we can accept the theory that friction results from the interlocking of slight bumps on two surfaces which are in contact with one another.

In Fig. 7-1 two such surfaces are shown as they might appear if enlarged under a powerful microscope so that the roughness on each is visible. If you try to slide one over the other, the grip of the tiny projections on each surface resists the motion. Extra force is necessary to overcome the obstruction of the bumps. If the surfaces are very smooth, indeed so smooth that their molecules are close enough to adhere, then extra force is needed to overcome the effects of adhesion as well. This resistance to movement is called **friction.** Since friction may be defined as the force that opposes the rolling or sliding of one object over another, it is measured in units of force, namely pounds or grams.

2. Friction is useful

You might imagine that the job of pushing and pulling objects would be much easier if there were no friction. In reality it would be far more difficult.

If there were no friction, you could not walk along the ground because your feet would slip. You would not be getting any traction. Without friction all objects in a room would have to be fastened down; otherwise the slightest push would send them sliding around the floor as though they were on roller skates. If there were no friction, nails and screws would pull right out of wood.

It is the friction in brakes which stops a car. By pressing your foot on the brake pedal, you cause the rough brake lining to press against the drum of the wheel, thus increasing the friction and slowing down the motion of the wheel. In similar fashion the clutch plates in automobiles that have standard gear shift use friction to transfer the motion of the engine to the gears. In automatic transmissions the cohesion or friction between the molecules in a fluid accomplishes the same purpose.

3. Friction causes heat

Since friction is a force, work is required to overcome it. Try rubbing your hands together very hard and very fast. You are doing work to overcome friction. Remembering the Law of Conservation of Energy, you will realize that this work is not destroyed. What be-

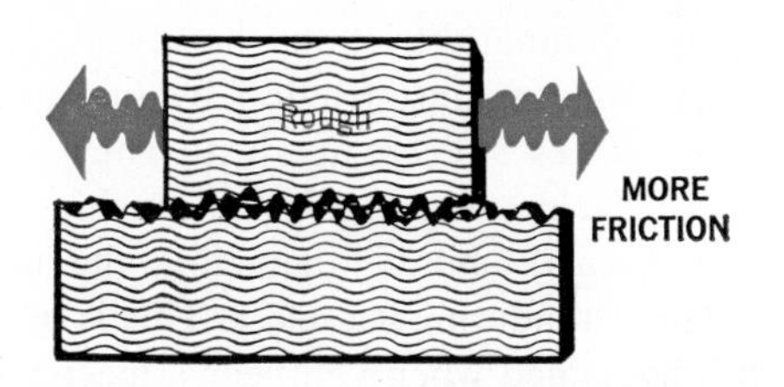

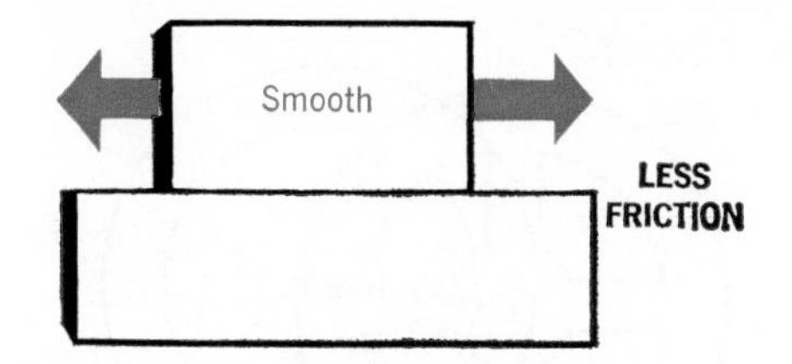

7-1 Friction is caused by surface irregularities. Friction is reduced by removing the irregularities and making the surfaces smooth.

comes of it? Your experiment will convince you that your work is turned into heat. Heat, we saw in Chapter 2, is energy of motion of molecules. When you do work against friction, you cause the molecules of the rubbing surfaces to move faster; hence the surfaces grow hotter.

If you keep up with recent developments in air travel, you have probably heard that at speeds of about 1,000 miles per hour airplanes and rockets encounter the heat barrier. If airplanes are going to fly for long at such immense speeds, the work done against friction creates so much heat that some method of cooling is required to prevent scorching. Your experiment with rubbing your hands together showed you a similar form of heat barrier; if you keep on rubbing your hands together hard, they will become unbearably hot.

The heat caused by friction also has its uses. For example, when you strike a match, it is the heat resulting from friction which sets the match on fire.

Reducing friction

A certain amount of friction, then, is absolutely necessary. Sometimes, as when we are bolting two pieces of steel together, we want to achieve the greatest amount of friction possible, so as to increase holding force. At other times we want to reduce friction (Fig. 7-2). There are several ways in which friction can be reduced.

4. Wheels reduce sliding friction

One of the greatest of all mechanical inventions is the wheel. Before the wheel was invented, friction made it very difficult to move large objects; everything that man wanted to move had to be dragged or carried. But as a wheel rolls forward, the spot that is touching the ground lifts up and another portion of the wheel comes down to take its place

New York Central System

7-2 Friction has been reduced so much that these men can move this several-hundred-ton locomotive!

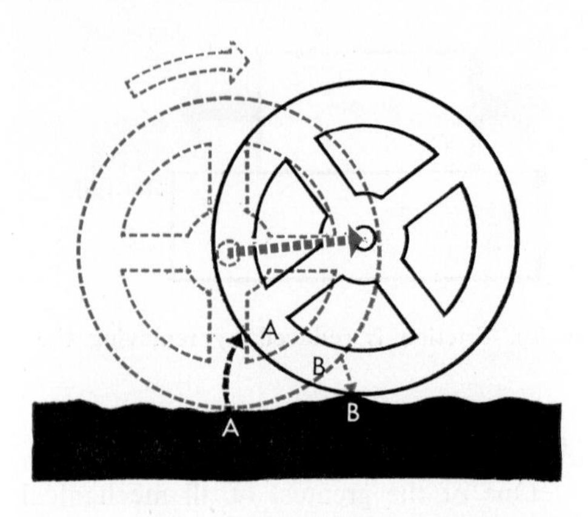

7-3 The wheel lifts up over the bumps instead of sliding across them. Rolling friction is much less than sliding friction.

(Fig. 7-3). Thus, most of the sliding is eliminated, and the friction is greatly reduced.

Friction at the axle surfaces can be reduced by inserting ball bearings. In this way, rolling friction is again substituted for sliding friction.

5. Smooth surfaces reduce friction

If we level out the irregularities in two surfaces, less effort will be needed to make them slide over each other. For this reason all moving parts of machines are fitted together as accurately as possible and the surfaces in contact are polished smooth.

6. An "antifriction" metal reduces friction

Certain alloys, formed by combining various metals, have been found to create very little friction when sliding on ordinary metals. They are called **antifriction metals. Babbitt metal,** an alloy of tin, copper, and antimony, is frequently used for bearings which hold a rotating steel shaft. Steel moving against Babbitt metal produces less friction than would steel against steel.

Even the use of such metals and the use of highly polished surfaces never entirely eliminate friction, however. The work done against friction by the moving surfaces still creates heat which can cause the moving parts of a machine to expand and "bind," damaging the machine or rendering it useless. This is what has happened when you hear a mechanic speak of "burning out a bearing."

7. Lubricants reduce friction

One of the best ways to reduce friction is by means of lubrication. The **lubricant** is usually a liquid such as oil which is placed between the two surfaces. The lubricant forms a liquid layer between the sliding surfaces and thus prevents them from rubbing directly against each other.

Although oil is the most common lubricant, it is far from being the only one. When a glass tube is inserted into a rubber stopper, water or glycerine may be the lubricant used. If a wooden window frame sticks, you might use wax. Soap is a good lubricant when you are driving a screw into wood. If the sliding catch of a door sticks, it is often helpful to rub on some graphite from a soft pencil.

8. Coefficient of friction

You have probably found that it is much easier to drag a box across the floor when it is empty than when it is loaded. If you were to weigh the box and measure the force necessary to pull it, you would find that the force needed to pull it is less than the weight of the box. Will this still be true if a weight is put inside the box?

If you were to try the experiment and double the weight of the box by putting a load inside it you would find that it now takes twice as much force to pull it. You would find that it takes three times as much force to pull it when its weight has been tripled, four times as much force to pull it when its weight is four times as great, and so on. That is, the ratio between the weight of the loaded box and the force needed to pull it remained constant. We can summarize all these observations in a formula:

$$\frac{\text{Force needed to overcome friction (F)}}{\text{Weight (W)}} = \text{a constant (k)}$$

or

$$\frac{F}{W} = k$$

This formula says that the force of friction is a constant fraction of an object's weight. The constant fraction, *k,* is called the **coefficient of friction.**

Notice that this *k* is *not* the same as the constant, *k,* in Hooke's Law.

Experiments show that the coefficient of friction (friction force divided by weight) is nearly constant for a given pair of surfaces no matter how great the weight pressing them together. Different kinds of surfaces have different coefficients. The more slippery the surface, the smaller the coefficient of friction. For example, you might find that it took a force of 40 pounds to drag a 100-pound box along a gravel driveway. The coefficient of friction is

$$k = \frac{40 \text{ lb}}{100 \text{ lb}} = 0.40$$

If you dragged the box along a wet asphalt road, you might find that it took only 25 pounds of force to drag it. The coefficient of friction, *k,* has been reduced to 25 lb/100 lb or 0.25.

Some useful coefficients of friction are listed in Table 3.

TABLE 3
COEFFICIENTS OF FRICTION

Wood on wood, dry	0.25 to 0.50
Wood on wood, soapy	0.20
Metal on metal, dry	0.15 to 0.20
Metal on metal, greased	0.03 to 0.05
Metal on metal, rolling	about 0.002
Rubber on concrete, dry	0.60 to 0.70

SAMPLE PROBLEM

A 3,000-pound car on a concrete road is being used to pull a second car out of a ditch. With how much force can the 3,000-pound car pull before its tires begin to slip on the road? Noting that the lowest coefficient of friction of rubber on concrete is 0.60, we write

$$\frac{F}{W} = k$$

$$\frac{F}{3{,}000 \text{ lb}} = 0.60$$

$$F = 0.60 \times 3{,}000 \text{ lb}$$

$$= 1{,}800 \text{ lb}$$

STUDENT PROBLEM

A boy and his bicycle together weigh 154 lb. How much force can the bicycle tires exert on a concrete road without beginning to slip? (Use the lowest value of *k.*) (ANS. 92.4 lb)

■ ■ ■ ■ ■ ■ ■ ■

Our formula applies to the common case of a weight, *W,* being pulled along a horizontal surface. But there are many instances of friction between surfaces that are not horizontal (a sled coasting down a steep hill, for example). There are even situations in which the force holding the surfaces together is not a weight (for example, the force of friction between a board and the nail being pulled out of it). The same formula will handle these situations too. In the preceding examples the weight, *W,* was a force acting perpendicularly to the ground and thus holding the two sliding surfaces together. When the surfaces are not horizontal, we continue to put in the denominator the force that acts perpendicular to the two surfaces, thus holding

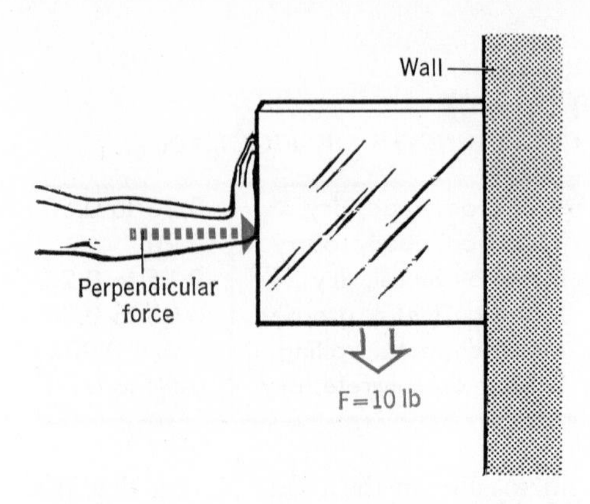

7-4 Even in this unusual situation the coefficient of friction is the force, *F*, needed to cause slipping divided by the force holding the surfaces together.

them together. On sloping surfaces, of course, this force is not at right angles to the horizontal surface of the earth—it is at right angles to the sloping surface. In Chapter 8 we shall learn more about how to compute the force when the surfaces are not horizontal.

SAMPLE PROBLEM

How hard must you push a 10-pound wooden box horizontally against a wooden fence in order to keep it from slipping to the ground? Coefficient of friction = 0.40 (Fig. 7-4). Here the perpendicular force holding the two surfaces together is certainly not the weight of the box. This force is unknown. The force, *F*, causing the box to slip is the weight of the box. Hence

$$\frac{\text{F}}{\text{Perpendicular force}} = \text{k}$$

$$\frac{\text{10 lb}}{\text{Perpendicular force}} = 0.40$$

$$\text{Perpendicular force} = \frac{\text{10 lb}}{0.40} = \text{25 lb}$$

STUDENT PROBLEM

If the coefficient of friction between a pencil and your skin is 0.3, how hard must you squeeze a 10-gm pencil between your fingers when you hold it vertically? (ANS. 33.3 gm)

9. Fluid friction

The resistance air offers to a falling parachute is an example of **fluid friction.** The friction between fluids (liquids or gases) and solid objects acts quite differently from friction between sliding solid objects. Sliding friction hardly varies at all with velocity, whereas the friction encountered by a solid moving through a fluid (such as a parachutist or an airplane moving through air, or a boat moving through water) depends greatly upon the speed with which it travels. A simple equation first worked out by Sir Isaac Newton gives the relation between fluid friction, F_f, and velocity, v:

$$F_f = kv^2$$

Here k is a constant whose numerical value depends upon the size, shape, and roughness of surface of the moving object as well as the fluid through which the object is moving. F_f is the force of friction due to fluid resistance.

The equation tells us that fluid friction increases very rapidly with velocity. Doubling the speed, v, of an object moving in a fluid makes fluid friction, F_f, four times as great. Fluid friction on an object moving in liquids such as water is very great, even at relatively low speeds, but in gases such as air it is usually unimportant until velocities over 60 miles per hour are reached. The reason is, as you learned in Chapter 2, that there are far fewer molecules in a given volume of air than in the same volume of water. Automobiles and especially aircraft are **streamlined** to reduce the effects of fluid friction (Fig. 7-5). The fact that resistance due to fluid friction increases so rapidly with speed explains why high-speed planes must have such powerful engines.

It makes no difference whether an object is moving through the air or whether the air (in the form of wind) is moving past the object. In either case, fluid friction (air resistance) acts in the same way. The force of the wind against objects and buildings is nothing more than the effect of fluid friction. You can see from the data in Fig. 7-6 (calculated from the equation for fluid friction, $F_f = kv^2$) why winds of hurricane velocity cause so much destruction. The data is based on the assumption that a 10-mile-per-hour wind exerts a force of 10 pounds on a windowpane whose area is 5.8 square feet, and that doubling the velocity of the wind quadruples the force.

SAMPLE PROBLEM

Let us apply the equation for fluid friction to a practical situation. If the wind force on a billboard 10 feet high and 20 feet long is 115 pounds when the wind velocity is 1 mile per hour, what would be the force in tons on the billboard in a hurricane with a wind velocity of 100 miles per hour?

The equation for fluid friction states that the force is proportional to the square of the wind velocity. We solve the problem by applying the formula to each of the two situations mentioned. Thus at the first wind velocity, v, the force exerted can be written as F_1, and the formula becomes

$$F_1 = kv_1^2$$

Again, at the second wind velocity, v_2, the force exerted can be written as F_2, and

$$F_2 = kv_2^2$$

When we divide the first equation by the second, the k's cancel and we get:

$$\frac{F_1}{F_2} = \frac{v_1^2}{v_2^2}$$

Substituting the given data:

$$\frac{115}{F_2} = \frac{1^2}{100^2}$$

$$F_2 = \frac{115 \times 100 \times 100}{1}$$

$$= 115 \times 10^4 \text{ lb}$$

Or, in tons,

$$\frac{115 \times 10^4}{2 \times 10^3} = 575 \text{ tons}$$

STUDENT PROBLEM

What would be the force on the billboard in the sample problem above in a tornado in which the wind velocity is 400 mi/hr? (ANS. 9,200 tons)

Efficiency

Since all machines have friction, there must always be a certain amount of energy wasted when they are used to do work. Engineers are continually striving to reduce this wasted energy to a minimum. In so doing, they increase the **efficiency** of the machine. Efficiency is

General Motors Corp.

7-5 Thanks to streamlining, a modern car does far less work against air resistance when traveling at high speed.

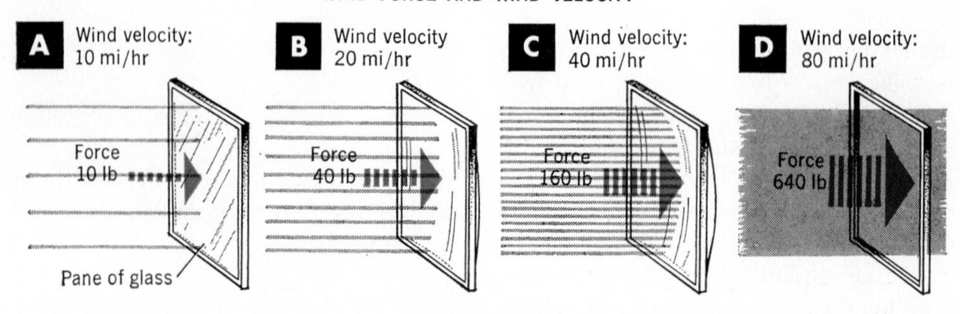

7-6 The force of the wind on a plate glass window increases as the square of the wind velocity. In D the force of the wind has shattered the glass. Wind velocities of 80 mi/hr often occur in hurricanes.

measured by comparing the amount of useful work we get out of a machine to the work, or energy, we put into it.

10. Perpetual motion?

It would be very nice if we could make a **perpetual motion machine.** Such a machine, once started, would continue operating and doing work by itself without any outside source of energy. Many men have tried to make such machines, but all attempts have failed (Fig. 7-7).

The person who tries to invent a perpetual motion machine is trying to get around the Law of Conservation of Energy. This law states that the most we can get out of an ideal or frictionless machine is exactly the amount of work we put into it. Since friction is always present in actual machines, the useful work we will get out will always be less than the work we put in. But with a perpetual motion machine we would be getting out more work than we put in.

7-7 As the wheel turns, the balls on the right will always be farther from the center and should therefore keep the wheel spinning because of greater leverage. All seems logical. But why won't this perpetual motion machine work?

Scientific laws have often been proved wrong in the past, and we should not be dogmatic even about this one. It just might be possible that a perpetual motion machine will be invented some day. But scientists are practically certain that such a machine is impossible.

11. Calculating efficiency

We have said that we find the efficiency of any given machine by comparing useful work output to work input. Thus:

$$\text{Efficiency} = \frac{\text{Useful work output}}{\text{Work input}}$$

Suppose we want to find the efficiency of the inclined plane shown in Fig. 7-8. The plane is 10 feet long and 2 feet high, and the weight the boy is pulling up is 500 pounds. The force he must exert on the rope, however, is 125 pounds. By the method we learned on page 76 we find that

$$\text{M.A.} = \frac{L}{h} = \frac{10\text{ ft}}{2\text{ ft}} = 5$$

Since this figure is the M.A. for an ideal or frictionless plane we call it the **ideal M.A.** or **I.M.A.** (All the M.A.'s we have calculated up until now have been ideal M.A.'s.)

Ideally, then, he should need only $\frac{1}{5}$ of 500 pounds, or 100 pounds of effort, to slide the weight up to the top of the incline. But when he actually does the job, he finds it takes more effort than our calculation showed on paper. He needs 125 pounds instead of 100. We call this 125 pounds the **actual effort.**

The useful work output is the work accomplished by raising the 500-pound weight to a height of 2 feet. Or:

$$\begin{aligned}\text{Useful work output} &= \text{Force} \times \text{Distance}\\ &= 500 \text{ lb} \times 2 \text{ ft}\\ &= 1{,}000 \text{ ft-lb}\end{aligned}$$

The work input is the result of the exertion of 125 pounds of effort for a distance of 10 feet along the incline. Or:

$$\begin{aligned}\text{Work input} &= \text{Force} \times \text{Distance}\\ &= 125 \text{ lb} \times 10 \text{ ft}\\ &= 1{,}250 \text{ ft-lb}\end{aligned}$$

Using these in our formula for efficiency, we find:

$$\begin{aligned}\text{Efficiency} &= \frac{\text{Useful work output}}{\text{Work input}}\\ &= \frac{1{,}000 \text{ ft-lb}}{1{,}250 \text{ ft-lb}} = 80\%\end{aligned}$$

Thus we obtain a useful work output of only 1,000 foot-pounds, although our work input was 1,250 foot-pounds. The difference of 250 foot-pounds has been converted into heat all along the incline; in other words, this energy has been wasted in doing work to overcome the force of friction. We say "wasted" because this extra energy was not useful to us in lifting the resistance.

The efficiency of any device is always less than 100 per cent because the numerator (useful work output) is always less than the denominator (work input). An efficiency of more than 100 per cent would mean that the useful work output (numerator) was greater than the work input (denominator), and this would mean that the device would keep running forever once it had been started; that is, it would be a perpetual motion machine. An efficiency greater than 100 per cent is therefore impossible.

STUDENT PROBLEM

An inclined plane used to slide a 200-lb desk from the ground into a truck is 15 ft long and 4 ft high. A force of 90 lb is needed to slide it up the plane. What is the efficiency of the plane? (ANS. 59.2%)

12. Efficiency when mechanical energy is used

For machines using mechanical energy, we can find the efficiency in other ways. One way is to compare the ideal

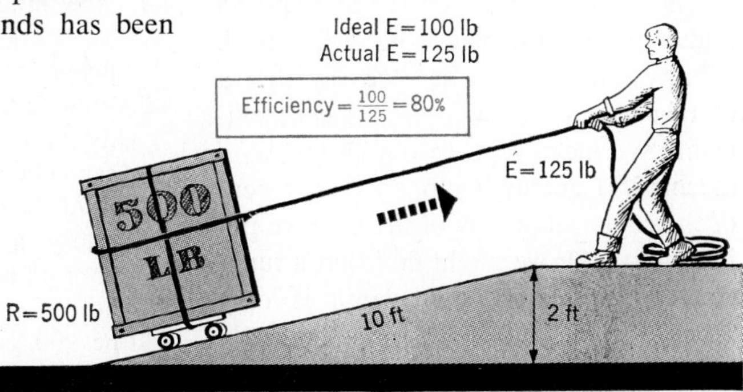

7-8 Figuring the efficiency of an inclined plane.

effort (neglecting friction) with the actual effort. In the situation we have just been considering, this would be:

$$\text{Efficiency} = \frac{\text{Ideal effort}}{\text{Actual effort}}$$

$$\text{Efficiency} = \frac{100\ \text{lb}}{125\ \text{lb}} = \frac{4}{5} = 80\%$$

We can also find the efficiency by comparing the actual M.A. with the ideal M.A. Thus:

$$\text{Efficiency} = \frac{\text{Actual M.A.}}{\text{Ideal M.A.}}$$

$$\text{Efficiency} = \frac{4}{5} = 80\%$$

Our actual M.A. is 4 because we need an actual effort of 125 pounds to lift the weight of 500 pounds.

$$\text{Actual M.A.} = \frac{\text{Resistance}}{\text{Actual effort}} = \frac{500\ \text{lb}}{125\ \text{lb}} = 4$$

STUDENT PROBLEM

1. A force of 30 lb is needed to lift a certain load with a block and tackle. If there were no friction, a force of 25 lb would be required. What is the efficiency of the block and tackle? (ANS. 83%)

2. An inclined plane has an ideal M.A. of 3.4 and an actual M.A. of 2.6. What is its efficiency? (ANS. 76%)

13. Efficiency when other forms of energy are used

The idea of efficiency is the same no matter what form of energy is involved. What do we mean when we say that a certain steam engine is 5 per cent efficient? We mean that we obtain in *useful* mechanical energy (output) 5 per cent of the chemical energy of the coal we put in (input). Or we might find that a tungsten (wolfram) filament lamp is only 2 per cent efficient. This means that only 2 per cent of the electrical energy we put in (input) appears in the *useful* form of light energy (output). Most of the remaining 98 per cent appears as heat, which is wasted energy for our purpose of providing light.

Going Ahead

THE VOCABULARY OF PHYSICS

Define each of the following:

friction	efficiency
lubricant	actual effort
perpetual motion machine	ideal M.A.
	actual M.A.
Babbitt metal	streamlined
coefficient of friction	fluid friction
antifriction metals	

THE PRINCIPLES WE USE

Explain why

1. Rolling friction is less than sliding friction.
2. The maximum efficiency of a machine is 100%.
3. Friction is often useful.

PROBLEMS IN FRICTION AND EFFICIENCY

For coefficient of friction use the lowest value listed in Table 3 on page 91.

1. If it requires 8 lb to drag a 40-lb block across the floor, what is the coefficient of friction?
2. How much horizontal force can a 2-ton tow truck exert on dry concrete?
3. How much pull can a 40-ton locomotive exert?
4. How much force is required to drag a 2-kg wooden block across a wooden floor?
5. What is the minimum force with which the vertical jaws of a wood-lined vise can support a 25-lb block of wood?
6. A 7.9-kg object is dragged along a laboratory table by means of a spring balance attached to one side. The pull needed to keep it moving is found to be 900 gm. What is the coefficient of friction between object and table?

7. What is the weight of an object if a force of 20 lb is needed to pull it across the floor, and the coefficient of friction is 0.35?

8–12. Calculate the missing information in the table at the foot of the page.

13. A machine with an ideal M.A. of 5 is used to lift a 300-lb weight. How much effort will actually be needed if the efficiency is 75%?

14. 2,000 ft-lb of work must be put into a machine in order to accomplish 1,500 ft-lb of work. What is the efficiency of the machine?

15. A block and tackle has an efficiency of 80%. What work must be put into the block and tackle to lift a 300-lb load a vertical distance of 20 ft?

16. What is the ideal mechanical advantage of an inclined plane if its actual mechanical advantage is 3.6 and its efficiency is 40%?

17. What actual effort must be applied to a block and tackle that is 80% efficient to lift a 100-lb load if the ideal effort is 25 lb?

18. A block and tackle has four ropes directly supporting the load, a 200-lb crate. An effort of 60 lb is needed to lift the crate 20 ft. Find (a) the ideal M.A., (b) the effort distance, (c) the work output, (d) the work input, and (e) the efficiency.

19. In Problem 18, what is: (a) the work done against friction, (b) the force needed to overcome friction?

20. Theoretically a gallon of gasoline has energy sufficient to do about 100 million ft-lb of work. If a 3,000-lb car climbs a mountain 4,000 ft high using 1 gallon of gasoline, what is the overall efficiency of the car?

21. A submarine cruising beneath the surface at 2 mi/hr increases its speed to 7 mi/hr. How much greater is the force the propellers must exert at this increased speed?

22. An inclined plane has an efficiency of 40%. (a) How does the force of friction compare with the ideal effort? (b) Will the object on the plane slide back down if no effort is applied? (c) What general statement can you make about machines when efficiency is less than 50%? (d) Give some examples of such machines. (e) Would they be "better" machines if their efficiency were greater than 50%?

	Effort distance	*Ideal effort force*	*Actual effort force*	*Resistance force*	*Resistance distance*	*Ideal M.A.*	*Actual M.A.*	*Efficiency*	*Work wasted overcoming friction*
8.	20 ft	?	20 lb	150 lb	2 ft	?	?	?	?
9.	?	?	120 lb	250 lb	2 ft	$2\frac{1}{2}$	?	?	?
10.	7 ft	30 lb	?	70 lb	3 ft	?	?	75%	?
11.	16 cm	100 gm	?	?	?	8	?	?	800 gm–cm
12.	90 cm	?	?	450 gm	50 cm	?	$1\frac{1}{2}$	?	?

CHAPTER

8

Combining Forces

If you were flying an airplane, which information would you consider more important: knowledge of the speed of the plane or knowledge of the direction in which it is traveling? A slight error in the speed will simply mean getting there a bit sooner or later than expected. A slight error in measurement or calculation of the direction may take the plane far off its course. Of course it is more important to know the direction in which it is going.

Combining motions

When you are measuring motion, it is important to know not only "how much" motion but also the direction of the motion. A quantity like velocity, which involves *both* amount and direction, is called a **vector** quantity. Forces are also vector quantities. For instance, when we say that a certain object weighs 10 pounds, we have only partly described what is happening. For a complete description we should say that it is being pulled downward toward the center of the earth (direction) with a force of 10 pounds (amount). Here the force of gravity (10 lb) is a vector quantity, since it tells us both the amount of the force and the direction in which it is acting. No vector quantity is completely described unless both its *amount* (magnitude) and its *direction* are given.

1. Addition of vectors

If a boat travels at 20 miles per hour on a river that flows at 5 miles per hour, the velocity of the boat will be the result of these two velocities combined. Experiment shows that if the boat is traveling downstream and hence both velocities are in the same direction, then the total velocity will be 20 + 5, or 25 miles per hour downstream. This velocity is the result of adding the two velocities.

But what if the two velocities are in exactly opposite directions? Then, experiment shows that the resulting velocity will be 20 − 5, or 15 miles per hour, in the direction of the larger velocity.

Frequently several velocities are acting at an angle to one another. If we know the direction and amount of each velocity, we can predict the resulting motion. We can understand the method by considering how wind affects the flight of a plane.

2. Action of wind on an airplane

Suppose a plane is traveling 100 miles per hour due north. A tail wind from the south is blowing 30 miles per hour. If the wind and the plane are traveling in exactly the same direction, we find the total velocity of the plane by adding the two velocities, 100 + 30, or 130 miles per hour (Fig. 8-2A). The direction is north. We call this vector of 130 miles per hour north the **resultant.**

A resultant is the vector formed by combining two or more vectors.

But what if the wind is blowing from the east instead of from the south? In 1 hour it would blow the plane 30 miles off course to A, as shown in the scale drawing, Fig. 8-2B. By measuring on the scale drawing, we find that the distance PA which the plane covers is about 104 miles. The angle P is about 17 degrees, which shows us that the plane is 17 degrees off course. Our resultant is the vector PA, which is the actual path the plane takes.

A head wind from the north of 30 miles per hour would slow the plane down. Then the plane would travel 100 − 30, or 70 miles in one hour. The resultant vector would be 70 miles per hour north (Fig. 8-2C).

If the 30-mile-per-hour wind were blowing from the southeast, it would carry the plane off its course to A, as shown in the scale drawing, Fig. 8-2D. The plane would be off its course by the angle P, which we find by measurement to be 11 degrees. Our new resultant is the vector PA. This shows us that the plane would travel 125 miles per hour in a direction 11 degrees west of north as a result of the two motions.

It does not matter which velocity we measure first—that of the plane or that of the wind. We would have obtained the same result if we had marked off the wind velocity first at PB (Fig. 8-2D) and then had marked off the velocity of the plane from B to A.

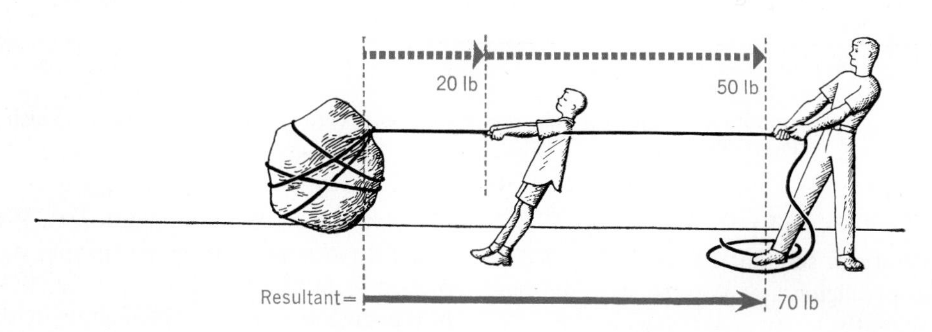

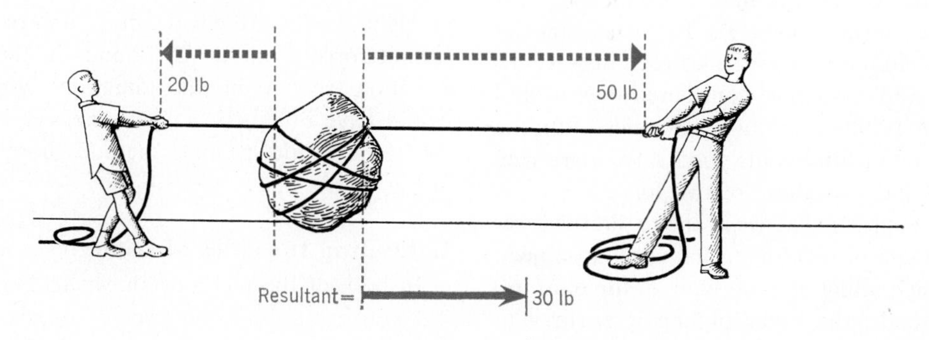

8-1 **An experimental tug of war to find the resultant of two forces. In which position is the rock more likely to move?**

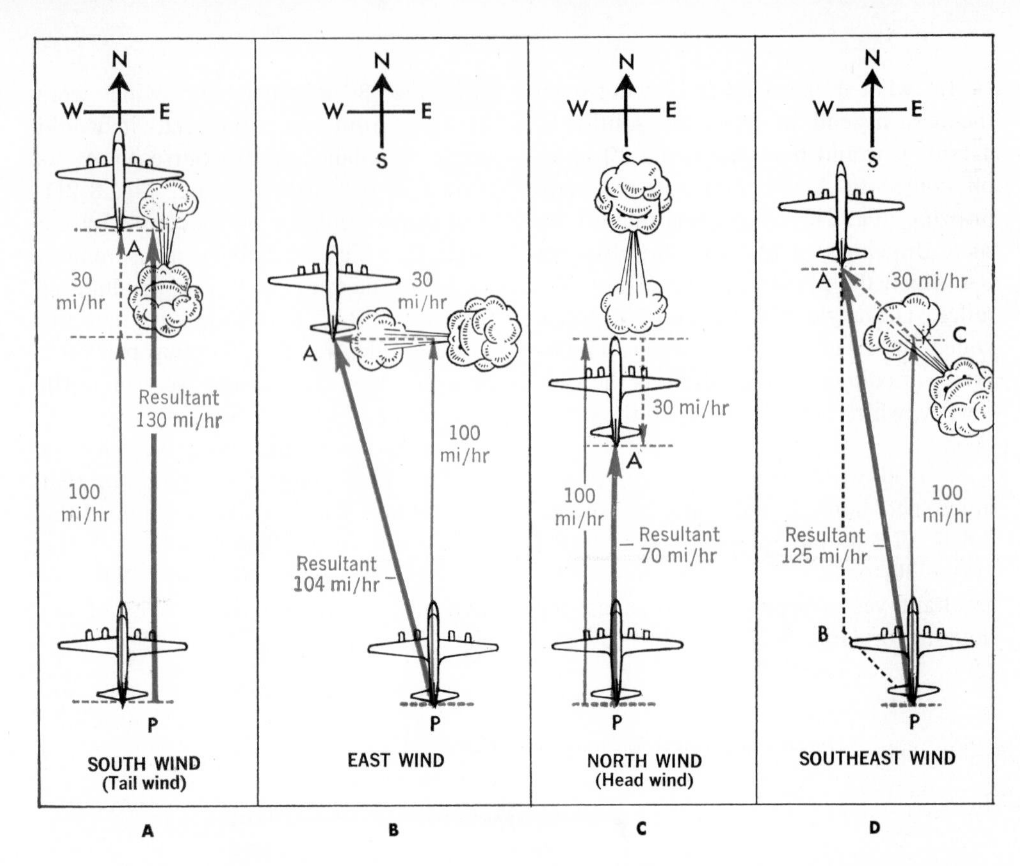

8-2 The resultant, PA, of two velocities will differ according to their directions. Which of these situations does the pilot like best?

3. Vectors and parallelograms

You can see that PBAC in Fig. 8-2D is a parallelogram. We can describe our resultant velocity as the diagonal, PA, of a parallelogram whose sides represent the two velocities we are trying to combine. We always take the diagonal from the starting point P. The diagonal CB would not give us the correct answer.

So far we have considered how to find the resultant of two velocities. But the same method holds good if we are trying to find a resultant of two forces.

Suppose that we want to find the resultant of two forces acting on an object, one pulling it east with a force of 60 pounds, the other pulling it north with a force of 40 pounds (Fig. 8-3). We proceed in the same way as with airplane velocities. We draw the forces, PA and PB, to scale, and in the proper direction from point P. Our resultant, PC, is the diagonal of a parallelogram with the two forces as sides (Fig. 8-3B). By measurement we find that when two forces of 60 pounds and 40 pounds act at right angles to each other, the resultant force is about 72 pounds in the direction shown in the diagram. We would have obtained the same answer if we had used the triangle method shown in Fig. 8-3A.

4. Finding the sides of triangles

In both of the methods shown in Fig. 8-3 you must draw the two forces to a suitable scale. Then you can calculate the resultant by measuring your diagonal and converting it to miles per hour or pounds, using the scale you have chosen. There is another and more accurate method of finding the resultant, provided the figure contains a right angle.

In Fig. 8-3B the diagonal PC is the hypotenuse of a right triangle, PAC. The Pythagorean Theorem in plane geometry states that the square of the hypotenuse is equal to the sum of the squares of the other two sides. Hence,

$$\overline{PC}^2 = \overline{PA}^2 + \overline{AC}^2$$

Since PA = 60 lb and AC = 40 lb,

$$\begin{aligned}\overline{PC}^2 &= (60)^2 + (40)^2\\ &= 3{,}600 + 1{,}600\\ &= 5{,}200\\ PC &= 72 \text{ lb}\end{aligned}$$

Many vector diagrams contain right angles. Often the Pythagorean Theorem will help you to find the unknown sides.

If the angle opposite the resultant is not 90°, the triangle may be solved by drawing a scale diagram. Students who have a knowledge of trigonometry will be able to find the unknown sides by the Law of Cosines. (See page 674.)

STUDENT PROBLEM

A 90-lb force and a 120-lb force act at right angles. What is their resultant? (Solve by means of a diagram drawn to scale, and check by the Pythagorean Theorem.) (ANS. 150 lb E of NE)

---

5. The equilibrant

You will remember from your study of equilibrium (page 63) that an object can be in equilibrium only if the sum of the forces acting on it in one direction is equal to the sum of the forces acting on it in the opposite direction.

If PB and PA are the only forces acting on an object at point P (Fig. 8-3B), then the object cannot possibly be in equilibrium. It will begin to move in the direction of the resultant force, PC.

The object at P can be in equilibrium only if a balancing force of 72 pounds acts opposite to PC. The vector PE represents the single force necessary to balance PC and hence to produce equilibrium. A balancing force which produces equilibrium is called an **equilibrant.** According to this definition PE is the equil-

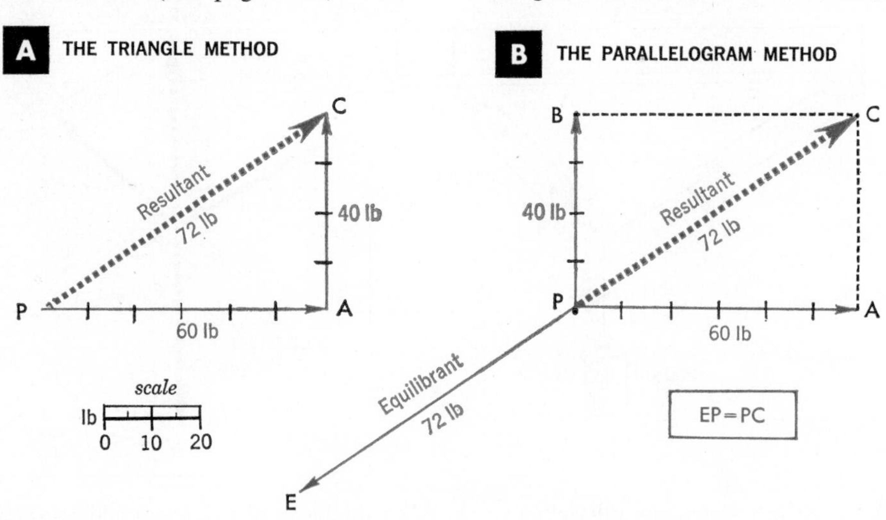

8-3 Two ways of combining forces to obtain a resultant. A force of 60 pounds east and one of 40 pounds north combine to form a resultant of 72 pounds in the direction PC.

ibrant of PC; in the same way, PC is also the equilibrant of PE.

A simple example of a force and its equilibrant is a man holding up a suitcase. Consider the handle of the suitcase. The weight of the suitcase pulls downward on the handle. But the handle is in equilibrium. Hence we know that the man must be exerting an equal upward force on the handle—the equilibrant. Can you draw the vector diagram to show this?

Another example of a force and its equilibrant is provided by your trying to push a car that is stuck in the mud. If the car is in equilibrium—that is, if it won't budge—then your force on it in one direction must be precisely balanced by an equal and opposite force which acts as an equilibrant. The equilibrant is the force of friction between the car and the mud.

These two examples are particularly simple because they involve only two forces acting along the same straight line. But many familiar situations involve three or more forces in equilibrium, not all of them acting along the same straight line.

6. Three forces in equilibrium

Let us try an experiment. As shown in Fig. 8-4, we suspend a 200-gram weight from an overhead support by means of two strings at an angle, with a spring balance attached to each string. We observe that the 200-gram weight is held up by two forces; one of 147 grams, the other of 180 grams.

Consider the forces acting on point P. P is in equilibrium. Therefore, the sum of all the downward forces on P must equal the sum of all the upward forces on P. We can see at once that the downward force is 200 grams. Hence, there must be a force acting vertically upward equal to 200 grams. This upward force is the equilibrant of PW. In this case the equilibrant is the resultant of the two forces exerted by the spring balances.

To show that the resultant of the two upward forces is 200 grams, we draw to

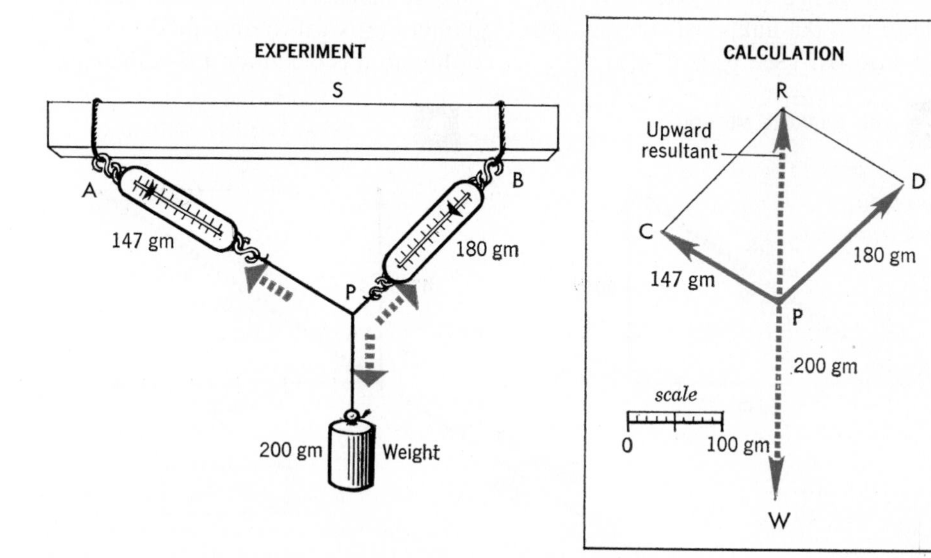

8-4 Measurement and calculation check. When the forces of 147 grams and 180 grams are marked off in proper directions and to scale, the resultant (PR) turns out to be 200 grams upward. This upward resultant of 200 grams balances the weight of 200 grams and keeps it from moving downward.

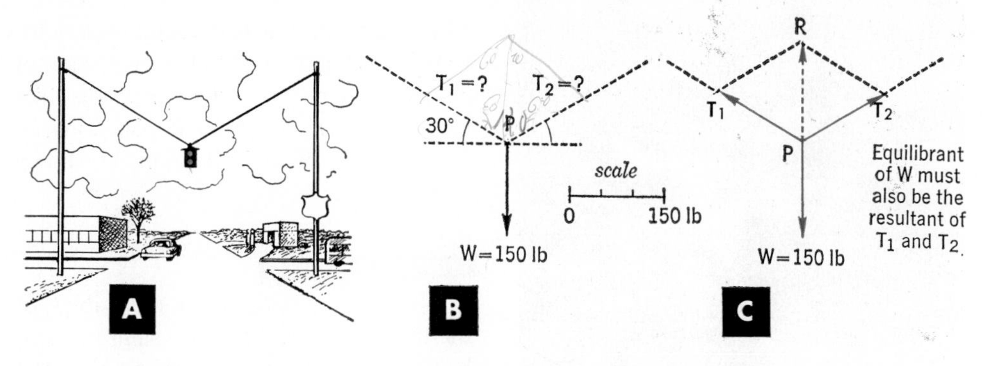

8-5 To find the tension on the supporting cables a vector diagram (B) is drawn showing as completely as possible all the forces acting at the point of suspension. (C) Because we know the traffic light is in equilibrium we know how to complete the diagram, drawing the tensions, T, of such a length that their resultant is equal and opposite to W.

scale a careful vector diagram showing all the forces on P. We do this as shown in Fig. 8-4. Notice that the three vectors, PW, PD, and PC, have their *tails,* not their heads, at P, and that each arrow points in the direction of one of the three forces acting on P. Notice that their lengths are all drawn to the same scale. Whenever you draw a vector diagram, you must draw it in this manner if you are to get accurate answers.

Now we are in a position to decide whether force PW, acting as equilibrant, is equal and opposite to the resultant of the two forces, PC and PD. We can decide by constructing a parallelogram. We first draw through C and D lines parallel to PC and PD. Then we draw the line PR through the point R where they meet. We note that this diagonal PR extends vertically upward and is precisely equal and opposite to PW. Hence, the forces PC, PD, and PW are in equilibrium. (Since PC and PD do not form a right angle, you cannot use the Pythagorean Theorem to find the length PR in this problem.)

In the same manner you can show from the same vector diagram that force PC is exactly balanced by the resultant of forces PD and PW. Construct the parallelogram and draw the resultant. See if it is equal and opposite to PC. Likewise PD is the equilibrant of the resultant of PC and PW.

SAMPLE PROBLEM

A traffic light weighing 150 pounds is supported above the street by two cables (Fig. 8-5). The cables sag at an angle of 30° below the horizontal. What is the tension in the cables?

First draw a sketch of the situation, like Fig. 8-5B, showing all the important lengths and forces. Second, find a point on which all the important forces act. In our problem this will be the point where the two supporting wires are attached to the light. Label this point P. Third, draw a vector diagram like Fig. 8-5C, showing as completely as possible all the forces acting on P. This must be a separate diagram; it should not be drawn on top of your first sketch. In your first diagram all your lines were actual ropes, poles, etc., but in your vector diagram all your lines are vectors (in our present problem, forces). When you draw your vector diagram, be sure to draw all three forces with their tails at P. Be sure, too, that each is drawn in the proper direction, and, if you know it, to the proper length.

You will notice that both the direction and the length of the downward vector

Thomas Muir from Frederic Lewis

8-6 **The forward component of the force (OF) makes the lawnmower move forward, while the downward component (OA) pushes it into the ground.**

(the weight of the traffic light) are known, but only the *directions* of the two tension vectors pulling on the light are known. Our problem is to find their lengths.

The problem is quickly solved if you remember that the resultant of the two tension vectors must be exactly equal and opposite to the downward weight vector, W, since the point P is in equilibrium. Thus we can draw the equilibrant of W pointing upward, and from its tip draw in the sides of a parallelogram. Where the sides meet T_1 and T_2 we locate the ends of the tension vectors. By measurement of the lengths of T_1 and T_2, using the scale, we find the unknown tension on each cable to be 150 pounds. You can also find the lengths of vectors T_1 and T_2 with the help of elementary geometry. (HINT. Triangles whose angles are all equal are equilateral triangles.)

We have drawn two vector diagrams, Figs. 8-5B and 8-5C, for the sake of clarity. When you solve problems like this one it will be easier if you draw all your vectors on one diagram, as shown in Fig. 8-5C.

STUDENT PROBLEM

What is the tension in the cables of the traffic light in the above sample problem if the cables sag at an angle of 45° below the horizontal? (ANS. 106 lb)

Components of forces

It often happens that a single force has more than one effect. For example, if you push on a pole, upward and at an angle, to close a window that is out of reach, the window tends to move up; at the same time it also tends to break outward. When you push on a lawnmower (Fig. 8-6), it tends to move forward, and at the same time dig downward into the earth.

7. Dividing a force into parts

You are pulling someone on a sled with 30 pounds of force. As shown in Fig. 8-7A, if you pull upward, the front of the sled tends to rise, but this upward pull has no effect in moving the sled forward. If you were to get down near the ground and pull horizontally, as shown in Fig. 8-7B, then the upward pull would be 0, and all of your 30 pounds of effort would be used in moving the sled ahead. When you pull the sled in a normal way, with the rope at an angle as shown in Fig. 8-7C, your force has both upward and forward effects.

How shall we find the actual upward force and forward force your 30 pounds of force are exerting? We might put the question this way: What two forces, one upward and one forward, will produce the same effect as the single force of 30 pounds?

Look at parallelogram OARB (Fig. 8-7D), which has a resultant equal to the 30-pound force OR. Then forces OA (upward) and OB (forward) will have OR as their resultant. OA and OB thus can be considered equivalent to OR. These two forces, OA and OB, one upward and the other forward, are called the **components** (kŏm·pō′nĕnts) of force OR. They can be considered to be the parts of which OR is made. Together they produce the same effect as OR alone. OR is said to be **resolved** into the two components OA and OB.

Finding these components is not difficult. Using OR as a diagonal, we work backward and reconstruct the parallelogram (Fig. 8-7D). As you see, the figure happens to be a rectangle because the upward and forward components are at right angles to each other. All we need do to find the components is to drop perpendiculars from the end of the force (at R) to the vertical and horizontal lines. If OR is marked off to scale and the angles are drawn properly, we can get the amount of each component by measuring with a ruler (Fig. 8-7D).

STUDENT PROBLEM

If the boy in Fig. 8-7C pulls upward at an angle of 45° with a force of 28.3 lb, what are the vertical and horizontal components of his force? (ANS. 20 lb each way)

---

8. Using components

Figure 8-8 shows a movie marquee jutting far out over the street. Unless it is supported, the great weight of the structure will cause it to break and fall. This is prevented by the cable OA, which comes down at an angle from the wall to the edge of the marquee.

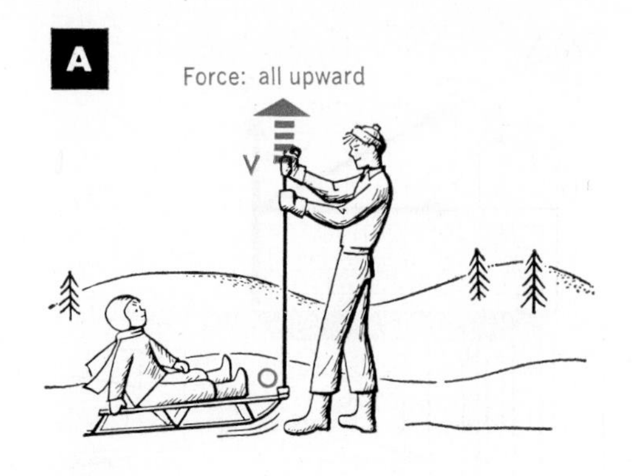

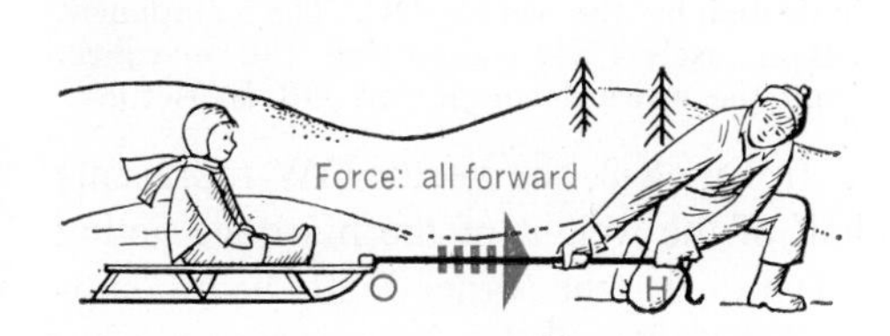

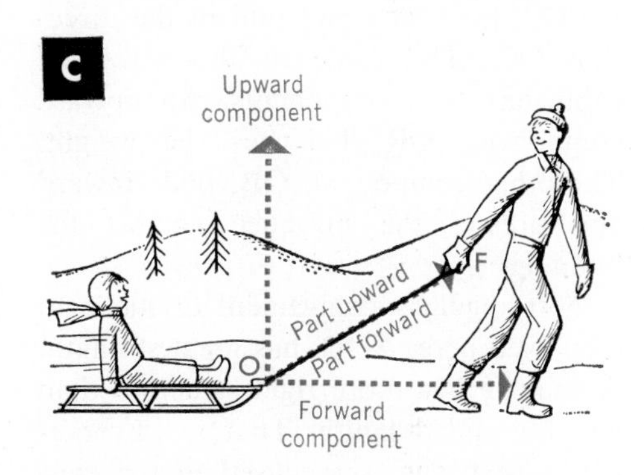

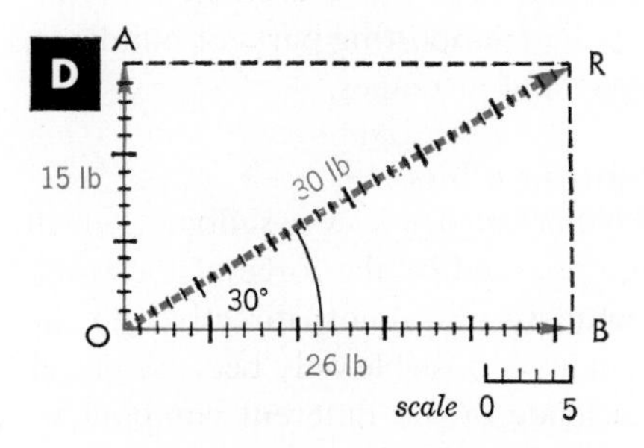

8-7 When the rope is at an angle to the ground, the force has upward and forward components. Why is it best to pull the sled at an angle?

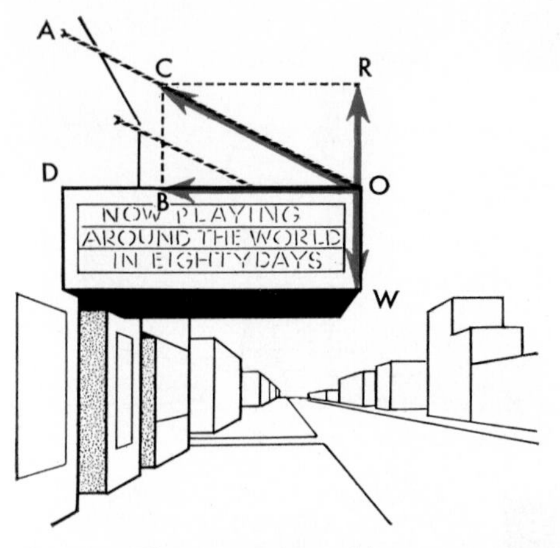

8-8 The supporting cable exerts a force indicated by the vector OC. The component OB is useless for supporting the marquee. Only the upward component OR is useful.

In Fig. 8-8, the vector OW represents half of the weight of the marquee acting downward (the other half being supported by the wall). This causes the cable OA to stretch and pull in the direction OC. The force in this stretched cable has two components. An upward component, OR, balances the weight. The other component, OB, pulls inward and presses the marquee against the building.

A triangular arrangement of the kind employed here, AOD, has great strength. A triangle is a more rigid structure than any four-sided figure. That is why networks of triangles are used in the construction of supporting parts of buildings, bridges, and airplanes.

9. Sailing a boat

It is hard to see how a sailboat, which is being pushed by the force of the wind, can actually sail almost directly into the wind. This is possible only because clever use is made of the different components of the wind's force.

Suppose we experiment with a simple sail, a flat piece of cardboard, C, and place it near the wind from a fan as shown in Fig. 8-9. Instead of being blown in the direction of the wind, OR, the flat cardboard moves in the direction OB, perpendicular to its surface. This happens because the force of the wind, OR, has two effects. The component OA, parallel to the surface, has little effect because it simply slides off the smooth almost frictionless surface of the cardboard. The only component which does affect the cardboard is OB, which is at right angles to it. The sail therefore is pushed in that direction.

Now let us consider this same surface as the sail on a boat, and fasten it to the mast M (Fig. 8-10). MS represents the sail. The force of the wind striking the sail has a component which pushes at right angles to the sail in the direction OB. But the boat is constructed with a long, narrow keel so that sideward motion is difficult. Also the boat is streamlined so that it moves easily in a forward direction. Therefore, we may break down the force on the sail, OB, into two components: one, OF, acts in the direction in which the boat can move, and the other, OL, in a direction at right angles to the boat. Thus it is the component OF which acts to push the boat forward (Fig. 8-10). Because of the keel, component OL cannot move the boat sideways to any extent; instead it causes the boat to tip while sailing (Fig. 5-6).

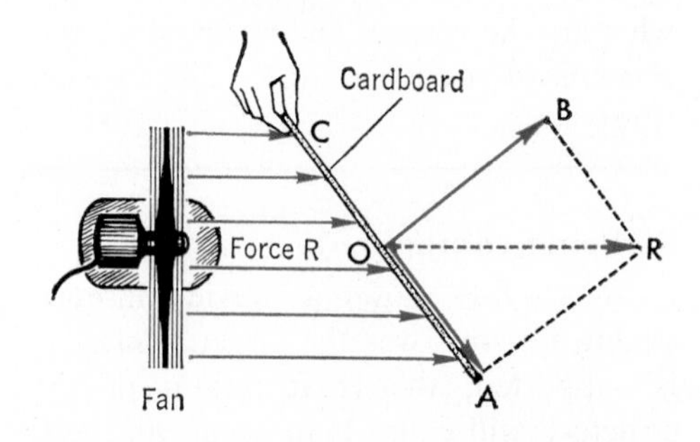

8-9 The wind from a fan exerts a force at right angles to the cardboard. Can you see the connection between this experiment and sailing a boat?

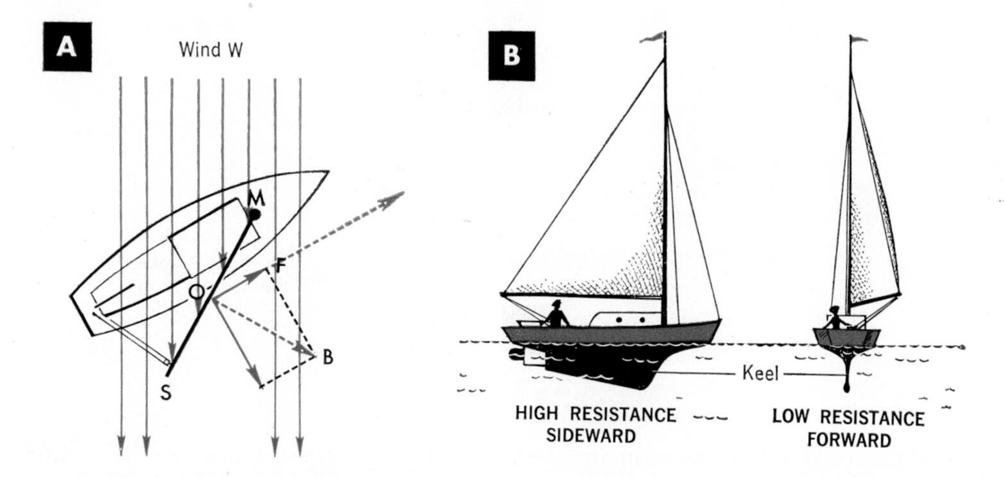

8-10 **Components of wind forces, when properly used, can make a boat sail into the wind at an angle. Note how the sailboat is built to move forward but not sideward. What effect does sideward component OL have?**

We might say, then, that a boat can sail almost directly into the wind because the designer of the craft gives all possible assistance to those components of force which are helpful in pushing it ahead, while hindering those forces that tend to push the boat in other directions.

10. Two practical problems

1. What force must you exert on a 200-pound cart in order to roll it up a plank into the back of a truck, if the plank makes an angle of 30° with the ground?

First, sketch the situation, as in Fig. 8-11A, showing all the forces on the cart and their directions. N is the force the plank exerts on the cart at right angles to the plank. Next, select a suitable scale and construct the vector diagram, as shown in Fig. 8-11B, making angle OPN = 30°, OPF = 60°, and FPW = 120°. Since the cart is in equilibrium as you hold it on the plank, we can assume that the 200-pound force, W, will be equal and opposite to the resultant of N and F. Hence we can draw PO upward as a 200-pound vector and from its tip construct the sides of the parallelogram. Thus we find the lengths of vectors F and N and can calculate their values from the scale.

The problem can also be solved analytically as follows. Since the plank forms a 30°-60° right triangle, triangle PFO in the vector diagram is also a 30°-60° right triangle. (For useful information about the lengths of the sides of such triangles, see Mathematics for Physics, page 673.) Since PO is 200 pounds, PF (the force you must exert parallel to the plane) is half of PO, or 100 pounds. The force PN with which the plank pushes up on the cart is $\sqrt{3}$ times PF, or 173 pounds.

2. A 300-pound street light is suspended over the sidewalk at the end of a beam. The beam is supported by a guy wire that makes a 45° angle with the beam (Fig. 8-12A). Find the tension on the guy wire. (Neglect the weight of the beam.)

The weight of the light, W, the tension of the guy wire, T, and the outward thrust, F, are the three forces acting on point P in the diagram. Draw a vector diagram showing these forces and their directions (Fig. 8-12B). Since P is in equilibrium, we can assume that the

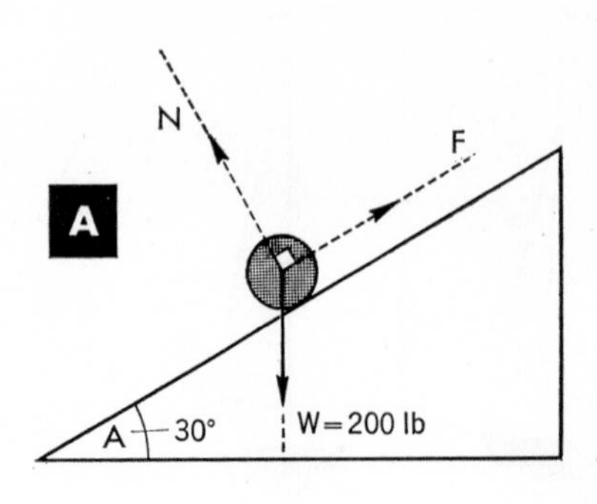

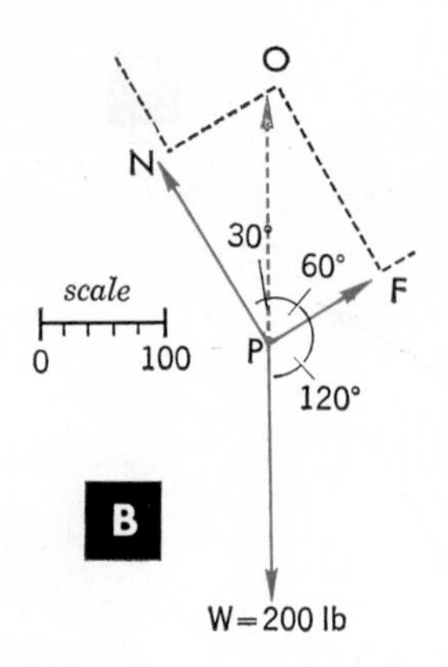

8-11 Note that, by geometry, angle OPN must be drawn equal to angle A; that is, 30°.

equilibrant of W will be the resultant of T and F. (We might equally well assume that the equilibrant of T is the resultant of W and F, or again, that the equilibrant of F is the resultant of W and T. In each case we would draw a different equilibrant, but in all cases the values of W, F, and T turn out to be the same.) Draw O equal to W upward as shown, and from its tip construct the sides of the parallelogram. Thus the ends of vectors F and T are located. By measurement the length of T can be found to be equivalent to a force of 423 pounds.

You can solve the problem analytically in the following manner. Observe that since TOP is a 45°-45° right triangle, its hypotenuse is 1.41 times the length of its sides. Therefore the tension vector, TP, is 1.41 × PO, or TP = 1.41 × 300 = 423 lb.

STUDENT PROBLEM

What would be the tension in the guy wire in the sample problem above if the angle between the beam and the guy wire were 30°? (ANS. 600 lb)

---

11. Friction on an inclined plane

How high must you raise one end of a plank so that a block will just barely slide down the plank at constant speed when given a slight push?

We draw the forces as in Fig. 8-13. PN is perpendicular to the board, XZ,

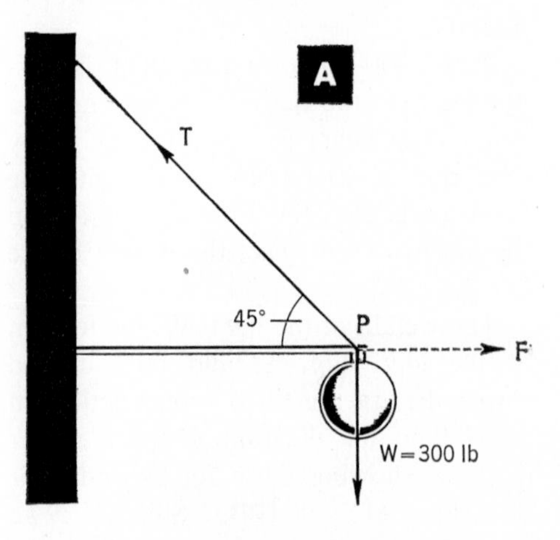

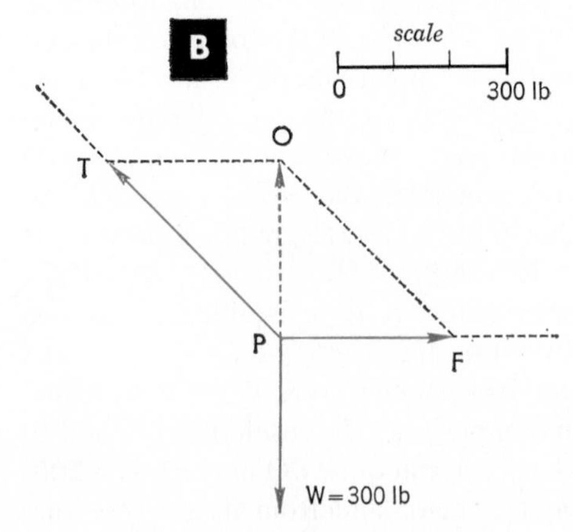

8-12 Forces on a street light at the end of a beam supported by a guy wire.

since it is the perpendicular force the board exerts against the block. Likewise PW is perpendicular to the ground, XY, since it is the force of gravity acting downward on the block. PF is the force of friction. Since the block is in equilibrium, PW, PF, and PN are in equilibrium. Hence PR, the resultant of PF and PN, is the equilibrant of PW.

Since PN is perpendicular to XZ and PR is perpendicular to XY, the angles between them must be equal. These angles are labeled *a*. Hence right triangles PNR and XYZ must be similar triangles. (Similar triangles are triangles which have the same shape; each of the angles of one triangle equals one of the angles of the other.)

Because the triangles are similar, the ratio between any two sides of one triangle is the same as the ratio between the corresponding two sides of the other. Thus

$$\frac{PF}{PN} = \frac{\text{Height of plane (YZ)}}{\text{Base of plane (XY)}}$$

But PF is the force of friction preventing the block from sliding down the plane, and PN is the force holding the surfaces together. Hence PF/PN is the coefficient of friction, *k* (which is explained on page 91).

$$k = \frac{\text{Height of plane}}{\text{Base of plane}}$$

or, from trigonometry

$$k = \tan a$$

This is a convenient and practical way of measuring the coefficient of sliding friction.

Thus, if you know the coefficient of friction, *k,* the first formula above tells you how high you must raise one end of a plank so that the block will just slide when given a slight push.

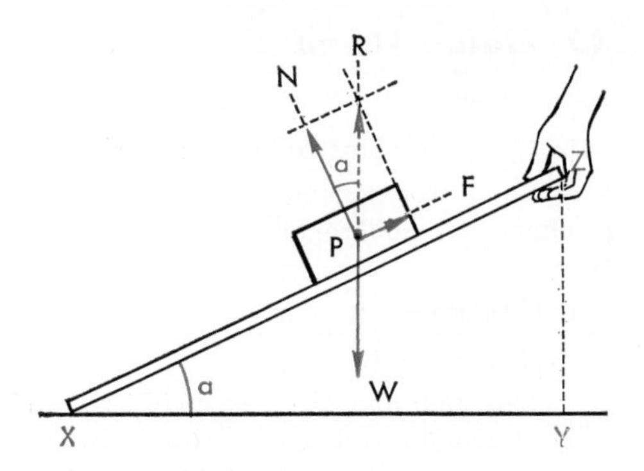

8-13 How high must you raise the end of a plank so that a block will slide down at constant speed when given a slight push?

SAMPLE PROBLEM

If you have to raise one end of a 6.00-foot plank 2.50 feet off the ground in order to make a box slide down it at a constant speed when given a slight push to start it, what is the coefficient of friction between the box and the plank?

We are given the height of the plane, 2.50 feet, but to find the base of the plane it is necessary to use the Pythagorean Theorem.

$$(6)^2 = (2.50)^2 + (\text{Base})^2$$
$$(\text{Base})^2 = 29.75$$
$$\text{Base} = 5.45 \text{ ft}$$

Hence

$$k = \frac{2.50}{5.45} = 0.458$$

Notice that the above relation for the coefficient of friction is true *only* for the inclination of the plane when the object just slides down at constant speed. It is *not* true for any other inclination of the plane.

STUDENT PROBLEM

A box will slide at constant speed down an inclined plane whose length is 2.5 ft and whose height is 1.5 ft. What do you find to be the coefficient of friction? (ANS. 0.75)

Going Ahead

THE VOCABULARY OF PHYSICS

Define each of the following:

vector — component

resultant — equilibrant

THE PRINCIPLES WE USE

Explain why

1. Forces and velocities are considered to be vector quantities.
2. In pushing up a window with a window pole, it is best to keep the pole nearly vertical.
3. The coefficient of friction of an object on a plane is equal to the height of the plane divided by the base of the plane *only* if the object slides down the plane at *constant* speed.
4. It is possible for a sailboat to sail into a wind.

SOLVING VECTOR PROBLEMS

Calculate both the size and the direction of the resultant for each situation in the table that follows:

	FORCE #1		FORCE #2	
	Size	*Direction*	*Size*	*Direction*
1.	50 lb	S	120 lb	W
2.	51 kg	E	68 kg	S
3.	500 gm	NW	500 gm	NE
4.	100 mi/hr	W	60 mi/hr	N
5.	30 lb	N	40 lb	NE

6. A man pushes along the handle of a lawnmower with a force of 100 lb. If the handle is held at 45° to the ground, how much is the force that (a) urges the mower forward, (b) pushes the mower downward?
7. A boy pulls along the rope of a sled with a force of 25 lb. If the rope makes an angle of 60° to the ground, how much force is urging the sled forward?
8. How much force is required to push a 500-lb cart up an inclined plane that makes an angle of 30° with the ground? (NOTE. Check your answer, using the Principle of Work.)
9. A 200-lb man sits in the middle of a hammock. If the two halves of the hammock make an angle of 120° with each other, what is the tension in the ropes supporting the hammock?
10. A 20-lb picture is hung from a nail by a single wire attached to each side of the picture. If the two halves of the wire make an angle of 90° with each other at the nail, what is the tension in each half of the wire?
11. A rope mooring an advertising balloon makes an angle of 30° with the vertical and has a tension of 300 lb. (a) What is the lift on the balloon? (b) What is the horizontal force of the wind on the balloon?
12. A guy wire holds up an 86.5-lb street light on the end of a horizontal beam. If the guy wire makes an angle of 60° with the beam, what is the tension on the guy wire? (Neglect the weight of the beam.)

---

13. A horizontal cargo boom supports a 1,500-lb load from its end. A guy wire to the end of the boom makes an angle of 35° with the end of the boom. (a) What is the tension in the guy wire? (b) What is the outward push exerted by the boom? (Neglect the weight of the cargo boom.)
14. A sinker on the end of a fishline is being towed behind a boat. The sinker weighs 0.3 lb in water, and the fishline makes an angle of 25° with the horizontal. (a) What is the force of the water against the sinker? (b) What is the tension in the fishline?
15. Solve Problem 12 assuming that the beam is uniform and weighs 150 lb.
16. Solve Problem 13 assuming that the cargo boom is uniform and weighs 800 lb.
17. What is the coefficient of friction if a block of wood will just slide down an inclined plane that makes an angle of 30° with the horizontal?
18. A steel ball weighing 100 gm rests in a V-shaped groove whose sides make an

angle of 45° with the horizontal. What is the perpendicular force the ball exerts against the sides of the groove?

19. What is the least force necessary to roll a 100-lb barrel up an inclined plane 50 ft long that rises 30 ft?

20. A 20-lb weight is supported from a single movable pulley. The rope through the pulley is 20 ft long, and its ends are attached to the ceiling from hooks 10 ft apart. What is the tension in the rope?

21. What would be the tension in each section of the rope in Problem 20, if the rope were 14 ft long and the weight were *tied* to the rope at a point 6 ft from one end?

PROJECT: TO STUDY THE FORCES ON A CRANE

You can easily study the forces acting on a weight supported by a simple crane. What you will need is a ringstand, a stick (or dowel) about a foot long, some strong cord, one or two spring balances, and a weight of 500 gm or 1 kg.

With a little work you will find that you can hang the weight on your crane model in the same way that the street light is supported at the end of the horizontal beam in Fig. 8-12.

You now need to measure all three forces acting on the point corresponding to P in Fig. 8-12 in order to see if the sum of any two of them really *is* equal and opposite to the third as stated in Section 10.

You already know the force corresponding to W, since it is the known weight you have hung on the end of your crane. The tension, T, in the supporting cord is found by tying the top end of the cord to one of the spring balances and finding what tension is needed to support the arm in its horizontal position. The third force, F, is the force the beam exerts on P horizontally out from the ringstand. This is found by tying a spring balance to P and pulling horizontally on it until the beam is pulled just clear of the ringstand. The force exerted by your spring balance is now substituting for the force exerted previously by the beam. Its reading is F.

Now you are in a position to draw these forces in a vector diagram similar to that in Fig. 8-12B, being careful to draw each one to the proper length and in the proper direction. A protractor will help you get the angles right.

Try adding up any two of the forces in your diagram and see if their resultant is not just equal and opposite to the third. Remember that there are three possible combinations. It is worth trying all three to convince yourself that within the accuracy of your measurements the forces on P obey the rules described in the preceding chapter.

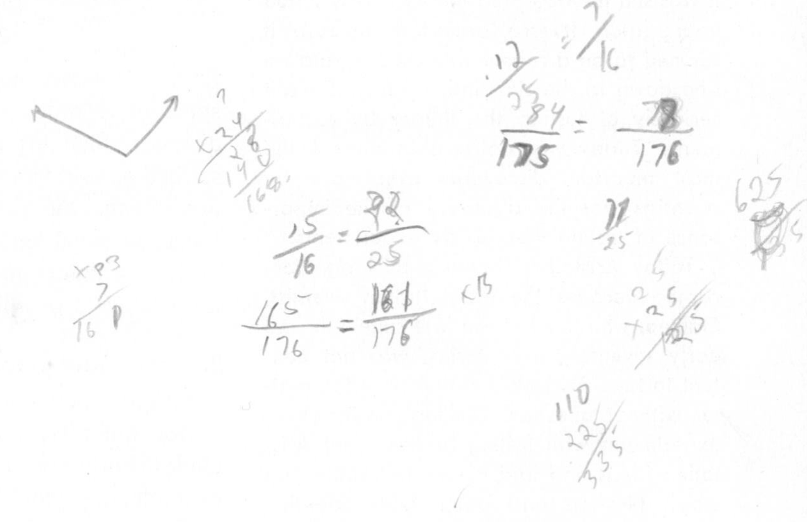

CHAPTER

9

Accelerated Motion

Which falls to the ground more quickly —a heavy object or a light one?

Take a feather and a coin and drop them at the same time from the same height. The coin falls much faster than the feather. From this we might conclude that the rate of fall is determined by the weight of the object. This was, in fact, the theory put forward by the Greek philosopher, Aristotle, several centuries before the Christian era. It was still the accepted theory nearly 2,000 years later. It was accepted because it *seemed* to be a reasonable theory, and no one down to the sixteenth century thought seriously of testing the theory by experiments. Today we realize that one of the most important discoveries ever made by scientists was the discovery of the importance of testing theories by experiments.

Today Aristotle's theory is no longer accepted because the great Italian scientist, Galileo, who lived in the late sixteenth and early seventeenth centuries, was not content to take Aristotle's authority in the matter without question. Galileo's painstaking experiments with falling bodies upset Aristotle's ideas and laid a sure foundation on which Newton and other later scientists were able to build.

Speed, velocity, and acceleration

In order to understand Galileo's discoveries we must first be sure that we know the scientific meaning of three important words: speed, velocity, and acceleration. The meaning of the first two is made clear by a study of uniform motion.

1. Uniform motion

On entering the New Jersey Turnpike a motorist is handed a card with the time of entry stamped on it. As he leaves the turnpike the toll officer stamps the leaving time on it. A quick calculation tells the officer whether the motorist has exceeded the speed limit. Suppose, for example, the motorist enters the turnpike at 2 P.M. and leaves by an exit 150 miles away at 4 P.M. If the speed limit is 60 miles per hour, the officer can be sure that the motorist must have broken the law, because covering 150 miles in 2 hours means that he must have averaged 75 miles per hour. The officer's calculation follows:

$$\text{Average velocity } (v_{av}) = \frac{\text{Distance (s)}}{\text{Time (t)}}$$

$$v_{av} = \frac{150 \text{ mi}}{2 \text{ hr}} = 75 \text{ mi/hr}$$

If the motorist maintained a constant speed of 75 miles per hour for the entire distance, then his motion might be described as **uniform motion.** Uniform motion means the covering of equal distances in equal lengths of time. The motion of escalators and marchers on parade are examples of uniform motion.

2. Speed and velocity

Strictly speaking, since velocity is a vector quantity, the word **velocity** includes both speed and direction. For example, we should speak of a velocity of 75 miles per hour *south* or 75 miles per hour *east*. Thus, velocity can be represented by a vector. Speed, on the other hand, is only the numerical part of velocity. (A numerical quantity having no direction is called a **scalar quantity.** Speed is, therefore, a scalar quantity.) The two speeds mentioned above are identical (75 mi/hr), but the velocities differ because of the difference in direction (south and east). This distinction in meaning can be disregarded, and the words used interchangeably, when the direction of motion either is of no concern or is obvious.

Common units of speed are miles per hour (mi/hr) and kilometers per hour (km/hr). In many problems it is more convenient to use feet per second (ft/sec) or centimeters per sec (cm/sec).

To convert miles per hour to feet per second, one multiplies by 5,280 (the number of feet in a mile) and then divides by 3,600 (the number of seconds in an hour). A convenient relationship to remember is that 60 mi/hr is the same as 88 ft/sec. For example, to convert 15 mi/hr to ft/sec:

$$15 \text{ mi/hr} = \frac{15}{60} \times 88 \text{ ft/sec} = 22 \text{ ft/sec}$$

3. Average speed and instantaneous speed

Our erring motorist on the turnpike may have learned that he had better take $2\frac{1}{2}$ hours for that 150-mile trip. Then the toll officer's calculation for speed would have been

$$v_{av} = \frac{s}{t}$$

$$v_{av} = \frac{150 \text{ mi}}{2.5 \text{ hr}} = 60 \text{ mi/hr}$$

But it is entirely possible that the motorist covered the first 90 miles in 1 hour, stopped for $\frac{1}{2}$ hour to eat lunch, and then proceeded to cover the remaining 60 miles in 1 hour. The toll officer would know only that the motorist covered 150 miles at an **average speed** of 60 mi/hr, and he would be unaware that the motorist had exceeded the 60-mi/hr limit.

The speed that the speedometer shows at any moment is called the **instantaneous speed.** Although our driver averaged 60 mi/hr, there were times when his instantaneous speed exceeded 60 mi/hr, and at such moments he was breaking the law. In order to spot cases of excessive instantaneous speed, police patrol cars are necessary.

Now that we understand the meaning of speed and velocity, we are ready to look at acceleration.

4. What is acceleration?

When the speedometer hand moves up or down the dial, it means that the car is changing speed. The rate at which the speed increases is called **acceleration.** The rate at which speed decreases is called negative acceleration or **deceleration.** Every time a car starts up from rest it is accelerating, since its speed is increasing. A falling object is another example of acceleration. The slowing down of a car as the brake is applied is an example of deceleration; so is the slowing down of a ball hurled up into the air.

Since acceleration is the rate at which your speed is changing,

$$\text{Acceleration} = \frac{\text{Change in speed}}{\text{Time}}$$

For example, if you are traveling at 30 mi/hr and wish to pass a car ahead of you, it may be necessary to increase your speed to 50 mi/hr in 10 seconds. Your change in speed is $50 - 30 = 20$ mi/hr. Hence your acceleration is

$$a = \frac{20}{10} = 2 \text{ miles per hour per second (mi/hr/sec)}$$

Notice that acceleration involves two units of time, one to express the change in velocity and the other to express the time in which the change took place.

The formula for acceleration is

$$a = \frac{\text{Change in velocity}}{\text{Time taken to change}} = \frac{v_f - v_o}{t}$$

Here v_0 is the starting velocity; v_f is the final velocity. In the example given above $v_f = 50$ mi/hr and $v_0 = 30$ mi/hr. Thus

$$a = \frac{50 - 30}{10} = 2 \text{ mi/hr/sec}$$

SAMPLE PROBLEM

When the traffic light turns green, your velocity is 0 miles per hour. 10 seconds later you may have accelerated to 45 miles per hour. How great is your acceleration? Here

$$a = \frac{45 - 0}{10} = 4.5 \text{ mi/hr/sec}$$

Can you show that your acceleration could also be written as 6.6 feet per second per second (ft/sec/sec)?

STUDENT PROBLEM

What is the average acceleration of a rocket that starts from rest and attains a speed of 1,500 mi/hr in 20 sec? (ANS. 75 mi/hr/sec)

If you are interested only in objects starting from rest, such as your car at the traffic light, then $v_0 = 0$, and the change in velocity is simply v_f. Then our formula can be simplified to

$$v_f = at$$

When the acceleration is constant, then the average velocity, v_{av}, of an accelerating object is midway between the starting velocity and the final velocity. Thus for an object starting from rest,

$$v_{av} = \frac{v_f}{2}$$

In the problem above, your car during "pickup" had an average velocity

$$v_{av} = \frac{45}{2} = 22.5 \text{ mi/hr (or 33 ft/sec)}$$

Accelerated motion

Now we are in a position to understand some of the discoveries Galileo made when he studied acceleration.

Galileo, in about 1590, was one of the first scientists to study accelerated motion by means of experiments and careful measurements. In one of his experiments he observed the motion of a polished brass ball rolling down an inclined trough 12 yards long. After many careful measurements he came to the conclusion that

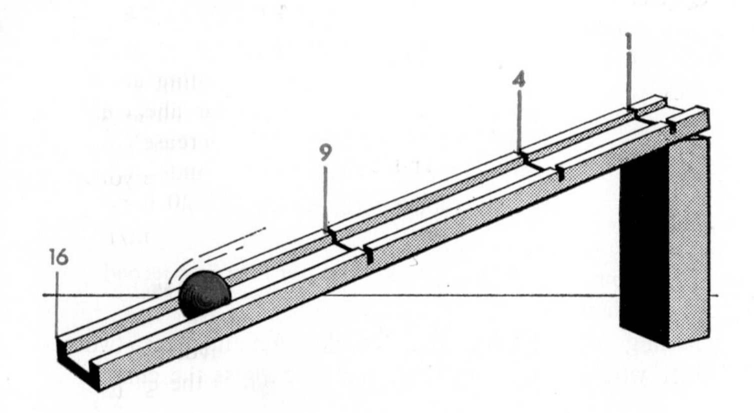

9-1 Here is one of Galileo's key experiments. What does this experiment show?

the velocity of his accelerating ball was proportional to the time it had been rolling. Thus, he was probably the first man to study the formula, $v_f = at$, that we discussed on the opposite page.

5. Distance covered by an accelerating object

By means of further careful measurements Galileo was able to show that the distance his accelerating ball traveled was proportional to the *square* of the time it had been rolling. We can repeat this experiment with a long tilted plank as shown in Fig. 9-1. We mark the plank off into 16 equal parts. Then we cut a small notch at the first, fourth, ninth, and sixteenth intervals so that the ball will click as it passes each notch. As the ball picks up speed, it covers the distance between the notches (0 to 1; 1 to 4; 4 to 9; 9 to 16) in equal periods of time, as shown by a series of regular clicks (Fig. 9-1). This fact can be checked by means of accurate timing devices.

Notice that in Galileo's experiment the distance increases as the square of the time. During the first second the ball covers the distance 0 to 1. During two seconds the *total* distance it covers is 2^2, or 4 times the distance covered in the first second; during three seconds the distance is 3^2, or 9 times as great; during four seconds it is 4^2, or 16 times as great as the distance during the first second.

The formula for distance, *s,* is

$$s = \frac{at^2}{2}$$

This formula not only describes the behavior of Galileo's rolling ball; it describes the motion of *any* object gaining or losing velocity with a constant acceleration. Moreover, it makes it possible for us to find the distance the accelerating object has traveled without our having first to find the average velocity.

6. Proving that $s = at^2/2$

We can work out the above formula from the formulas we already have. Remembering that they all apply only to a body starting from rest and moving with a *constant* acceleration, we write them down once more:

$$(1) \qquad v_f = at$$

$$(2) \qquad v_{av} = \frac{v_f}{2}$$

$$(3) \qquad s = v_{av}t$$

By substituting the value of v_f from equation (1) into equation (2), and then putting the resulting expression for v_{av} into equation (3), we obtain

$$(4) \qquad s = \frac{at^2}{2}$$

also for a body starting from rest.

7. A third acceleration formula

We can rewrite equation (1) to read

$$t = \frac{v_f}{a}$$

Substituting this expression for t into $s = at^2/2$ leads to another useful formula:

$$(5) \qquad v_f^2 = 2as$$

Thus suppose your brakes can slow down your car (decelerate it) at the rate of 11 ft/sec/sec. Within what distance can you stop if you are traveling 30 mi/hr? 30 mi/hr is 44 ft/sec. Hence,

$$(44)^2 = 2 \times 11s$$
$$s = 88 \text{ ft}$$

But if you are going twice as fast (88 ft/sec) your stopping distance is

$$(88)^2 = 2 \times 11s$$
$$s = 352 \text{ ft}$$

which is *four* times as great a distance. Thus, doubling your speed multiplies your stopping distance by four, a fact every driver ought to know.

If you now examine equations (1), (4), and (5), you will see that we now have an equation for each pair of quantities, *v* and *t*, *s* and *t*, and *v* and *s*. Thus, we are now in a position to solve problems involving accelerated motion in which we are given or wish to find *v*, *s*, *t*, or *a*. It will save you time in solving problems if you memorize equations (1), (4), and (5).

8. Applying the laws of acceleration

To illustrate the use of these equations we shall solve some practical problems dealing with acceleration.

1. If the upward acceleration of an elevator cannot be greater than 15 ft/sec/sec without causing the passengers' knees to buckle, how long will it take an elevator to attain a speed of 25 ft/sec?

Here we are given *a* and *v* and are asked to find *t*. The equation relating these three quantities is $v_f = at$. Hence we have

$$25 = 15t$$
$$t = 1.67 \text{ sec}$$

STUDENT PROBLEM

How long will it take a boy on a bicycle to reach a speed of 30 ft/sec if he accelerates at a rate of 4 ft/sec/sec? (ANS. 7.5 sec)

2. If a hockey puck is slowed down 5 ft/sec/sec as it slides across the ice, how far will it go when you give it a speed of 60 ft/sec?

Here the object is slowing down instead of gaining speed. However our formulas work equally well. The answer to the problem would be the same if the puck *gains* speed at 5 ft/sec/sec until it reaches 60 ft/sec. In either instance we are given the values for *v* and *a* and asked to find *s*. The equation relating these quantities is $v^2 = 2as$. Hence

$$(60)^2 = 2 \times 5s$$
$$s = \frac{3{,}600}{10} = 360 \text{ ft}$$

STUDENT PROBLEM

A chair sliding across the floor decelerates at a rate of 3.5 ft/sec/sec. How far will it slide if it is given an initial speed of 8 ft/sec? (ANS. 9.15 ft)

3. Your car may be able to accelerate 5 mi/hr/sec (7.34 ft/sec/sec). If the car starts from rest, can it cover a distance of 100 yards in 10 seconds? Here we must be careful that the acceleration, *a*, involves the same units of measurement as the distance, *s*, and the time, *t*. Thus, we must write $s = 300$ ft, *not* 100 yd, and we must write $a = 7.34$ ft/sec/sec, *not* 5 mi/hr/sec. In all physics problems you must be careful to use consistent units in your equations. Substituting in the equation

$$s = \frac{at^2}{2}$$
$$300 = \frac{7.34 \times t^2}{2}$$
$$t^2 = 81.8$$
$$t = 9.04 \text{ sec}$$

Hence, your car can indeed cover the 100 yards in less than 10 seconds.

STUDENT PROBLEM

A falling object accelerates at a rate of 32 ft/sec/sec. How far will it fall from rest in 3 sec? (ANS. 144 ft)

Falling motion

Which falls to the ground more quickly—a heavy object or a light one? You will remember that for nearly 2,000 years people agreed with the statement of Aris-

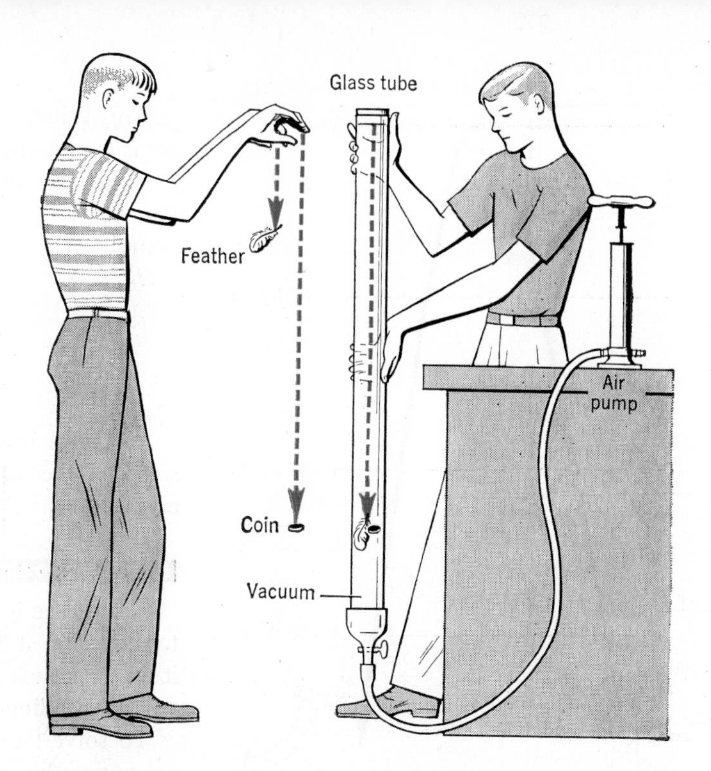

9-2 Why does the feather fall as fast as the coin does in the vacuum tube? And why does the coin fall faster than the feather in air?

totle that a heavy object always fell faster. After all, why shouldn't it? If you try dropping a coin and a feather together, you will find that Aristotle seems to be right. We say "seems" because further experiment shows that this result is due to air resistance slowing down the feather (Fig. 9-2).

Galileo had no vacuum pump, but he was able to reduce the effect of air resistance by using two objects of different weight made of a dense material. He is said to have dropped at the same instant a heavy weight and a light weight from the famous Leaning Tower of Pisa—with the result that both hit the ground at the same time. Whether or not he really did use the Leaning Tower for this purpose, he did change our ideas on the subject and show Aristotle to be wrong. Apparently, if we can neglect air resistance, the weight of a body has nothing to do with the speed at which it falls. A simple experiment may help to convince us of this fact.

9. Weight and acceleration

Suppose we take two identical objects, each weighing 1 pound, and drop them separately but side by side from the same height. Each will take the same time to reach the ground. If we tie the two objects together—making a 2-pound bundle—the rate of fall will not change. Nor will the rate of fall change if we make a bundle of ten 1-pound objects. Therefore, we conclude that objects of different weight, dropped from a given height, reach the ground in the same time, and with the same acceleration (neglecting air resistance).

10. Acceleration of freely falling bodies

We have seen that all falling objects gain speed at a uniform rate, and that this rate is not affected by the weight of the object. But what actually *is* the acceleration?

Imagine that you are riding on a freely falling body, carrying a speedometer and

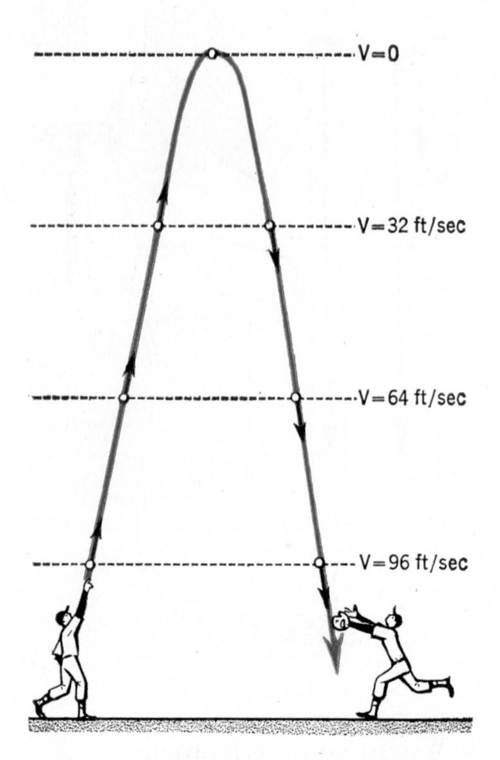

9-3 At each level on the way down, the velocity of a baseball is the same as it was at that level on the way up.

a watch. Starting from rest, the object on which you are mounted gains speed at a uniform rate. One second after the start you find you are traveling downward at a speed of 32 feet per second. Two seconds after the start you are going 64 feet per second (32 + 32). After three seconds your speed is 96 feet per second (64 + 32). After four seconds it is 128 feet per second (96 + 32). After ten seconds you are traveling 320 feet per second (32 + 32 + 32 + 32 . . . up to 10 times). Thus, by experiment you have found that a freely falling body (in the absence of air resistance) accelerates at a constant rate of 32 feet per second per second (32 ft/sec/sec). This is also equal to 980 cm/sec/sec. We shall see in a later section of this chapter why this figure varies slightly from place to place around the earth.

This value of the **acceleration of gravity** is commonly represented by the letter *g*. Thus, for objects falling freely from rest our three acceleration formulas become:

$$v_f = gt$$

$$s = \frac{gt^2}{2}$$

$$v_f^2 = 2gs$$

Problems involving falling objects are solved in the same way as those discussed earlier in this chapter.

SAMPLE PROBLEM

A package is dropped from a helicopter hovering at 2,000 feet. How long will it take to reach the ground, and how fast will it be falling when it lands?

To solve the first part of the problem we notice that we are given *s* and asked to find *t*. We know a = g = 32 ft/sec/sec. Hence, we use the formula:

$$s = \frac{gt^2}{2}$$

$$2{,}000 = \frac{32\,t^2}{2}$$

$$t^2 = \frac{2{,}000}{16} = 125$$

$$t = 11.2 \text{ sec}$$

In the second part of the problem we wish to find *v*, having been given the value of *s* and, of course, *g*. The formula that relates these three quantities is:

$$v_f^2 = 2gs$$

$$v_f^2 = 2 \times 32 \times 2{,}000 = 128{,}000$$

$$v_f = 358 \text{ ft/sec}$$

STUDENT PROBLEM

A boy drops a stone from a bridge 64 ft above the water. (a) How long will it take the stone to reach the water? (b) How fast will the stone be moving when it hits the water? [ANS. (a) 2.00 sec, (b) 64 ft/sec]

SAMPLE PROBLEM

If you throw a ball straight up in the air and catch it again 6 seconds later (a) how fast did you throw it, and (b) how high did it go?

This problem is slightly harder because there are two parts to the ball's flight—up and down. Obviously the upward flight covers exactly as many feet as the downward flight. Moreover, during the upward flight it loses speed at the rate of 32 ft/sec/sec, while on the way down it gains speed at the same rate. Hence we can see that it arrives back in your hands at the same speed with which it left them. In fact the two halves of the ball's flight are identical in duration, and at each level the upward and downward velocities are the same, though in opposite directions (Fig. 9-3).

To solve part (a) of the problem we need only consider the second half of the ball's flight. It lasts for 3 seconds. The speed of the ball, v_f, the time of fall, t, and the acceleration, g, are related by the formula:

$$v_f = gt = 32 \times 3 = 96 \text{ ft/sec}$$

This must also be the speed at which you threw it upward.

To solve part (b) we need merely ask how far an object falls from rest in 3 seconds. This involves the quantities s, t, and g. Hence our formula is:

$$s = \frac{gt^2}{2}$$
$$= \frac{32 \times (3)^2}{2} = 144 \text{ ft}$$

STUDENT PROBLEM

If the ball in the above sample problem had been in the air for 4 sec, what would have been (a) its initial upward velocity, (b) the distance it rose? [ANS. (a) 64 ft/sec, (b) 64 ft]

American Museum of Natural History

9-4 **The "flying" squirrel is a living parachute. He spreads out his legs and tail and increases the air resistance so much that his speed is checked. As a result, he can jump from considerable heights without injury.**

11. Parachuting to earth

As you remember, our formulas apply only when we neglect the resistance of the air. Actually, unless we are dropping objects in a vacuum, the upward force exerted by air resistance is always present and increases as the object falls faster and faster. Eventually this resistance becomes equal to the force of gravity which is pulling the object down (the weight of the object). At this speed the forces on the object are balanced; its motion is no longer accelerated but becomes uniform. Notice that the falling object is now in equilibrium, since the vector sum of all the forces on it is zero.

If the surface of the object is spread out, like the surface of a feather or a piece of paper, air resistance becomes equal to the weight of the object when a very low speed has been attained. A man who drops straight out of a plane without a parachute reaches this steady speed—called the **terminal velocity**—when he is falling about 120 miles per hour (176 ft/sec). An iron ball, much more compact than a man, continues to accelerate until it reaches a much greater speed. A squirrel, whose bushy tail and outspread legs offer a large surface to resist the air, has a low terminal velocity and can therefore safely jump from a height (Fig. 9-4).

Air resistance explains why an open parachute permits a flyer to make a safe jump. The larger the parachute, the lower the terminal velocity—and the safer the fall.

Projectile Motion

A bullet in flight differs in only one respect from an object falling freely straight down. The bullet has been given a component of horizontal velocity by the gun. But the horizontal velocity of the bullet has no effect whatever upon the downward motion, and hence a bullet projected horizontally from, say, the top of a cliff will hit the ground below at precisely the same moment as if it had been dropped straight down from rest. The same would be true for a golf ball driven from the cliff.

You can verify this quite simply by snapping two pieces of chalk off a table, as shown in Fig. 9-5, and listening to the click each makes as it hits the floor. One piece of chalk flies some distance across the room, but the other barely topples off the table and falls straight down. As they strike the floor the two clicks sound together as one.

12. Throwing a baseball

Suppose you throw a baseball horizontally from a tall building at a speed of 50 feet per second. The ball will travel horizontally 50 feet every second (neglecting air resistance). Let us assume that gravity is not acting. In that case line ABCD in Fig. 9-6B would show us the distance the ball travels every second. One second after being thrown it would be 50 feet away (at A); in two seconds it would be 100 feet away (at B); in three seconds, 150 feet (at C); in four seconds, 200 feet (at D), etc.

But, as we know, gravity is acting to pull the ball downward. This downward pull is the same whether the ball is moving or not. Therefore, the ball drops 16 feet in 1 second, 64 feet in 2 seconds, 144 feet in 3 seconds, etc. (see Fig. 9-6A), just as if it had been dropped in-

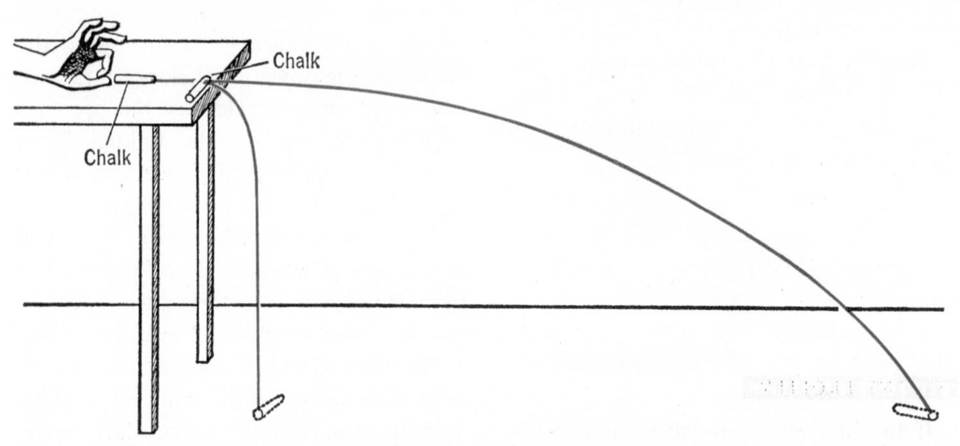

9-5 If you are careful (and lucky), the chalk you snap with your fingers will travel at high speed off the table top and will just topple the second piece off the table with negligible forward speed. Yet both will hit the floor at the same instant, as you can clearly hear. This experiment therefore shows that the horizontal speed of an object does not affect its vertical acceleration.

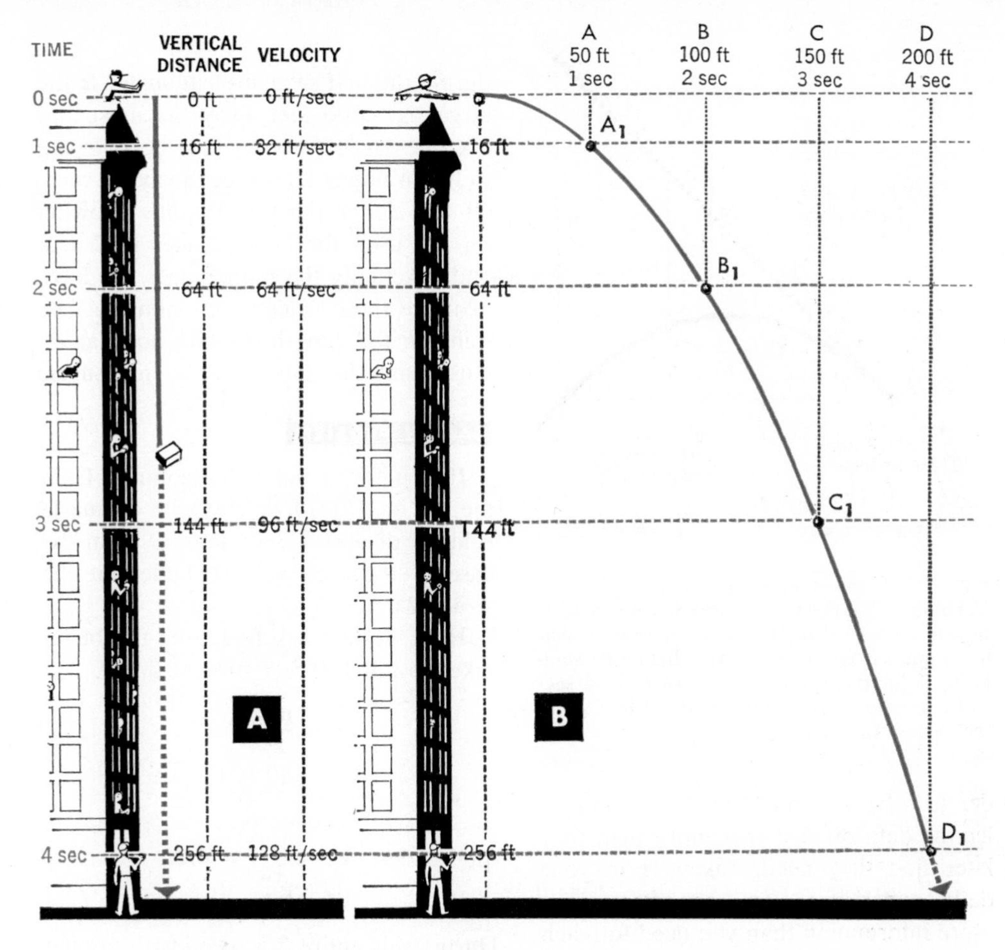

9-6 The forward and downward motions on an object thrown outward horizontally are independent of each other. The object will fall 16 feet in one second, 64 feet in 2 seconds, etc., just as though it had no forward motion.

stead of thrown. The forward motion and the downward motion can be accounted for separately; one has absolutely no effect upon the other. Figure 9-6B shows how you can calculate where the ball will be at any given time. After 1 second it will be at A_1, 50 feet forward and 16 feet down; in 2 seconds it will be at B_1, 100 feet forward and 64 feet down; in 3 seconds at C_1, 150 feet forward and 144 feet down.

Figure 9-7 shows how similar calculations are made for objects thrown upward at an angle. The shape of the curved paths in Figs. 9-6B and 9-7 is known as a **parabola.**

SAMPLE PROBLEM

If a fielder's 180-foot throw to third base is in the air for 2.0 seconds, how high in the air does the ball rise above the fielder's hand?

The ball was in the air 2.0 seconds. It must have spent half the time rising to the top of its path and half the time curving down again to the third baseman's mitt. Since it spent, therefore, 1.0 second in falling, it must have dropped

$$s = \frac{gt^2}{2} = \frac{32 \times (1)^2}{2} = 16 \text{ ft}$$

Notice that you do not need to know the horizontal distance, 180 feet, in or-

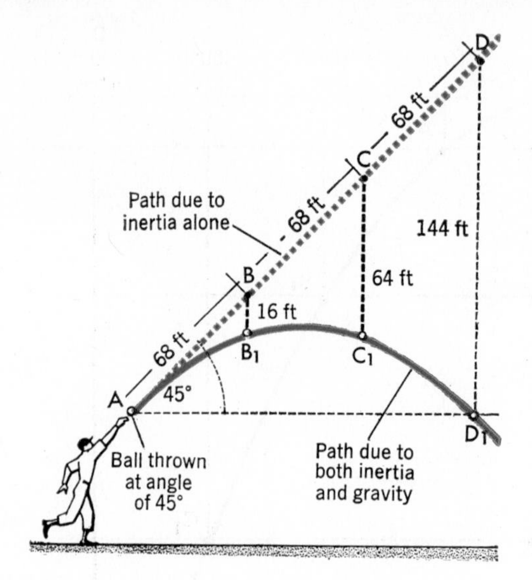

9-7 If the straight line motion of an object due to inertia would carry it from A to B in one second, then the effect of gravity will be to make it fall to B_1, 16 feet below B. Similarly, if inertia would carry it to C in 2 seconds, gravity would make it drop to C_1, 64 feet below C.

der to solve this problem. Scientists collecting data often gather more than they later find they need. Likewise in your daily experience you are often given more information than you need to reach decisions. It is important to learn how to select the information you need in order to solve a problem.

STUDENT PROBLEM

If the fielder's throw in the above sample problem had been 150 ft and the ball had been in the air for 6 sec, how high would you find the ball to have risen? (ANS. 144 ft)

13. Shooting a rifle

You can see how important parabolic motion is to a hunter when he uses his gun. Suppose the bullet he is using travels at a speed of 1,000 feet per second. He knows that he must allow for the force of gravity pulling the bullet down 16 feet during the first second of its flight. If the target is 1,000 feet away, he must aim his rifle 16 feet above it. If he is trying to hit an object 2,000 feet away, he aims 64 feet above the target—thus allowing for the drop the bullet makes in 2 seconds. Actually the hunter does not have to make these calculations himself. The sight on the gun shows him how far he must raise the gun for different ranges.

SAMPLE PROBLEM

If you fire a bullet horizontally from the top of a 100-foot cliff with a muzzle velocity of 900 ft/sec, how far from the base of the cliff will the bullet hit the ground?

Here we first ask how long the bullet takes to fall 100 feet straight down.

$$s = \frac{at^2}{2}$$

$$100 = \frac{32 \times t^2}{2}$$

$$t^2 = \frac{100}{16} = 6.25$$

$$t = 2.5 \text{ sec}$$

During this entire 2.5 seconds the bullet is also traveling horizontally across the ground at a speed of 900 ft/sec. Hence it will travel a horizontal distance

$$s = 900 \times 2.5 = 2{,}250 \text{ ft}$$

STUDENT PROBLEM

What would be the range of the bullet in the above sample problem if the bullet's velocity were 1,200 ft/sec and the cliff were 64 ft high? (ANS. 2,400 ft)

■ ■ ■ ■ ■ ■ ■

The study of moving bodies has been of immense importance in creating our modern industrial world. It is not too much to say that industry as we know it would be entirely different if it were not for the studies of motion conducted by the pioneers Galileo and Newton. The laws of gravitation and of acceleration of freely falling bodies apply not only to our knowledge of our own planet, but to our understanding of the universe that surrounds us. The motion of earth satellites through the sky and the path of projectiles of all sorts fired out into space are only two familiar kinds of objects to which these laws apply.

The pendulum

14. Motion of the pendulum

The motion of a pendulum is an illustration of positive and negative acceleration (deceleration). As you watch a pendulum swing back and forth, you will notice at once that at the end of its swing its velocity is zero. Then as it swings downward from A to B in Fig. 9-8, its speed increases, and its acceleration is positive. As soon as it passes the bottom point, B, however, it begins to slow down. From B to C its acceleration is therefore negative.

The vectors in Fig. 9-8 show why the pendulum moves in this way. At B there are two equal forces acting on the pendulum in opposite directions: T, the tension on the cord, and W, the force of gravity pulling downward on the pendulum bob. But at A and C *these two forces are neither equal nor opposite. Therefore, at any point but B, the pendulum is not in equilibrium.* At A the two forces, T and W, have a resultant, F, which urges the pendulum bob down toward B. Under the action of this force, F, the pendulum gathers speed.

When the pendulum bob reaches B, the forces T and W are now in exactly opposite directions; hence the pendulum is in equilibrium for an instant. As it begins to climb the curve toward C, the two forces acting on it, T and W, are no longer opposite in direction. Their re-

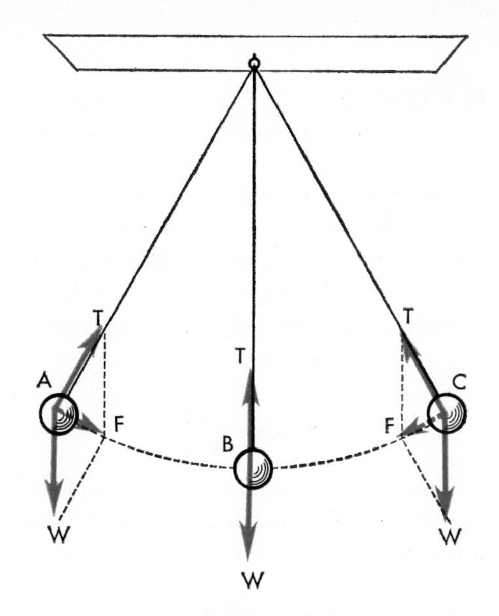

9-8 Vector diagrams of a pendulum at the beginning, middle, and end of its swing. Can you explain the direction in which the forces operate?

sultant, F, now points in a direction *opposite* to the motion of the pendulum. Since F opposes the motion, the pendulum slows down (negative acceleration).

15. The period of a pendulum

The time for one complete round trip (from A to C and back to A, for example) is called the **period of the pendulum.** You can easily prove for yourself that the period of the pendulum has nothing to do with the weight of the pendulum bob nor with the length of the swings, provided the swings are small. How would you prove it? These rather surprising facts were first discovered by Galileo, according to a well-known story, as he watched the gently swinging chandeliers in the cathedral at Pisa.

With further experiments you can show that the period of a pendulum depends only on the square root of its length. Thus, if you make the length of a pendulum four times as long, you will make its period only twice as long.

It has also been found that the period

of a pendulum depends on the acceleration of gravity, g. Indeed, the most accurate measurements of g are made, not by timing falling objects, but by carefully timing the swings of a pendulum. You too can make a rough measurement of g very simply by substituting your measurements of the period, T, and the length, L, into the formula.*

$$T = 2\pi\sqrt{\frac{L}{g}}$$

Thus, a rock on the end of a thread 1.0 meter long will swing back and forth with a period

$$T = 2 \times 3.14\sqrt{\frac{100}{980}}$$

$$T^2 = \frac{(6.28)^2 \times 100}{980}$$

$$T = 2.0 \text{ sec}$$

Notice that since $g = 980$ cm/sec/sec, L must be in centimeters, not meters.

SAMPLE PROBLEM

How long is the pendulum whose period is 1.00 seconds?

$$T = 2\pi\sqrt{\frac{L}{g}}$$

$$1.00 \text{ sec} = 2\pi\sqrt{\frac{L}{32}}$$

Squaring both sides:

$$1.00 = \frac{4\pi^2 L}{32}$$

$$L = \frac{32}{4\pi^2} = \frac{32}{39.5} = 0.810 \text{ ft}$$

STUDENT PROBLEM

A boy swinging on the end of a rope hung from the ceiling of the school gymnasium notices that his period of swing is about 7 sec. About how high is the ceiling? (ANS. about 40 ft)

* L must be measured from the point of support to the *center of gravity* of the pendulum bob.

Going Ahead

THE VOCABULARY OF PHYSICS

Define each of the following:

average speed	terminal velocity
instantaneous speed	uniform motion
velocity	parabola
scalar quantity	period of a pendulum
acceleration	deceleration
	acceleration of gravity

THE PRINCIPLES WE USE

Explain why

1. An object dropped from a great height will reach a terminal velocity.
2. A feather falls more slowly in air than a coin does.
3. A parachute permits a man to jump out of an airplane safely.
4. A hunter raises the barrel of his rifle when aiming at distant targets.
5. Supplies must be dropped from an airplane a considerable distance before the plane reaches the spot where they are supposed to land.

SOLVING PROBLEMS IN ACCELERATED MOTION

Neglect air resistance in *all* the following problems.

Calculate the unknown quantity in each problem below for objects accelerating at constant rates.

	a	$v_f - v_o$	t
1.	?	30 ft/sec	1.5 sec
2.	?	30 cm/sec	2.1 sec
3.	980 cm/sec/sec	?	5 sec
4.	32 ft/sec/sec	60 ft/sec	?
5.	5 mi/hr/sec	15 mi/hr	?

	s	a	t
6.	?	10 ft/sec/sec	3 sec
7.	100 cm	?	2.5 sec
8.	256 ft	32 ft/sec/sec	?
9.	1,960 cm	?	2 sec

	v	a	s
10.	?	10 ft/sec/sec	5 ft
11.	32 ft/sec	980 cm/sec/sec	?
12.	20 ft/sec	?	200 ft

13. If a ball starting from rest rolls 1 ft down an incline in 1 sec, how far will it roll in (a) 2 sec, (b) 5 sec, (c) 10 sec?

14. What upward velocity must be given to a ball if it is to (a) rise 128 ft, (b) be in the air for 5 sec?

15. How far horizontally from the base of a 128-ft cliff will a stone hit the ground if given a horizontal velocity of 50 ft/sec?

16. An airplane flying at a speed of 200 ft/sec at an altitude of 1,600 ft drops a package. (a) How long does it take the package to hit the ground? (b) How far will the package move horizontally during its fall? (c) Draw a graph showing the actual path of the package through the air.

17. How high is a building if a boy finds that when he throws a ball horizontally from its top at 30 ft/sec it hits the ground 120 ft from the building?

18. A bullet fired horizontally from a 100-ft cliff hits the ground 0.25 mi from the cliff. What was the muzzle velocity of the bullet?

19. A powerful rocket can reach a height of 200 mi. (a) How long will it take to fall straight down from this height? (b) With what speed will it hit the ground?

---

20. Find the period of a pendulum whose length is (a) 6 in, (b) 4 ft, (c) 30 cm, (d) 10 m.

21. A bomber flying north at 300 mi/hr and at an altitude of 10,000 ft drops a bomb on a destroyer steaming north at 30 mi/hr. How far (horizontally) from the destroyer was the bomber when it released the bomb?

22. A stone is thrown vertically upward with a velocity of 64 ft/sec. (a) Where is it at the end of the third second?

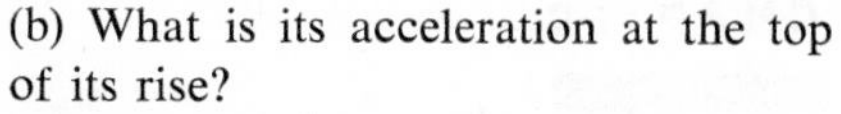

(b) What is its acceleration at the top of its rise?

23. A bullet is fired at an angle of 45° with the horizontal with a muzzle velocity of 2,000 ft/sec. What is its range? (HINT. Find the vertical and horizontal components of its initial velocity.)

24. What is the direction and magnitude of the velocity of the bullet in Problem 23 at the end of 20 sec?

25. How far will a stone fall during the sixth second after it is dropped?

26. Derive a formula for the distance an object falls during the nth second.

PROJECT: TO DEMONSTRATE THE PATH OF A PROJECTILE

At the top of a tall laboratory ringstand, clamp a meter stick horizontally, its zero end held in the clamp. With a piece of thread hang a machine nut from the 10-cm mark, its center just 1 cm below the bottom edge of the meter stick. From the 20-cm mark in the same way hang a nut 4 cm below the meter stick. From the 30-cm mark hang a nut 9 cm below the stick, at 40 cm another nut 16 cm below the stick, etc. Notice that the vertical position of each nut is proportional to the *square* of its horizontal distance from the end of the stick.

Now arrange a jet of water at the top of the ringstand. A glass tube through a cork can be clamped horizontally, one end at the 0-cm position on the meter stick, the other end connected by rubber tubing to the water supply.

When the water is turned on the stream will be projected in a parabola (like the baseball in Fig. 9-6B). The vertical distance of fall varies with the square of the time of travel ($s = at^2/2$), and hence when the water-jet speed is properly adjusted the water will follow the parabolic path defined by the nuts.

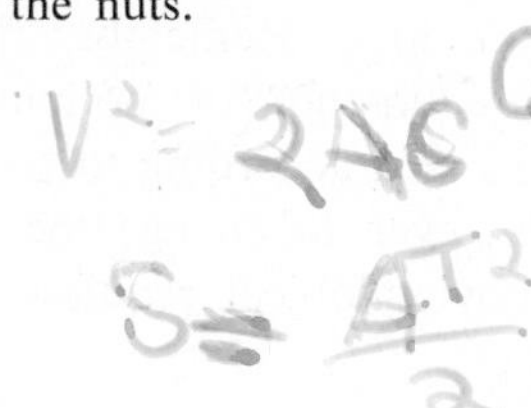

CHAPTER

10

The Laws of Motion

In the seventeenth century the great English scientist, Sir Isaac Newton, was particularly interested in the way heavenly bodies behave. His observations of objects in motion on the earth eventually led him to formulate his three Laws of Motion, which continue today to be among the most important laws of physics.

Newton's First Law

1. Inertia

Newton's First Law of Motion is also called the **Law of Inertia.** It states:

An object at rest remains at rest, or if in motion remains in motion in a straight line and at the same speed, unless acted upon by an unbalanced force.

In other words, you cannot change the velocity of an object unless you apply an unbalanced force to it. Remember that velocity involves direction of motion as well as speed, so that according to Newton's First Law, even to change the direction of a moving object requires an unbalanced force.

What do we mean by an **unbalanced force?** If we have two opposite forces of 10 pounds pulling on an object, as in an absolutely equal tug of war, each force balances the other and there is no motion toward either side. But if one force is changed to 20 pounds, this force, being greater than the other, is not balanced by it. Hence motion occurs in the direction of the greater (or "unbalanced") force.

Let us see how the Law of Inertia applies to your experiences on a moving vehicle such as a bus. You pay your fare. At the moment the bus starts to move, you remain at rest with respect to the ground. ("An object at rest remains at rest . . .") Hence *you* seem to move toward the back of the bus but only because the bus has started to drive out from under you.

Once the bus is moving at a steady speed you, too, travel at that same speed. You can walk around and do anything you wish just as though you were on solid ground. Aside from vibrations, you feel no forces, because the bus is traveling at a constant speed in a straight line, and so are you (". . . and when in motion remains in motion in a straight line and at the same speed . . ."). In other words, you are traveling in the same direction and at the same speed as the bus.

But as soon as the bus changes either the rate or the direction of its motion, you immediately become aware of that change. If the bus rounds a curve to the left, you *seem* to move to the right—because you tend to maintain the same direction you had before. If the bus slows down, you feel yourself traveling toward the front of the bus—because you tend to keep the same speed you had before. For the same reason, when the bus speeds up you feel yourself traveling toward the rear of the bus.

This tendency of an object to maintain

whatever motion it has originally is called **inertia** (ĭn·ûr′shȧ) (Fig. 10-1). Let us see how inertia acts.

2. Inertia and the pilot

The Law of Inertia governs any activity in which motion occurs, whether the activity is driving a car, running a locomotive, or flying a plane. Let us consider the plane as an example. But first try an experiment.

Hold up a piece of chalk, then drop it. As the chalk falls, it picks up speed (accelerates). There must be an unbalanced force to make it fall with increasing speed. Newton called this force which the earth exerts on all objects the **force of gravity.** When acting on objects, this force of gravity is called **weight.** When we say that an object weighs 10 pounds, we mean simply that it is being pulled downward toward the earth's center with 10 pounds of force. If the object is unsupported (dropped), then the object is pulled downward with an **unbalanced force** of 10 pounds, which is responsible for causing it to fall with accelerated motion.

To make an airplane fly above the earth, we must first balance its weight, or the downward force of gravity, with an equal and opposite upward force. In other words, we need an upward **lift** equal to the weight of the plane (Fig. 10-2).

This lift is provided by the motion of the air over the wings as the plane is driven through the air by the propeller or by a jet. As the plane gains speed, this lift increases and soon is equal to the plane's weight. At this point the downward force of gravity is balanced by the upward force of lift, and the plane can remain in the air without falling.

At first the unbalanced forward force of the propeller—called the **thrust**—causes the plane to accelerate forward. As it does so, the air resistance increases. This force of air resistance is called the **drag.** When the drag becomes equal to the thrust, the two forces balance each other, and a constant speed is reached (Fig. 10-2B), as Newton's Law requires.

When the pilot no longer wants to travel at constant speed in a straight line, he changes one of the four forces

New York *Mirror*

10-1 The driver of this truck forgot about inertia. He started up too fast and left his load of pipes behind.

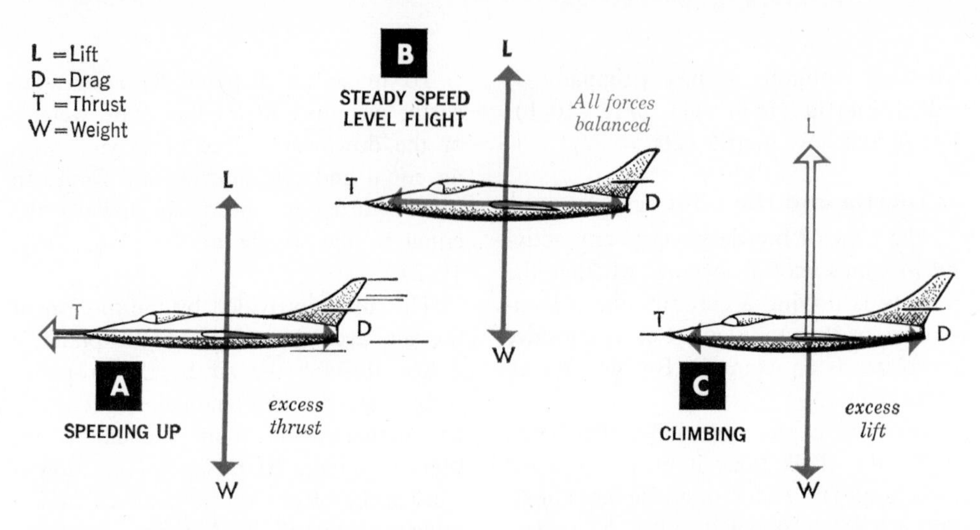

10-2 The problem of flying an airplane is really one of controlling the interaction of four forces.

acting on the plane. Thus, he causes an unbalanced force to act on the plane. Now, according to Newton's Law, the plane will no longer travel at constant speed in a straight line. For example, the plane climbs when the upward force (the lift) is made greater than the weight. It glides down when the lift is made less than the weight. When the pilot wants to gain speed, he can decrease the drag on the plane by reducing the angle at which the air strikes the wing. He can also make the propellers turn faster, thus increasing the thrust.

3. Inertia and equilibrium

Read the Law of Inertia again carefully (page 126). It describes an object on which no unbalanced forces are acting—an object for which the sum of the forces pushing in one direction is equal to the sum of the forces pushing in the opposite direction. From Section 10 of Chapter 5, you will recognize that Newton's Law of Inertia describes the behavior of an object in equilibrium.

We might even restate the Law of Inertia to read: *An object in equilibrium may be either at rest or moving at constant speed in a straight line.*

At once you will be able to think of a number of objects that are obeying Newton's Law of Inertia:

1. The chair you are sitting on.
2. You, seated in the motionless bus.
3. You, seated in the bus as it travels in a straight line at constant speed.
4. A parachutist falling at his terminal velocity (constant speed).
5. A traffic light suspended above a street intersection.

Notice that in none of these situations is there an unbalanced force acting on the object.

Now let us observe what happens when an *unbalanced* force acts on these objects.

1. If the legs of your chair collapse, the upward force exerted by the floor upon the chair no longer is as great as the downward force exerted by you on the chair. The chair begins to *change* its velocity—to accelerate downward.

2. When the motionless bus starts up, you are an object in equilibrium tending to remain at rest. Only when an unbalanced force is exerted on you (by the support you grabbed at) do you cease to be in equilibrium—that is, you begin to accelerate.

3. When the bus you are seated in swerves or stops suddenly, your inertia causes you to continue moving in a straight line at a constant speed until the force of a seat or a fellow passenger (an unbalanced force) causes your speed to change.

4. When the parachutist hits the ground, he tends to continue his downward motion, due to his inertia. However, the ground exerts on him an unbalanced upward force. At once he ceases to be an object in equilibrium. His constant downward speed changes rapidly to zero. That is, he undergoes negative acceleration as a result of the unbalanced upward force exerted on him by the ground. (Of course he is very quickly in equilibrium once more when the upward force on the ground has brought him to rest.)

5. When the guy wire supporting the traffic light breaks, the traffic light is no longer acted on by equal and opposite forces. Though inertia makes it tend to remain at rest, there is an unbalanced force acting on it—the force of gravity. At once the traffic light begins to accelerate downward.

Newton's Second Law

We have seen that when an object is in equilibrium, its velocity remains constant (often zero). We have also seen that its velocity changes only when an unbalanced force acts on it, upsetting its equilibrium. Now we must examine the connection between the unbalanced force on an object and the acceleration it produces.

4. What determines acceleration?

Suppose that we arrange a small cart so that it can be pulled by a falling weight which we attach to it by a string (Fig. 10-3). We put a heavy mass in the cart. If the suspended weight balances the maximum force of friction on the cart, the cart remains stationary. However, we can cause the cart to accelerate by hanging more weights from the string over the edge of the table. These additional weights provide the unbalanced force that produces acceleration. If we use small pulling weights, the car accelerates slowly; the rate of gain in speed (acceleration) is small. The greater the force we put on the end of the string, the greater the acceleration. Careful measurements show that the acceleration increases exactly in proportion to the unbalanced force applied to the string. Twice the unbalanced force produces twice the acceleration; an unbalanced force five times as great increases our acceleration five times.

Now let us change the mass of the cart. If we increase the mass, keeping the same weight on the string, we notice that the acceleration is less. Measurements show that if we double the mass of the cart, we have to double the pulling force on the string to maintain our rate of acceleration. If we triple the mass of the cart, we have to use three times as much force on the string to maintain the acceleration we had originally. Thus we find that the force required to produce a certain acceleration is exactly proportional to the mass of the object.

Newton combined both these facts in-

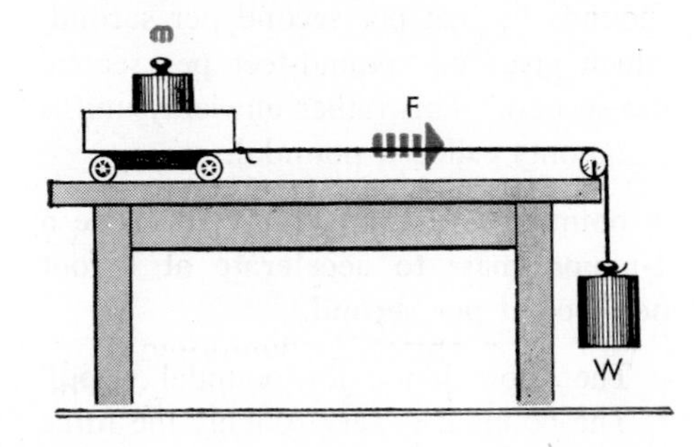

10-3 **Experimenting with Newton's Second Law of Motion.**

to his **Second Law of Motion.** This law states that:

The force applied to accelerate an object is proportional to the mass and to the acceleration produced.

Or, written as a formula:

Force applied = mass × acceleration produced

$$F = ma$$

5. Units of force

Let us suppose that the cart, together with the mass it contains, has a mass of 1 pound. We now adjust the pulling force, *F,* until measurements show an acceleration of 1 foot per second per second. How great is the accelerating force, *F?*

Newton's Law says:

Unbalanced force causing acceleration = mass moved × acceleration produced

Force = 1 lb × 1 ft/sec/sec

Force = 1 × 1 what?

You might suppose that the answer would be 1 pound. But this cannot be right since we have already chosen our unit of mass as the pound. A mass in pounds multiplied by feet per second per second obviously cannot yield an answer whose unit is still pounds.

One way out of the difficulty is to use the units formed by multiplying the unit of mass by the unit of acceleration (pounds by feet per second per second) which gives us "pound-feet per second per second." This rather unwieldy unit is commonly called a **poundal.**

A poundal is a force that will cause a 1-pound mass to accelerate at 1 foot per second per second.

The abbreviation for poundal is **pdl.**

The poundal is very roughly the force with which a silver half-dollar presses down on your hand due to gravity.

SAMPLE PROBLEM

In Fig. 10-3 the cart and the mass it contains have together a mass of 6.3 pounds. How rapidly will the cart gain speed if the tension on the string is 46 poundals?

Applying Newton's Second Law:

$$F = ma$$

$$46 \text{ pdl} = 6.3 \text{ lb} \times a$$

$$a = \frac{46}{6.3} = 7.3 \text{ ft/sec/sec}$$

STUDENT PROBLEM

An automobile with a mass of 3,000 lb is accelerated by a force of 9,000 pdl. What is its acceleration? (ANS. 3 ft/sec/sec)

■ ■ ■ ■ ■ ■ ■

In a similar manner we can define the unit of force in the metric system. If our cart and its contents (Fig. 10-3) have a mass of 1 gram and are made to accelerate 1 centimeter per second per second, then the accelerating force is defined as 1 **dyne.**

A dyne is the force that will cause a 1-gram mass to accelerate at 1 centimeter per second per second.

The dyne is a very tiny unit of force. When you hold a five-cent piece in the palm of your hand, the force of gravity pulls downward on it with a force of 5,000 dynes. A mosquito resting on your hand presses down with a force considerably more than 1 dyne.

6. The force exerted by gravity

As you know, it is the force of gravity on an object that makes it accelerate when it falls. This force of gravity, called the object's weight, is commonly measured in pounds. But it does not seem logical to measure a force in pounds, since the pound is the unit used for mass.

Force and mass are, after all, quite different quantities. As we saw in the previous section, a force is a push or a pull applied to an object, whereas mass is simply a measure of inertia—an object's resistance to acceleration when a force is applied to it.

Why is it, then, that we commonly say someone's *weight* is 120 *pounds,* when weight is a force and should be measured in *poundals?*

This is the result of the historical growth of our knowledge of force and mass. Before Newton's Laws of Motion were widely understood, people thought the mass of an object and the force of gravity acting on it were the same thing. After all, the larger the mass, the greater the force of gravity, and so the same unit, pounds, was used as a measure of both. Once this use of the word "pound" became a habit, it was impossible to break; and so today we are confronted with the confusion of measuring weight and other forces in pounds as well as measuring mass in pounds. Scientists prefer to avoid such confusion and use only poundals and dynes to measure force, but we must get along in a practical world where pounds and grams are still used to measure both force and mass.

We can easily find out the relation between the two units of force, the pound and the poundal, by considering a simple problem. Suppose a rock whose mass is 1 pound falls off a cliff. The force of gravity pulling it downward is found by:

$$\begin{aligned} F &= ma \\ &= 1.00 \text{ lb} \times 32 \text{ ft/sec/sec} \\ &= 32 \text{ pdl} \end{aligned}$$

Thus, on the rock whose mass is 1 pound, gravity pulls downward with a force of 32 poundals. Or, in other words, its weight is 32 poundals. But since a rock whose mass is 1 pound is commonly said to "weigh 1 pound" we see that **a force of 32 poundals is equivalent to a force of 1 pound.**

Similarly, if a pebble whose mass is 1 gram falls off the cliff, the force of gravity is

$$\begin{aligned} F &= ma \\ &= 1.00 \text{ gm} \times 980 \text{ cm/sec/sec} \\ &= 980 \text{ dynes} \end{aligned}$$

and we see that in the metric system a **force of 980 dynes is equivalent to a force of 1 gram.**

We too have used pounds and grams as units of force even in this physics book, and we shall continue to do so. This is permissible as long as we keep clearly in mind a very important point:

In using equations involving force and acceleration, if the force is expressed in pounds (or grams), we must change it to poundals (or dynes) before substituting.

SAMPLE PROBLEM

How fast will a 4-pound rock accelerate if a force of 2 pounds is applied?

$$\begin{aligned} F &= ma \\ 2 \times 32 \text{ pdl} &= 4a \\ a &= \frac{2 \times 32}{4} = 16 \text{ ft/sec/sec} \end{aligned}$$

STUDENT PROBLEM

How fast will a 10-lb ball accelerate if a force of 20 lb is applied to it? (ANS. 64 ft/sec/sec)

SAMPLE PROBLEM

What is the mass of an object if a force of 10 grams gives it an acceleration of 60 cm/sec/sec?

$$\begin{aligned} F &= ma \\ 10 \times 980 \text{ dynes} &= m \times 60 \\ m &= \frac{10 \times 980}{60} = 163 \text{ gm} \end{aligned}$$

STUDENT PROBLEM

What is the mass of the object in the above sample problem if the accelerating force is 20 gm and the acceleration is 100 cm/sec/sec? (ANS. 196 gm)

. . . .

There is more to the difference between mass and weight than we have considered here. Other differences will be discussed in Chapter 11.

7. Some practical problems

As a result of the preceding discussion it should be clear that the *weight* of an object on the surface of the earth *given in pounds* is *numerically* equal to the *mass* of the object in pounds. Either value can be used for m in Newton's Law.

Many practical situations require the use of Newton's Second Law together with one of the acceleration formulas.

AN ALTERNATIVE PROCEDURE

In other textbooks you will occasionally find a different procedure for solving problems in English units. Instead of multiplying the force in pounds by 32 to convert to poundals, it is perfectly possible instead to *divide* the mass in pounds by 32.

Thus instead of writing the solution of the first sample problem on page 131 as

$$F = ma$$

$$\underbrace{2 \times 32}_{\text{poundals}} = \underbrace{4a}_{\text{pounds}}$$

it would be just as accurate to write

$$\underbrace{2}_{\text{pounds}} = \underbrace{\frac{4a}{32}}_{\text{slugs}}$$

Thus the pound is treated as a unit of **force.** The mass, when divided by g, is now in units called "slugs." The slug is a mass to which a force of 1 pound will give an acceleration of 1 ft/sec/sec, and hence equivalent to a mass of 32 pounds.

If, in $F = ma$, we write mass as pounds/32 or W/g, then Newton's Law becomes

$$F = \frac{W}{g} a$$

where both F and W are in pounds.

A locomotive exerts a force of 8,000 pounds on a train weighing 320,000 pounds. How rapidly does the train gather speed?

Here $F = 8{,}000$ lb and $W/g = \frac{320{,}000}{32}$ slugs. Hence, we find the acceleration of the train by

$$F = \frac{W}{g} a$$

$$8{,}000 = \frac{320{,}000}{32} a$$

$$a = 0.8 \text{ ft/sec/sec}$$

What will be the acceleration of a 1-lb hockey puck which a player on a hockey rink shoves with a force of 20 lb? (ANS. 640 ft/sec/sec)

. . . .

The existence of the slug as a unit of mass only confuses the understanding of mass and force. Although all problems can be solved both by the use of slugs and by the procedure of using poundals or dynes as developed in Section 6—indeed engineers use slugs very commonly—it is never *necessary* to use slugs in the solution of a physics problem. Throughout the book we shall follow the procedure presented in Section 6, using poundals and dynes.

Let us solve three problems involving Newton's Second Law.

1. A car weighing 3,000 pounds is traveling at 30 mi/hr when it runs out of gas. It coasts to a stop in 11 seconds. How great was the retarding force of friction?

We would like to apply Newton's Second Law in order to find the force of friction decelerating the car, but we do not know the deceleration. Hence we must first find a from

$$v = at$$

Remember that 30 mi/hr = 44 ft/sec. Hence

$$44 = a \times 11$$
$$a = \frac{44}{11} = 4 \text{ ft/sec/sec}$$

Now, remembering that the weight in pounds is numerically equal to the mass in pounds, we can use

$$F = ma$$
$$= 3{,}000 \times 4 = 12{,}000 \text{ pdl}$$

Since 32 poundals are equivalent to 1 pound, this force is equivalent to $\frac{12{,}000}{32}$ pounds, or 375 pounds.

STUDENT PROBLEM

How great is the force of friction on a bicycle and rider weighing 150 lb if they coast to a stop in 18 sec from a speed of 24 mi/hr? (ANS. 9.17 lb)

. . . . .

2. A 5-gram bullet traveling 300 meters per second imbeds itself 4 centimeters in a wall. What force did it exert on the wall?

Here as before we must first find the acceleration (negative) as the bullet is brought to rest in the wall. Only then can we apply Newton's Second Law to find the force. Converting 300 meters/sec to cm/sec and using the appropriate acceleration formula,

$$v^2 = 2as$$
$$(30{,}000)^2 = 2 \times a \times 4$$
$$a = \frac{(3 \times 10^4)^2}{2 \times 4} = \frac{9 \times 10^8}{8}$$
$$= 1.13 \times 10^8 \text{ cm/sec/sec}^*$$

Now we are in a position to apply Newton's Second Law to find the force exerted on the wall.

$$F = ma$$
$$= 5 \times 1.13 \times 10^8 = 5.65 \times 10^8 \text{ dynes}$$

This is equal to

$$\frac{5.65 \times 10^8}{9.80 \times 10^2} = 5.77 \times 10^5 \text{ gm}$$

Notice how much easier it is to use powers of 10 to handle these large numbers.

STUDENT PROBLEM

A boy moves his baseball mitt six inches "with the ball" while catching a 5-oz baseball thrown at a velocity of 40 ft/sec. What is the force of the ball on his mitt? (ANS. 15.6 lb)

. . . .

3. (This is a more difficult one to solve.) A tractor is dragging a loaded supply sled weighing 2,800 pounds across the snow. The coefficient of friction between the sled and the snow is 0.25. What force must the tractor exert in order to get the sled up to 15 mi/hr within a distance of 121 feet?

The acceleration required here is found from

$$v^2 = 2as$$

Since 15 mi/hr = 22 ft/sec

$$(22)^2 = 2 \times a \times 121$$
$$a = \frac{484}{2 \times 121} = 2 \text{ ft/sec/sec}$$

The force needed to get the sled up to the required velocity must do two things.

* For the use of powers of 10 notation, see page 666.

First, it must overcome friction. The force, F, needed to overcome friction is

$$F = kW$$
$$= 0.25 \times 2{,}800 = 700 \text{ lb}$$

Second, the force must cause the sled to accelerate at 2 ft/sec/sec. The force needed to do this is

$$F = ma$$
$$= 2{,}800 \times 2 = 5{,}600 \text{ pdl}$$

The total force required is the sum of these two (700 lb plus 5,600 pdl). Since we cannot add pounds and poundals, we must change 5,600 poundals into pounds (or 700 pounds into poundals).

$$5{,}600 \text{ pdl} = \frac{5{,}600}{32} = 175 \text{ lb}$$

Hence the total force the tractor has to exert in order to accelerate the sled is

$$F = 700 \text{ lb} + 175 \text{ lb} = 875 \text{ lb}$$

From what you know about equilibrium you will see that as soon as the tractor and sled have speeded up to 15 mi/hr, the tractor need exert only 700 pounds of force to maintain this speed. This force is just equal to the force of friction, and hence the sled will be in equilibrium (forward and backward forces are equal). Thus the sled will travel at constant speed. In order to slow down the sled, the tractor must exert less than 700 pounds of force. The forward force exerted by the tractor will then be less than the force of friction. Equilibrium no longer exists and the sled decelerates. Indeed, if the tow rope should break, the sled will decelerate at the following rate:

$$F = ma$$
$$\underset{\text{(friction)}}{(32 \times 700) \text{ pdl}} = 2{,}800 \text{ lb} \times a$$
$$a = \frac{32 \times 700}{2{,}800}$$
$$= 8.00 \text{ ft/sec/sec}$$

and will, therefore, come to rest from a velocity of 22 feet per second in a distance, s, such that:

$$v^2 = 2as$$
$$(22)^2 = 2 \times 8 \times s$$
$$s = \frac{484}{2 \times 8} = 30.3 \text{ ft}$$

STUDENT PROBLEM

If the coefficient of friction between sled and snow in the above problem is 0.20 and the loaded sled weighs 3,200 lb, (a) what force must the tractor exert to get the sled up to 15 mi/hr in 121 ft? (b) how far will it coast if the tow rope breaks? [ANS. (a) 840 lb, (b) 37.8 ft]

Newton's Third Law

When you start to run, what exerts the force that makes you pick up speed? The answer may surprise you. You accelerate only because the ground pushes you forward. Similarly, automobiles, bicycles, and trains can accelerate only because the earth pushes them forward. The reason is that, according to Newton's Second Law, acceleration can occur only if an unbalanced force pushes them forward. Can you imagine any other force driving you forward as you start to run?

This may seem quite absurd to you. "After all," you say, "the only force I exerted was backward against the ground." This is, of course, exactly right. The only force *you* exerted was a force directly backward. The ground in turn exerted an exactly equal and opposite force on you—forward. It was this equal and opposite force exerted on you by the ground that made you accelerate. Forces always occur in pairs.

8. Equal and opposite forces

When you jump upward, you push downward on the earth with a force that

10-4 This steam reaction vehicle was designed by Isaac Newton. Can you explain how it works?

exceeds your weight. The equal and opposite force the earth exerts on you, being more than your weight, provides the unbalanced force that accelerates you upward until your feet leave the ground.

When the propeller of a motorboat pushes the water back, the water exerts an equal and opposite force on the boat, which drives the boat forward.

In a jet-propelled plane or a rocket, a mass of gas is hurled backward with tremendous force. But the mass of gas exerts an exactly equal and opposite force on the plane or rocket, driving it forward.

When you try to run on ice, you exert as much force backward on the ice as the coefficient of friction allows. But, because ice is so slippery, you cannot exert a very large force before your feet begin to slide. Hence, ice cannot be made to exert a very large force to accelerate you forward.

Try as you will, you will be unable to find any force that is not opposed by an exactly equal and opposite force. Newton expressed this fact in his **Third Law of Motion:**

Every force (action) is opposed by an exactly equal and opposite force (reaction).

This does not mean that every force acting on an object is opposed by an exactly equal and opposite force acting on the *same* object. If such were the case, it would be impossible for an unbalanced force to exist, and hence impossible for any object to accelerate. Remember, the Third Law of Motion involves two forces and two objects (each exerting a force on the other), not two forces and one object.

9. Applying the Third Law: rockets and jets

Suppose you were out in space with no ground, water, or air to push against. How could you move forward? You would have to push against something. In this case you would have to carry the material you wish to push against along with you. A rocket does precisely this. The material it pushes against is the burning exhaust gases it drives out behind it.

It helps to understand how a rocket operates if we think of it as a kind of gun (Fig. 10-5). When we fire a bullet, the gases which are formed inside the barrel of the gun force the accelerating bullet out the front end. At the same time, these gases force the barrel of the gun backward; this is the backward "kick" or recoil of the gun. The body of the rocket can be likened to the barrel of the gun; the hot gases coming out the open end at the back of the rocket can be compared to the bullet. The force of the recoil drives the rocket forward.

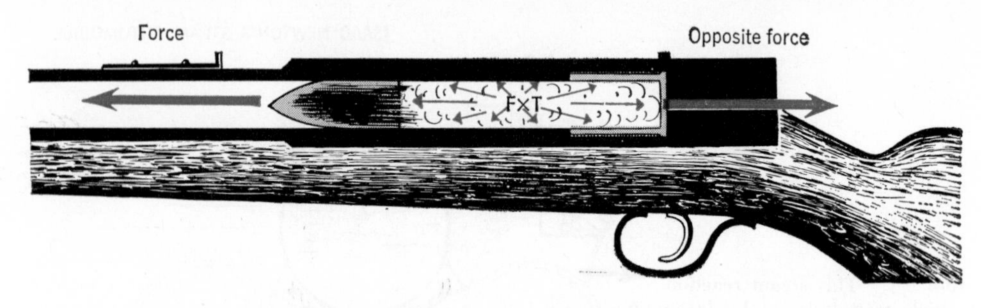

10-5 The explosion in a gun exerts forward on the bullet a force as great as the force with which the gun is kicked backward. But notice that the bullet with its smaller mass is given a much larger acceleration than the heavy gun, while the heavy gun is given a small acceleration in the opposite direction.

Clearly, if we attached a gun to a vehicle, we could make the vehicle go forward by firing the gun to the rear. As long as we keep firing, the vehicle will continue to accelerate. This actually happens when a machine gun is fired from a plane. The firing of a rear gun gives the plane a greater speed forward.

The same principle applies when auxiliary rockets are used to assist the take-off of a plane. The explosive charge shoots high-speed gas out of the back of the rockets. The recoil from the blast increases the forward speed of the plane. The "bullets," in this case, are the high-speed molecules of gas.

Because most rockets are designed to operate where there is little or no air, they have to carry along their own oxygen supply with which to burn their fuel (Fig. 10-7). A rocket can therefore travel through the vacuum of outer space. A jet-propelled plane, on the other hand, is designed to get its supply of air for burning fuel through an opening in the front of the plane. Jet planes are therefore limited to flight in the earth's atmosphere.

U.S. Navy

10-6 This machine exerts a *downward* force on the air, just as a flying bird does. The downward force is equal to the weight of the man and machine; hence the opposite force upward is too, and the man and machine are motionless in mid-air!

Rotary motion

How does it happen that we can tie a can of water to a rope and whirl it around in a circle without spilling the water? To answer this you need to know about centripetal and centrifugal forces.

As you whirl the can, you notice that you have to keep pulling inward on the rope (Fig. 10-8). This inward force which is required to make a whirling object move in a circle is called the **centripetal** (sĕn·trĭp′e·*tăl*) **force.**

If the speed of rotation is increased, the centripetal force must also increase. Thus, if you whirl a can of water around you at high speed, the centripetal force

Douglas Aircraft Co., Inc.

10-7 **Newton's Third Law at work. This Thor missile is driven skyward by a force opposite to the immense force that drives the exhaust gases downward.**

may become greater than the weight of the water in the can. As a result, the water does not fall out even though the can may be upside down (Fig. 10-8).

10. Applying centripetal force

When a wheel is spinning rapidly, whatever is on the wheel tends to be hurled off with great force. Wet clothes in the rotating drum of a washing machine can be partly dried because the loosely held water tends to travel in a straight line through holes in the drum.

In a cream separator, the heavier portion of the milk (the "water portion") is separated from the lighter (cream) portion by whirling action. The dense water portion, having more density than the lighter cream, has a greater tendency to travel in a straight line and hence moves to the outside of the container. The cream concentrates near the center and can be readily collected.

The **centrifuge** is an important application of this principle. Liquids placed in whirling containers separate out with the denser parts moving to the outside.

11. Effects of centripetal force

You have noticed that a runner leans inward (banks) on a turn. A pilot banks the wings of his plane to make a turn without sideslipping. Highway turns are usually banked to eliminate skidding.

In Fig. 10-10A we see the forces acting on a car as it travels down a road in a straight line at constant speed. You will see at once that the car is in equilibrium. If the driver wants to turn a corner, there must be exerted on the car a centripetal force, C, such as is shown in Fig. 10-10B. This unbalanced force is really the force of friction exerted sideways by the road against his tires as he turns the front wheels to the left. You will remember from Chapter 7 that the force of friction can never be greater than the coefficient of friction, represented by R in Fig. 10-10, times the

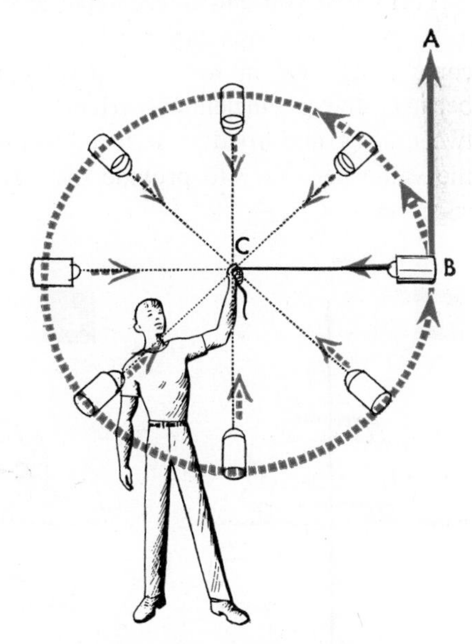

10-8 **The inward centripetal force BC causes an object to swerve away from the straight line path BA and travel in a curve.**

Clay-Adams, Inc.

10-9 This centrifuge is an application of Newton's Laws of Motion. As it whirls mixtures of various materials in the test tubes, the denser material in each tube has a greater tendency to travel in a straight line than the less dense material does and therefore is separated by being driven outward to the bottom of the test tube.

weight, W. To turn left very sharply the driver needs a relatively large sideways force, C. If it is more than the force of friction can provide, the car will skid.

Turns that you can safely make at 30 miles per hour on a dry road may become dangerous in wet or icy weather because the coefficient of friction between tires and road is reduced below the value necessary to provide the sideways force, C.

Banking makes a curve safer even in wet weather. Imagine that the car in Fig. 10-10C is rounding a curve to the left. The three forces, W, C, and R, are acting on it, just as in Fig. 10-10B. Due to the banking of the curve, however, the force, R, exerted on the car by the road is not vertical. You can see that R has a component, C, toward the left. This force is exactly the same size as the force of friction in the preceding figure, and hence it is just what is required to force the car around the curve. The vertical component of R (not drawn) is, of course, just equal to the weight of the car. By **banking the road** at the correct angle for the speed of the car and the sharpness of the curve, we have found a substitute for the force of friction. We have replaced it by the leftward component of the force, R, that the road exerts on the car (Fig. 10-10).

12. Calculating centripetal force

We can calculate centripetal force, F, by means of the following formula, in which m is the mass of the moving object, v is its velocity, and r is the radius of the curve it is following.

$$F = \frac{mv^2}{r}$$

In this formula F is in poundals when m

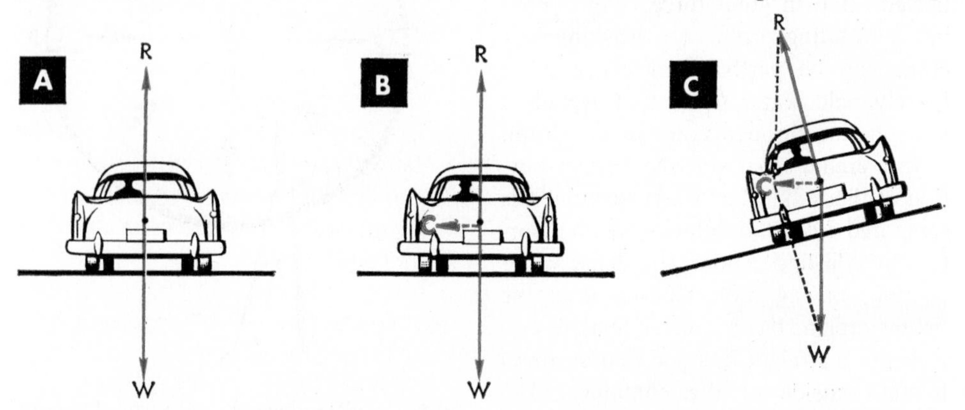

10-10 Banking the road enables the car to make the turn at high speed without skidding or turning over. Can you explain how the forces act in each of the drawings?

New York State Dept. of Commerce

10-11 This bobsled is being forced into a curved path by the centripetal force exerted on it, towards your left, by the sloping wall. If the wall had no slope, friction would have to serve the same purpose.

is in pounds, v is in feet per second, and r is in feet. F, M, v, and r may also be in dynes, grams, centimeters per second, and centimeters, respectively.

SAMPLE PROBLEM

If your 3,000-pound car is traveling at 30 mi/hr (44 ft/sec) around an unbanked curve whose radius is 100 feet, what will be the centripetal force against the tires?

$$F = \frac{3{,}000 \times (44)^2}{100}$$

$$= 58{,}100 \text{ pdl (or } 58{,}100/32 \text{ lb)}$$

Since the coefficient of friction of rubber tires on the road is approximately 0.65, the road cannot exert a sideways force on the car greater than

$$F = 0.65 \times 3{,}000 = 1{,}950 \text{ lb}$$

Since this is more than the centripetal force that is needed, the car will not quite start to skid.

STUDENT PROBLEM

Can the car in the above sample problem go as fast as 20 mi/hr around the unbanked curve without skidding on an icy day when the coefficient of friction is 0.12? (ANS. No. Maximum force of friction is 360 lb; centripetal force needed is 805 lb)

• • • • • • •

You have probably heard the expression "9g." In testing a plane, the pilot noses it down and goes into a steep dive. The plane accelerates and reaches a very high speed. Then the pilot suddenly pulls back his control stick, forcing the plane to curve upward out of its almost vertical downward path. At this moment centripetal force presses the pilot toward the center of the curve with a force equal to as much as 9 times his normal weight. This force exerted on him by the seat of the plane may equal 9 times the force which gravity normally exerts on his body. He accelerates toward the center of the curve with an acceleration of 9g—nine times his acceleration in free fall. It is impossible for a human being to stand such a force for long.

These applications of the laws of motion show us that a "law" in science is far from being just an abstract thing. Once a scientific law is established we have a tool we can use to solve the most practical problems in our daily lives.

Going Ahead

THE VOCABULARY OF PHYSICS

Define each of the following:

unbalanced force	inertia
drag	centripetal force
force of gravity	centrifuge
thrust	poundal
lift	dyne
banked road	*g*
weight	mass

THE PRINCIPLES WE USE

Explain why

1. You fall backward in a bus when it starts.
2. You fall forward in a bus when it stops.
3. You tend to strike the left side of a bus when it swerves to the right.
4. You finally reach a constant speed on a bicycle even though you continue to push on the pedals.
5. A bicycle accelerates when you exert force on the pedals starting from rest.
6. You find it difficult to run on ice.
7. A rapidly spinning wheel tends to break and fly apart.

APPLYING THE LAWS OF MOTION

1. State Newton's three Laws of Motion and give a practical example of each.
2. Using Newton's First Law, explain how the four forces of lift, weight, thrust, and drag can produce level flight at constant speed at one time and acceleration, climbing, or gliding at other times.
3. Show how Newton's three Laws of Motion are applied to the design of guns.
4. Show how Newton's Third Law is illustrated in walking, jumping, rowing, accelerating a car, driving a boat, and flying a plane.
5. Explain the similarities in the operation of rockets and jet planes. In what way does their operation differ?
6. Explain how banking a road helps provide centripetal force to help make a turn.
7. Explain why the *g* of a turn measures the force on a swerving object.

SOLVING PROBLEMS OF MOTION

Find the missing quantity in each problem in the table:

	Force	*Mass*	*Acceleration*
1.	5 lb	20 lb	?
2.	100 gm	350 gm	?
3.	10 gm	?	60 cm/sec/sec
4.	? lb	15 lb	20 ft/sec/sec
5.	? gm	5 kg	2,000 cm/sec/sec

6. A car has a mass of 3,000 lb. What force is required to give it an acceleration of 5 ft/sec/sec?
7. A sled has a mass of 500 kg. What force is needed to give it an acceleration of 150 cm/sec/sec?
8. What acceleration will an unbalanced force of 300 pdl give a mass of 15 lb?
9. What acceleration will an unbalanced force of 5,000 dynes give a mass of 250 gm?
10. What is the mass of an object that is accelerated at a rate of 8 ft/sec/sec by an unbalanced force of 272 pdl?
11. What centripetal force is needed to swing a mass of 5 kg on the end of a 1-m string at a speed of 10 cm/sec?
12. A 1-lb mass is swung at the end of a 1-ft string. How fast must the mass travel in order to exert a force of 10 lb on the string?

---

13. A force of 20 gm is applied to a 200-gm mass. How far will it go before getting up to a speed of 600 cm/sec?
14. What force must be applied by the brakes of a 3,000-lb automobile to bring it to a stop from 60 mi/hr in 8 sec?
15. An unbalanced force of 30 lb is applied to a 150-lb wagon for 10 sec. How far does it go while the force is being applied?
16. A 4,000-lb car is traveling at 45 mi/hr around an unbanked curve of radius 200 ft. What is the centripetal force against the tires?
17. If the coefficient of friction of the tires against the road is 0.65, will the car in Problem 16 take the turn without skidding?

18. A boy weighing 150 lb rides in a car of "the whip" at an amusement park. If the car is whirled through an arc of 25-ft radius at a speed of 15 mi/hr, what force does the car exert against the boy?

---

19. If the coefficient of friction between a hockey puck and the ice is 0.1, how far will a puck coast that is given a speed of 40 ft/sec?

20. If the car and load in Fig. 10-3 weigh 1 kg and the weight suspended over the pulley weighs 500 gm, what will be (a) the acceleration of the car, (b) the tension in the string? (HINT. What is the total mass being accelerated?)

21. Find the acceleration of the car in Problem 20 (Fig. 10-3) if the coefficient of friction between the car and the table is 0.2.

22. A block slides down a frictionless inclined plane 100 cm long and 60 cm high. What is its acceleration?

23. What is the acceleration of the block in Problem 22 if the coefficient of friction is 0.2?

24. At what angle must a road be banked if a car is to travel around a curve of 500-ft radius at 60 mi/hr without skidding and without any centripetal force of friction being applied to the tires?

25. How fast must a plane be flying if it is to do an inside loop of 500-ft radius and have the pilot "weightless" at the top of the loop?

26. How many g's does a pilot experience as he pulls his plane out in a curve of 1.0-mi radius at the bottom of a vertical dive at 600 mi/hr?

27. A rope passes over a single fixed pulley. A 10-lb monkey holding onto the rope on one side is exactly balanced against a 10-lb weight on the other side. What will happen to the weight if the monkey starts to climb? Assume that the rope is weightless and the pulley is frictionless.

PROJECT: VERIFYING NEWTON'S SECOND LAW

Newton's Second Law is such a fundamental law of physics that it is well worth your taking the trouble to verify it carefully for yourself.

For apparatus use two masses, m_1 and m_2, joined by a light cord which passes over a pulley as nearly frictionless as possible. If one of these masses, m_1, is heavier than the other it will fall when it is released, raising the other weight. The *difference* between the two masses is proportional to the accelerating force, F, and the *sum* of the two masses is equal to the mass being accelerated.

Two small buckets of sand are easily adjustable masses. In all cases their difference, $m_1 - m_2$, should exceed 30 grams in order to reduce the importance of pulley friction.

The procedure consists in adjusting the masses over the pulley, pulling the smaller one down to the floor, releasing it, and timing the fall of the larger mass over a measured distance.

The points to be established are: (1) With the same mass subject to varying forces, is the acceleration *directly* proportional to the force acting? (2) With the same force acting on a variety of masses, is the acceleration *inversely* proportional to the mass?

Measure at least three different combinations of mass for both (1) and (2) and graph the data. Note that some data from (1) may be used for (2). Interpret your graphs and account for any failure of your graphs to pass through the origin. Have you verified Newton's Second Law?

8. F = ma
300 = 15 A
A = 20. (ft./sec²)

9. 5000 = 250 A
A = 20 cm/sec²

10. 272 = M 8
m = 38 lb

CHAPTER

11

Gravitation

Newton wondered why the moon continues to travel around the earth. He reasoned that the moon's inertia *should* carry it off into space in a straight line. To hold the moon to its approximately circular path around our earth, an inward centripetal force must be acting. Newton offered his theory of gravity as an explanation. Since his time we have accumulated so much evidence in support of this theory that we no longer call it a theory at all but refer to it as the **Law of Gravitation.** This law, in simplified form, states that all objects in the universe attract all other objects with a force called gravitation.

Attraction between objects and the earth

Notice that the Law of Gravitation applies to *all* objects. That means that you too are exerting a force of gravitation on everything around you. You are pulling upward on the earth with as much force as the earth pulls you down. As you know, the force of gravitation of the earth on objects is called weight. Say you weigh 150 pounds. That means you are being pulled toward the earth with a force of 150 pounds. And, according to Newton's Third Law, you are pulling upward on the earth with the same force.

The force of gravitation which ordinary objects exert on one another is so slight that it is hardly noticeable. But an enormous object like the earth exerts an attraction so great that it affects the motions of other planets and the moon.

1. Measuring the force of gravitation

Laboratory experiments give us direct evidence of the truth of the Law of Gravitation. Figure 11-1 is an example. Here a large lead ball, M, of mass about 5 tons, is placed under one of the pans of a very sensitive balance which contains a spherical flask of mercury, m, whose mass is about 10 pounds. The attraction of the large mass for the smaller one pulls the balance down slightly. The weight we must add to the left-hand pan to restore balance is our measure of the attraction of gravity the lead ball exerts on the mercury.

2. Newton's equation for the Law of Gravitation

Newton discovered that the force of gravitation between two objects depends upon the product of the masses of the two attracting objects. Thus if you hold a 3-pound mass in your hand, the gravitational force between you, m_1, and the mass, m_2, depends upon the product of the masses, or m_1m_2. If you replace the 3-pound mass by a 6-pound mass, the product m_1m_2 is now twice as great; the gravitational force is doubled.

Newton also stated as part of his theory of gravitation that the force grows less as the distance between masses is increased. At twice the distance the attraction of gravitation is, not $\frac{1}{2}$, but $(\frac{1}{2})^2$ or $\frac{1}{4}$ as great (Fig. 11-2). At three times the distance the force of gravitation is $(\frac{1}{3})^2$ or $\frac{1}{9}$. In other words, Newton said that the force of gravitation varies *inversely* with the *square* of the distance. (The

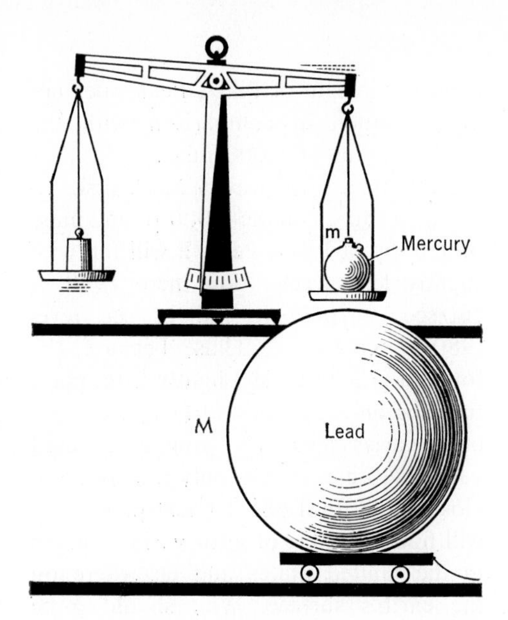

11-1 A vivid demonstration of gravitational attraction.

word "inversely" means that two quantities so depend on each other that when one increases the other decreases.)

All of the facts we have just described are summarized in Newton's equation for the Law of Gravitation. The force of gravitation in dynes, F, between two objects of masses m_1 and m_2 grams that are a distance r centimeters apart is

$$F = \frac{Gm_1m_2}{r^2}$$

where G in the equation is a constant, that is $G = 6.67 \times 10^{-8}$.

SAMPLE PROBLEM

If you are sitting 1 meter (100 cm) from one of your classmates, and you both weigh 60 kilograms (6×10^4 gm), what is the force of gravity pulling you toward each other?

$$F = \frac{(6.67 \times 10^{-8}) \times (6 \times 10^4) \times (6 \times 10^4)}{(100)^2}$$

$$= 0.024 \text{ dynes}$$

This force is comparable to the weight of the ink in the period at the end of this sentence.

Why is the force of gravity between you and the earth (your weight) so much greater than this tiny force between you and your classmate? The answer is that the mass of the earth is so very large (5.98×10^{27} gm).

STUDENT PROBLEM

Using Newton's Law of Gravitation, find the force of gravity at the Equator pulling a 1-gm mass toward the center of the earth. (Radius of earth at the Equator = 6.38×10^8 cm; mass of the earth = 5.98×10^{27} gm.) (ANS. 978 dynes)

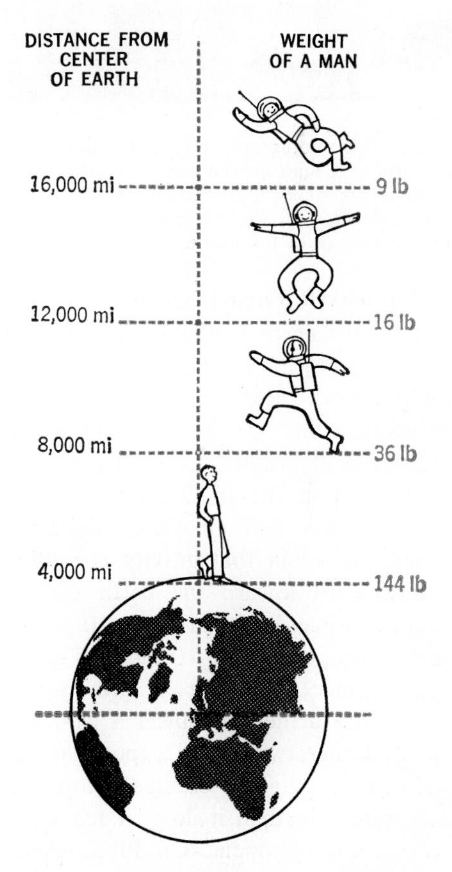

11-2 The force of gravity (weight) decreases as one goes up away from the earth. At a distance of 16,000 miles from the center of the earth, a 144-pound man would weigh only 9 pounds.

George Pal Productions, Inc.

11-3 **This photograph is from the film, *Destination Moon.* A man on the surface of the moon would jump six times higher than on the earth. Gravitational force on the moon is only one-sixth that on the surface of the earth.**

3. The earth's gravitation

Since the earth is almost spherical in shape, all objects on the earth's surface are practically the same distance from its center—that is, about 4,000 miles. The force of gravity, as we have seen, depends upon the mass of the object. When we say that an object weighs 1 pound, we mean that gravity is pulling it downward toward the earth with 1 pound of force. In the case of a 10-pound object, this pull will be 10 times as great.

But, since the force of gravity decreases the farther the object is from the earth's center, we would expect an object to weigh slightly less at the top of a high mountain than it does at sea level. This actually happens. On Pike's Peak, at an altitude of 14,000 feet, a 1,000-pound object weighs about 1 pound less than it does at sea level in the same latitude. The change in weight is small because the height of Pike's Peak does not matter much in comparison with the earth's radius of 4,000 miles.

Moreover, since gravity exerts a slightly smaller force on the 1,000-pound mass on the top of Pike's Peak, it will fall with slightly less acceleration there (978.53 cm/sec^2) * than it will at sea level (980.07 cm/sec^2). Thus, because the force of gravity varies slightly from place to place on the earth's surface, the value of *g* varies also. The commonly used value, 980 cm/sec^2, is only a convenient close value. In Table 4 (next page) you will find the value of *g* for various places in the United States and elsewhere on the earth's surface. Why should *g* be larger at the North Pole than at the Equator? (You will find the answer if you read on.)

Variations in the force of gravity on the earth's surface can be measured by a delicate instrument called a **gravimeter** (grȧ·vĭm′ē·tẽr). These measurements are often of great practical use. For instance, oil is usually found under domed caps of rock which are hidden below ground. These rock caps increase the pull of gravity slightly at the points where they occur. Hence their presence can often be detected by the gravimeter.

Although the earth is spherical in shape, it is not a perfect sphere. The earth's diameter is actually 27 miles more at the Equator than it is between the North Pole and the South Pole. Hence the same object weighs a little more at the North Pole, since it is nearer the center of the earth, than it would at the Equator. Measurements with gravimeters have shown that an object which weighs 1,000 pounds at sea level close to the North Pole weighs about 5 pounds less at the Equator. There is an

* The unit cm/sec^2 is the same as cm/sec/sec. We shall use sec^2 as meaning second per second from this point on.

additional reason for this change in weight. At the Equator the object is rotating with the earth at high speed. Hence it is subject to centripetal force which reduces the apparent weight.

The planets differ greatly among themselves in mass and diameter. We realize from our knowledge of the Law of Gravitation that the weight of an object would vary on different planets and their moons. A man who weighs 180 pounds on earth would weigh only 30 pounds on our moon, and could jump over a two-story house there (Fig 11-3).

4. Tides

The moon and the sun exert a gravitational pull on the earth which causes the tides. The net result of the attraction of the moon, and to a lesser extent that of the sun, is to lift the water in the oceans several feet. The earth rotates beneath the bulges of water and thus at most places around the earth the tides rise and fall twice a day.

Objects in motion

5. Weight and mass

We are now in a position to understand more fully the difference between "weight" and "mass."

We have just seen that weight is the downward pull of gravity on an object, and that it varies slightly from place to place on the earth's surface. Out in empty space, far from the earth or any other object, you would have practically no weight.

Imagine yourself in a space suit, floating weightless in outer space. Just in front of you floats a brick, also weightless, of course. Now, would you be willing to kick that weightless brick with all your strength? You would probably hesitate, remembering from your study of physics that the brick still has inertia—the ability to resist an accelerating force. Yes, you would bruise your toe on that weightless brick just as badly as if you and the brick were still on the school playground. The brick still has mass.

TABLE 4
VALUES OF g IN CM/SEC^2

North Pole (sea level, 90° latitude)	983.217
New York City (sea level, 41° latitude)	980.267
Pike's Peak (14,000 ft, 39° latitude)	980.068
New Orleans (sea level, 30° latitude)	979.324
Panama Canal (sea level, 9° latitude)	978.243

Thus the *mass* of an object is the same wherever it is, on the earth or off it, while the *weight* of an object changes from place to place, depending on the force of gravity. In other words, mass is the measure of an object's inertia, whereas weight is a force.

For practical purposes it is convenient, as long as we stay on the surface of the earth, to measure the mass of an object and its weight in the same units. After all, the weight of an object does not change very much from one place to another on the surface of the earth.

The connection between weight and mass is given by Newton's Second Law. When we drop an object whose mass is m, its acceleration will be g (980 cm/sec^2) as a result of the force of gravity, W, acting on it. Hence, substituting in $F = ma$, we get

$$W = mg$$

Thus on a mass of 1 gram, gravity exerts a force $W = 1 \times 980 = 980$ dynes. This is why a force of 980 dynes can be spoken of as a force of 1 gram. Similarly we can show that 32 poundals of force can be spoken of as a force of 1 pound.

In many situations we find it useful to use grams and pounds as units of force instead of mass. We are justified in doing so by the relationship.

$$W = mg \text{ (or } m = W/g)$$

where m is in grams (or pounds), W is in dynes (or poundals), and g is 980 cm/sec^2 (or 32 ft/sec^2).

6. Weight and falling motion

You will remember that Galileo is said to have experimented with falling bodies by dropping two objects of different weights off the Leaning Tower of Pisa. He was interested in finding out by means of an experiment which object would reach the ground faster. If he had known what we now know about force and gravitation, he could have predicted what would happen before he tried the experiment. He could have done it in this way.

Popular Science Monthly

11-4 Most satellites have a reflecting metal surface, but in this composite picture we see from a few feet away how a satellite might look in its orbit if it had a transparent cover. Its battery-powered instruments are recording a variety of data about the atmosphere a few hundred miles up and radioing them to receiving stations on the earth below.

He could have said that the force of gravity, F, pulling down on one of his objects of mass m_b would be

$$F = \frac{Gm_em_b}{r^2}$$

where m_e is the mass of the earth and r is the distance to the center of the earth. Then he would have written down the law that describes how his mass, m_b, would accelerate when the force of gravity, F, pulls downward on it;

$$F = m_ba$$

He would observe that in each of these equations F is the force of gravity on the ball. Hence

$$\frac{Gm_em_b}{r^2} = m_ba$$

The quantity m_b appears on both sides of the equation, and so dividing both sides by m_b gives the acceleration of his falling ball as

$$a = \frac{Gm_e}{r^2}$$

This equation contains the answer to his question about falling bodies. It says that the acceleration of a falling object depends *only* on the mass of the earth and on the distance to the center of the earth. Since m_b does not appear in the equation (you will remember that it canceled out), *the acceleration of a falling object does not depend on the object's mass at all.* Thus (neglecting air resistance), all objects have the same rate of fall regardless of mass.

7. Earth satellites

Of course you have read about, and perhaps even seen, the earth satellites, tiny man-made moons that revolve around the earth in the outer fringes of the earth's atmosphere. Why don't they fly off in a straight line into outer space? Or why don't they fall to the ground at once?

Of course Newton's First Law says that they *should* fly off into outer space in a straight line at constant speed unless they are acted upon by some unbalanced outside force. The satellites are apparently acted upon by an unbalanced outside force that is just sufficient to keep pulling them into a circular path. From what you know about objects moving in circular paths you recognize that this is a centripetal force, and that its strength is given by the formula

$$F = \frac{mv^2}{r}$$

Here m is the mass of the satellite, v is its velocity, and r is its distance from the center of the earth (the center of its circular path).

You will recognize at once that the centripetal force is provided by the force of gravity. We know that the force of gravity on an object is given by

$$F = \frac{Gm_em}{r^2}$$

where r is the distance between the satellite and the center of the earth (Fig. 11-5), and hence we can write this value for F in the centripetal force equation:

$$\frac{Gm_em}{r^2} = \frac{mv^2}{r}$$

Canceling m and r and cross-multiplying we get

$$v^2r = Gm_e$$

Here G and m_e are fixed numbers * whose product is 3.98×10^{20} dyne-cm²/gm or, in English units, 1.40×10^{16} pdl-ft²/lb.

* $G = 6.67 \times 10^{-8}$ dyne-cm²/gm², or 1.06×10^{-9} pdl-ft²/lb², and $m_e = 5.98 \times 10^{27}$ gm, or 1.32×10^{25} lb.

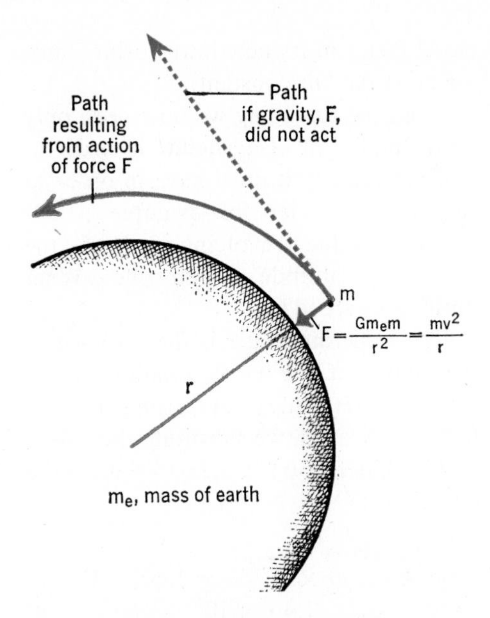

11-5 If a satellite is to remain in a circular orbit, the centripetal force, mv^2/r, required to keep it in its orbit must be supplied by the force of gravity, Gm_em/r^2. When these two forces are equal, $v^2r = Gm_e$, which means that the velocity and radius of the orbit do not depend upon the mass, m, of the satellite.

The equation states that for any satellite the product of v^2 and r must always equal a fixed number. Thus if the satellite is to revolve in an orbit close to the earth (r), then v^2 must be large. But if the satellite is to revolve in a distant orbit, say as far away as the moon, then the satellite's velocity, v, must be smaller in order that the product, v^2r, remain the same as before.

Notice carefully that m, the mass of the satellite, has canceled out. This proves the interesting fact that the weight of the satellite has nothing to do with the velocity, v, and the distance, r, of the orbit. At the same speed, heavy and lightweight satellites will revolve in precisely the same orbits.

It also follows from our equation $Gm_e = v^2r$ that if an earth satellite loses energy and hence descends to a lower orbit (that is, if r decreases), then it must travel faster in its new lower orbit, since v^2r must remain constant.

From the foregoing we now need only know the height of a satellite in order to find the velocity it must have in order to stay up. Consider for example a 50-pound satellite revolving around the earth at an altitude of 250 miles. What must be its speed?

Remembering that r is the distance to the center of the earth (radius of the earth = 3,950 miles) and using units of feet and seconds, we substitute the value, $r = (3{,}950 + 250) \times 5{,}280$ into the equation

$$v^2r = Gm_e$$
$$v^2 \times 4{,}200 \times 5{,}280 = 1.40 \times 10^{16}$$
$$v^2 = \frac{1.40 \times 10^{16}}{2.22 \times 10^7}$$
$$= 6.30 \times 10^8$$

Hence
$$v = 2.51 \times 10^4 \text{ ft/sec}$$
$$= \tfrac{60}{88} \times 2.51 \times 10^4 \text{ mi/hr}$$
$$= 17{,}100 \text{ mi/hr}$$

The given weight of 50 pounds has no bearing on the solution to the problem.

The moon itself is a similar satellite. Its motion is governed by precisely the same equation.

STUDENT PROBLEM

Using the same procedure, find the velocity with which the moon travels in its orbit 239,000 mi from the center of the earth. (ANS. 2,290 mi/hr)

You can check your answer by substituting the data into the formula $v = s/t$, making use of the fact that to travel the distance, s, all the way around the earth the moon takes a time, t, of 27.3 days.

▪ ▪ ▪ ▪ ▪ ▪ ▪

The motion of the planets around the sun is governed by the same equation too, though here the value Gm_e, where m_e is the mass of the earth, must be replaced by Gm_s, where m_s is the mass of the sun ($m_s = 329{,}390\ m_e$).

Thus the more distant planets travel more slowly. The earth, for example (93,000,000 miles from the sun), travels around the sun at a speed of 18.5 mi/sec, while Neptune, which is one of the farthest planets from the sun, travels only 3.4 mi/sec.

Going Ahead

THE VOCABULARY OF PHYSICS

Define each of the following:

gravitation	gravimeter
inversely	G
weight	*g*
mass	

THE PRINCIPLES WE USE

Explain why

1. People on the other side of the earth do not fall off into space.
2. The moon revolves around the earth instead of traveling in a straight line off into space.
3. An object will vary slightly in weight if taken to different places on the earth.
4. If you were on the moon, you could jump over a house.
5. If you weigh 110 lb, you are pulling *up* on the earth with a force of 110 lb.
6. At the center of the earth your weight would be zero.
7. The tides rise and fall, usually twice a day.
8. The acceleration of a falling object does not depend on its mass.
9. Mass is not the same thing as weight.
10. The earth satellites neither fall to earth nor fly off into space.

SOLVING GRAVITATION PROBLEMS

1. How much will a 1-lb mass weigh at an altitude above the earth's surface equal to four times the earth's radius?
2. What is the gravitational force between the moon and the earth? (Moon's mass = 7.35×10^{25} gm; moon's distance = 3.84×10^{10} cm.)

3. What is the velocity of the moon along its orbit? Use the answer from Problem 2.
4. From the answer to Problem 3 calculate the period of the moon's motion (time of one revolution) around the earth.
5. What is the centripetal force on a 1-gm mass at the Equator due to the earth's rotation? What bearing has your answer on the term "weight of an object"? (See the answer to the student problem on p. 143.)
6. How fast must an earth satellite travel if it is to remain indefinitely in an orbit 800 mi above the earth's surface?
7. The planet Mercury has only 0.05 the mass of the earth and revolves around the sun at a distance of only 36 million miles. How fast must it travel in its orbit?
8. At what altitude above the earth's surface would an artificial satellite have to be placed if it is to travel parallel to the Equator in the direction of the earth's rotation, with a time of revolution of 24 hr? Such a moon would always remain *directly over* some point on the Equator.
9. Why couldn't a satellite be made to remain motionless over any part of the earth except the Equator?

HOW AN ASTRONOMER "WEIGHS" A STAR

When astronomers find a pair of stars revolving around one another (a common situation) they can apply Newton's Second Law and the Law of Gravitation to find the mass of each star. For example the mass of Plaskett's Star, the most massive known, has been found by essentially the following method.

From the Doppler effect (see page 586) the star's velocity around its invisible companion star is found to be 250 km/sec. It is partially eclipsed by its companion every 1.24×10^6 sec, which must be the time it takes to complete its orbit. Hence its orbit must have a circumference of 250 km/sec $\times$ 1.24×10^6 sec $= 3.1 \times 10^8$ km, and a radius, $r = 5.0 \times 10^7$ km.

In the centripetal force equation, $F = m(v^2/r)$, the expression v^2/r is the centripetal acceleration—the acceleration towards the center of rotation. For Plaskett's Star, where $v = 2.5 \times 10^7$ cm/sec and $r = 5 \times 10^{12}$ cm, the acceleration towards the center of the orbit is $a = \frac{v^2}{r} = \frac{(2.50 \times 10^7)^2}{5 \times 10^{12}} = 125$ cm/sec^2. This is the value of a in $F = ma$. Now we must find F from the Law of Gravitation.

Since both stars revolve around their common center of gravity, how do we know that they have the same mass? If they don't have the same mass, they won't be equidistant from the center of rotation. Hence, the two stars must be $2 \times 5 \times 10^{12}$ cm or 10^{13} cm apart. Let us call m_p the mass of Plaskett's Star and m_c the mass of its companion. Then, as on page 146, we can write $\frac{Gm_pm_c}{r^2} = m_ca$. Cancelling m_c and substituting for the quantities which we now know, we get for the mass of Plaskett's Star $m_p = 1.8 \times 10^{35}$ gm. The mass of our own sun is 2×10^{33} gm, so this most massive of all known stars has 90 times the mass of our sun.

This treatment has made several assumptions in order to keep the derivation reasonably simple. Can you find what they are?

For a more detailed treatment of the problem you should look up the very clear and elementary article by Otto Struve on page 18 of the magazine *Sky and Telescope,* November, 1957.

CHAPTER

12

Energy and Momentum

So far in physics you have spent most of your time studying forces. From Newton's Laws you have learned how the motion of an object is changed by the forces acting on it. For example, you have seen how

—the force of gravity acting on a falling object makes it gain speed.

—centripetal force acting on a satellite changes its velocity by forcing it into a curve.

—the force of an explosion gives a bullet a high velocity.

—the force of friction slows down a moving car.

—the force driving exhaust gases backward speeds a rocket forward at take-off.

Conservation of energy

In all of these situations you have seen how forces bring about a change in velocity. They also do something else. They bring about a change in momentum and a change in energy. Energy and momentum are two of the most important things to know about a falling object, a swerving airplane, a whizzing bullet, a skidding car, or an accelerating rocket. In this chapter we shall see what ideas these two words represent and why they are so useful.

1. Kinetic energy

You will remember that energy comes in two different forms: potential energy (or stored-up energy) and kinetic energy (or energy due to motion) (page 3). You have already learned how to compute the potential energy acquired by an object that has been lifted upward through a distance, *s*, against the force of gravity, *F* (Potential energy = force × distance, or E = Fs) (page 49). This formula describes the amount of potential energy stored in the object as it rests motionless at a height, *s*, above the ground. Now we should learn how to compute the energy possessed by an object when it is moving—its **kinetic energy** (Fig. 12-1).

Suppose we fire a bullet of mass *m* out of a gun barrel *s* feet long. Let us assume that a force, *F*, acts on the bullet while it is traveling down the barrel. What will be the kinetic energy of the bullet when it comes out of the gun?

To compute the energy we compute the work done on the bullet in getting it up to speed. The work done, *E*,* is

$$E = Fs$$

where the bullet is being accelerated by the force *F*. According to Newton's Second Law

$$F = ma$$

Substituting this value of *F* into the first equation we get

$$E = mas$$

We know from our formulas for accelerated motion that the velocity of the

* We shall find that work is equivalent to energy; hence, since this chapter deals with energy, *E* is substituted for *W* in the formula developed in Chapter 4. The letter *s* is commonly used for distance.

bullet as it leaves the gun is $v^2 = 2as$, and hence that

$$as = \frac{v^2}{2}$$

When we put this value of *as* into the last formula we get

$$E = \frac{mv^2}{2} \text{ ft-pdl or dyne-cm}$$

This is the energy given to the bullet, and since this energy takes the form of motion it is kinetic energy.

Notice that in the formula for Newton's Second Law, $F = ma$, the force is either in poundals or dynes, and hence the formula $E = mv^2/2$ gives the kinetic energy in foot-poundals or dyne-centimeters (the latter are more often called **ergs**). If we wish to find the kinetic energy in foot-pounds or gram-centimeters, we shall have to divide by 32 ft/sec² or 980 cm/sec², the values of *g* in the English and metric systems, respectively. For most practical situations it is easier to use foot-pounds or gram-centimeters, and so our equation becomes

$$E = \frac{mv^2}{2g}$$

where *E* is in foot-pounds or gram-centimeters.

SAMPLE PROBLEM

If you weigh 120 pounds and are running at 10 ft/sec, what is your kinetic energy?

$$E = \frac{mv^2}{2g}$$

$$= \frac{120 \times (10)^2}{2 \times 32}$$

$$= \frac{12{,}000}{64} = 187 \text{ ft-lb}$$

Here we are using English units, and hence $g = 32$ ft/sec².

SAMPLE PROBLEM

An earth satellite weighing 3.00 kilograms is traveling with a speed of 7,000 meters/sec. What is its kinetic energy?

U.S. Air Force

12-1 The force driving the rocket exhaust gases towards the right must be matched by an opposite force that gives the rocket kinetic energy towards the left.

Since this problem involves metric units and the unit of length is meters, g = 9.80 meters/sec² and

$$E = \frac{mv^2}{2g}$$

$$= \frac{3.00 \times (7{,}000)^2}{2 \times 9.8} = 7{,}500{,}000 \text{ kg-m}$$

STUDENT PROBLEM

What is the kinetic energy of a 1-oz bullet moving 1,000 ft/sec? (ANS. 977 ft-lb)

---

2. Conservation of energy

How does the Law of Conservation of Energy (Chapter 1) apply to a falling penny?

Before the penny is released all its energy is in the form of potential energy. It has no kinetic energy, $mv^2/2g$, since $v = 0$. But when the penny is released, its height, *s*, begins to decrease, and hence its potential energy, *Fs*, steadily decreases. On the other hand, as it falls it travels faster and faster, and hence its kinetic energy steadily increases. At the instant it hits the ground, $s = 0$, and hence *Fs*, its potential energy, is all gone; it has all been turned into kinetic energy, $mv^2/2g$. When it strikes the ground, the energy is not destroyed; it does work in compressing the ground and in creating heat.

In a similar way the energy of a pendulum changes from potential energy (at A in Fig. 1-8) into kinetic energy which reaches a maximum at the bottom of its swing. As the pendulum swings upward once more its kinetic energy is converted again into potential energy. You may wonder whether energy is lost when the pendulum coasts to a stop. It is not! As the pendulum runs down, the energy is not being destroyed; it is merely being turned into heat energy as the pendulum does work against friction with the air. Thus molecules of air are made to move faster and their energy is increased at the expense of the energy of the pendulum. A coasting automobile or a speedboat or a whizzing bullet likewise turn their energy of motion into heat energy as friction brings them to a stop. In the ordinary world around you energy is never destroyed; it is merely changed from one form into another.

3. Applying the Law of Conservation of Energy

You can use the Law of Conservation of Energy to solve many practical problems.

SAMPLE PROBLEM

How much force do you exert on an 0.4-pound baseball if you give it a speed of 100 feet per second when your arm swings through 3 feet in throwing it? When you throw a baseball you exert a force, *F*, on it through the distance your arm moves, *s*. In other words, you do work on it equal to *Fs*. Since energy is not destroyed this energy is exactly equal to the kinetic energy of motion as the ball leaves your hand.

Work done by ball player = Energy acquired by ball

$$Fs = \frac{mv^2}{2g}$$

$$F \times 3 = \frac{0.4 \times (100)^2}{2 \times 32}$$

$$F = 20.8 \text{ lb (or } 20.8 \times 32 = 666 \text{ pdl)}$$

SAMPLE PROBLEM

Over a dam on a small stream 800 cubic feet of water a minute fall 20 feet. Find the kinetic energy of 1 cubic foot of water as it hits the bottom of the waterfall.

The quickest way to find the kinetic energy, $mv^2/2g$, is to remember that it equals the potential energy, Fs, the water had at the top of the dam. Hence the kinetic energy of a cubic foot of water (which weighs 62.4 lb) will be $62.4 \times 20 = 1{,}250$ ft-lb.

You could have found the kinetic energy also by figuring out the value of $mv^2/2g$. To do this, though, you need to know v^2. From our study of accelerated motion we know $v^2 = 2gs$. Substituting this expression for v^2 into the kinetic energy equation and simplifying, we get

$$E = \frac{mv^2}{2g} = ms$$

Hence

$$E = 62.4 \times 20 = 1{,}250 \text{ ft-lb}$$

as before.

STUDENT PROBLEM

What is the kinetic energy of 1 ft^3 of crushed stone (density 180 lb/ft^3) falling from a chute into a freight car 20 ft below, if 80 ft^3 per minute pass out of the chute? (ANS. 3,600 ft-lb)

■ ■ ■ ■ ■ ■ ■

There are many situations in which kinetic energy is used up doing work against friction, thereby producing heat.

SAMPLE PROBLEM

Suppose you take a running start and slide across an icy sidewalk. You might expect to get up to a speed of 10 miles per hour (this is equivalent to $(\frac{10}{60} \times 88 = 14.7$ ft/sec). You weigh 120 pounds, and your shoes have a coefficient of friction, k, of 0.10 with the ice. How far will you slide?

Here your kinetic energy, $mv^2/2g$, is all used up doing work against friction by exerting a force, F, through a distance, s.

$$\frac{mv^2}{2g} = Fs$$

but $F = kW$, so

$$\frac{mv^2}{2g} = kWs$$

$$\frac{120 \times (14.7)^2}{2 \times 32} = 0.10 \times 120 \times s$$

Canceling 120 from both sides of the equation and solving for s,

$$s = \frac{(14.7)^2}{2 \times 32 \times 0.10} = 33.8 \text{ ft}$$

It is interesting to note that since your weight cancels out, your weight has no effect on the distance you slide. Thus a row of your friends can line up and run together toward the ice with the same speed. Provided they all wear the same kind of shoes, they will all slide the same distance on the ice regardless of their size.

STUDENT PROBLEM

How far will a sled coast on the level if it is moving 16 ft/sec and the coefficient of friction is 0.18? (ANS. 22.2 ft)

Momentum

4. Changing momentum

If you have ever tried to get a stalled car started on a level road by getting out and pushing it (Fig. 12-2), you probably know that the longer you continue to push it, the faster it moves. You also know that the heavier the car the more slowly it gathers speed, so that it is easier to start it rolling if all the passengers get out. Both of these facts are summed up in an important formula:

$$Ft = \text{Change in } mv$$

or

$$Ft = \Delta mv$$

where (the Greek letter “delta”) means “change in.”

Hays from Monkmeyer

12-2 The gain in momentum of this car depends on the *unbalanced* force, F, exerted on it and on the length of time it is exerted.

The formula states that if you wish to change the velocity of an object, whose mass is m, by an amount v, you exert a force, F, on it for a period of time t. You can exert a small force for a long time, or (perhaps with the help of a towing truck) you can exert a large force for a shorter time. Whichever you do, if the product of F and t is the same, the resulting change in the momentum of the object, Δmv, will be the same. If the mass of the object does not change, the resulting change in velocity will be the same in either case.

The product, Ft, is called the **impulse,** and the product, Δmv, is the **change in momentum** that it produces. There are many everyday situations in which you apply an impulse to an object. Whenever you throw a baseball, hit a tennis ball, swing a club against a golf ball, fire a bullet from a rifle, bring a moving hammer to rest against the head of a nail, or bring a car to a halt by applying the brakes, you are changing the velocity and hence the momentum of an object by exerting a force on it for a period of time. The impulse that you apply to the object equals the resulting *change* in momentum.

Our units in the above equation are as follows: F is in poundals or dynes, t is in seconds, m is in pounds or grams, and v is in feet per second or centimeters per second. For practical applications where we express force in pounds or grams we must rewrite our equation, dividing the mass by g:

$$Ft = \Delta \frac{mv}{g}$$

where g, you will remember, is 32 ft/sec^2 or 980 cm/sec^2.

Using the above formula we can calculate how much *time* it will take for a force to accelerate an object to a given speed. (Compare this formula with $Fs = mv^2/2g$, which is used to find how much *distance* is needed for a force to accelerate an object to a given speed.)

SAMPLE PROBLEM

How long will it take you to get a 3,000-pound car up to 5 miles per hour by pushing, if you can exert on it a steady unbalanced force of 50 pounds?

Being careful to write our velocity of 5 mi/hr as $\frac{5}{60} \times 88 = 7.33$ ft/sec, we find

$$Ft = \frac{mv}{g}$$

$$50t = \frac{3{,}000 \times 7.33}{32}$$

$$t = \frac{3{,}000 \times 7.33}{50 \times 32} = 13.7 \text{ sec}$$

STUDENT PROBLEM

How long will it take to stop a 3.12-lb ball moving 64 ft/sec if the catcher's hands exert a retarding force of 20 lb on the ball? (ANS. 0.312 sec)

5. Deriving the formula for momentum

We do not need to depend on our intuition to be sure that our formula is correct. We can derive it from Newton's Second Law of Motion. You will remember the formula for acceleration, $a = v/t$. Substitute this expression for *a* into the formula

$$F = ma$$

which gives

$$F = \frac{mv}{t}$$

or

$$Ft = mv$$

for force, *F*, in poundals, or

$$Ft = \frac{mv}{g}$$

when *F* is written in pounds.

6. The Law of Conservation of Momentum

Figure 10-5 is a diagram of what happens inside a gun as a bullet is fired. The explosion inside the gun barrel exerts a tremendous force, F_b, on the bullet, but it also exerts, as you know, an equal and opposite force, $-F_g$ backward on the gun. Hence $F_b = -F_g$. Moreover the explosion exerts its force on the bullet for a time, t_b, which is exactly the same as the time, t_g, for which the recoil force of the explosion pushes back on the gun. Hence, $F_b t_b = -F_g t_g$, or the impulses on the gun and bullet are equal and opposite. Since the impulses are equal, and since both gun and bullet start from rest, the change in momentum of the gun and bullet must also be equal and opposite. Thus

$$\Delta m_b v_b = -\Delta m_g v_g$$

This equation says that the forward momentum given to the bullet is exactly equal to the backward momentum given to the gun. A similar statement can be made about every situation in which

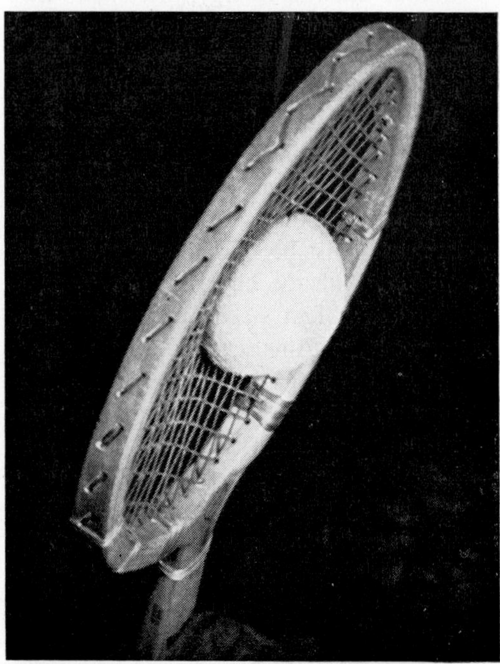

Harold E. Edgerton

12-3 This tennis racquet is in contact with the ball for only a very brief time, *t*. Hence it must exert a very large force, *F*, on the ball in order that its impulse, *Ft*, can give the ball a large change in momentum, Δ/mv. What evidence is there of the large force on the ball?

equal and opposite forces produce changes in velocity. In other words:

Whenever the momentum of one object is changed, the momentum of some other object must change also by exactly the same amount and in the opposite direction.

This is known as the **Law of Conservation of Momentum.**

Such statements as these *cannot* be made about kinetic energy. For instance, in a collision between a car and a massive truck kinetic energy is not conserved. The total momentum before and after the collision remains the same, but the large kinetic energy of motion has become zero. All of it is turned into other forms of energy, such as heat and sound. Thus, while there is a Law of Conservation of Energy, there is no Law of Conservation of Kinetic Energy.

When a rocket is fired, the momentum given to the rocket is precisely equal to the momentum given to the exhaust gases in the same length of time. Thus, to give a rocket the highest possible velocity and hence momentum, the available mass of fuel, m, must be burned and ejected to the rear with the highest possible velocity, v. The first stage of the Vanguard, one of the American rockets used to launch an earth satellite, burns approximately 7 tons of fuel in 70 seconds, giving the rocket a thrust of 38,000 pounds. This is sufficient to give the rocket a forward velocity of 5,200 mi/hr in the same length of time. The energy of motion of the exhaust gases turns very quickly into energy of heat and sound (Fig. 12-4).

SAMPLE PROBLEM

In a freight yard an empty freight car weighing 10 tons rolls at 3 ft/sec along a level track and collides with a loaded freight car weighing 20 tons, which is standing at rest with the brakes released (Fig. 12-4). If the two cars couple together, find their velocity after the collision.

Here the object that loses momentum in the collision is the rolling empty freight car, since it will be slowed down by the collision. The object that gains that momentum will be the combined freight cars, coupled together as one mass after the collision. Since momentum is never destroyed, the momentum of the moving car before collision, mv_1/g, must equal the momentum of the two cars coupled together after the collision, $(m + M)v_2/g$. Thus

$$\frac{mv_1}{g} = \frac{(m + M)v_2}{g}$$

$$\frac{10}{32} \times 3 = \frac{(10 + 20)}{32} \times v_2$$

$$v_2 = \frac{30}{30} = 1 \text{ ft/sec}$$

It is interesting to notice that kinetic energy *has* disappeared in this collision. Before the collision the kinetic energy of the moving car was

$$\frac{mv_1^2}{2g} = \frac{10 \times (3)^2}{2 \times 32} = 1.40 \text{ ft-tons}$$

After the collision the two cars together share the kinetic energy

$$\frac{(m + M)v_2^2}{2g} = \frac{(10 + 20) \times (1)^2}{2 \times 32} = 0.47 \text{ ft-ton}$$

Thus $1.40 - 0.47 = 0.93$ ft-ton of kinetic energy turned into heat energy and sound energy as the cars coupled together. In a later chapter we shall learn how to measure this amount of heat energy.

STUDENT PROBLEM

A boy weighing 120 lb steps vertically into a 150-lb canoe that is moving at a speed of 3 ft/sec. What is the velocity of the canoe after he steps aboard? (ANS. 1.67 ft/sec)

12-4 CONSERVATION OF MOMENTUM. In all three of these events we see objects possessing energy that are either colliding with one another or being driven apart. Energy of motion is turned into heat energy and sound energy in each one, but the total momentum always remains unchanged.

Olin Mathiesen Chemical Corp

Pennsylvania Railroad

Department of Defense

Going Ahead

YOUR PHYSICS VOCABULARY

Define each of the following:

kinetic energy
erg
momentum
conservation of momentum
impulse

THE PRINCIPLES WE USE

Explain why

1. There is no such thing as a Law of Conservation of Kinetic Energy.
2. The heavier an object is, the harder it is to get it moving at a particular speed.
3. The flying fragments of an exploding firecracker have the same total momentum as the firecracker just before the explosion.
4. Kinetic energy is a scalar quantity, but momentum is a vector quantity.

SOLVING PROBLEMS OF MOMENTUM

Use energy and momentum equations to solve these problems. For the present purpose, do *not* use Newton's Second Law and equations of accelerated motion.

1. A nickel weighing 5.00 gm falls and hits the ground with a speed of 200 cm/sec. What is its kinetic energy?
2. From what height did the nickel of the preceding problem fall?
3. What is the kinetic energy of a 3,500-lb car traveling at 60 mi/hr?
4. What is the kinetic energy of a 1-oz bullet traveling at a speed of 1,200 ft/sec?
5. How much kinetic energy is acquired by a 5-oz baseball dropped from the top of the Empire State Building (1,472 ft high)?
6. A 10-lb weight falls a distance of 35 ft. How much kinetic energy does it acquire?
7. What speed will a 5-oz baseball acquire if it is hit with a 100-lb force for an interval of 0.5 sec?
8. What average force will a 5,000-lb boat which is moving at 5 mi/hr exert on a piling if the piling bends for an interval of 2 sec?
9. A golfer hits a ball of mass 1.5 oz, giving it a speed of 100 ft/sec. How long should he "follow through" if the force he applies to the ball is 150 lb? Is this practical?
10. A 5-kg weight is dropped from a height of 2 m. How much kinetic energy will it have on striking the ground?
11. What is the force that propels a rocket if 1,000 lb of exhaust gases are emitted each second at a speed of 10,000 ft/sec?
12. How far will a hockey puck slide across the ice if it is given a velocity of 80 ft/sec and the coefficient of friction is 0.15?
13. How much energy is needed to get an 80-ton locomotive up to a speed of 60 mi/hr?
14. What is the "kick" of a 15-lb gun when it fires a 2-oz bullet with a muzzle velocity of 800 ft/sec?

---

15. Do you think it is possible to catch a 5-oz baseball dropped from the top of the Empire State Building—a height of 1,472 ft? Assume that the ball meets with no air resistance and the catcher pulls his hands back a distance of 3 ft as he absorbs the kinetic energy of the ball.
16. An archer exerts an average force of 12 lb as he draws the arrow back 3 ft preparatory to shooting. If the arrow weighs 2 oz, how much kinetic energy will it have after it is released?
17. What will be the horizontal speed of the arrow in Problem 16?
18. A 10,000-lb truck moving 20 mi/hr south collides with a 3,000-lb automobile moving 60 mi/hr north. If they become locked together and friction is negligible (an icy road), what will be their resultant velocity?
19. In Problem 18, how much kinetic energy is lost? What happens to it?
20. Assume that the truck and automobile in Problem 18 collide at right angles and solve the problem again. (HINT. Notice that momentum is a vector quantity.)

21. What force must be applied to a 400-lb sled to get it up to 15 mi/hr in 5 sec if the coefficient of friction is 0.2?
22. Two tons of coal fall from a chute into a 5-ton car moving slowly (2 mi/hr) beneath the chute. What is the *change* in velocity of the car?
23. Two satellites are moving in circular orbits around the earth. One of them is twice as far from the center of the earth as the other. How do their velocities compare?
24. Mars has two tiny moons: Deimos, which takes 1 day 6 hr 18 min to revolve around it; and Phobos, which takes only 7 hr 39 min. Phobos is 5,800 mi from the center of Mars. How far is Deimos?

PROJECT: VERIFYING THE LAW OF CONSERVATION OF MOMENTUM

The Law of Conservation of Momentum is such an important law that it is well worth your taking the time to verify it for yourself.

Support a smooth board 1½ to 4 feet long at an angle to the floor or table top. Cover it and the table top for a foot or two beyond its base with a sheet of smooth *unwrinkled* paper, wax paper, or foil. At the base of the plane the paper should curve smoothly out onto the horizontal table. Where the paper first becomes horizontal draw a "starting line" parallel to the bottom of the plane.

When a heavy coin (preferably a half dollar) slides down the inclined plane and crosses the "starting line" at the bottom it has a kinetic energy $m_1v_1^2/2g$. As it slides to a stop in a distance, s_1, beyond the starting line it is converting its kinetic energy into work against the constant force of friction, F. Therefore $Fs_1 = m_1v_1^2/2g$. Rewriting, $v_1^2 = 2Fs_1g/m_1$, or, since $2F$, g, and m_1 can all be lumped together in a single constant, k: $v_1 = k\sqrt{s_1}$.

This last equation says that the velocity of the coin as it crosses the starting line is proportional to the square root of the distance within which it slides to a stop.

If the coin is now allowed to collide with a second identical coin, m_2, originally motionless at the starting line, then by the Law of Conservation of Momentum it should be true that $m_1v_1 = m_1v_1' + m_2v_2'$. Dividing to eliminate the m's (since $m_1 = m_2$), $v_1 = v_1' + v_2'$ or $\sqrt{s_1} = \sqrt{s_1'} + \sqrt{s_2'}$.

If this relationship can be shown to hold, then the conservation of momentum has been verified.

Slide coin m_1 down the plane, and observe how far it coasts beyond the starting line, s_1. Then from the same height release it so as to collide with coin m_2 on the starting line. Measure the two distances, s_1' and s_2', that they move after the collision and substitute these values into the formula.

Some important precautions are: (a) the paper must be smooth and free from grit; (b) the collisions must be nearly head-on, otherwise the coins will acquire rotational momentum; (c) the target coin does not have the same starting point as the projectile at the moment of collision.

Is momentum conserved? Is kinetic energy conserved?

CHAPTER

13

Pressure in Liquids

Have you ever seen a car come to a screeching halt as a child runs into its path? By jamming on his brakes, the driver has saved the child's life. Did you know that the brake, like many other machines, depends on the action of fluids under pressure? (When we say **fluid** here, we mean any material—either liquid or gas—that will flow.)

The fact that fluids can exert pressure is of the greatest practical importance in our daily lives. For example, just consider the automobile.

When you step on the brakes of your car, you are applying force to a liquid which, in turn, transmits it to the wheels. The body of the car itself was probably stamped out by a gigantic press operated by a liquid under pressure. When you pump up a flat tire, you are actually lifting the car off the ground by forcing a fluid (air) into the tire. And when your car is sent to the garage to be greased, the mechanic raises it off the ground by means of a lift operated by a liquid under pressure and forces grease into the grease cups by the pressure he exerts with his grease gun. We can understand these devices better if we look a little more closely into the nature of pressure.

Concentration of force

1. Pressure and surface area

Suppose we have a block that measures 5 inches by 4 inches by 1 inch and weighs 10 pounds (Fig. 13-1). If we set this block on a table on its largest side (5 in × 4 in), all of its weight is resting on 20 square inches of the table. Since 10 pounds are resting on 20 square inches, we can say that each single square inch holds up $\frac{1}{2}$ pound, or, to put it in other words, that the pressure on the table due to the weight of the block is $\frac{1}{2}$ pound per square inch. The pressure tells us how the force is concentrated or distributed over an area.

Now let us turn the block over on another side (5 in × 1 in). Now the 10 pounds are resting on 5 square inches. In this position each square inch supports 2 pounds. The pressure on the table

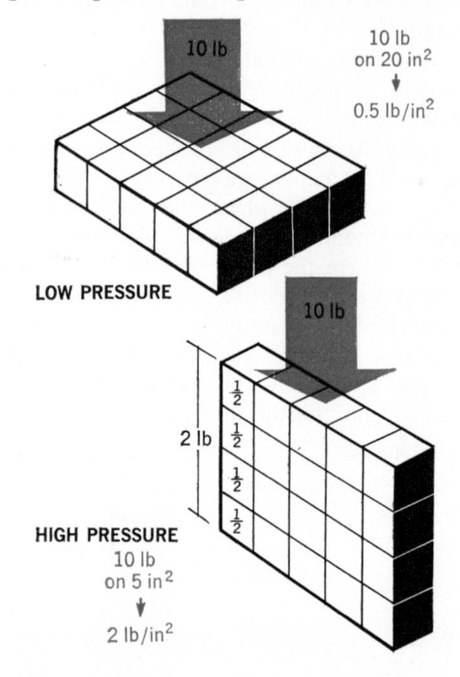

13-1 When the block is resting on its edge, the pressure it exerts on the table top is greater than when it is flat. The pressure is greater because the same weight is concentrated on a smaller area.

now is 2 pounds per square inch (2 lb force on 1 in² area). Note that the force in both positions is the same (10 lb) even though the pressures differ.

You see now that **pressure** is the force acting on a unit of area (ft^2, in^2, cm^2, or any other square unit). We find the amount of the pressure by dividing the force by the area on which it acts, thus:

$$\text{Pressure (P)} = \frac{\text{Force (F)}}{\text{Area (A)}}$$

SAMPLE PROBLEM

Suppose we balance this same block on the point of a thumbtack. Its weight will probably press the point of the tack into the wood. This happens mainly because the concentration of force (as measured by the pressure) is very great. If the area of the point of the tack is $\frac{1}{100,000}$ of a square inch, we can find the pressure thus:

$$P = \frac{F}{A} = \frac{10 \text{ lb}}{\frac{1}{100,000} \text{ in}^2}$$

$$= \frac{10 \times 100,000}{1}$$

$$= 1,000,000 \text{ lb/in}^2$$

When we say that the pressure on the point of the tack is 1,000,000 pounds per square inch, we mean that 10 pounds resting on this small area of the head of the tack produce as much effect on the point of the tack as 1,000,000 pounds would on an area of 1 square inch. This great concentration of force upon a small area, together with the mechanical advantage of the point of the tack (a wedge), causes the tack to penetrate the wood.

STUDENT PROBLEM

What is the pressure exerted on the ground by a pair of crutches supporting a 150-lb boy if the area of the bottom of *each* crutch is 0.75 in²? (ANS. 100 lb/in²)

American Museum of Natural History

13-2 **The thick steel walls and spherical shape of this bathysphere resist the pressure of water half a mile down. Recently a self-propelled deep-water submarine called a "bathyscaphe" has been developed by Prof. Auguste Piccard which can safely carry men much deeper.**

Similarly, if the force is spread over a large area the pressure is reduced. A good example of this is the surface area of skis. Without skis, a 150-lb man whose feet have a total area of $\frac{1}{4}$ ft² would exert a pressure of 600 lb/ft². Such a large pressure would cause him to sink in soft snow. However, skis provide an area of about 3 ft², which reduces the pressure to only 50 lb/ft².

Pressure under liquids

You may have heard of the bathysphere (băth'ĭ·sfēr), a hollow ball about five feet in diameter developed by the American scientist William Beebe to explore the ocean depths. The walls of this bathysphere were of solid steel, 1½ inches thick, and its specially constructed windows were 6 inches thick (Fig. 13-2).

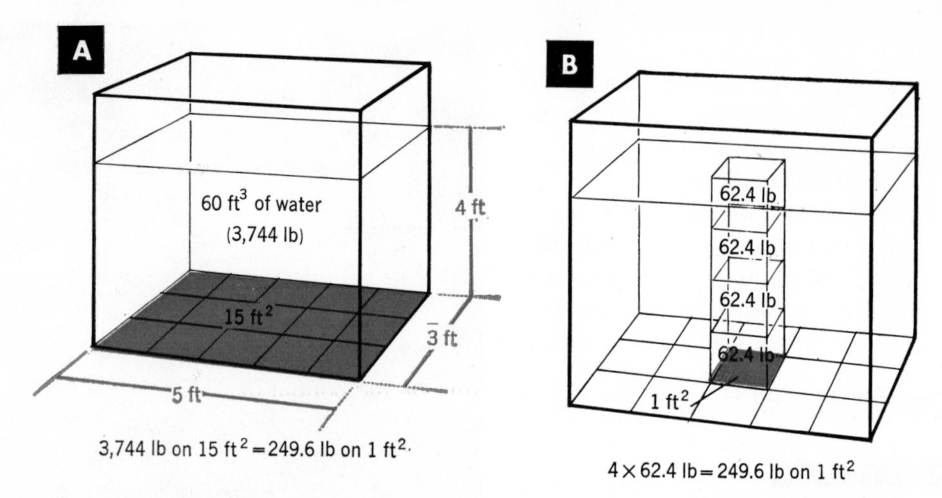

13-3 The pressure (pounds per square foot) on the bottom of this tank can be calculated by dividing the total weight by the total area. Or we can simplify the calculation by finding the weight of water just on one square foot of the bottom.

2. Pressure and the bathysphere

Beebe constructed the bathysphere because he wanted to observe ocean life more than a half mile below the surface of the sea. The pressure of water increases so rapidly with depth that a diver in a specially constructed suit can safely go down only 500 feet at most. With a bathysphere, the much greater depth of 3,000 feet can be reached. At this depth the pressure on the bathysphere amounts to almost 100 tons for every square foot of surface!

An example will reveal why the pressure on Beebe's bathysphere was so great. Suppose we calculate the pressure of the water in a tank 5 feet long, 3 feet wide, and 4 feet high (Fig. 13-3.) This tank contains 60 cubic feet of water ($5 \times 4 \times 3$). We know that each cubic foot of water weighs 62.4 pounds; hence 60 cubic feet will weigh 60×62.4 pounds, or 3,744 pounds. This total weight is resting on the bottom of the tank, which measures 5 feet by 3 feet (15 ft² of surface) (Fig. 13-3A). Hence, the pressure (force per unit of area) on the bottom is:

$$P = \frac{F}{A} = \frac{3{,}744 \text{ lb}}{15 \text{ ft}^2} = 250 \text{ lb/ft}^2$$

We can also calculate this pressure by considering just the weight of the water directly above a square foot. For example, in the tank in Fig. 13-3B we have a column of water 4 feet high and 1 foot square at the base. This column has a volume of 4 cubic feet ($1 \times 1 \times 4$). Since 1 cubic foot of water weighs 62.4 pounds, the total weight of this column resting on a square foot will be 4×62.4, or 250 pounds. The pressure is therefore 250 pounds per square foot—the same answer that was obtained previously by a different method.

So we find the formula for calculating pressure in water as follows:

$$\text{Pressure} = 4 \text{ ft} \times 62.4 \text{ lb/ft}^3$$
$$= 250 \text{ lb/ft}^2$$
$$\text{Pressure} = \text{Height} \times \text{Density}$$

Similarly beneath a height, h, of any other liquid whose density is D the pressure, P, is

$$P = hD$$

SAMPLE PROBLEM

What, then, would be the pressure on the outside of Beebe's bathysphere at a depth of 3,000 feet in salt water (density 64 lb/ft³)?

$$P = hD$$
$$= 3{,}000 \text{ ft} \times 64 \text{ lb/ft}^3$$
$$= 192{,}000 \text{ lb/ft}^2$$

At a depth of 3,000 feet, the pressure on the bathysphere would be 192,000 pounds per square foot! No wonder that water seeped into the bathysphere during test dives at this great depth.

STUDENT PROBLEM

What is the pressure at the bottom of a pond that is 25 ft deep? (ANS. 1,560 lb/ft^2)

3. Pressure and water levels

If we have a number of connecting tubes of different shapes and pour a liquid into one of them, the liquid will flow into the others until it is at the same level in all. This is so because the pressure caused by the weight of the water in any one tube is transmitted to the connecting tubes (such as BC in Fig. 13-4). If the pressures are unequal in the different tubes, the liquid flows from the points of higher pressure to points of lower pressure until the pressures at the bottom of each tube are the same. Since the height of the water in all the tubes is the same when the water comes to rest, we conclude that the pressure at the bottom of every tube is the same. Apparently the pressure at the bottom of a tube is not affected in any way by its shape.

You might think that the water pressure on the bottom of a lake a mile wide and 10 feet deep would be greater than that at the bottom of a thin water pipe 10 feet high. But remember that pressure means force per unit area. Every square inch at the bottom of the lake supports the weight of the 10 feet of water above it, just as every square inch at the bottom of the pipe carries the same weight directly above it. The total weight of the water in the lake is much greater than that in the pipe, but *per square inch* the force on the bottom of both the lake and the pipe is the same.

4. Water tanks

It frequently happens that the water supply of a town comes from a source at a lower level than the town itself. In

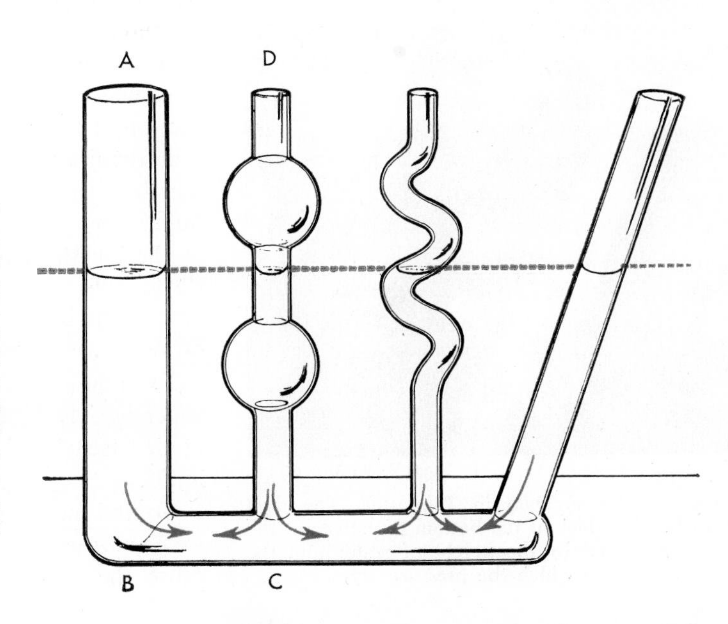

13-4 When water is poured into this vessel with tubes of different shapes, the pressures at the bottom equalize and the liquid finally comes to rest at the same level in each tube. This experiment indicates that the shape of the container has no effect on the pressure at the bottom. Pressure in a fluid is determined only by height and density.

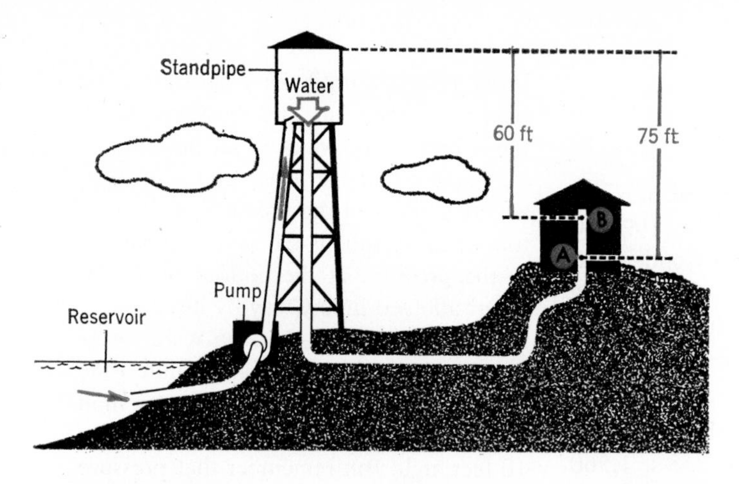

13-5 The standpipe here supplies pressure to homes because it contains water at a height. The water pressure in the home at A and at B is determined solely by the difference in levels. Why is there more pressure at A than at B?

this case the water has first to be pumped into raised tanks called **standpipes.**

Figure 13-5 shows us how a standpipe supplies water to two faucets in a building. The pressure at faucets A and B can be calculated solely from the difference in level between the faucets and the water at the top of the tank (disregarding the loss in pressure due to friction as the water flows through the pipe). As we have seen, the shape of the pipe does not affect our calculation of pressure. Therefore, to get the pressure of the water at faucet B, 60 feet below the surface of the water in the tank, we use the formula:

$$\begin{aligned} P &= hD \\ &= 60 \text{ ft} \times 62.4 \text{ lb/ft}^3 \\ &= 3{,}744 \text{ lb/ft}^2 \end{aligned}$$

If we want the pressure in pounds per square inch, we divide this answer by 144, since there are 144 square inches to 1 square foot. Thus we have a pressure of $\frac{3{,}744}{144}$, or about 26 pounds per square inch.

Calculation of pressure and total force becomes very important in the construction of a dam. Note that the bottom of the dam in Fig. 13-6 is much thicker than the top. The greater pressure at the bottom requires more concrete to prevent the dam from bursting. In some of the larger dams the bottom may be as thick as the length of a long city block (600 ft).

Trans World Airlines, Inc.

13-6 Norris Dam in Tennessee. The dam must be very thick at the bottom to withstand the pressure caused by the height of water and the large area upon which the pressure acts.

5. Force against a vertical wall

Even on a small dam behind which the water is only 10 feet deep, the pressure at the base of the dam is $10 \times 62.4 = 624$ lb/ft^2.

To find the *total* force against this dam we *cannot* use the formula $P = F/A$, because the pressure is different at each level behind the dam, varying from 0 pounds per square foot at the water surface to 624 pounds per square foot at its base. It would appear that there is no single value of pressure which we can substitute into the formula. It is possible, though, to find an average value.

Since the pressure varies uniformly from 0 to 624 pounds per square foot, the **average pressure** is halfway between these two pressures, or

$$\frac{624 + 0}{2} = 312 \text{ lb/ft}^2$$

The total force against the dam will be the number of square feet of surface, A, multiplied by the *average* pressure, P_{av}, on each square foot. If our dam is 20 feet long, its area is $10 \times 20 = 200 \text{ ft}^2$. We have already found that the average pressure on each square foot of the dam is 312 pounds per square foot. Hence the force, F, on the dam is:

$$\text{Force} = 312 \text{ lb/ft}^2 \times 200 \text{ ft}^2 = 62{,}400 \text{ lb}$$

Expressed as a formula:

$$F = P_{av}A$$

We can solve the problem in a slightly different way and obtain another useful formula. Notice that the average pressure behind our small dam is found at a depth, h, given by

$$h = \frac{P}{D} = \frac{312}{62.4} = 5 \text{ ft}$$

This is a depth halfway between the base of the liquid and the surface. Hence to find the average pressure we need merely find the pressure halfway down in the liquid, that is, the pressure at the average depth, h_{av}. Thus

$$P_{av} = h_{av} D$$

Combining this with the formula above we have

$$F = h_{av} DA$$

SAMPLE PROBLEM

Find the force against the side of a water-filled swimming pool 30 feet long and 12 feet deep. Here the average depth is $\frac{12}{2} = 6$ ft and the area of the wall is $30 \times 12 = 360 \text{ ft}^2$. Using the formula,

$$\begin{aligned} F &= h_{av} DA \\ &= 6 \text{ ft} \times 62.4 \text{ lb/ft}^3 \times 360 \text{ ft}^2 \\ &= 135{,}000 \text{ lb} \end{aligned}$$

STUDENT PROBLEM

Find the force against a dam 40 ft long behind which the water is 14 ft deep. (ANS. 245,000 lb)

Fluids and the hydraulic press

6. Pascal's Law

If you examine Fig. 13-7, you will notice that the force applied to the small **piston** is exerted downward. Yet the fluid confined in the box transmits the pressure upward against the large

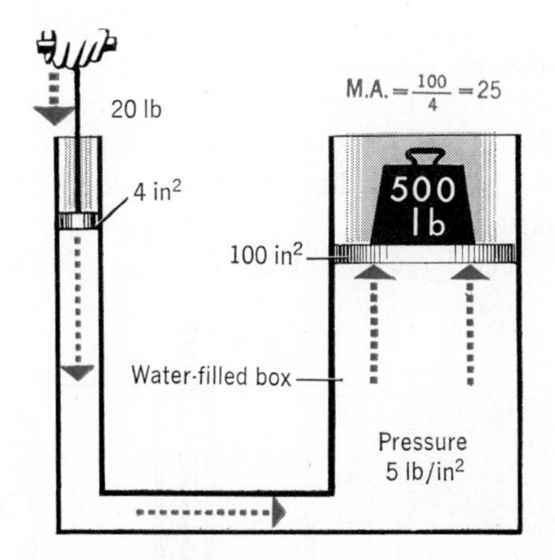

13-7 Twenty pounds of force on the 4-square-inch area of the small piston cause a pressure of 5 pounds per square inch. This pressure is transmitted to the large piston, where 5 pounds on each square inch cause a total force of 500 pounds on the 100-square-inch area.

piston as well as sideways against the walls of the box. When pressure is applied to a liquid in a closed container, the liquid transmits that pressure equally to all parts of the container and in all directions. It would make no difference if the pistons were horizontal or if they were connected together by a long coiled tube. It would still be true that:

Pressure applied to a fluid is transmitted equally in all directions.

This information about fluid pressure was first discovered by the French scientist, Blaise Pascal, in 1653, and it is known as **Pascal's Law.**

7. The hydraulic press

Figure 13-7 illustrates an interesting fact. An effort force of 20 pounds exerted on the small piston is lifting a load (or resistance) of 500 pounds resting on the large piston. This increase in force is a direct consequence of Pascal's Law.

The small piston has an area of 4 square inches, and the force on it is 20 pounds. Hence its bottom surface exerts a pressure of $\frac{20}{4} = 5$ pounds per square inch upon the liquid. According to Pascal's Law this pressure is transmitted by the liquid in all directions to all parts of the box—including the base of the large piston.

The large piston has an area of 100 square inches, however; and hence the total force, F, on it is 100 in^2 × 5 lb/in^2 = 500 lb.

The force on the small piston has been multiplied by $\frac{500}{20} = 25$. In other words the device shown in Fig. 13-7 has an M.A. of 25. You will also notice that the ratio of the areas of the pistons is $\frac{100}{4} = 25$. Hence the M.A. of this device can be found by taking the ratio of the piston areas. Thus

$$\text{M.A.} = \frac{\text{Area of large piston}}{\text{Area of small piston}}$$

This device is called a **hydraulic press.** Hydraulic presses are exceedingly useful devices because they can be made with a very large M.A.

STUDENT PROBLEM

Find the M.A. if the area of the large piston is 70 in^2 and the area of the small piston is 5 in^2. (ANS. M.A. = 14)

8. Some uses of the hydraulic press

Hydraulic presses can be made with an M.A. so great that they can stamp steel into the shape of an automobile body. Similarly, scrapped cars are compressed into blocks by such presses before being shipped as scrap iron. Cotton bales are pressed into compact shape for transportation in the same manner. In

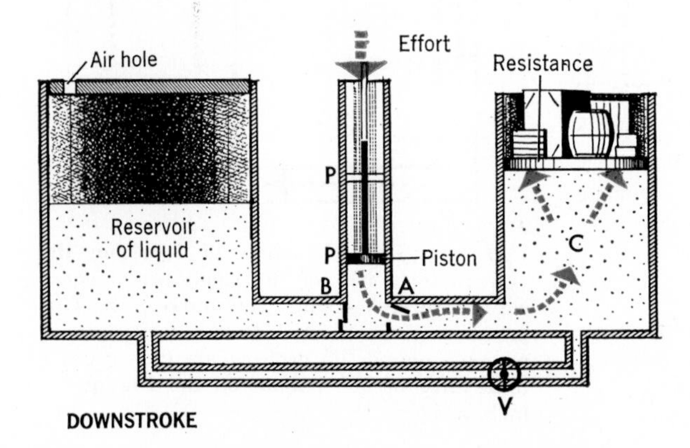

13-8 A practical hydraulic press must have valves and a reservoir. When the piston P is pushed down, valve A opens and lets oil be pushed into the large cylinder C. But valve B closes and prevents any oil from flowing back into the reservoir. On the upstroke, valve B opens and lets fluid in while A closes. When the job is completed, valve V can be opened by hand to permit the load to drop and push the oil back into the reservoir.

U. S. Air Force

13-9 Using the pressure of air to lift the wing of a test missile off the ground so that repairs can be made. Flat rubber bags are placed under the wing and air is pumped in. A small pressure distributed over a wide area can exert enough force to lift an entire airplane weighing many tons.

service stations, the hydraulic press is often used to raise cars off the floor. You can recognize a lift of this type by the large cylindrical piston that rises out of the floor.

In a garage lift, there is no effort piston at all. Compressed air is applied directly to a reservoir of oil, pushing the oil into the large cylinder. Suppose that the piston that lifts the car has an area of 100 square inches. If air which is under pressure of 50 pounds per square inch is applied to the fluid under the piston, that pressure is transmitted equally to the piston. The total weight that can be lifted is 5,000 pounds (50 lb/in^2 $\times$ 100 in^2).

Important use is made of Pascal's Law in hydraulic brakes (Fig. 13-10). When you step on the pedal, P, the pressure is transmitted equally to all four wheels of the car, and so gives equal braking on each side. The pressure acts on the piston in each wheel brake to push the brake lining up against a steel drum on the wheel. Friction then stops the car. It is better to have slightly more braking force at the front wheels. This is easily accomplished by making the front brake pistons a bit larger in area.

Still other hydraulic devices that operate in the same manner are the barber's and dentist's chairs, the hydraulic jack, and the controls on large airplanes.

9. A hydraulic-press problem

A hydraulic press has a small-piston diameter of 3 inches. The diameter of the large piston is 16 inches. What is its M.A?

Having been given the diameters of the two pistons, we must first find the areas of the pistons to substitute into the formula. To find the area of a circle we use either of the formulas

$$A = \pi r^2 = \frac{\pi d^2}{4}$$

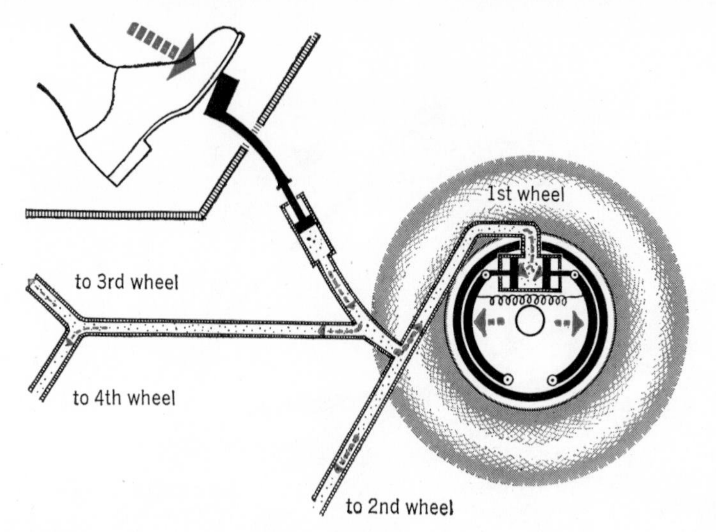

13-10 When you step on the brake pedal of a car, Pascal's Law goes to work. You get even braking on each side of the car. Without such brakes, it would be dangerous to drive a modern car.

Hence the area of the large piston is $\pi \times (16)^2/4$ sq. in, and the area of the small piston is $\pi \times (3)^2/4$ sq. in. When we put these values into the formula for the M.A. of the press we have

$$\text{M.A.} = \frac{\pi \times (16)^2/4}{\pi \times (3)^2/4}$$

or, since the π's and 4's cancel,

$$\text{M.A.} = \frac{(16)^2}{(3)^2} = \left(\frac{16}{3}\right)^2 = (5.33)^2 = 28.4$$

Notice that the ratio of the areas of two circles (A/a) is the ratio of the *squares* of their diameters (D^2/d^2) or the ratio of the *squares* of their radii (R^2/r^2).

STUDENT PROBLEM

Find the M.A. of a hydraulic press whose large piston has a diameter of 20 inches and whose small piston has a diameter of 4 in. (ANS. 25)

10. Let's look ahead

There is a very close connection between what we have studied in this chapter and the floating of a large ocean liner or a giant dirigible. In the next chapter we shall see how the facts about pressure can be used to explain why objects float or sink. Using this knowledge, we shall then go on to study the operation of floating dry docks, life preservers, and submarines.

Going Ahead

THE VOCABULARY OF PHYSICS

Define each of the following:

pressure	standpipe
piston	fluid
hydraulic press	average pressure

THE PRINCIPLES WE USE

Explain why

1. You can lift a heavy car by inflating a tire with a simple air pump.
2. A 5-lb force applied to the head of a thumbtack can push the thumbtack into wood, while a man weighing 150 lb standing on that same piece of wood has little effect.
3. A tire becomes flatter if the pressure in the tire is reduced.
4. The pressure at the bottom of a lake 5 mi wide and 5 ft deep is the same as the pressure in a tank 10 ft wide and 5 ft deep.
5. In a hydraulic press whose M.A. is 30, the small piston must move through 30 times the distance of the large piston.

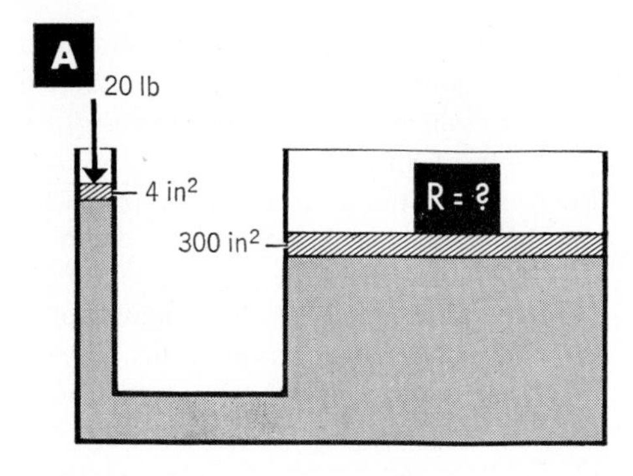

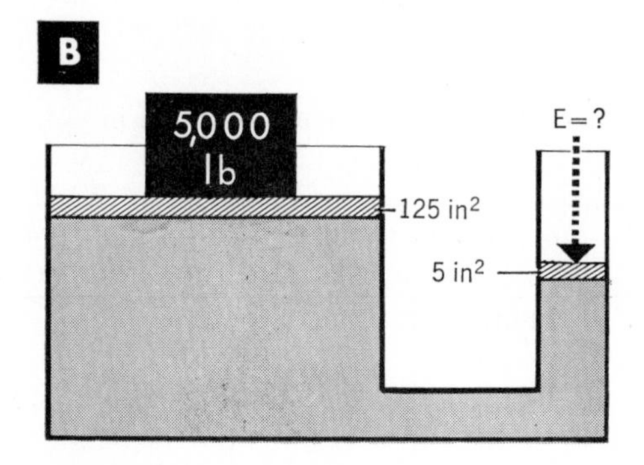

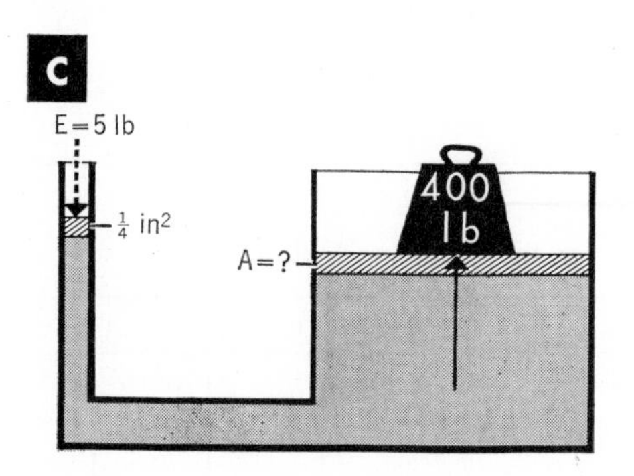

APPLYING PRINCIPLES OF PRESSURE

1. Draw a diagram of a hydraulic press and explain its operation.
2. Mention three practical hydraulic devices.
3. State Pascal's Law and mention one illustration of the law.
4. The water level in large tanks is often measured by a simple gauge consisting of a vertical glass tube along the side of the tank. How does this gauge work?

SOLVING PRESSURE PROBLEMS

1. What pressure is exerted by a 300-lb force acting on an area of 22 in^2?
2. What is the pressure in lb/ft^2 at a depth of 36 ft in a lake? In lb/in^2?
3. The pressure at the bottom of a tank of water is 480 lb/ft^2. How deep is the water?
4. What is the pressure 20 cm below the surface of a tank of mercury?
5. The pressure at the bottom of a tank of water is 600 gm/cm^2. What is the force on the bottom if its dimensions are 60 cm by 20 cm?
6. What is the M.A. of a hydraulic press whose pistons have areas of 10 in^2 and 3 in^2?
7. An effort of 2 lb is applied to a hydraulic press whose pistons have areas of 3 in^2 and 40 in^2. What resistance is being lifted?
8. Find the unknown quantities in Fig. 13-11.

---

9. The base of a tank of water measures 3 ft by 4 ft, and the water is 5 ft deep. What is the force on the bottom of the tank?

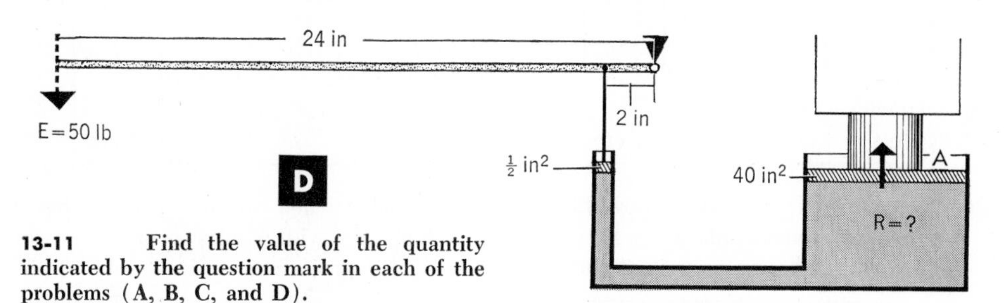

13-11 Find the value of the quantity indicated by the question mark in each of the problems (A, B, C, and D).

10. What would be the force on the bottom of the tank in the preceding problem if the liquid were alcohol?

11. The density of sea water is slightly greater than that of fresh water—about 64 lb/ft^3. What will be the pressure on a diver 200 ft under the surface of the ocean?

12. A submarine is built to withstand a pressure of approximately 10,000 lb/ft^2 before damage results. What is the maximum depth to which it can safely go in sea water (density 64 lb/ft^3)?

13. What is the pressure in lb/ft^2 at a depth of 20 ft in sea water?

14. What is the force against one end of a rectangular tank 60 cm long and 20 cm wide which is filled with mercury to a depth of 45 cm?

15. Calculate the pressure of the ocean on the body of a skin-diver at a depth of 150 ft. Calculate the pressure at 200 ft.

16. Suppose that one of the windows of a bathysphere has an area of 60 in^2. What is the force of the water against this window at a depth of 100 ft?

17. Suppose the faucet at A in Fig. 13-5 were 40 ft below the level of water in the standpipe. What would be the pressure at the faucet in lb/in^2?

18–22. Calculate the missing information for the rectangular tanks of liquid in the table that follows.

	Length	*Width*	*Depth*	*Density of liquid*	*Pressure on tank bottom*		*Total force on tank bottom*
18.	20 ft	25 ft	15 ft	64.0 lb/ft^3 sea water	? lb/ft^2	? lb/in^2	?
19.	35 ft	20 ft	30 ft	49.4 lb/ft^3 alcohol	? lb/ft^2	? lb/in^2	?
20.	100 ft	50 ft	?	62.4 lb/ft^3 water	? lb/ft^2	14.7 lb/in^2	?
21.	15 cm	15 cm	20 cm	1.8 gm/cm^3 sulfuric acid	? gm/cm^2		?
22.	50 cm	40 cm	30 cm	13.6 gm/cm^3 mercury	? gm/cm^2		?

23–27. Calculate the missing information for the following hydraulic devices:

	SMALL PISTON					LARGE PISTON			
	Piston dimension	*Effort force*	*Effort distance*	*Pressure*	*MA*	*Piston dimension*	*Resistance force*	*Resistance distance*	*Pressure*
23.	Area = $\frac{1}{2}$ in^2	80 lb	1 in	?	?	Area = 100 in^2	?	?	?
24.	Radius = 2 in	50 lb	10 in	?	?	Radius = 10 in	?	?	?
25.	Area = ?	?	75 in	?	150	Area = 600 in^2	1,500 lb	?	?
26.	Diameter = 10 cm	5 kg	?	?	?	Diameter = 200 cm	?	20 cm	?
27.	Radius = 1 cm	500 gm	1 cm	?	?	?	?	75 cm	?

28. Neglecting air resistance, how high a column of water could a fire hose produce with a pressure of 50 lb/in^2?

29. A cubical box 40 cm on a side has a long tube with a cross-sectional area of 1 cm^2 projecting vertically out of the top. The whole apparatus is filled with

water to a level in the tube 50 cm above the top of the box. Find (a) the force on the bottom of the box, (b) the force on the top of the box, (c) the force on one side of the box.

30. Into the tube described in the preceding problem there is poured another 10 cm^3 of water (10 gm). By how many grams is the force on the bottom of the box increased? Why is this added force so much greater than 10 gm?

31. Derive the formula for the M.A. of a hydraulic press by finding the ratio of the effort distance to the resistance distance without computing pressures and without using Pascal's Law.

PROJECTS: THE EFFECT OF FLUID PRESSURE

There are many interesting experiments you can do to illustrate the effects of fluid pressure. Here are a few.

1. Demonstrate Pascal's Law with an old unwanted gallon jug. Fill the jug with water to the very top, and insert a cork. Give the cork a sharp tap with your fist, and the bottom will burst out of the jug. It is well to wear heavy gloves and to stand back as you do this! How much force was applied to the bottom of the jug if your blow exerted a force of 15 pounds on the cork? How does this illustrate Pascal's Law?

2. Find the water pressure at your home with the help of a garden hose. With the hose you can direct the water vertically upward, and by measuring the height of the stream you can compute the pressure from the formula, P = hD. An interesting (but harder) further problem is to measure the greatest horizontal distance the water can be projected (hose at an angle of about 45°) and then, using what you know about the motion of falling bodies (Chapter 9), to deduce the "muzzle velocity" of the water and from this the water pressure.

3. Support 20 pounds of books with your breath! To a rubber hot-water bottle attach a one-holed rubber stopper through which passes a short length of glass tubing. Attach to it a few feet of rubber tubing. Pile a number of heavy books on the hot-water bottle, and while a friend balances them blow gently into the rubber tube. Thanks again to Pascal's Law, the books are raised easily. How does this arrangement compare with the hydraulic lift in a garage?

4. A Perpetual Motion Device? Now that you have learned what is meant by the Law of Conservation of Energy (page 5), by the idea of capillary action (page 19), and by the idea of pressure at the base of a column of liquid (page 162), you may be able to decide whether the following perpetual motion scheme would work. In a very narrow-bore open glass tube, capillary action forces water to a height of 3 inches. This tube is now immersed in a jar of water with only 2 inches of tube vertically above the water surface, and of course capillary action *easily* raises water to the open top end with an inch to spare. If the water raised in this way now overflows the top of the tube and returns to the jar, the action should continue indefinitely (barring evaporation). Do we have a perpetual motion device? If not, why not?

CHAPTER

14

Buoyancy

The principle underlying the facts of buoyancy was first discovered by the ancient Greek scientist, Archimedes. According to legend, Archimedes is supposed to have made his discovery while taking a bath and pondering a problem the king had given him. He was to discover whether a certain crown that had been presented to the king was made of pure gold or alloyed with silver. While taking his bath, he noticed that he seemed to lose weight as he displaced more water in the tub. He figured out a connection between this fact and a method of solving his problem. According to the story, he was so excited by his discovery that he forgot to dress and ran through the streets shouting, "Eureka" ("I have found it").

We cannot vouch for the accuracy of this story, but we do know that his discovery enables us to understand such things as why a steel battleship weighing thousand of tons will float on water while a 1-ounce pebble will sink, why a big dirigible rises while a toy balloon full of air falls to the ground, and why you can float on the surface of a swimming pool although a rock thrown into the pool will sink immediately.

Archimedes' Principle is another excellent example of how a general principle in science describes an enormous number of situations, all apparently very different from one another. How many other principles of physics can you name that apply to a wide variety of situations?

Archimedes' Principle

1. Buoyancy is upward force

We know that the weight of an object is a downward force which tends to make the object fall. If an object floats in a liquid, it must be that its weight is balanced by an equal force which is pushing it upward. This upward force, called **buoyancy**, is supplied by the liquid, as illustrated by the experiment shown in Fig. 14-1.

The glass plate at the bottom of the hollow cylinder is held in place by the pressure of the water. This pressure, caused by the height of the water outside the cylinder, pushes upward on the glass plate, and so keeps it from falling.

How much is this upward pressure on the glass? Pour some water into the hollow cylinder. The glass plate does not fall off until the water inside the cylinder is at the same level as the water outside. Since the plate will not fall until the upward pressure on it is equalized by the downward pressure of the water in the cylinder, this experiment shows that:

The upward pressure on a surface under water is equal to the downward pressure at that point.

Let us use this information to figure out the buoyancy in an actual case.

2. Calculating buoyancy

Suppose that we have a solid cube which measures exactly 1 foot on each edge. Each side has an area of exactly 1 square foot, and the volume of the object is 1 cubic foot. Let us place this cube so that its top is 2 feet below the surface of a tank of water (Fig. 14-2). At this depth the downward pressure of the water on the top of the cube is

$$P = hD = 2.00 \text{ ft} \times 62.4 \text{ lb/ft}^3 = 124.8 \text{ lb/ft}^2$$

Since the area is 1 square foot, the force pushing down on the cube is 124.8 pounds.

Now if the top of the cube is 2 feet below the surface of the water, then its bottom is 3 feet below. We find the upward pressure on the bottom thus:

$$P = hD = 3.00 \text{ ft} \times 62.4 \text{ lb/ft}^3$$
$$= 187.2 \text{ lb/ft}^2$$

and the force pressing up on the cube is 187.2 pounds.

We see that the upward force of the water on the cube exceeds the downward force of the water on the cube by 62.4 pounds (187.2 lb − 124.8 lb). Therefore we say that the resulting upward force (buoyancy) on the cube is 62.4 pounds.

Mind you, we are not saying that the cube will float. We will consider that in a few moments. We are simply saying that the water exerts a buoyant force of 62.4 pounds on the cube because that is the difference between the upward and downward forces caused by the water pressures upon it.

This difference in upward and downward pressures remains the same no matter how far below the surface of the water we place the cube. If the cube were lowered in the water, the downward pressure on the top would increase. But at the same time the upward pressure on the bottom would also increase by exactly the same amount. For every increase in downward pressure there is a corresponding increase in upward pressure, so the difference in pressures which causes the buoyancy remains the same. Clearly, the buoyancy of an object in a liquid is the same regardless of depth.

Note that the buoyancy of this object, whose volume is 1 cubic foot, is exactly 62.4 pounds. This is the weight of 1 cubic foot of water. The cube has a volume of 1 cubic foot and hence displaces this same volume of water. Thus the buoyancy of this cube is equal to the weight of the water it displaces.

For convenience in calculation, we have taken an object with a volume of 1 cubic foot and considered it as being

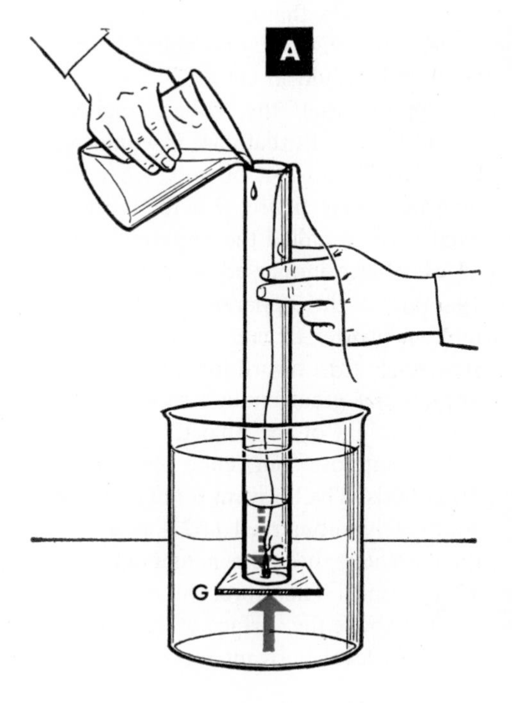

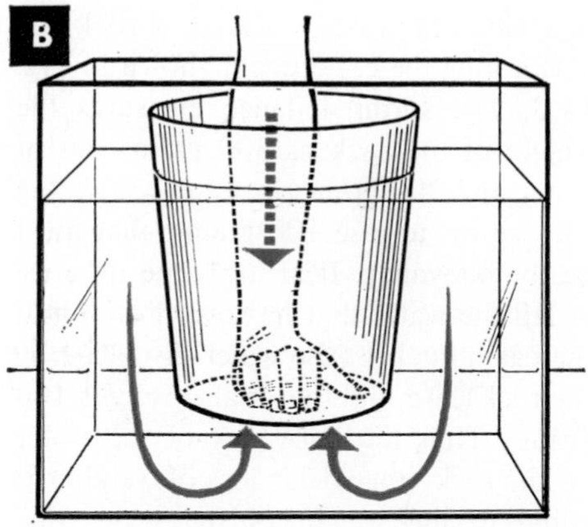

14-1 The buoyant force which keeps the glass plate from falling in A and pushes the pail up in B is due to the pressure of water acting upward. In A, the glass does not fall off until the water level in the tube is the same as in the jar, thus showing that upward pressure of water at G is the same as downward pressure, if water heights are the same. Why is it harder to push the pail deeper?

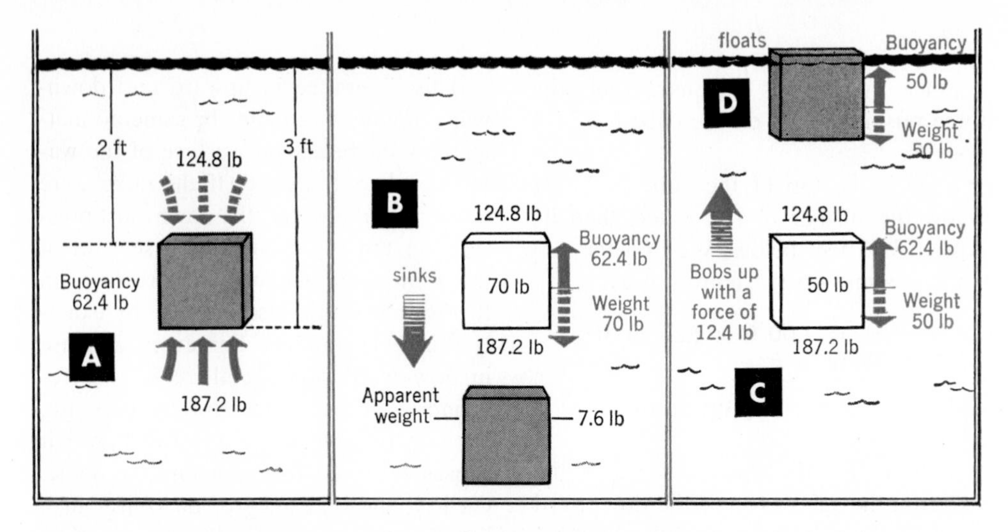

14-2 The buoyant force on a one-cubic-foot object, as shown in A, is exactly equal to the weight of the one cubic foot of water displaced (62.4 pounds). In B the object sinks because its weight of 70 pounds is more than the buoyancy of 62.4 pounds. In C the object bobs up because its buoyancy of 62.4 pounds is greater than its weight of 50 pounds. At D it floats because the buoyancy is only 50 pounds, which is exactly equal to the weight of the cube.

immersed in water. But when we make similar calculations for any other object, we find that, no matter what the dimensions of the object and no matter what the liquid, the buoyancy of the object is equal to the weight of the fluid it displaces.

We can check this fact for an object that sinks in water, such as a rock, by performing the experiment shown in Fig. 14-3. The spring balance measures the weight of the rock as 300 grams out of water and 200 grams in water. Since the rock seems to lose 100 grams, that must be the buoyancy. If at the same time we weigh the water that has overflowed into the catch bucket (the water displaced by the rock), we find that it also weighs 100 grams. Thus the buoyancy of the water on the rock (the rock's loss of weight) is equal to the weight of the water displaced. Other objects and other fluids give the same result.

From such experiments we have verified **Archimedes' Principle:**

An object is buoyed up by a force equal to the weight of the fluid it displaces.

3. Sink or float?

We still have not decided whether the cube in Fig. 14-2 will float. It will help us to understand why one object sinks and another does not if we recall that buoyancy causes an apparent loss of weight. If the buoyant force on this cube is 62.4 pounds, that means, in effect, that it seems to weigh 62.4 pounds *less* in water than it would in air.

Suppose that the cube weighs 70 pounds in air. In that case it would seem to weigh 70 − 62.4, or only 7.6 pounds in water. This figure, 7.6 lb, represents what is often called the **apparent weight** of an object immersed in a fluid (Fig. 14-2B). If the apparent weight is any number greater than zero, the object will sink, since the upward force due to buoyancy is not great enough to support the actual weight of the object.

But suppose our cube weighs only 50 pounds. The buoyant force on it when completely submerged (62.4 pounds) is then greater than the actual weight by 12.4 pounds. Therefore, the cube will be pushed above the surface of the water. At a certain level buoyant force and weight are both equal to 50 pounds, and the cube will not rise any farther (Fig. 14-2D). This occurs when 50/62.4 or about $\frac{5}{6}$ of the volume of the cube is under water.

If we immerse our cube in a liquid other than water, the liquid will exert a different buoyant force on it, but our principle continues to apply in the same way, as follows.

Suppose we submerge a cube whose volume is 1 cubic foot in a liquid which has a density of only 50 pounds per cubic foot. Now the immersed cube displaces only 50 pounds of liquid, and the buoyant force on it is only 50 pounds. If the cube itself weighs more than 50 pounds, it will sink; if less, it will float. But if the density of the liquid were 75 pounds per cubic foot, our cube would sink if it weighed more than 75 pounds and float if it weighed less. So we see that the greater the density of the liquid, the greater the buoyant force that it exerts upon objects. A floating object will float higher in a denser liquid (Fig. 14-8).

To summarize what we have discovered so far about buoyancy:

1. The buoyant force on an object immersed in a fluid depends upon the volume of fluid it displaces. The buoyant force also depends upon the density of the fluid, the force being greater in denser fluids. Combining these two ideas we can say that the buoyant force is equal to the volume of the object times the density of the fluid—or, in other words, to the weight of the displaced fluid. This is Archimedes' Principle.

2. Objects float in liquids which are denser than the objects and sink in fluids which are less dense. If an object floats, it does so at the level where the buoyant force equals the weight of the floating object.

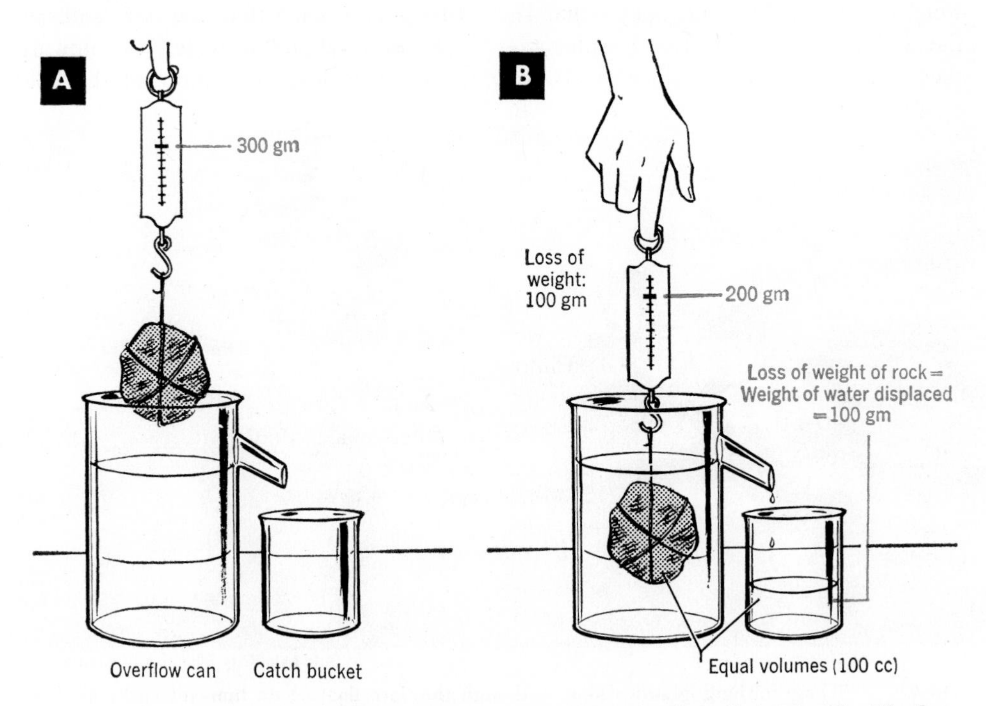

14-3 An experiment providing evidence of the truth of Archimedes' Principle. The loss of weight of the rock is equal to the weight of the water displaced.

Applications of Archimedes' Principle

Ships made of steel float in water, and balloons made of rubber float in air. If an object is made of a material denser than the fluid it is submerged in, there are two things we can do to make it float. First, we can decrease its weight so that less buoyancy is needed to support it. Hollowing it out will accomplish this. Second, we can increase the buoyant force on it by increasing its volume so that it displaces more fluid. Shaping it into a hollow hull will do this.

4. Why do iron ships float?

Suppose we obtain a block of iron whose volume is 10 cubic centimeters. Such a block weighs about 77 grams. As shown in Fig. 14-5A, when placed in water it displaces 10 cubic centimeters of water. This displaced water weighs 10 grams, since 1 cubic centimeter of water weighs 1 gram. The buoyancy—that is, the weight of the displaced water—is therefore only 10 grams. Since this is not enough to support the 77 grams of iron, the iron block will sink, and have an apparent weight in water of 67 grams (77 − 10).

But now suppose that we hammer the iron into a thin sheet and then shape it into a hollow form like a boat (Fig. 14-5B). This new shape has far greater volume and displacement than our original block of iron. When we put our "boat" into water, we find that its volume is so great that when only partly submerged it displaces 77 cubic centimeters of water. This means that the buoyant force on it is 77 grams. Since this force equals the weight of the iron, the "boat" will float.

5. Adding cargo to a boat

As we add cargo to a boat, each 62.4 pounds of additional weight causes the boat to sink slightly, displacing another cubic foot (62.4 lb) of water (Fig. 14-6). And each time the boat sinks to a lower level in the water, the upward pressure against the bottom of the boat

Salt Lake City Chamber of Commerce

14-4 These bathing beauties look as though they are floating on buoyant cushions, but they aren't. They are floating in Great Salt Lake, Utah, which is 27 per cent salt. This water is much denser than fresh water or sea water and therefore exerts a greater buoyant effect.

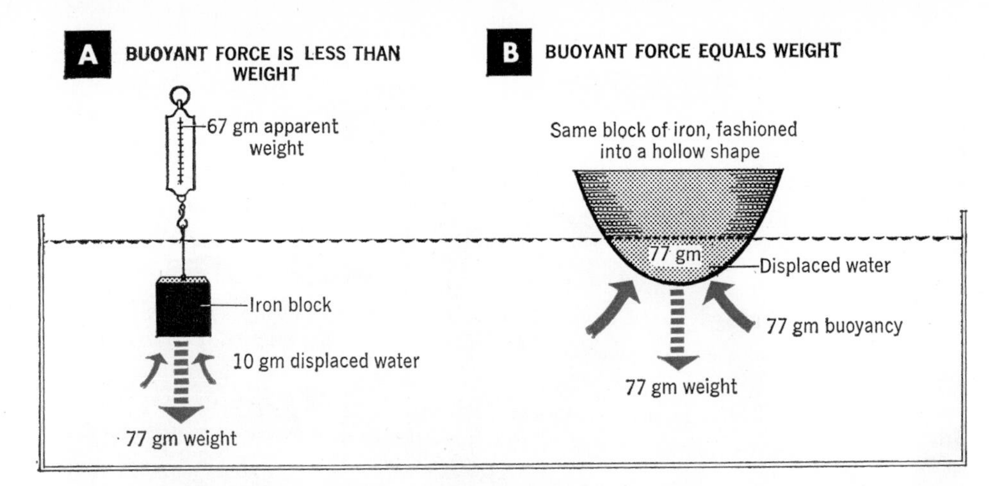

14-5 Shaping a piece of iron to make it float. The hollowed-out shape will float when it displaces a weight of water exactly equal to its own weight.

increases, giving greater buoyancy to balance the new cargo.

Boats that have been loaded in an ocean port will sink deeper when they enter a fresh-water lake or river. This happens because salt water has a greater density than fresh water (about 64 lb/ft^3 instead of 62.4 lb/ft^3). The ship need not sink as deep in the denser salt water in order to displace its own weight. For the same reason swimming is easier in salt water (Fig. 14-4).

6. Life preservers

Life preservers obtain high lifting force by having low weight and large displacement. That is why they are made of materials of low density like cork or kapok. A life preserver that weighs 10 pounds and has a volume of 1 cubic foot, when completely immersed in water, will displace 1 cubic foot of water—or 62.4 pounds. This is its buoyancy. The net lifting force of the life preserver will be 62.4 − 10, or 52.4 pounds.

An interesting use of this principle is illustrated by the "marsh buggy" in Fig. 14-7. The large volume of the superballoon tires enables the buggy to float on water, and the four wheels enable it to ride on either land or water. This buggy also illustrates the principles of pressure. Since the weight of the buggy is distributed over a large area, the pressure on

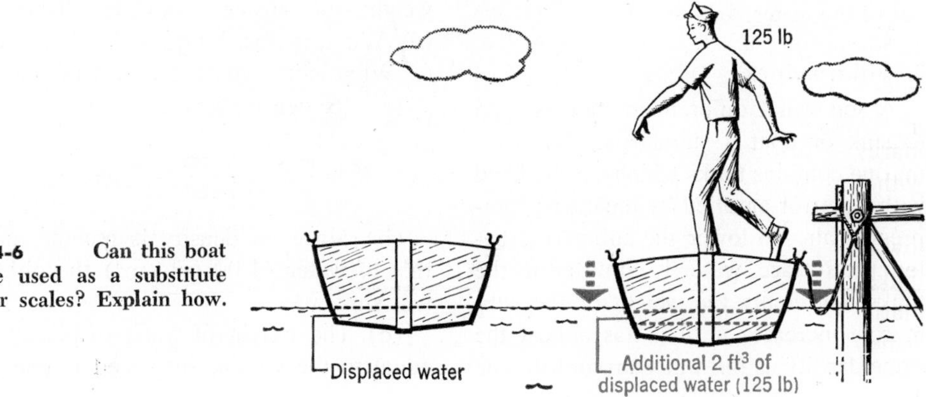

14-6 Can this boat be used as a substitute for scales? Explain how.

Gulf Oil Corp.

14-7 Archimedes' Principle is put to use in this "marsh buggy," a vehicle employed for seeking oil in swampy regions. This strange contraption can ride on land, over swampy areas, or in water. Can you find a resemblance between this "marsh buggy" and snowshoes?

the ground is small and the buggy does not sink far into soft earth.

A swimmer doing the "dead man's float" can reduce his buoyancy so as to sink in water by breathing air out of his lungs, thus decreasing the volume of his body and the amount of water it displaces. On the other hand, if he breathes in before putting his head in the water, he fills his lungs with air and increases his volume, so that his body floats higher out of the water.

7. Submarines

A somewhat different method is used to sink or float a submarine. The submarine contains tanks which can be filled with water or emptied by means of compressed air. To lower the submarine below the surface, water is admitted to the tanks. Gradually the weight of the submarine increases until it has almost the same density as the water around it. The submarine can then navigate under water by means of a rudder and an elevator similar in principle of operation to those used in an airplane.

SAMPLE PROBLEM

An object having a weight of 50 grams and a volume of 10 cubic centimeters is immersed in water. (a) What is its density? (b) Will it float or sink? (c) What weight of water does it displace? (d) What is the buoyant force on it? (e) What is its apparent weight in water?

(a) Its density is:

$$D = \frac{m}{V} = \frac{50 \text{ gm}}{10 \text{ cm}^3} = 5 \text{ gm/cm}^3$$

(b) Since its density is greater than that of water (1.00 gm/cm^3) the object will sink.

(c) The weight of water displaced is equal to the volume displaced (same as the volume of the object) times the density of water:

$$\text{Weight displaced} = 10 \text{ cm}^3 \times 1 \text{ gm/cm}^3 = 10 \text{ gm}$$

(d) The buoyant force, by Archimedes' Principle, is equal to the weight of water displaced, 10 gm.

(e) The apparent weight in water is the weight of the object in air minus the buoyant force:

$$\text{Apparent weight} = 50 - 10 = 40 \text{ gm}$$

STUDENT PROBLEM

Answer questions (a) through (e) in the sample problem on the preceding page for an object weighing 180 gm whose volume is 100 cm^3. [ANS. (a) 1.80 gm/cm^3; (b) it will sink; (c) 100 gm; (d) 100 gm; (e) 80 gm]

---

8. Specific gravity

Up to this point we have described density in terms of particular units. Water, for example, has a density of 1 gram per cubic centimeter; it also has a density of 62.4 pounds per cubic foot. The numerical value depends upon the units we are using. For instance, the density of water in pounds per cubic yard would be 1,680.

It would be simpler if we could describe density in terms of a number alone, without any unit. We can do this by using the idea of **specific gravity.**

The specific gravity of any material is the mass (or weight) of any given volume of it compared to the mass (or weight) of an equal volume of water; that is,

$$\textbf{Specific gravity} = \frac{\textbf{Mass of object}}{\textbf{Mass of equal volume of water}}$$

Since specific gravity is a *ratio* it has no units.

SAMPLE PROBLEM

Iron weighs about 480 pounds per cubic foot, while water weighs 62.4 pounds per cubic foot. What is the specific gravity of iron?

$$\text{Specific gravity} = \frac{\text{Mass of 1 ft}^3 \text{ of iron}}{\text{Mass of equal volume of water}} = \frac{\text{density of iron}}{\text{density of water}} = \frac{480}{62.4} = 7.7$$

The fact that the specfic gravity of iron is 7.7 means that iron is 7.7 times as heavy as water, volume for volume. If we say that a certain material has a specific gravity of 5, we mean that it is 5 times as heavy as an equal volume of water. Another way of looking at it is that the material is 5 times as dense as water.

STUDENT PROBLEM

What is the specific gravity of a metal if 1 ft^3 weighs 168 lb? (ANS. 2.7)

. . . .

Zinc has a specific gravity of 7.1. What is the weight of 5 cubic feet of zinc? If water weighs 62.4 pounds per cubic foot, zinc must weigh 7.1 times as much, or about 443 pounds per cubic foot (7.1×62.4). In that case 5 cubic feet of zinc must weigh 5 times 443, or about 2,215 pounds.

We know that mercury has a density of 13.6 grams per cubic centimeter. What is the specific gravity of mercury? The density of water is 1 gram per cubic centimeter. Comparing the densities of mercury and water, we get:

$$\text{Specific gravity} = \frac{13.6 \text{ gm/cm}^3}{1 \text{ gm/cm}^3} = 13.6$$

Notice that the figure for the specific gravity of mercury is the same as its den-

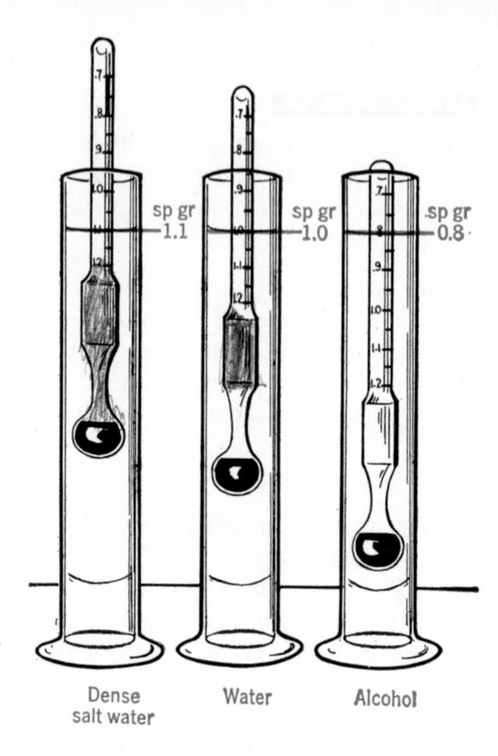

14-8 A hydrometer is a floating tube which measures the specific gravity of the liquid in which it floats. The denser the liquid, the higher the hydrometer floats.

sity **in grams per cubic centimeter,** 13.6. This is so for any material, because the density of water is 1 gram per cubic centimeter. A material with a specific gravity of 5 is 5 times as dense as water and therefore has a density of 5 grams per cubic centimeter.

9. The hydrometer

Hydrometers (hī·drŏm′ē·tẽrz) are instruments that measure the specific gravity of liquids by making use of Archimedes' Principle. They are usually made of glass and weighted so as to float upright in the liquid.

Figure 14-8 shows hydrometers floating in three different liquids: salt water, fresh water, and alcohol. In each liquid the hydrometer, like all floating objects, comes to rest when it has sunk deep enough to displace its own weight. Salt water is denser than fresh water; hence a hydrometer does not have to sink as far into it to displace its own weight. But alcohol is less dense than either, so that a hydrometer sinks farther into alcohol to displace its own weight.

A calibrated scale (a scale marked off properly) inside the glass tube enables us to read the specific gravity of each of these liquids (Fig. 14-8). The higher the specific gravity, the nearer to the bottom of the hydrometer is the number indicating it.

One use for the hydrometer is to measure the amount of cream in a given amount of milk. Cream is less dense than the water portion of the milk (as shown by the fact that cream rises to the top of a milk bottle). The more cream in the milk, the less dense is the entire mixture. A special hydrometer placed in the milk after it has been thoroughly shaken tells its specific gravity—and from this it is possible to tell how much cream the milk contains. Hydrometers are also used in testing automobile batteries and antifreeze solutions in radiators.

10. Archimedes' Principle problems

1. A certain stone weighs 500 grams in air and 400 grams when it is immersed in water. In alcohol it weighs 420 grams. How would we go about finding (a) the volume, (b) the density, and (c) the specific gravity of the stone, and (d) the specific gravity of the alcohol? Let us consider these problems in the order named.

(a) Volume of the stone. We know that an object completely immersed in a fluid always displaces its own volume of the fluid. We will know the volume of the stone if we can find the volume of the fluid it displaces. The volume of the displaced fluid is not given, but its weight can be found, and from that we can find its volume. The stone weighs 500 grams in air and 400 grams in water. Therefore its loss of weight (or buoyancy) is 100 grams. Now:

Buoyancy (loss of weight) = Weight of water displaced
= 100 gm of water displaced
= 100 cm^3 of water displaced

So the volume of the stone is 100 cm^3.

(b) Density of the stone. We find this in the following way:

$$D = \frac{m}{V} = \frac{500 \text{ gm}}{100 \text{ cm}^3} = 5 \text{ gm/cm}^3$$

(c) Specific gravity of the stone. We find the specific gravity of any object by comparing its weight with the weight of an equal volume of water. Since 1 cubic centimeter of this stone weighs 5 grams, whereas 1 cubic centimeter of water weighs only 1 gram, the stone is 5 times as dense as water. Hence its specific gravity is 5. We could have found the specific gravity in another way:

$$\text{Specific gravity} = \frac{\text{Weight of object}}{\text{Weight of equal volume of water}}$$

The weight of an equal volume of water is the weight of the water the stone displaces. We know what this is from the buoyancy of the stone (its loss of weight in water, 500 gm − 400 gm). So we can rewrite our formula for finding specific gravity:

$$\text{Specific gravity} = \frac{\text{Weight of object}}{\text{Loss of weight in water}}$$

$$\text{Specific gravity} = \frac{500 \text{ gm}}{100 \text{ gm}} = 5$$

(d) Specific gravity of the alcohol. The stone, whose volume is 100 cubic centimeters, lost only 80 grams of weight in the alcohol (500 gm − 420 gm). Now:

Buoyancy = Weight of fluid displaced
80 gm = Weight of alcohol displaced
80 gm = Weight of 100 cm^3 of alcohol

Since 100 cubic centimeters of alcohol weigh only 80 grams, while an equal volume of water weighs 100 grams, we find the specific gravity of the alcohol by comparison.

$$\text{Specific gravity} = \frac{80 \text{ gm}}{100 \text{ gm}}$$

Specific gravity of alcohol = 0.8

STUDENT PROBLEM

A piece of metal weighs 30 gm. When immersed in water its apparent weight is 27 gm. In an unknown liquid its apparent weight is 24 gm. What is (a) the volume of metal, (b) the density of the metal, (c) its specific gravity, (d) the specific gravity of the unknown liquid? [ANS. (a) 3 cm^3, (b) 10 gm/cm^3, (c) 10, (d) 2]

■ ■ ■ ■ ■ ■ ■

2. A cylindrical stick 30 centimeters long has a specific gravity of 0.83. When the stick is floating vertically in water, how many centimeters of its length will be submerged?

The situation is pictured in Fig. 14-9. Archimedes' Principle states that the weight of the water displaced by the submerged part of the stick (the shaded volume) is equal to the buoyant force on the stick. Since the stick floats, this buoyant force must also equal the weight of the stick. In other words, 30 centimeters of stick weigh the same as the water displaced by L centimeters of stick. Since these two weights are equal, the stick is only $L/30$ times as dense as water. But the ratio between the density of the stick and the density of water is just the specific gravity of the stick, and this was given as 0.83. Hence

$$\frac{L}{30} = 0.83$$

$$L = 30 \times 0.83 = 24.9 \text{ cm}$$

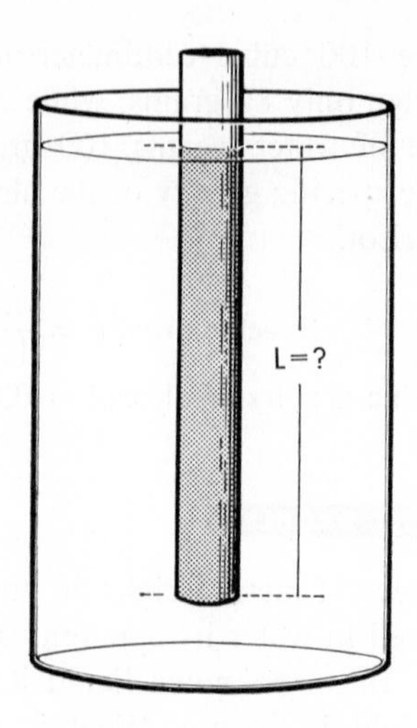

14-9 The water displaced by this floating stick (shaded gray) has the same weight as the entire stick.

Since 0.83 of the stick is under water, and 0.83 is also the specific gravity of the stick, you can see that the fraction of a floating object that lies beneath the fluid is simply the ratio between the two densities (or specific gravities). Thus

$$\frac{\text{Volume under fluid}}{\text{Total volume}} = \frac{\text{Specific gravity of object}}{\text{Specific gravity of fluid}}$$

When we say that an ice cube floats with 0.917 of its volume under water, we know at once that its specific gravity is 0.917. If a block of wood whose specific gravity is 0.70 is put in kerosene, whose specific gravity is 0.80, you can expect it to float with $0.70/0.80 = 0.875$ of its volume submerged.

The same principle is employed in marking off the scale of a hydrometer.

STUDENT PROBLEM

What length of a vertically floating stick will be under water if the stick has a specific gravity of 0.65 and a length of 20 cm? (ANS. 13.0 cm)

Floating in the air ocean

At first glance there would seem to be little connection between an ocean liner floating in the ocean and a balloon floating in the air. Actually balloons float in an ocean of air for the same reason that the liner floats in an ocean of water. Archimedes' Principle explains why.

11. Why a balloon floats

A balloon must displace its own weight of fluid—in this case, air. Or we may say that a balloon floats if it has the same average density as air. In order to give the balloon low density, it is made as light as possible and with a volume as great as possible.

Ideally, if the inside of the balloon were a vacuum, it would have the least possible weight. But the envelope of the balloon would then have to be of enormous strength to prevent the pressure of the outer air from crushing it. And this strength could be obtained only by using heavy rigid materials which would make it too heavy to rise. In actual practice a balloon is a lightweight bag filled with a light gas. This gas is less dense than air but keeps the pressure on the inside of the envelope the same as the pressure on the outside so that the balloon retains its shape.

The two lightest gases are hydrogen (which weighs about 0.1 oz/ft^3) and helium (which weighs about 0.2 oz/ft^3). Air weighs about 1.2 ounces per cubic foot. Thus the net lifting force of hydrogen (the difference between its weight and the weight of air) is 1.1 ounces per cubic foot $(1.2 - 0.1)$. By the same calculation we find that the net lifting force of helium is 1.0 ounce per cubic foot $(1.2 - 0.2)$. If helium were used, the lifting force would be slightly less than for hydrogen, but helium is safer because it does not burn.

12. Limits of balloon flight

At high altitudes air pressure is much less than it is at sea level. As a result, the air at these altitudes is not compressed as it is at sea level, and its density is much less. At a certain point in its ascent, a balloon reaches a level where the density of the air is the same as the average density of the balloon. At this level the balloon will cease to rise, and remain floating. Balloons carrying people have gone as high as 19 miles above sea level. Balloons containing weather instruments for measuring temperature, pressure, and humidity have gone up more than 25 miles.

Weather balloons are often equipped with **radiosondes** (rā'dĭ·ô·sŏndz'), tiny radio transmitters which send messages back to earth giving readings of pressure, temperature, and relative humidity at different altitudes. These weather balloons are often tied together in clusters. As they rise, the air pressure outside the balloons becomes less and less, allowing the gas inside the balloons to expand. When a balloon bursts, the radiosonde floats down to earth on a small parachute; it can be used again should it be found.

U.S. Air Force

14-10 **This high-altitude balloon is headed for the stratosphere, 10 miles up and more.**

Going Ahead

THE VOCABULARY OF PHYSICS

Define each of the following:

buoyancy
apparent weight
hydrometer
specific gravity
radiosonde

THE PRINCIPLES WE USE

Explain why

1. It is difficult to push a large rubber ball under water.
2. The buoyancy of an object with a volume of 1 ft^3 immersed in water is exactly 62.4 lb.
3. The buoyancy is the same no matter how deep an object is immersed in a liquid (provided the density of the liquid does not change).
4. An object floats in a liquid if it is less dense than the liquid.
5. An object sinks in a liquid if it is more dense than the liquid.
6. A person swimming in water can make himself sink by breathing out.
7. A boat sinks deeper as cargo is placed in it.
8. A boat sinks deeper as it travels from the ocean into a fresh-water river.

9. Describe an experiment to prove that:
   (a) Archimedes' Principle holds true for a stone.
   (b) A floating object displaces its own weight of liquid.
10. Explain how Archimedes might have applied his principle to test the king's golden crown for its gold content.

APPLYING PRINCIPLES OF BUOYANCY

1. Explain why it is possible for an iron boat to float.
2. Explain how a submarine can be made to sink or float at will.
3. Explain how it is possible for a hydrometer to measure the specific gravity of liquids.
4. Describe two uses of hydrometers.

SOLVING BUOYANCY PROBLEMS

1. What is the buoyant force on a 300-cm^3 piece of metal submerged in water?
2. A block of stone whose weight is 2,100 gm and whose volume is 700 cm^3 is submerged in water. (a) What volume of water is displaced? (b) What is the weight of water displaced? (c) What is the buoyant force on the stone? (d) What is the apparent weight of the stone in water?
3. A piece of material weighs 680 gm and has a volume of 763 cm^3. Will it float or sink in water? Explain.
4. What is the buoyant force on a 10-gm piece of aluminum (density 2.7 gm/cm^3) submerged in water?
5. A 100-gm brass weight of density 8.4 gm/cm^3 is immersed in water. (a) Will it float or sink? Why? (b) What is its volume? (c) How many grams of water does it displace? (d) What is the buoyant force on it? (e) What is its apparent weight in water?

---

6. Repeat Problem 5 using alcohol of density 0.8 gm/cm^3 instead of water.
7. Repeat Problem 5 using mercury of density 13.6 gm/cm^3 instead of water.
8. A block of wood with a volume of 2 ft^3 weighs 100 lb. (a) Why will it float? (b) What weight of water does it displace? (c) What volume of water does it displace? Assume specific gravity = 0.7. (d) What is the apparent weight of the block?
9. What force must be applied to the block of wood of Problem 8 to submerge it completely in water?
10. An object weighs 80 gm in air and 60 gm in water. Find (a) the weight of the water displaced, (b) the specific gravity of the object, (c) the volume of the object, (d) the density of the object in metric units, (e) the density of the object in English units.
11. A block of metal weighs 100 gm in air, 70 gm in water, and 60 gm in a liquid. Find (a) the specific gravity of the metal, (b) the volume of the metal, (c) the specific gravity of the liquid.
12. A bottle weighs 40 gm when empty, 150 gm when filled with salt water, and 140 gm when filled with pure water. What is the specific gravity of the salt water?
13. The horizontal dimensions of a block of wood are 2 ft by 3 ft, and its vertical dimension is 1 ft. It floats with 9 inches submerged under water. (a) What is the weight of the wood? (b) What is its specific gravity?
14. A certain rock weighed 500 lb in air and 375 lb in water. (a) What was its volume? (b) What was the density of the rock? (c) What was its specific gravity?
15. A 500-lb cylindrical pontoon, 4 ft in diameter and 5 ft high, is to be used to raise a small boat that has sunk 30 ft under water. The pontoon is submerged and then filled with air. What lifting force will it exert on the boat?
16. What will be the apparent weight of a 100-gm block of metal of specific gravity 5 when immersed in alcohol of specific gravity 0.8?
17. A barge weighing 100 tons is 20 ft wide, 80 ft long, and 15 ft high. How much deeper will it sink when filled with cargo weighing 250 tons?

---

18. An iron anchor has an apparent weight of 85 lb when submerged in water. What is its actual weight?
19. A stone weighs 800 gm in air, 600 gm in water, and 580 gm in a certain salt solution. (a) What is the volume of the stone? (b) What is its density? (c) What is its specific gravity? (d) What is the specific gravity of the salt solution? (e) What would the stone weigh in alcohol of specific gravity 0.8?
20. A block of iron of specific gravity 7.7, measuring 10 cm on each edge, is placed in mercury of specific gravity 13.6. Prove that the iron will float. What fraction of the volume of the iron will be immersed in the mercury?
21. A corked bottle weighs 50 gm and has a volume of 200 cm^3. It is tied to 20 cm^3 of iron of specific gravity 7.7. Will the combination sink or float? Explain.
22. Show by calculation of pressures that a rectangular object of length, width, and height a, b, c, with its top a distance h below a liquid of density D, will be buoyed by a force equal to the weight of the fluid displaced.
23. A small piece of wood (specific gravity 0.7) floats in a jar whose bottom half is filled with water and whose top half is filled with oil (specific gravity 0.6). What fraction of the wood's volume is in water?
24. How many pounds of passengers and cargo can be carried in a balloon that weighs 800 lb, has a gas capacity of 150,000 ft^3, and contains helium?
25. A cork is floating in water in a closed container. If the air above the water is pumped out of the container, will the cork sink deeper, float higher, or remain at exactly the same level? Why?

PROJECTS: APPLYING ARCHIMEDES' PRINCIPLE

1. Your friends may have a hard time explaining how this diver works. Fill a bottle to the top with water. Float in it a medicine dropper filled with water until the dropper just *barely* floats. Now insert a cork in the bottle, and press firmly on the cork. The diver descends. As you release the pressure the diver rises once more. Having learned how a submarine works (Section 7) you will know why the diver's density changes.
2. Into a small wide-mouth bottle put some mercury, some gasoline, and some water. Since none of these liquids mix and their densities differ, they will form three separate layers. Now obtain an iron nut or bolt, a small block of wood, and a cork. *Before you drop them into the bottle* use Archimedes' Principle and Table 2 (page 33) to predict what will happen. Then drop them in and see if you were right.
3. Making Your Own Hydrometer. Drive a nail with a few iron washers under its head into one end of a 6-inch length of wooden dowel. The dowel should now float vertically in water. Mark on the dowel the level at which it floats. Do the same with other liquids of known specific gravity. You now have a "calibrated" hydrometer which you can use to identify unknown fluids and find the strength of various water solutions. Explain why the divisions on your scale marking off equal differences in specific gravity are *not* equal distances apart.

CHAPTER

15

Air Pressure

What would you say about the sample of handwriting shown in Fig. 15-1? You might think it was the scribble of a very young child. Actually, however, it is the writing of a full-grown, intelligent man. You would write the same way under the same conditions—if you were in a plane without an oxygen mask (or a pressurized cabin) at an altitude of 23,000 feet.

Man lives at the bottom of an ocean of air. When he ascends in an airplane, the rapid drop in pressure reduces the amount of oxygen he takes in at each breath and causes failure of his body mechanisms. The man who wrote this illegible scribble was almost unconscious at an altitude of 23,000 feet. Had he remained at that altitude without an oxygen mask for more than a short time, he would have perished.

Mountain climbers who climb as high as 29,000 feet can only survive as a result of long preliminary training.

Pressure and weight

1. Air pressure compared to water pressure

As you know, when we go under water, the weight of the water above us creates a pressure. The deeper we go, the greater the pressure. On the other hand, as we rise toward the surface, the weight of the water decreases and the pressure becomes less.

In the same way, the weight of the air above us produces pressure on the earth's surface, the bottom of the **air ocean** where we live. The higher we go in this air ocean, the less the weight of the air above us and the lower the **air pressure.** At very high altitudes, air pressure becomes so low that effective breathing becomes difficult.

The air ocean, then, is very much like an ocean of water. But while there is a dividing line at the surface of the sea where water ends and air begins, the air ocean thins out gradually so that there is no sharp boundary line between it and the airless spaces beyond.

2. The weight of air

Volume for volume, air has much less weight than water. We can weigh it and find its density by using a flask, as shown in Fig. 15-2. When we pump air out of the flask, the weight of the flask decreases by an amount equal to the weight of the air removed. Knowing the volume of the

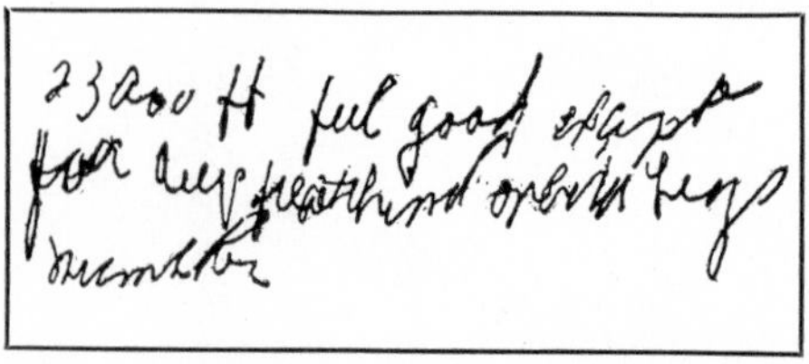

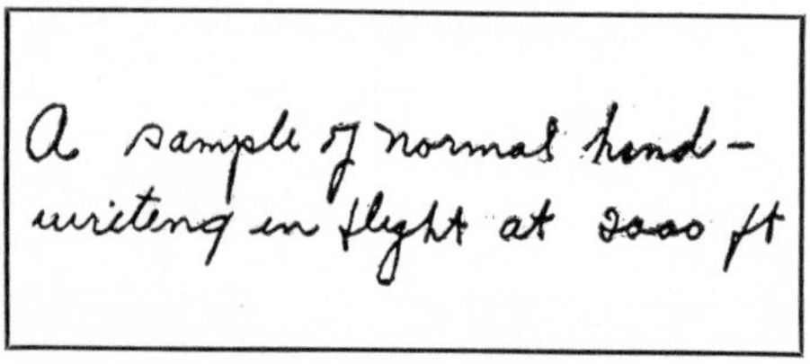

15-1 Compare this sample of handwriting at an altitude of 23,000 feet (left) with the same person's normal handwriting (right). The low air pressure makes breathing very difficult, and death would follow if a person remained too long under these conditions.

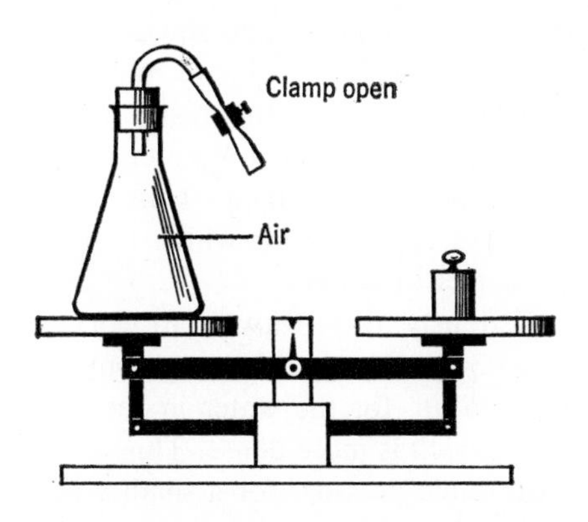

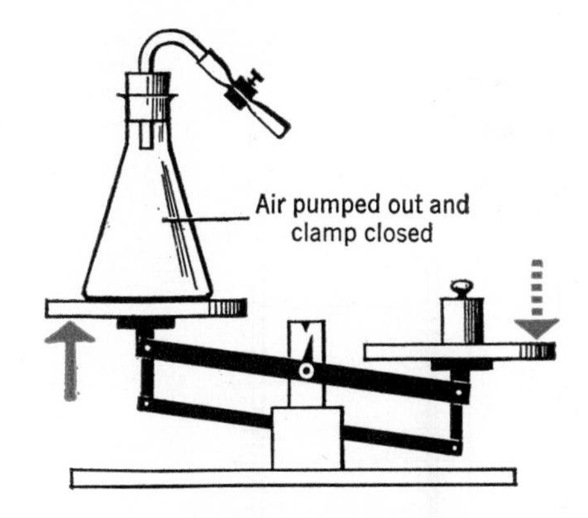

15-2 As the air is pumped out of the flask, the platform rises, showing loss of weight. The amount of weight needed to restore balance is equal to the weight of the air removed from the flask.

flask, we can calculate the density of the air. We find that air, at sea level and room temperature, weighs about 1.2 ounces per cubic foot. Water, as we know, weighs much more—62.4 pounds per cubic foot.

Thus water is more than 800 times denser than air. From this we see that if we go down 800 feet in the elevator of a skyscraper, we experience an increase in pressure approximately the same as if we were to go 1 foot under water.

As a result of the great height of air, the air pressure at sea level is considerable. Figure 15-3 shows two experiments which provide evidence of this great pressure.

Barometers

The experiment illustrated in Fig. 15-3B provides the clue to a method of measuring air pressure. Suppose we have a long glass tube filled with water. We invert the tube, and cover the bottom with a piece of cardboard. The cardboard will support the water as long as the upward pressure of air on it is greater than the downward pressure of water. By using tubes of different lengths, we find that when the water height is close to 34 feet, its downward pressure becomes equal to the air pressure, and the cardboard falls off. Do this experiment over the sink!

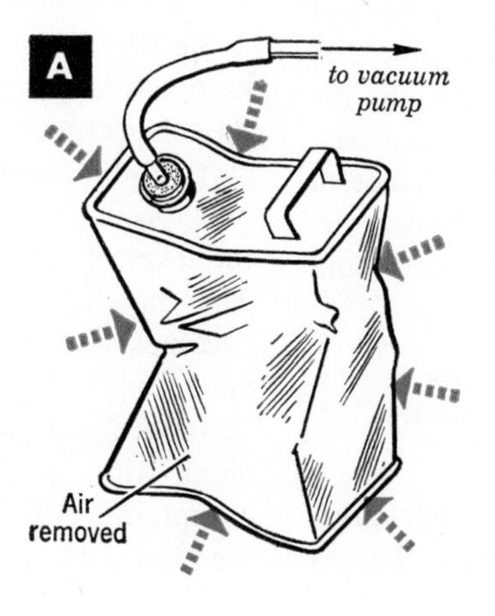

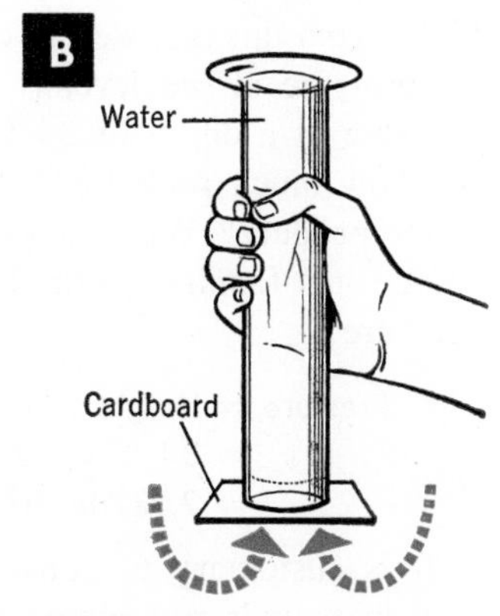

15-3 Two ways of demonstrating air pressure. In A a tin can has been crushed by the outside air pressure after the air inside the can has been removed. In B a card has been placed on a tall glass tube filled with water which has then been inverted. Air pressure holds up the column of water and the card. With how tall a cylinder could you get this to work?

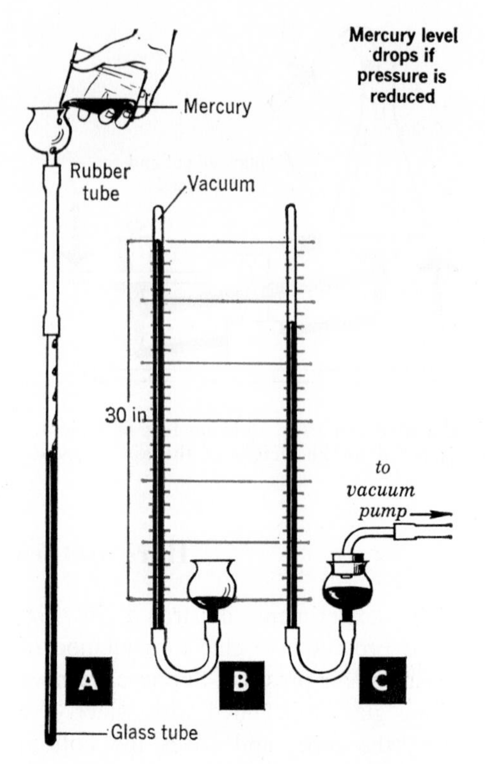

15-4 You can make a barometer by filling a long glass tube (about 36 inches) with mercury. When the tube is inverted as shown in B, the mercury level falls slightly, but remains 30 inches higher than the level in the wide tube. Raising or lowering the pressure with pumps, as shown in C, causes the mercury level to rise or fall.

3. Measuring air pressure

Early in the seventeenth century it was discovered that air pressure at sea level will hold up a column of water 34 feet high. From this fact we can conclude that air pressure at sea level is equal to the pressure at a depth of 34 feet in water, but how much pressure does 34 feet of water produce? We find the answer by using our familiar formula for liquid pressure:

$$\begin{aligned} \text{Pressure} &= \text{Height} \times \text{Density} \\ &= 34 \text{ ft} \times 62.4 \text{ lb/ft}^3 \\ &= 2{,}122 \text{ lb/ft}^2 \end{aligned}$$

It is customary to consider air pressure in pounds per square inch. To express 2,122 pounds per square foot as pounds per square inch, we must divide by 144. We obtain the result that the pressure of 34 feet of water is 14.7 pounds per square inch. This, then, is the pressure of the air which is holding up 34 feet of water.

Obviously, it is awkward to work with a tube 34 feet long. It is more convenient to substitute for the water in the tube a liquid which is more dense. This will give us the same pressure for a smaller height of liquid, so that we can use a much shorter tube. Mercury is a very dense liquid with a specific gravity of 13.6. This means that a column of mercury that exerts the same pressure as 34 feet of water need be only $\frac{34}{13.6}$ ft $= 2\frac{1}{2}$ ft, or 30 in (76 cm) high.

A glass tube, sealed at the top, in which air pressure holds up a column of mercury, was first made by Galileo's pupil, Torricelli (tôr·rê·chĕl'lê). This device, called a **barometer,** provides an easy method for measuring air pressure (Fig. 15-4). If air pressure is lowered for any reason, it can no longer hold up the normal 30-inch height of mercury, and the mercury drops to a new level in the tube. This drop in the mercury level is our measure of the drop in air pressure.

4. Using a barometer

Shortly after Torricelli invented the barometer, the French scientist Blaise Pascal arranged to have the instrument carried up a mountain. He found, just as he expected, that as one climbed higher, the level of the mercury fell in the tube. Thus he showed that the pressure of air becomes less as one goes higher in the air ocean.

Suppose you were to take a barometer along with you as you climbed a high mountain. You would find that, during the first 10,000 feet of the ascent, the mercury level in the barometer would drop approximately 1 inch for each 1,000 feet you climbed. At about 18,000 feet (this is a *real* mountain you are climbing!), it would have dropped to 15 inches—or half the height of the column of mercury at sea level. So we can say that at an altitude of 18,000 feet, air pressure is only about half as great as it is at sea level.

Barometers are also useful in forecasting a change in the weather. A drop in the mercury level—a falling barometer—means the approach of a low-pressure area, usually bringing bad weather with it, while a rising barometer is a good sign that the weather will clear.

5. The aneroid barometer

Torricelli's mercury barometer is convenient enough for a weather station, where it can be attached to a wall. But it is too clumsy for an airplane. Hence, a more convenient substitute—called the **aneroid barometer**—has been developed. (The word "aneroid" means "not moist"—or "without liquid.")

The aneroid barometer measures changes in air pressure by the motion of a springy metal box from which most of the air has been removed. When the pressure of the outer air increases, the top and bottom of the box are squeezed together slightly. A combination of machines (consisting of gears and levers) is used to magnify this slight motion and to rotate a pointer. A calibrated dial gives the reading for pressure in inches or centimeters of mercury. Altitude above sea level can be measured by a barometer which has been marked off properly. Such a barometer, used for measuring altitude, is called an **altimeter.**

If a continuous record of changes in pressure is needed, the pointer of the aneroid barometer can be made to make a continuous mark on a rotating paper roll (see Fig. 15-5). The paper roll is rotated by a clock mechanism so that it makes one complete revolution every week. The pointer responds to changes in pressure by moving up and down, while the clock rotates the paper slowly. This device, known as a **barograph,** records changes in pressure that occur during the week.

6. Units of air pressure

Pressure is the force acting on a unit of area. Strictly speaking, we should talk about air pressure in units such as pounds per square inch or grams per square centimeter. But it is customary to measure atmospheric pressure by the height of the mercury column in a barometer. Therefore, you will usually find atmospheric pressure referred to as, say, 29.54 inches, or perhaps 750 millimeters.

Since weather affects air pressure, actual barometer readings at sea level can vary from hour to hour. Therefore, a standard sea-level pressure of 76 centimeters of mercury has been established. This pressure is equivalent to 760 millimeters; can you show that it is also equal to 29.92 inches? This is also known as a pressure of 1 atmosphere.

The Weather Bureau also uses another unit for measuring atmospheric pressure, called the **millibar,** which is 1,000 dynes/cm^2. Standard atmospheric pressure at sea level is 1,013.2 millibars. This is equivalent to the pressure caused by 29.92 inches (or 760 millimeters) of mercury in a barometer tube.

A law governing air pressure

Figure 15-6 shows a famous scientific experiment which took place in the year 1654, a time when little was known about air pressure. The German scientist

Taylor Instrument Co.

15-5 **Photograph of a device to record barometric pressure on a revolving drum. This device is known as a barograph. Note how the pressure dropped from 29.9 inches to 29.5 inches in one day, indicating that a storm was approaching. Can you see why the captain of a ship carefully watches a falling barometer?**

Otto von Guericke (gā′rĭ·kĕ) arranged this experiment as a sort of tug of war to prove his point. The strength of twenty men was pitted against the pressure of the atmosphere.

A rope attached to the piston in the pump branched out into twenty separate strands, each of them held by a man pulling with all his strength. When von Guericke opened a small valve, the piston shot downward, hurling all twenty men forward.

Von Guericke was making use of a fact on which all air-pressure devices are based. If a connecting valve is opened between two tanks of air, each at a different pressure, the air in the region of higher pressure pushes toward the region of lower pressure. If a piston is located between the two tanks, then the air from the high-pressure side exerts force on the piston. The greater the difference in pressure, the greater the force. Also, the larger the area of the piston, the greater the force will be.

As Fig. 15-6 shows, von Guericke pumped the air out of a tank, T. When the valve, V, was opened, the higher-pressure air outside the pump piston, P, pushed toward the lower-pressure region in the tank. If the area of the piston was 100 square inches, and the pressure in the tank was almost zero, the total force pulling against the twenty men would be about 15 $\text{lb/in}^2 \times 100\ \text{in}^2$, approximately 1,500 lb.

If the atmospheric pressure in von Guericke's experiment could overcome the resistance of twenty men, it is obvious that it can also be made to push and pull objects and do useful work. There is a simple law that describes how the pressure exerted by a confined gas is related to its volume.

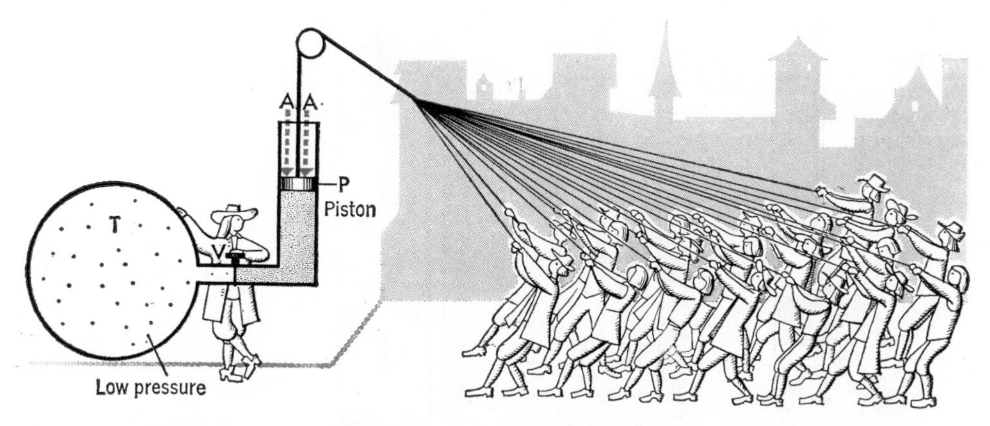

15-6 Scientific "magic" in the year 1654. An empty tank outpulled 20 men. When the valve was opened, air at A pushed inward toward the low pressure in the tank T.

7. Boyle's Law

Suppose we have a sliding piston in an enclosed cylinder. A ruler alongside tells us how far the piston moves up or down (Fig. 15-7). When the air in the cylinder is at normal pressure, we say its pressure is equal to 1 atmosphere. Now we slowly push the piston halfway down. Our pressure is now twice as great, or 2 atmospheres. We continue to exert force on the piston until the air is compressed to $\frac{1}{3}$ of its normal volume. Air pressure in the cylinder is now 3 times normal, or 3 atmospheres. At $\frac{1}{4}$ of its original volume, air pressure is 4 times normal.

This fact—that the pressure increases in exactly the same proportion as the volume decreases—was first discovered by the English scientist Robert Boyle in 1662. **Boyle's Law** says that:

The pressure in a mass of gas is inversely proportional to its volume provided the temperature remains constant.

In the experiment just described, we increased pressure by making the volume of the gas less. If we pull the piston up to a point where the volume of the gas is doubled, the pressure of the air is just $\frac{1}{2}$ what it was. If we increase the volume of the gas 3 times, then our pressure is only $\frac{1}{3}$ what it was originally (Fig. 15-8).

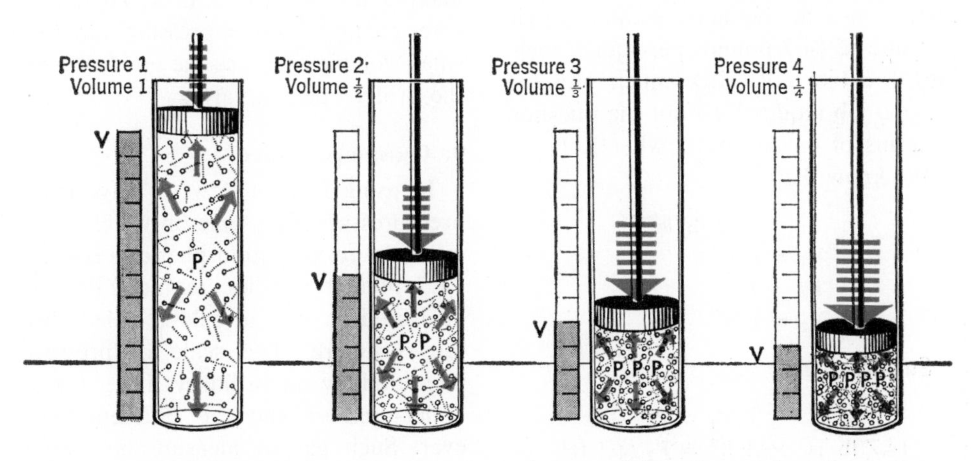

15-7 Boyle's Law. High pressure can always be created by reducing the volume of gas.

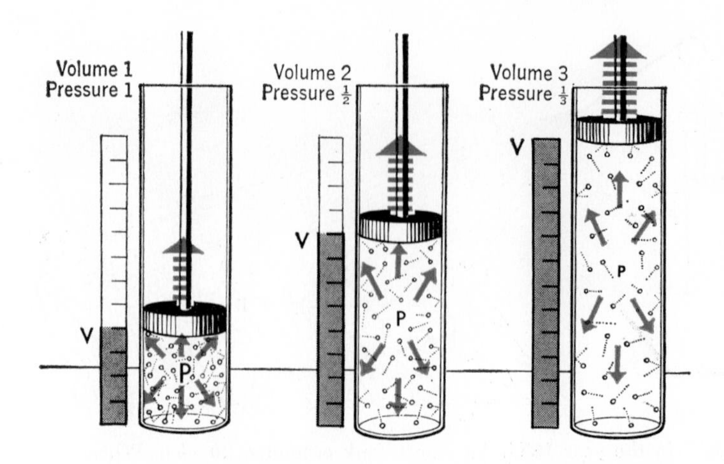

15-8 Low pressure can be created in a gas by increasing the volume. Air outside tends to push the piston back, thus making it more difficult to pull up the piston.

8. Applying Boyle's Law

A formula summarizes these facts:

$$P_1V_1 = P_2V_2$$

Here P_1 and V_1 represent, respectively, the pressure and volume of a certain parcel of gas under one set of conditions. P_2 and V_2 represent the pressure and volume of the same parcel of gas under different conditions. Any units of pressure will do, of course—pounds per square inch, centimeters of mercury, feet of water, or atmospheres—as long as the units of P_1 and P_2 are the same.

SAMPLE PROBLEM

Suppose we want to compress 4 cubic feet of gas, at the normal atmospheric pressure of 14.7 pounds per square inch, into 1 cubic foot. How much pressure will the job require? To put the question in terms of our formula, what is P_2?

We know that:

$$P_1 = 14.7 \text{ lb/in}^2$$
$$V_1 = 4 \text{ ft}^3$$
$$P_2 = ?$$
$$V_2 = 1 \text{ ft}^3$$

Substituting:

$$P_1V_1 = P_2V_2$$
$$14.7 \text{ lb/in}^2 \times 4 \text{ ft}^3 = P_2 \times 1 \text{ ft}^3$$
$$P_2 = 58.8 \text{ lb/in}^2$$

We see again that pressure varies inversely with volume, as stated by Boyle's Law. In order to reduce our volume to $\frac{1}{4}$ (from 4 ft^3 to 1 ft^3), we have to increase our pressure 4 times (from 14.7 lb/in^2 to 58.8 lb/in^2).

STUDENT PROBLEM

300 ft^3 of gas at normal atmospheric pressure are compressed into 5 ft^3. What pressure is needed? (ANS. 882 lb/in^2)

In the centuries since Boyle discovered his law many practical applications have been made. The man operating the compressed air drill in Fig. 15-9 is making use of Boyle's Law. High pressure created by compressing air to a small volume operates the air drill when the driller opens a valve.

9. Gauge pressure

In using Boyle's Law we have measured pressure on a scale on which a vacuum is zero and atmospheric pressure is 14.7 pounds per square inch. Pressure measured on such a scale is called **absolute pressure.** The scale of pressures measured by an ordinary tire gauge or boiler-pressure gauge is different, however. Such gauges measure how much the pressure of the gas is above atmospheric pressure. For example, a tire gauge, when put on a flat tire, will indicate zero, because the air pressure in the tire is atmospheric pressure. Yet we know that the pressure in the tire is not *really* zero, but 14.7 pounds per square inch. The pressure read on a gauge such as a tire gauge is called **gauge pressure.**

To change gauge pressure to absolute pressure, add 14.7 pounds per square inch to the pressure indicated by the gauge. Thus, if a tire gauge indicates 30 pounds per square inch, we know that the absolute pressure is 44.7 pounds per square inch. It is important to remember that the pressures used with Boyle's Law *must always* be absolute pressure if the formula is to work.

10. Explaining Boyle's Law

It has been found that all gases obey Boyle's Law, with only slight variations. Boyle's Law provides an important piece of evidence for the theory that all matter is made up of tiny particles, called **molecules,** which are in constant motion. According to the kinetic molecular theory, molecules of gas in a container are in constant motion and bounce off each other and off the walls of the container. The pressure we have observed in the experiments with air pressure is caused by billions of molecules a second bouncing against the walls of a container.

Now suppose we compress a given quantity of molecules of air into half the space they occupied before. There are now twice as many molecules per cubic inch as before. Each square inch of surface of the container is now being bombarded by twice as many molecules as before. Hence the pressure will be twice as great. If we squeeze the enclosed air into $\frac{1}{4}$ of its original volume, then 4 times as many molecules will be striking each square inch of the container—and pressure will be 4 times as great.

But this rule does not apply if the temperature changes. Temperature affects the speed at which the molecules move and changes the push they exert against the walls.

Making use of air pressure

The application of Boyle's Law to pumps can be made clearer by studying an actual situation such as that shown in Fig. 15-10.

11. High and low pressure at the same time

Suppose you have a pump with the inlet valve, I, attached to a closed can and the outlet valve, O, attached to a football, as shown in Fig. 15-10. Each of

Lewsen from Black Star

15-9 The energy contained in compressed air can be used to shatter solid rock. Discoveries by men such as Robert Boyle and Otto von Guericke, who lived centuries ago, made the modern compressed air drill possible.

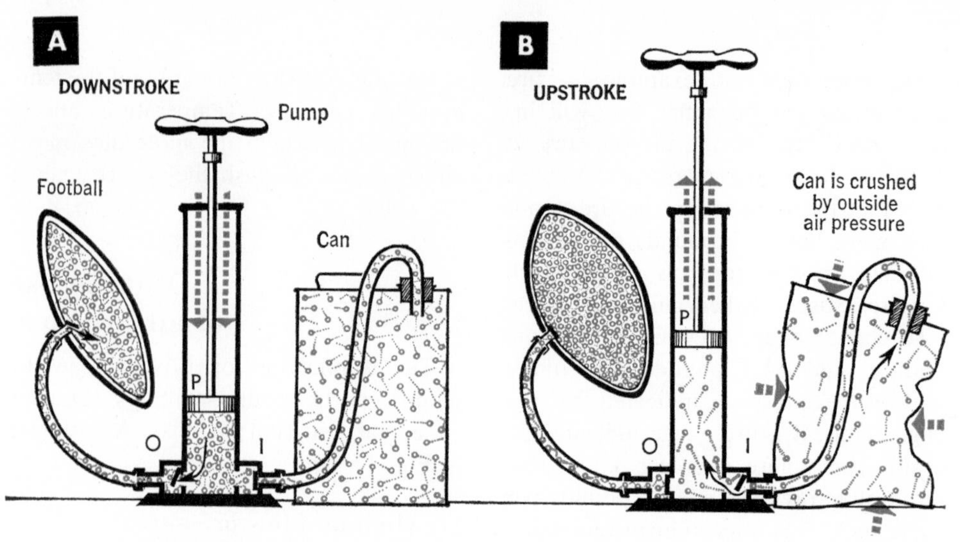

15-10 Every downstroke decreases volume, raises the pressure, and pumps up the football. At every upstroke volume is increased, pressure is reduced, air rushes from the can into the pump, the pressure in the can is lowered, and the outside air makes the can collapse. Thus the same pump creates both high pressure and low pressure.

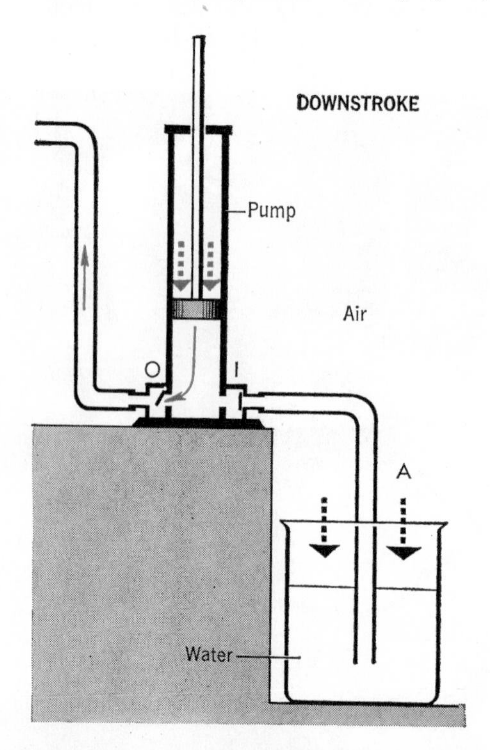

15-11 A simplified water pump. If the inlet side of a pump is attached to a tube inserted into water, then the outside air pressure will push water ahead of it into the low pressure in the pump. On the downstroke the inlet valve I is closed, and the water is forced through the outlet valve O up the tube.

these valves is a one-way door, permitting the air to flow in one direction only. The inlet valve lets air in only; the outlet valve lets it out only.

At the start of your experiment, the pressure of air is the same in the pump, the can, and the football (14.7 lb/in^2). But if you pull the pump piston upward, the volume of air in the pump is increased (Fig. 15-10B). Hence, as we would expect from Boyle's Law, the pressure of air in the pump cylinder is reduced by the increased volume so that it is less than the pressure in either the can or the football. Now air from both the football and the can tends to rush into the cylinder.

But the outlet valve, O, jams shut so that air cannot pass from the football into the cylinder (Fig. 15-10B). At the same time the inlet valve, I, to the can is opened by the air rushing from the can into the cylinder. Now the pressure in the can is lowered. If the inside pressure is lowered sufficiently, the pressure of the atmosphere outside the can will crush it, provided that the can is not too strong.

When the piston is pushed down again, the new air that has entered the cylinder is compressed into a smaller volume (Fig. 15-10A). This, as we know, means higher pressure. The air in the cylinder now moves toward the lower-pressure regions in the can and football. But valve I now jams shut and valve O is pushed open. Thus the air is pushed into the football and raises its pressure.

So we see that on every upstroke an increase in volume in the pump results in lower pressure, and air rushes into the pump from the can by way of the inlet valve. At every downstroke this new amount of air is compressed in volume; as a result its pressure is increased, and it pushes its way through the outlet valve and into the football.

This illustrates a use of Boyle's Law to create high or low pressure at will so that we can transfer air from one place to another.

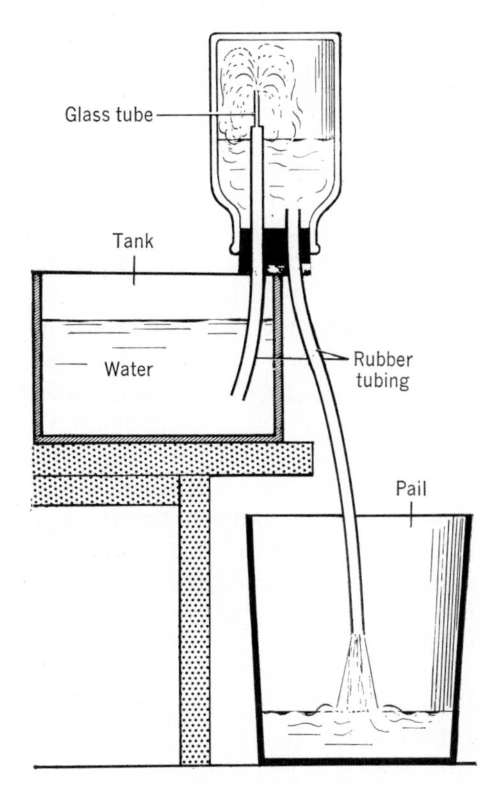

15-12 Perpetual motion? Why does the water spray upward by itself?

12. Using air to push water

We can use this same arrangement to pump water. If we attach a pipe from the inlet side, I, to the source of water, we have a simple kind of water pump (Fig. 15-11). The upstroke produces low pressure in the cylinder, just as it did before. The outside air at A tries to rush in. But now the water is in the way. If we have created a difference in air pressure that is great enough, the air from the outside will push the water into the cylinder. After a few strokes the cylinder is filled with water. On the downstroke the pressure of the piston on the water then pushes it up a tube to any desired height. The higher we want the water to go, the greater the pressure we must exert on the piston.

The pump will not work if the piston is more than 34 feet above the level of the water, since it is the atmospheric pressure that pushes the water up into the pump. You will recall that normal air pressure can hold up 34 feet of water, at the most.

13. Other uses of air pressure

Compressed air has countless uses in our modern world. Air under pressure is used to operate air brakes on trains, trucks, and buses in a manner similar to that shown in Fig. 13-10. Riveting guns, pneumatic drills, paint sprayers, and air rifles are all operated by compressed air. Compressed air opens and shuts doors on trains and buses. Your dentist uses compressed air to run a machine in cleaning your teeth. And last but not least, a cow on a modern farm is very likely to be milked by machines that use low air pressure!

The great variety of jobs performed by air pressure has given rise to many different kinds of pumps, as shown in Fig. 15-13. See if you can explain how each pump operates.

The airplane

In the year 1500, the great artist and scientist Leonardo da Vinci speculated on the possibility of flying machines. But it was beyond the limits of even Leonardo's imagination to think that some day man would fly faster than the speed of sound—over 750 miles an hour. Yet this modern achievement is no miracle. It is the result of the use of scientific principles arrived at through the years by painstaking research. We have mentioned some of these principles before, but let us take time for a closer look at them here.

14. Lift

Everyone is familiar with the way a kite is held aloft by the wind. As the wind increases in velocity, the kite can fly higher and can carry more weight. The force, or lift, that accomplishes this is created partly by the air striking the bottom of the kite.

As shown in Fig. 15-14A, the kite is held in such a position that the moving air strikes the bottom at an angle. The kite deflects this air downward, changing its path from Q to R. At the same time the equal and opposite force of the air upon the kite forces the kite in an upward and backward direction, represented by force F in Fig. 15-14A.

The wing of an airplane acts like a kite. As the propeller pulls the plane through the air, the motion of the air against the underside of the wing exerts an upward and backward force. The upward component of this force is responsible for part of the **lift.** But this desired effect can be achieved only when the wing is tilted at an angle to the air stream. The angle, *a,* which the bottom of the wing makes with the air stream is called the **angle of attack** (Fig. 15-14A).

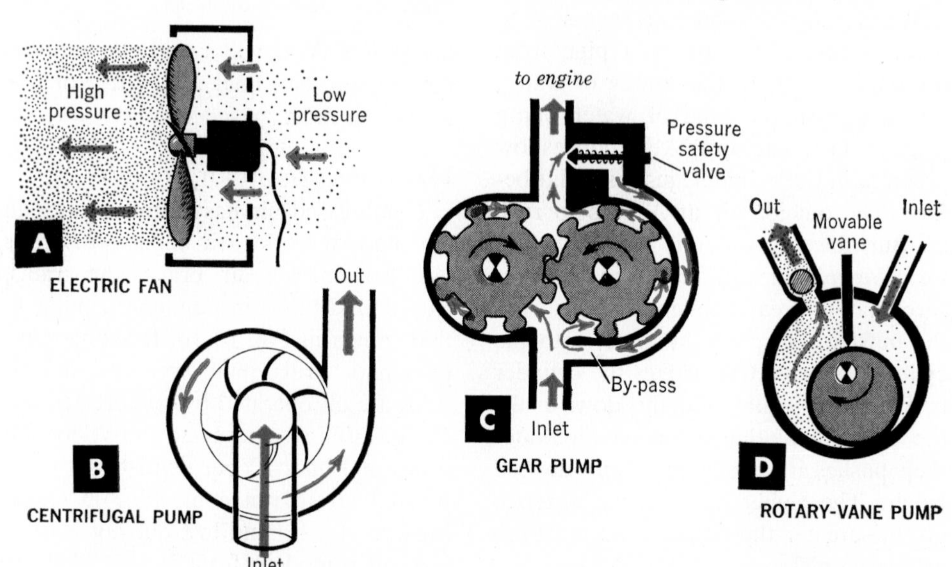

15-13 There are many kinds of pumps for fluids. A shows the fan type used to create low or high pressure in air, as used in the vacuum cleaner. B is a centrifugal pump, as used in the automobile engine to pump water through the cooling system. C is a gear pump as used in some engines to pump oil. D is a rotary vane pump used to obtain very low air pressures in making radio tubes.

15-14 Airplanes are held aloft by two forces. In A we see the "kite effect" in which air striking the bottom of the inclined surface pushes the kite (or the bottom of an airplane wing) upward. In B we see how the fast-moving air above the paper (or the upper surface of a wing) creates a low pressure that makes the paper (or wing) rise. Of course, this latter effect applies to A as well.

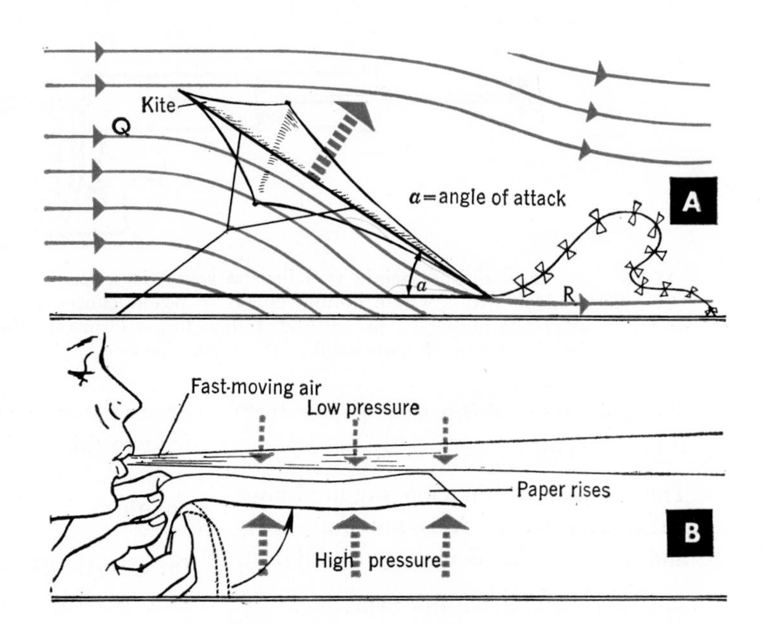

15. Lift on the upper surface

The force of the moving air against the underside of the tilted wing accounts for only part of the lift on the wing of a modern plane. Most of the lift is caused by reduced pressure on top of the wing as air travels over the curved upper surface.

You can observe this action for yourself by holding a strip of paper in your fingers and blowing over the top of it, as shown in Fig. 15-14B. The motion of the air over the top causes the paper to rise.

It will help us understand why this happens if we take a pipe filled with a fluid like air and make it narrow at one part, B, as shown in Fig. 15-15. If the fluid is to travel through the pipe, it must speed up in the narrow part and slow down in the wider part. This is similar to the way water flows in a river. The narrow parts of the river are the "rapids." The water in the wide parts moves very slowly.

But when a slowly moving fluid is suddenly speeded up, there must be extra force behind it to cause the acceleration. Thus there must be greater pressure behind the fluid in the wide part of the tube A than in the narrow part B. If the two pressures were equal, it would be impossible to increase the speed of the fluid in the narrow part.

Similarly in Fig. 15-15, the pressure in the wide part C has to be greater than that in the narrow part B, exerting a backward force that slows down the fluid.

BERNOULLI'S PRINCIPLE

v V v A B C p T P Atmospheric pressure P Mercury

15-15 Pressure drops when the fluid speeds up in the narrow part of the tube. Rising mercury in the tube T shows that the pressure at B is less than normal.

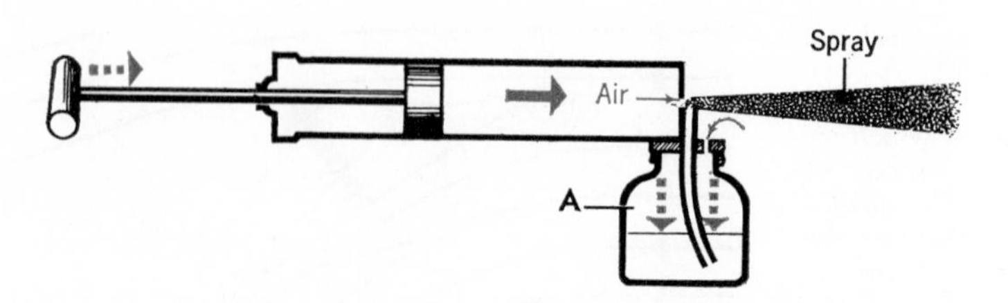

15-16 Bernoulli's Principle explains the action in a spray gun. A stream of air passing the open end of a tube whose bottom end is in liquid causes a low pressure because of its high velocity. The higher air pressure at A then forces liquid up the tube. A spray is formed as it is struck by the moving air passing above the open end.

The fluid slows down as it passes from B into C. Thus:

The pressure within a moving fluid must be greater where it is moving slowly and less where it is moving rapidly.

This is called **Bernoulli's Principle.**

In an airplane, the curved upper surface of the wing forces the air to travel up over the top. Thus the curved surface acts as one side of a narrowing tube, while the still air above the wing acts as the other side (Fig. 15-17). The air is jammed into the narrowing space between these two sides and is forced to move faster. As we would expect from Bernoulli's Principle, this increase in speed results in a loss of pressure. This lower pressure above the wing accounts for two thirds of the lift, the higher pressure underneath giving the remaining one third.

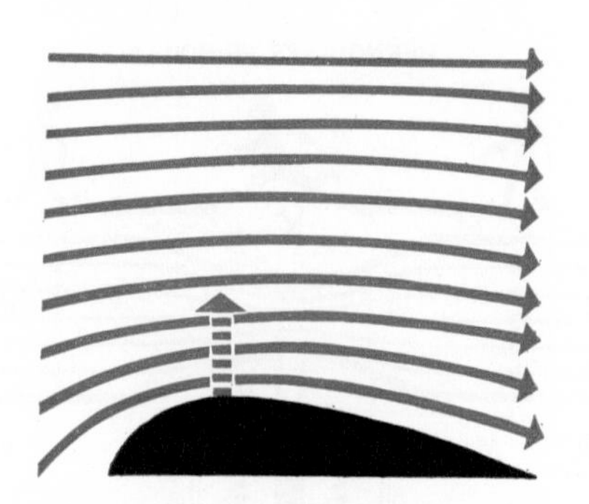

15-17 We can consider a real airplane wing as the lower side of a narrowing tube of air, with the air far above the wing making the upper side of the tube. The low pressure on the upper surface of the wing, resulting from the flow of air, causes the airplane wing to be lifted upward.

16. Controlling the lift

A pilot makes his plane rise or fall mainly by controlling the lift. One of his methods is to change the angle of attack (Fig. 15-18). At a low angle of attack, when the bottom of the wing is almost parallel to the motion of the air, the lift is small. This is the situation at high speed. The higher impact of the air increases lift and compensates for the reduced lift caused by the low angle of attack (Fig. 15-18A).

Up to a certain point an increase in the angle of attack means an increase in lift; the pilot can then reduce his speed and still remain aloft (Fig. 15-18B). But beyond a certain angle, the air breaks away from the upper surface of the wing, and whirlpools of air at the back edge of the wing result in a great loss of lift (Fig. 15-18C). The angle at which this occurs is called the **stalling angle.** As the name suggests, if a wing is at or near this angle, the plane (not the engine, of course) may stall in mid-air and fall. Therefore, the pilot must be careful not to exceed this stalling angle of attack.

The pilot controls the angle of attack by pushing the **stick** forward or backward. This stick is connected to the elevators in the back of the plane. When the pilot pushes the stick forward, the elevators move down (Fig. 15-19). The air which strikes the lower part of the elevator surfaces then pushes upward, lifting the tail of the plane. In this way the pilot decreases the angle of attack.

If the stick is pulled backward, the elevators move up. Air now strikes the upper part of the elevator surface and pushes the tail of the plane down. The angle of attack is now greater; the lift is increased and the plane climbs. But the pilot must watch his plane carefully when he pulls the stick back so as to avoid stalling the plane.

The pilot also controls lift by means of the **throttle**, which regulates the amount of gasoline the engine burns. When he "opens up the throttle," he causes the engine and propeller to turn faster so that they pull the plane through the air at greater speed. As speed increases, the air passes more rapidly across the wing, thus increasing lift and making the plane climb. In actual practice the pilot uses both throttle and stick to climb, level off, or glide, as he chooses.

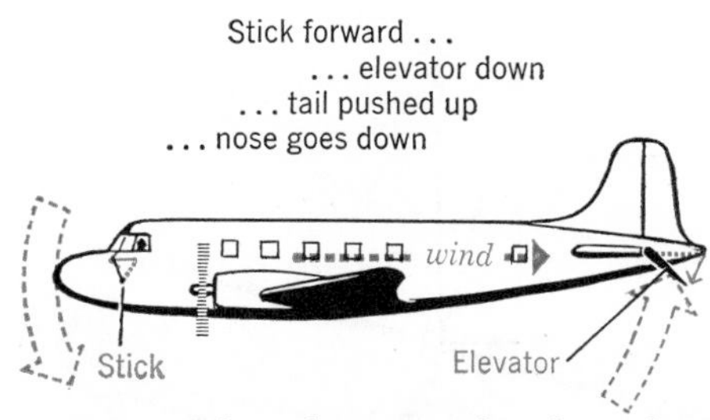

15-19 The pilot noses the plane down by pushing the stick forward. This motion pulls a cable which is connected to the elevator in back of the plane and causes it to move downward. Air striking the bottom of the elevator now pushes the tail of the plane upward, and the plane then noses downward. Pulling the stick backward causes the reverse action, and the plane noses upward.

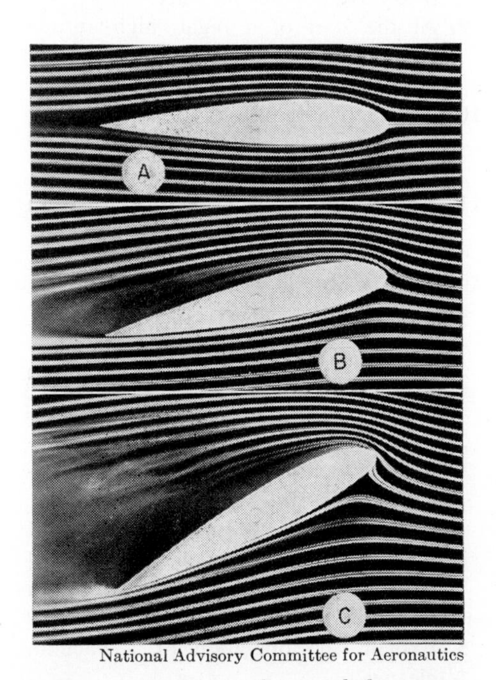

National Advisory Committee for Aeronautics

15-18 Smoke streaks reveal the manner in which air travels past a wing. A shows the air motion when the plane is moving at high speed. B shows the motion of air during slow speed of the plane when the wing makes a large angle of attack with the air. In C the angle has been made too great and the plane will stall (start to fall) because of the drag created by the turbulence of the air.

As the plane rises, the density of the surrounding air decreases. This lower density reduces the impact of the air against the wing—with the result that lift is also reduced. This means that if the pilot wants to continue flying level at a high altitude, he must increase his speed to maintain the necessary lift. Of course, at a certain altitude, known as the **ceiling**, the density of the air is so low that it requires the top speed of the plane just to fly level. The plane can climb no higher if it depends on the force of air on the wings to give it its lift.

Of course a rocket-propelled plane can fly entirely out of the earth's atmosphere, depending on the action of the forces defined by Newton's Third Law to propel it and to steer it.

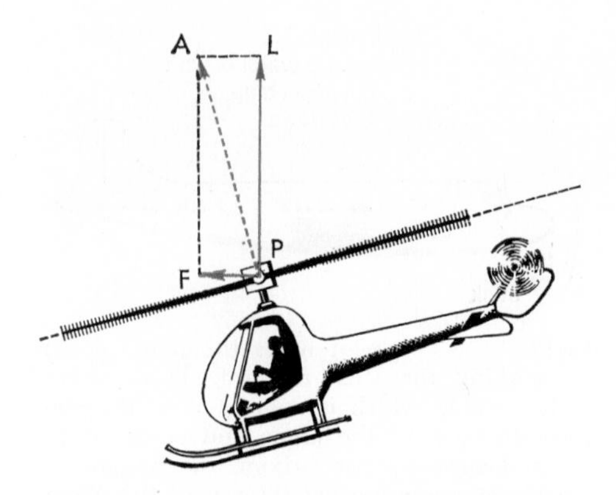

15-20 The pilot of a helicopter can make his plane move forward by causing the blades to tilt forward at an angle. The force of the helicopter blades is now exerted in the direction PA. The vertical component of this force PL causes lift. The horizontal component PF causes the helicopter to move forward.

17. The helicopter

In the helicopter a large propeller (called a **rotor**), rotating horizontally, provides the lift that holds up the plane. When lift is exactly equal to weight, the helicopter hovers motionless. If the pilot wants to go forward or backward, he changes the angle of the rotating blades. In the position shown in Fig. 15-20, the pull of the blades is no longer vertical. There is now a forward component PF which moves the plane forward, while the upward component PL provides lift. If the blades were tilted backward, the helicopter would move backward.

Going Ahead

THE VOCABULARY OF PHYSICS

Define each of the following:

air pressure	altimeter
barograph	lift
absolute pressure	aneroid barometer
millibar	stick
air ocean	angle of attack
gauge pressure	throttle
barometer	stalling angle
ceiling	

THE PRINCIPLES WE USE

Explain why

1. Air pressure becomes less as you go up in the air.
2. It takes a height of several hundred miles of air to produce the same pressure as 30 inches of mercury.
3. A tin can is crushed when the air is pumped out of it, but a television picture tube is not.
4. The water in the inverted cylinder of Fig. 15-3 does not fall.
5. Good weather is usually accompanied by a rising barometer.
6. Air is removed from the inside of the box of an aneroid barometer.
7. A balloon reaches a certain height and can go no higher.
8. Increasing the volume of a gas causes it to have a lower pressure.
9. At the start of a flight, high-altitude balloons may contain less than $\frac{1}{10}$ the normal volume of gas (Fig. 14-10).
10. A pump whose piston is more than 34 ft above the level of the water it is pumping will not operate.
11. You breathe in when the diaphragm in your body moves down.

APPLYING LAWS OF AIR PRESSURE

1. Explain (with a diagram) the operation of an air pump.
2. Diagram the operation of a pump for liquids and explain its operation.
3. Find out how a siphon works, and explain its action with the help of a diagram.
4. Explain how air pressure is used to fill a fountain pen. Why does a half-full pen sometimes leak if carried in an airplane, while a full pen does not?
5. Explain the operation of a vacuum cleaner.
6. Draw a diagram of a mercury barometer and explain how it works.
7. Compare the advantages and disadvantages of aneroid and mercury barometers.
8. State two factors that can cause a change in the reading of a barometer.
9. Why is it possible for a balloon to float

in air? How can the lift of the balloon be increased?

10. How is Bernoulli's Principle applied in the flight of an airplane?
11. How is Bernoulli's Principle applied to a spray gun?
12. Explain how the pilot uses the stick to glide downward or climb upward.
13. Explain four factors that affect the lift of an airplane.

EXPERIMENTS WITH AIR PRESSURE

Describe an experiment to

1. Show that air exerts pressure.
2. Measure air pressure.
3. Find the weight of a liter of air.
4. Illustrate Bernoulli's Principle.

SOLVING AIR-PRESSURE PROBLEMS

NOTE. To simplify the arithmetic, assume in these problems that atmospheric pressure at sea level is 15 lb/in².

1. 1 ft³ of air in a pump at atmospheric pressure is compressed by a piston to $\frac{1}{8}$ ft³. What is the final pressure on the compressed air in (a) lb/in², (b) cm of mercury, (c) atmospheres, (d) ft of water?
2. 2 liters of air at a presure of 76 cm of mercury are compressed to a volume of 0.3 liters. What pressure was applied to compress the gas?
3. If the absolute pressure on the gas in a cylinder is 15 lb/in² and the volume of the gas is 5 ft³, what is the volume when the pressure is increased to 45 lb/in²?
4. 500 cm³ of gas are under a pressure of 76 cm of mercury. What volume will the gas occupy if the pressure becomes 570 cm of mercury?

---

5. If 20 ft³ of air at atmospheric pressure are squeezed into 2 ft³, what is the resulting (a) absolute pressure, (b) gauge pressure?
6. The gauge put to a tire reads 30 lb/in². The volume of air in the tire is 3 ft³. What volume will the air occupy after a blowout?
7. How many cm³ of oxygen at normal atmospheric pressure are required to fill a tank at 100-lb/in² gauge pressure if the tank has a volume of 20 liters?
8. Some air at a gauge pressure of 200 lb/in² expands to 6.7 times its original volume. What does its gauge pressure become?
9. How many ft³ of air at atmospheric pressure can be released from a 10 ft³ tank containing air at 300 lb/in² gauge pressure?
10. A rubber suction cup 3 inches in diameter attaches a baggage carrier to the top of a car. If the cup is perfectly sealed, how much force is required to

15-21 You would think that the air blown under the paper would blow the paper off. Instead it collapses downward. Do you know why?

pull it away from the top of the car?

11. To a tire pump whose handle has a 15-in stroke there is attached a tire gauge. The gauge reads zero when the handle is raised as far as possible. What will the gauge read when the handle has pushed the piston to 3 in from the bottom?

---

12. How many liters of air at atmospheric pressure must be pumped into a 10-liter tank of air at atmospheric pressure in order to raise the gauge pressure to 100 lb/in^2?
13. An inverted container is pushed under water until the air in it is compressed to 0.9 of its original volume. How far down has the container been pushed?
14. What is the maximum height to which atmospheric pressure can force alcohol (specific gravity 0.8) in a tube?
15. Two tanks of air are connected by a pipe fitted with a valve. The valve is initially closed, and the air in one tank (volume 200 ft^3) is at an absolute pressure of 500 lb/in^2. The air in the other tank (volume 50 ft^3) is at an absolute pressure of 800 lb/in^2. What will be the pressure in the two tanks after the valve has been opened?
16. A bubble of marsh gas (methane) at the bottom of a pond has a volume of 0.5 in^3. When it gets to the surface it has a volume of 0.85 in^3. What is the depth of the pond?
17. What weight of air at atmospheric pressure can be released from a 200-liter tank of air at a gauge pressure of 300 lb/in^2?
18. A piece of glass tubing 28 cm long and open at both ends is lowered vertically into a dish of mercury until its bottom is 15 cm below the surface. The top is now closed off and it is pulled upward a distance of 12 cm. (a) What is the pressure of the air in the tube? (b) How high is the level of the mercury in the tube above the level of the mercury in the dish?
19. A vacuum pump removes $\frac{1}{3}$ of the air in a flask with every stroke. What is the density of the air in the flask after 6 strokes?
20. A McLeod gauge is a very useful laboratory instrument for measuring low gas pressures. Find out how it works.

PROJECT: THROWING BALLS IN A CURVE

A great many people argue that it is impossible for a spinning ball to curve sideways when it is thrown—that the effect is really an optical illusion.

You can settle the matter by a simple experiment. Obtain a 15-inch length of cardboard mailing tube 2 inches in inside diameter and line it with sandpaper. Simply roll up the sandpaper and slide it in. This tube is a very effective throwing tool for projecting ping-pong balls with lots of spin. Depending on the way the balls are thrown from the tube, they may be given a spin in any direction. With a little practice you will be able to throw them well enough to settle the curved-ball question for yourself.

UNIT THREE

Heat Energy—The Motion of Molecules

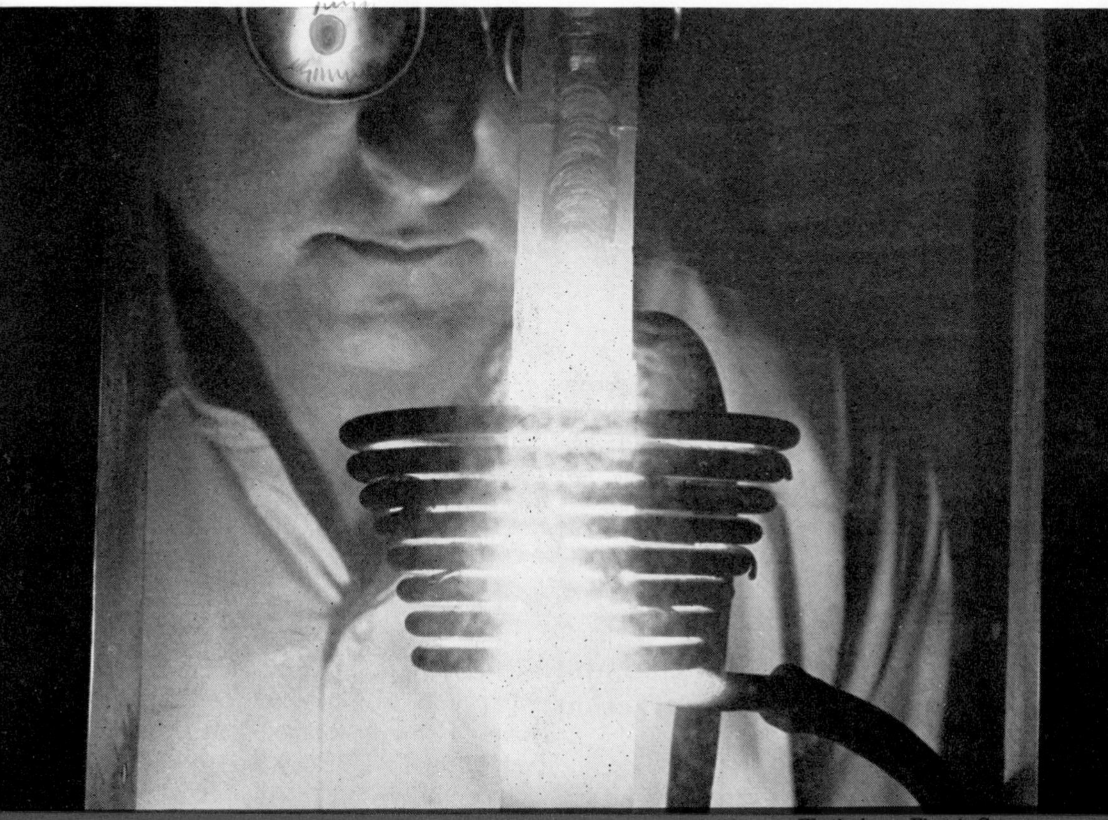

Westinghouse Electric Corp.

You have of course lit a fire. You have many times turned your face to the sun and felt its heat. You probably have had your temperature taken. In the illustration heat from a high-frequency alternating current in the coil is being used to purify a bar of the rare metal, niobium. All these events have one thing in common—heat energy.

Since earliest times man has sought to understand heat energy; increasingly he has brought it under control. In this unit you will study the evidence that the motion of molecules plays a role in the production of heat. You will also see the ways heat energy is produced and how it is measured and put to work to drive ocean liners, airplanes, and automobiles and operate many other devices important in your daily life.

CHAPTER

16

Expansion by Heat

Many simple experiments, like those shown in Fig. 16-1, illustrate the following important fact: objects tend to expand when heated and to contract when cooled. With very few exceptions this is true of all gases or liquids or solids. Even our weather is partly caused by warmed air expanding and hence (by Archimedes' Principle) rising.

Expansion and contraction

1. Expansion can be harmful

Every driver of a car knows that a steaming radiator is a danger signal he cannot afford to ignore. It means that his engine is too hot. In other words, the close-fitting moving parts of the engine are operating at a higher temperature than that for which they were designed. If the driver does not remedy the situation, the parts will expand, grip, and become scratched, warped, or broken as they move.

All of us, whether or not we drive a car, can think of examples of the harmful effects of expansion. We know that if we pour boiling water into a chilled glass, the glass may crack (Fig. 16-2A). This happens because a strain is set up as the cool outer surface keeps the heated inner surface from expanding. In laying railroad tracks, space is left between adjacent rails to allow for expansion in hot weather. The center span of a large steel bridge is made in a separate section and moves on large rollers so that expansion will not warp the bridge (Fig. 16-2C). The parts of a clock expand in warm weather; the clock does not give correct time unless constructed to compensate for this. Even the apparently everlasting rocks of the earth are gradually broken up by continual expansion and contraction due to temperature changes. During the day the surfaces of the rocks expand under the hot sun; at night they contract when the temperature falls. The gradual effect is to flake thin sheets off the rock; these sheets are further broken down by the action of freezing water (see page 248) and are finally carried away by wind and rain.

2. Explaining expansion and contraction

All of the foregoing examples of heat expansion are the result of the fact that heat is the motion of molecules. When a material is heated, its molecules vibrate more rapidly than they did before. Each molecule pushes its neighbors harder. Despite the force of cohesion drawing molecules toward one another, the increasing violence of their motion drives them farther and farther apart. The increased distances between molecules account for the expansion of the material.

On the other hand, as an object cools, the molecules of which it is composed move more and more slowly. Hence they push against their neighbors with less and less force, and allow the immense

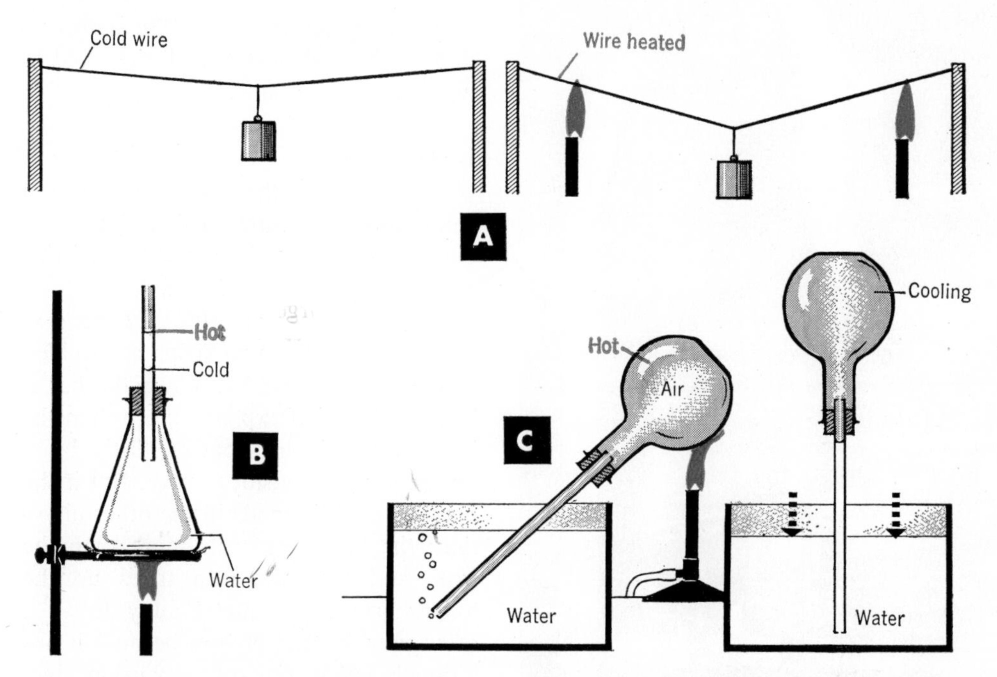

16-1 Heat usually causes expansion. In A the hot wire sags more because it has expanded and become longer. In B the heated liquid has expanded and risen in the narrow tube. In C the heated air expands and bubbles outward while the cooling air contracts and permits the outside air to push water up the tube.

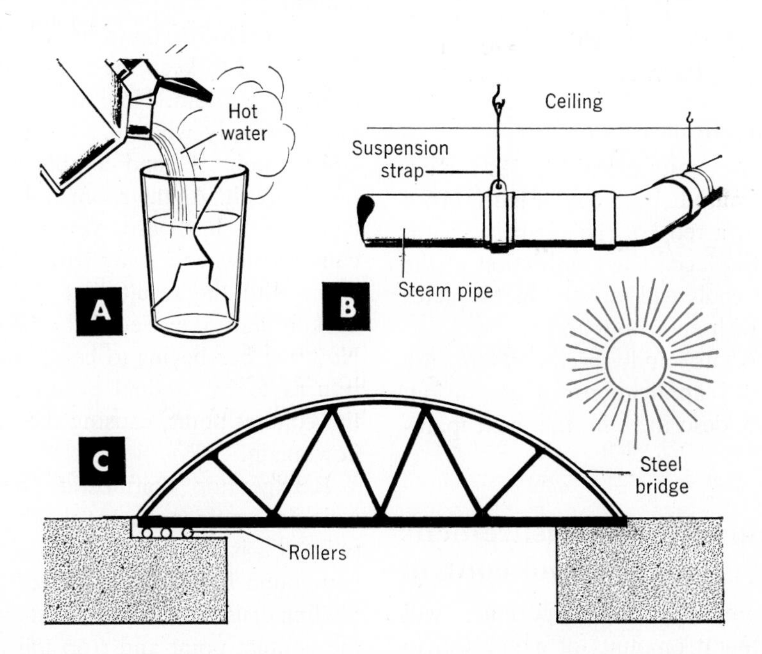

16-2 The cracking of the glass when hot water is poured in, the loose suspension of pipes from a ceiling, the rollers under a bridge are all connected with expansion caused by heat. Can you explain each one?

Standard Oil Co. (N J.)

16-3 Temperature changes that can make a straight pipe expand and contract several inches will not crack this one, thanks to the flexible expansion loop.

force of cohesion (which is always present) to draw them closer together.

3. Expansion can be useful

There are many practical uses for expansion caused by heat. Figure 16-4 shows how a red-hot rivet is being hammered into place. The contraction of the rivet as it cools will pull the metal plates together with enormous force.

Thermostats and thermometers are other useful applications of expansion, which are described in the next paragraphs.

Temperature measurement and control

The same rise in temperature will cause different amounts of expansion in different materials. Brass, for example, will expand more than iron. This fact is put to practical use in the construction of the **thermostat,** a device used to regulate temperatures.

4. Thermostats

One form of the thermostat uses a **compound bar** (often referred to as a bimetallic strip). It is composed of strips of two metals, such as brass and iron, which have different rates of expansion when heated. The two strips are welded together (Fig. 16-5A). When the bar is heated, the brass expands more than the iron, so that the bar bends with the brass on the longer (outside) curve. But if the bar is cooled, it bends in the other direction; the brass contracts more than the iron, and now the iron takes up the longer curve. The metal alloy Invar is often used instead of iron because it has a much lower rate of expansion than iron. The greater difference in expansion causes more bending of the brass-Invar combination than the brass-iron combination.

Figure 16-5B shows how a compound bar is used to keep the temperature constant in a room. Note that the current can flow only when the bar touches the contact point, C, and completes the electrical circuit. As the room warms up, the compound bar also warms up. This causes it to bend away from the contact point with the result that the circuit is broken and the electricity is shut off. Now the bar begins to cool, and soon it returns to its original position to touch the contact point, causing the current to flow again.

Refrigerator thermostats operate in a similar manner, except that the contact point, C, is on the opposite side of the compound bar (Fig. 16-5C). Too much cooling causes the bar to bend away from the contact point and stop the motor.

Temperature regulators in houses operate in the same way, except that the current controls the operation of an oil or gas burner. Automatic irons, ovens, waffle irons, and other controlled heating devices use thermostats in the same way. In a car, a thermostat of different design controls the cooling flow of water through the radiator, thus keeping the temperature of the water in the motor at a fixed point set by the engineers who designed the car (usually between 160° and 180°F). If the temperature is too high, the thermostat opens up the water passageway to the radiator and thus allows more cooling to occur. If the temperature is too low, the radiator is closed off until the engine warms up.

5. Measuring temperature by expansion

Among the many effects caused by heat are changes in length, pressure, color, and electrical characteristics. All

United Press International

16-4 Rivets hold this ship together. Can you explain how expansion and contraction by heat are put to use here?

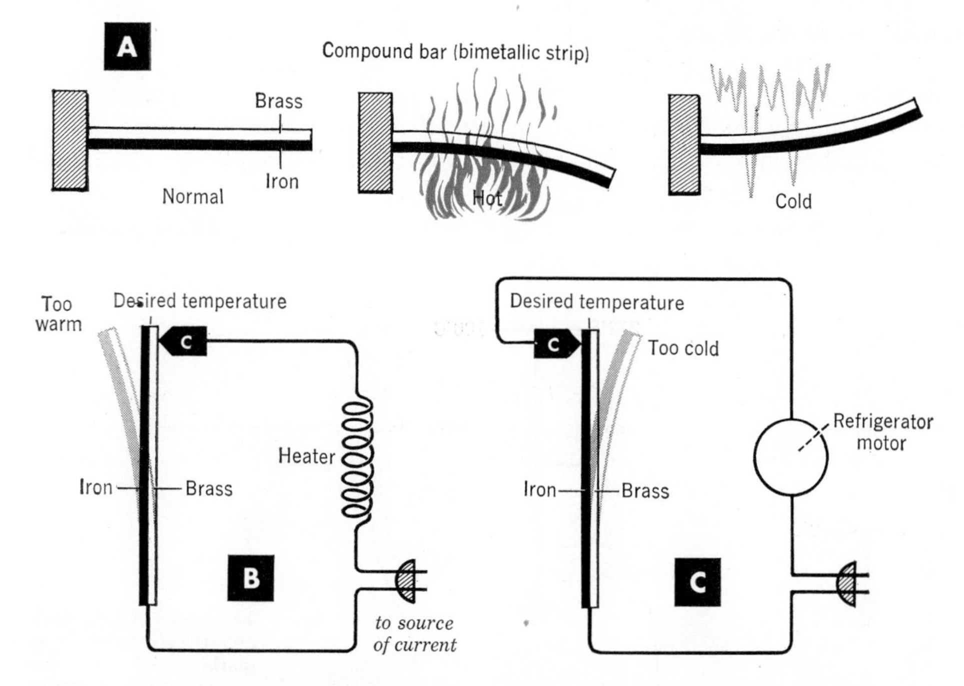

16-5 The thermostat makes use of the fact that metals expand at different rates. Can you explain the bending of the bar when heated and when cooled? Can you explain how the thermostat controls the heater in B and the refrigerator motor in C?

of these changes can be used to measure temperature, but most **thermometers** make use of expansion.

In a metallic thermometer, heat expansion causes a coiled compound bar to bend. The motion of the bar turns a pointer to indicate the correct temperature on a dial, which is calibrated (that is, marked off) in degrees. Oven thermometers are usually of this type.

The most common thermometers make use of liquid expansion. In most of them the liquid is either mercury or alcohol, whose expansion forces it up a very narrow tube, thus permitting very slight temperature changes to be measured.

6. Standards of temperature

Just as we need a standard of length, so we need a standard of temperature. Our judgment of hot and cold is a very unsure guide, as you can see by performing a simple experiment. Obtain three jars. Put hot water in one, water at room temperature in the second, and ice water in the third. Place one hand in the hot water, the other in the cold. Keep them there for about a minute. Now place both hands together into the jar containing water at room temperature. The hand which was formerly in the hot water feels cool; the hand which was in the ice water feels warm.

Standard fixed temperatures are needed which will be the same for all people at all times. For this purpose the freezing and boiling points of water at an atmospheric pressure of 76 centimeters of mercury have been found to be excellent standards for calibrating thermometers. On the Fahrenheit scale the freezing point of water is labeled 32°, the boiling point 212°. The metric system uses the centigrade scale (also called the Celsius scale), setting 100° as the boiling point of water, 0° as the freezing point. The selection of these numbers is quite arbitrary; any other numbers could be chosen for freezing and boiling, and from them a workable temperature scale could be made.

7. Converting from one temperature scale to the other

Just as we must sometimes convert feet into meters, so it is often necessary to convert a Fahrenheit temperature to a centigrade, and vice versa. The simplest

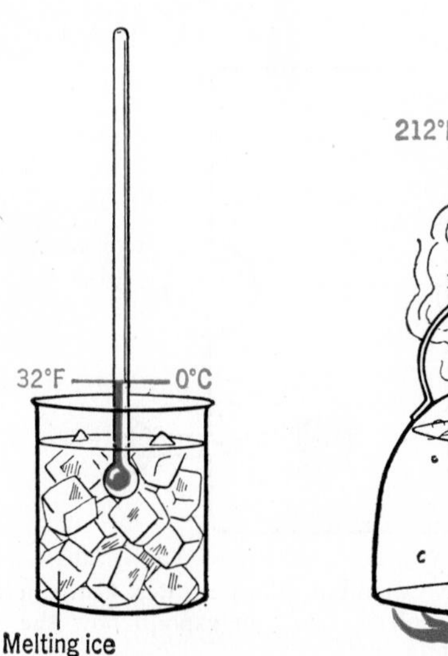

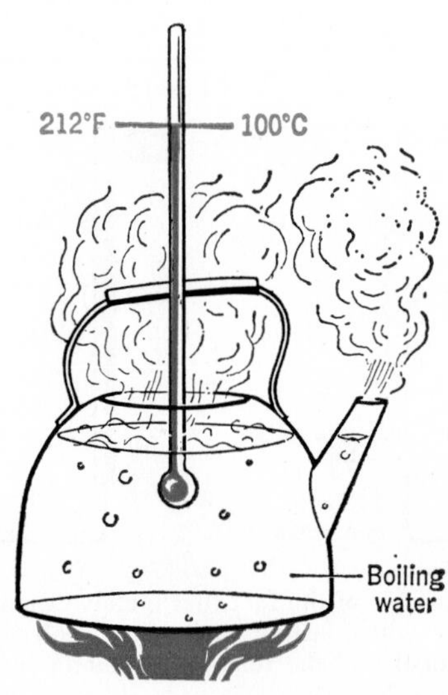

16-6 A thermometer is calibrated by first marking off the fixed points (the freezing and boiling points of water).

way is to use a thermometer scale in which the two readings are set down side by side. Many thermometers for scientific work are made with both scales. For example, the room temperature which gives the greatest comfort is usually considered to be about 68°F. As we see from the double-scale thermometer in Fig. 16-7, this is the equivalent of 20°C.

Lacking a scale which gives us both readings at once, the following formula is useful to convert back and forth from centigrade to Fahrenheit:

$$C = \frac{5}{9}(F - 32)$$

In the formula you notice that you first subtract 32° from the Fahrenheit temperature. This tells you the number of degrees above freezing on the Fahrenheit scale. Now if you look at Fig. 16-7, you see that the centigrade degree is a larger unit than the Fahrenheit degree. To be precise, 5 degrees C are equal to 9 degrees F. For this reason it takes fewer centigrade degrees than Fahrenheit degrees to express the same temperature change. Therefore you take $\frac{5}{9}$ of (F − 32) to get the equivalent centigrade temperature.

SAMPLE PROBLEM

A cookbook says that to cook a certain dish the oven temperature should be 400°F. What temperature would this be on an oven thermometer calibrated in centigrade degrees? Substituting in the equation:

$$\begin{aligned} C &= \tfrac{5}{9}(F - 32) \\ &= \tfrac{5}{9}(400 - 32) \\ &= \tfrac{5}{9}(368) = 204°C \end{aligned}$$

STUDENT PROBLEM

What temperature would a centigrade oven thermometer indicate in an oven at a temperature of 300°F? (ANS. 149°C)

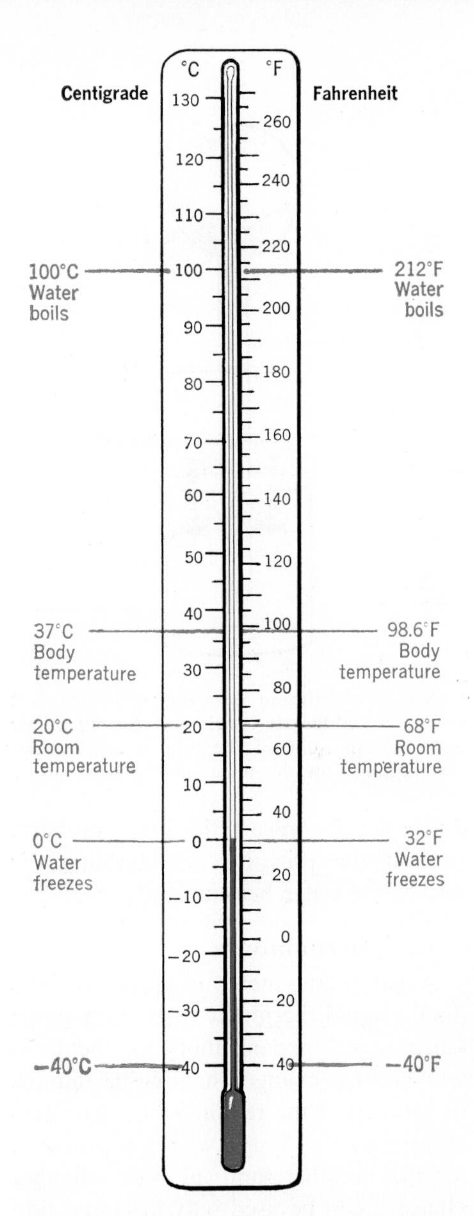

16-7 Centigrade and Fahrenheit scales. Can you discover from this chart which is larger, a centigrade degree or a Fahrenheit degree? How much larger?

Expansion of gases

The first thermometer, invented by Galileo, made use of the expansion of air. When the bulb, B, in Fig. 16-8 is heated, the air inside expands and forces the water down the tube. When the air in the bulb cools, the reduced speed of

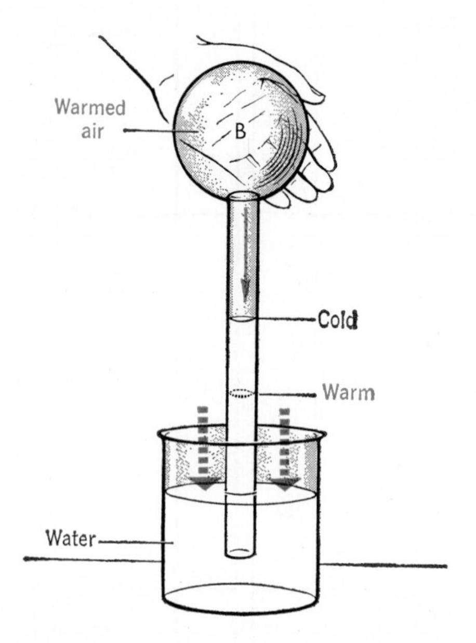

16-8 A simple air thermometer, first made by Galileo. It consists of a bulb of air with the open end of the tube in water. Can you explain how the water got into the tube?

the molecules lowers the inner pressure so that the pressure of the outer air pushes the water back up the tube.

8. Gas thermometers

A gas thermometer is more sensitive than a liquid thermometer because gases (such as air) expand more for the same temperature change than either liquids or solids. But the readings of a gas thermometer will change with air pressure, as the weather and altitude change. Hence it can be used only in connection with special apparatus which adjusts each reading to standard sea-level pressure.

9. Absolute zero

The French scientist J. A. C. Charles measured the expansion of gases as they were heated. He found that if any gas is heated 1 degree, it will expand $\frac{1}{273}$ of its volume at 0°C. Thus if 1 liter of any gas were heated from 0°C to 273°C, it would double in volume to 2 liters.

What would happen if we cooled a gas 273 degrees? If the gas were to keep up the same rate of contraction of $\frac{1}{273}$ for each degree, then it would have contracted $\frac{273}{273}$, or its full volume, at 273° below 0° on the centigrade scale. In other words, the gas would vanish! This provided scientists with the clue to the fact that the lowest possible temperature is −273°C. This temperature is called **absolute zero** (Fig. 16-9). (Actually, no gas ever disappears at −273°C because if a gas is cooled enough, it will first condense into a liquid and finally change into a solid, and the rate of contraction changes.) At absolute zero practically all motion of molecules and parts of molecules would cease, and an object would have lost almost all its heat. No lower temperature is possible.

Scientists often use a temperature scale based on absolute zero. This scale is known as the **absolute scale** or the **Kelvin scale.** A temperature of 20°C is 20° above the freezing temperature of water. But freezing water is 273° above absolute zero. So 20°C is equivalent to 293°K (273 + 20). Thus to change centigrade degrees to Kelvin degrees, we simply add 273 to the centigrade reading ($K = 273 + C$).

10. Charles' Law

The discoveries of Charles about the expansion of gases can best be put in the form of a mathematical law called **Charles' Law.** Let us examine a graph of the change in volume of a gas with temperature (Fig. 16-10). From what you learned about the motion of molecules on page 204, you know that as the temperature falls the molecules move more slowly. When molecules move more slowly the impulse they give to the walls of their container decreases. (Remember the formula for impulse, $Ft = \Delta mv$.) The force and hence the pressure they exert on the walls of their container becomes less. If the pressure is to be maintained constant, the volume occupied by the gas must be reduced. In other words, as the graph shows us, the gas contracts in direct proportion to the decrease in temperature. You can see that the volume would become zero at −273°C if the gas did not liquefy. Hence the volume is proportional to the absolute temperature (centigrade temperature + 273°). We can state Charles' Law as follows:

The volume of a mass of gas is directly proportional to the absolute temperature if the pressure remains constant.

We can write this law in the form of an equation. If we start with a gas whose volume is V_1 and whose temperature is T_1, and change its temperature to T_2, its volume will become V_2, and:

$$\frac{V_1}{V_2} = \frac{T_1}{T_2}$$

If you write the law in the form

$$\frac{V_1}{T_1} = \frac{V_2}{T_2}$$

and compare these two ratios with the graph below you may see that they can be made to represent the ratio between corresponding sides of similar triangles.

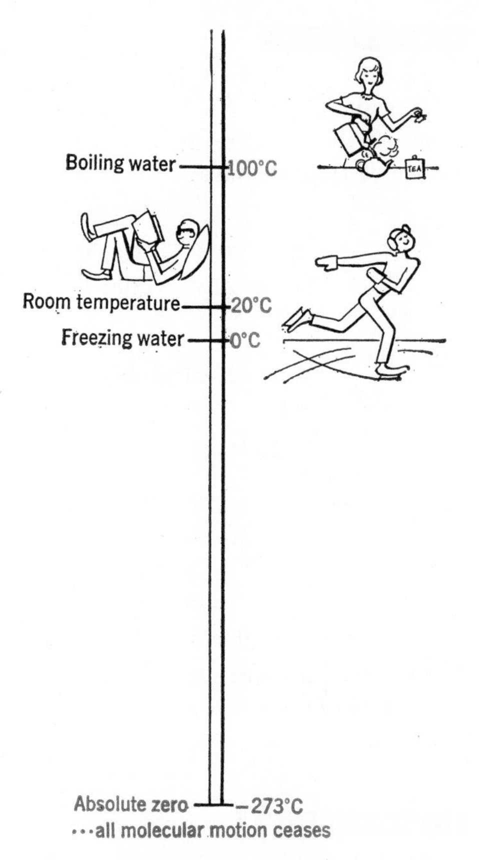

16-9 The lowest, coldest temperature is −273°C (or −459°F). At this *absolute zero* practically all molecular motion stops, all gases are condensed, and materials have strange properties. Many scientists are at work today on the study of materials at temperatures near absolute zero.

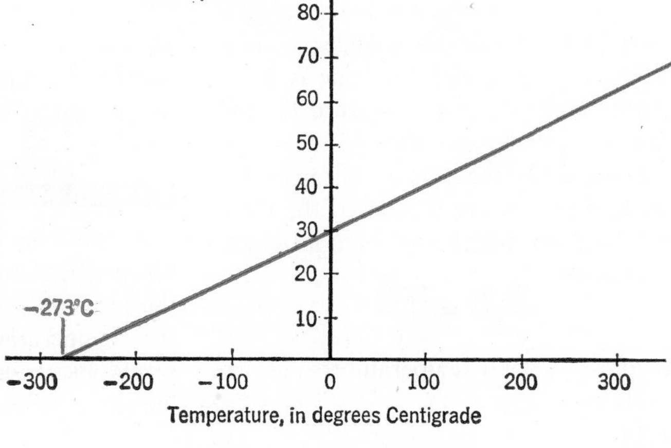

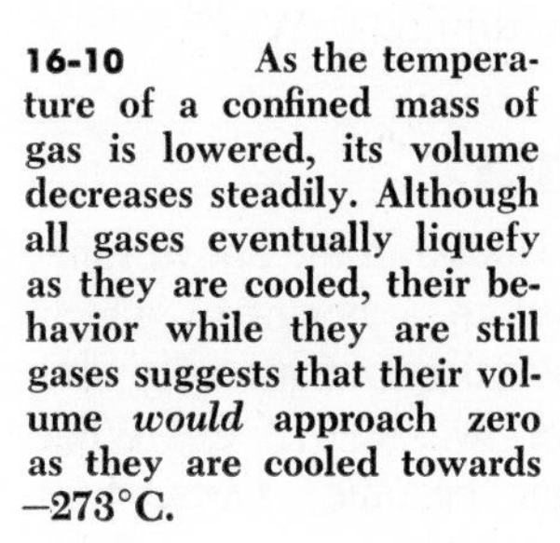

16-10 As the temperature of a confined mass of gas is lowered, its volume decreases steadily. Although all gases eventually liquefy as they are cooled, their behavior while they are still gases suggests that their volume *would* approach zero as they are cooled towards −273°C.

SAMPLE PROBLEM

Suppose we have a liter flask filled with air at 227°C. What will be its volume if it is cooled to 27°C at constant pressure? Here we have:

$$V_1 = 1.0 \text{ liter}$$
$$T_1 = 227° + 273° = 500°K$$
$$T_2 = 27° + 273° = 300°K$$
$$V_2 = ?$$

Substituting in the equation for Charles' Law we have:

$$\frac{V_1}{V_2} = \frac{T_1}{T_2}$$
$$\frac{1.0 \text{ liter}}{V_2} = \frac{500°K}{300°K}$$
$$V_2 = \frac{1 \times 300}{500} \text{ liters}$$
$$= 0.600 \text{ liter}$$

Notice that *you must always express the temperature in absolute (Kelvin) degrees* when using this law.

STUDENT PROBLEM

300 cm³ of air at 77°C are heated to 427°C. What is the volume of the heated air? (ANS. 600 cm³)

11. The General Gas Law

Boyle's Law (page 191) and Charles' Law may be combined into one law to solve problems in which volume, temperature, and pressure all change at once. If we have a gas whose volume is V_1 at a temperature T_1 and a pressure P_1, and we change the temperature to T_2 and the pressure to P_2, the volume becomes V_2. These quantities are related by the **General Gas Law**, which applies to all gases.

$$\frac{P_1V_1}{T_1} = \frac{P_2V_2}{T_2}$$

All pressures and temperatures must be absolute.

Notice that if the pressure remains constant, $P_1 = P_2$, P_1 and P_2 cancel out, and the equation reduces to Charles' Law. Similarly if there is no change in temperature, the equation reduces to the formula for Boyle's Law ($P_1V_1 = P_2V_2$). And if the volume is kept constant we get

$$\frac{P_1}{T_1} = \frac{P_2}{T_2}$$

SAMPLE PROBLEM

Suppose a toy balloon is filled with helium at 27°C until its volume is 3 liters at a pressure of 76 centimeters of mercury. If it rises up through the atmosphere to an altitude where the pressure is 38 centimeters of mercury and the temperature is −50°C, what will its volume become?

$$V_1 = 3 \text{ liters}$$
$$P_1 = 76 \text{ cm of mercury}$$
$$T_1 = 27 + 273 = 300°K$$
$$T_2 = -50 + 273 = 223°K$$
$$P_2 = 38 \text{ cm of mercury}$$

Substituting in the General Gas Law:

$$\frac{P_1V_1}{T_1} = \frac{P_2V_2}{T_2}$$
$$\frac{76 \times 3}{300} = \frac{38\ V_2}{223}$$
$$V_2 = \frac{76 \times 3 \times 223}{38 \times 300} = 4.46 \text{ liters}$$

Notice that it is not necessary to convert the pressures from cm of mercury to lb/in² before substituting. Why?

STUDENT PROBLEM

Suppose the balloon in the above sample problem rose to a height at which the pressure was 50 cm of mercury and the temperature was − 10°C. What would its volume become (ANS. 4.00 liters)

16-11 Measuring coefficient of expansion. Hot water passing through the tube causes expansion which is measured by the motion of the pivoted lever.

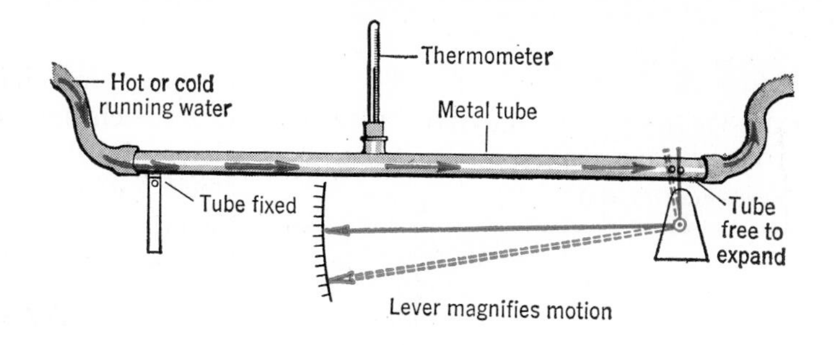

Expansion of liquids and solids

All gases expand at practically the same rate because there is almost no cohesive attraction between the molecules to interfere with their movement. Hence the expansion of a gas depends only upon the energy of motion of its molecules, and this depends only on the temperature of the gas (provided the pressure is kept the same).

But it is quite otherwise with solids and liquids. Here molecules are so much closer than in gases that molecular attraction (cohesion) prevents the material from expanding as much as in the gaseous state. Since the force of cohesion differs in different materials, each liquid and solid has its own rate of expansion, which must be measured by experiment (Fig. 16-11).

12. Coefficient of expansion of solids

The rate of expansion, as we have said, will vary from one solid to another. But it has been discovered that the dimensions (length, width, and height) of any one solid increase approximately the same amount for each degree rise in temperature. If we know the fraction by which the length of a given material increases for each degree, we can calculate the total expansion of the material for any temperature rise. The number that expresses this fractional increase in length for a temperature rise of 1 degree is called the **coefficient of linear expansion.** Table 5 gives the coefficients of linear expansion for a number of important materials.

We can measure the coefficient of linear expansion as shown in Fig. 16-11. The change in length that occurs in a tube when hot water replaces cold water is measured. Temperatures are also carefully measured. The change in length per degree can then be calculated.

Suppose an engineer wants to know how much space he must leave for expansion between the 50-foot iron rails of a railroad track. Let us say that the temperature varies from −10°C in the winter to about 35°C in the summer. From Table 5 it appears that each iron rail will increase 0.000011 of its 50-foot length for each degree centigrade rise in temperature. Now, each rail will be sub-

U.S. Forest Service

16-12 Because of linear expansion, steel rails may buckle under extreme heat.

TABLE 5

COEFFICIENTS OF LINEAR EXPANSION (Fractional increase in length per degree centigrade)

Aluminum	0.000025
Brass	0.000019
Copper	0.000017
Glass (ordinary)	0.000009
Glass (Pyrex)	0.000004
Invar	0.0000009
Iron and Steel	0.000011
Lead	0.000028
Platinum	0.000009
Zinc	0.000026

jected to a temperature rise of 45°C (from −10°C in winter to 35°C in summer). So the total expansion of the rail will be 45 times as much as for 1 degree centigrade. The estimated increase of length for each rail is, then, 0.000011 × 50 feet × 45°C. This is 0.0248 foot (about $\frac{1}{3}$ inch).

From these figures we can see that the increase in length is given by:

Change in length =
Coefficient of expansion × Original length × Change in temperature

Using the Greek letter "delta" (Δ) to mean "change in," and k_e for the coefficient of expansion, our formula becomes:

$$\Delta L = k_e L \Delta T$$

13. Coefficient of cubical expansion

It is meaningless to talk about the linear expansion of a liquid, since it takes the dimensions of the container it occupies. However, liquids do expand, and we can easily measure their change in *volume*. We do this in exactly the same way that we measure linear expansion except that we use the **coefficient of cubical expansion,** which is **the fractional change in volume of a substance per degree change in temperature.** The change in volume, ΔV, equals the coefficient of cubical expansion, k_c, times the original volume, V, times the change in temperature, ΔT:

$$\Delta V = k_c V \Delta T$$

This formula applies to the change in volume of solids as well as liquids. Notice the close similarity between this formula and the formula for linear expansion just above. Table 6 gives the coefficient of cubical expansion for some common substances. You will see by comparison with Table 5 that the coefficient of cubical expansion of a given solid is approximately three times its coefficient of linear expansion.

SAMPLE PROBLEM

A tanker is loaded with 80,000 gallons of oil at a port in the Red Sea where the temperature is 110°F. How many gallons will be unloaded when the ship arrives in New Jersey on a winter day when the temperature is 10°F? Assume that there has been no oil lost by evaporation.

First we must convert our temperatures to the centigrade scale to agree with the units given in Table 6 for the coefficients of cubical expansion. Actually, we don't have to convert each temperature separately, since we are interested only in the *change* or difference in temperature—not the actual temperatures. The temperature change of the oil is 110 − 10 = 100 Fahrenheit degrees. This is equal to $\frac{5}{9} \times 100 = 55.6$ centigrade degrees. This is our value for ΔT to be substituted in the formula. Our value for V is 80,000 gallons, and we find from Table 6 that the coefficient of cubical expansion for oil is 0.000899 per degree centigrade. Substituting:

$$\Delta V = k_c V \Delta T$$
$$= 0.000899 \times 80{,}000 \times 55.6$$
$$= 3{,}998 \text{ gal}$$

This is the *decrease* in volume due to the drop in temperature, and so the volume of oil unloaded in New Jersey will be:

$$80{,}000 - 3{,}998 = 76{,}002 \text{ gal}$$

STUDENT PROBLEM

What would be the volume of oil unloaded in New Jersey in the sample problem above if 50,000 gal were loaded in the Red Sea at 110°F and the temperature in New Jersey at the time of unloading were 50°F? (ANS. 48,303 gal)

TABLE 6

COEFFICIENTS OF CUBICAL EXPANSION (Fractional increase in volume per degree centigrade)

Alcohol	0.00113
Aluminum	0.000075
Glass (ordinary)	0.000027
Iron	0.000034
Mercury	0.000182
Oil (petroleum)	0.000899

14. Irregular expansion

Not all materials expand regularly when heated. In fact, some of them actually contract with heating at certain temperatures. For example, when ice is melted by heating it contracts about 10 per cent in volume. And this same water continues to contract as the temperature rises from the freezing point at 0°C to 4°C. Beyond 4°C it begins to expand. Thus water is most compact and has its greatest density at 4°C.

This fact about water means the difference between life and death to fish and other creatures which inhabit frozen lakes in winter. Since water at 4°C is at its greatest density, it sinks to the bottom of a lake that is about to freeze. Water that is colder than 4°C is less dense than the 4-degree water below, so it floats on top and freezes there. Thus the lake freezes from the top down. The water at the bottom may remain unfrozen at 4°C—giving living space to the fish (Fig. 16-13).

The expansion of water as it freezes can do a great deal of damage. It can burst the water pipes in a house or crack the motors and radiators of automobile engines. This is one reason why you use an antifreeze solution in your car in winter. These solutions have a very low freezing point; when mixed with the water in your engine, they lower the point at which the water would normally freeze.

You might ask whether this irregular expansion of water—the fact that it expands between 4°C and 0°C as it freezes—does not form an exception to our explanation of expansion as the result of increased motion of molecules with heat. It is a good question to ask. When an exception is found to a scientific theory, the exception must be explained if the theory is to stand.

Investigation has shown that molecules of water occur in groups rather than singly. We have good evidence that as water approaches the freezing point—that is, the solid state of ice—the molecules in these groups reorganize themselves in such a way as to require more volume.

Scientists are always alert to catch facts which seem to contradict their theories. Following up these exceptions often leads to new and exciting discoveries. Radium, for example, was found to behave in a way that seemed to contradict the normal behavior of atoms. The investigation of this exception resulted finally in the discovery of how to release atomic energy.

Going Ahead

THE VOCABULARY OF PHYSICS

Define each of the following:

thermometer	compound bar (bimetallic strip)
absolute zero	centigrade scale
coefficient of linear expansion	coefficient of cubical expansion
thermostat	absolute scale
Kelvin scale	

THE PRINCIPLES WE USE

Explain why

1. Most materials expand when heated.
2. A glass may crack when hot water is poured into it.
3. A bottle made of very thin glass will not crack when hot water is poured into it.
4. A compound bar bends when heated.
5. The temperature −273°C was selected as absolute zero.
6. Ice forms at the top of a lake first.

EXPERIMENTS IN EXPANSION

Describe an experiment which gives evidence to show that

1. Solids expand when heated.
2. Liquids expand when heated.
3. Gases expand when heated.

APPLYING PRINCIPLES OF EXPANSION

1. Mention four instances of harm caused by expansion due to heat.
2. Explain how expansion and contraction caused by changes in temperature could be used to fasten two steel beams together by means of rivets.
3. Show by a diagram how the compound-bar type of thermostat can be used in regulating the temperature of a home. Use another diagram to show how this might be changed to regulate the temperature of a refrigerator.
4. Describe how a thermometer can be calibrated.
5. Scientists are always examining data that seem to contradict their theories. Why do they do so? From everyday life, can you give one example of an error that results from failure to examine the exceptions to our pet beliefs or prejudices?
6. Refer to Table 5 and answer these questions: (a) What combination of metals would make a compound bar that would bend to a greater extent than one of brass and iron for the same change in temperature? (b) What material would be best for the pendulum of a clock? (c) Why does Pyrex glass break less readily when heated than ordinary glass? (d) Electric light bulbs have wires that pass through the glass. What metal might be suitable for this purpose?
7. Careful observation shows that when mercury thermometers are placed in hot water, some of them show a slight fall in the mercury level before it starts to rise. Explain why this happens. Predict what would happen to the mercury level of such thermometers if they were placed in ice water.

SOLVING PROBLEMS OF EXPANSION

1. Convert 20°C into °F.
2. Convert 30°C into °K (Kelvin).
3. The melting point of a certain alloy is 890°F. What is its melting point in degrees centigrade?
4. What is the melting point of the alloy in Problem 3 in degrees absolute?

---

5. Convert (a) 86°F into °C, (b) −15°C into °F, (c) −40°F into °C.
6. What will be the amount of expansion in (a) a 10-inch piece of aluminum heated from 10°C to 60°C, (b) a 50-ft copper wire heated from 5°C to 30°C, (c) an iron bridge 300 ft long heated from −10°C in winter to 30°C in summer?
7. How much does a 100-ft copper wire shrink when the temperature drops from 35°C to 5°C?
8. What is the coefficient of expansion of aluminum if a 4-ft aluminum rod at 20°C measures 4.00832 ft at 100°C?
9. A 20-inch rod of aluminum is heated from the freezing point of water to the boiling point of water. How much does the rod expand?
10. 300 cm^3 of hydrogen at 0°C are cooled

16-13 Can you explain this picture? Why doesn't the water at the bottom of the pond freeze and why don't the fish die?

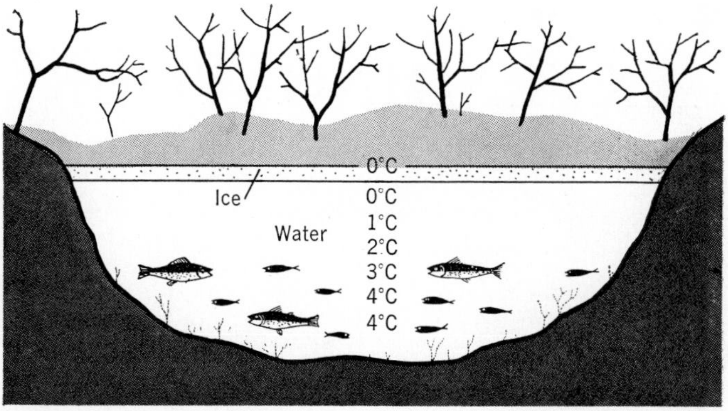

until the volume is 150 cm^3. What is the temperature of the gas in degrees centigrade? Assume pressure to be constant.

11. 5 ft^3 of air at 100°C and a pressure of 76 cm of mercury are cooled to 20°C and the pressure raised to 228 cm of mercury. What volume does the air now occupy?

12. When the piston of a pump is pushed down, the volume of air in the cylinder is decreased to 0.05 of its original volume and the temperature rises from 20°C to 160°C. If the original air pressure was atmospheric pressure, what is the pressure of the compressed gas?

13. A certain town has a yearly range of temperature of 90°F. What is the range of temperature in centigrade degrees?

---

14. What will be the change in length of a 20-ft iron girder after it has been heated 200°C?

15. How much will a 50-ft aluminum girder expand if it is heated from 60°F to 160°F?

16. A tank contains 5,000 gal of oil at 20°C. What volume will it occupy when heated to 40°C?

17. A flask contains 1 liter of mercury at 68°F. What volume will it occupy at −10°F?

18. A 100-ft steel tape is accurate at 20°C. When used to measure the length of a steel bridge when the temperature is −20°C, it gives a reading of 55.000 ft. What is the true length of the bridge at −20°C? $k_e = 0.000011$ for steel.

PROJECT: TO SHOW THE IRREGULAR CONTRACTION OF WATER

You can easily demonstrate the expansion of water as it cools close to its freezing point. Fill a small bottle or flask with water colored with a few drops of ink. Seal the mouth with a stopper through which pass a few inches of glass tubing. Pressing the stopper in place should force the water an inch or two up the tube.

Now immerse the bottle in a container full of a mixture of salt and crushed ice. As your bottle of water cools, watch the level of water in the glass tube. After it has fallen it will suddenly start to rise once more. The water will continue to expand until it has frozen—and burst the bottle.

Compare the results of this experiment with the facts illustrated in Fig. 16-13 above. Is Archimedes' Principle involved in the situation in any way?

EXTENDING YOUR STUDY

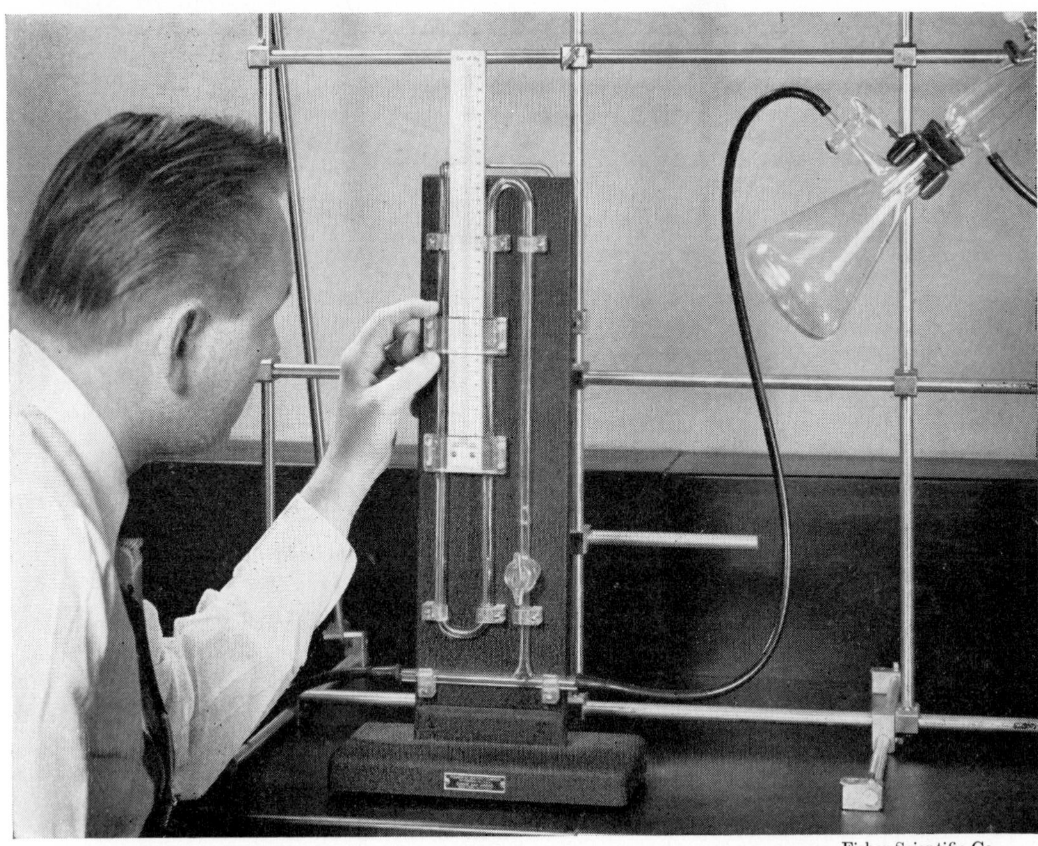

Fisher Scientific Co.

We begin, strangely enough, by going back to 1662, when Robert Boyle investigated the relation of pressure to volume of a gas. His apparatus was simple: mercury poured into the open end of a *J* tube was permitted to compress air trapped in the other end. As the volume of the air decreased, its pressure increased according to a definite relationship of which you are now well aware (p. 191). But Boyle was in at the beginning (when Nature was thought to abhor a vacuum) and this simple experiment had not been done before. Here is his conclusion as stated by him nearly 300 years ago. Note especially his last sentence. If he had only had a "model of a gas."

"It is evident, that as common air, when reduced to half its wonted extent, obtained near about twice as forcible a spring as it had before; so this thus compress air being further thrust into half this narrow room, obtained thereby a spring about as strong again as that it last had, and consequently four times as strong as that of the common air. And there is no cause to doubt, that if we had been here furnished with a greater quantity of quicksilver and a very strong tube, we might, by a further compression of the included air, have made it counterbalance the pressure of a far taller and heavier cylinder of mercury. For no man perhaps yet knows, how near to an infinite compression the air may be capable of, if the compressing force be competently increased." *

And now in the 20th century we go on with the work Boyle himself began.

Predicting the Gas Laws from the Molecular Theory

In this unit we have built up, piece by piece, a mental picture or theory of what a gas is really like. We have seen that this theory assumes that a gas is made up of tiny molecules, far apart compared to their size, which move about at very high speeds in a random manner. Some have extremely high speeds, others relatively low speeds, but most have speeds close to the average speed of all the molecules in the sample. We have seen that both Charles' Law and Boyle's Law seem to fit this picture. But using our knowledge of the laws of mechanics, can we start with this picture of a gas and derive the formulas for Boyle's and Charles' Laws? If we can do this, it would indeed be strong evidence for the correctness of our kinetic theory of gas. Let us try.

The force of gas molecules

Suppose we consider the motion of a single molecule moving in a small cubical box the length of whose side is L. For simplicity we will assume the molecule is moving parallel to one side, so that it

* "New *Experiments* Physico-Mechanicall, Touching The Spring of the Air, and its Effects (Made, for the most part, in a New *Pneumatical Engine*), Written by way of letter to the Right Honorable Charles Lord Viscount of Dungarvan Eldest Son to the Earl of Corke. Part II of an Appendix to the second edition of this book titled: Wherein the Adversaires Funicular Hypothesis is Examined."

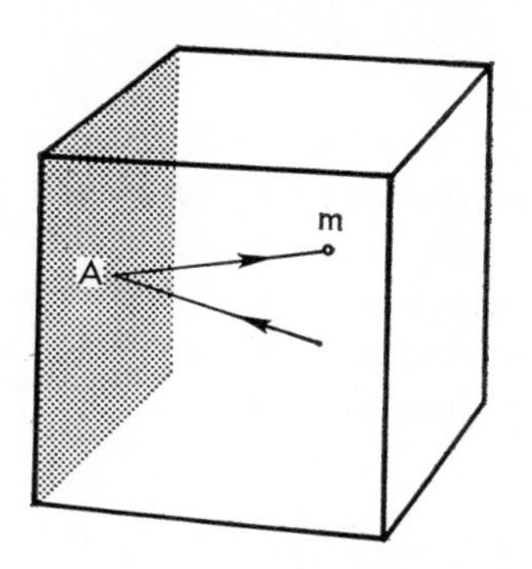

16-14 Diagram showing how a molecule of mass m rebounding from wall A of a cube exerts a force on the wall.

will strike and rebound perpendicularly from wall A (Fig. 16-14). The collision of the molecule with the wall will produce a force. How great is this force? If we can find the change in momentum and the time for this change, then we can use the equation

$$\text{Impulse} = \text{Change in momentum}$$
$$Ft = \Delta mv$$

to find the force. The force will then be

$$F = \frac{\Delta mv}{t}$$

or the change in momentum per unit time.

We evaluate $\Delta mv/t$ in the following way. If the molecule strikes the wall with a velocity, v, it will rebound in the opposite direction with an equal velocity if we assume that the collision is perfectly elastic and the molecule loses no energy to the wall. If the rebound velocity is equal to v, then the momentum of the rebounding molecule is of the same magnitude (but in the opposite direction) as the momentum before collision with the wall. The *change* in momentum must therefore be (since momentum is a vector quantity)

$$mv - (-mv) = 2mv$$

To find the change in momentum per unit time, which is the force, we must find the time between collisions of the molecule with the wall, A. Since the molecule is moving perpendicular to the wall, it must travel a distance $2L$, from A to the opposite wall, B, and back again, between collisions. Since distance equals velocity times time,

$$2L = vt$$

$$t = \frac{2L}{v}$$

Having now found the change in momentum, $2mv$, and the time within which it occurs, $2L/v$, we can find the force, F, that a single molecule exerts on a wall of the box. It is

$$F = \frac{\text{Change in momentum}}{\text{Time in which it occurs}} = \frac{2mv}{2L/v} = \frac{mv^2}{L}$$

What we have so far calculated is the force produced by one molecule. This force is a sudden push against the wall, repeated at regular intervals. In any container of gas, there are a tremendous number of pushes per second, but their average effect, even when the moving molecules collide with each other and strike the walls with different velocities at various angles, is a steady push.

A careful calculation of all the changes in momenta at different angles would show that we can consider one third of all the molecules to be moving in the simple way we have calculated between walls A and B at the average velocity, while one third can be considered to move in a similar fashion between the top and bottom walls and the remaining third between the front and back walls. The mathematics needed to show this is more complex than we can include in this book, but we shall make use of this simplification and say that the total force, F_t, against the wall, A, is exerted, in effect by $\frac{1}{3}$ of the N molecules in the box and hence

$$F_t = \frac{N}{3} \times \frac{mv^2}{L}$$

Molecular pressure and Boyle's Law

It is now a simple matter to calculate the pressure on wall A. The pressure is the force acting on a unit area, so

$$P = \frac{F_t}{L^2} = \frac{N}{3} \times \frac{mv^2}{L^3}$$

But L^3 is the volume, V, of our box, so

$$P = \frac{N}{3} \times \frac{mv^2}{V}$$

or

$$PV = \frac{Nmv^2}{3}$$

Let us consider this equation in the light of what happens when we take a sample of gas containing N molecules and change its pressure while assuming that the molecular velocity stays constant. This means that N and v stay constant while pressure and volume change from P_1 and V_1 to P_2 and V_2:

$$P_1V_1 = \frac{Nmv^2}{3}$$

$$P_2V_2 = \frac{Nmv^2}{3}$$

Dividing one equation by the other, we have

$$\frac{P_1V_1}{P_2V_2} = \frac{Nmv^2/3}{Nmv^2/3} = 1$$

$$P_1V_1 = P_2V_2$$

This is just Boyle's Law—*the pressure of a mass of gas at constant temperature is inversely proportional to the volume.*

Not only have we derived Boyle's Law from our theory of a gas by using our knowledge of the laws of mechanics, but we have also learned what assumptions it is necessary to make if Boyle's Law is to hold. Boyle's Law is true for "a mass of gas"—the mass must remain constant. We derived our above equation by assuming that N remains constant. This, of course, is equivalent to keeping the mass constant, since the total mass of gas is the total number of molecules, N, times the mass of an individual molecule, m. Also in Boyle's Law we assumed constant temperature. Our other assumption in deriving the above equation was that the average molecular velocity, v, stayed constant. Thus it seems that there must be some connection between molecular velocity and temperature. Let us investigate this connection further.

Molecular velocity and Charles' Law

Suppose we keep pressure, P, and mass, Nm, constant, and see how volume changes with v, the molecular velocity.

$$PV_1 = \frac{Nmv_1^2}{3}$$

$$PV_2 = \frac{Nmv_2^2}{3}$$

Dividing:

$$\frac{V_1}{V_2} = \frac{v_1^2}{v_2^2}$$

Here, the volume of a mass of gas at constant pressure is seen to be directly proportional to the *square* of the average molecular velocity. This seems to indicate that absolute temperature is proportional to the square of molecular velocity, since if we replace v_1^2/v_2^2 by T_1/T_2 we have Charles' Law,

$$\frac{V_1}{V_2} = \frac{T_1}{T_2}$$

This result, that absolute temperature is proportional to the *square* of the average molecular speed, is the basis for our statement earlier in this book that temperature rises when molecular speed increases.

A mechanical definition of temperature and heat

Let us now return to the equations we used to derive Charles' Law.

$$PV_1 = \frac{Nmv_1^2}{3}$$

$$PV_2 = \frac{Nmv_2^2}{3}$$

and put them in a different form that will tell us more about the true meaning of temperature. First we shall multiply numerator and denominator of both equations by 2:

$$PV_1 = \frac{2}{3}N \times \frac{mv_1^2}{2}$$

$$PV_2 = \frac{2}{3}N \times \frac{mv_2^2}{2}$$

Now, dividing the first equation by the second, we have

$$\frac{V_1}{V_2} = \frac{mv_1^2/2}{mv_2^2/2}$$

But $mv_1^2/2$ and $mv_2^2/2$ are just the two different *average kinetic energies* of the molecules. We can therefore say

$$\frac{V_1}{V_2} = \frac{(KE)_1}{(KE)_2}$$

If we compare this with Charles' Law,

$$\frac{V_1}{V_2} = \frac{T_1}{T_2}$$

we see immediately that

$$\frac{T_1}{T_2} = \frac{(KE)_1}{(KE)_2}$$

This tells us that:

Absolute temperature is a measure of the average kinetic energy of the molecules of a gas.

We see that the Molecular Theory of Gases gives us a definition of temperature based on the simple laws of mechanics. Furthermore, we can now see that the *total* kinetic energy of the molecules in a sample of gas will be proportional to the temperature—the *average* kinetic energy of the molecules. The heat in a gas, therefore, which is also proportional to the temperature, must be just the total kinetic energy of the molecules. Thus we see that heat is just another form of mechanical energy—the random kinetic energy of molecules. This view of the nature of heat energy is called **The Kinetic Molecular Theory** (or simply the Kinetic Theory).

The problem of designing gas engines is the problem of converting this random energy of molecules moving every which way in a gas into directed energy in which the molecules move in one direction so that their energy can be used to produce useful forces for useful work.

The General Gas Law

Returning to our equation on page 220, $PV = Nmv^2/3$, we now see that since temperature is proportional to v^2, we can write $PV = NkT$ by combining all constants into one constant, k. Then, allowing P, V, and T all to vary,

$$P_1V_1 = NkT_1 \text{ and } P_2V_2 = NkT_2$$

$$\frac{P_1V_1}{P_2V_2} = \frac{T_1}{T_2}$$

or

$$\frac{P_1V_1}{T_1} = \frac{P_2V_2}{T_2}$$

which is the General Gas Law.

The power of a scientific theory

In the preceding sections we have seen a simple picture or theory of the nature of a gas (often called a *model* by physicists). It has led, through the simple laws of Newton's mechanics, to a derivation of the gas laws and a simple explanation of the nature of temperature and heat. This same Kinetic Theory can tell us much about the properties of matter. We have only made a beginning here. There are many advanced physics books that carry the predictions of the theory further, explaining mathematically and in detail such things as heat conduction, mean free path, specific heat, the efficiency of heat engines, and even the nature of galaxies and the stars, and the future history of the universe. When you learn more mathematics and physics you will be able to read and understand such books and perhaps add to the theory.

Every theory has its limitations. It is constantly being changed and improved and is sometimes completely discarded in favor of a better theory or model—a better picture of the physical world. In physics a theory is the best description or model of matter and energy that we have at any particular moment. As our observations improve, with more ingenious and precise experimental apparatus, and are applied to new and unfamiliar situations, they confirm much of what is predicted by a good theory; but eventually our experiments are likely to come up with a contradiction. The theory has to be changed. Thus the Kinetic Theory of Gases, useful as it is, has been modified from time to time to explain better what is observed. Each modification has led to further predictions by the new modified theory, but like all theories it is never complete—never completely successful for long.

There appears to be little likelihood that man will ever have a complete and perfect set of theories that will explain all that happens in the universe. The Kinetic Theory is only one of many theories attempting to explain the nature of matter and energy. Other theories attempt to explain light and radiation, the nature of space, the fundamental nature of energy, and many other things. But they are always being improved as man, by careful experiment, with ingenious tools, learns more about the universe he inhabits. There is always more to learn, better theories to build to give a more complete and satisfying picture of the universe. This is the fascination of science to those who make it their life work.

CHAPTER

17

Transporting Heat

Not only our comfort but our very existence depends upon our being able to control temperature in order to avoid extremes of hot and cold. To control temperature, to raise or lower it at will, requires a knowledge of how heat is transported from one place to another. There are only three ways by which heat can be transported—by conduction, by convection, and by radiation. In this chapter we shall consider each of them in turn.

Conduction

1. Passing heat through a solid

Suppose an aluminum pan is resting on one hand and you pour boiling water into it. In a few seconds the pan is too hot to hold. If the pan were made of iron, it would take longer for you to feel the heat. And if you were using a Pyrex baking dish, you would feel the heat much later than you would with either of the metal containers. Everyday experiences like these illustrate two facts about heat:

1. Heat can flow through materials from molecule to molecule, or, as we say, can be transported by **conduction.**

2. Some materials are better conductors of heat than others (Fig. 17-1).

Table 7 shows that metals are generally the best conductors of heat.

2. Selecting the proper conductor

Our choice of material from Table 7 would depend upon our purpose. Thus for cooking utensils, where we want heat to pass quickly from the fire through the pan to the food, we could use silver, copper, or aluminum. But since cost is a factor, we would rule out silver and choose copper or aluminum. Copper

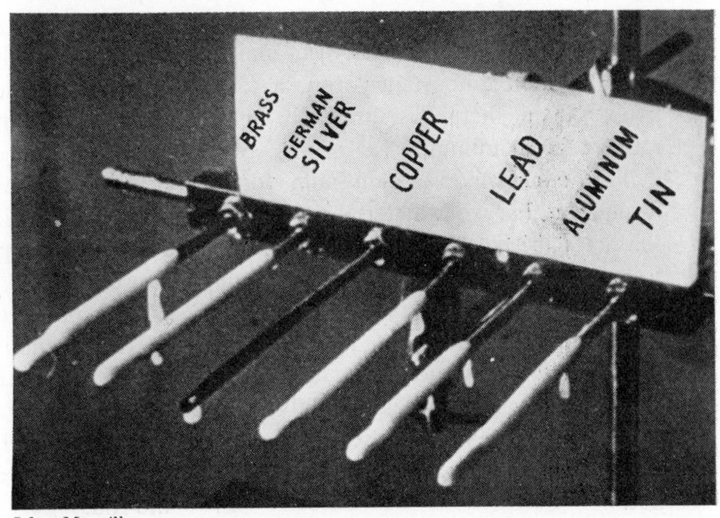

Johns-Manville

17-1 The different rates of melting of the wax on these different rods reveal which are the best conductors. List the metals in the order of their heat conductivity as shown by this experiment.

TABLE 7

HEAT CONDUCTIVITY

(Air is taken as the standard, with a conductivity of 1. Thus silver conducts 19,300 times as well as an equal thickness of air. These values are approximate.)

GOOD CONDUCTORS	
silver	19,300
copper	17,400
aluminum	8,800
iron	2,900
POOR CONDUCTORS	
glass (window)	44.0
concrete	35.0
water	25.0
wood	7.0
asbestos	4.8
linen	3.8
glass wool	2.6
cotton	2.4
cork board	1.9
rock wool	1.8
air	1.0

utensils, although more expensive than aluminum, are favored by first-rate chefs, since the flavor of food often depends on its staying as short a time over the fire as possible. Copper conducts heat so rapidly that the cooking time is short. In making large tanks, boilers, and radiators—where the size of the object makes cost still more important—iron, a relatively cheap metal, is generally used rather than aluminum.

You might choose aluminum for a frying pan, but what material would you select for the handle? As you may know from sad experience, an iron or aluminum handle can conduct heat so rapidly that it may be too hot to touch. A wooden handle is much better because it conducts heat much less rapidly than aluminum; thus you can still grasp the handle even though the frying pan is very hot.

3. Insulation

Often, as for the handle of the frying pan, we want a material which is a poor conductor of heat. Such materials are called heat **insulators.** They serve a twofold purpose: not only is the heat kept from passing on to something else, but it is retained by the original object. Thus when asbestos is wrapped around hot water or steam pipes, it reduces the rate at which they lose heat. At the same time, it prevents the surroundings from heating up. The clothes we wear in winter should be good insulators in order to reduce the rate at which our body heat escapes.

Whenever we want good conduction of heat, the conducting material should be as thin as possible. On the other hand, when we want to decrease heat conduction, we thicken the insulator. Aluminum cooking pans are made thin. But the walls and roof of a house are often packed with thick insulating material to keep the heat inside in winter and outside in summer.

4. Why do materials conduct heat?

Let's see how our molecular theory explains the fact that heat can pass through solid objects.

We know that heat is the energy of motion of molecules, and that all materials, solid, liquid, or gas, are made of molecules. When we heat one end of a metal rod, the molecules at the heated end begin to vibrate more rapidly. These rapidly moving molecules bombard the slower-moving ones next to them, giving them more energy of motion, and they in turn bombard their neighbors. This transfer of motion (or heat energy) takes place all the way down the length of the rod until the heat energy put into one end of the rod arrives at the other end of the rod.

5. Gases are the best insulators

We see from Table 7 that air is one of the poorest of the common heat conductors—and one of the best heat insulators. In fact, all gases are better insulators than solids or liquids, and the molecular theory explains why. In a gas, such as air, the molecules are much farther apart than the molecules of a solid or a liquid. Hence, there is less chance of a fast-moving molecule bombarding a slower one nearby. Thus the transfer of molecular motion, or heat energy, from one part of the substance to another proceeds at a slower rate.

As we can see from this discussion, the best possible heat insulator—or the worst heat conductor—would be a vacuum. A vacuum contains no molecules of anything—and heat transfer by conduction depends on the bombardment of molecules. The vacuum bottle makes use of a vacuum to cut down heat loss by conduction (Fig. 17-3).

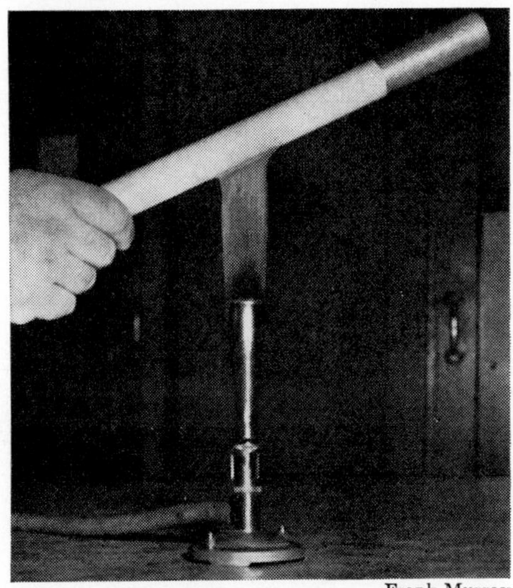

Frank Murray

17-2 Why doesn't the paper burn? It is wrapped around aluminum (an excellent heat conductor) which conducts the heat away so rapidly that the paper cannot reach its kindling temperature.

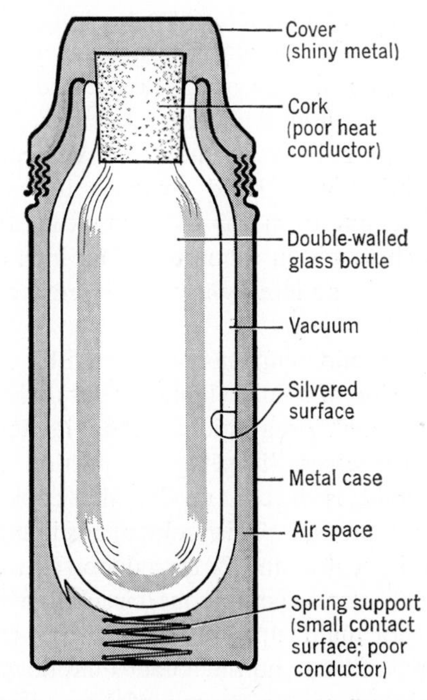

17-3 The vacuum bottle uses a vacuum between the double walls of a bottle to eliminate most of the conduction and convection of heat. The cork prevents convection and reduces conduction through the top. Silvered surfaces stop transfer of heat by radiation.

Convection currents

Suppose you want to transport heat from a furnace in the basement of your house to the rooms upstairs. You could conduct the heat through solid rods of aluminum or copper. But it would take such a long time for the heat to travel in this way that the system would not be practical. There is a much better way.

6. Currents caused by heat

Take a look at the rectangular glass tube shown in Fig. 17-4A. Put some colored crystals in the water and heat the tube at the bottom. By following the movement of the color, we see that the water moves in the direction of the arrows. It goes up the tube, across to the other side, then down, and returns to the starting point. This circulation goes on as long as we continue to heat the water.

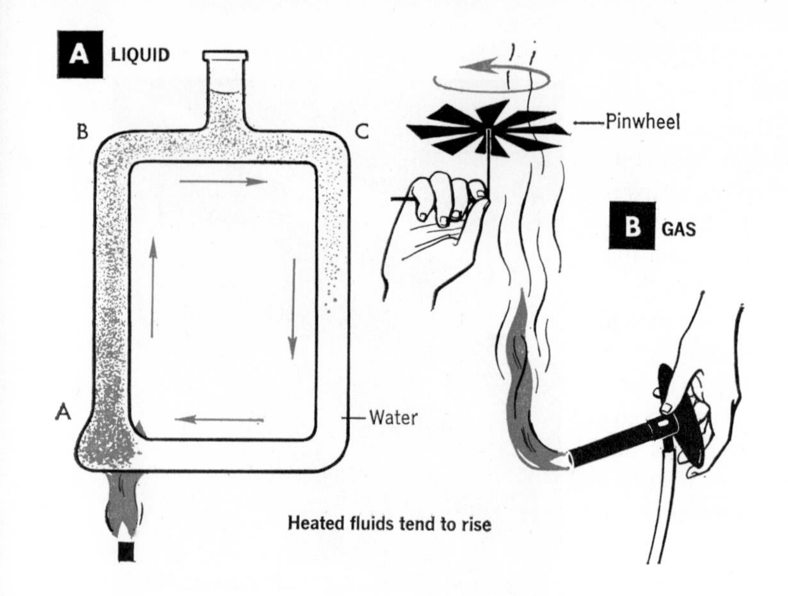

17-4 The difference in density between expanded warm fluid and contracted cold fluid causes convection. Can you explain each of these experiments? The red dots in the water are coloring matter.

Heat at point A has been transported to B; when we touch the tube at B it feels warm. We have here a **convection current.** This is the name given to the flow of a liquid or gas as it transfers heat.

Convection currents also take place in the air. Take a Bunsen burner and point the flame downward. The heated air curves upward and rises. Because hot gases are rising from the flame, you cannot bring your hand close to it from above. But from the side you can approach to within a quarter of an inch of the flame and feel very little sensation of heat.

Convection currents are caused by the expansion of a fluid (liquid or gas) when it is heated. As a fluid expands, its density decreases. This lighter portion of the fluid is forced up by the colder, denser portion. This colder portion in turn is heated and is forced up. In this way we get a continuous circulation of the hot fluid upward away from the source of heat and the colder fluid downward toward the source of heat. Both motions occur at the same time.

You can see from this why convection currents do not occur in solids. The material in a solid cannot flow. Convection currents occur only in liquids and gases. On the other hand, conduction—the transfer of heat from molecule to molecule—can occur in all states of matter, solid, liquid, and gas.

7. Convection currents and the sea breeze

A convection current is set up if there is a difference in temperature between any two parts of a fluid. Air is a fluid, and convection currents can occur if part of it is heated or cooled (Fig. 17-5). Winds are caused in this way. For example, the shore of a lake is usually warmer than the water on a hot day. As the air over the land is heated, its density decreases. Obeying Archimedes' Principle, this warmer air rises and moves out at high levels over the water. At the same time the cooler air from the water moves in underneath. A person on the shore describes this colder air moving in from the water as a cool sea breeze (Fig. 17-6).

8. Dead-air spaces

Materials that are often used to prevent the flow of heat are linen, cork, cotton, glass wool, sawdust, paper, rock wool, and asbestos. All of these are simi-

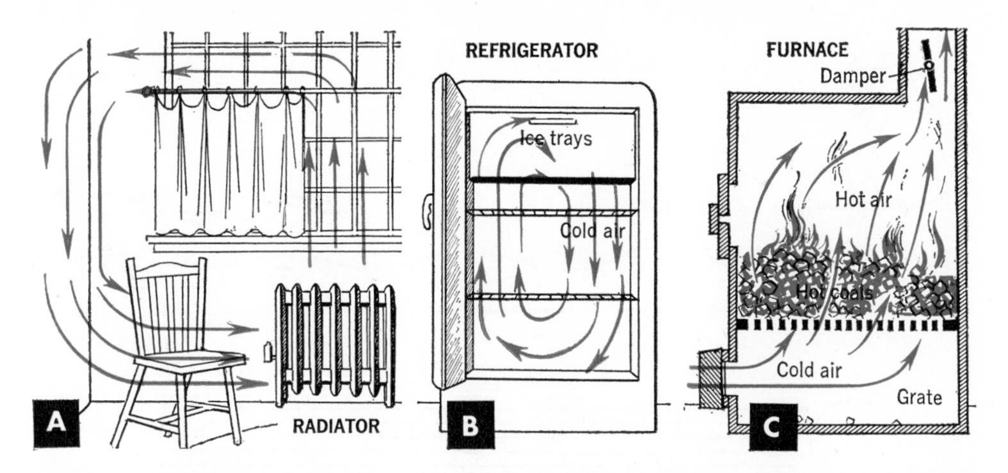

17-5 Convection currents have many uses. In A the convection currents from a radiator heat up the distant parts of a room. In B a cold convection current caused by placing the ice trays at the top of a refrigerator keeps all parts of the box cold. In C a convection current caused by the heated air creates a draft which maintains the air supply for the fire.

lar in that they are made up of fibers, or of separate bits, or have pores. In other words, their composition is such that they contain a good deal of air. Air, as Table 7 shows, is a poor conductor of heat. Provided it is prevented from moving and hence transporting heat by convection, we should expect air to be a good insulator. Porous or fibrous materials are good insulators because they can trap quantities of air and keep it motionless. These spaces of trapped air are called **dead-air spaces.**

A lady who buys a fur coat almost always buys one with the fur on the outside. If her sole object were to keep warm, she would wear the fur on the inside and have an airtight material like leather on the outside. In this way there would be dead-air spaces between the fur and the leather. When fur is on the outside, cold winds blow the warm air out of the spaces between the hairs, and thus the body heat is dissipated. Explorers and Eskimos in the Arctic regions, whose lives depend upon keeping warm, have learned this lesson and therefore wear clothing with the fur on the inside.

It has also been found that several layers of thin clothing offer better insu-

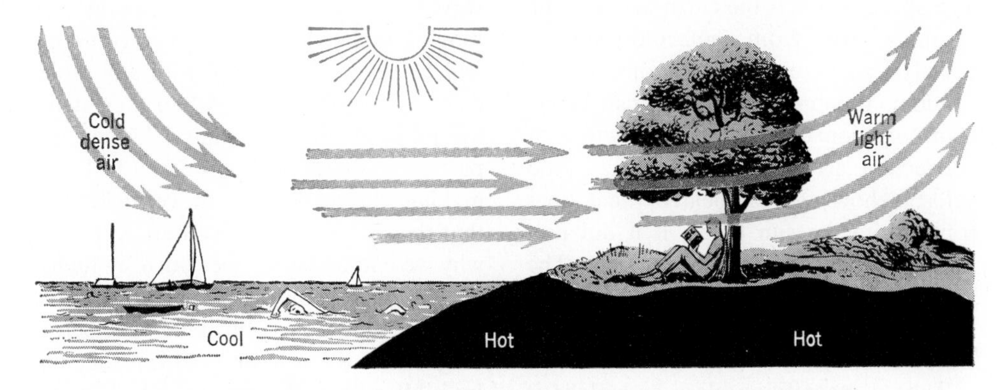

17-6 Ah! a cooling breeze. Convection currents caused by the difference in temperature between cool air over the water and warm air over the land make the seashore a nice place to relax in the summertime.

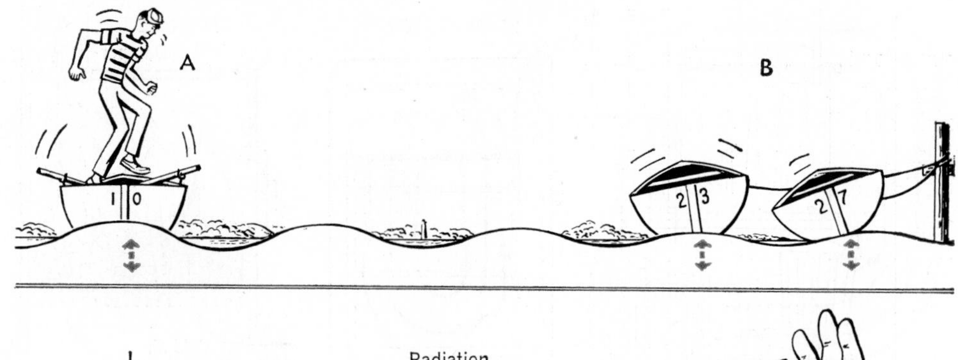

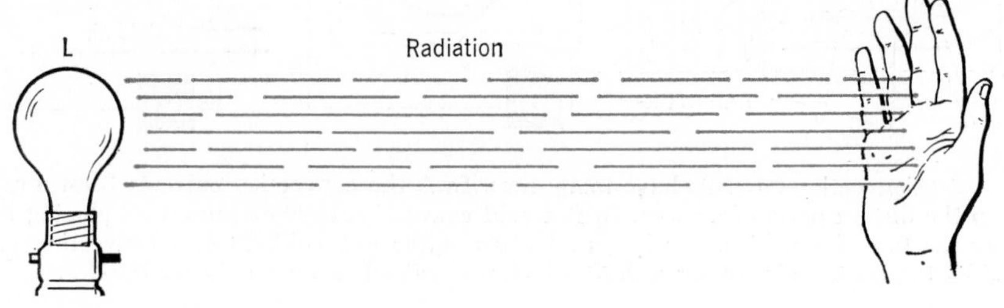

17-7 Motion of the boat at A causes wave motion in the water which in turn causes motion of the boats at B. Similarly, molecular motion in the hot filament of the lamp (L) causes infrared waves which in turn cause molecular motion (heat) when they strike the hand.

lation than one layer of the same combined thickness. The space between the two thin layers provides dead-air spaces and thereby increases insulation. A coat lining acts in the same way to provide additional dead-air space to reduce conduction of heat and keep the wearer warm.

Radiation

How does heat reach us from the sun? The space between the earth and the sun is almost free from molecules; for all practical purposes it can be considered a vacuum. It follows that there can be no transportation of heat by conduction or convection across this vast space. And yet somehow a great deal of heat does travel from the sun to us; if it did not, life on earth would not be possible. The heat reaches us by **radiation.**

9. Heat from the sun

Take a magnifying glass and place it in the path of the sun's light. Then focus these rays of light in such a way that they converge to a small bright spot on a sheet of paper. The paper soon starts to burn.

As we shall see in Chapter 43, light is a form of energy which travels as "electromagnetic waves." Radio waves and infrared rays are also examples of such waves. Electromagnetic waves are caused in the sun by the violent movement of atoms and electrons. All electromagnetic waves travel at a speed of 186,000 miles per second. They are able to pass through a vacuum. Although electromagnetic waves do not themselves contain any moving molecules, they do carry energy, which they are able to give to any molecules they strike. Thus, when they strike the molecules of any material, they impart some or all of their energy to them, making the molecules move faster, so that the material becomes hotter. This is why we can make a paper burn by focusing the sun's radiation upon it.

Light is not the only kind of electromagnetic wave that we get from the sun. Invisible **ultraviolet waves,** which cause sunburn and kill germs, also reach us from the sun in this fashion. Invisible

American Motors Corp.

17-8 This battery of heating lamps dries the paint on these cars in a very short time. Why aren't the car bodies left to dry overnight? Why is it better to use radiation than conduction or convection?

infrared waves, however, are responsible for most of the heat energy we get from the sun and hence are sometimes called "heat rays." These rays are invisible because they have no effect on the nerves in the eye when they enter the eye. Figure 17-7 shows how wave motion can carry energy from one place to another. Figure 17-8 shows one way in which infrared rays are put to use.

Just as the sun radiates heat by infrared rays, so a hot flatiron cools off by radiating the same invisible infrared rays. Its violently moving molecules turn some of their energy of motion into infrared radiation, which travels away from the hot object. The violently moving molecules move more slowly. Although at ordinary temperatures we cannot see the infrared radiation from the flatiron with the naked eye, we can feel the effect of heat waves as they give their energy to the molecules of our hands near by. We can even photograph the iron in total darkness with special film (Fig. 17-9).

In a similar way the molecules of hot

Myles J. Adler

17-9 This photograph of a hot iron was taken in *total darkness* with no visible light radiating from the iron. All hot objects emit invisible infrared rays which can be recorded by special infrared film. In this case the radiation was so great that it illuminated a plastic bird salt and pepper shaker placed about three inches in front of the iron.

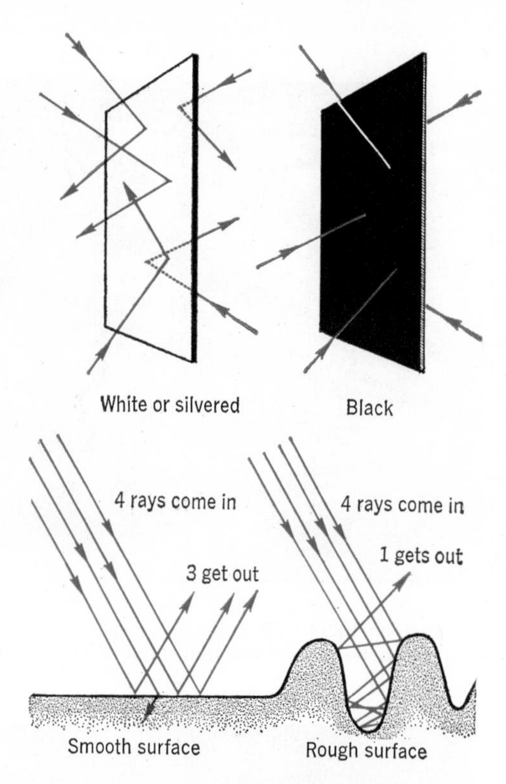

17-10 Radiation is reflected by light colors and absorbed by dark colors. Radiation is also absorbed more readily if the surface is rough.

radiators, warm automobile engines, and even people give off enough infrared radiation for them to be detected in total darkness. Indeed, any object warmer than its surroundings gives off infrared radiation at a greater rate than it receives it, and hence can be observed and even have its temperature measured by appropriate instruments. Thus astronomers know the temperature of the surface of the moon and of the planet Mars, and photographers can take a picture of a blacked-out city at night.

10. Black and white

On a sunny day, touch the fender of a black car and then one of the chromium-plated parts. You will find that the black fender feels hot, while the white, mirror-like parts of the car are not as hot.

Black objects become hotter than white objects in sunlight because they absorb most of the sun's radiation that strikes them. This causes an increased movement of molecules, so the objects warm up. White or shiny objects, however, reflect a great deal of this radiant energy, and the heating is less.

The simplest way to prevent an object from being heated by radiation is to paint it white or to silver it (Fig. 17-10). Then most of the heat rays (infrared rays) as well as the light rays that strike the object will be reflected rather than absorbed.

Similarly, if the radiation of heat is coming from inside the object, then a silvered surface outside will reduce the rate at which heat is lost by radiation. Thus aluminum foil lining the walls of houses reflects infrared radiation back into the house in winter, conserving heat. This may be illustrated by a simple experiment. Take two similar containers, one white or shiny (such as a tin can) and the other black. Fill both with hot water at the same temperature and set them aside away from radiators or direct sunlight. After a while test the temperature of both containers. Though conduction and convection has been identical in both cases, the water in the black container is not as warm as in the white or shiny one, thus showing that the black surface has radiated heat more rapidly.

The vacuum bottle makes use of the fact that a silvered surface keeps infrared rays from getting in or out. Both inner glass walls of the vacuum bottle are silvered to keep infrared rays from passing into or out of the bottle (Fig. 17-3).

So we see that a silvered surface can be used to keep heat from getting into an object by radiation, or to keep heat in the object from escaping by radiation. On the other hand, a black surface is effective if we want to absorb radiation from the outside or if we want to let the heat in an object escape by radiation. Figure 17-11 shows an important application of the fact that white or shiny objects reflect infrared rays.

11. Effect of roughness and area on radiation

The roughness of a surface is important in both absorption and radiation of heat. When heat rays approach a smooth, white, or silvered surface that reflects light like a mirror, most of the rays simply bounce off (Fig. 17-10). The surface absorbs only a small part of the energy which could be converted into heat. But the irregularities of a rough surface cause some rays to reflect back and forth from different parts of the surface which absorb a little more radiation at each reflection (Fig. 17-10).

A large surface also gives off more heat by radiation than a small one because it has more radiation area. So roughness of surface and amount of surface are important considerations if we want a good radiator or good absorber of heat.

The cylinder of an air-cooled airplane engine, designed to get rid of heat rapidly, is made with many fins in order to increase the radiating surface. It is also painted a dull black to increase the loss of heat by radiation. For the same reason automobile radiators are painted a dull black and provided with a large radiating surface. The giant radio tubes shown in Fig. 17-12 make use of the same principles in carrying away the heat developed during their operation.

12. Stefan's Law

From your experience with radiators and bonfires you know that the hotter an object is, the more rapidly it radiates heat energy. Actually the rate at which

Standard Oil Co. (N.J.)

17-11 **Danger—flammable liquids! The possibility of fire and explosion is reduced by painting these tanks in Baytown, Texas, a smooth silvery color, thus reflecting infrared rays from the sun and keeping down the temperature.**

it radiates heat energy depends upon the *absolute* temperature. To be precise:

The rate at which heat is radiated from an object is proportional to the fourth power of its absolute temperature.

This is **Stefan's Law**; it can be written:

$$R = kT^4$$

where R is the rate at which heat energy is radiated and k is a constant that depends upon the type of surface radiating the heat and on its area. Since T is the absolute temperature, it follows that any object above absolute zero radiates heat.

SAMPLE PROBLEM

Let us compare the rate at which heat is radiated from a piece of iron at a room temperature of 27°C with that for the same piece of iron when it is red hot (927°C). Converting our temperature to the absolute scale gives us 27 + 273 = 300°K for the temperature of the room and 927 + 273 = 1,200°K for red heat. The higher temperature is three times the lower. If the rate of radiation were

Westinghouse Electric Corp.

17-12 These giant 25,000-watt radio transmission tubes would melt unless cooled. Heat is conducted away by a copper-finned "radiator." The large amount of surface area radiates the heat into the air. The heat liberated in this manner is sufficient to keep a small house warm in winter.

proportional to the temperature, then heat would be radiated three times as fast. If the rate of radiation were proportional to the *square* of the temperature, then heat would be radiated $(3)^2$ or nine times as fast. But Stefan's Law says the rate of radiation is proportional to the *fourth* power of the temperature. Hence the rate heat is radiated by the red-hot iron is $3^4 = 3 \times 3 \times 3 \times 3$ or 81 times as fast as for the iron at room temperature.

The above sample problem shows why very hot objects lose heat so fast by radiation. If the radiators in your classroom are heated up from 60°C to 123°C they will give out heat twice as fast.

STUDENT PROBLEM

How many times faster does molten iron at 2,127°C radiate heat than red-hot iron at 927°C? (ANS. 16 times)

Heating the home

Heating the home is one of the most important practical applications of heat transmission. Conduction, convection currents, and radiation are all involved.

13. Heating with air

In the hot-air system, cool air is taken into large pipes, or ducts, which pass around the furnace. The heat in the furnace is transmitted through the metal firebox by conduction, thus heating the air inside the pipes. This heated air rises through the pipes to the rooms upstairs and enters them through grates in the wall or floor. The warm air rises by convection in each room and circulates, part of it returning to the furnace by another duct once it has given up its heat (Fig. 17-13). In some installations the circulation is speeded up by means of an electric fan.

14. Hot-water heat

In the hot-water system, water instead of air fills coils which are run into the firebox. Again heat from the furnace passes by conduction through the metal

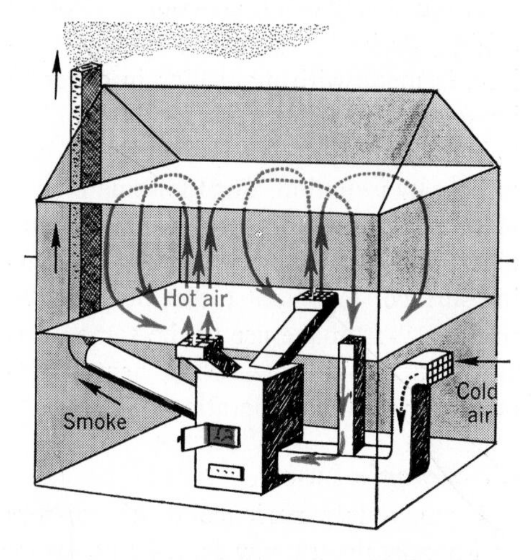

17-13 The simple hot-air heating system uses convection currents to circulate the heated air through the house.

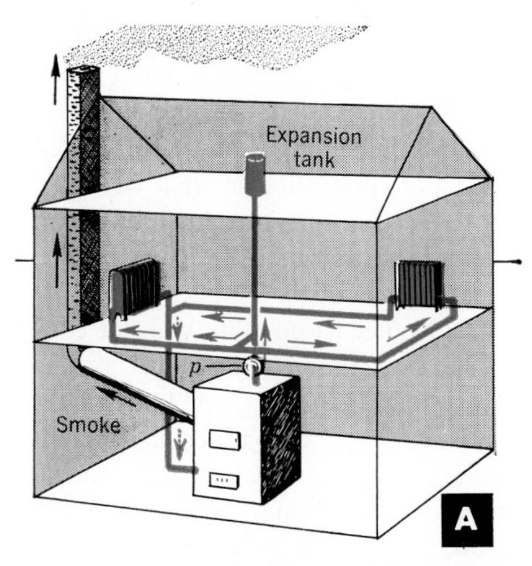

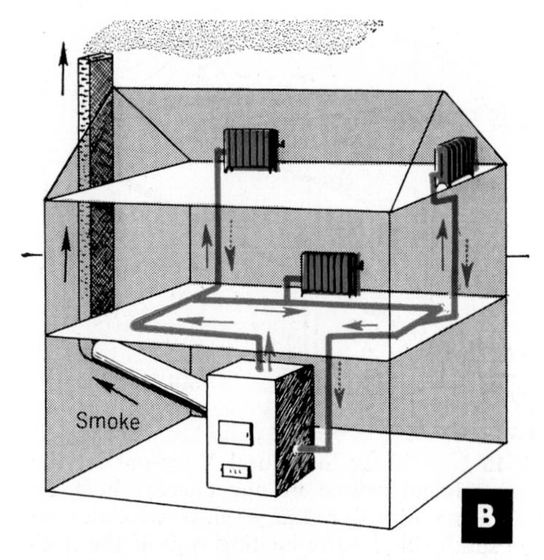

17-14 Both the hot-water heating system (A) and the steam-heating system (B) transport the heat through pipes to the radiators. In many hot-water systems a pump (*p*) is used to increase the circulation of water. Why is an expansion tank used? Why does the steam system use only one pipe to each radiator? What are the advantages and disadvantages of each system?

coils to the water. This heated water is pumped through the pipes to the radiators. The (hot) water in the radiators then gives off its heat by radiation and by conduction to the surrounding air. The water, somewhat cooled, travels back to the furnace, where it is reheated to be returned to the rooms (Fig. 17-14A).

To improve radiation, the radiator is constructed in a number of separate sections which increase the total area of its surface.

15. Steam heat

Steam, as well as hot air and hot water, is often used to bring heat from the furnace to the rooms upstairs. If the system is of the one-pipe variety, water is heated to the boiling point, and the steam forces its way up the pipe and into the radiators. In the radiator the steam condenses back into water, giving up its heat to the room. Since this condensed water occupies less space than the steam did, it can drip back inside the pipe to be reheated by the furnace and still leave room in the pipe for more steam to come up (Fig. 17-14B).

16. Heating in the modern manner

A number of new ideas in heating systems have recently been developed. One such modern system is **radiant heating.** Here pipes are set into the floors or walls of a building and warm water is pumped through them. The walls or floors heat up moderately (from 80°F to 115°F) by conduction, and because of the large surfaces involved there is good radiation to all parts of the room. Since waves of radiant energy can pass through air, the people in a room heated in this way can be warm even though the air itself is slightly cool (Fig. 17-16).

Modern architects are beginning to make use of the simplest and cheapest heating system of all—the direct heat of the sun. If **solar heating** is desired, houses are built with large windows on the southern side and large eaves which overhang the windows (Fig. 17-15). In summer, when the sun is high in the sky most of the day, the eaves cast a shadow over the southern windows and keep the midday heat out of the house. In winter, when the sun is low, the sun's radiation

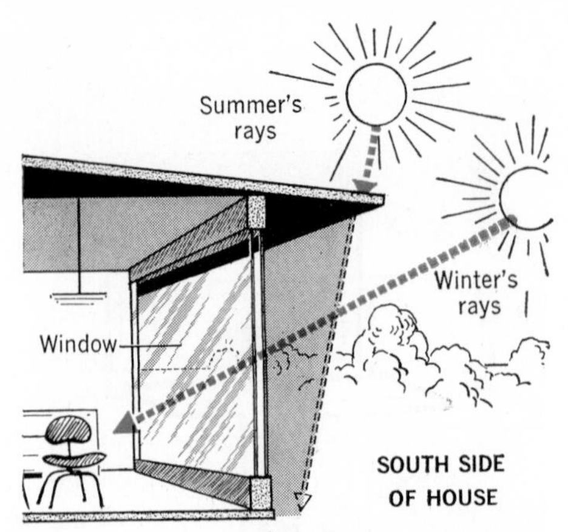

17-15 Back to nature! Why not let the sun, original source of our energy, heat up the home directly? Large glass windows on the south side let the heating rays of the mid-winter sun enter the house. The overhanging eaves keep out the sun's rays in summertime.

enters the large windows to warm the house. Such solar radiation can save a large part of the fuel bill. In one experimental modern house in southern New England, heat from the sun alone was sufficient to keep the house at a healthful temperature of 70°F all winter long without fuel.

17. Conserving heat in the home

All of us have probably had experience with houses that are not properly heated. Much heat loss can be eliminated by a proper application of what we know about heat transmission. Fuel bills can be reduced if we make sure that heat is not leaking out of the house.

First, we need walls that will not conduct heat to the outside. This means walls containing good insulating materials. Rock wool and other lightweight materials full of dead-air spaces can be inserted between inner and outer walls and under the roof. A double wall containing insulation will keep the heat from passing through it better than a solid brick or concrete wall of the same thickness because of the insulating properties of dead-air spaces.

Windows account for much of the heat leakage. We can reduce this loss by using storm windows (an extra window over the regular one). In this way an insulating space of air is provided between the

U.S. Steel Corp.

17-16 Radiant heating pipes being installed in a factory. Concrete will be poured around these pipes to form the floor. Warm water circulating through the pipes will heat the floor and radiate heat upward.

The Eagle-Picher Sales Co.

17-17 Which house has better insulation, the one with little snow on the roof, or the one all covered with snow? Before you make a snap decision, remember that the snow has melted from the roof on the right but not from the one at the left.

two windows. Another application of the storm-window idea is a double windowpane, made of two separate pieces of glass with an air space between (Fig. 17-15).

We can reduce the radiation of heat from a house to the outside by including thin, mirror-like sheets of aluminum in the walls and roof. These sheets are placed in such a way as to reflect heat rays back into the house. As we know, a house which is painted white on the outside, or which has an aluminum roof, tends to keep its heat in winter. In summer the same roof and walls tend to keep the house cool since they reflect the radiation from the sun.

Going Ahead

THE VOCABULARY OF PHYSICS

Define each of the following:

conduction	ultraviolet waves
insulator	infrared waves
convection current	radiant heating
dead-air space	solar heating
radiation	

THE PRINCIPLES WE USE

Explain why

1. Heat can be conducted through objects.
2. Air is a poor conductor of heat.
3. A vacuum is a perfect insulator and yet heat reaches us from the sun through a vacuum.
4. Sea breezes blow on hot days.
5. Good insulators are often porous.
6. White objects heat up more slowly in the sun than black objects.

EXPERIMENTS WITH HEAT TRANSPORTATION

Describe an experiment to show that

1. Materials differ in their rates of conduction of heat.
2. Black objects absorb heat more readily than white objects.
3. Black objects radiate heat more readily than white.
4. Suspend a piece of wire gauze several inches above a Bunsen burner and turn on the gas. If a lighted match is held below the gauze, the burner flame will light below the gauze but not above it. If the lighted match is placed above the gauze, then the gas catches fire above the gauze but not below it. Explain.

Show how this experiment with wire gauze applies to a miner's safety lamp, in which a metal screen around a burning candle keeps explosive gases in the air from igniting.

APPLYING PRINCIPLES OF HEAT TRANSPORTATION

Explain

1. The operation of a furnace.
2. Good ventilation in a room.
3. How a vacuum bottle keeps liquids hot or cold.
4. Why ice cream is sometimes wrapped in a silvered bag.
5. Hot-air heating.
6. Hot-water heating.
7. Steam heating.
8. Advantages and disadvantages of steam, hot-water, and hot-air methods of home heating.
9. An ideal air-conditioning system.
10. The cooling system of an automobile.
11. How it is possible for dogs in a cold northern climate to keep warm by digging themselves into the snow.

---

12. Compare the rate of heat loss by radiation from a jet engine at 327°C with the rate of heat loss from the same engine at 1,227°C.

PROJECTS

1. To Illustrate the Poor Heat Conduction of Water. In your bare fingers hold a test tube $\frac{2}{3}$ full of water by its base, tipped in such a position that the water in the upper part of the tube is exposed to a Bunsen flame. The water in the upper part of the tube will boil briskly while your fingers (if you are careful!) will remain quite cool. Why doesn't convection carry the heated water downward in the tube?
2. To Illustrate the Absorption of Heat by Dark and by Shiny Surfaces. Obtain a clean tin can 4 or 5 inches in diameter. Paint half of one side flat black from top to bottom, both inside and outside. Leave the other half shiny. Now with the can mouth upwards support a 100-watt or 150-watt bulb vertically inside it.

   Hold one hand firmly against the shiny outside surface of the can, your other hand against the black surface. You will be amazed at the difference in the two sides.

CHAPTER

18

Measuring Heat

It is natural to assume that heat and temperature are the same thing. For instance, you might think that a thimbleful of boiling water contained more heat than a glassful of cool water. Certainly if you placed a thermometer in the boiling water, the mercury column would rise higher than it would in the cool water.

But actually temperature and amount of heat are not the same. A simple experiment will illustrate this. Take the thimbleful of boiling water and pour it over an ice cube in a glass. This small amount of boiling water has little effect on the cube. Now repeat the experiment, pouring the glass of cool water over a fresh ice cube. The cube melts much more. The large amount of cool water in the glass melts more ice than the small amount of hot water in the thimble (Fig. 18-1).

From this experiment we conclude that the glassful of cool water contains more heat than the small thimbleful of boiling water. In other words, the *amount* of material, as well as its temperature, is important in determining the amount of heat it contains.

To understand why this is true we must find out what heat and temperature really are.

Heat units

Suppose we could observe the molecules of an object in action. We would notice that some of them are moving rapidly, while others are moving slowly. Because of its motion, each molecule has some kinetic energy (energy due to motion). By considering the motion of all the molecules in the object, we could arrive at an average kinetic energy per molecule for the object as a whole. This average for the molecules in the thimbleful of boiling water would be higher than the average for the molecules in the cool water. A thermometer would register this fact. The higher the **temperature,** the higher the **average kinetic energy** of the molecules (Fig. 18-2).

But when we refer to the **heat** in a glassful of water, we mean the **total kinetic energy** of *all* its molecules. Because there are many more molecules of water in the glass than in the thimble, their total kinetic energy is greater, even though the average kinetic energy of the molecules is less.

1. Units of heat energy

Scientists and engineers use two units of heat energy. One of them, the **British Thermal Unit (Btu),** is defined as the amount of heat required to raise the temperature of 1 pound of water 1 Fahrenheit degree. This heat unit is used by engineers in making calculations of heating requirements in homes, office buildings, and factories.

The other unit, the **calorie,** belongs to the metric system. A calorie is defined as the amount of heat needed to raise the temperature of 1 gram of water 1 centigrade degree.

Since there are 454 grams in 1 pound and a Fahrenheit degree is $\frac{5}{9}$ of a centigrade degree,

$$\textbf{1 Btu} = \textbf{454} \times \tfrac{5}{9} = \textbf{252 cal}$$

18-1 Which has more heat, a large glass of cool water or a thimbleful of hot water? Wrong again! The glass of cool water has more heat, as shown by its ability to melt more ice. The fast-moving high-temperature molecules of the small mass of hot water can be compared to the small speedboat, and the many slow-moving molecules of the cool water to a slow, heavy ocean liner. The slower-moving ocean liner has more energy than the speedboat because of its enormously greater weight.

2. Calculating the amount of heat

How much heat would we need to raise the temperature of 50 grams of water from 20°C to 80°C?

We know that to heat 1 gram of water 1 centigrade degree requires 1 calorie of heat. Therefore, to raise a single gram of water from 20°C to 80°C would require 60 calories, 1 calorie for each degree of temperature rise. Our 50 grams, then, would take 50 × 60, or 3,000 calories. From these figures we can derive a formula for finding the total heat added:

$$3{,}000 \text{ cal} = 50 \text{ gm} \times (80 - 20)°\text{C}$$

or

Heat added or lost = Mass × Temperature change

$$\Delta H = m\Delta T$$

(Here we are again using Δ, the Greek letter "delta," as the symbol for "change in"; thus ΔT means "change in temperature," and ΔH means "change in heat.")

Note that in this calculation we assumed that it takes the same amount of heat to raise a gram of water from 20°C to 21°C as it does to raise the same amount of water from 70°C to 71°C. Very accurate heat measurements show slight differences, but for our purposes we can assume the heat required to be the same.

STUDENT PROBLEM

How much heat is needed to raise the temperature of 150 gm of water from 30°C to 100°C? (ANS. 10,500 cal)

3. Exchanging heat

We have seen that heat measurements involve more than the use of a thermometer. Since the quantity of the material involved is important, a balance is also necessary. The principle behind all heat measurement is illustrated by an experiment.

Take a certain quantity of cold water—say, 100 grams at a temperature of 10°C. Mix this with a certain quantity of hot water—say, 50 grams at 40°C. The hot water loses its heat to the cold water until all the water in the mixture is at the same temperature; measurement with a thermometer shows the final temperature of the mixture to be 20°C (Fig. 18-3).

U.S. Steel Corp.

18-2 High temperature means high average kinetic energy—and fast motion of molecules. Here a steelmaker is making use of the fact that at high speeds molecules (and atoms) give off light energy. The kind of thermometer used here is known as an *optical pyrometer* and measures temperature by comparing the color of the furnace with that of an electrically heated hot wire.

We can calculate the heat gained by the 100 grams of cold water in the following way, remembering that the temperature of the cool water rose from 10°C to 20°C.

$$\Delta H = m\Delta T = 100 \text{ gm} \times (20 - 10)°C$$
$$= 1{,}000 \text{ cal}$$

Using the same formula, we find the amount of heat lost by the 50 grams of hot water as it cooled from 40°C to 20°C.

$$\Delta H = m\Delta T = 50 \text{ gm} \times (40 - 20)°C$$
$$= 1{,}000 \text{ cal}$$

We see that the heat gained by the cold water is exactly equal to the heat lost by the hot water. This fact, confirmed by many similar experiments, is stated as the **Law of Heat Exchange:**

The heat lost by an object is equal to the heat gained by the objects to which the heat is transferred.

As the exchange of heat occurs, the speed of the molecules in the hot water is reduced and the speed of the molecules in the cold water is increased until the average kinetic energy of motion of all the molecules in the mixture is the same.

We would expect this to be so from our knowledge of the Law of Conservation of Energy. Since heat is a form of

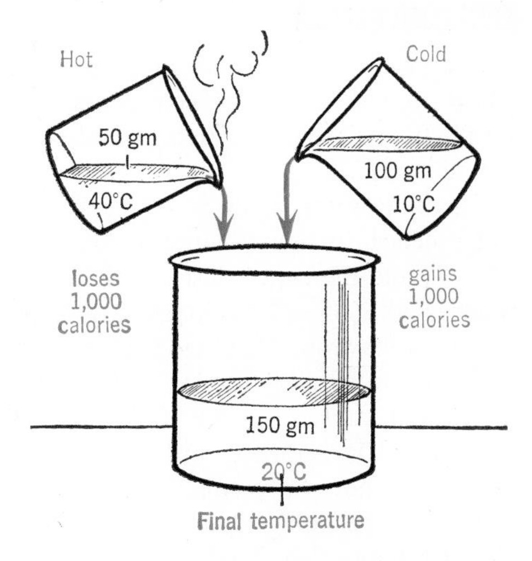

18-3 One thousand calories of heat lost by the hot water are absorbed by the cool water. When mixed, the hot water cools, the cool water warms up, and both end up at 20°C. ***Heat flows from hot to cold; heat lost equals heat gained.***

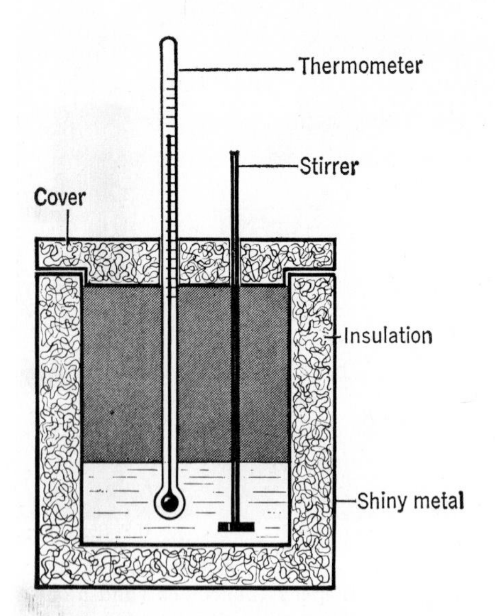

18-4 The *calorimeter,* used in measuring heat, is simply an insulated container with a thermometer. Loss or gain of heat is calculated from the rise in temperature of a given mass of water. Why is the container insulated?

energy, it cannot be created or destroyed by ordinary means. If any object loses heat energy, that same energy must show up elsewhere in other objects.

SAMPLE PROBLEM

300 grams of hot water at 65°C are added to 800 grams of cold water at 10°C. What is the final temperature of the mixture?

The hot water cools from 65°C to T°C; hence its temperature change is $65 - T$, and the heat it loses to the cold water is $300\ (65 - T)$ calories. The cold water warms from 10°C to T°C; hence its change in temperature is $T - 10$, and the heat it gains from the hot water is $800\ (T - 10)$ calories. Setting the heat lost by the hot water equal to the heat gained by the cold water (the Law of Heat Exchange) we have:

Heat lost by hot water =
Heat gained by cold water

$$300\ (65 - T) = 800\ (T - 10)$$
$$T = 25.0°C$$

STUDENT PROBLEM

What is the final temperature of a mixture of 200 gm of water at 80°C and 150 gm of water at 20°C? (ANS. 54.3°C)

* * * * * * *

The container used for a heat-exchange experiment is generally insulated, made of shiny metal, and covered, so as to reduce transfer of heat from or to the outside by conduction, convection, and radiation (Fig. 18-4). Such a container designed for making heat measurements is called a **calorimeter** (kăl′ō̂·rĭm′ê·tẽr).

4. Heat of burning fuels

The value of any fuel depends upon the amount of heat it can produce. To measure this heat a carefully measured weight of fuel is burned in a small oxygen-filled sealed container inside a special calorimeter. Surrounding the container is a known weight of water, which is heated by the burning fuel. From the rise in temperature of the water we can calculate the energy value of a fuel, using the Law of Heat Exchange.

The energy value of different foods is measured in the same way. The energy value of any food is simply its fuel value when used inside the body. To measure fuel value, the food is first dried, then burned in a container placed in the calorimeter. Temperature measurements are made and the amount of heat is calculated.

For consistency all the values in Table 8 are listed in calories/gram—called **small calories.** Food energy values, however, are usually listed in terms of **large calories.** A large calorie is equal to 1,000 small calories. Hence a large calorie is the amount of heat which will raise 1 *kilogram* of water 1 centigrade degree. When we say that a slice of bread has 100 calories, we are speaking of large calories; the heat value in the slice of bread is equal to 100,000 small calories. Throughout this book the word "calories" will always refer to *small* calories unless otherwise stated.

SAMPLE PROBLEM

How much heat energy is released when 150 grams of gasoline are burned in an automobile engine? Since gasoline gives 11,500 calories per gram when completely burned, we have:

$$\text{Heat released} = 11{,}500 \times \text{Mass in grams}$$
$$H = 11{,}500 \times 150$$
$$= 1{,}725{,}000 \text{ cal}$$

STUDENT PROBLEM

How much heat is released by burning 1 kg of soft coal? (ANS. 6,000,000 cal)

TABLE 8

HEAT RELEASED BY BURNING OF FUEL (These figures are approximate)

	cal/gm	Btu/lb
Hydrogen	34,000	61,200
Gasoline	11,500	20,750
Butter	9,200	16,500
Coke	8,000	14,400
Coal, hard	7,400	13,300
Ethyl alcohol	6,500	11,620
Coal, soft	6,000	10,800
Oak wood	4,000	7,200
Sugar	4,000	7,200
Bread, white	2,600	4,800
Potatoes	1,000	1,800
Milk	720	1,300
Apples	630	1,130
Spinach	575	1,030
String beans	430	775

Specific heat

Suppose we take a large cake of ice and place on it, one after the other, different substances which are all at the same temperature and have the same mass, say 1 kilogram. Assume that the substances we use are water, aluminum, iron, and lead, and that they have all been heated to a temperature of 100°C. Figure 18-5 shows how much of the ice each of these substances would be able to melt. We see that the aluminum melts about $\frac{1}{5}$ as much ice as the hot water, the iron melts $\frac{1}{8}$ as much as the water, and the lead about $\frac{1}{30}$ as much.

We conclude from this that aluminum releases about $\frac{1}{5}$ as much heat as an equal mass of water going through the same temperature change. This number, $\frac{1}{5}$ (or 0.22, to be more precise), is called the **specific heat** of aluminum. The specific heat of any substance is the ratio of the heat it gives out to the heat given out by an equal mass of water, the temperature changes being the same. As we have seen, the specific heat of iron is about $\frac{1}{8}$; that of lead about $\frac{1}{30}$. According to the definition, the specific heat of water is 1.

We can also think of specific heat as the number of calories needed to raise the temperature of 1 gram of a substance 1 degree centigrade (or the number of Btu needed to raise 1 pound of a substance 1°F). Thus 1 gram of aluminum,

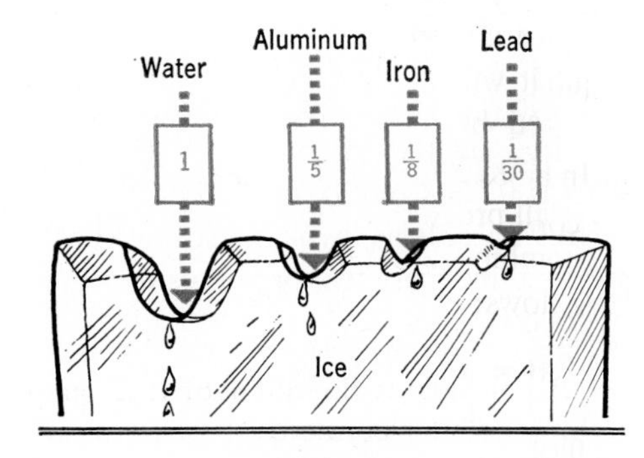

18-5 The same masses of hot water, aluminum, iron, and lead—at the same temperature—melt different amounts of ice and thus have different amounts of heat. Water has a much higher *specific heat* than most common materials.

TABLE 9

SPECIFIC HEATS

Water	1.00
Ethyl alcohol	0.60
Ice	0.50
Steam	0.50
Aluminum	0.22
Glass	0.16
Iron	0.12
Copper	0.092
Lead	0.030

with a specific heat of 0.22, can be raised in temperature 1 degree centigrade by only 0.22 calorie, whereas 1 gram of water, with a specific heat of 1, will require 1 calorie. The reason that the specific heat of water is 1, you will remember, is that it takes 1 calorie to heat 1 gram of water 1 centigrade degree.

5. Calculations using specific heat

How much heat would we need to raise the temperature of 100 pounds of aluminum from 40°F to 100°F?

First, we can calculate how much heat would be required to heat water by this amount:

$$\Delta H = m\Delta T = 100 \text{ lb} \times (100 - 40)°\text{F} = 6{,}000 \text{ Btu}$$

Since aluminum has a specific heat of 0.22, it will require 0.22 times the heat required by the same weight of water. This is 0.22 × 6,000, or 1,320 Btu.

In all problems involving specific heats we combine these two processes into one, as follows:

$$\Delta H = \text{Mass} \times \text{Specific heat} \times \text{Change in temperature}$$

or more simply

$$\Delta H = mS\Delta T$$

$$\Delta H = 100 \times 0.22 \times (100 - 40) = 1{,}320 \text{ Btu}$$

STUDENT PROBLEM

How much heat would be needed to raise the temperature of 25 lb of aluminum from 60°F to 150°F? (ANS. 495 Btu)

Note that water has the highest specific heat in Table 9. Water has, in fact, one of the highest specific heats known, and is therefore an excellent material to use for transporting heat in cooling and heating systems.

6. Heating the ocean

You may think that the high specific heat of water is of no great practical importance to you. Actually, however, it is of enormous importance, since the waters of the earth have a very great influence on climate.

Because of its high specific heat, water requires more heat and hence more time than other materials to reach the same temperature. Thus, in summer the oceans and large lakes warm up more slowly than the land around them and help to keep the surrounding land cooler. For the same reason, water cools off more slowly than the land in winter, and keeps the land from getting as cold as it otherwise would.

Large bodies of water serve to keep down extremes of temperature on land. In summer the winds that blow over oceans and large lakes are cooled, and carry this cooler temperature hundreds of miles inland. In winter the cold winds are warmed by the water, and nearby land is therefore not as cold as it would otherwise have been.

7. Finding temperatures by the method of mixtures

Suppose we place a 200-gram piece of aluminum at 100°C in 100 grams of water at 20°C. What, exactly, will happen?

The aluminum will lose heat to the water and cool off. The water will gain heat from the aluminum until both reach the same temperature. What will this temperature be? If we assume that no heat is lost to or gained from the surroundings, the Law of Heat Exchange tells us that the heat lost by the aluminum exactly equals the heat gained by the water. If we call the final temperature of the mixture T, we can express the Law of Heat Exchange in the form:

Heat lost by aluminum = Heat gained by water

$$200 \times 0.22 \times (100 - T) = 100 \times 1 \times (T - 20)$$
$$T = 44.4°C$$

This method of finding the final temperature of a mixture of materials by the use of the Law of Heat Exchange is called **the method of mixtures.**

STUDENT PROBLEM

What would have been the final temperature in the above sample problem if the weight of aluminum had been 100 gm and the weight of the water 75 gm? Assume again that no heat is gained from or lost to the surroundings. (ANS. 38.2°C)

8. The method of mixtures—taking the calorimeter into account

If we actually performed the experiment described in the previous section, we would use a calorimeter to hold the water. Hence our calculation of the final temperature would be more accurate if we took into account the heat absorbed by the calorimeter can. To do this, we solve the problem in the same way except that we must add a term to the "heat gained" side of the equation to take care of the heat absorbed by the calorimeter.

SAMPLE PROBLEM

Assume that the inner can of the calorimeter (which is the only part that absorbs any significant amount of heat) is made of aluminum and weighs 50 grams. Find the final temperature.

Since the calorimeter can is in direct contact with the water it contains, it is always at the same temperature as the water. Therefore, in the equation below, its temperature change is the same as that of the water. Our mixture equation now becomes:

Heat lost by aluminum = Heat gained by water + Heat gained by calorimeter

$$200 \times 0.22 \times (100 - T) = 100 \times 1 \times (T - 20) + 50 \times 0.22 \times (T - 20)$$
$$T = 42.7°C$$

Since there is a big difference between the result obtained by taking the calorimeter into account (42.7°C) and the result obtained by neglecting it (44.4°C), it is necessary to consider the heat absorbed by the calorimeter if we are to be accurate.

STUDENT PROBLEM

What would be the answer to the student problem at the end of Section 7 if the aluminum calorimeter can (weight 50 gm) were taken into account? (ANS. 36.2°C)

9. Measuring specific heat by the method of mixtures

We can use the method of mixtures to measure the specific heat of a substance.

SAMPLE PROBLEM

To find the specific heat of copper an experimenter heats a 100-gram piece of copper to 100°C in boiling water and then drops it into a 50-gram aluminum calorimeter can containing 200 grams of water at 20°C. He observes the final temperature of the mixture to be 23.3°C. What is the specific heat of the copper? His data are:

Weight of copper	100 gm
Initial temperature of copper	100°C
Specific heat of copper	S
Weight of water	200 gm
Specific heat of water	1.00
Weight of calorimeter	50 gm
Specific heat of calorimeter	0.22
Initial temperature of water and calorimeter	20°C
Final temperature of water, calorimeter, and copper	23.3°C

Using the method of mixtures,

Heat lost by copper = Heat gained by water + Heat gained by calorimeter

$$100 \times S \times (100 - 23.3) = 200 \times 1 \times (23.3 - 20) + 50 \times 0.22 \times (23.3 - 20)$$

$$S = 0.0908$$

STUDENT PROBLEM

200 gm of metal at 100°C are placed in a 50-gm aluminum calorimeter containing 100 gm of water at 20°C. The final temperature is 31.4°C. What is the specific heat of the metal? (ANS. 0.0921)

Heat and mechanical energy

In science we are continually coming back to the Law of Conservation of Energy. Important use of this law was made by the English physicist James Prescott Joule, who experimented with heat energy in the middle of the last century.

10. Mechanical equivalent of heat

Joule wanted to find out how much heat energy could be developed by a given amount of mechanical energy. He caused a falling weight to turn a paddle wheel in a container of water. As the

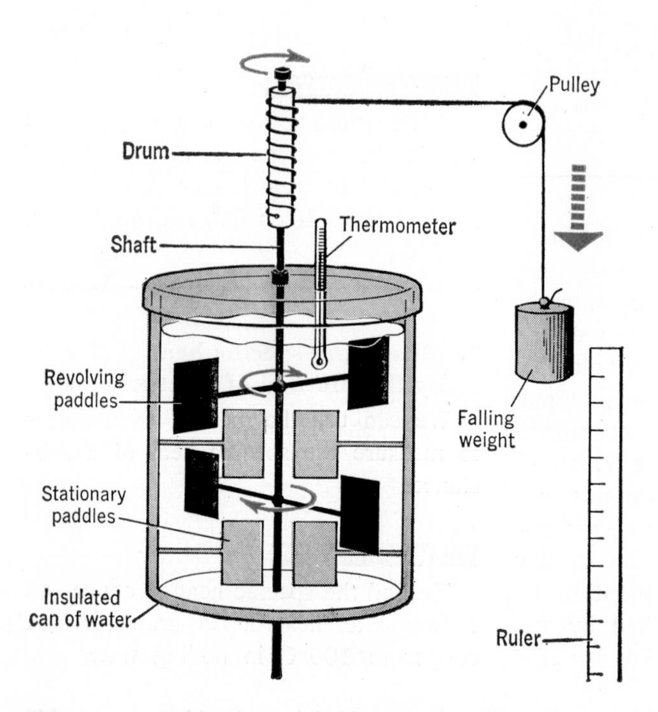

18-6 This paddle-wheel contraption takes advantage of one of the most important laws in science—the Law of Conservation of Energy. The falling weight loses potential mechanical energy, and the paddle wheel converts it into heat energy in the water. In this manner Joule discovered that 1 Btu = 778 foot-pounds.

weight slowly fell, the paddle wheel began to turn. The paddles churning the water caused friction, which heated the water (Fig. 18-6).

As we know, the potential energy of the falling weight is converted into the heat energy gained by the water. The potential energy of the weight can be found from the formula: Energy = Weight × Height. The heat energy gained by the water can be found from the formula: $\Delta H = m\Delta T$. Using the Law of Conservation of Energy, we can say that the mechanical energy lost by the falling weight is equal to the heat energy gained by the water—assuming, of course, that no energy is lost in other forms.

As a result of his experiment, Joule showed that **778 foot-pounds of mechanical energy are equivalent to 1 Btu of heat energy** (and 42,700 gm-cm = 1 cal).

SAMPLE PROBLEM

How much heat energy in foot-pounds is released by burning 2 pounds of gasoline?

Table 8 (page 241) gives the heat of combustion of gasoline as 20,750 Btu/lb. Hence the heat energy in 2 pounds of gasoline is

$$\text{Energy} = 2\ \text{lb} \times 20{,}750\ \frac{\text{Btu}}{\text{lb}} \times 778\ \frac{\text{ft-lb}}{\text{Btu}} = 32{,}300{,}000$$

STUDENT PROBLEM

How much heat energy in ft-lb is released by burning 25 lb of coke? (ANS. 280,000,000 ft-lb)

■ ■ ■ ■ ■ ■ ■

This does not mean that when we burn a fuel like gasoline in an engine we get 778 foot-pounds of useful work for every Btu of heat produced. As we know, machines and engines are never 100 per cent efficient, and some of the energy produced is always wasted by being converted into energy that cannot be used. In a gasoline engine with an efficiency of 25 per cent, we would get 25 per cent of 778, or about 195 foot-pounds of useful work, for each Btu of heat produced. The remaining 75 per cent of the heat energy is wasted as heat lost to the surroundings.

Joule's measurement of the **mechanical equivalent of heat** was one of the first and led to many others. We can say without exaggeration that it was one of the most important scientific experiments ever performed, partly because it was the first really convincing demonstration that heat was simply another form of energy.

Going Ahead

THE VOCABULARY OF PHYSICS

Define each of the following:

temperature	large calorie
heat	specific heat
Btu	mechanical equivalent of heat
calorie	
calorimeter	

THE PRINCIPLES WE USE

Explain why

1. There is a difference between heat and temperature.
2. There is a connection between the Law of Heat Exchange and the Law of Conservation of Energy.
3. Large bodies of water change in temperature much more slowly than the land.

EXPERIMENTS IN MEASURING HEAT

Describe an experiment to

1. Find the caloric value of a slice of bread.
2. Measure the mechanical equivalent of heat.
3. Verify the Law of Heat Exchange.

SOLVING PROBLEMS OF HEAT EXCHANGE

1. How much heat is needed to raise the temperature of 365 gm of water from 35°C to 86°C?
2. How much heat is given off by 365 gm of water in cooling from 86°C to 35°C?
3. How much heat is needed to raise the temperature of 36.5 lb of water from 68°F to 200°F?
4. 65 gm of water give off 195 cal of heat when cooled from 35°C. To what temperature did the water cool?
5. By how many Fahrenheit degrees will 300 gm of water be heated if they absorb 6,000 cal of heat?
6. How much heat is released when 2.5 gm of hard coal are burned?
7. How many gm of coke must be burned to produce 64,000 cal of heat?
8. How much energy, in heat units, can be obtained from eating 20 gm of sugar?
9. How much heat is released by the burning of 300 gm of hydrogen?

---

10. 120 gm of water at 72° are added to 200 gm of water at 40°C. What is the final temperature of the mixture?
11. 450 gm of cool water at 39°C are added to 110 gm of hot water. If the mixture acquires a temperature of 50°C, what was the starting temperature of the hot water?
12. 500 lb of water at 97°F are added to 300 lb of water at 17°F. What is the final temperature of the mixture?
13. 200 lb of hot water at 35°F are added to 100 lb of cold water. If the mixture temperature becomes 25°F, what was the starting temperature of the cold water?
14. 150 gm of hot water at 89°C are added to a mass of cold water at 4°C. If the final temperature of the mixture is 19°C, what is the mass of the cold water?
15. How much heat is given off by 30 lb of iron in cooling from 600°F to 70°F?
16. 60,000 cal of heat are absorbed by a 400-gm piece of copper at 20°C. How hot does it become?
17. 75 gm of iron at 90°C are added to 36 gm of water at 65°C. What is the final temperature?
18. What is the specific heat of lead if 60 Btu can warm 20 lb of lead from 20°F to 120°F?

---

19. What is the final temperature in Problem 17 if a 50-gm aluminum calorimeter can is used to hold the water?
20. A 2,100-gm sample of a certain metal at 100°C is placed in a copper calorimeter can containing 1600 gm of water at 45°C. The calorimeter can weighs 100 gm. The final temperature of the mixture is 60°C. What is the specific heat of the sample? What might the sample of metal be?
21. A 50-gm piece of iron is taken from a hot furnace and placed in a 50-gm copper calorimeter containing 100 gm of water at 20°C. The water and calorimeter are heated to 62.5°C. What is the temperature of the furnace?
22. How many Btu of heat are produced by 6,000 ft-lb of work?
23. Assuming 50% efficiency of a furnace, how many kg of coke would be needed to heat 400 kg of water from 30°C to boiling?
24. A piece of iron falls 200 ft, and half its kinetic energy is converted into heat. How many Fahrenheit degrees does its temperature rise?
25. How many Btu of heat are produced in the brakes of a 3,200-lb car when it is brought to a stop from a speed of 60 mi/hr?
26. What speed must a lead bullet have for its temperature to be raised 200 Fahrenheit degrees when it hits a steel plate, if half its energy is converted into heat?

CHAPTER

19

Freezing and Boiling

Processes such as boiling and freezing are called **changes of state.** In boiling, a material changes from a liquid into a gaseous state. In freezing, it changes from a liquid to a solid. The most common changes of state are:

1. melting (fusion)—a solid becomes a liquid
2. freezing (solidification)—a liquid becomes a solid
3. boiling and evaporation—a liquid becomes a gas
4. condensing—a gas becomes a liquid

The three states of matter—solid, liquid, and gas—and their molecular pictures are shown in Fig. 19-1. Note that it is the increased motion of molecules (heating) which frees the molecules from a state of "bondage" in the solid, and gives them partial freedom in the liquid, and complete freedom in the gaseous state.

The molecules of a material move most slowly when it is a solid. They are therefore closest together when it is a solid, and the forces of cohesion can operate most strongly. For these reasons a solid is able to keep its shape—unlike liquids and gases.

Freezing

We know that water commonly freezes at 0°C. But this freezing point temperature can be changed by dissolving materials in the water or by applying pressure to it. These two ways of changing the freezing point of water have many practical consequences in the everyday world.

1. Dissolved materials lower the freezing point

In wintertime you may have melted the ice on your front sidewalk by scattering salt on it. You can do this because salt, like all materials that dissolve in water, lowers the freezing point of water. You have created a mixture (salt water) which no longer freezes at the freezing temperature of pure water (0°C) but at a lower temperature. Because the ocean contains salt, it does not freeze during the winter in temperate climates. But if the temperature falls low enough, as it commonly does in polar regions, even salt water will freeze.

You can perform an easy experiment to prove that a mixture of salt and water freezes at a lower temperature than pure water. Put two ice-cube trays in the ice compartment of your refrigerator, one filled with fresh water, the other with salt water. Which will take longer to freeze? If the concentration of salt is high enough, salt water will not freeze at all. A practical application of this is to put salt under the ice-cube trays. The salt keeps the water underneath the trays from freezing them into place and the trays are easily removed.

2. Pressure and the freezing point of water

If you grasp a strong, sharp needle with a pair of pliers, you can slowly push the needle all the way through an ice cube without cracking the ice. (Keep your hand out of the path of the needle!) The ice directly under the needle melts into water and flows away to freeze again at a spot where there is no pressure.

Why does pressure melt ice? Water has an odd property, one that we have mentioned before. Most materials contract when they pass from the liquid into the solid state during freezing, but water expands when it passes into the solid form of ice. It does so because its molecules become rearranged and require more space as ice than they did as water. When water is under pressure, it is more difficult for it to expand into ice. A temperature lower than its normal freezing point of 0°C is needed to cause the molecules to rearrange themselves and the water to freeze. In other words, pressure lowers the freezing point of water. The greater the pressure the lower the freezing point will be. For this reason, as the needle was pushed through the ice cube, the freezing point of the ice at the point of contact was lowered by the pressure, and hence the ice melted.

For example, if the block of ice were at *minus* 2°C, two degrees lower than the temperature at which water freezes under normal air pressure, and if the pressure on the ice lowered its freezing point to −5°, our temperature of −2° would not be low enough to freeze water. Because of the increased pressure, the ice would melt.

When we remove this pressure, the water freezes into ice again, because at −2°C it is colder than the normal freezing point of 0°C. Note that the actual temperature of the ice (−2°C) is not altered—it is the freezing point that is shifted from 0°C to −5°C and back to 0°C again. This process of melting un-

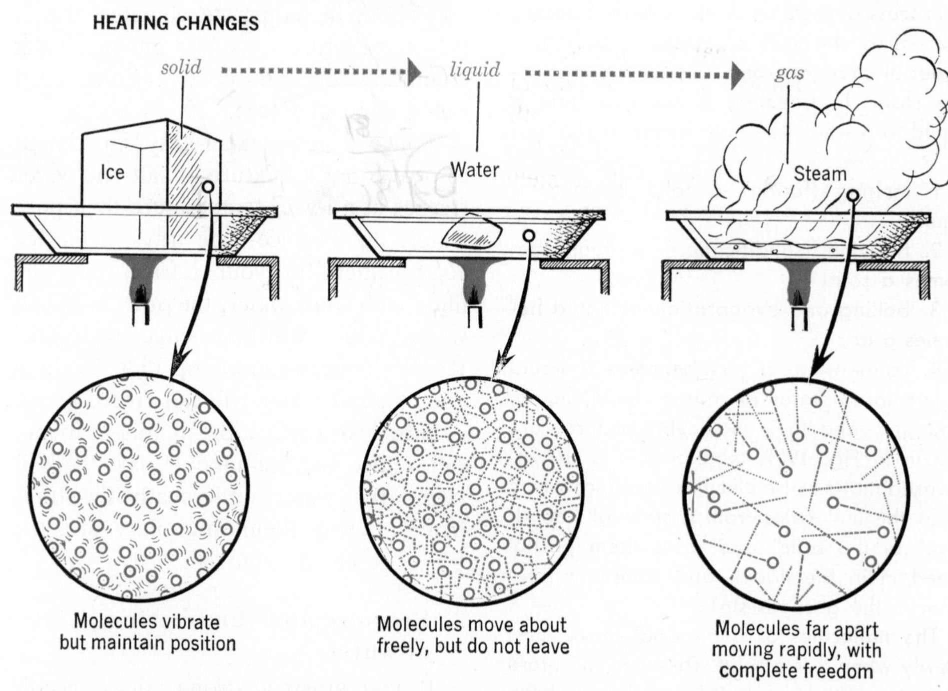

19-1 Slow-moving molecules of water are trapped by the cohesion of fellow molecules and stick together to form a solid. With increasing motion, they acquire more freedom as the ice becomes liquid. But with enough molecular speed caused by heating, they escape from the liquid to complete freedom.

der pressure and then refreezing when the pressure is released is called **regelation.**

The opposite occurs with most liquids. They take up less room as solids than they do as liquids; hence pressure brings them nearer to the solid state. In other words, pressure raises the freezing point of most liquids other than water.

The behavior of ice under pressure accounts for the movement of glaciers. The pressure of snow and ice against projecting rocks on the ground causes the snow and ice to melt. The resulting water flows around the rocks and refreezes as soon as the pressure on it is released. This continuous melting of ice by pressure where it touches the ground allows the glacier to move slowly down and around the side of the mountain into the valley below (Fig. 19-2).

3. Ice-skating

You may think that we are able to skate on ice because the ice is so smooth, but this is not so. Glass is just as smooth, but we cannot skate on it. Again we find the explanation in the fact that ice melts under pressure. In ice-skating, your weight on the small area of the sharp edges of the skates produces a very high pressure and melts the ice. This melted ice water acts as a lubricant over which the blades can glide and the groove in the ice keeps your skate from slipping sideways. In very cold weather, when the temperature of the ice is so low that the pressure of the blades is not sufficient to melt it, the blades of the skates will not slide, and skating is difficult because there is no groove for your skate to push against.

Regelation enables you to make snowballs, too. The pressure of your hands causes some of the snow to melt. It refreezes again when the pressure is released.

U.S. Coast and Geodetic Survey

19-2 **A river of ice! Glaciers like this one in Alaska could not flow around bends unless pressure lowered the freezing point of water.**

Hidden heat

4. Heat of fusion

Pack some melting ice around a thermometer in a pan. In a short while the thermometer will read 0°C. Now hold the pan full of melting ice over a flame and heat it strongly, stirring gently and keeping the thermometer from actual contact with the bottom of the pan. Even at the end of a minute or two the thermometer still reads 0°C (Fig. 19-3). We notice that the flame has melted a good deal of ice, yet the temperature of the ice

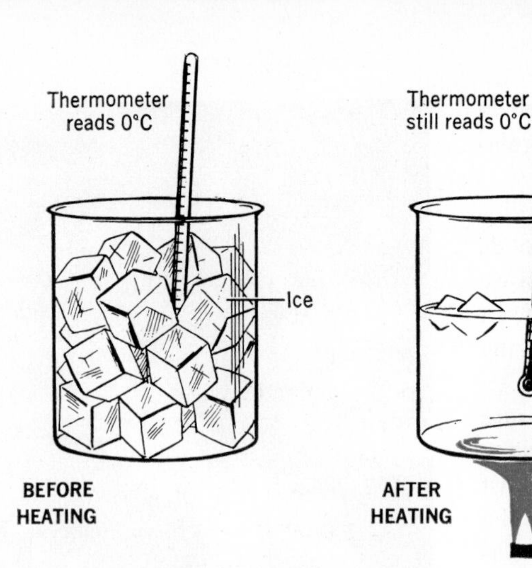

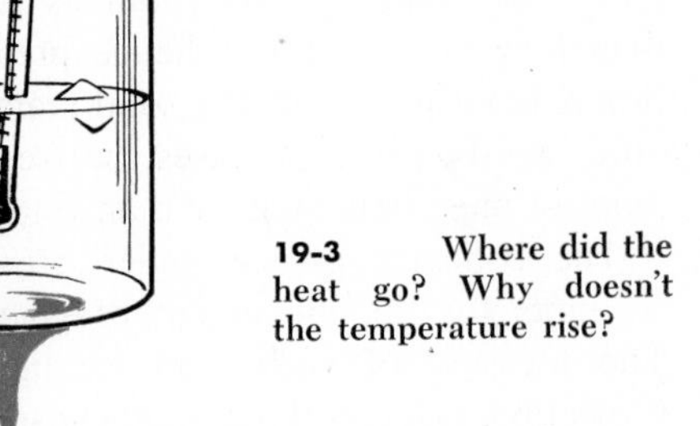

19-3 Where did the heat go? Why doesn't the temperature rise?

water is also 0°C. How can this be? We have added heat but we have not increased temperature. Since we know that energy is never destroyed, we cannot conclude that the heat has simply disappeared.

You may have already guessed what happened to this heat which apparently "disappeared." It was being used to accomplish a change of state. The change was from solid (ice) to liquid (water). Heat which is used to change a material at its freezing point from solid to liquid (or liquid to solid) is called **heat of fusion** (or **latent heat of fusion**—"latent" meaning "hidden").

5. Calculations using heat of fusion

Measurements show that it takes 80 calories to change 1 gram of ice at 0°C to water at 0°C. This quantity, 80 calories per gram, is the heat of fusion of water.

It is obvious that if 80 calories are required to melt 1 gram of ice, 160 calories (2×80) will be needed to melt 2 grams of ice, and 240 calories (3×80) to melt 3 grams of ice. We can put this in the form of a simple equation:

$$\Delta H = mL$$

where H is the number of calories needed to melt m grams of a substance whose heat of fusion is L calories per gram. Conversely, H is the amount of heat that must be given off when m grams of the substance freeze. Notice that the specific heat does not appear in the above equation. This is to be expected since *no change in temperature* of the substance is involved in melting or freezing. Specific heat has to do only with heat absorbed or liberated when there is a change in temperature.

SAMPLE PROBLEM

How much heat is needed to melt 1 pound of ice? Since 1 pound equals 454 grams, and the heat of fusion of ice is 80 calories per gram, we can say:

$$\Delta H = mL$$
$$\Delta H = 454 \times 80 \text{ cal} = 36{,}320 \text{ cal}$$

STUDENT PROBLEM

How much heat is absorbed by 4.5 lb of ice in melting? (ANS. 163,440 cal)

---

6. Measuring heat of fusion

Let us examine several problems involving the heat of fusion.

1. Suppose we wish to find out how much heat is extracted from 10 grams of water at 20°C placed in an ice tray in a

refrigerator as it freezes into an ice cube whose temperature is −10°C. We must think of this problem as having three parts, which we solve in order.

First, the water (whose specific heat, S_w, is 1) is cooled from 20°C to 0°C, and the heat extracted is:

$$\Delta H_1 = mS_w\Delta T$$
$$= 10 \times 1 \times (20 - 0) = 200 \text{ cal}$$

Second, we must calculate the heat extracted as the water freezes:

$$\Delta H_2 = mL$$
$$= 10 \times 80 = 800 \text{ cal}$$

Third, we must find out how much heat is released when the resulting ice cube cools to −10°C. Recalling that the specific heat of ice, S_i, is 0.50 (p. 242), we substitute in the formula:

$$\Delta H_3 = mS_i\Delta T'$$
$$= 10 \times 0.50 \times 10 = 50 \text{ cal}$$

Adding all these together, we have for the total heat, *H*, given off:

$$H = \Delta H_1 + \Delta H_2 + \Delta H_3$$
$$= 200 + 800 + 50 = 1{,}050 \text{ cal}$$

Instead of doing all these calculations separately, we can put them all together in one calculation:

Total heat given off = Heat from cooling water + Heat from change of state + Heat from cooling ice

$$H = 10 \times 1 \times (20 - 0) + 10 \times 80 + 10 \times 0.50 \times 10$$
$$= 1{,}050 \text{ cal}$$

STUDENT PROBLEM

How much heat is lost by 25 gm of water at 24°C in changing to ice at −12°C? (ANS. 2,750 cal)

■ ■ ■ ■ ■ ■ ■

How can we measure the heat of fusion of ice in the laboratory? We can do it with a calorimeter and by applying the method of mixtures.

2. Let us assume that we place a known weight of ice, say 40.0 grams, at 0°C in a calorimeter can containing warm water. When all the ice is melted, we measure the temperature of the water in the can. Our data might be:

Weight of ice	40.0 gm
Weight of calorimeter	50.0 gm
Specific heat of calorimeter can (aluminum)	0.22
Weight of warm water	200 gm
Initial temperature of water	20.0°C
Final temperature of water and water from ice	4.2°C
Heat of fusion of ice	L cal/gm

We can now calculate the heat of fusion of ice by setting the heat lost by the calorimeter can and water equal to the heat gained by the ice:

Heat lost by water + Heat lost by calorimeter can = Heat gained by melting ice at 0°C + Heat gained by melted ice (water)

$$200 \times 1 \times (20.0 - 4.2) + 50.0 \times 0.22 \times (20.0 - 4.2) = 40.0 \times L + 40.0 \times 1 \times (4.2 - 0)$$
$$L = 79.1 \text{ cal/gm}$$

STUDENT PROBLEM

100 gm of ice at 0°C are placed in a copper calorimeter weighing 100 gm, which contains 200 gm of water at 40°C. The final temperature is found to be 1.2°C. Find the heat of fusion of ice. (ANS. 80.0 cal/gm)

Evaporation

If you wet the back of your hand with water and wave it back and forth, your hand feels cool as the water evaporates. If alcohol or ether are used, the effect is even greater. The cooling effect of evaporation depends upon the **heat of vaporization** absorbed by a liquid when it vaporizes (changes into a gas). This heat

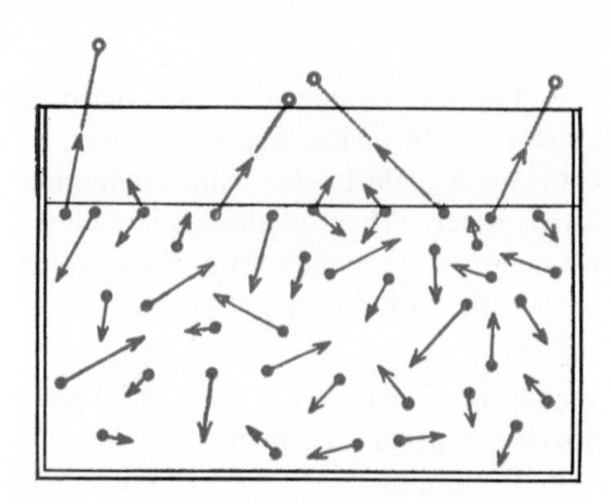

19-4 Molecules near the surface of a liquid that are moving fast enough in the right direction are able to break away from their neighbors and become vapor molecules. This process is called *evaporation.*

of vaporization also plays a part in explaining why you can't possibly make an uncovered pan of boiling water hotter than about 100°C no matter how high you make the temperature under it. To understand what heat of vaporization is and how these processes take place, we must consider the motion of molecules in liquids and gases.

7. Evaporation and heat of vaporization

If we place a shallow dish of water in the sun, we find that in a few hours the water has disappeared. It has changed state from a liquid (water) to an invisible gas (water vapor), or, more simply, it has evaporated. Why does this change of state occur? As you learned in Chapter 16, the molecules in a liquid are moving around in a random manner. Some molecules are moving very fast, some very slowly; their average speed depends on their temperature. Molecules moving toward the surface of the liquid at speeds great enough to overcome the attraction of neighboring molecules will leave the liquid and become vapor molecules (Fig. 19-4). Since only the fastest molecules get away, the remaining molecules have an *average* speed lower than the average speed before evaporation took place; in other words, the temperature of the remaining liquid falls. (You will remember from Chapter 18 that temperature depends on the average kinetic energy—energy of motion—of the molecules in a substance.) This is why evaporation is a cooling process.

If we wish to keep the temperature of the water in the shallow dish from falling while the water is evaporating, we must add heat to the liquid. In other words, heat is required to change a liquid to a gas without raising its temperature. When water is at its boiling point, 540 calories of heat must be supplied to 1 gram of water to evaporate it without changing its temperature. This heat is called **heat of vaporization.** We therefore say that the heat of vaporization of water is 540 calories per gram.

Conversely, if the Law of Conservation of Energy is to hold true, when water vapor condenses into liquid water, 540 calories of heat are *released* for every gram of water vapor that changes to liquid water. This is why a burn from steam at 100°C is much worse than one from water at the same temperature. Much more heat is released by the condensation of 1 gram of steam and the cooling of the resulting water on the skin than by the mere cooling of an equal amount of water from 100°C—ten times as much, in fact, if we assume that the temperature of the skin is 40°C. Can you prove this?

8. Some useful effects of evaporation

In desert regions, drinking water is often kept in porous pottery jugs or porous bags. The water that oozes through to the outside surface evaporates and absorbs heat from the jug and the remaining water, keeping the water cool. Metal canteens carried by Boy Scouts and soldiers are covered with canvas, which, if kept wet, keeps the contents cool by evaporation.

In a similar manner the evaporation of perspiration from our skin keeps us cool in hot weather. By an intricate biological process, sweat glands near the skin's surface release perspiration in hot weather. The evaporating liquid absorbs heat from the skin, keeping the body cool. When the humidity is high, evaporation is slow, and we feel hot and sticky—hot because evaporation is too slow to keep us cool and sticky because the perspiration that does not evaporate wets our clothes.

Most people think of an electric fan as creating a cool breeze. Actually, however, the fan has no cooling effect on the air at all. If anything, the hot motor tends to warm up the room. The cooling effect of the fan depends entirely on evaporation. For a flow of air tends to carry the escaping molecules away and replace the moist air by drier air. The rate of evaporation is increased, and our bodies feel the cooling result. If the air into which water molecules are evaporating is stationary, many molecules return to the liquid state instead of flying off into the air.

The process of evaporation is important for reasons other than its cooling effect. We depend upon evaporation to obtain salt from sea water; when the water evaporates, the salt is left behind. The odor of perfume reaches the nose as a result of evaporation. Liquid kerosene in a kerosene lamp vaporizes as it leaves the wick, and in this way burns more vigorously.

Liquids that evaporate easily are said to be **volatile.** Alcohol, ether, and gasoline are quite volatile. Water is fairly volatile, while mercury is not.

A gas also can be formed directly from the solid state by a process called **sublimation.** Dry ice (solid carbon dioxide) offers an example of sublimation. The solid carbon dioxide does not melt into liquid but passes directly into the gaseous state. Moth balls act in a somewhat similar way. They do not melt but change directly into a gas.

9. Increasing the rate of evaporation

Our previous discussion shows how a wind or breeze will make the molecules of a liquid evaporate faster (A in Fig. 19-5).

Increasing the temperature also helps, since it provides the energy necessary for escape (B). A third way of increasing the rate of evaporation is to reduce the air pressure above the liquid. This lowers the concentration of air molecules and so permits more molecules of liquid to escape (C). Finally, the greater the surface of the liquid exposed to the air, the greater the chance for its molecules to escape and the more rapid the evaporation (D).

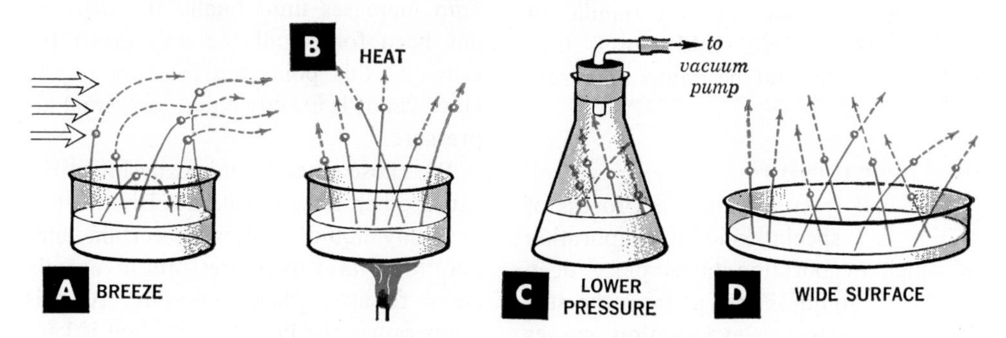

19-5 **Four ways to speed up evaporation.**

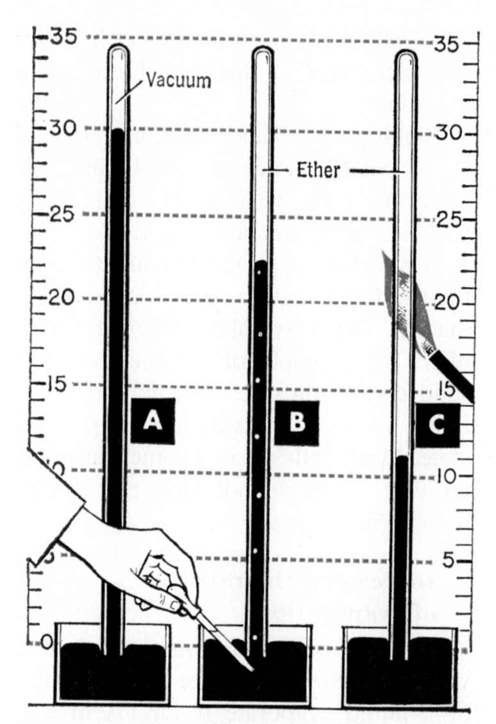

19-6 The pressure of ether vapor in B pushes the level down considerably. Very gentle heating of the ether vapor (C) increases its vapor pressure and makes the mercury drop even more. Can you estimate the vapor pressure of the ether in B and C? Don't try this experiment without supervision! The ether could explode if the glass cracked.

We make use of three of these facts in drying clothes. We prefer to hang the clothes in a breeze. We spread them out to provide the largest possible surface for evaporation. We put them in the sun so they will be heated. On the other hand, if we were to throw the wet bundle of clothes into a dark, cold closet, they would probably still be damp after several weeks.

10. Vapor pressure

We shall soon see that the process of boiling is a special kind of evaporation in which evaporation takes place deep within the liquid as well as from the surface. This internal evaporation causes the vigorous bubbling that we observe in a boiling liquid, and it is the result of vapor pressure.

What is vapor pressure? Let us study this problem by performing a simple experiment with a mercury barometer (Fig. 19-6). By means of a medicine dropper, we insert some liquid ether into the open end of the barometer tube (the bottom). Almost immediately the mercury level drops.

When the liquid ether rises to the top of the tube, its molecules evaporate and become vapor, or gas. Molecules of liquid ether evaporate quickly because they have little cohesion to begin with. These molecules of ether vapor jump out into the vacuum at the top of the barometer tube and exert a pressure which pushes the mercury column down. We can measure the amount of pressure exerted by the vapor—called the **vapor pressure**—by the fall in the mercury. If the mercury drops one fourth of the way down the tube, we see that the vapor pressure of the ether is one fourth that of normal air pressure.

Now let us warm the tube. (Warm it gently, because of the danger of the glass cracking and the ether exploding. Be sure that no open container of ether is anywhere near the flame!) The mercury level drops still further, indicating that the vapor pressure has increased. As we continue to heat the tube, the ether vapor pressure on the top of the mercury column increases until finally the mercury has been forced all the way down the tube. At this point the pressure of the ether vapor is exactly equal to normal air pressure.

We chose ether for this experiment because ether is very volatile. But we can heat any liquid until, at a certain temperature, its vapor pressure is exactly equal to atmospheric pressure. At this temperature the liquid would boil if kept in an open container. Hence, the boiling point of a liquid is the temperature at which its vapor pressure is equal to air pressure.

11. Saturated vapor pressure

You will remember that when ether was first introduced into the barometer tube the level of the mercury fell rapidly. In a few moments the mercury level fell more slowly, and finally it became stationary.

At the beginning there were no ether molecules at all in the near-vacuum above the mercury. Ether molecules could evaporate, but there were no vapor molecules to return to the liquid state.

The more vapor there was present, however, the more rapidly molecules returned to the liquid state. Finally there was so much vapor present that the rate of condensation was equal to the rate of evaporation. The vapor and the liquid were then in equilibrium. The space above the ether could contain no more ether vapor.

When a liquid is in equilibrium with its vapor it is said to be **saturated.** The pressure the vapor exerts on the walls of the container is called the **saturated vapor pressure.** In other words:

The saturated vapor pressure of a liquid is the pressure of the vapor when vapor and liquid are in equilibrium.

If we raise the temperature of the liquid, the molecules in the liquid move faster, and more of them can move fast enough to escape. This means that the liquid evaporates more rapidly, and more vapor (greater vapor pressure) must be present before the rate of condensation becomes great enough to equal the new, increased rate of evaporation and produce equilibrium. So we see that the saturated vapor pressure of a liquid increases with temperature. The relation between saturated vapor pressure and

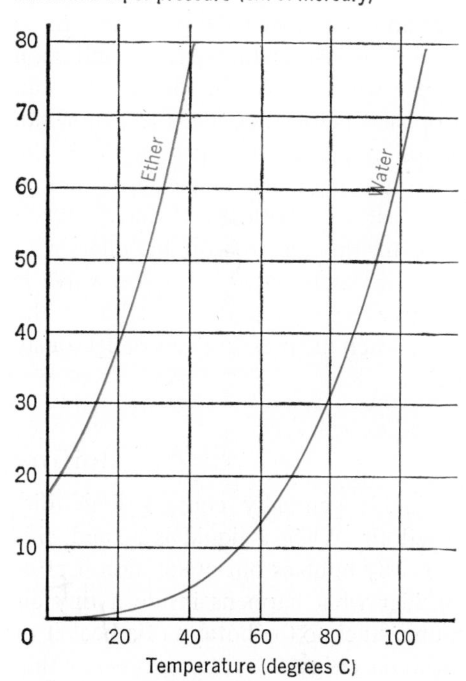

19-7 The saturated vapor pressure of liquids rises at an ever-increasing rate as their temperature increases.

temperature can best be shown by a graph (Fig. 19-7). Here we see the saturated vapor-pressure curves of both water and ether. You will note two things that will be of interest to us as we examine the process of boiling in detail: (1) the saturated vapor pressure of water at 100°C is 76 centimeters of mercury (normal atmospheric pressure) and (2) water has a measurable vapor pressure at 0°C (its freezing point).

12. Some effects of vapor pressure

Have you ever tried to heat a can of beans over an open flame without first opening the can? Don't try it! It is very dangerous. As the temperature rises, the vapor pressure inside the can also rises. At 100°C it is 1 atmosphere, and, as you can see from Fig. 19-7, the pressure rises at an ever-increasing rate as the temperature goes up. At 200°C (easily reached over an open flame) the pressure will be more than just 2 atmospheres, and soon the can will explode. Of course if the can is opened before heating, the vapor can escape as fast as it is produced, so the pressure cannot rise dangerously high.

Disastrous explosions have taken place, for the same reason, in boilers with defective safety valves (a safety valve is a device set to release steam from the boiler when the pressure gets dangerously high).

Boiling

Liquids generally contain some dissolved air. When a liquid is heated, this air slowly bubbles out of solution. Let us consider what happens inside a tiny air bubble stuck to the bottom of a beaker of hot water (Fig. 19-8). Let us assume that the temperature is about 70°C. Water evaporates into the bubble until the air in the bubble is saturated with water vapor. This raises the pressure in the bubble so that it expands somewhat until the total pressure it exerts outward equals the total pressure (of atmosphere and water) acting inward upon it. If the water temperature is increased to 85°C, the bubble expands some more until the total pressure inside it is again equal to the pressure exerted by the atmosphere and water above it. But if the temperature rises to just above 100°C, the pressure inside the bubble becomes greater than atmospheric pressure (Fig. 19-8) and will remain so, no matter how much more the bubble expands. So at 100°C bubbles of water vapor grow very rapidly and rise to the surface. The water is now boiling. If we continue to supply heat to the water, more bubbles will form and grow. The heat we supply goes into changing the state of the water and not into raising its temperature. This is why the temperature of boiling water remains at 100°C until all the water boils away.

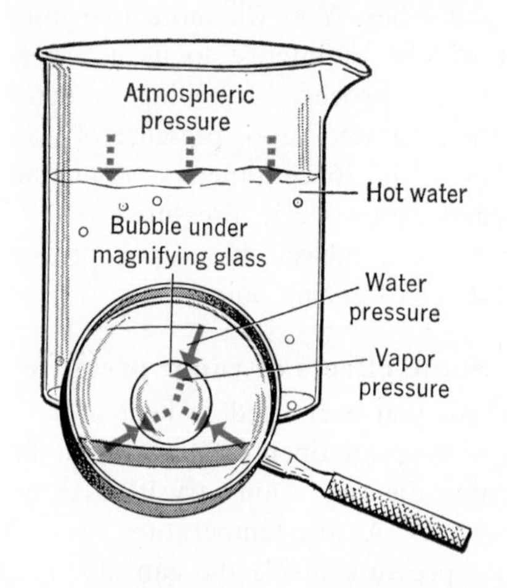

19-8 **Water boils when its saturated vapor pressure equals the pressure on its surface. When this happens, internal evaporation causes the tiny bubbles to grow rapidly.**

From the above discussion of the process of boiling we can now give a precise definition of the **boiling point.**

The boiling point of a liquid is the temperature at which the saturated vapor pressure is equal to the pressure on the surface of the liquid.

From the vapor-pressure curve of water (Fig. 19-7) we can see that if the pressure on the surface of the water is raised, the boiling point will be raised. Conversely, a lowering of the surface pressure reduces the boiling point. Pure water boils at 100°C when the air pressure is 760 millimeters of mercury (normal sea-level atmospheric pressure).

13. Reducing air pressure lowers the boiling point

Let us show by an experiment that reducing air pressure lowers the boiling point.

Suppose we put some liquid ether in a flask connected to a vacuum pump, and begin pumping out the air (Fig. 19-9A).

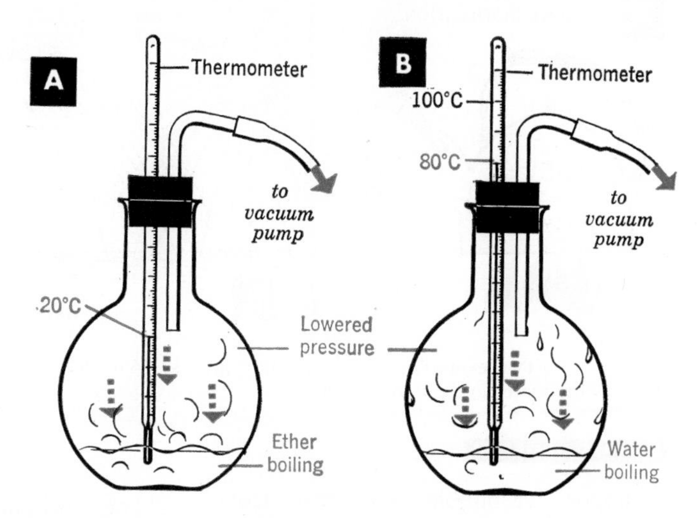

19-9 What's the boiling temperature of water? of ether? It all depends—if you lower the air pressure sufficiently, the vapor pressure of ether can be made equal to the lowered air pressure, and it boils by itself at room temperature! In the same way water can be made to boil at temperatures far below 80°C.

Almost immediately the ether begins to boil. Here it is true that we have not added any heat. We have simply reduced the pressure of air on the liquid. As a result, the vapor pressure of the ether at room temperature is soon equaled by the reduced air pressure, and the ether boils. In other words, we have lowered the boiling point by lowering the air pressure.

We can lower the boiling point of water, just as we did for ether, by lowering the air pressure on its surface. In order to do this, we use a flame to boil some water in a flask. When we remove the flame, the water stops boiling. But by pumping out the air above it (Fig. 19-9B) we can again bring the water to a boil. Thermometer readings show us that under reduced pressure water will boil at temperatures far below the normal boiling point. For instance, at the top of Pike's Peak (about 14,000 feet above sea level), air pressure is only a little more than half what it is at sea level; at this altitude water will boil at 186°F instead of its normal (sea-level) boiling point of 212°F.

14. The pressure cooker

If we were to raise air pressure above normal, it would take a higher temperature to make water boil. Raising cooking temperature by raising the air pressure is the principle applied in a pressure cooker. The pot is made of strong aluminum or steel and is sealed so tightly that steam cannot escape. In a pressure cooker water may be made to boil at a temperature of 250°F. At such a high temperature, the food in the container can be cooked in about one third the usual time.

15. Distillation

The facts about boiling are put to important use in the process of distillation pictured in Fig. 19-10A. Water can be obtained in pure form (distilled water) by boiling it, then cooling and condensing it in a separate container. Nature's "water cycle," which gives us rain, is an example of distillation (Fig. 19-10B).

16. Measuring heat of vaporization

As we have seen on page 252, 540 calories of heat are needed to change 1 gram of water into water vapor at 100°C without any change in temperature. How can we measure this heat of vaporization? It can be done with a calorimeter in the same way that we measured the heat of fusion—by the method of mixtures.

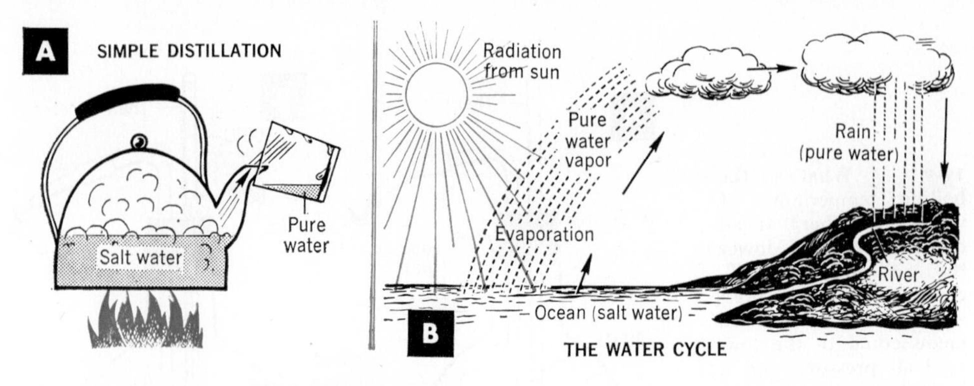

19-10 Two examples of the distilling process. What becomes of the salt? Of what value to us is the water cycle shown in B?

SAMPLE PROBLEM

Suppose we bubble some steam from a boiler into a calorimeter containing water at room temperature. Our data are:

Weight of steam	10.0 gm
Weight of calorimeter can	50.0 gm
Specific heat of aluminum calorimeter can	0.22
Weight of water	200 gm
Initial temperature of cold water	20.0°C
Final temperature of cold water and water from steam	48.1°C
Heat of vaporization of steam	L cal/gm

Assuming that the heat lost by the steam equals the heat gained by the calorimeter can and water:

Heat lost by condensing steam + Heat lost by water resulting from condensing steam = Heat gained by cold water + Heat gained by calorimeter can

$$10L + 10 \times 1 \times (100 - 48.1) = 200 \times 1 \times (48.1 - 20) + 50 \times 0.22 \times (48.1 - 20)$$

$$L = 541 \text{ cal/gm}$$

STUDENT PROBLEM

20 gm of steam are bubbled into 200 gm of water at 10°C contained in a 100-gm aluminum calorimeter. The final temperature is 62°C. Calculate the heat of vaporization of water. (ANS. 539 cal/gm)

Refrigeration

The mechanical refrigerator (Fig. 19-11) offers a most important application of the cooling effect of vaporization.

17. A refrigerator

In Fig. 19-11 the tank, T, contains a volatile liquid called the **refrigerant**, which evaporates easily. As the refrigerant evaporates through the small hole at A it takes heat from its surroundings, which become cooler. If the evaporating refrigerant surrounds a freezing compartment containing water (B in Fig. 19-11), the water loses heat to the evaporating refrigerant and cools off. But it would be wasteful—and possibly dangerous—to let refrigerant escape into the air. To avoid this, we must get the refrigerant gas back to tank T while changing it back into a liquid again.

We begin this change from gas back to liquid by means of pressure. Compression brings the molecules of a gas closer together so that their cohesive attraction for one another is greater. As soon as pressure has increased enough, the cohesive attraction causes the gas to begin to liquefy. In the refrigerator, a high-compression pump at C (Fig. 19-11) compresses the refrigerant gas back into a liquid state after it has done its cooling.

But compression not only brings the molecules of a substance closer together;

it also speeds them up. This increased speed makes the gas hotter. To counteract the heating effect of compression, the gas is passed through the coils of a small radiator, D, called the **condenser,** and cooled by a flow of air from a fan, F. This cooling, combined with pressure, brings the gas back into a liquid state at D. As the gas liquefies, the latent heat it took from the water in the ice-cube tray is given up to the air flowing past the condenser and transferred to the outside air by means of the fan and condenser. Now the pump forces the liquefied refrigerant into the tank, T, where it originally started, and the cooling process begins all over again.

Unless there were some boundary line between the high-pressure and low-pressure sections of the piping, the liquid would simply be pumped around continuously without ever evaporating. So the tube is narrowed to a pinhole at A. This opening is called the expansion jet. It permits the liquid to spray out into the low-pressure side of the tube connected to the pump, C. This lower pressure makes it easier for the liquid to evaporate.

18. Choosing a refrigerant

In selecting the refrigerant, we want a gas that can be easily liquefied by the combination of pressure and cooling we have just described. There are many materials far more suitable than ether for this purpose. Ammonia is excellent, because it is a gas at room temperature but liquefies under moderate pressure. Sulfur dioxide has the same qualities. But both these gases, although they are often used, share a disadvantage: they are poison-

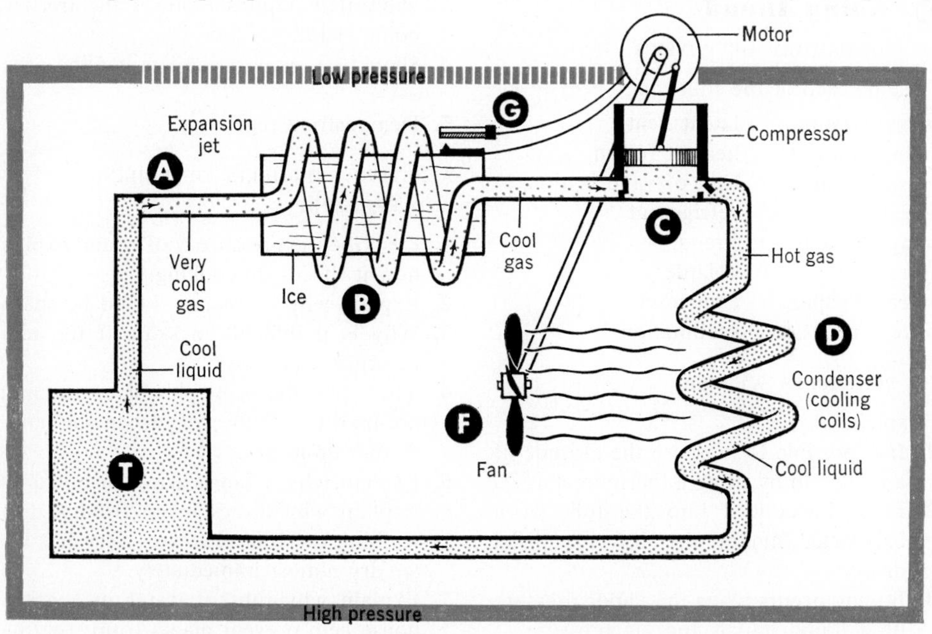

19-11 The refrigerator uses evaporation to cause cooling. A pump at C creates low pressure at A and causes the liquid in T to evaporate. The evaporation causes cooling and the cold vapor flowing in pipes around trays of water can freeze the water in B. Now the vapor is compressed once more and driven onward by the pump at C. The compression heats the gas, which is then cooled by the fan at F. The combination of pressure and cooling changes the gas back to liquid in D. The motor operates the pump and fan. It is controlled by the thermostat at G. The narrow expansion jet at A serves as a boundary between the high pressure in T and the low pressure in A.

ous and can cause trouble if they escape through leakage. Hence most modern refrigerators use a material known as **Freon.** This gas liquefies easily under pressure, has a high heat of vaporization, and is not poisonous.

The refrigerating processes described in this chapter have been used to produce temperatures so low that they approach absolute zero within a fraction of a degree. In one interesting experiment at low temperatures, pieces of solid air at 340°F below zero were dropped into liquid helium at 452°F below zero. Since the cold solid air had more heat than the liquid helium, it supplied enough heat to cause the helium to boil!

Going Ahead

THE VOCABULARY OF PHYSICS

Define each of the following:

vapor pressure	latent heat
boiling point	heat of fusion
pressure cooker	heat of vaporization
regelation	refrigerant
change of state	Freon
saturated	volatile
saturated vapor pressure	condenser
	sublimation

THE PRINCIPLES WE USE

Explain why

1. It is possible to measure the altitude of a mountain by using a thermometer.
2. Ether introduced into the tube of a barometer makes the mercury level drop.
3. Boiling occurs when the vapor pressure of a liquid equals the air pressure.
4. Lowering the pressure on a liquid lowers the boiling point.
5. Hard-boiled eggs take longer to prepare at the top of Pike's Peak.
6. The ocean does not freeze at 32°F.
7. Pressure can cause ice to melt.
8. A glacier can flow around a bend.
9. Heat seems to disappear when a solid is melted or a liquid is boiled.
10. Lakes take a long time to freeze or to thaw.
11. It takes much longer to boil off the water in a pan than to start it boiling.
12. A burn by live steam is more severe than one from boiling water.
13. Evaporation is a cooling process.
14. Pressure helps liquefy a gas.
15. Increasing pressure causes a gas to heat up.
16. Evaporation can be increased by heat, by circulation of air, by lowering the pressure, and by increasing the surface.

EXPERIMENTS WITH HEATING AND COOLING

Describe an experiment to

1. Measure the vapor pressure of a liquid.
2. Show that lowering the pressure lowers the boiling point.
3. Show that impurities lower the freezing point of ice.
4. Show that evaporation is a cooling process.
5. Demonstrate regelation.

APPLYING PRINCIPLES OF BOILING AND FREEZING

1. Describe the pressure cooker and explain how it speeds up cooking.
2. Explain why ice skates should be sharp.
3. Why is it difficult to skate if the temperature is too low?
4. Show how the principles of evaporation are used by a housewife when she hangs clothes up to dry.
5. Explain why a fan is used on hot days.
6. Explain why dishes taken from a dishwashing machine that uses very hot water dry almost immediately.
7. Explain why tubs of water in a greenhouse help prevent plants from freezing.
8. Without looking at Fig. 19-11, draw a simple labeled diagram of a refrigerator and explain how it causes the water in the ice tray to freeze. Explain the purpose of pump, motor, condenser, expansion jet, refrigerant, and thermostat. Label the hottest part and the coldest part.

SOLVING PROBLEMS OF TEMPERATURE CHANGES

1. How much heat is needed to melt 60 gm of ice?
2. How much heat is needed to melt 30 lb of ice?
3. What is the heat of fusion of a substance if 150 gm of it absorbs 3,000 cal in melting?
4. How much heat is needed to change 20 gm of water at 100°C to steam at 100°C?
5. 3,000 cal of heat are released by a certain mass of steam in condensing. How much steam condensed?

---

6. How much heat is required to change 20 gm of ice at −30°C to water at 0°C?
7. How much heat is needed to change 45 gm of ice at −30°C to steam at 200°C?
8. How much water at 20°C must be slowly poured over a large cake of ice at 0°C to melt 3 gm of ice?
9. How much heat is released in a room per hour if 10 lb of steam condense in the radiators and cool to 30°C every hour?
10. What is the boiling point of water when the pressure on its surface is (a) 38 cm of mercury, (b) $\frac{1}{4}$ atmospheric pressure, (c) 5 mm of mercury?
11. From the following data find the heat of fusion of water when ice is dropped into warm water in a calorimeter:

Weight of copper calorimeter	90 gm
Weight of calorimeter and warm water	400 gm
Temperature of warm water	30°C
Temperature of calorimeter and water when ice is melted	10°C
Weight of calorimeter and cold water at end of the experiment	471 gm

12. From the following data find the heat of vaporization of water when steam is condensed in a calorimeter containing cold water:

Weight of copper calorimeter	80 gm
Weight of calorimeter and cold water	500 gm
Temperature of cold water	2°C
Final temperature of mixture	42°C
Weight of calorimeter and mixture	528.8 gm

13. Approximately how much will a pan that contains 2,000 gm of water cool if 3 gm of water evaporate? Assume no heat is lost by conduction, convection, or radiation.
14. 15 gm of ice are placed in a 50-gm aluminum calorimeter containing 150 gm of water at 20°C. What is the final temperature?
15. How many grams of ice must be put into an 80-gm copper calorimeter containing 250 gm of water at 60°C in order to lower the temperature to 10°C?
16. How many grams of steam must be bubbled into 5 kg of water to raise the temperature from 20°C to 80°C?
17. 60 gm of ice are mixed with 200 gm of water at 20°C. What is the final temperature?

---

18. 10 gm of steam at 150°C are bubbled into a 50-gm aluminum calorimeter containing 200 gm of water and 40 gm of ice. What is the final temperature?
19. A steam heating plant with an efficiency of 60% burns 50 kg of coal per hour. Water is fed to the boilers at a temperature of 20°C. How much steam at 100°C is produced per hour?
20. How fast must a lead bullet travel if it is to be completely melted when it hits a wall? Assume that half of its energy is converted into heat, and that its temperature before collision is 20°C. (Melting point of lead = 326°C; heat of fusion of lead = 5.6 cal/gm; specific heat of lead = 0.031.)
21. How much snow is melted when a 300-lb sled is dragged 200 ft across snow at 0°C if the coefficient of friction is 0.2?

CHAPTER

20

Humidity and Weather

A complete description of the weather at any given time and place must contain details about wind, about temperature, and about the condition of moisture in the air (water vapor, clouds, fog, rain, snow, hail, and sleet). In this chapter we shall consider mainly the water content of the air.

Humidity

To understand why clouds, fog, and other forms of water appear in the air, we must first understand humidity.

1. Absolute humidity and saturation

The atmosphere is never completely dry, even in the Sahara Desert. There is always some invisible water vapor present. What do you think would be the simplest way to describe the amount of water vapor in the air? Since water vapor is a gas, its density is very low, and we describe the air's water vapor content as the number of grams of water vapor in a cubic meter of air (gm/m^3). This is called the **absolute humidity.**

Is there any limit to the amount of water vapor the air can hold? You will remember from the previous chapter that water can be evaporated into a given space (whether air is present or not) until the rate of condensation equals the rate of evaporation. Then no more water can be evaporated, and we say the space is **saturated.** In other words, there is an upper limit to the absolute humidity of the air. This limit—the maximum amount of water vapor the air can hold—increases with temperature. Thus the warmer the air, the higher its absolute humidity can be. Table 10 gives the absolute humidity of *saturated* air at different temperatures.

2. Relative humidity

Air at a given temperature can have any value of absolute humidity up to the maximum value (the saturation value) shown in the table. Usually the amount of water vapor in the air is less than the maximum. If the air contains 80 per cent of the maximum amount it can hold, we say the **relative humidity** is 80 per cent. In other words:

The relative humidity of air at a given temperature is the ratio of the amount of water vapor in the air to the maximum amount the air can hold at the given temperature.

$$\text{Relative humidity} = \frac{\text{Actual absolute humidity of the air}}{\text{Absolute humidity of saturated air}}$$

SAMPLE PROBLEM

Suppose the absolute humidity of the air is 13 gm/m^3 on a day when the temperature is 70°F. What is the relative humidity? The absolute humidity of the air at 70°F is found from Table 10 to be 18.2 g/m^3. Hence

$$\text{Relative humidity} = \frac{13.0}{18.2} = 0.715$$
$$= 71.5\%$$

STUDENT PROBLEM

If there are 4 gm of water vapor in a cubic meter of air at 66°F, what is the relative humidity? (ANS. 25%)

3. Measuring relative humidity

Measure the temperature of your room with a thermometer. Now put a wad of cotton around the bulb of the thermometer and wet it with water at room temperature. As the water on the cotton starts to evaporate, the thermometer cools. Do you remember why?

If the air in the room were saturated, there would be no evaporation and the thermometer would not show any drop in temperature. If the humidity were not at the saturation point but were still high, the rate of evaporation of water from the cotton would be slight and the temperature would drop only a few degrees. On the other hand, if the air were dry, evaporation would be rapid and there would be a considerable temperature drop. Thus, by using a thermometer with a wet bulb, you can determine whether the air is saturated, moist, or dry by the drop in temperature.

TABLE **10**
ABSOLUTE HUMIDITY OF SATURATED AIR

Temperature (°F)	Absolute Humidity (gm/m³)	Temperature (°F)	Absolute Humidity (gm/m³)
26	3.7	60	13.0
28	4.0	62	14.0
30	4.4	64	15.0
32	4.8	66	16.0
34	5.2	68	17.1
36	5.6	70	18.2
38	6.0	72	19.4
40	6.4	74	20.6
42	6.9	76	22.0
44	7.5	78	23.5
46	8.1	80	25.0
48	8.7	82	26.5
50	9.3	84	28.2
52	10.0	86	30.0
54	10.7	88	32.0
56	11.4	90	34.0
58	12.2	92	36.0

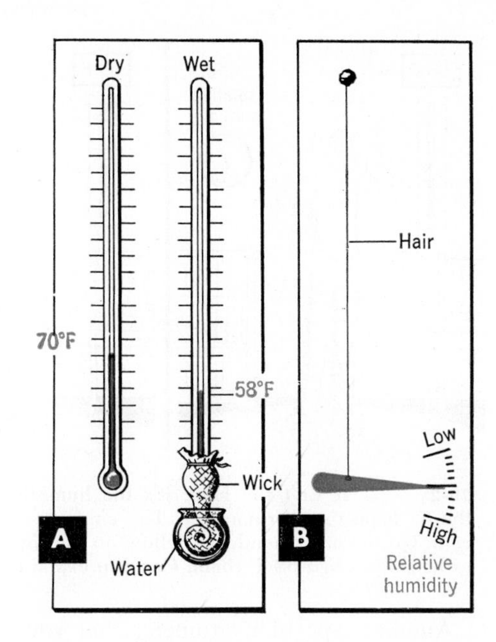

20-1 Two ways of measuring humidity.

An instrument which measures relative humidity is called a **hygrometer** (hī-grŏm′ē·tẽr). One type of hygrometer, called the wet-and-dry-bulb hygrometer, uses a dry-bulb thermometer next to a wet-bulb thermometer. The wet-bulb readings drop most in dry rooms and least in moist rooms (Fig. 20-1A). In another type the wet and dry thermometers are whirled around by a handle. The whirling creates a breeze which speeds up evaporation and cuts down the time for the temperature change to occur. This particular type of hygrometer is called a **psychrometer** (sī·krŏm′ē·tĕr).

Both hygrometers must be used in conjunction with humidity tables. These tables are based on careful measurements made under controlled conditions. (See page 680.)

20-2 "It isn't the heat, it's the humidity." So says the boy on the left, as he looks at the 68°F on the thermometer. The air feels warm and sticky because the humidity is 90 per cent. On the other hand, the fellow on the right says, "It isn't the cold, it's the low humidity," as he shivers in a 68°F room. Can you explain why?

Another type of hygrometer that you can make (Fig. 20-1B) makes use of the fact that a human hair gets longer when the air is moist and contracts when it is dry. In this hygrometer the expansion or contraction of a hair with changes in relative humidity causes a pointer to move along a dial. The dial has been marked off so that relative humidity readings can be obtained directly in per cent.

4. Relative humidity and comfort

We are usually more interested in the relative humidity of the air than we are in the absolute humidity, since the rate of evaporation of water depends directly on the relative humidity. The closer the air is to saturation (100% relative humidity), the nearer the rate of condensation of water vapor is to the rate of evaporation, and so water evaporates more slowly. Hence evaporation slows up when the relative humidity is high and increases when the relative humidity is low.

Evaporation, as we know, is a cooling process, and we feel cooler when the humidity is low enough to allow perspiration to evaporate freely. We can walk into a room where the temperature is 68°F and the humidity very low and feel chilly. In another room at the same temperature, where humidity is high, we can feel too warm because we are not being cooled by evaporation. (Fig. 20-2.)

Have you ever wondered how people can stay alive at temperatures of 120°F or more in desert regions? Air in the desert is dry, and the rapid evaporation of perspiration into the dry air can cool the body temperature to normal. The man in Fig. 20-3 in the special hot-air tank is still alive at a temperature of 160°F!

We are most comfortable when the temperature is around 68°F and the humidity about 50 per cent. For perfect air conditioning, these conditions should be maintained all the time—summer and winter.

The cold air of winter contains little moisture; consequently, when this air is heated in the house, its relative humidity drops to a very low value. We may then feel chilly at a temperature of 70°F, and have to adjust the thermostat to 75°. But this dry air is not healthful because it tends to increase the evaporation of moisture from nose and throat passages. We need this moisture to ward off colds.

To add moisture to the air and avoid unhealthful dryness, you may put pans of water on the radiators. But pans are not very effective, since the average room requires the evaporation of several quarts of water before the humidity is improved. In some air-conditioning systems, the air is blown through a fine spray of water to increase the humidity before it reaches the rooms of the house.

In summer the problem is to get the excess moisture out of the air. This can be done by passing the humid air over cold pipes which condense some of the moisture out of it.

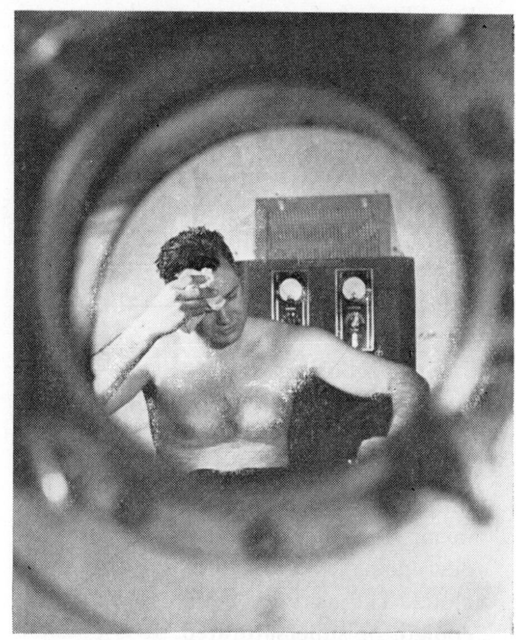

Kurt Severin from Black Star

20-3 Still alive at 160°F! This hothouse experimental tank is used to test the effects of heat and humidity on men and equipment. Some men have withstood air temperatures in this tank above 200°F. The cooling effect of evaporation of perspiration into dry air enables these men to live.

5. The dew point

What happens to saturated air (100% relative humidity) if it is cooled? From Table 10, on page 263, we see that as the temperature of air is lowered, it can hold less and less water vapor. Hence the cooling of saturated air results in its condensation as dew, frost, or fog. The more the cooling, the greater the condensation.

Consider what happens to unsaturated air when we cool it. Consider air at 70°F whose relative humidity is 60 per cent. From Table 10 we see its absolute humidity *if saturated* would be 18.2 gm/m^3, and hence its absolute humidity when only 60% saturated is $0.6 \times 18.2 = 10.9$ gm/m^3. As we cool it, the amount of water vapor remains constant at 10.9 grams while the capacity drops to 16.0 grams. Thus the relative humidity rises to $\frac{10.9}{16.0} = 0.681 = 68.1\%$. Eventually, when the air is cooled to 54°F, the air becomes saturated, since from the table we see that air at this temperature can hold only 10.7 grams of vapor per cubic meter of air (the approximate absolute humidity of our air). 54°F is the **dew point** of our air.

The dew point is the temperature to which a given sample of air must be cooled to become saturated.

If the air is cooled below this temperature, condensation (as dew, frost, fog, or clouds) is produced.

SAMPLE PROBLEM

Suppose we have some air at 38°F whose relative humidity is 66.7%. What is its dew point?

We find from Table 10 that its absolute humidity will be $0.667 \times 6.0 = 4.0$ gm/m^3. Now according to the table, 4 grams of moisture will saturate a cubic meter of air at 28°F. Hence 28°F is the dew point for air that has an absolute humidity of 4 grams per cubic meter. This is below freezing, and if condensation occurs it will be in the form of ice crystals or frost.

STUDENT PROBLEM

What is the dew point of air at 58°F if its relative humidity is 52.5%? (ANS. 40°F)

---

6. Supercooling

For water vapor to condense into droplets, there must be small dust particles or ice particles around which the drops can form. If these are not present in air, then it can be cooled below the dew point without forming dew or a cloud. The air is then said to be **supercooled,** or **supersaturated.** It is possible to "trigger" precipitation from this supercooled air by seeding it with the necessary particles.

In 1947 Dr. Vincent Schaefer performed a historic experiment in which he made many tons of snow fall from a supercooled cloud by seeding it with particles of dry ice from an airplane. The seeding resulted in snow instead of rain because the temperature of the air at the time was below freezing. In 1950 during New York City's water shortage and in 1957 during Boston's shortage, scientists were called in to try to increase the rainfall over the watershed areas by such a seeding method.

7. Measuring the dew point

You have probably noticed how the outside of a pitcher of ice water or a cold pipe is often covered with moisture—particularly if the day is warm and humid. The pitcher or pipe has cooled the nearby air below the dew point, the temperature at which the air is saturated, and water vapor in the air has condensed into drops of water. The same thing happens when the inside surface of windows becomes covered with water or ice on a cold day. We can observe it again in the refrigerator when frost forms on the ice trays. All these observations provide us with a clue to designing an instrument for measuring the dew point.

Fill a shiny metal can with water. Now lower the temperature of the can gradually by putting ice in it. By means of a thermometer, keep track of the temperature of the water. A temperature is finally reached at which water droplets begin to appear on the outside of the can. This temperature is the dew point. A thermometer reading taken at the moment the outside of the can begins to cloud over is the dew-point temperature you are looking for. For greater accuracy you might permit the can to warm up slowly and notice the temperature at which this outside moisture disappears. The dew point is the average of your two readings.

It is such information that enables the meteorologist to predict how much the air will have to cool to form clouds and rain. From other measurements he can also judge whether the necessary degree of cooling is likely to take place. He then makes his predictions as to the possibility of rain or snow.

Clouds and rain

We have seen that water vapor in the air can be made to condense into dew, frost, or fog by cooling the air below the dew point. Accordingly, we would expect clouds to form if the air is cooled sufficiently.

We have also seen (Chapter 18) that when water is heated, it evaporates and changes into water vapor. Therefore, a cloud—which consists of tiny droplets of water floating in the air—could be made to disappear by heating.

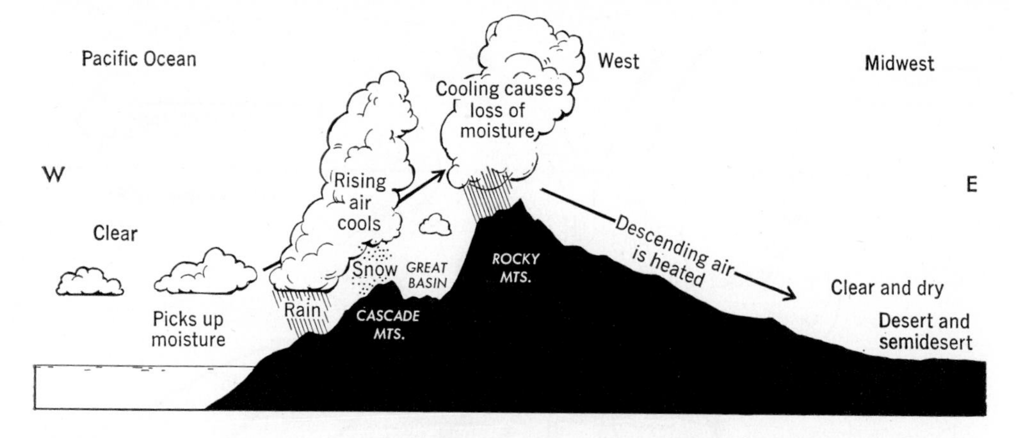

20-4 Geography plays an important part in the climate of an area. The cooling by expansion of moist ocean air as it rises up the West Coast mountains of the United States causes rain and fertile valleys on the coast. But the heating by compression of the descending air on the other side creates the dry and desert regions of Nevada and Arizona.

Anything which cools the air tends to result in clouds and possibly rain or snow. Thus rainy weather is caused by a cooling of the air. Anything which warms the air tends to make clouds disappear. Hence fair weather tends to result from a warming of the air.

Let us consider here what causes the air to cool and thus bring clouds and rain.

8. Rising air

In summer, people who live in hot, "steamy" cities can, if they are lucky, escape the heat by going to the mountains. The higher the mountains, the cooler the air becomes. On the average there is a temperature drop of 3.3°F for every 1,000 feet of altitude. Thus at the top of a 10,000-foot peak, the temperature would average about 33°F lower than at sea level. There might be snow at such a height even in summertime. If the air were actually rising up the mountainside, the cooling would be even greater because of the expansion of air due to the lower air pressure found at high altitudes. In that case the cooling might amount to as much as 5.5°F for every 1,000 feet of rise.

So we see that if air can rise to a great enough height, it will be cooled to its dew point by **expansion.** Condensation then occurs and may produce rain.

What causes air to rise? Figure 20-4 shows us one cause. When moist air from the Pacific Ocean strikes the mountain ranges on our western coast, the air is pushed upward. As it ascends on the coast side of the mountains, the air cools below the dew point and produces heavy rainfall.

But now the air passes over to the other side of the coastal range. It has already lost a great deal of its moisture through precipitation. Furthermore, the air has warmed up as a result of the heat released by condensation, and as it descends to lower and warmer altitudes, increasing air pressure warms it still more. This heat evaporates whatever water droplets were left in the air after it crossed the mountains. Rain is therefore not likely. This pattern accounts for the deserts and parched areas on the Nevada side of the Sierra Nevada range.

9. Thunderstorms

Convection currents can also bring about the process of rising, cooling, and condensing. For instance, on a hot summer day convection currents (see Chap-

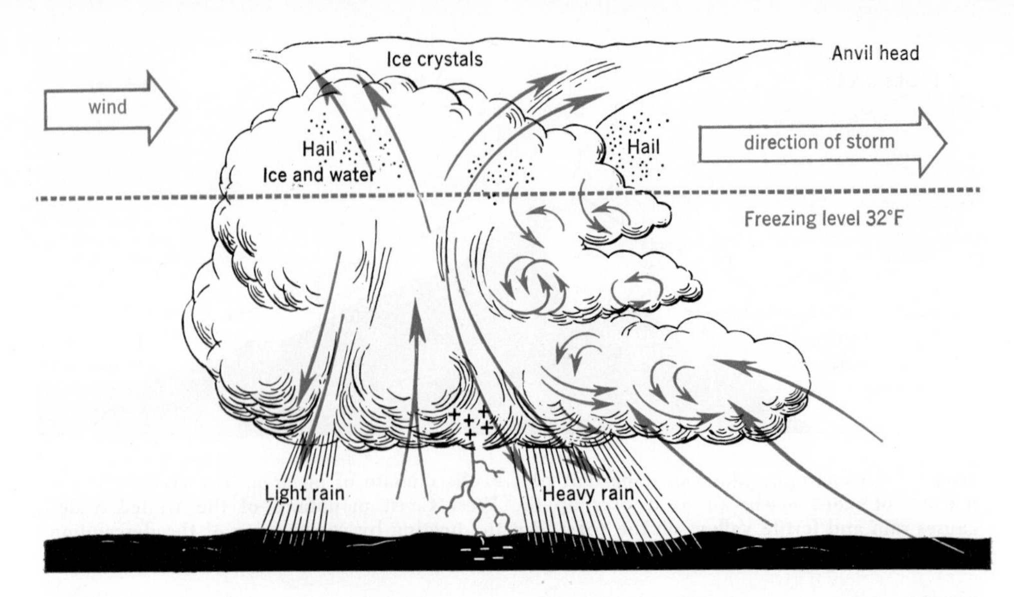

20-5 On a hot summer afternoon, rising convection currents cause the air to cool and condense. Some of the rising droplets freeze, fall and gather moisture, rise in another upward current, and repeat to grow into hailstones. Finally the hailstones fall, along with torrents of condensed water in the familiar thunderstorm.

ter 16) may cause the air to rise as high as 20,000 feet (Fig. 20-5). Such a rise might cool the air by as much as 110°F (5.5×20)! Most of the water vapor in the air would certainly condense as the temperature dropped far below the dew point.

Convection currents of this sort cause thunderstorms. The condensation of the water in the air heats the air, which makes the upward convection currents become more powerful. Water condenses still more rapidly, and soon torrents of rain are falling. This process is pictured in Fig. 20-5. Sometimes the convection currents in such storms are so violent that they can break an airplane in two in mid-air.

10. Vertical "hurricanes"

The velocity of the updrafts in a thunderstorm have been measured by observing the size of hailstones. The larger a hailstone, the greater its terminal velocity (page 119). As a hailstone forms, it falls but is caught by the updrafts in a cloud which carries it upward again where another layer of moisture freezes on the outside of it. The speed of the updrafts in a cloud must be slightly greater than a hailstone's terminal velocity, in order to lift it above the freezing level to freeze on its last layer of ice (Fig. 20-5). The terminal velocity of a hailstone can be determined from its weight and density (by a formula known as Stokes' Law). The largest stone on record was 6 inches long; computations show that its terminal velocity, and therefore the vertical speed of convection currents in the storm from which it fell, must have been in excess of 200 miles per hour!

11. Fronts

Most rain and snow in temperate climates result from the lifting and cooling of air along **fronts.** A front is the dividing surface between two large masses of air of different temperatures, called **warm** and **cold air masses.** Warm air masses usually originate in warm latitudes, and cold air masses come from Arctic regions. Air masses may be several thousand miles in diameter; they move slowly across the earth's surface from their place of origin. When a warm air mass is riding up over and pushing a cold air mass before it, the boundary between the two is called a **warm front.** When a cold air mass is pushing under a warm air mass as it drives the warm air before it, a **cold front** results. Since cold air is more dense than warm air, the warm air is lifted and expands and cools as it rises, producing rain if the humidity of the warm air is great enough.

12. The future of weather control

You can see how our understanding of the "forces of nature" has made weather predicting more reliable. Actually the air ocean is simply a gigantic "humidity box"; air is cooled and heated in this box according to the same principles we have studied in this chapter. How can man not only predict the weather but also control it? What is needed is more complete knowledge of the actual condition of the air at every moment. In the meantime, experiments like Dr. Schaefer's "triggered" snowfall add to our knowledge of weather control. It is possible that such experiments may one day lead to much greater control over the weather than we now have.

Going Ahead

THE VOCABULARY OF PHYSICS

Define each of the following:

relative humidity	psychrometer
dew point	front
absolute humidity	warm front
supercooled	cold front
hygrometer	

THE PRINCIPLES WE USE

Explain why

1. Water vapor in the air tends to condense when cooled.
2. Clouds tend to disappear when air is heated.
3. Air cools as it rises to high altitudes.
4. Air heats up as it descends from high to low altitudes.
5. Water evaporates in a closed box until saturation is reached.
6. Relative humidity increases as the temperature in a closed room decreases.
7. The dew point is low if the relative humidity is low.

APPLYING PRINCIPLES

1. Describe two different ways by which the water vapor in the air in the earth's atmosphere can be cooled and condensed to form clouds.
2. Describe and explain the weather that is likely to result from cold air passing over warm ground.
3. Describe and explain the weather that is likely to result from warm air passing over cold ground.
4. What kind of weather would you expect to find in the Appalachian Mountains as warm air from the Gulf of Mexico passes over the mountains? Explain your answer.
5. Why can you feel cold at one time when the air is 68°F and yet warm at another time at the same temperature?
6. How is it possible for people to live when the temperature of the air is greater than 98.6°F—the temperature of the body?
7. Explain Dr. Schaefer's method of producing rain or snow from a cloud.
8. Describe two ways of measuring the relative humidity of the air.
9. Describe a method of measuring the dew point.
10. How does the defroster of a car work?

SOLVING PROBLEMS OF HUMIDITY AND MOISTURE

1. What is the relative humidity of air at 68°F that contains 10.0 gm of water vapor per m^3?
2. How much water vapor is there in air at 80°F with a relative humidity of 34%?
3. How much water vapor is there in saturated air at 32°F?

4. What is the dew point of air at 68°F with a relative humidity of 43.9%?
5. What is the dew point of air with an absolute humidity of 4.2 gm/m³?
6. What is the dew point of air at 78°F whose relative humidity is 40.4%?
7. What is the relative humidity of air at 72°F if the wet-bulb is 5°F lower? (See page 680.)
8. What will be the wet-bulb temperature of air at 80°F with a relative humidity of 50%?

---

9. As air rises it expands and cools at a rate of 5.5°F for every 1,000-ft rise in altitude. At the same time the dew point drops 1°F for every 1,000-ft rise. If the temperature on the ground is 70°F and the dew point is 34°F, at what altitude will clouds form from the rising air?
10. How much heat is released by condensation when air at 80°F, relative humidity 70%, is cooled to 32°F?
11. Hurricanes get their energy from the evaporation of sea water as a result of the sun's heat. This latent heat of vaporization is later released as heat energy when the resulting water vapor condenses at high altitude. (a) How much energy (in foot-pounds) is fed to a hurricane as a result of the evaporation of 1 mm depth of water from a square meter of ocean surface? (b) Why do hurricanes rapidly die out when they move inland? (c) Why do hurricanes become weaker as they approach northern latitudes?

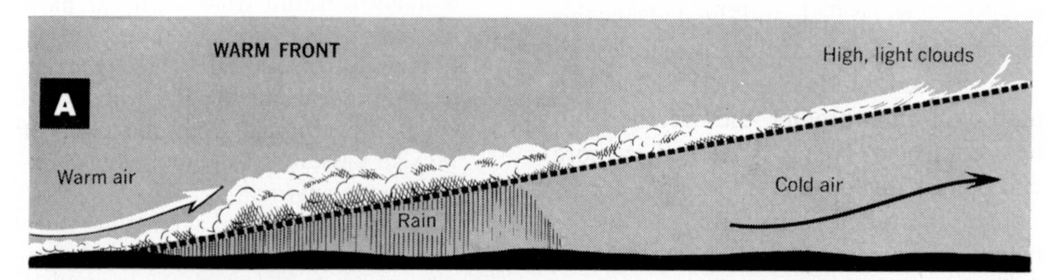

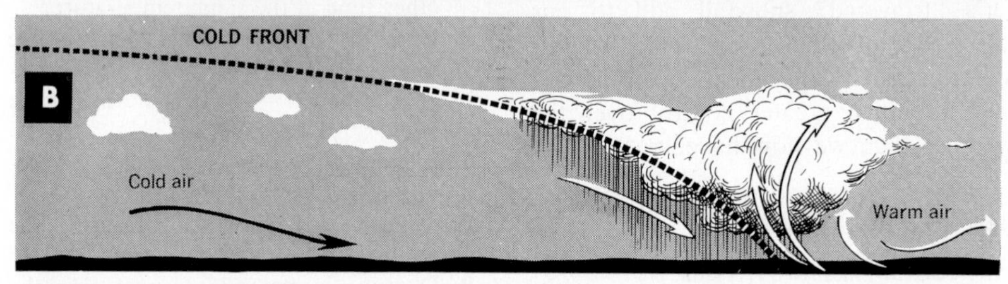

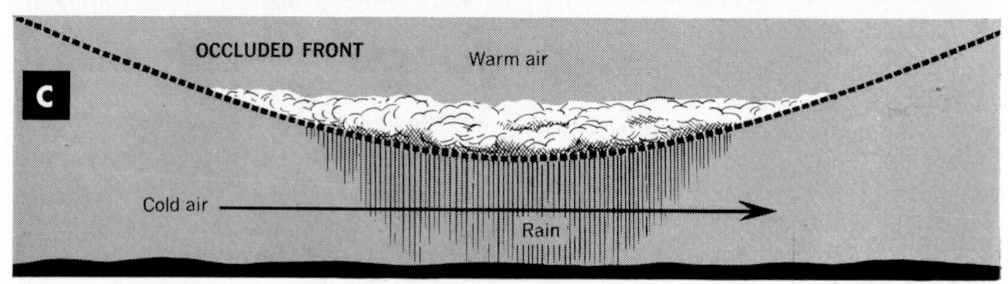

20-6 Most bad weather occurs along boundaries between cold and warm air masses known as "fronts." In A, light warm air rides up an incline of heavy cold air. The warm air is cooled, condenses, and forms clouds, and probably rain. In B the heavy cold air pushes under light warm air. The rising warm air cools, condenses, and forms thunderstorms. In C the cold air has lifted the warm air up, causing it to cool, condense, and fall as cold rain.

CHAPTER

21

Heat Engines

Today jet-propulsion and rocket engines rule the skies overhead. When do you suppose the first such engine was invented? 1932? 1920?

Actually the first such engine was invented 2,000 years ago, around 100 B.C., by Hero of Alexandria in Egypt. Hero's device consisted of a boiler in which steam was generated. The steam issued from a hollow sphere in two jets arranged like the curved arms of a lawn sprinkler (Fig. 21-1). The reaction to the motion of the steam as it shot out of the jets caused the sphere to turn. Here is true jet propulsion in operation. But Hero's engine was considered in his time to be mainly an interesting toy.

Early engines

Today not only do jet-propulsion and rocket motors run airplanes; there are also gasoline engines for cars and planes; diesel engines for trucks, boats, and trains; steam turbines to generate electricity and propel boats; and reciprocating steam engines to run boats and locomotives. But all of these engines make use of the same basic principle that operated Hero's toy. This is the principle: Heat gives increased motion to molecules and causes expansion of gases.

1. Watt's steam engine

The first usable engine, which made use of the heat obtained from burning fuel, was invented by an Englishman named Thomas Newcomen (nū'kô·mĕn) in 1705 (Fig. 21-2). It is interesting to notice why Newcomen's engine was never very efficient. The steam that drove the piston, P, upward (Fig. 21-2) is removed by condensing it *in the cylinder*. To do this the entire cylinder has to be cooled by water entering at B. Hence when steam is injected into the cold cylinder through valve A, much of it is uselessly condensed on the cool cylinder walls before they are heated again and steam pressure once more exerted. A better, more efficient engine was obviously needed.

About a half century after Newcomen's engine first appeared, it was great-

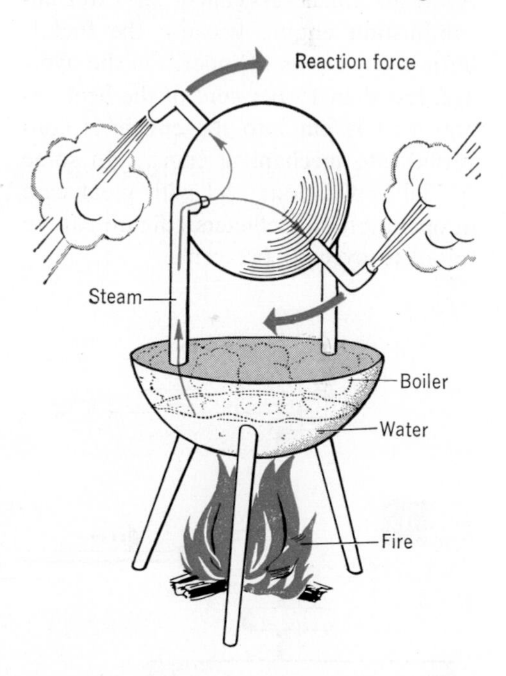

21-1 Jet propulsion in 100 B.C. This first heat engine was operated by a reaction force caused by a jet of steam. This engine is very similar to the lawn sprinkler, which rotates by reaction to a stream of water instead of steam.

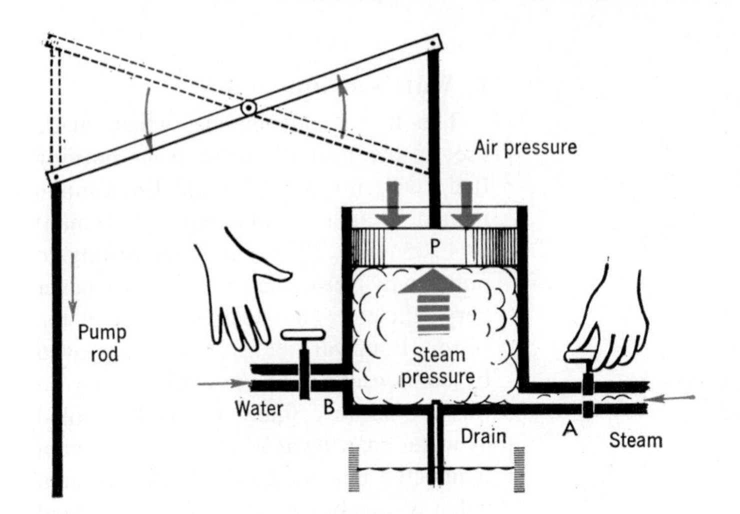

21-2 In the 1700's boys were hired at very low pay to turn the valves of this Newcomen steam engine. Steam pressure from A made the piston rise. Then the valve at A was shut, and the water valve at B was opened to condense the steam and make a vacuum. Air pressure above P then pushed the piston back to repeat the cycle.

ly improved by James Watt. Because Watt was the first to make the steam engine really practical, he is given the chief credit for its invention. Figure 21-3 shows how Watt's steam engine works. A steam engine is called an **external-combustion engine** because the fuel is burned outside its cylinder. On the average, less than 15 per cent of the heat energy that is put into the engine is converted into mechanical energy. In some special applications and with great care in operation, this efficiency figure can be raised to over 30 per cent.

Modern engines

One reason for the low efficiency of steam engines is that they burn their fuel outside the cylinder. Much of the heat escapes from the bulky equipment that produces the steam. But if the heat is created inside the cylinder, this cause of loss of energy is removed; we then have an **internal-combustion engine.** The gasoline engine is one familiar example.

2. The gasoline engine

In the gasoline engine the burning of fuel takes place inside the cylinder. Fig-

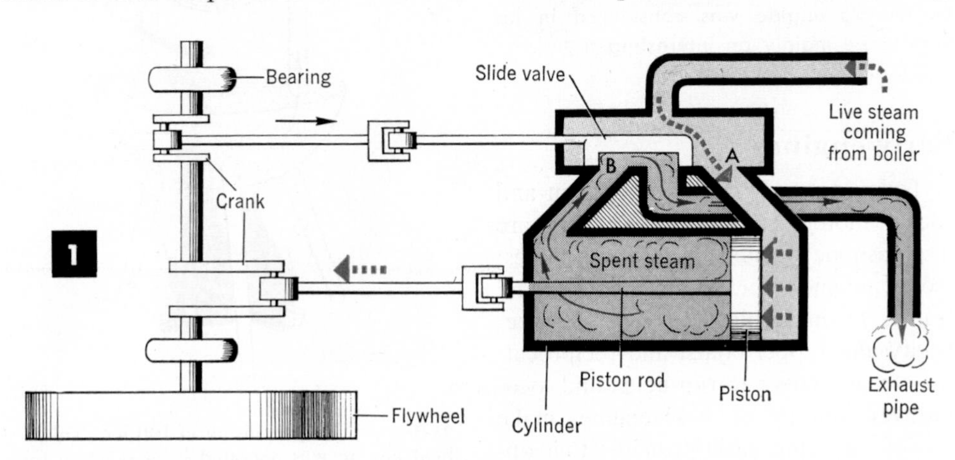

21-3 Watt gets the credit for inventing the steam engine because he made the valve operation automatic and thus created a practical engine. In diagram 1 steam from the boiler enters through port A and pushes the piston to the left. Spent steam from the previous cycle goes out the exhaust through port B. The moving piston pushes the rod and causes the crank to turn the heavy flywheel. The crank on the shaft of the flywhel now pushes the slide

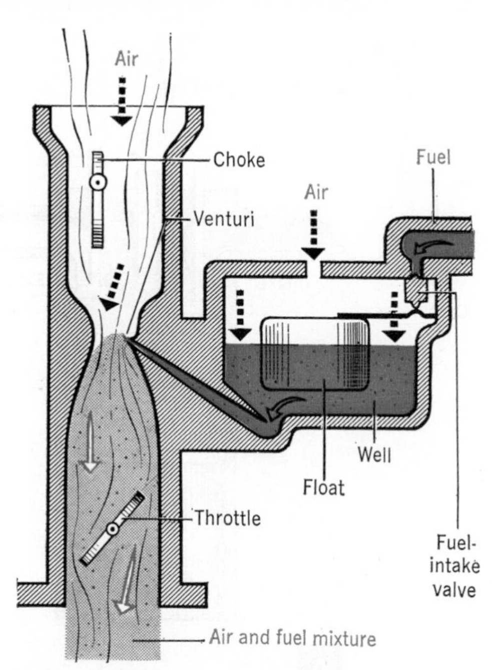

21-4 The carburetor is an essential part of the gasoline engine. Air rushing through the narrow "venturi" tube creates a low pressure (see p. 197), and air above the well forces gasoline to spray into the air stream, forming an explosive mixture which goes to the cylinders. The float keeps the gasoline in the well at the same level all the time. As gasoline leaves, the float drops, causes the fuel-intake valve to open, and lets more gasoline in. This raises the float and stops more gas from entering. Turning the throttle controls the amount of mixture that gets into the cylinders and thus controls the speed of the engine. For cold-weather starting, the choke is turned to choke off the air supply and get more gasoline into the cylinders.

ures 21-4 and 21-5 show how this engine operates. The explosion of the fuel produces hot, expanding gases which force the piston to move. Figure 21-5 shows the four strokes of the piston that are required to complete one cycle of operation—intake, compression, power, and exhaust. Such an internal-combustion engine is called a **four-stroke engine**; that is, four strokes occur in one complete cycle of operation of the engine. A common arrangement of cylinders in an automobile engine appears in Fig. 21-6. Gasoline engines generally have many small cylinders rather than one large one to reduce vibration and to enable the engine to run more slowly without stalling.

A four-stroke engine in an automobile or airplane may make more than 1,800 revolutions per minute (1,800 rpm). This amounts to 30 revolutions per second! The pistons race up and down, the valves open and close, and the sparks

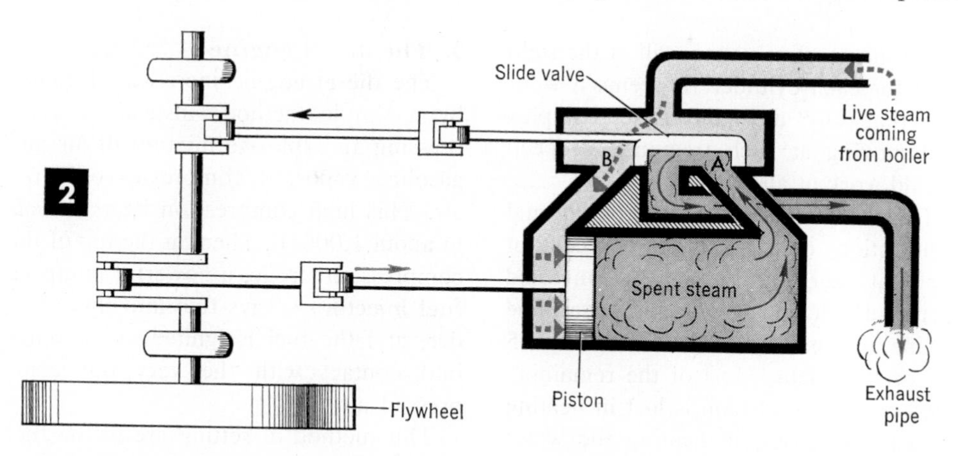

valve over the right to close port A and open port B (diagram 2). By this time the piston is over at the left of the cylinder. Steam now enters port B and pushes the piston back to the right. Spent steam on the right goes out the exhaust through A. The flywheel turns some more, pulling the slide valve back again as shown in diagram 1. This process is repeated to keep the engine going. What is the purpose of the flywheel?

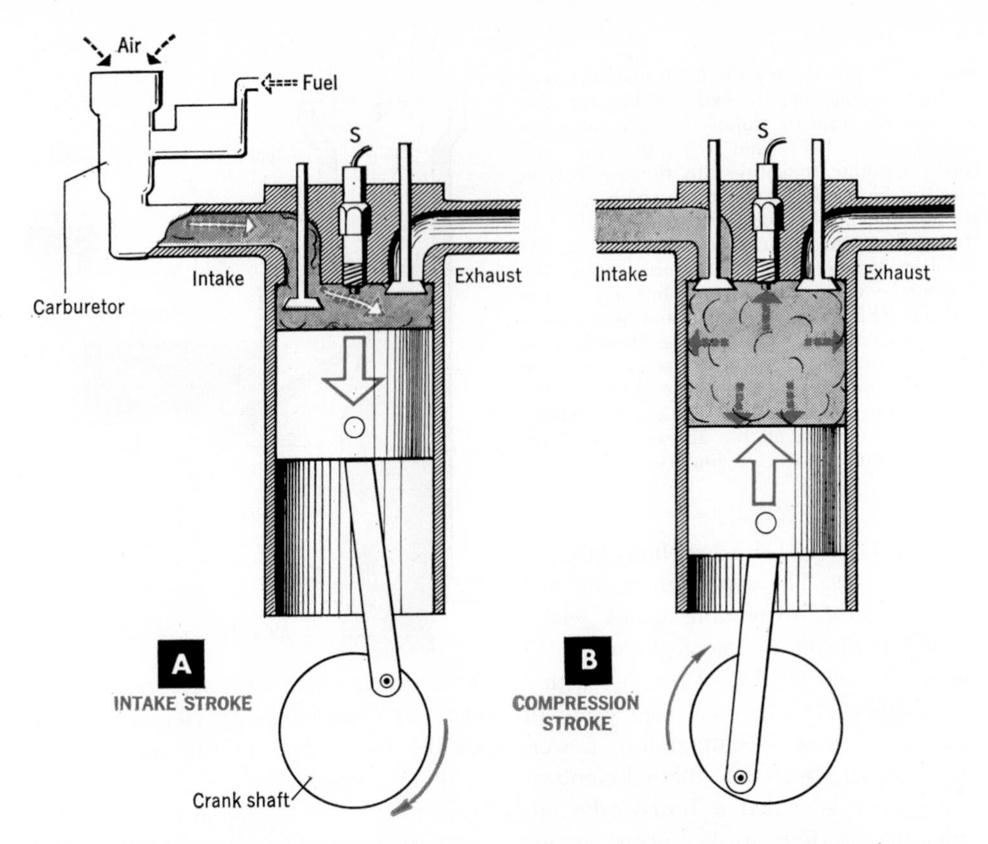

21-5 There are four strokes (up or down motions) in one cycle of operation of a gasoline engine: (A) *Intake stroke.* The piston moving down creates a partial vacuum above it which causes air to rush into the cylinder through the open intake valve on the left. On the way through the carburetor the air picks up a proper amount of gasoline to form an explosive mixture. (B) *Compression stroke.* The intake valve closes and seals in the explosive mixture. The upward-moving piston now compresses the gasoline-air mixture to a smaller volume.

ignite the gasoline vapor—all at the right time for each cylinder! It seems a wonder that any mechanism so complex and moving at such tremendous speed should work at all.

In spite of the advantages of internal combustion, only about 25 per cent of the heat energy in the gas is converted to useful mechanical energy. A gasoline engine is therefore said to be only 25 per cent efficient. Most of the remaining energy of the explosion is lost in heating the exhaust gases, in heating the water that keeps the cylinders cool, and in overcoming the friction of the moving parts inside the engine. None of these can be easily reduced.

3. The diesel engine

The **diesel engine** increases efficiency by a simple method: instead of compressing an explosive mixture of air and gasoline vapor, it compresses only the air. This high compression heats the air to about 1,000°F. Then, at the top of the compression stroke, a powerful pump (a **fuel injector**) sprays fuel into the cylinder, and the fuel is ignited as it comes into contact with the very hot compressed air.

This method of setting fire to the fuel requires no electrical ignition system of the kind found in gasoline engines. Furthermore, cheap oil can be used in the diesel engine instead of expensive gaso-

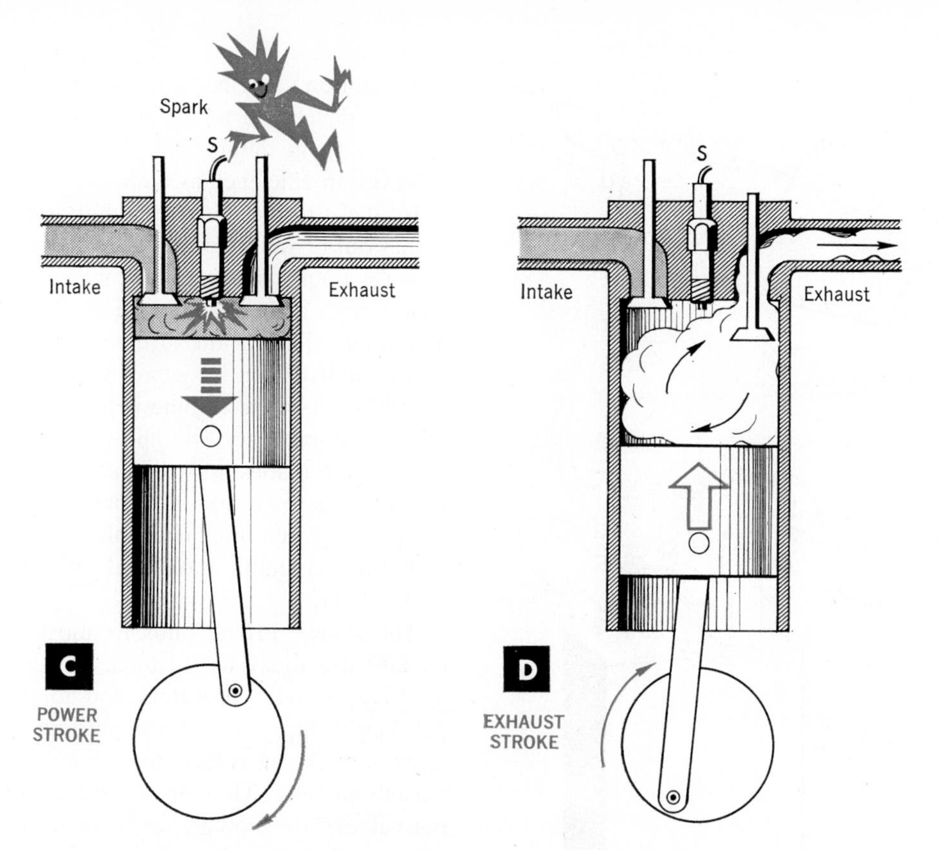

(C) *Power stroke.* Both valves still closed. An electric spark goes off just as the piston reaches the top and ignites the mixture, causing it to burn. The piston is now forced downward by the high pressure of the hot gases. (D) *Exhaust stroke.* The exhaust valve on the right opens and the upward-moving piston pushes the waste gases out. At the top of the stroke the cylinder is ready for a new cycle, starting with the next intake stroke. In each one of the four strokes the gases in the cylinder obey the General Gas Law (page 212).

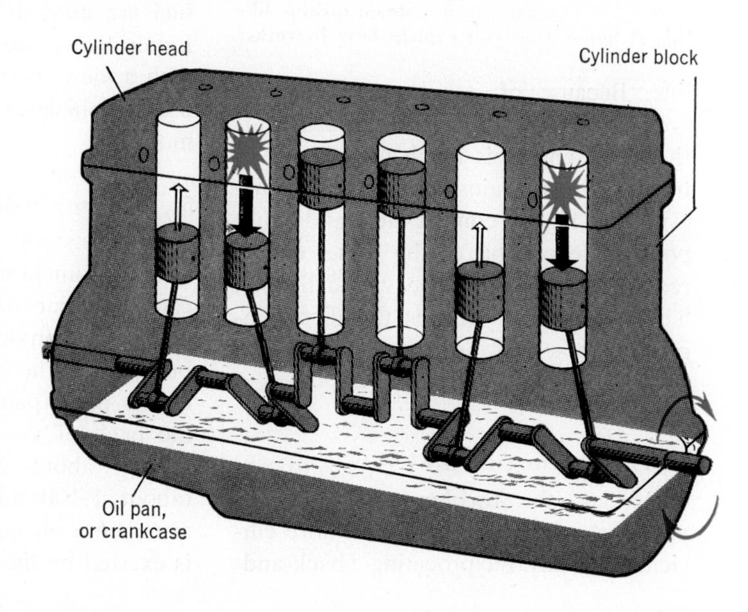

21-6 All automobile and airplane gasoline engines use many cylinders rather than one. As a result, the flow of power is smoother, there is less vibration, and the chance of stalling is reduced. The arrangement shown here is "six cylinders in line." Another good arrangement is the V-8, in which two groups of four cylinders are placed at an angle to each other to make a V shape.

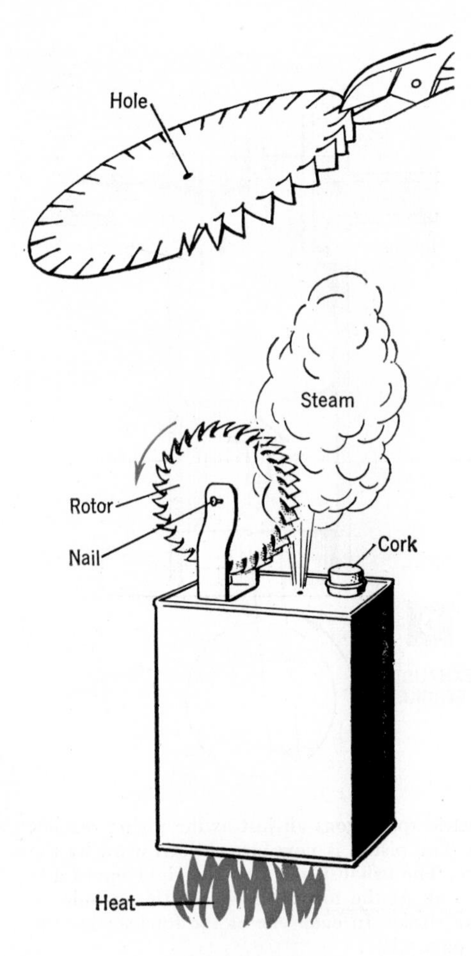

21-7 You can make a steam turbine like this at home. Can you explain how it works?

line. Because of its cheaper fuel and greater efficiency (30 to 40 per cent), the diesel engine is much used in heavy trucks, buses, trains, and boats. However, the higher compression and more powerful explosion of the diesel engine requires stronger and heavier cylinders. This extra weight makes the diesel engine less practical for light passenger cars and airplanes.

4. The steam turbine

Because the **turbine** eliminates the piston, it is much simpler and more efficient than any reciprocating (back-and-forth motion) engine, and some day it may be developed to the point where it replaces them altogether. The turbine achieves an efficiency as high as 50%.

In the turbine the expanding steam rushes out of small nozzles at high speed. This gas then strikes a bladed wheel (the **rotor**) and makes it turn (Fig. 21-7). You might think of the turbine as a kind of windmill driven by jets of high-speed gas. Of course, the turbine is not as simple as the ordinary windmill; the positions of the blades must be carefully designed so as to convert the energy of the expanding gases into mechanical energy with the maximum force and efficiency (Fig. 21-8).

The steam turbine finds its most important use today as a source of power to drive electric generators and to propel large ships. The large turbines in powerhouses can rotate steadily for long periods of time. They do not have moving valves, spark plugs, carburetors, and other parts which in other engines frequently require repair.

Recently gas turbines have been developed. In gas turbines the expanding hot gases which result from burning the fuel are used to turn the rotor directly. Experiments are now being conducted with a view to adapting gas turbines for use in vehicles such as trains and automobiles.

5. Jet-propulsion engines

In your study of the Law of Conservation of Momentum (Chapter 12) you became familiar with the principle of **jet-propulsion engines.** In a jet plane the burning of the fuel (usually kerosene) causes the expanding gases to shoot out the back of the plane with a speed of usually about 2,000 feet per second (about 1,400 miles per hour). At the same time an equal and opposite force is exerted by the expanding gases, which pushes the plane forward (Fig. 21-9).

Jet engines are efficient only if the gases are very hot. Thus the metal chamber in which burning takes place must be constructed of special alloys so that it does not melt or burn under the tremendous heat—in the neighborhood of 900°C. Only at very high speeds is such an engine more efficient for an airplane than an ordinary gasoline engine.

To increase the efficiency of jet engines, a powerful, high-speed blower is used to compress the air before it enters the combustion chamber. As shown in Fig. 21-9, this compressor is driven by a turbine which is rotated by the hot expanding gases coming out of the back of the engine. This type of engine is called a **turbojet** (Fig. 21-10). Of course, some energy is used to rotate the turbine and compressor, but the increase in efficiency resulting from extra compression more than makes up for this energy loss.

Rocket engines differ from jets mainly in that they carry their own oxygen supply. This makes it possible for them to burn fuel even when they are traveling above the earth's atmosphere. Expanding gases from the burning fuel push the rocket forward in much the same way as in the jet-propulsion engine.

Westinghouse Electric Corp.

21-8 A giant turbine used to turn an electric generator is exposed to view as it is being installed. Note the series of bladed wheels, which are pushed by the jets of steam. Such a steam turbine can supply a small city with electricity.

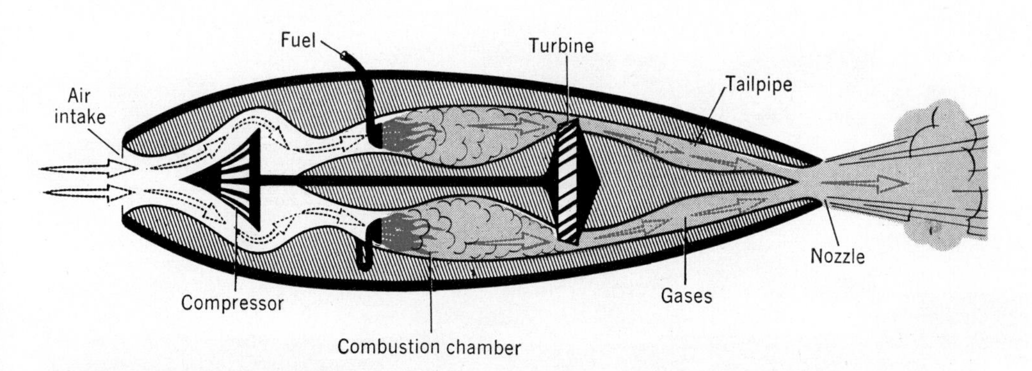

21-9 The jet engine gulps in air at the front. The compressor jams the air into the combustion chamber where it is mixed with fuel and burned. The hot expanding gases rush out the back, and the reaction kicks the engine forward. On the way out the gases turn the turbine which operates the compressor at the front.

6. Atomic energy and engines of the future

Atomic energy is the most concentrated source of energy yet known. In all atomic power plants in use in 1959 the heat released by the splitting of atoms in the heart of the reactor was used for one purpose only. That one purpose was to turn water to steam. The steam was used to drive conventional turbines for ship propulsion or for the generation of electric energy (Fig. 21-11).

In Chapter 46 we shall learn how atomic energy is released and how the heat it produces is put to work.

The power of engines

Could a mouse lift a grand piano to the top of a five-story building? Yes, theoretically it could! Arrange an apparatus in which the mouse runs on a treadmill. Then attach the treadmill to a simple machine having a mechanical advantage of several million, sufficient to lift a grand piano. Because of this high M.A., the mouse would have to run several million times as far as the piano rises. It might take a team of mice months to accomplish the job.

On the other hand, a man would get the job done in less than an hour. A horse requires even less time. The horse might get the piano to the top of the building in a few minutes.

When the piano has been hoisted to the top of the building, there is no way of telling whether the job was done by a mouse, a man, or a horse. Each would have accomplished the same amount of work. The only difference would lie in the time it took to complete the job.

7. What is horsepower?

It was not until James Watt was faced with the problem of producing steam engines which could replace horses for work that it became necessary to compare the rate at which men, horses, and engines can work. This quantity, the rate at which work is done, is called **power.**

For instance, a team of two horses may have been required to operate a pump which lifted water out of a mine. To do the same amount of work in the same time Watt needed a two-horsepower engine.

Watt measured a horsepower by harnessing a horse to lift weights over a period of time and then measuring the work done (Fig. 21-12). Suppose it was found that a horse could lift—on the av-

General Electric Co.

21-10 The inside of a turbojet engine. Can you locate the compressor, the combustion chamber, and the turbine?

U.S. Navy

21-11 The *Nautilus,* first United States atomic submarine. An atomic reactor provides heat to turn water into the steam used to propel the submarine.

erage—a 100-pound weight to a height of 55 feet in 10 seconds. Then the work done would be:

$$\text{Work} = \text{Force} \times \text{Distance}$$
$$W = FD = 100 \text{ lb} \times 55 \text{ ft}$$
$$= 5{,}500 \text{ ft-lb}$$

Since this amount of work was done in 10 seconds, the work done per second would be:

$$\frac{5{,}500 \text{ ft-lb}}{10 \text{ sec}} = 550 \text{ ft-lb/sec}$$

As a matter of fact, Watt decided to take this figure—550 foot-pounds per second—as representing the power (rate of doing work) of the horse, or 1 **horse-power.**

A man might take 10 times as long—or 100 seconds—to do the same job. In this case the power would be:

$$1 \text{ manpower} = \frac{5{,}500 \text{ ft-lb}}{100 \text{ sec}}$$
$$= 55 \text{ ft-lb/sec}$$

From these calculations we see that the formula for obtaining power is:

$$\textbf{Power} = \frac{\textbf{Work}}{\textbf{Time}}$$

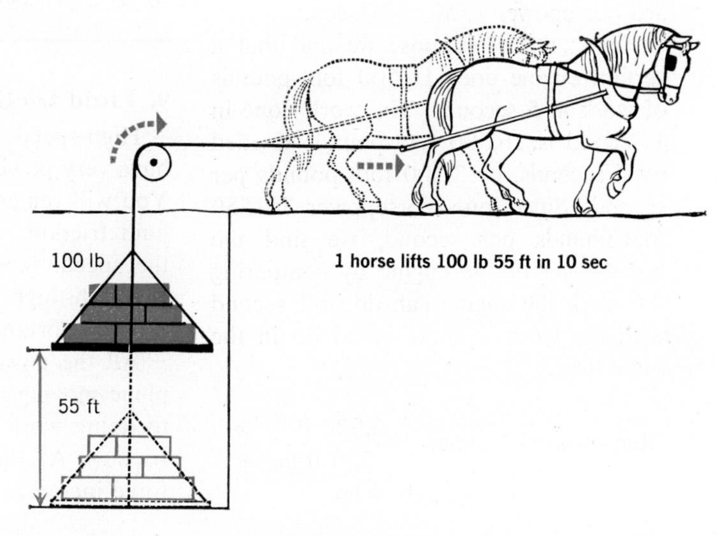

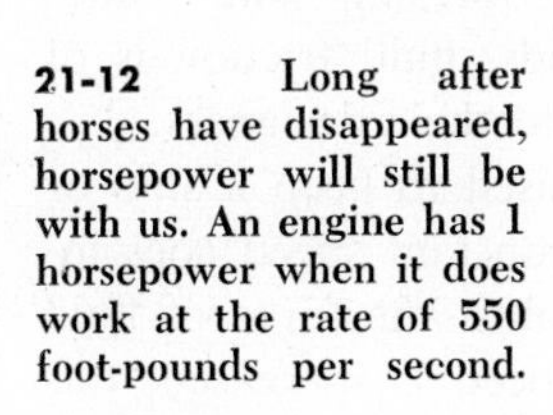

21-12 Long after horses have disappeared, horsepower will still be with us. An engine has 1 horsepower when it does work at the rate of 550 foot-pounds per second.

We note also that the approximate rate at which a man can do physical work (about 55 ft-lb/sec) is $\frac{1}{10}$ the rate at which a horse does physical work. Actually, this figure applies to work that is done steadily over a period of time. For brief spurts of activity a man can exert more than 1 horsepower. But he can't keep it up for more than a few seconds.

SAMPLE PROBLEM

What is the power of an electric motor that can do 3,000 foot-pounds of work in 30 seconds?

Using the formula for power:

$$P = \frac{W}{t} = \frac{3{,}000 \text{ ft-lb}}{30 \text{ sec}} = 100 \text{ ft-lb/sec}$$

STUDENT PROBLEM

What is the power of an electric motor that can do 1,480 ft-lb of work in 20 sec? (ANS. 74.0 ft-lb/sec)

8. Calculating horsepower

In most situations we want a direct comparison of the power of an engine with the power of a horse. To get this we compare our engine with Watt's standard horsepower (550 ft-lb/sec).

For example, suppose we find that a certain engine does 16,500 foot-pounds of work in 5 seconds. The work done in 1 second is 16,500 foot-pounds divided by 5 seconds, or 3,300 foot-pounds per second. Since one horsepower is 550 foot-pounds per second, we find the horsepower of the engine by comparing the work the engine can do in 1 second with the work a horse could do in the same time.

$$\text{Horsepower of engine} = \frac{3{,}300 \text{ ft-lb/sec}}{550 \text{ ft-lb/sec}} = 6 \text{ hp}$$

This calculation could be made in one step:

$$\textbf{Horsepower} = \frac{\textbf{Work done by engine (ft-lb)}}{\textbf{550} \times \textbf{time (sec)}} = \frac{16{,}500 \text{ ft-lb}}{550 \text{ ft-lb/sec} \times 5 \text{ sec}} = 6 \text{ hp}$$

Sixty times as much work can be done in 1 minute as in 1 second. So 1 horsepower is also 550 foot-pounds per second × 60, or 33,000 foot-pounds per minute.

Let us work out the horsepower in an actual problem.

SAMPLE PROBLEM

A 3,000-pound elevator is to be lifted 110 feet in 10 seconds. Neglecting losses of energy, what is the horsepower of the motor needed for the job?

$$\text{Horsepower} = \frac{\text{Work done (ft-lb)}}{550 \times \text{time (sec)}} = \frac{3{,}000 \text{ lb} \times 110 \text{ ft}}{550 \times 10 \text{ sec}} = 60 \text{ hp}$$

STUDENT PROBLEM

What is the horsepower of the motor needed to lift a 4,000-lb elevator 220 ft in 20 sec? (ANS. 80 hp)

9. Fluid friction and horsepower

High-speed ships and aircraft must have very powerful engines. Why is this? You will remember from Chapter 7 that fluid friction increases as the square of the speed. Since both ships and planes move through fluids, fluid friction is of great importance to their designers.

All the power used to keep a ship or plane moving at constant speed goes into doing work against the force of fluid friction. As the speed is increased, this force increases and more power must be expended. Let us see exactly how the power needed depends upon the speed. The fluid friction, F_f (of water or air), depends upon the shape of the particular object and the nature of the fluid in a way expressed by a constant, k, and on the velocity, v (see page 92):

$$F_f = kv^2$$

but we saw on page 279 that power is:

$$P = \frac{W}{t}$$

and since W = FD

$$P = \frac{FD}{t}$$

Here F is the force of fluid friction, F_f, and D/t is, of course, the velocity, v, of the ship or plane. Substituting, we get:

$$P = F_f \frac{D}{t} = kv^2 \frac{D}{t} = kv^3$$

Thus we see that the power needed depends upon the *cube* of the velocity! This means that to increase the speed of a plane or a ship from v to $2v$, the power needed increases from kv^3 to $k(2v)^3$ which is $8kv^3$—just 8 times as much as before! It is no wonder that high-speed destroyers and fighter planes must have very powerful engines. If a plane needs a 200-horsepower engine to travel 150 miles per hour, it would need a 1,600-horsepower (200×2^3) engine to travel 300 miles per hour (it takes 8 times the horsepower to travel twice as fast)!

There is another aspect to the problem of power and fluid friction that is also of great importance. It has to do with fuel consumption. Suppose a plane flies from New York to London at a speed of 200 miles per hour. The work done by the engines (and consequently the amount of fuel used) is equal to the force of fluid friction times the distance covered. If the plane doubles its speed to make the trip in half the time, the fluid friction is four times greater, and four times as much fuel is used. Thus to cut the time of a plane trip in half requires not twice, but four times as much fuel—which is usually more than a plane can afford to carry.

Going Ahead

THE VOCABULARY OF PHYSICS

Define each of the following:

external combustion engine
internal combustion engine
four-stroke engine
power
carburetor
diesel engine
jet propulsion
fuel injector
turbine
turbojet
horsepower

THE PRINCIPLES WE USE

Explain why

1. An internal combustion engine such as the gasoline engine is more efficient than an external combustion engine such as the steam engine.
2. Automobile engines are built with a number of small cylinders in the engine rather than one large cylinder.
3. Diesel engines do not need spark plugs.

APPLYING PRINCIPLES OF HEAT ENGINES

1. Using simplified diagrams, explain the operation of
   (a) the four-stroke engine
   (b) Watt's steam engine
   (c) the diesel engine
   (d) the steam turbine
   (e) the jet engine
   (f) the rocket engine
2. Compare the above types of engines as to advantages, disadvantages, and uses.

SOLVING PROBLEMS IN POWER

1. A 300-lb weight is lifted 40 ft by a crane in 8 sec. (a) How much work is done? (b) What is the power of the crane in ft-lb/sec?

2. How much weight can be lifted to a height of 30 ft in 20 sec by a gasoline engine whose power is 36,000 ft-lb/sec?
3. How long will it take a diesel engine whose power is 10,000 ft-lb/sec to lift 500 lb a distance of 8 ft?
4. What is the power of a motor needed for a hoist to lift 5,500 lb 33 ft in 30 sec?
5. What is the power of a motor needed for a pump which raises 100 ft^3 of water a minute to a tank 20 ft high?
6. How high could a 10-hp motor lift a 300-lb weight in 20 sec?
7. How long would it take for a 3-hp motor to raise a 1,000-lb machine a height of 44 ft?
8. How high could a 3-hp motor lift a 200-lb weight in 1 min?
9. A locomotive exerts 3,300 lb in pulling a train a distance of 100 ft. (a) How much work is done in pulling the train? (b) How much power (in ft-lb/sec) does the locomotive develop if it takes 30 sec to do the job? (c) What horsepower does the locomotive develop doing this job?
10. How long will it take a 20-hp motor to raise 330 lb a height of 250 ft?
11. How high can a 50-hp motor lift a 200-lb weight in 20 sec?
12. A plane is flying at 90 mi/hr. How many pounds of thrust are being exerted by the propeller if it is developing 100 hp?

---

13. What is the power input to an 80%-efficient electric motor that lifts a 3,000-lb elevator 110 ft in 60 sec?
14. A pulley system with an ideal M.A. of 6 is used with an effort of 10 lb to lift a load 60 ft in 15 sec. The pulley system is 80% efficient. What horsepower is obtained from the system?
15. A steam engine burns 200 lb of hard coal per hour and has an over-all efficiency of 20%. What is the useful power output in horsepower?
16. A boy pulls a sled at a velocity of 5 mi/hr. The sled weighs 150 lb and the coefficient of friction is 0.2. What horsepower must the boy exert?
17. A power of 0.1 hp is needed to row a boat at a speed of 2 mi/hr. What horsepower is needed to row the boat at a speed of 4 mi/hr?
18. A jet engine with a thrust of 1,000 lb drives a plane at a velocity of 750 mi/hr. What is its horsepower?

FURTHER READING

Anyone interested in the machines described in this chapter and how they work will surely enjoy one or more of the books and pamphlets listed below.

The Boy Mechanic. Published by Popular Mechanics Company and distributed by Simon and Schuster. More than 500 projects of all sorts for the young home craftsman. Fully illustrated.

James Watt and the History of Steam Power. Ivor B. Hart. Henry Schuman Co., New York, 1949. A very readable account of the history and development of steam engines.

Automotive Mechanics. William H. Crouse. McGraw-Hill Book Co., 1946. 673 pages. A thorough technical manual for anyone seriously interested in the details of all aspects of automobile mechanics.

A Power Primer. An excellent introduction to the internal combustion engine, distributed free of charge by the Department of Public Relations, General Motors Corp., Detroit, Michigan. This elementary and well-illustrated booklet is one of a series that also includes *Diesel, the Modern Power* and *Power Goes to Work* (which explains power transmission systems in cars, planes, and ships).

EXTENDING YOUR STUDY

Project Matterhorn, Princeton

The importance of the statement Newton once made, that he "stood on the shoulders of others," is reinforced by the fact that the statements in your text are based on the work of many men. For instance, the quotation which follows goes back about 200 years to the work of Joseph Black—who in his turn referred to other workers.

Liquefaction

"Our experience of freezing of liquids when exposed to more or less powerful degrees of cold is almost universal. The exceptions are very few. The strongest spirit of wine (alcohol) and a few subtle and volatile oils are the only substances that have not yet been solidified by any degree of cold hitherto known. As these, however, are so few in number, it appears unreasonable to believe them to be so different in nature and constitution from all other bodies that liquidity is in them an essential quality, of which they cannot be deprived by any diminution of their heat. We have no certain knowledge of what is the lowest possible temperature, but, on the contrary, shall have reason hereafter to be persuaded that the most violent cold which has yet been observed is very far short of the most extreme degree. So it is reasonable to suppose that

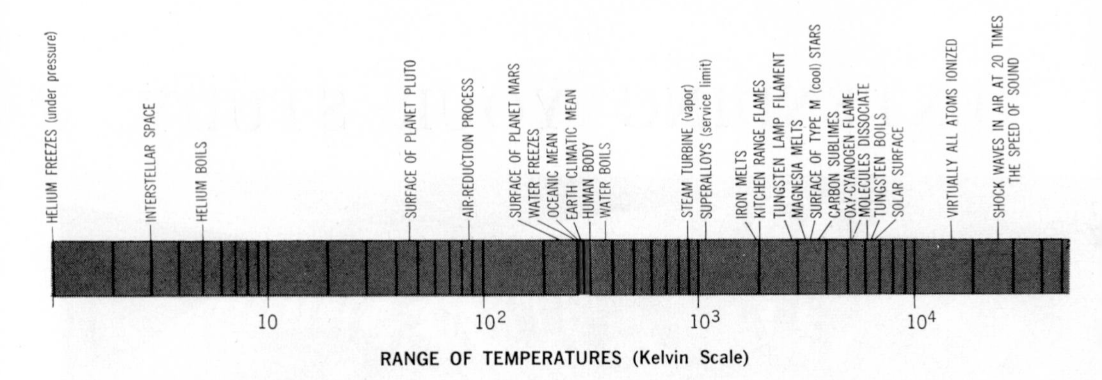

21-13 The range of temperatures runs from absolute zero to those encountered in the interior of the hottest stars. These temperatures are shown on the Kelvin scale in powers of 10.

these few substances differ from others only in having a much greater *disposition* to liquidity, so that we have never yet known a degree of cold sufficient for solidifying them; but that they would undoubtedly freeze, like other liquids, were they exposed to a sufficiently low temperature." *

Extremes of Temperature

Over and over again in this unit we have reminded ourselves that temperature is a measure of the average energy of motion of atoms and molecules. We have seen that this "kinetic molecular theory" of temperature leads to the conclusion that there is a lowest possible temperature at which all molecular motion ceases. It is meaningless to discuss temperatures lower than this "absolute zero," just as it is meaningless to talk about speeds slower than 0 mi/hr.

While there is a definite *lower* limit to temperature, our kinetic theory of heat sets no upper limit to temperature at all. Temperatures of thousands of millions of degrees have been observed. As we look at the universe around us we find represented every temperature in this enormous range.

In Fig. 21-13 you see displayed the range of temperatures in our universe. The temperatures are of course in Kelvin degrees, beginning with 1°K and written in powers of 10 along the bottom of the chart. For example, 10^2 is 100°K. The first line to the right of 10^2 is 200°K, the next 300°K, and so on until we come to 10^3, which is 1,000°K. Notice that the boiling point of water is 373°K.

The advantage of a scale marked in powers of 10 like this one is that an enormous range of numbers can be compressed clearly into a very small space. It is called a **logarithmic scale.**

When you examine the chart carefully you will find a number of interesting facts. We shall describe a few of them in the following paragraphs.

At 3°K you will notice the words "Interstellar Space"—the space between the stars (Fig. 21-13). Astronomers estimate that approximately half of all the atoms in the universe are in interstellar space. Hence half of all the atoms that

* "(Excerpts from Volume I of) *Lectures on the Elements of Chemistry* Delivered in the University of Edinburgh by the late Joseph Black, M.D. Professor of Chemistry in that University, Physician to His Majesty for Scotland; Member of the Royal Society of Edinburgh, of the Royal Academy of Sciences at Paris, and the Imperial Academy of Sciences at St. Petersburgh. Now published from his Manuscripts by John Robison, L.L.D. Professor of Natural Philosophy in the University of Edinburgh 1803."

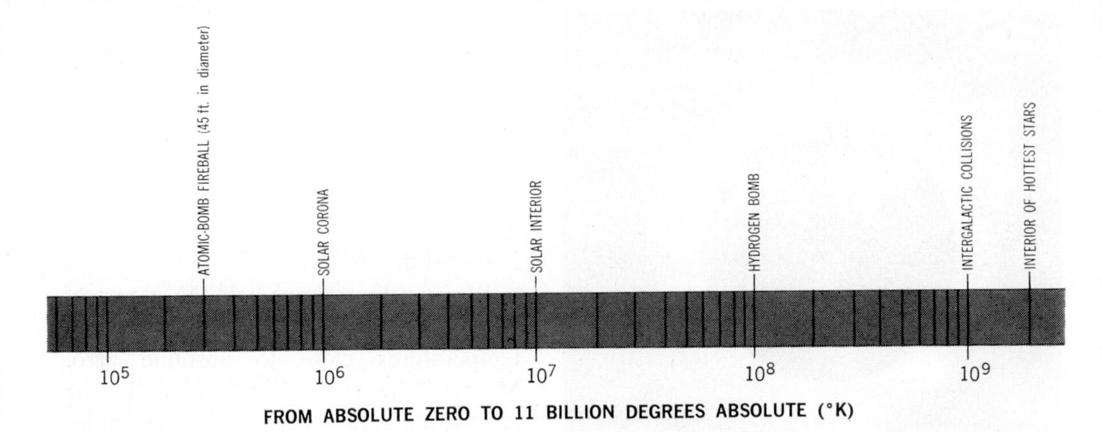

Thus on the Centigrade scale the freezing point of water is slightly more than 273° Kelvin. What is the boiling point on the Kelvin scale? (Redrawn from *Scientific American.*)

exist are at 3°K, a temperature 270° below the freezing point of water.

Even at the temperature of the surface of Pluto, the outermost planet of our solar system, temperatures are so low that the gases that might have formed Pluto's atmosphere probably lie frozen solid to its lifeless rocks at 45°K. You will notice that the gases in the atmosphere liquefy at about 77°K, which they are commonly made to do in the commercial production of oxygen and other gases.

One of the most remarkable regions of the temperature range lies between about 273°K and 300°K. This region, shown in red, is the range of reasonably comfortable temperatures for living creatures on the surface of the earth. Notice how narrow this region is. Outside it, living creatures carry on their life processes only with extreme difficulty. Indeed, some biologists consider it doubtful that any life at all would survive if our earth's surface temperature were suddenly raised or lowered only 20°C. And a change of only 1 per cent in the sun's radiation would more than suffice to bring this about.

In this life-supporting region temperatures are high enough to produce the molecular motions necessary to sustain the chemistry of life, and yet are not so high that the chemicals involved in life processes become unstable. Since the laws of chemistry are presumably the same throughout the universe, we may be reasonably sure that planets supporting life like ours elsewhere in the universe are not much hotter or cooler than ours.

What do we mean by hot? The answer depends upon whom you ask. For a cook it may be 100°C, the boiling point of

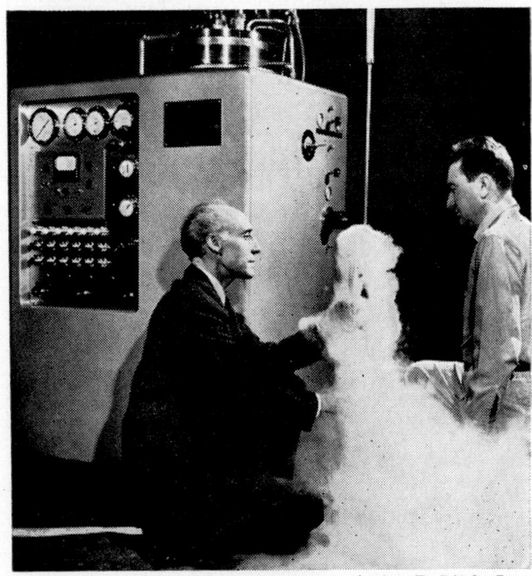

Arthur D. Little, Inc.

21-14 **This device, known as a cryostat, produces liquid helium and liquid hydrogen by the liter at a temperature within a few degrees of 0°K. Approximately half of all the atoms in the universe are this cold.**

American Museum of Natural History

21-15 **The gaseous "atmosphere" or corona of the sun, shown here during a total eclipse, is composed of atoms whose temperature is close to 1,000,000°C.**

water; for the designer of an airplane gas turbine it may be 1,100°C; for a metallurgist making alloys in an electric furnace it may be 2,000°C; and for a rocket-designer it may even be as high as 3,000°C.

Above 2,000°C very few chemical molecules remain intact; most have broken apart into their constituent atoms. Above 6,000°C molecules and ordinary chemical reactions cease to exist. All matter is in the form of atoms or ions. This temperature of 6,000°C is the temperature of the surface of the sun.

Temperatures higher than this cannot in general be produced on earth for any appreciable length of time. We may create a temperature of 10 million degrees in an atomic-bomb explosion, attaining for a brief moment the temperature of the interior of the sun. Indeed, when such an explosion is used to set off a hydrogen fusion bomb we may even achieve 100 million degrees. At such temperatures atoms move at speeds of thousands of miles a second, and the force of their collisions with one another strips off most of the outer electrons. Yet the *controlled* and sustained release of useful energy from hydrogen fusion requires that we somehow produce a temperature such as this and find a container in which to maintain it. No wonder that the production of useful power from this source is still in the future.

We saw that about half the atoms of the universe are in interstellar space at temperatures close to absolute zero. Most of the remaining half are in the stars at temperatures of a million or more degrees.

Thus only a tiny fraction of all that you see as you look up at a starry sky is at a temperature that can support life. And as you look at the living world around you, you should reflect on the delicacy of balance between the extremely high and the extremely low temperatures that can be experienced close at hand all around you.

Rocketdyne

21-16 **The temperature of the gases roaring out of this rocket engine on a test stand actually approaches the temperature of the surface of the sun. Why is such a high temperature important to rocket propulsion?**

UNIT FOUR

Sound Energy— The Motion of Vibrations

Bell Telephone Laboratories

Physicists studying sound are conducting experiments in this soundproof room. The odd appearance of walls, ceiling, and floor is due to the "honeycombs" of glass wool insulation which has been formed into five-foot wedges to absorb virtually all sound made in the room. Experiments such as these may someday play an important part in your life, just as countless experiments with sound in the past have already improved the conditions of our lives.

In this unit you will learn the nature of sound, how it is produced, and how it is transmitted. You will learn how different sounds are made, how musical instruments produce music, and how that music is recorded and how you hear it. Then, when you sit down at the piano or finger your saxophone or turn the knob of a radio, you will know more about the vibrations you make and hear. And you will also begin to understand the reasons behind experiments with sound.

CHAPTER

22

Sound Waves

Any sound, anywhere, has its origin in the **vibration** of something. The musical sound of a violin, for example, is caused by the vibration (back-and-forth motion) of a string. Examine a string or tuning fork as it vibrates to produce sound, and you will notice that each appears blurred because it is moving so rapidly; if you touch the string or tuning fork, you can feel the vibrations. The sound of a clarinet comes from the vibration of air in the hollow tube and of a flexible reed in the mouthpiece. You can feel the vibration of a radio loudspeaker when your hand is placed on the front of it. The sound of a bell comes from the vibration of the metal when it is struck.

If we stop the vibration, the sound stops. Without vibration there can be no sound.

How is sound transmitted?

You have possibly noticed that when you are swimming with your head under water, you can hear the engine noises of distant motorboats. The Indians used to put their ears to the ground to catch the pounding of horses' hoofs; from this we get our expression "putting your ear to the ground."

From such experiences and from the additional experiments shown in Figs. 22-1 and 22-2, we see that sounds can be transmitted through liquids and solids, as well as through gases.

1. No sound in a vacuum

Air is the most important of all materials for transmitting sound—for the simple reason that it is usually air which brings sounds to our ears. (If you are swimming under water, the water carries the sounds.) The fact that we need a transmitting material like air if we are to hear any sound can be demonstrated by the experiment in Fig. 22-2. A ringing bell is suspended in a jar. As air is pumped out of the jar, the sound becomes fainter and fainter until it can no longer be heard. When the valve is reopened and air is permitted to rush back into the jar, the sound grows louder and louder until it has regained its normal loudness. Thus it is demonstrated that a vacuum will not transmit sound.

Unlike light, sound can be transmitted only through matter. The vibrations of the object causing the sound set up vibrations in the material surrounding the object. It is these vibrations in the surrounding material which travel outward, carrying the sound to our ears.

Certain materials carry sound much better than others. Iron, water, and air, as we have seen, are all excellent transmitters of sound. When a bar of iron is struck at one end, the motion is easily transmitted from one molecule to the next down the length of the bar.

In general, materials that are good conductors of heat are also good conductors of sound and for the same reasons. On the other hand, soft, fluffy objects, such as pillows, rugs, and drapes, absorb any motion that is applied to them and therefore transmit sound (as well as heat) very poorly. Hence such objects are sound insulators.

2. The speed of sound

Although you may not have thought about it at the time, you have undoubtedly had experiences which show that sound travels at a measurable speed. Perhaps you have seen the puff of steam from the whistle of a far-off locomotive many seconds before you heard the whistle. Indeed, the lapse of time between the steam puff and the whistle may have been so great that you did not connect the two events at all. Lightning and thunder offer another example. Most of us know that thunder and lightning are somehow related, but many people do not realize that it is the lightning that causes the sound of thunder. They do not connect the two events because they see the lightning as much as half a minute before they hear the thunder.

The speed of light is so great that it can travel more than 7 times the distance around the earth in 1 second (Chapter 37). Since light travels with such enormous speed, we see something happening several miles away in less than one ten-thousandth of a second after it happens. For practical purposes we can say we are seeing it at the very instant it occurs.

We can make use of the tremendous speed of light to help measure the speed of sound. Say, for example, that we measure off a distance of 3,850 feet. At one end of the distance a man stands with a gun containing blank cartridges. At the other end another man holds a stop watch. As soon as the man with the stop watch sees the smoke of the gun, he starts the watch. When he hears the sound of the explosion, he stops it. Suppose it has taken 3.5 seconds for the sound of the gun to reach his ears over

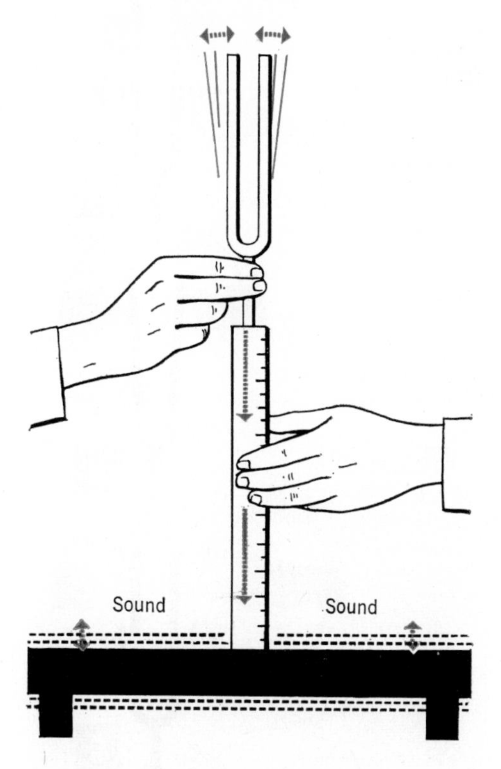

22-1 The sound is transmitted through the ruler and sets the desk vibrating. As a result the sound becomes much louder.

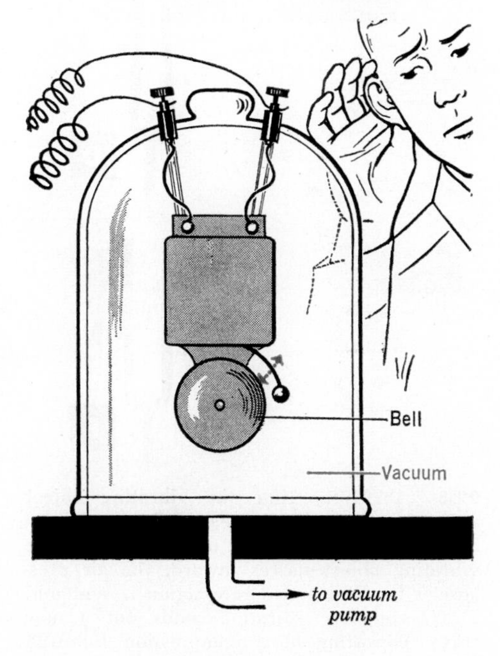

22-2 Can a bell in a vacuum be heard?

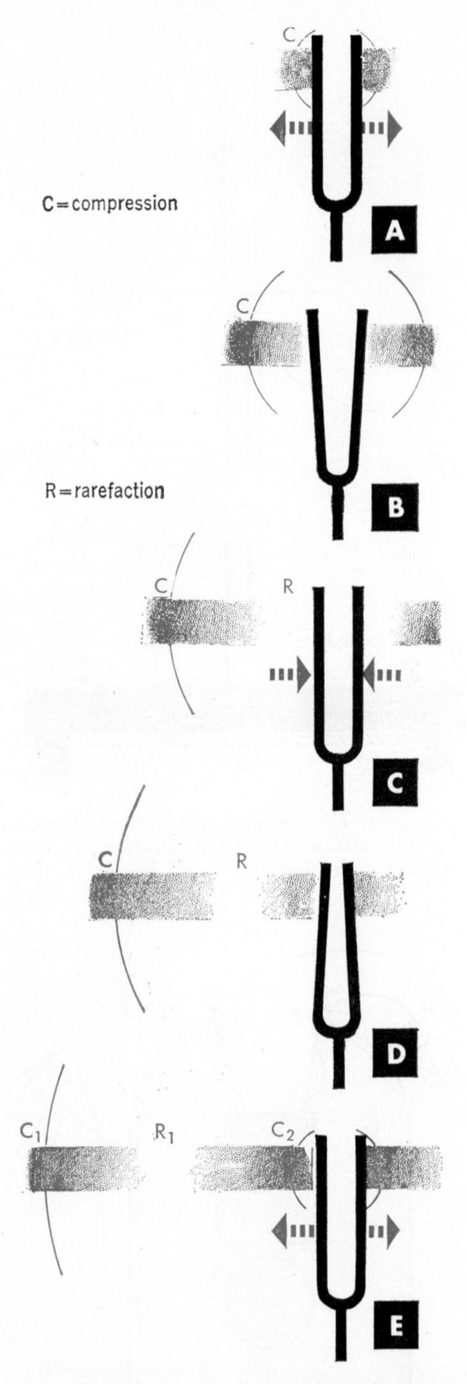

22-3 Every time the vibrating object moves outward, the air near by is compressed and a compression is sent out. Every time the vibrating object moves inward, the air pressure is lowered and a rarefaction is sent out. Every complete vibration sends out a new wave, consisting of a compression followed by a rarefaction.

the distance of 3,850 feet. He then finds the velocity of the sound according to the following formula:

$$\text{Velocity} = \frac{\text{Distance}}{\text{Time}} = \frac{3{,}850 \text{ ft}}{3.5 \text{ sec}} = 1{,}100 \text{ ft/sec}$$

This is a speed of approximately 1 mile in 5 seconds. The speed is the same for all sounds regardless of pitch or loudness.

Careful measurements made in this way have revealed an interesting fact about sound: it travels slightly faster in warm air than in cold air. At 0°C sound travels about 1,090 feet per second. With every centigrade degree of temperature rise this speed increases 2 feet per second. Thus at 10°C sound would travel 20 feet per second faster than at 0°C; that is, the speed of sound at 10°C would be 1,090 + 20, or 1,110 feet per second.

The speed of sound varies according to the material transmitting it. Air, as we have seen, carries sound at the rate of approximately 1,100 feet (about $\frac{1}{5}$ of a mile) in 1 second. In water (fresh or salt) sound travels about 4,700 feet (almost a mile) in 1 second. In iron the speed is even greater, 16,400 feet or about 3 miles per second.

3. The sound barrier

The speed at which sound travels through air has taken on new significance with the development of high-speed airplanes. It has been found that when a plane approaches the speed of sound (about 750 miles an hour at sea level), its motion through the air creates highly energetic **shock waves.** Their creation requires so much energy that the airplane engine has to exert a very much greater thrust as the plane reaches the speed of sound. This increasing resistance to further speed is called the **sound barrier.**

Once a plane is through the barrier and is therefore traveling faster than the speed of sound, the sound waves are left behind, and to a man in the airplane flight is silent.

In returning to a speed less than that of sound, the plane experiences the shock waves again. The shock waves are a hazard which must be taken into account in designing a plane, lest it be wrenched apart by them.

What is sound?

Let us examine in detail how sound travels through a material.

4. Compressions and rarefactions

Picture in your mind a tuning fork vibrating back and forth. High-speed photographs of the vibrating fork reveal that the two prongs move outward and inward together, as shown in Fig. 22-3. When a prong moves outward, it pushes against the neighboring molecules of air and creates a high-pressure area at C (Fig. 22-3A and B). As a result the air next to the prong is pushed outward. This air in turn pushes against the air next to it. Thus a pressure wave, or **compression,** travels outward from the prong. Once the pressure wave, or compression, has left the prong, it no longer has any connection with the tuning fork. It is now on its own and travels outward from the prong.

Meanwhile, in moving out beyond its normal position, the prong has been stretched. Hence it springs back. This leaves a low-pressure area (Fig. 22-3C and D), called a **rarefaction** (rarefied air), which also moves outward from the fork. It travels at the same speed as the compression, the exact speed depending on the material.

This formation of compression and rarefaction is repeated every time the prong moves back and forth. One compression and one rarefaction are sent out for each complete **cycle** (back-and-forth motion) of the prong. Together they form one complete **sound wave.** If the tuning fork makes 200 complete cycles (vibrations) each second, then 200 complete waves, each consisting of a compression and a rarefaction, are sent forth into the surrounding air. Once a wave has started, it keeps on going even though the vibrations of the fork itself may stop.

Figure 22-4 shows one way of picturing the wave. Compressions are represented by the colored areas, rarefactions by the white areas. Since sound waves travel outward in all directions

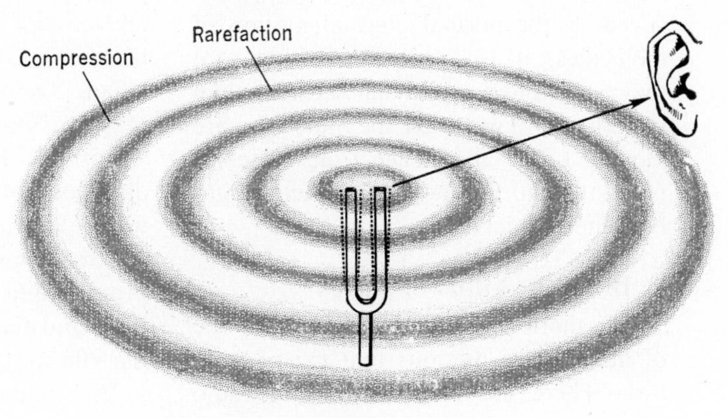

22-4 A sound wave pictured. It spreads like an expanding ball outward from the vibrating object. The wave consists of compressions (squeezed air) followed by rarefactions (low-pressure air) in rapid succession.

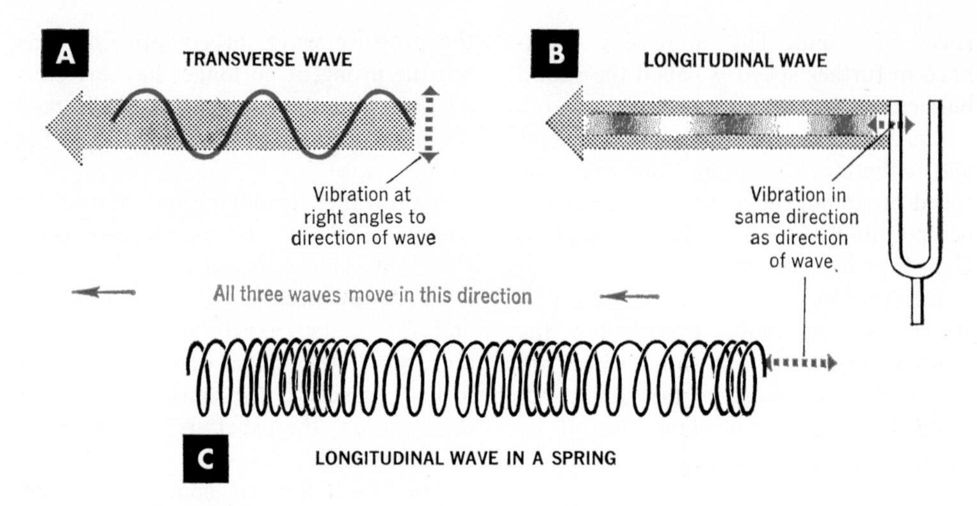

22-5 Transverse (crosswise) waves move at right angles to the motion causing them. The wave in the rope in A is of this kind. However, in longitudinal (lengthwise) waves, the direction of wave motion is in the same direction as the vibration causing it. Sound waves (B) are of this type. Longitudinal waves can also be set up in a long spring (C).

with equal speed, the shape of the wave is spherical, like a ball. As the wave spreads outward from the fork, the pressure becomes less and the amounts of compression and rarefaction decrease. This explains why sound grows fainter as it moves farther from its source.

From this explanation we can understand why sound waves travel about 1,100 feet per second. This is approximately the speed of molecules of air at 0°C and hence the speed with which a molecule that has been pushed by the tuning-fork prong can communicate the compression to its neighbors.

Moreover, since sound waves are conveyed by the normal thermal motion of molecules, it is hardly surprising that they travel more rapidly in warm air than in cold air. In warm air the molecules are moving faster and hence transmit compressions and rarefactions to their neighbors more quickly.

5. Different kinds of waves

The motion of sound waves is but one of many wave motions that exist in nature. There are light waves and radio waves. There are earthquake waves that travel in the earth and waves made by water.

All these waves can be classified into two main types. To illustrate one type, take a rope about 10 to 20 feet long and attach one end of it to a wall. Now hold the rope tight at the other end and move your hand rapidly up and down once. A wave travels out along the rope, strikes the wall, and is reflected back. While your hand is moving up and down in a vertical direction, the wave is traveling along the rope horizontally (Fig. 22-5A). In other words, the vibrations of the parts of the rope that cause the wave are moving at right angles to the motion of the wave itself. The prefix *trans-* means "across." Thus we call this kind of wave motion **transverse** because the vibrations of your hand move across the direction of the wave motion. Light waves and radio waves are of this kind.

Sound waves are of the other type. The prongs of a tuning fork, as they move back and forth, send out waves in which the vibrating particles (molecules) move in the same direction as the prongs (Fig.

22-5B), and hence the motion of the vibration is *along* the direction in which the wave is moving. This type of wave is called a **compressional** or a **longitudinal wave.**

Frequently a vibration sets up waves of both types. Thus there are always two sets of waves in the ground whenever an earthquake occurs. The first to arrive is the longitudinal wave; this brings the rumbling sound but results in little damage. Following it, sometimes minutes later, come the transverse waves, which shake buildings up and down and can cause enormous destruction.

Reflecting sound waves

If you stand about 100 feet away from a large wall or cliff and give a good-sized bellow, the sound is reflected back to you as an echo. As you go farther away from the cliff, the time between the original sound and the echo becomes longer. In fact, you can measure the distance between yourself and the cliff by the time it takes the echo to return.

The sound wave in an echo is being bounced back to you by **reflection** in much the same way that a ball bounces back from a wall.

6. Sonar

Various devices for measuring distance make use of sound reflections. One of these is the **sonar** system, similar to radar but using sound waves instead of radio waves. Sonar has been developed to detect the presence of unseen underwater objects near a boat. A sound is sent out under water from the boat at regular intervals, and the time required for the sound to be returned by reflection through the water is measured. Knowing the speed at which sound travels through water, a person on the boat can measure the distance of the reflecting objects from the boat. This method is also used to find the depth of the ocean bottom.

Socony Mobil Oil Co., Inc.

22-6 Sound reflection is used by oil prospectors. The sound of a dynamite explosion passes down through the earth and is reflected from the hard rock dome which covers the oil deposit. The reflection is timed and the position of the dome under the earth is calculated.

Johns-Manville

22-7 This ceiling is lined with soundproofing material that reduces the noise in an office or classroom by absorbing most of the sound waves and reflecting very few. Working in a noisy room can be very tiring.

The bat, though almost blind, is able to fly because it is guided by a system similar to sonar. It sends out extremely high-pitched sounds, above the range of human hearing, and senses the presence of nearby objects by the reflections of the sounds. This method is so effective that the bat can catch insects in flight in the dark. Blind people can also learn to sense nearby objects by tapping a cane and listening for slight changes in sound that occur because of reflection.

7. Improving acoustics

In small rooms the reflection of sound from the walls is usually not noticed. The sound strikes the wall and bounces back in so short a time that the reflection (the echo) and the original sound seem like one. But in large rooms and auditoriums, the time interval is greater, and the original sound and the echo will strike the ear at different times.

In a room 55 feet long, how long will it take a sound to travel from the front of the room to the back and then be reflected to the front again? The total distance traveled is 110 feet, so the time is $\frac{110}{1,100}$ or $\frac{1}{10}$ second, and two separate sounds will be heard. The same sound wave may bounce back and forth several times before being finally absorbed. Multiple echoes of this kind are called **reverberations.** An auditorium in which hearing is difficult because of too much reverberation is said to have poor **acoustics** (*ȧ*·kōōs′tĭks).

Poor acoustics may be improved by reducing the reverberation. Reverberation can be reduced by covering walls and ceiling with a special inelastic soundproofing material which absorbs most of the sound waves (Fig. 22-7). Drapes on the walls and rugs on the floor are also good sound absorbers.

Houses can be soundproofed by having thick walls that contain sound-absorbent materials. Since a material conducts heat in precisely the same way it conducts sound—by passing energy from moving molecule to moving molecule by means of collisions between them—it is easy to see that good soundproofing material is usually good heat insulation as well. Walls that absorb sound are also likely to conduct less heat.

Pitch and frequency

Close your eyes for a moment and listen to the sounds about you. Perhaps there is an automobile horn in the street, a telephone ringing in the house next door, shouts from the baseball game in the corner lot, or the whirr of the lawnmower as your brother mows the lawn.

All these sounds result from vibrations. Yet each sound is quite distinct from the others. You can identify them all. So there must be differences among these vibrations that strike your ear. These differences are three in number: pitch, loudness, and quality.

8. Pitch

One of the most easily recognizable differences between sounds is the difference between high and low tones. This difference is called **pitch.**

A fire engine's siren can reach quite a high pitch. What causes this bloodcurdling cry that can rouse you out of a deep sleep in the middle of the night? The siren consists of a rotating wheel with evenly spaced holes punched in it. Air under pressure rushes out of a jet (or a series of jets), alternately striking a hole and then a metal space between holes. When the air strikes a hole, it rushes through; when it strikes a metal space, it is blocked. Thus there is a rapid series of blocking and unblocking, blocking and unblocking. Each puff of air that shoots through a hole causes a compression to travel outward; each blocking of air creates a rarefaction. It is this series of compressions and rarefactions which we hear as the siren sound.

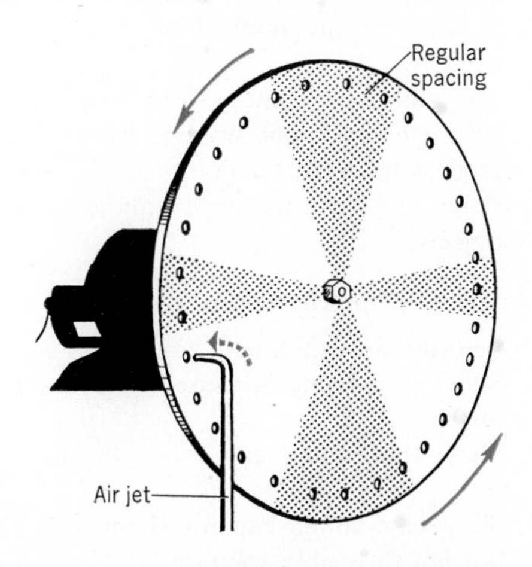

22-8 As the wheel spins past the air jet, the air alternately strikes a hole and then a metal space between holes, producing a sound whose pitch depends on the speed of the wheel.

Now let us experiment with a laboratory model of the siren (Fig. 22-8). We notice that if we turn the wheel slowly in front of the jet, the musical sound that comes out is low-pitched—very different from the shrill, high-pitched sound we associate with sirens. But as we speed up the wheel of the siren, the pitch rises. If we increase and then decrease the speed, the pitch rises and falls, and we have the familiar wail of the siren.

This experiment would seem to indicate that pitch (the highness or lowness of tone) depends upon the number of compressions and rarefactions (waves) per second. If the siren wheel is turning fast, there are more waves per second, and the pitch is high. As we slow the wheel down, the number of waves per second decreases, and the pitch is lowered.

We can show the same thing with two tuning forks of different pitches (Fig. 22-9). Each has a pin attached to one prong. If both forks are vibrating and are moved at the same speed across a plate of soot-blackened glass, the pins will leave wavy tracks behind them.

You will notice that there is a distinct difference between the wavy traces produced by the sideways vibrations of the two forks. There are many more waves in the trace made by the high-pitched fork than in the same length of trace made by the low-pitched one. An interesting example of how pitch increases with more rapid vibration is shown in Fig. 22-10.

9. Frequency

We have said that pitch depends upon the number of waves per second in a tone. But that is just another way of saying that pitch depends upon the **fre-**

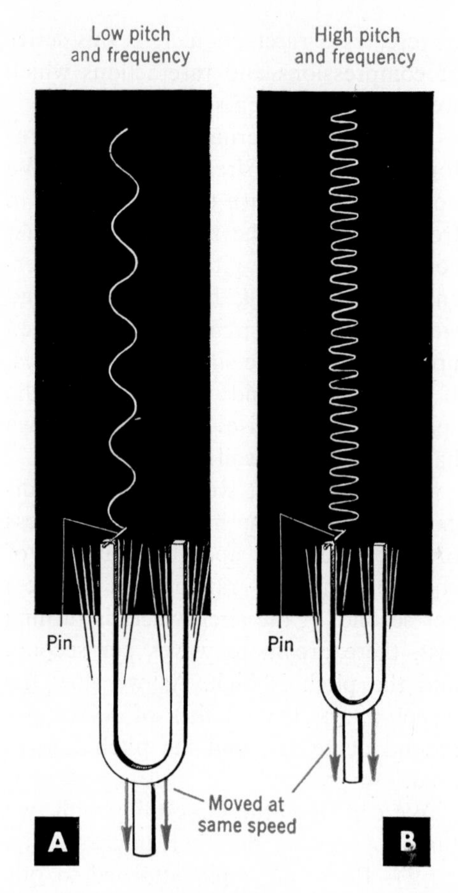

22-9 If the pin on the moving tuning fork is kept in contact with the soot-blackened glass plate, waves are traced on the glass. The higher the pitch or frequency, the more waves will be made on the tracing.

quency of the vibrations. The more frequent the vibrations, the higher the pitch. When we speak of the frequency of any musical tone, we mean the number of vibrations per second (vps) or cycles per second required to produce it.

How would we measure the frequency of a sound? Let us take the siren wheel as an example. Suppose that the wheel contains 80 holes and is making 10 revolutions per second. Then 80×10, or 800 holes will pass the air jet every second, so our frequency is 800 vibrations per second (vps).

Suppose that we rotate a siren so that it produces a tone whose frequency is 400 vps. If we sound a 400-vps tuning fork, we shall find that the tones are much the same. A musician would say that both tones have the same pitch. A scientist would explain the similarity by saying that the frequencies of vibration are the same. If the frequencies were different, then the pitches would be different—with the higher frequency producing the higher pitch.

Going Ahead

THE VOCABULARY OF PHYSICS

Define each of the following:

vibration	reverberations
shock waves	ultrasonic
frequency	sonar
compression	transverse wave
pitch	echo
rarefaction	longitudinal wave
acoustics	sound barrier
cycle	

THE PRINCIPLES WE USE

Explain why

1. Sound does not travel through a vacuum.
2. Sound is easily muffled by drapes and rugs.
3. The vibration of the prongs of a tuning fork causes a sound wave to travel away from it through the air.
4. A bat can fly in the dark without hitting objects.

EXPERIMENT WITH SOUND

Describe an experiment to

1. Show that sound is caused by vibrating objects.
2. Show that sound can travel through a solid.
3. Show that sound can travel through air but not through a vacuum.
4. Measure the speed of sound.
5. Show that the higher the frequency of a tone the higher the pitch.
6. Measure the frequency of a tuning fork.

APPLYING PRINCIPLES OF SOUND

1. Describe two practical applications of reflection of sound.
2. What causes poor acoustics in an auditorium? Mention two ways of improving acoustics.
3. How is soundproofing accomplished in homes?
4. Find out how an oscilloscope is used to analyze sound.

SOLVING SOUND-WAVE PROBLEMS

1. What is the speed of sound at 0°C? What is the increase in speed per degree centigrade?
2. How far away is a boat if the sound of a whistle is heard 6.5 sec after the steam puff from the whistle is seen? The temperature of the air is 20°C.
3. How long will it take for the sound of thunder to reach a point 3 mi from the place where lightning struck if the temperature is 15°C?
4. What is the temperature of the air if sound on a certain day travels 4,534 ft in 4.100 sec?
5. How far away is a cliff that returns an echo in 6.0 sec when the temperature of the air is 15°C?
6. What is the depth of the ocean at a point where a sound takes 0.6 sec to travel to the bottom, to be reflected and to return?
7. A hunter shoots a rifle, and 6 sec later he hears the sound echoed from a cliff. If the temperature of the air is 10°C, how far away is the cliff?
8. What is the temperature of the air if sound travels 8,960 ft in 8.000 sec?
9. Three seconds after he sees a puff of steam from a locomotive whistle, a man hears the sound. If the temperature is 10°C, how far away is the train?
10. At a ball game the crack of the bat is heard $\frac{3}{4}$ sec after the batter hits the ball. How far away is the batter if the temperature is 20°?

Irving Subway Grating Co., Inc.

22-10 **This road hums! When the car wheels ride over this grating, a loud humming sound results from the regular vibrations of the tire against the grating. The pitch of the sound increases with the speed of the car and tells the motorist to "watch out."**

11. A mountain climber drops a rock over the edge of a high cliff; 15 sec later he hears the rock hit the ground below. How high is the cliff? (Assume the velocity of sound is 1,000 ft/sec and there is no air resistance.)
12. A boy struck a railroad track with a hammer. Another boy stationed a distance down the track heard two sounds, one coming through the rail and the other through the air. If the interval between the two sounds was 1.4 sec, calculate the distance between the boys. Can you figure out how this method could be used by scientists to measure how far away an earthquake had originated? (Keep in mind that there are two distinct kinds of waves produced by earthquakes, each with a different speed.)

CHAPTER

23

When Sound Waves Meet

Have you ever seen a marimba? This interesting musical instrument is a large xylophone with a series of hollow tubes under the bars, one tube for each bar (Fig. 23-1). Notice that the tubes are of different lengths, with the longest tubes under the bars that produce the tones of lowest pitch. What is the purpose of these tubes?

Wave measurements

The purpose of marimba tubes can be demonstrated by a simple experiment. Suppose you hold a card under one of the bars of the marimba so as to block the opening to the hollow tube below, and then strike the bar. Instead of the usual loud sound, there is only a faint tone. As soon as the card is removed, the sound becomes loud again. Apparently one purpose of the hollow tube is to increase the loudness of the sound. This increase in sound due to a hollow tube is called **resonance.** The tube (actually the *air* in the tube) is said to resonate to the sound produced by the bar.

Another experiment shows something else about this increase in the loudness of a tone. Notice that the bars of the marimba and the tubes under them differ in length (Fig. 23-1). Suppose you change the position of two of the tubes, putting a short one in place of a long one and vice versa. Now when you strike the bars above the two tubes, you will find that the sounds are not loud but faint. Apparently each bar on the marimba requires a tube whose length is specially measured for the pitch produced by that bar.

1. Wave Length

Before examining this connection between the pitch of a tone and the length of its resonating tube, you should know some more facts about sound waves. Comparison with water waves will help.

Figure 23-2 shows two different water waves. Although their general shape is the same, they differ in one important

J. C. Deagan, Inc.

23-1 The hollow tubes under each bar of a marimba are designed so that they increase the loudness of the sound many times. Why are the tubes of different length?

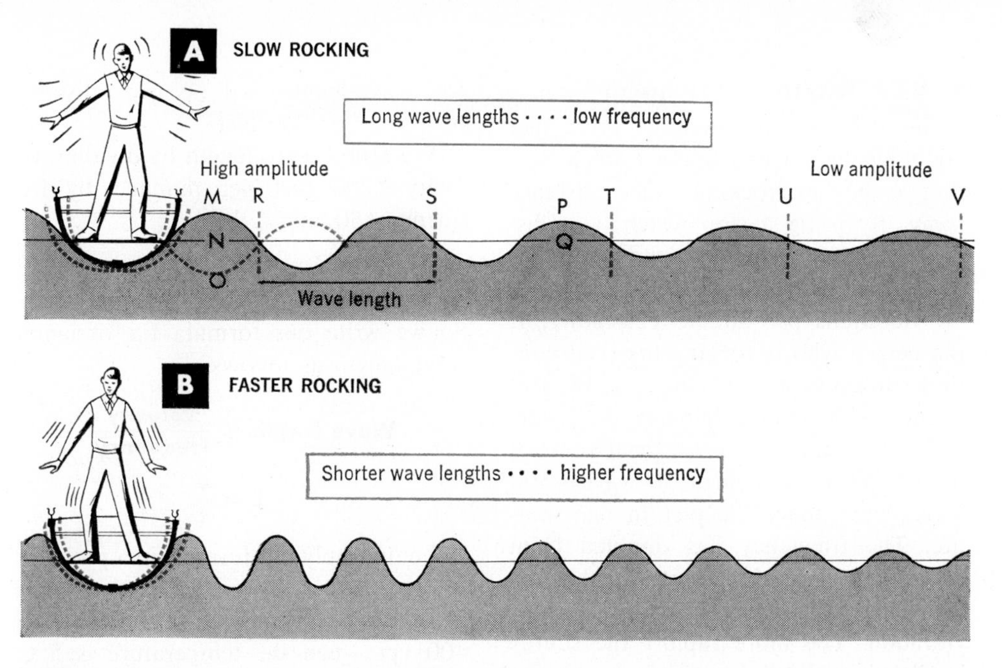

23-2 **The more rapid the rate of vibration, the higher the frequency and the shorter the wave length.**

respect. In wave A the successive crests are spaced farther apart than in wave B. Or, as we say, the **wave length** of A is greater than that of B. Wave length is the distance between the crest of one wave and the crest of the next wave. Or we could measure it as the distance between the bottom of one wave trough and the bottom of the next. **Wave length** may be defined as **the distance from any point on a wave to the corresponding point on the next wave,** as from R to S in Fig. 23-2.

2. Amplitude

Have you ever observed the wave caused by a boy rocking a rowboat? As the wave travels out from the boat, its height gets less and less until eventually it disappears. Near the boat, where the wave is high, it causes a twig on the water to bob violently up and down. If the twig were farther from the boat, its up-and-down motion due to the wave would be comparatively slight.

The word **amplitude** is used to describe the size of a vibrating motion such as that of the twig. Large amplitude means that the vibration is intense; small amplitude means that it is feeble. Amplitude is the distance a vibrating particle travels from its normal position before returning. Thus, in Fig. 23-2, a twig lying on the water at N is moved up to the top of a wave crest at M, so distance MN is the amplitude of the twig's movement. We could also measure the amplitude from N to O, the farthest point the twig reaches in traveling downward. As the wave travels outward from the source, it spreads its energy over a wider and wider circle, and as a result its amplitude decreases.

The boy in the rowboat could also regulate the amplitude of a wave by rocking the boat harder or more gently.

We note in Fig. 23-2A another important fact about waves: the wave length does not change as the amplitude changes. The distance RS (the wave length at R) is the same as UV (the wave length farther out at U).

3. Wave length and frequency

Suppose you are standing on a pier counting the waves coming from a boat that a boy is rocking some distance away. By using a stop watch, you discover that 50 waves pass under the pier in one minute. That is, the frequency is 50 vibrations per minute (50/min). If the person who is rocking the boat continues to do so at the same rate, the frequency of vibration remains the same.

But now the boy in the boat rocks it more rapidly. Instead of 50 waves, 100 waves pass under the pier in one minute. The frequency has doubled from 50/min to 100/min. So we see that the frequency is determined by the source of vibration. The more rapidly the source is vibrating, the higher the frequency.

Compare the spacing of the waves in Fig. 23-2A and B. The more waves that are sent out per minute the shorter the space (wave length) between them. So we see that an increase in frequency is accompanied by a decrease in wave length.

4. Wave length, frequency, and velocity

Since frequency and wave length depend upon each other, the relationship can be expressed in a formula. Suppose that we measure the velocity of the wave produced by the rocking boat in Fig. 23-2 and find that it is 200 feet per minute. Let us assume that the boat is rocking at the rate of 50 vibrations per minute.

In one minute the boat has sent out 50 waves. And at the end of one minute the first wave is 200 feet away from the boat. We can find the distance between these waves (their wave length) if we divide 200 feet into 50 equal parts. Thus our wave length is $\frac{200}{50}$, or 4 feet. We can check this answer by noting that 50 waves which are spaced 4 feet apart (the wave length) will occupy a distance of 4 × 50 or 200 feet.

We found wave length by dividing velocity (200 feet per minute) by frequency (50/min), thus:

$$\text{Wave length} = \frac{200 \text{ ft/min}}{50/\text{min}} = 4 \text{ ft}$$

So we write our formula for obtaining wave length as follows:

$$\textbf{Wave length} = \frac{\textbf{Velocity}}{\textbf{Frequency}}$$

$$L = \frac{v}{f}$$

Let us apply this formula in a problem.

Find the wave length of the sound given off by a tuning fork of frequency 400 vps when the temperature is 5°C.

Using the relationship between wave length, frequency, and velocity, we find that

$$L = \frac{v}{f}$$

$$L = \frac{1{,}100 \text{ ft/sec}}{400/\text{sec}} = 2.75 \text{ ft}$$

Note that the units for frequency are vibrations/sec (sometimes written vps) or vibrations/min (sometimes written vpm).

STUDENT PROBLEM

What is the wave length of a sound wave whose frequency is 22 vps at 5°C? (ANS. 50 ft)

▪ ▪ ▪ ▪ ▪ ▪ ▪

The formula in the above problems and the terms frequency and wave length apply to all types of waves. For example, a radio station is said to be broadcasting on 400 meters. This means that its radio signals have a wave length of 400 meters. Or we may say that the frequency of this broadcast is 750 **kilocycles.** The prefix "kilo-" means "thousand," and hence a frequency of 750

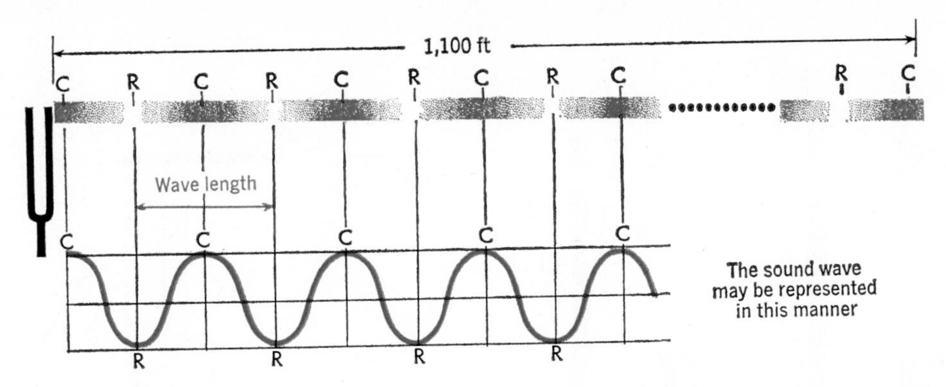

23-3 It is more convenient to graph a sound wave in a drawing as a transverse wave (bottom) than to draw it as a longitudinal wave (top). The distance from one C to the next (or from R to R) is the wave length. How could you calculate the wave length from the velocity and frequency?

kilocycles (750,000 cycles) means 750,000 vibrations per second.

In Fig. 23-3 we see that a sound wave can be represented by drawing a graph of it. The graph, you will notice, resembles a transverse wave in *appearance,* with the crests representing the compressions and the troughs representing the rarefactions. It must be remembered, however, that the actual sound is a longitudinal wave, and what appears to be a transverse wave is only a graph of the air pressure along a series of sound waves. A device for drawing the graph of a sound automatically is the oscilloscope (Fig. 23-4).

Allan B. du Mont Laboratories, Inc.

23-4 The wavy line on the screen of this oscilloscope is a faithful graph of the sound waves recorded on the record. Oscilloscopes are valuable laboratory tools for many other purposes where a visible display of a changing electrical signal is needed. Even your TV set is a kind of oscilloscope.

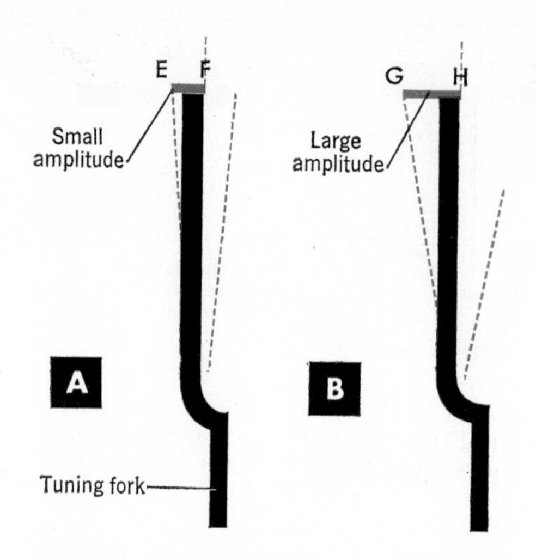

23-5 The amplitude of vibration of a tuning fork is the maximum distance that the prong moves from its normal position. The greater the amplitude, the louder the sound.

5. Amplitude and loudness

By amplitude of sound we mean the maximum distance that air molecules are displaced by the sound wave. If we produce the sound wave by striking a tuning fork, the amplitude depends on how hard the tuning fork is struck. The distance the prong moves when we strike the fork hard (GH in Fig. 23-5B) is greater than when we strike it gently (EF in Fig. 23-5A). A violent blow to the fork thus causes the air particles near the prong to move outward a greater distance than they would if the blow were gentle. In other words, the harder the blow the greater the amplitude of the wave that is set up and the more energy it contains. Our ears detect this increased amplitude as an increase in the loudness of the sound.

As the sound wave moves away from the tuning fork, its energy is spread out in all directions. This means that the amplitude of vibration of the air particles is less at a distance than it is at the source. Because of this decrease in amplitude, the sound grows weaker as we go farther from the tuning fork.

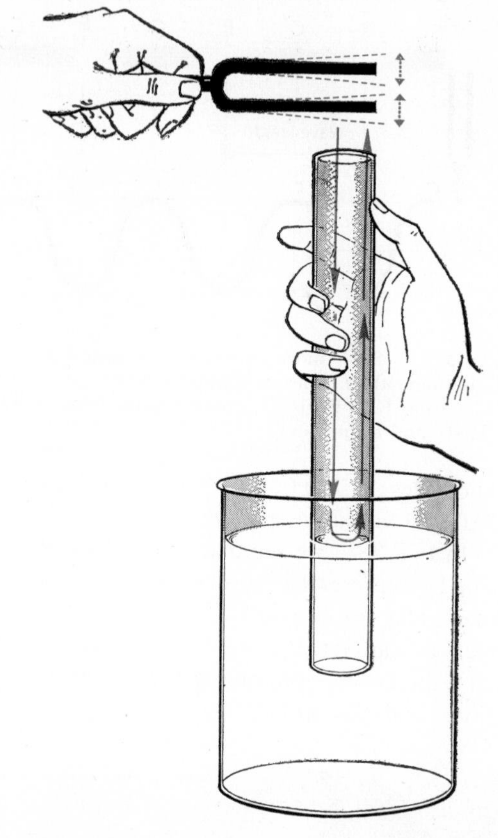

23-6 At a certain position of the hollow tube in water, the air in the tube resonates with the sound from the tuning fork and the sound becomes louder. Why?

Resonance

We are now in a better position to take up a question we raised at the beginning of this chapter: why do the different tones on a marimba require resonating tubes of different lengths?

6. Resonance and wave length

We can study the effects of resonance more closely by using the apparatus shown in Fig. 23-6. Take a hollow glass tube and place one end in water. Strike a tuning fork on a rubber stopper to make it vibrate, and place it above the open end of the hollow tube, so that a vibrating prong sends sound waves down the tube. Now raise and lower the tuning fork and tube together. At a certain position the tuning fork and air column resonate with maximum loudness; if we raise or lower the tube to other positions, the sound becomes faint. Apparently we get a resonating effect only when there is a certain definite length of hollow tube between the vibrating fork and the surface of the water. The length is exactly $\frac{1}{4}$ the wave length of the sound coming from the tuning fork. In other words, the wave length of the sound waves produced by the tuning fork is 4 times the length of the hollow tube required to resonate to it. The reason for this number is as follows.

Remember that one complete sound vibration consists of two halves, a compression and a rarefaction. The compression is sent out when the prong is moving downward, and the rarefaction is sent out when the prong is moving upward. In the experiment pictured in Fig. 23-7 we see the tuning fork at the start of a vibration: the prong is moving downward from its normal position and sends a compression, C, down the hollow glass tube.

This compression travels at a speed of 1,100 feet per second (the speed of sound). When it reaches the water at the bottom of the tube (Fig. 23-7B), it is reflected back to the prong again (Fig. 23-7C). During this time the compression has traveled two lengths of the tube (down and up). If this returning compression is to reinforce the amplitude of the sound, it must help the motion of the prong move upward. If the upward-moving compression approached a downward-moving prong, the motion of the prong would be opposed and the amplitude of the vibration would be decreased.

Therefore, for resonance the length of the tube must be such that the compression travels two lengths of the tube (down and up) in the same time that the prong completes half a vibration, sending out half a wave. Since half a wave extends two tube lengths during half a complete cycle of the tuning fork, a whole wave must extend four tube lengths during a whole cycle. Thus, at resonance, one wave length equals four tube lengths. This fact gives us an easy way of measuring the wave length of a sound.

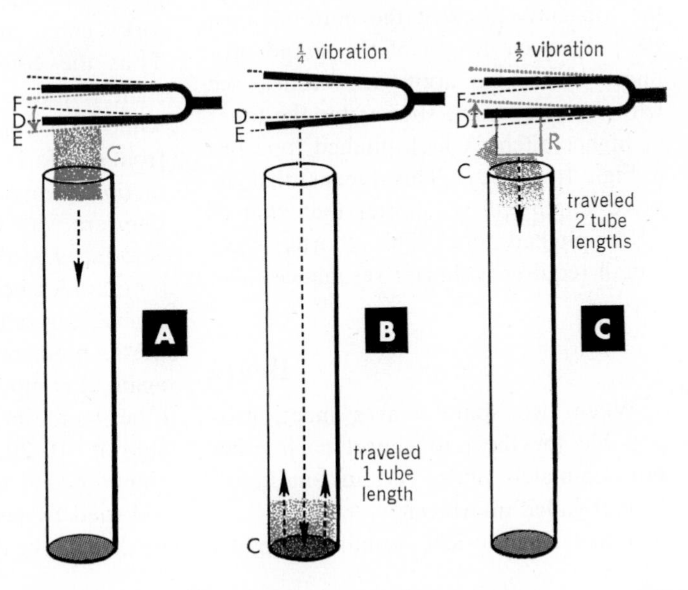

23-7 Resonance occurs when the length of the hollow tube is such that the reflected compression comes back exactly in time to join the next compression made by the fork itself. The two compressions reinforce each other, resulting in increased loudness.

SAMPLE PROBLEM

What is the frequency of the note produced by the bar on a marimba which has under it a $\frac{5}{8}$-foot closed resonance tube?

Since the resonance tube is closed, its length must be $\frac{1}{4}$ of the wave length it reinforces. Hence

$$\text{Wave length } L = 4 \times \frac{5}{8} = 2.5 \text{ ft}$$

The bar over the resonance tube should now be adjusted to give out this wave length. Using the relationship among frequency, velocity, wave length:

$$v = Lf$$
$$1{,}100 \text{ ft/sec} = 2.5 \text{ ft} \times f$$
$$f = \frac{1{,}100 \text{ ft/sec}}{2.5 \text{ ft}}$$
$$= 440/\text{sec (or vps)}$$

STUDENT PROBLEM

What is the resonant frequency of a closed tube whose length is 10 ft? (ANS. 27.5 vps)

▪ ▪ ▪ ▪ ▪ ▪ ▪

We see now why the hollow tubes under the bars of the marimba have to be measured carefully. Each tube, open at the top and closed at the bottom, must be $\frac{1}{4}$ the wave length of the sound produced by the bar above it. We also see why these tubes are shorter for the tones of higher pitch. A high-pitched tone has a high frequency. This means that its wave length will be shorter than that of a low-pitched tone. This shorter wave length requires a shorter resonance tube.

Beats

When two sound waves meet, it is possible for them to cancel each other out completely under certain conditions. This is called **interference.**

When two musical sounds of slightly different frequency alternately reinforce and interfere with each other, we get the effect known as **beats.** We can demonstrate beats by using two tuning forks of the same frequency and a heavy rubber band.

7. Producing beats

If we place the rubber band on the prong of one of the forks and then sound both of the forks together, we notice a peculiar throbbing or beating sound. The throbbing becomes more rapid the higher the band is placed on the prong.

What has happened is that the weight of the rubber band has slowed down the vibrations of one of the tuning forks and reduced its frequency. Now the compressions and rarefactions of the two forks are out of step. They cancel each other out at some moments and help each other at other moments. Suppose the two forks have the same amplitude and their frequencies are 30 vps and 27 vps respectively. Figure 23-8 shows all the compressions each of the two waves will produce in one second: 30 for one, 27 for the other.

Let us assume that at first the compressions and rarefactions from the two forks enter the ear at the same time. Thus the compressions at the extreme right reinforce each other. But a short while later compression 5 enters the ear from one fork at the same time that rarefaction 5 enters from the other. The two interfere with each other, and the sound is reduced. If the interference is complete (amplitudes being equal), no sound will be heard at all for a moment. Note that these moments of interference occur again at compressions 15 and 25. On the other hand, the compressions are back in step at 10, 20, and 30. Hence during a single second we get 3 periods of silence followed by periods of loudness; that is, we have 3 beats per second.

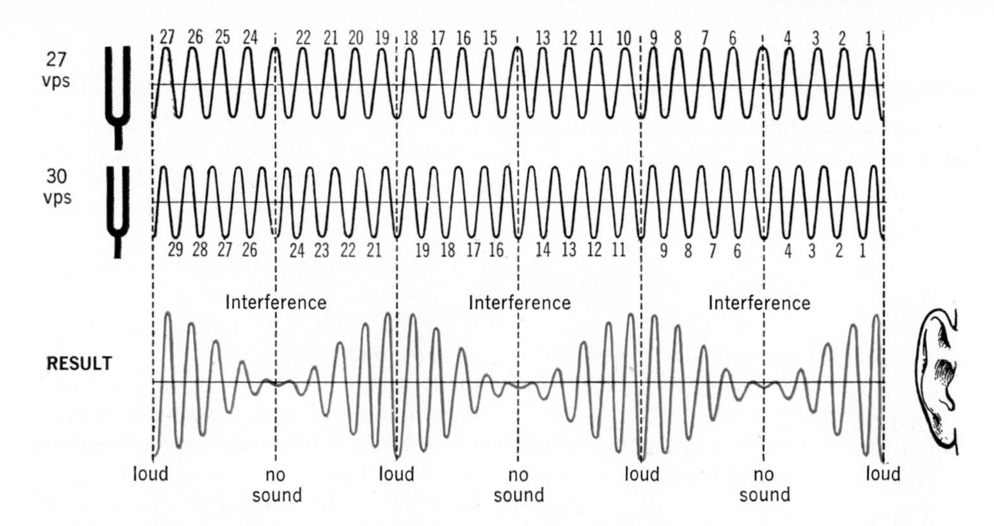

23-8 **Beats occur when two vibrations of slightly different frequency reach the ear together. Sometimes compressions from both reach the ear together and the sound is loud. At other times the rarefaction from one interferes with the compression from the other and little or no sound is heard.**

We see that the number of beats per second, 3, is equal to the difference between the frequencies of the two notes, 30 − 27. The number of beats per second can always be found in this way. Thus two notes with frequencies of 256 vps and 258 vps will have two beats per second (258 − 256). Notes with frequencies of 517 vps and 512 vps will have 5 beats per second, and so on. The throbbing you hear sometimes in a propeller-driven airplane is an example of beats, due to a very slight difference in the number of revolutions per minute of the propellers.

8. Looking ahead

In this chapter we have studied some of the important facts about sound waves. But many problems still remain. How do we detect sound waves when they strike our ears? And how is it possible for us to distinguish further between the many different sounds that we hear about us?

In the next chapter we shall consider these questions and show how our marvelous hearing mechanism enables us to distinguish one sound from another.

Going Ahead

THE VOCABULARY OF PHYSICS

Define each of the following:

resonance
beats
wave length
amplitude
kilocycles
interference
reinforcement

THE PRINCIPLES WE USE

Explain why

1. A note of high frequency has a short wave length.
2. The low tones of a marimba require long resonating tubes while the high tones have short ones.
3. Beats are heard when a rubber band is placed on the prong of a tuning fork and the tuning fork is sounded at the same time with another tuning fork of the same pitch.

EXPERIMENTS WITH SOUND WAVES

Describe an experiment to

1. Measure the wave length of the sound produced by a tuning fork.
2. Find out if two tuning forks of the same rated pitch are exactly the same.
3. Find the speed of sound in a room, using a tuning fork of known frequency and a resonance tube.

SOLVING SOUND-WAVE PROBLEMS

1. Calculate the unknown quantities in the following table:

Frequency	Wave length	Velocity
(a) 400 vps	?	1,100 ft/sec
(b) ?	10 ft	1,100 ft/sec
(c) 200 vps	5.6 ft	?

2. What is the approximate frequency of a sound that resonates to a tube 8 in long and closed at one end?
3. What is the frequency of the sound that resonates to a closed tube that is 6 ft long?
4. Find the wave length of "low C" (f = 128 vps) when the temperature is 31°C.
5. What is the velocity of sound in a room if a tuning fork of frequency 880 vps produces a 1.25-ft wave?
6. What is the temperature of the room in Problem 5?
7. What is the frequency of a bell that produces 20-ft waves when the temperature is 10°C?
8. What is the frequency of a drum that produces 3-ft waves when the temperature is 10°C?
9. What is the frequency of the note produced by a certain bar on the marimba if the bar has a $2\frac{1}{2}$-ft closed resonant tube under it?
10. A tuning fork is in resonance with a closed air column $\frac{7}{8}$ ft long. If the temperature is 10°C, what is the frequency of the fork?
11. What is the velocity of sound in the laboratory if a 512-vps tuning fork is in resonance with a closed air column $\frac{1}{2}$ ft long?
12. What is the frequency of a tuning fork in resonance with a 10-in closed air column if the room temperature is 16°C?
13. What is the velocity of the radio waves mentioned on page 300?

CHAPTER

24

How We Hear Music

Have you ever seen a conductor rehearsing a symphony orchestra? He may suddenly rap his baton for a halt in the midst of a passage and turn to the bassoon player. "That note was wrong," he says. "It should have been A flat!"

It seems incredible that out of such a complexity of sound the conductor can notice a single wrong note from one instrument. Yet he has simply developed to a high point an ability that all of us have to some degree. In this chapter let us look further into this ability to distinguish one sound from another.

Quality of a sound

All materials, including the air, can carry many vibrations of different frequencies at the same time. We can see how this is done by considering water waves.

A large wave in the ocean may have a wave length of hundreds of feet. The frequency of this wave will be low; that is, only a few such waves will pass a certain point in a given time. But over these large waves a small motorboat is passing. The boat sends out its own waves, which have a wave length of, say, 3 feet. These travel over the surface of the larger wave. A breeze arises and sends tiny ripples across the surface of the motorboat waves. The ripples may have a wave length of only a few inches. All three vibrations can be detected on the water at the same time, as shown in Fig. 24-1. In very much the same way different vibrations travel simultaneously through the air.

1. Many different vibrations

The fact that different musical tones exist in the air at the same time can be shown by the resonance experiment we performed in Chapter 23. Select an aluminum tuning fork with a broad side and set it up as shown in Fig. 23-6. Strike the fork hard on a flat rubber stopper. Now hold it over a resonance

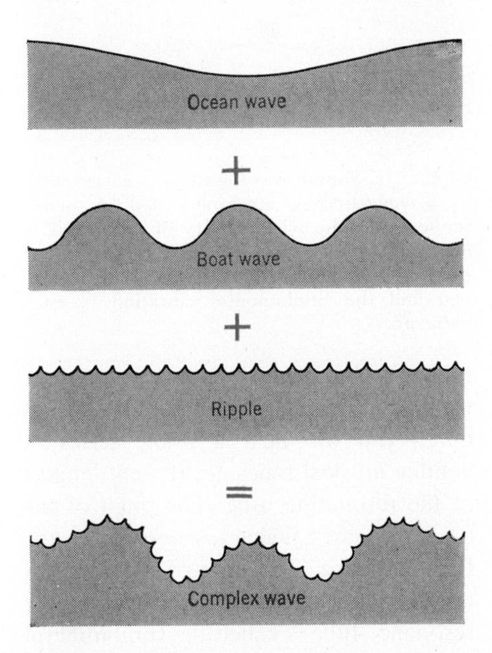

24-1 Look at the complex wave (bottom). Can you pick out the three waves of different frequency that formed the complex wave? In a similar way you can pick out the separate tones that make up the complex sound of an orchestra.

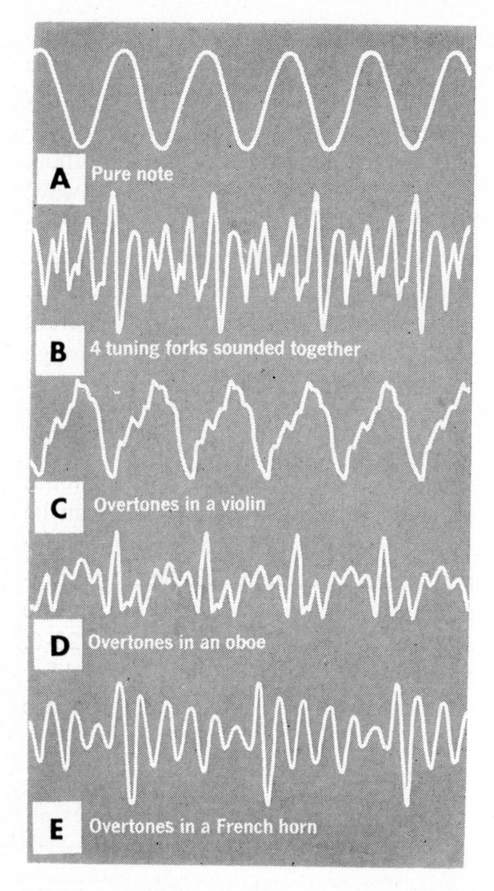

24-2 Sound wave patterns, as revealed by a wave-analysis machine called an oscilloscope, show the wide variety of vibrations in musical sounds. Each of the above vibration patterns is a sound of different quality. Can you find the fundamental vibration in each pattern?

tube and slowly move both tube and fork up and down together. If the room is quiet, you will hear a whole series of definite musical tones at different lengths of the resonating tube. The pitch of the tones becomes higher as you lower the fork and tube. The tone of lowest (and usually loudest) pitch produced by the resonance tube is called the **fundamental tone.**

So we see that the tuning fork is producing not only its fundamental vibration but vibrations of higher frequency as well.

2. Overtones

The higher tones which are produced in addition to the fundamental tone are called harmonics or **overtones.** Every musical instrument differs from every other in the frequency and relative intensity of the overtones it produces. The overtones of a violin are different from those of a tuning fork. The overtones of a piano are different again from those of a violin. If the instrument emphasizes the lower overtones, as a cello does, we say it has a mellow quality. If, like the oboe, it emphasizes the higher overtones, we say the quality is penetrating or brilliant. It is this difference in the overtones produced by various instruments that causes their difference in **tonal quality** and enables us to tell them apart. For the same reason we can identify people by the sound of their voices.

3. Duplicating tonal quality

By knowing the amplitude and frequency of the overtones of any instrument, we can produce artificially the tonal quality of that instrument.

Figure 24-2 shows an analysis of the vibrations of tones from various sources of sound. From it we can see that sounds can differ from each other in three ways: loudness, pitch, and quality.

The hearing mechanism

Your ears are amazing mechanisms which enable you to distinguish between different kinds of sound. To most people the visible part of the ear might seem the important part. This may be true for long-eared animals like the rabbit or donkey (where the large outer ear concentrates the sound), but it is not true for human beings. We could hear almost as well if we had no outer ears at all.

4. Inside the ear

A small, flexible membrane, the **eardrum** (Fig. 24-3), separates the middle ear from the outside air. The compressions and rarefactions of sounds reaching the flexible eardrum cause it to vibrate back and forth in time to the vibrations. A series of three bones, arranged as a system of levers, transmits these vibrations to the **cochlea** (kŏk′-lē·ȧ) in the **inner ear**, where they are sorted out. The cochlea is about an inch long and shaped like the coiled-up shell of a snail. It is filled with liquid and contains a flexible membrane along its entire length. All along this membrane, from the wide part of the cochlea to the tip, are numerous nerve endings which lead to the brain.

When sound vibrations reach the cochlea, the membrane begins to vibrate. Different parts of the membrane vibrate to the different frequencies that exist in the original sound. Thus one region responds to frequencies of 256 vps, another to 512 vps, and so on. Each of these parts of the membrane has its own nerves that send impulses to the brain. When a sound enters the cochlea, all the membranes resonant to the frequencies present in the sound are set into motion.

In this way the brain receives a sep-

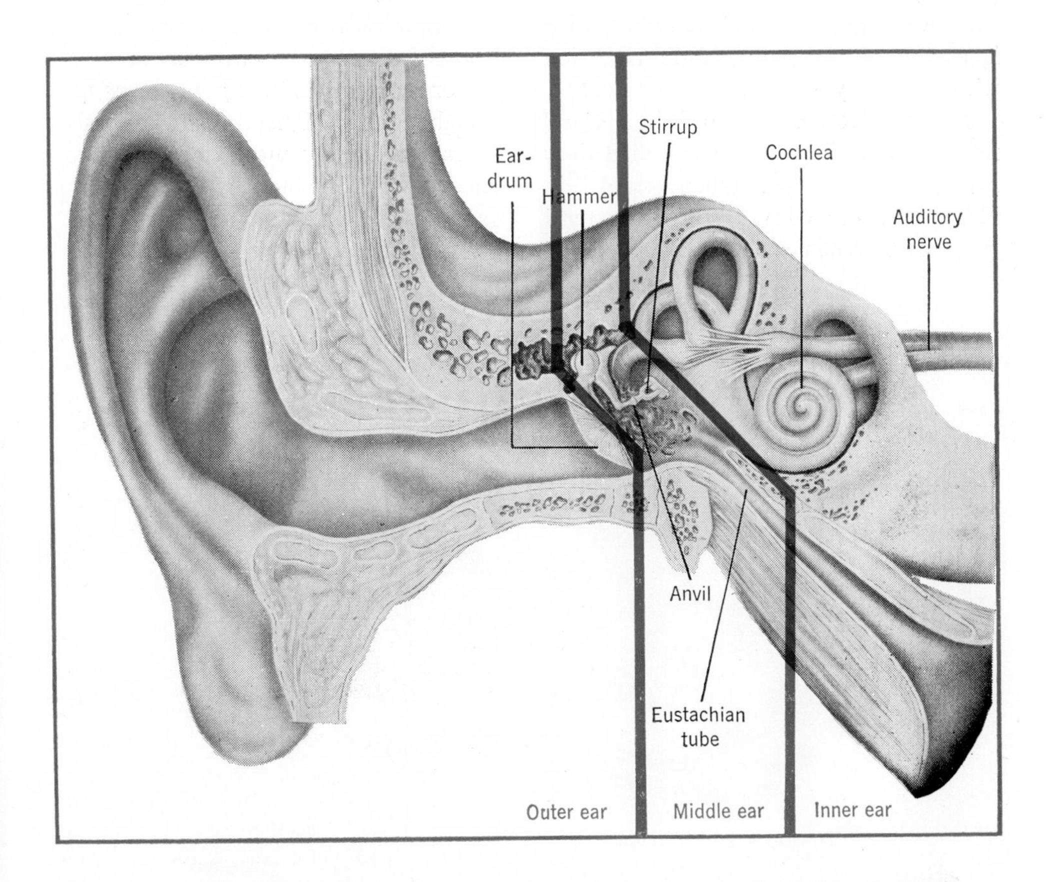

24-3 Sounds entering the outer ear set the eardrum into vibration. Then the bones of the middle ear transmit the sounds to the inner ear through a clever arrangement of movable bones—the hammer, the anvil, and the stirrup. Finally, the coiled cochlea of the inner ear sorts out the vibrations and sends nerve impulses to the brain, which interprets the sound.

arate combination of impulses for each vibration in the original sound. Thus it is possible to distinguish one sound from another having different overtones and a different fundamental tone.

5. Sympathetic vibrations

Open up a piano so that the strings are exposed. Step on the loud pedal, which lifts up the felt pad that muffles the strings and keeps them from vibrating freely. Now sing a tone into the strings. When you stop to listen, you will hear the same tone coming faintly back to you out of the piano. By feeling the strings, you can find the one that is vibrating; when you pluck it, it gives back the same tone you sang. The piano string is said to be in **sympathetic vibration** with the tone you sang.

Many objects have a definite *natural* frequency of vibration (also called their *resonant* frequency). When struck, an object tends to vibrate with its natural frequency. When sound waves of the same frequency reach the object from outside, it begins to vibrate. The piano string had the same natural frequency as the tone you sang. Had you tried another tone, a different string would have responded.

Although the hearing process in the cochlea is more complicated and is not yet fully understood, the action can be compared in a general way to the action of the piano strings. The different parts of the membrane seem to respond sympathetically only to vibrations of the same natural frequency.

6. Range of hearing

Marvelous as the human ear is, it has its limitations. We cannot hear sounds whose frequency is less than about 16 vps or more than 20,000 vps. Dogs have a higher range of hearing. It is possible to call a dog by means of a high-frequency whistle which has a sound that cannot be heard by the human ear.

The bat also hears sounds of a very

Edison Laboratory National Monument

24-4 Here is the first successful phonograph. The record was a sheet of tinfoil wrapped around the rotating drum. The sound was recorded by a needle scratching waves on the tinfoil; it was reproduced by the same needle vibrating as the wavy grooves were rotated under it.

Pickering and Co., Inc. *Hi-Fi and Music Review*

24-5 The grooves in the rotating LP record set the needle into vibration and produce the same sound that caused the grooves. Can you tell which parts of the record will give loud sounds? The wavy lines are a greatly magnified view of the actual grooves in a record.

high frequency, and emits them as well. It produces vibrations up to 80,000 vps, which are outside our range of hearing. We saw in Chapter 22 how the bat uses these sounds to guide itself in flight.

Many new experiments have been performed with very high-frequency (**ultrasonic**) sound waves of 20,000 vps and more. The energy of such waves can kill fish in a bowl and can heat up objects near by. Ultrasonic waves are being used to drill teeth painlessly, and a device has been developed which can use such waves to drill square holes.

7. The first phonograph record

Edison made the first phonograph record. He accomplished this feat by causing the vibrations from his voice to set a needle into the same kind of vibration. As the needle vibrated against a rotating drum, it scratched these vibrations into tin foil, producing a wavy groove which spiraled around the drum from one end to the other (Fig. 24-4). To play the record, the drum was then rotated with the needle sliding in the groove. The needle was thus forced to vibrate in the same manner as the original sound, producing waves in the air that were a reproduction of the original sound. More refined modern methods produce a comparable effect (Fig. 24-5).

Music

All musical instruments have one thing in common: they all make use of an elastic material that can vibrate easily. We make these vibrations by plucking, striking, or drawing a bow over taut strings, or blowing air into hollow tubes.

8. Kinds of instruments

Most musical instruments fall into three main groups: stringed instruments, wind instruments, and percussion instruments.

The stringed instruments include the violin, viola, cello, bass viol, mandolin, guitar, banjo, ukulele, and harp. Among the wind instruments are the flute, piccolo, oboe, trombone, French horn, tuba, trumpet, clarinet, and saxophone. The percussion instruments, in which vibration is achieved by striking, include the xylophone, marimba, bell, chime, triangle, and drum.

The piano may be considered a stringed instrument operated by percussion. The organ is a variety of wind instrument.

9. Stringed instruments

Examine a violin or any other stringed instrument. You will note from Fig. 24-7 that the violin has four strings of equal length, which produce different tones. The thick, heavy G string has the lowest pitch, while the thin, light E string has the highest. If you tighten a string, its pitch gets higher.

10. Controlling pitch

A violinist controls the pitch of a tone in three ways.

1. He has four strings to choose from in playing a note. Because a heavy string has greater inertia than a light one, it moves more slowly and produces fewer vibrations. The heavier the string, the lower the pitch.

2. He tunes each string to its correct pitch by tightening or loosening it, thus changing the tension of the string. A tight string produces more vibrations than a slack one because it springs back and forth more rapidly. The looser the string, the lower the pitch.

3. He shortens the vibrating part by pressing against the string with his finger. A short string vibrates more rapidly than a long one. The longer the string, the lower the pitch.

11. Making the sound louder

If we stretch a violin string between the ends of a board and pluck it, it gives out a rather faint, unimpressive sound. Some reinforcement is needed. This is provided in the violin by the hollow, curved wooden "box," which is forced

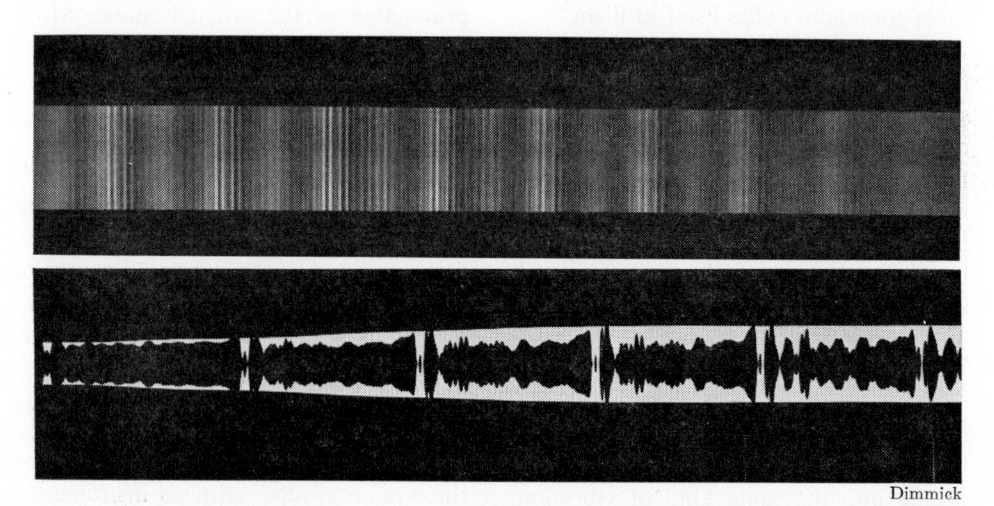

Dimmick

24-6 The wave nature of sound is clearly shown in these two types of movie sound track. The vibrations of the sound wave are converted into electrical vibrations by the microphone and then into changing light by a lamp, and are photographed on the movie film. When the film is run through the projector, light shining through the sound track is converted into electricity by a photoelectric cell and then back into sound in a loud-speaker. In the sound track in the bottom picture the width of the black and white area is varied. In the sound track at the top the tonal density of the film (blackness) is varied to produce the vibrations of the sound.

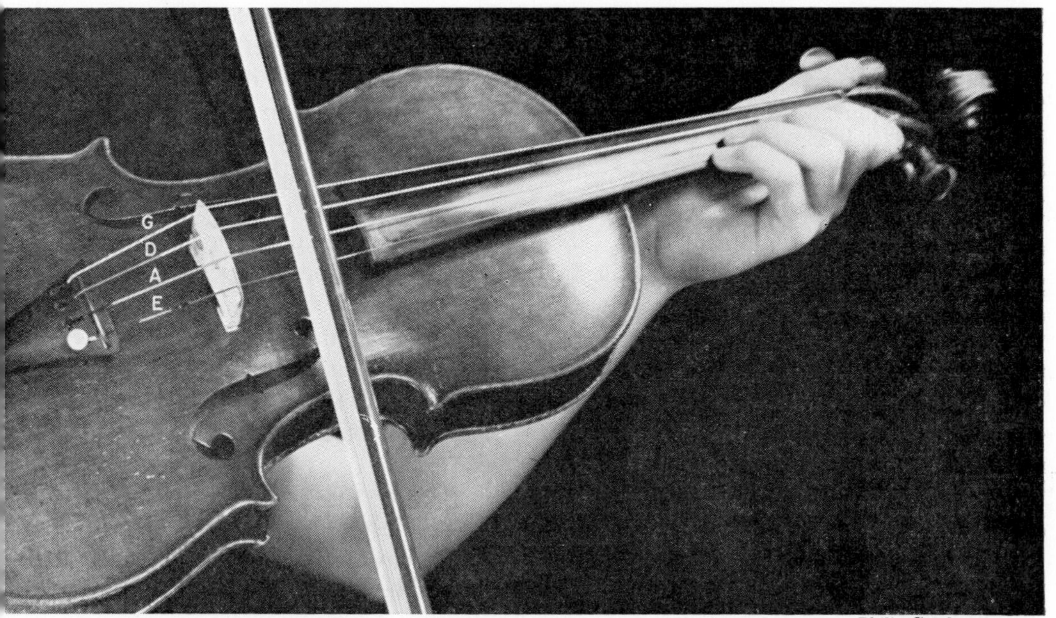

Philip Gendreau

24-7 A violin is a work of science as well as a work of art. What three methods are used by this violinist to change the pitch of the tone produced by the violin? Why does the hollow violin box have such a peculiar shape?

to vibrate at the frequency of the string. It is not only the vibration of the air in the hollow box that increases amplitude. The fact that the light wood of the box is set into vibration by the strings is also a great help. Because the whole surface of the box is vibrating, there is much more motion in the surrounding air than the string could produce alone, thus increasing the loudness.

You can illustrate this effect by placing a vibrating tuning fork against a large wooden desk (Fig. 22-1). The sound becomes much louder as the desk is set into vibration by the fork. This accounts for most of the amplitude of sound in instruments such as the violin and piano.

12. Wind instruments

All wind instruments use the principle of the **resonating air column.**

When you blow across the open end of the resonance tube pictured in Fig. 24-8, a compression starts down the

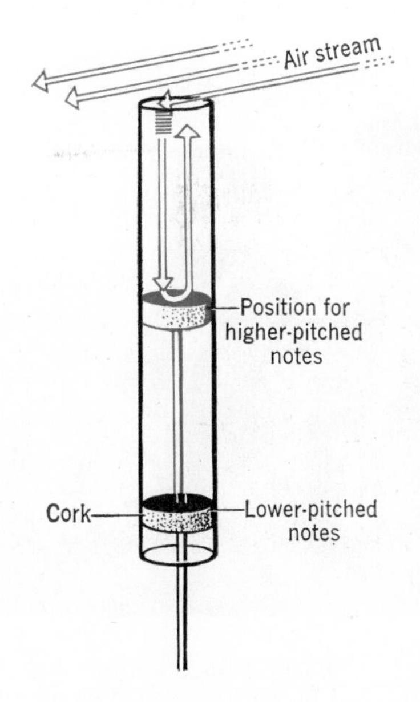

24-8 You can actually play a tune with this simple "trombone." Why not try it and see for yourself?

tube. It strikes the bottom, is reflected, and travels back up the tube. As this compression reaches the top, it deflects the air stream and pushes it away from the open end. As this air expands outward from the open end, a rarefaction starts down the tube, bounces off the bottom, and returns to the top. This rarefaction now deflects air back into the tube. In turn this air starts a second compression on its way down, and so on.

Thus the air in the tube is set into vibration with compression following rarefaction in rapid succession. The number of such vibrations per second depends only on the time it takes for a compression to travel down the tube and back again. If the tube is longer, the trip requires more time; there will be fewer vibrations per second and the note produced will have a lower pitch.

From our description of the manner in which the compression moves down and up the tube, you will recognize the resonance process described in Chapter 23. That is why a wind instrument of this type always produces the same tone that resonates with it.

However, most wind instruments have a tube which is open at both ends. The time required for a compression to get out in the air is only half what it is when the same tube is closed at the bottom end. This means that twice as many vibrations are produced per second, and the frequency of the tone is twice as great. We say that the tone produced in an open tube is an octave higher (twice the frequency) than that produced in a closed tube of the same length.

13. The human voice

The most remarkable of all musical instruments is the human voice itself. This complicated "instrument" operates on principles very similar to those we have just been describing.

We produce sounds by means of the **vocal cords,** which vibrate when air is blown through them. These cords are located in the **larynx** (Adam's apple) of the throat. They are not really cords, but

Concert Artists, Inc.

24-9 Each of these musical instruments is essentially an air column whose length can be changed by the musician. Vibrations are caused in each column by the air the musician blows against one end of the air column. The pitch of the resulting sound is determined by the length of the vibrating column, and the quality of the sound is determined by the shape of the instrument and the material of which it is made.

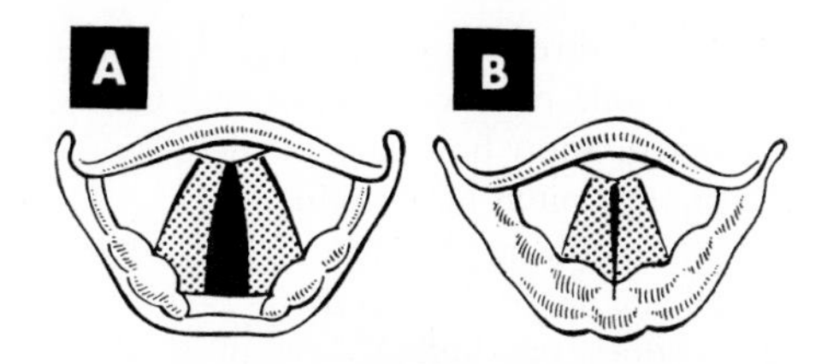

24-10 The wonderful variety of sounds (pleasant and unpleasant) that can be produced by a human being are the result of vibration of flexible membranes (vocal cords) in the windpipe. (A) The relaxed position of the cords during breathing. (B) The cords stretched for speaking or singing.

a pair of stretched membranes which can be tightened or loosened by muscular action. When the cords are taut and vibrate rapidly, a high-pitched tone is produced (Fig. 24-10B). When they are more relaxed, the vibration is slower and the pitch is low. Not all people have the same range of tones, because the size and tension of all vocal cords are not the same. Children, for example, have small vocal cords, and their voices are high-pitched.

The air spaces in the chest, head, nose, throat, and mouth passages are resonating air columns which amplify and change the quality of sounds besides giving each individual voice its special character. We make a variety of sounds in speech simply by changing the positions of the tongue, lips, cheeks, and vocal cords.

14. What does a phonograph do?

The widespread interest in music in this country and the development of high-quality records and playing equipment has created an entirely new hobby—"hi-fi." If you are a high fidelity fan you probably know a woofer from a tweeter, but if you are an "outsider" you may wish to know what all the talk is about. Perhaps the brief sketch below will help you.

The term *high fidelity* means the faithful recording and reproduction of all the details of a sound. It is fairly easy to invent equipment to record music and to play it back, but it is very difficult to design it to reproduce all the overtones of the original sound in the correct proportions. Every step in the process has to work perfectly to avoid distortion of the original sounds. The essential steps are as follows.

As the turntable bearing the record rotates beneath the pick-up arm, a needle mounted in a small container called a cartridge in the end of the arm follows the groove of the record (Fig. 24-5). The "needle" may be a tiny sapphire or a diamond. Inside the cartridge the vibrations of the needle are converted into a varying electric current which *should* be a faithful replica of the wiggles on the record and reproduce the original sound.

Current from the pick-up is so tiny that it often is fed into a preamplifier, an electrical circuit whose function is to boost the signal sufficiently to operate the next stage. The boosted electrical signal from the "preamp" is fed into the amplifier, an electrical circuit which enormously increases the strength of the current, again with the least possible distortion of its various frequencies.

The amplified current now flows to the loud-speaker system. Instead of one loud-speaker inside the carefully proportioned speaker cabinet, there may be two loud-speakers, one often called the tweeter, designed to reproduce the high-frequency sounds, the other often called the woofer, to reproduce the low-frequency sounds. The high and low frequency electrical signals from the amplifier must be sorted out and sent to the proper loud-speaker, which is done by an electrical circuit called a crossover network, mounted inside the loud-speaker cabinet.

Even if each piece of equipment we have named passes along the original sound with no distortion, you will still not hear music as it was originally played. When you are listening to a "live" orchestra, your two ears give you a sense of direction, a 3-D sensation, comparable to 3-D movies in which a slightly different image in each eye gives you a sense of depth. To create the same effect with a phonograph requires special records and two complete systems from preamplifiers to loud-speakers. The resulting stereophonic sound is one of the most dramatic improvements in high fidelity in recent years.

➤ *Going Ahead*

THE VOCABULARY OF PHYSICS

Define each of the following:

larynx	inner ear
fundamental tone	ultrasonic vibration
vocal cords	eardrum
overtones	sympathetic vibration
resonating air column	cochlea
	octave
tonal quality	harmonic

THE PRINCIPLES WE USE

Explain why

1. When a piano plays middle C (256 vps), it sounds different from the same note played on a violin.
2. It is possible for a dog to hear certain whistles which a man cannot hear.

EXPERIMENTS WITH VIBRATIONS

Describe an experiment to show

1. Sympathetic vibration with a piano.
2. That a vibrating tuning fork may emit many tones at the same time.

APPLYING PRINCIPLES OF SOUND WAVES

1. How are sounds of different pitch and quality produced by the human voice? Why do children have voices of higher pitch than adults?
2. Mention three ways in which the pitch of the tone produced by a string can be changed. Show how each of these is used to change pitch in a violin.
3. Explain how a phonograph record can duplicate speech and music.
4. Why does the hollow box of a violin make the sound louder?
5. Describe the mechanism of the ear and explain how we hear different sounds.
6. Explain how the musical sound of a wind instrument is produced.

SOLVING SOUND-WAVE PROBLEMS

1. What is the frequency of the tone obtained by blowing across the top of a test tube 6 in long? What tone would be obtained if the bottom of the tube were open?

---

2. The fundamental tone from a certain tuning fork resonates with a tube closed at one end and 9 in long. What is the frequency of the fundamental tone?
3. What are the frequencies of the main overtones which are heard to resonate to closed tubes $4\frac{1}{2}$, 3, $2\frac{1}{4}$, and $1\frac{1}{2}$ in long?
4. What are the answers to Problem 3 if the tubes are open tubes of the same lengths?
5. What is the range of sizes of closed tubes whose resonant frequencies include the full range of frequencies audible to the human ear?
6. How many beats would be heard if two open-pipe whistles were blown simultaneously? Their lengths are 8 and 9 cm respectively and the temperature of the air is 24°C.
7. If the above tubes produced their first overtones, how many beats would be heard?
8. If the ends of the tubes were closed, how many beats would be produced?
9. If the ends were closed and the tubes were blown to produce their first overtones, how many beats would be heard?

UNIT FIVE

Electrical Energy—The Motion of Electrons

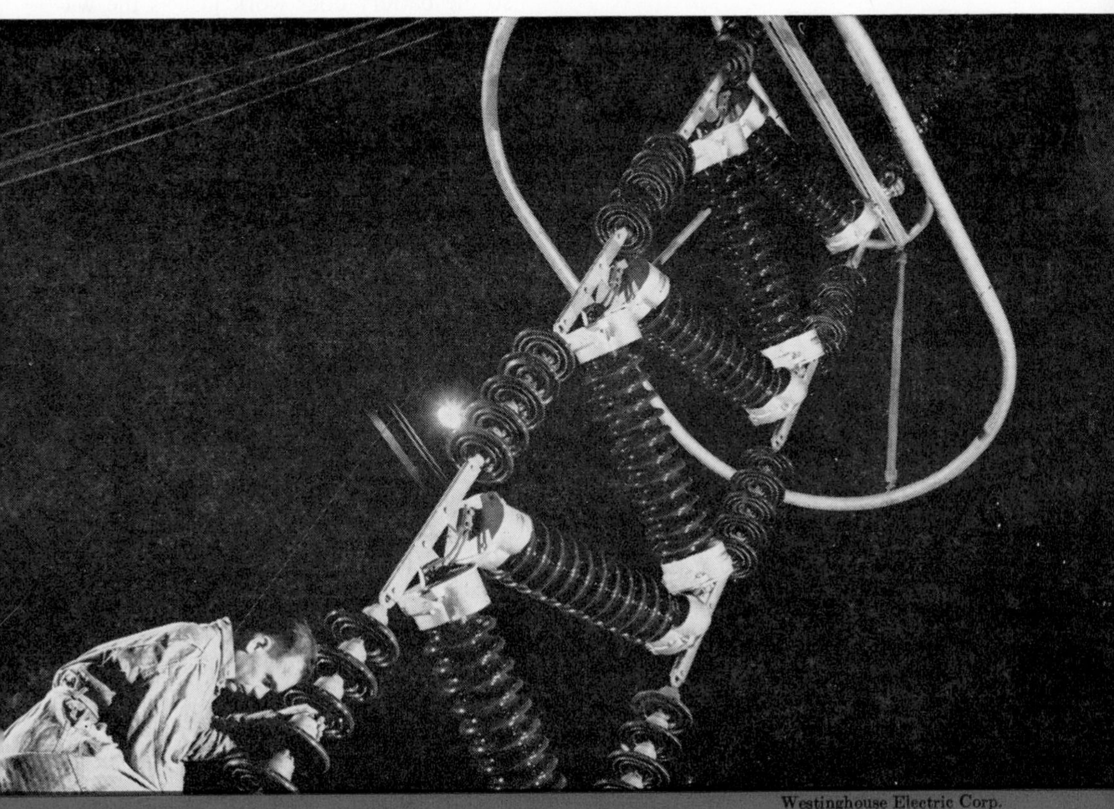

Westinghouse Electric Corp.

The development of electricity has given us the telephone, telegraph, radio, television, and electric lights and electric motors—to name only a few electrical marvels. Electricity can even operate giant "brains" that can multiply 2,864,928,253,837,490,496,107,668 by 395,012,893,129,804,211 and then divide the answer by 318,902,746,097,224,959,041 in less time than it takes to read this sentence.

Our expanding knowledge has made possible ever more powerful sources of electrical energy and more efficient devices for controlling it and using it. The equipment illustrated is a new lightning arrester to protect high-voltage power lines. The study of electricity leads us to a more thorough understanding of the nature of matter and energy, the very center of interest of modern physics.

CHAPTER

25

The Flow of Electricity

Beneath the streets of a large city lies a bewildering maze of pipes. Each one of them is a **conductor** of something, and together they make the life of a city possible. Some of the pipes conduct water, others gas; still others carry waste and sewage away. Then there are wires which carry electrical energy.

We can compare these electrical wires to pipes because electricity flows through them somewhat as water flows through a pipe.

The complete circuit

To understand how the flow of electricity resembles the flow of water, compare the action of a water circuit and an electrical circuit. Consider a water circuit that consists of a water pump, water pipes, and a water-wheel motor driven by the flow of water (Fig. 25-1A). To push molecules of water through the pipe and water wheel requires a force. The force is supplied by the water pump. Now compare this circuit with an electrical circuit in which a dry cell or battery operates a small electric lamp, using two wires as conductors (Fig. 25-1B). The wires are electrical pipes. The electrons in the copper atoms, of which the wire is made, may be considered "electrical fluid." The battery is the electricity pump that supplies the force necessary to push a current of electrons through the wires and the bulb. In doing so the battery does work just as the water pump does.

1. Water flow

Notice the resemblance between the conductors in the water circuit and the electrical circuit. The water pump has a pipe carrying water in and an outlet pipe carrying water away to the water-wheel motor. Similarly the battery, an "electricity pump," has one wire bringing electric current in and one carrying it out to the electric lamp.

The "round trip" of the water or the electric current is called a **complete cir-**

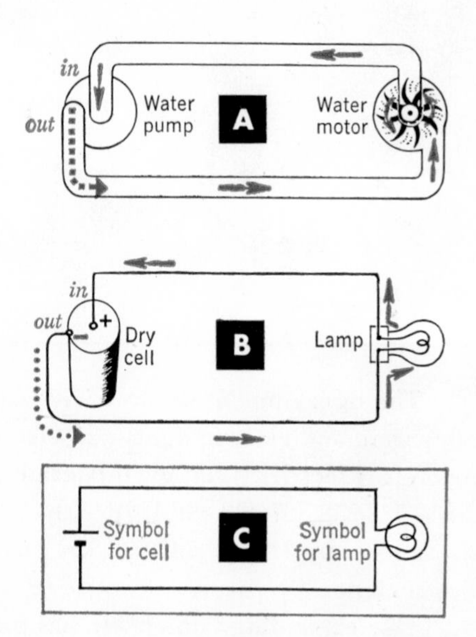

25-1 An electric current resembles the flow of water in a pipe. The dry cell is an "electric pump" and the wires are the "electric pipes." Each device needs an inlet and outlet pipe or wire. And the circuit must be "completed" by a return wire or pipe to the pump.

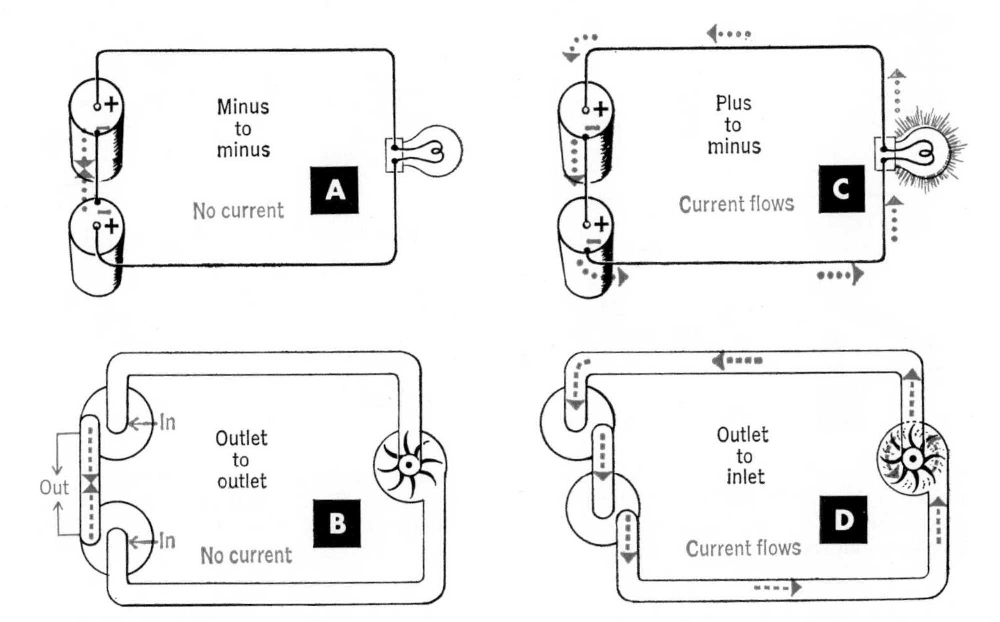

25-2 If two water pumps are connected outlet to outlet, no current will flow, because the pumps oppose each other. The connection must therefore be outlet to inlet. In the same way we get no current if cells are connected minus to minus (outlet to outlet). They must be connected minus to plus.

cuit. If the circuit is broken for any reason, or if the pump stops or the battery fails, the flow of water or electricity stops in the entire circuit.

2. Plus and minus

If you examine a dry cell, you will probably find the following information printed on it: "Negative terminal on outside. Positive terminal in center." These terms "positive" and "negative" come from an attempt to explain the nature of electricity made by Benjamin Franklin a long time ago. He assumed that the electrical fluid flowed out of one terminal of a cell, which he called the positive (plus) terminal, and in again at the negative (minus) terminal.

More recent information represents the electron flow as proceeding from minus through the external circuit and back to plus. We shall follow this modern convention throughout this book. The direction of electron flow is shown on the diagram by means of arrows drawn next to the wires.

3. Using two cells

Suppose we want to increase the flow of current in order to make a lamp give more light. Since two cells should cause more current to flow than one, we might try connecting them as shown in Fig. 25-2A. We have a complete circuit, but we get no current at all. Why? We have connected the minus terminal of one cell to the minus terminal of the other. But once we connect the plus of one cell to the minus of the other, we get the increased current we want in the circuit (Fig. 25-2C).

To see why this is so, let us turn again to the comparison with a water pump. In Fig. 25-2B we have connected the outlet of one pump to the outlet of another. Now the two pumps tend to send water through the pipe in *opposite* directions. But actually, because the push of

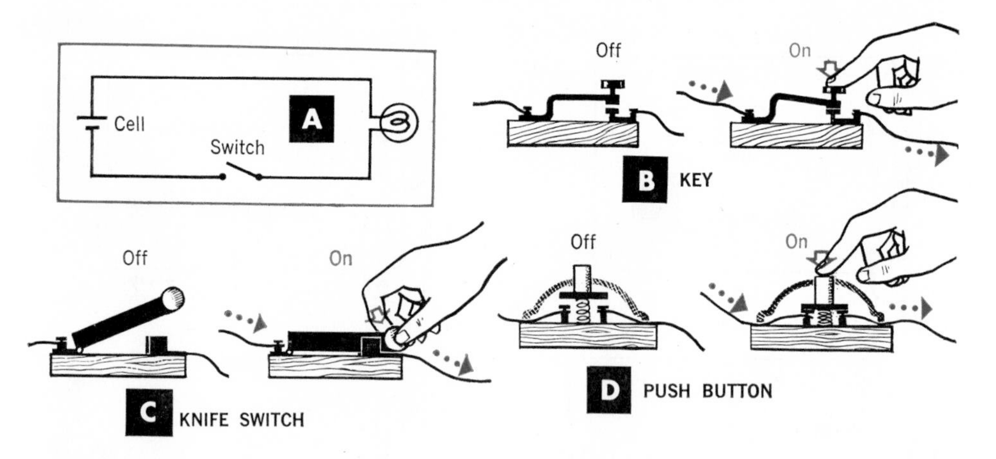

25-3 "On" or "off" in electrical devices is accomplished by switches. These are simply breaks in the circuit which are bridged by moving the switch. B, C, and D show simple types of switches for different purposes. A shows how a switch (or key) is connected in a circuit. The long and short line is the symbol for a cell.

one pump is counteracted by the equal and opposite push of the other, no water flows through the pipe at all. If we remedy the situation by connecting the outlet of one pump with the inlet of the other, then the two pumps aid each other and we get an increased flow of water (Fig. 25-2D). In other words, we have increased the force pushing water through the pipe.

4. Shutting off the current

Not all materials can be used as "pipes" to conduct electric current. For example, if we substitute string for metal wires, we find that the current stops flowing. A material like string, which does not permit electric current to flow, is called an **insulator.** Some commonly used insulators are air, glass, porcelain, plastics, paper, and oil. Some insulating materials used to cover wire are rubber, asbestos, lacquer, and plastics.

Actually, even the best insulator can be made to conduct some current. If there is sufficient electrical pressure, very small currents can be forced through a piece of string. But for all practical purposes we can think of such an insulator as blocking the flow of electrons.

Air is the most important of these insulators because it is the one we rely on most often. Only when there is a very high voltage (high electrical pressure), which gives a strong push to the flow of electrons, can they be made to jump the air gap from one wire to another. When this does happen, a "spark" jumps the gap.

Ordinarily even a small gap of air in the circuit is enough to cut off the current. This fact about air is useful in giving us a way to start or cut off the electric current in a wire at will. We arrange things so there is a break at some point in an otherwise complete circuit. When we bridge the gap with a piece of conducting metal, the circuit is completed and the current begins to flow (Fig. 25-3A). When we remove the bridge, the complete circuit is broken and the flow of electrons stops throughout the entire circuit. Switches, keys, and push buttons—shown in Fig. 25-3—are different types of "bridges" which we can use.

5. Parallel and series circuits

A dry cell can be used to operate several different devices at the same time. Figure 25-4A shows one way in which two devices can be connected to a single water pump. The water flows from the pump through device M_1 and then through device M_2 before returning to the pump. Figure 25-4B shows an electrical circuit of the same kind in which one dry cell is used to light two lamps. In the water circuit, a valve turns on or shuts off the flow; in the electric circuit a **switch** serves that purpose.

This arrangement is called a **series circuit.** You can see that in a series circuit, when you shut off the flow of electricity, all the devices in the circuit will go off together. Some Christmas-tree lights are arranged in this manner; all the lights in the circuit go out if a single bulb fails. A series circuit, however, has the big advantage of simplicity of wiring, and therefore costs less to make.

Obviously a series circuit would be a great disadvantage in the home, where we want to turn on different lamps, the radio, the fan, etc., independently of each other. Therefore, the circuits used in house wiring are of another kind, called **parallel** (branching) **circuits.** Figure 25-4D shows a parallel circuit operating with a water pump, and Fig. 25-4E shows a similar parallel electric circuit using a dry cell. You notice that each device that is being operated by the cur-

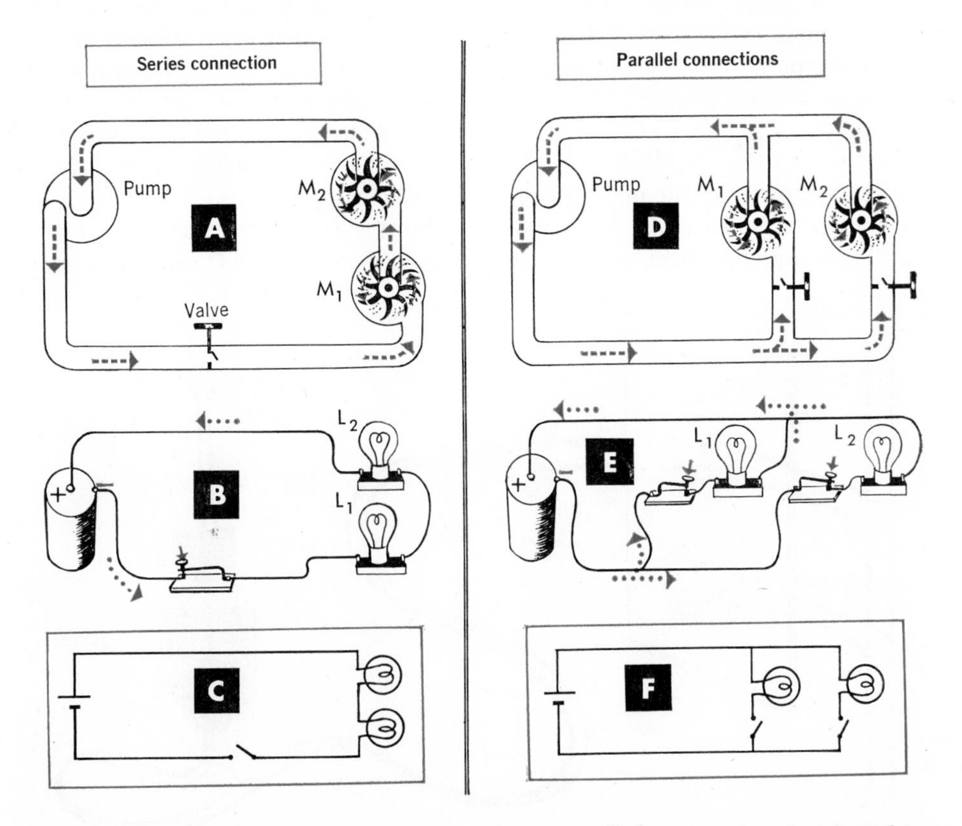

25-4 In the series (one-after-the-other) arrangement all the current passing through one device must pass through the others. If current is stopped in one, it is also stopped in the others. But in the parallel (branching) arrangement, each device gets current through a separate branch and so operates independently of the others.

rent can be shut off by separate valves in the water circuit and by separate switches in the electric circuit.

Figures 25-4C and 25-4F show a simplified way of drawing these devices in an electrical circuit.

A parallel circuit has another important advantage over a series circuit. You can see what this advantage is by arranging three lamps in series and another three in parallel. In the parallel circuit, you can turn one lamp on or off and the effect on the brightness of the other lamps is usually scarcely noticeable. But if the lamps are arranged in series, then every lamp that is added decreases the brightness of the others. Moreover the lamps that you would expect to be the brightest (the biggest lamps) turn out to be the dimmest. (The explanation for this peculiar property of series circuits will be discussed in Chapter 32.)

Practical wiring

Outlets are provided at many places throughout a house to make it possible to connect many appliances at the same time. So that these appliances can operate independently of each other, the outlets are connected in parallel. Each outlet contains two holes. Each hole leads to one of the wires of the main line. The cord connecting with the appliance (say, a lamp) is really two separate wires insulated from each other and bound in a single cord. At the end of this two-in-one cord is a two-pronged plug which fits into the wall outlet, as shown in Fig. 25-5. Each prong connects one of the wires from the lamp to the main line.

At the lamp end, the two wires of the connecting cord are attached, respectively, to the inlet and the outlet sides of the lamp. If we examine a light bulb before screwing it into the socket of the lamp, we see that the bottom of the bulb is made so as to contact these inlet and outlet wires in the socket (Fig. 25-5). Thus, when we put the plug in the outlet, we have a complete circuit, and the bulb lights up.

Most electrical systems use a type of current known as AC (alternating cur-

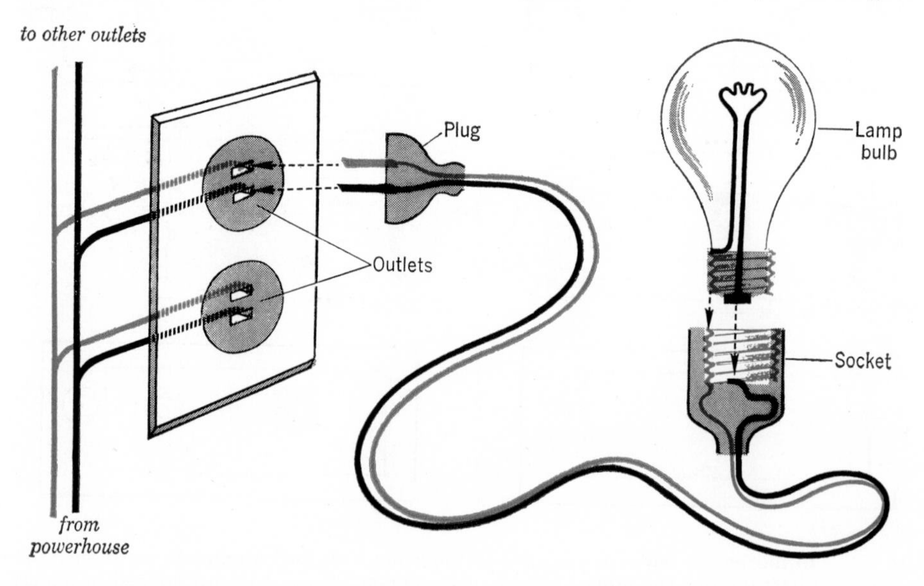

25-5 Connections are made to the double wires in the walls by means of two-pronged plugs and corresponding outlets.

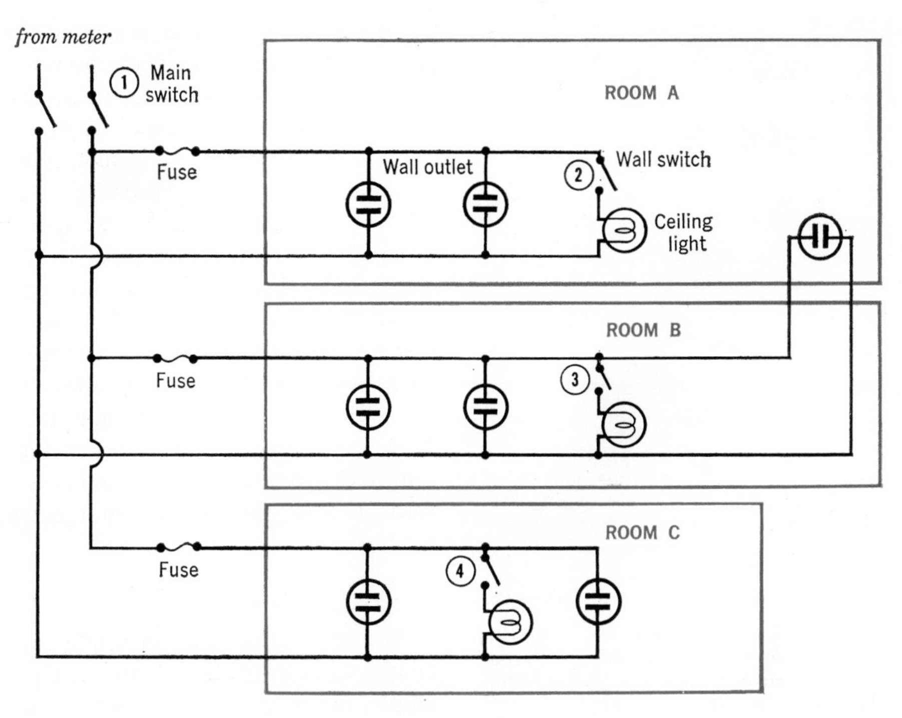

25-6 A simplified diagram of the parallel wiring in a 3-room house. What is the advantage of having an outlet in room A connected to the circuit of room B?

rent), in which the current reverses its direction many times a second. This makes no difference in the operation of lamps and heaters, since they work equally well whether the current is flowing in one direction or in the other. The reason for the use of alternating current will be explained in Chapter 35.

6. Wiring the home

Figure 25-6 shows a simplified wiring system for a house. The main cable (left) which carries current to each room branches off to different devices (lamp, radio, etc.) in each room. Each ceiling light can be turned on or off by a separate switch, as shown by 2, 3, and 4 in Fig. 25-6. All current in the house can be turned off by one switch where the main line enters the house at 1.

If you disconnect and cut open an electric cord going to a lamp, a radio, or any other device in the house, you

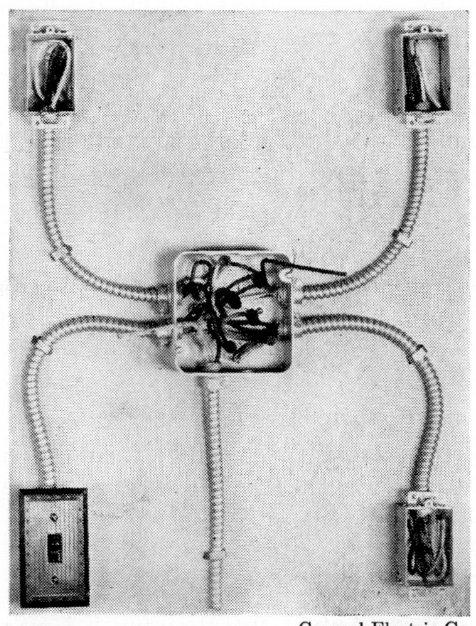

General Electric Co.

25-7 One type of house wiring in which all the wires are completely enclosed in metal cables and "junction boxes" (where connections are made to lamps, switches, and outlets).

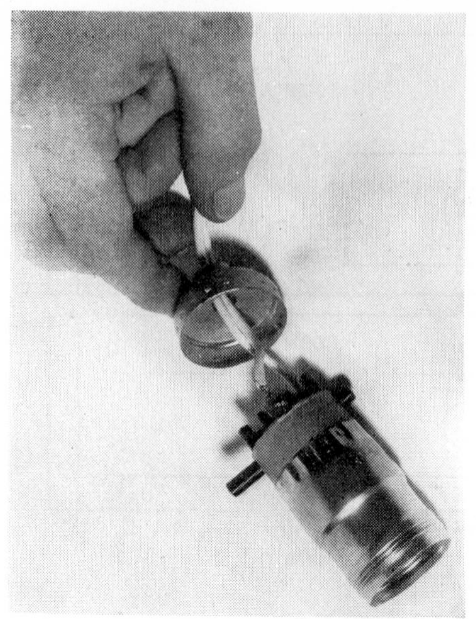

General Electric Co.

25-8 In this photograph we note that the electric wire is really double, with one wire carrying current into the socket while the other carries it out.

will be cutting not one wire but two (Figs. 25-7 and 25-8). If you do not disconnect the cord and these two cut wires should touch, you would have a **short circuit** (Fig. 25-9). This can be very dangerous, since a large amount of current then flows directly from one wire to the other. This causes sparks and heat which may ignite nearby objects. It is to prevent short circuits that each of the two wires is covered by several layers of insulating material—rubber, cloth, asbestos, etc. The wires should never be placed where they can be rubbed or stepped on, since that may wear the insulation away, cause the wires to touch, and create a short circuit.

7. Grounding a circuit

Often in an electrical circuit it is possible to save the expense of a return wire by **grounding** the circuit. Again we can illustrate this by comparison with a water circuit.

In Fig. 25-10A water is pumped from a well up to faucet F in the house. The waste pipe in the sink S then returns the waste water to the ground at O. Thus the water flows from the ground into the house and back to the ground again. We have a complete circuit, but we have not completed it by a return pipe. In effect, the ground itself serves as the return pipe. This does not mean that the sewage water necessarily returns to the well. The waste pipe is placed far enough from the well so that this does not happen. However, every gallon of water that is removed from the well eventually returns to the ground through the waste pipe.

8. Saving wires

The same grounding principle can be employed in an electrical circuit. For example, in an automobile one terminal of the battery (the "pump" in this case) is always connected to the metal body of the car, which corresponds to the ground in our water system (Fig. 25-10C). From the other terminal of the battery the current travels, by parallel circuits, to the various devices it operates—the

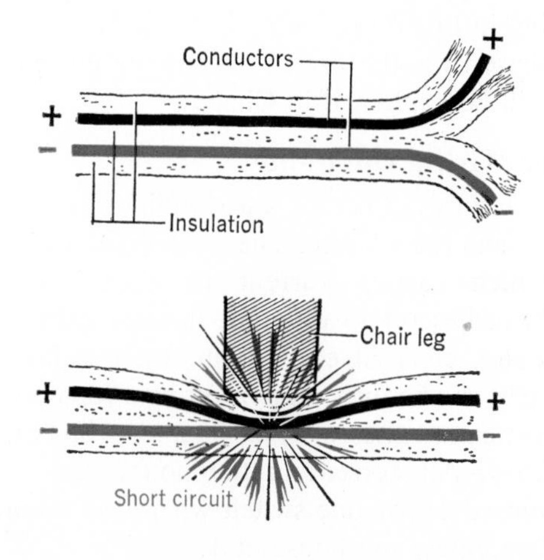

25-9 Keep that electric wire away from the floor! Stepping on it or putting objects on it may break or wear away the insulation between the two wires and cause a short circuit.

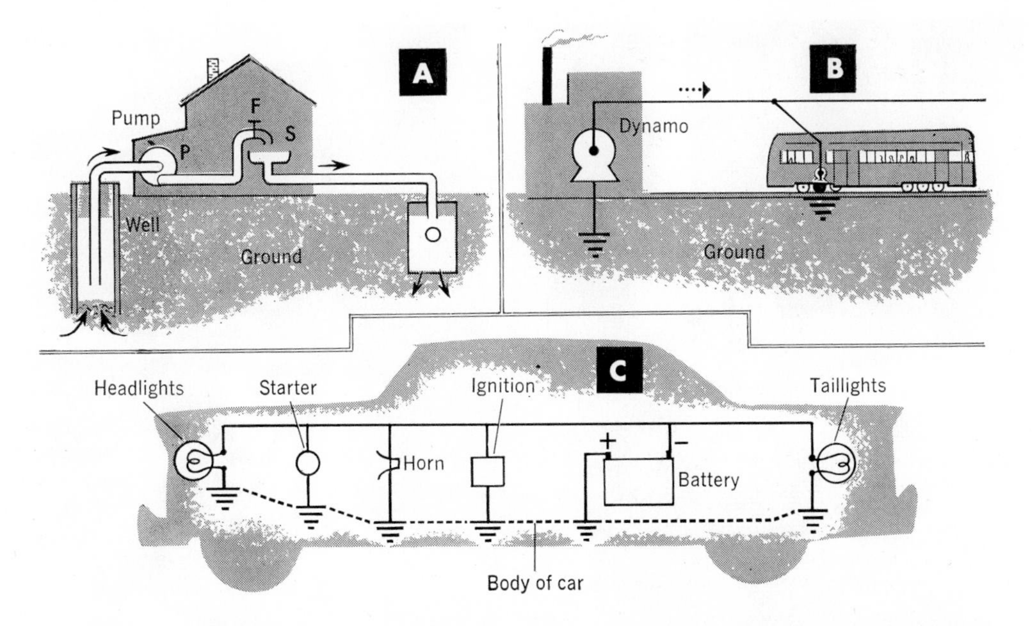

25-10 Wiring is reduced in electrical systems in the same way that return pipes are eliminated in water systems. What is the resemblance between A, B, and C? In some automobiles it is the negative wire that is grounded.

starter of the car, the headlights, etc. The circuit is completed when the current comes back to the battery through the metal car body. In a radio set the metal body of the set is the "chassis" which serves as a common ground for many of the parts. The same principle is used in a metal flashlight to save the expense of a return wire. Electric railways also use this system (Fig. 25-10B). In all of these examples, the metal body of the device serves as a return path for electrons.

The wiring principles you have just studied apply to all electrical devices. Of course, the wiring gets very complicated in some electrical apparatus such as TV sets or giant electronic "brains." Note the maze of wires in the electronic calculating machine shown in Fig. 25-11. Current cannot flow in any of these wires unless the circuits are completed.

In this chapter we have used a water circuit as an illustration in order to show that electricity, like water, involves a flow. But there are limits to the comparison. Electricity and water, being very different things, will not behave exactly alike in all situations. Electricity, unlike water, can flow through solids. It flows best through metals. Just why this is so we shall discover in the next two chapters.

Going Ahead

THE VOCABULARY OF PHYSICS

Define each of the following:

conductor	complete circuit
series circuit	parallel circuit
insulator	switch
short circuit	grounding

THE PRINCIPLES WE USE

Explain why

1. Electrical circuits must be "completed."
2. Current will not flow if two dry cells are connected minus to minus.
3. Electricity does not leak out of wires when the metal is exposed to the air.

Douglas Aircraft Co., Inc.

25-11 **Electronic computers in use in an aircraft plant. Computers are being used today for a wide variety of work in offices, factories, and even airplanes.**

4. All the devices in a series circuit stop operating when a break develops in any one of the devices.
5. One lamp in parallel with several others can be turned on or off independently of the others.
6. All electrical devices in the home are attached to plugs that have two prongs.
7. A short circuit may develop when a lamp cord is worn by rubbing.

APPLYING PRINCIPLES OF ELECTRICITY

1. Compare the advantages and disadvantages of series and parallel circuits in wiring houses.
2. Draw a diagram of three lamps in a house circuit, arranged so that each can be turned on or off independently.
3. Draw a diagram of three lamps that can be turned on or off all together and yet in which the burning out of one lamp will not affect the others.
4. Draw a diagram of a circuit arranged so that a bell can be rung by one push button located at the front door of a house or by another push button at the back door.
5. In a certain schoolroom six lamps are turned on or off by three switches at one place on the wall. Each switch controls two lamps. Draw the circuit that is used in this room.
6. In many homes a hall light is connected in such a way that the lamp can be turned on or off by either of two switches, one upstairs and another downstairs. Can you draw the circuit? Before you jump to conclusions, let us warn you that the situation is not the same as in Problem 4. The solution of this problem requires a special kind of switch that has not been discussed in this chapter. See if you can figure out what kind of switch would be needed.
7. Three-way lamp bulbs have two different filaments inside the lamp and a special switch that can connect three different circuits using these two filaments.

Work out the arrangement of the filaments and the nature of the switch. Draw the circuit.

FURTHER READING ABOUT ELECTRICITY

The following list of books and pamphlets covers most of the topics discussed in this unit on electricity. For more about atoms, see the reading list at the end of Chapter 46.

Fun With Electrons. Raymond F. Yates. Appleton-Century-Crofts, New York, 1945. 159 pp. Experiments you can do with simple equipment at home.

Exploring Electricity. Hugh H. Skilling. Ronald Press, New York, 1948. 277 pp. A human-interest account of the men who built up our knowledge of electricity and how they did it.

Electrical Things Boys Like to Make. Sherman R. Cook. Bruce Publishing Co., Milwaukee, 1949. 205 pp. Full of practical ideas, shop notes, and construction plans.

From Immigrant to Inventor. Michael Pupin. Charles Scribner's Sons, New York, 1925. 396 pp. The story of a boy who came to America as a nearly penniless immigrant and became a widely known physicist who is remembered today for his important inventions in electrical engineering.

House Wiring Made Easy. J. J. Smith. F. J. Drake Co., Chicago, 1950.

A First Electrical Book for Boys. Alfred Morgan. Charles Scribner's Sons, New York, 1951. 263 pp. An experienced popularizer describes and illustrates history, theory, and home experiments with electricity.

Radar and Other Electrical Inventions. Frank Ross, Jr. Lothrop, Lee, and Shepard Co., Inc., New York, 1954. 244 pp. A clear explanation for high school students and adults, describing the operation of television, radio, radar, and other modern electronic inventions.

Understanding Television. Orrin Dunlap. Harper & Bros., New York, 1948. 268 pp.

Radio Amateur's Handbook. Published every year by the American Radio Relay League, West Hartford, Conn. The standard manual for everyone interested in amateur radio. Over 500 pages, ranging from the most elementary principles of electricity to the description of advanced practical equipment.

Only a few useful pamphlets are listed below. Many other pamphlets and charts are available from electrical companies free of charge or at a low cost.

Fun with Dry Batteries. 10¢. National Carbon Company, 30 E. 42nd St., New York 17, N.Y.

The Story of Lightning. Free from the General Electric Company, 1 River Road, Schenectady 5, N.Y.

Transistors, Today and Tomorrow. Free from Radio Corporation of America, 30 Rockefeller Plaza, New York 20, N.Y.

How to Become a Radio Amateur. 50¢. American Radio Relay League, West Hartford, Conn.

Magic of Communications. Free from American Telephone and Telegraph Company, 195 Broadway, New York, N.Y.

Maintaining the Farm Wiring and Lighting System. 45¢ from the Coordinator's Office, Agricultural Engineering Department, University of Georgia, Athens, Ga.

Highways of Wire. A. C. Monteith. Free from the Westinghouse Electric Corporation, Box 868, Pittsburgh 30, Pa.

Birds, Bees and Capacitors. G. V. Peck. Free from P. R. Mallory and Co., Inc., 3031 E. Washington St., Indianapolis, Ind.

National Electrical Code. Free (4¢ postage) from the National Board of Fire Underwriters, 85 John St., New York, N.Y.

CHAPTER

26

Static Electricity

More than 2,000 years ago the Greek philosopher Thales observed that after a piece of amber had been rubbed with another material, such as wool, it attracted small, light objects to itself. This was the beginning of a long series of observations that led, after many centuries, to the discovery of electricity. In fact, the word "electricity" comes from the Greek word "electron," meaning amber.

Electricity at rest

What Thales was actually doing was "charging" his piece of amber with electricity. You can do the same thing if you rub a plastic fountain pen with wool and then hold the pen close to sawdust or small pieces of tissue paper. Particles of sawdust or tissue paper are now attracted to the pen. If the day is very dry, you may even hear faint crackling noises indicating that the pen is producing small sparks. We say that the pen has been charged with **static electricity.** "Static" means "stationary." Therefore static electricity is electricity that is at rest on an object. If you rub your hand over the pen, the electric charge is removed. Once the electricity flows, as in a wire, it is no longer static or stationary electricity but "current" or flowing electricity.

1. Static electricity

Tear off a strip of newspaper about 15 inches long and several inches wide, hold it against your woolen suit, and rub it back and forth with your hand. The paper is now electrically charged. Because of the charge, the paper will stick to a wall if the weather is fairly dry; the electrically charged paper "attracts" the wall (and is attracted by it) just as the fountain pen attracted the particles of sawdust. Now charge another piece of newspaper and hold the two strips end to end, as shown in Fig. 26-1. These two pieces of paper, which have been electrically charged in the same manner, repel each other and fly apart.

These electrical effects of **attraction**

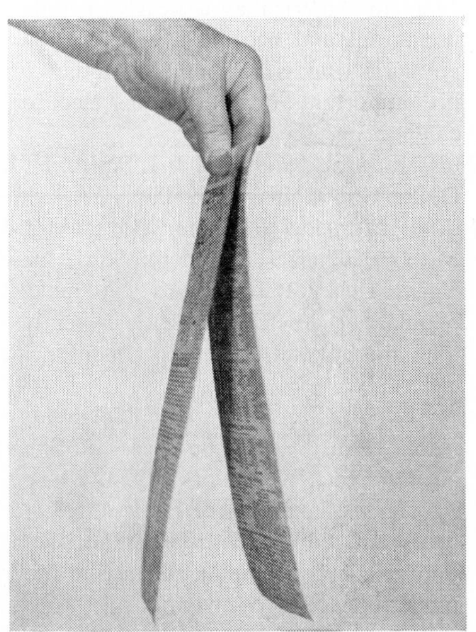

Frank Murray

26-1 Electrical repulsion of like charges keeps these newspaper strips apart. Do this experiment on a cold, dry day.

and **repulsion** can be observed when any objects of different material are rubbed against each other. When a comb is rubbed through dry hair, it produces a crackling noise and the hairs tend to "stand up." An automobile traveling in dry weather accumulates a static charge as it rubs against the air. Consequently it can give a considerable shock to anyone who comes up and touches it. That is why a small, flexible, vertical rod is placed in the center of the road at the approach to a toll booth. The rod permits the electricity to flow from the car into the ground so that the driver can hand his coin to the person in the booth without giving him a shock. A blimp can develop a terrific electric charge from traveling through the air. When the landing cable is thrown out from the blimp, it must be allowed to touch the ground before anyone handles it; otherwise the shock might be sufficiently great to kill a man.

The fact that friction between two different materials can charge an object with static electricity is put to use in some **electrostatic machines.** One form often used in science classrooms makes sparks more than an inch long. Larger machines can create sparks many feet long. Figure 26-2 shows a man sitting on the terminal of one such machine. Why does his hair stand on end? A particularly large machine used for atomic research is shown on page 617 at the head of Unit Seven.

Westinghouse Electric Corp.

26-2 **Why is this man's hair standing on end? Can you explain it? Static electricity provides the clue.**

2. Attraction or repulsion?

We noticed that when we charged two pieces of newspaper in the same way, they repelled each other. Let us perform a similar experiment with two Bakelite rods. Suspend one of them by a thread so that it is free to rotate. Rub it vigorously with fur to give it an electrical charge. Now bring another Bakelite rod, also charged by rubbing with fur, near the suspended one. We observe (Fig. 26-3A) that the rod is repelled by the similarly charged rod in the hand, just as the pieces of newspaper were repelled. The same thing happens if we use glass rods which have been rubbed with silk (Fig. 26-3B). In fact, when two identical objects are charged in the same way, they will always repel each other. Because the manner of charging is the same, we say they have like charges. Apparently **like charges repel.** We now see why the hairs of the man in Fig. 26-2 stand on end. The hairs are all charged alike and they repel each other so that they stand up straight.

Now let us bring the charged Bakelite rod which has been rubbed with fur close to the glass rod which has been charged by rubbing it with silk. We find that they attract each other (Fig. 26-3C). The charges on the two rods must be unlike, since we know that like charges re-

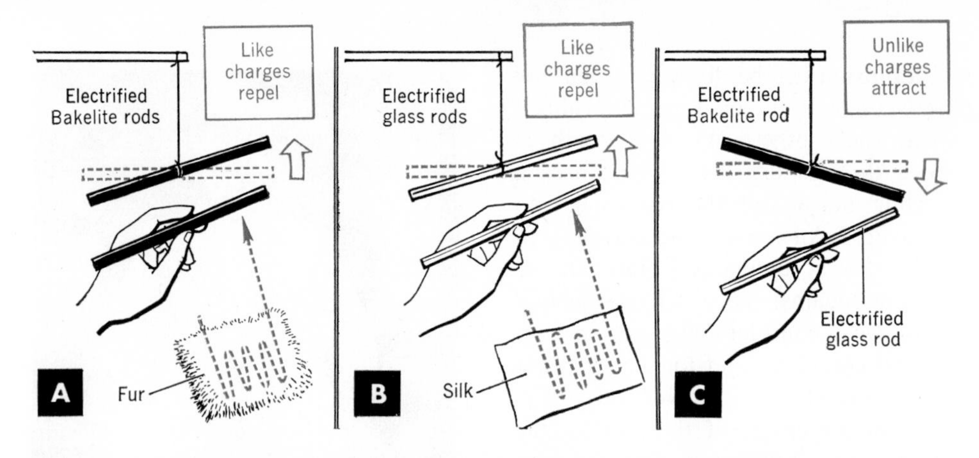

26-3 **A simple experiment with Bakelite rubbed on fur and glass rubbed on silk. Here is evidence that like charges repel and unlike charges attract. Hard rubber rods can also be used.**

pel. From this and many similar experiments, it can be shown that:

When two objects of unlike charge are brought close together, they attract each other, while like charges repel each other.

In performing these experiments you will find that the stronger the charges, the greater will be the force exerted between them. You will also find that the further the charges are from each other, the weaker the force between them will be. These ideas are summarized in **Coulomb's Law for Electric Charges** which states:

The force exerted by two charges on each other is directly proportional to the product of the charges and inversely proportional to the square of the distance between them.

The law may be expressed by the formula

$$F = \frac{kq_1q_2}{r^2}$$

F is the force, q_1 and q_2 are the charges, r is the distance between them, and k is a constant.

This is another example of an "inverse square law." For other examples see pages 146 and 367.

If we suspend a piece of flannel from a silk thread (which does not conduct electricity), and then charge a Bakelite rod by rotating it in the flannel, we find that we have charged the flannel as well as the rod. But the flannel and the rod attract each other. This must mean that the charge on the flannel is unlike that on the rod. Other experiments of the same kind with different materials always produce the same effect. We conclude that whenever two different insulating materials are rubbed against each other, they will become oppositely charged.

3. Franklin's contribution

Scientists of 200 years ago were much puzzled by the facts of attraction and repulsion. Some thought there must be two different kinds of electrical charge. Others, like Benjamin Franklin, believed there was only one kind. Franklin said there was a single "electrical fluid" which could pass from one object to another. Uncharged objects were those which possessed a certain normal, or neutral, amount of this fluid. A charged object was one which possessed either more or less than the normal amount. If the object possessed more, Franklin said it had a **positive** (+) **charge.** An object which possessed less than the normal amount of electrical fluid was said to have a **negative (−) charge.**

Of course, Franklin could not see the transfer of electrical fluid between two objects of unlike charge; therefore he could not be sure which object was positive and which was negative. For example, when a glass rod was charged by friction with silk, and then the silk and rod attracted each other, Franklin simply decided to call the charge on the rod positive and that on the silk negative. According to this explanation, the rod had acquired some of its electrical fluid from the silk. As a result of the transfer, the rod contained more fluid than a neutral, uncharged object—while the silk contained less.

Today we use Franklin's terms, positive and negative, to describe all electrical charges. Bakelite which has been charged by rubbing with fur attracts a positively charged rod. Because of this attraction, we say that the charge on the Bakelite is unlike that of the positive glass and is therefore negative. On the other hand, we find that the charged fur repels the positively charged rod. We conclude that the fur has a positive charge, since repulsion is caused by like charges.

Experiment has also shown that the two different kinds of charges are due to different kinds of electricity. We now know that every kind of material contains two main kinds of electrical particles: electrons and protons. The electrons are negative; the protons positive. In the next chapter we shall see how these particles were discovered and their properties investigated. But for the moment let us consider how scientists today explain the observed facts about electricity.

4. Electricity and the atom

According to modern theory, all the matter in the universe is composed of small particles called **molecules.** These molecules in turn are made up of still smaller particles called **atoms.** By 1959 one hundred and two distinctly different kinds of atoms (called elements) had been discovered. Atoms can combine in millions of different ways to form molecules of different materials.

Atoms are composed of three basic kinds of particles: a heavy positively charged particle called a **proton;** an equally heavy neutral particle called a **neutron;** and a very light negatively charged particle called an **electron** (Fig. 26-4). The charge of the electron is equal to that of the proton but opposite in kind. That is, one electron and one proton neutralize each other electrically and produce a neutral state (no electrical charge). The heavy protons and neutrons make up the nucleus of the atom. Around this nucleus the light electrons revolve like planets around the sun. The atom is normally electrically neutral; that is, it contains an equal number of negatively charged electrons and positively charged protons.

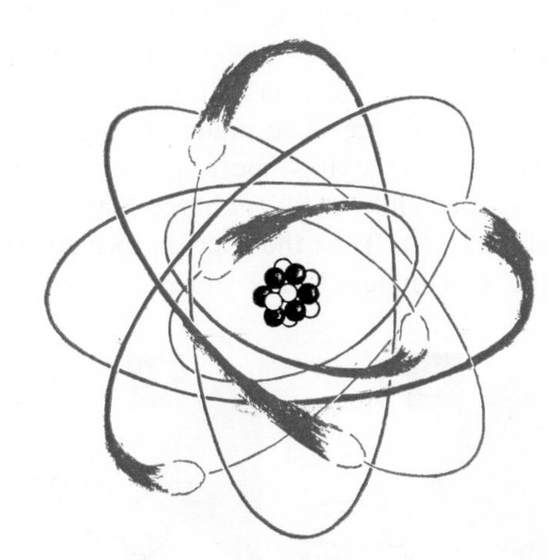

26-4 The atom is composed of protons (+), neutrons (no electrical charge), and electrons (−). The protons and neutrons are packed together in the center of the nucleus of the atom while the electrons whirl around the center. The arrangement in some ways resembles planets whirling around the sun.

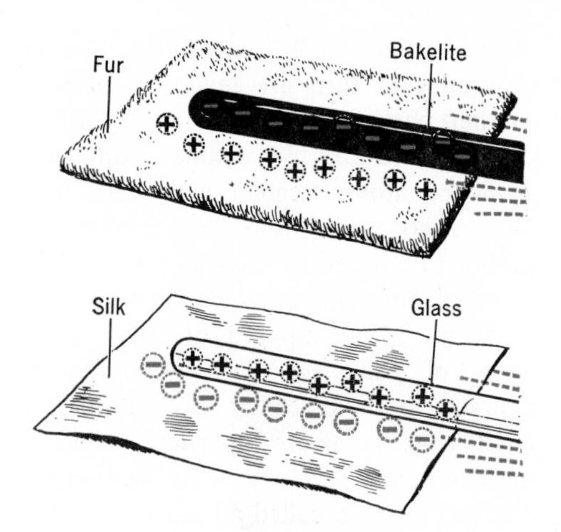

26-5 The process of rubbing to produce electric charges is really a borrowing-lending arrangement. One object gains electrons and becomes negatively charged, while the other loses electrons and becomes positively charged.

5. How a charge is acquired

When two different materials are rubbed together, it is only the electrons of each atom that are affected, since these are the outer particles that surround the heavy nucleus. If the contact between the two materials is close, a small number of electrons from the atoms of one material are transferred to the atoms of the other. The material which gains electrons thus acquires a negative charge. The material that loses electrons now has more protons than electrons and hence its charge is positive (Fig. 26-5). Thus the charges formed on the two materials are always opposite. It is important to remember that only the electrons are transferred from one solid material to another; the *protons remain stationary*.

What happens when we place a negatively charged rod in a heap of sawdust? We observe that many bits of sawdust cling to the rod. As we watch, each bit stands up on end and after a while shoots off the rod with some speed. Why do these things happen? Our theory explains these events very nicely.

The particles of sawdust are neutral to begin with; that is, each bit of sawdust contains as many electrons as protons. But the repulsion of the negatively charged rod drives some of the electrons in each bit of sawdust away from the rod. As shown in Fig. 26-6A, they move in the bit of sawdust from point R, near the rod, to point S, as far away as they can get. So the end of the sawdust at R becomes positive, since it has lost electrons and now has an excess of protons. At the same time the part at S, by gaining electrons, now has a negative charge. Though the charge has been redistributed, the particle of sawdust is still neutral as a whole.

So part R of the bit of sawdust is attracted to the negative rod, while part S is repelled. But since the positively charged part at R is nearer to the rod than is the negatively charged part at S, the attraction is greater than the repul-

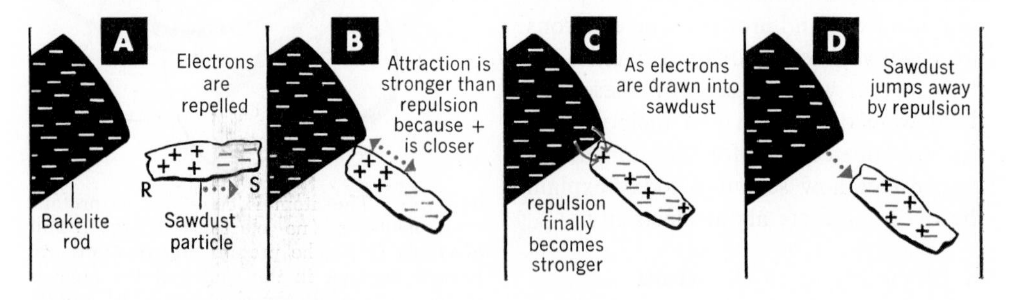

26-6 Here is an explanation of why sawdust is first attracted to and then repelled by a charged Bakelite rod.

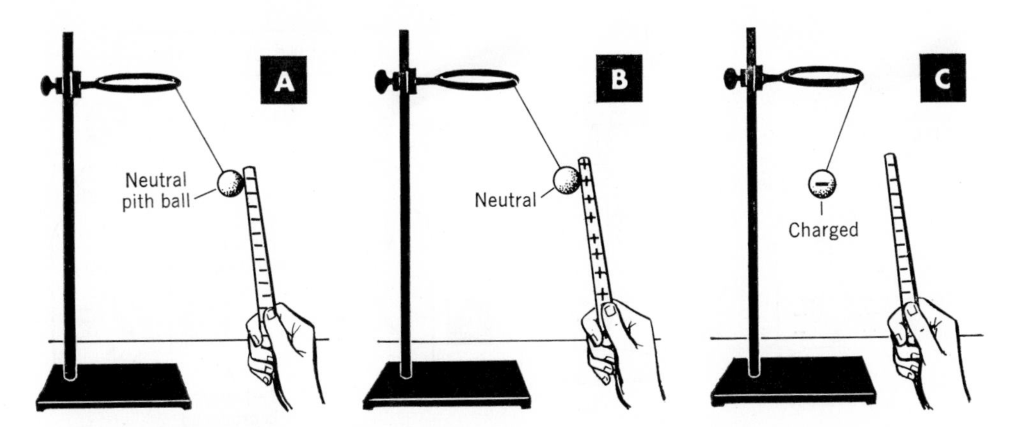

26-7 A neutral pith ball is attracted to both positive and negative rods (A) and (B). But when the pith ball is charged it is repelled by a rod carrying a similar charge.

sion, and the net effect is one of attraction. Hence sawdust clings to the rod (Fig. 26-6B).

But as soon as a bit of sawdust comes in contact with the negative rod, it begins to receive electrons from the rod, which has an excess of these negative particles. As these electrons leak into the piece of sawdust, it slowly builds up a negative charge (Fig. 26-6C). When a large enough negative charge has accumulated, the piece of sawdust stands up as the far end is repelled and finally leaps away from the rod; it now has a charge like that on the rod, and like charges repel (Fig. 26-6D).

The electroscope

6. How a charge is detected

If a particle of sawdust or a ball of light wood (such as one made of **pith**) is suspended by a thread from a ringstand, it can be used to detect whether an object brought near it is charged or neutral. Such a device is called a **pith-ball electroscope** (Fig. 26-7). If the ball is neutral, it will be attracted by either a positively or a negatively charged object. But suppose the ball is given a negative charge to begin with. (This can be accomplished by touching it with a negatively charged Bakelite rod.) Now if the ball is attracted by an object known to be charged, the charge on that object must be positive, but if the ball is repelled by an object, that object must be negative (Fig. 26-7).

Experiments with two pith-ball electroscopes help to verify Coulomb's Law for Electric Charges. Two identically charged pith balls (the charge may be either + or −) are touched together and are then hung from separate supports. If the balls are weighed and the angles of the supporting threads are measured, an accurately drawn vector diagram will show that the force between the balls is one quarter as great when the distance between them is doubled, one ninth as great when the distance is tripled, and so on. Thus **the force between two charges varies inversely with the square of the distance between them.**

A more sensitive instrument in the study of electricity is the **gold-leaf electroscope,** a small device used to test the presence of an electric charge. This instrument, simple to construct, has proved a valuable scientific tool. For instance, in 1910 in Zürich, Switzerland, a physicist conducted experiments with an electroscope high above the earth in a balloon. Little attention was paid to the

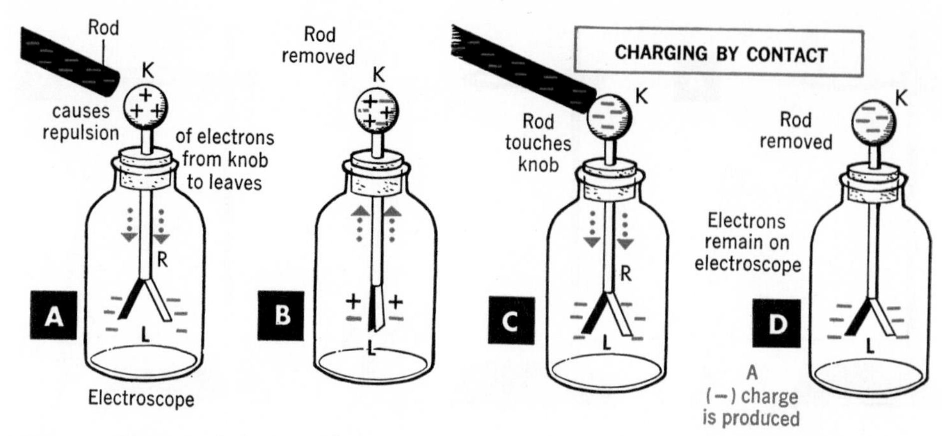

26-8 Electrons help to explain why the leaves of an electroscope move apart when a charged rod is brought near (A) and fall back when the rod is removed (B). Can you explain why the leaves remain spread apart in D?

flight at the time. But these experiments gave the first hint of the existence of cosmic rays (Chapter 43). Studies of these rays have yielded important information about the basic structure of the atom. And as late as 1932, Fermi in Italy used an electroscope in experiments that showed the properties of slow neutrons, experiments that eventually led him to design the first atomic reactor. Let us examine the instrument which made such discoveries possible.

The electroscope, as we have said, enables us to detect the presence of an electric charge. Figure 26-8A shows how we can make one. A strip of thin aluminum or gold foil to form the leaves, L, is folded over and hung from a metal rod, R. This metal rod is placed in a glass container and supported by a rubber stopper. At the top of the rod is a round metal knob, K.

7. Charging the electroscope

If we bring a charged rod near the knob without touching it, the two gold-foil leaves move apart (Fig. 26-8A). Why does this happen? If the rod is negatively charged, then as it approaches the knob some of the electrons in the knob are repelled down into the leaves, so that both leaves become negatively charged. As a result of this like charge, the leaves repel each other and spread apart, just as the like-charged strips of paper spread apart (Fig. 26-1). But suppose the rod we bring near the knob has a positive charge. Then electrons from the leaves are attracted upward to the knob. The leaves, having lost some electrons, are now charged positively. But the charges on the two leaves are again alike, so the leaves repel each other and spread apart. Thus the leaves spread apart whenever a charged object (either positive or negative) is brought near the knob.

When we remove the charged rod from the vicinity of the knob, the leaves in the electroscope return to their original position (Fig. 26-8B). This occurs because the electrons return to their original place in the leaves or knob, and the leaves become neutral again. You will note that throughout this entire experiment the electroscope *as a whole* has remained neutral or uncharged.

8. Charging by contact

So far we have not touched the knob of the electroscope with the charged rod. Suppose we touch the knob with a negatively charged rod (Fig. 26-8C). When we remove the rod, we find that some of the electrons on the rod have flowed onto the electroscope and remained there (Fig. 26-8D). We have charged the electroscope negatively by contact. If the rod is positive, electrons are attracted out of the knob and leaves. There is a deficiency of electrons in the electroscope, and the leaves acquire a positive charge. **We always produce the same charge as that of the rod, when charging by contact.** In either case the electroscope may be neutralized by touching (grounding) the knob with your finger. The reason for this is that if the electroscope is negative, the excess electrons flow into your body; if it is positive, the body supplies the missing electrons. In either case the electroscope becomes neutral.

These experiments show us that our bodies conduct electrons. If we touch a metal wire which is connected to the knob of the electroscope, we get the same results. This is because metal, as we know, is a good conductor of electrons. If, however, we touch a silk thread, one end of which is touching the knob of an electroscope, nothing happens. The silk is an insulator and a very poor conductor of electrons.

9. Charging by induction

You can produce an opposite charge to that of the rod if you charge the electroscope by **induction** (charging from a distance). Figure 26-9 shows how this is done. You ground the knob by touching it with your finger, at the same time bringing the rod near to the knob but *not touching it* (Fig. 26-9A). Now you remove your finger *first,* then the rod. The leaves of the electroscope stay apart, showing the presence of a charge (Fig. 26-9C).

If the rod was negative, it repelled electrons out of the knob into your body through your finger, making the knob positive. When you removed your finger, these electrons in your body could not return to the leaves, and the knob remained positive even after the rod had been removed.

When a positive rod is brought near the knob, electrons are attracted into the knob from your finger. When you take your finger away, the knob is left with an excess of electrons, or a negative charge, even after the rod is removed. So we see that charging by induction with a positively charged rod produces a negative charge on the leaves, and charging by induction with a negatively charged rod produces a positive charge

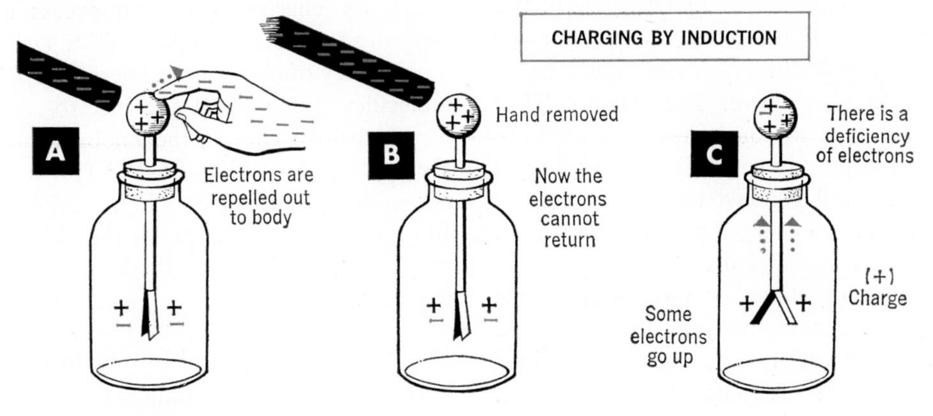

26-9 **Charging by induction produces an opposite charge. Explain the electron story in A, B, and C.**

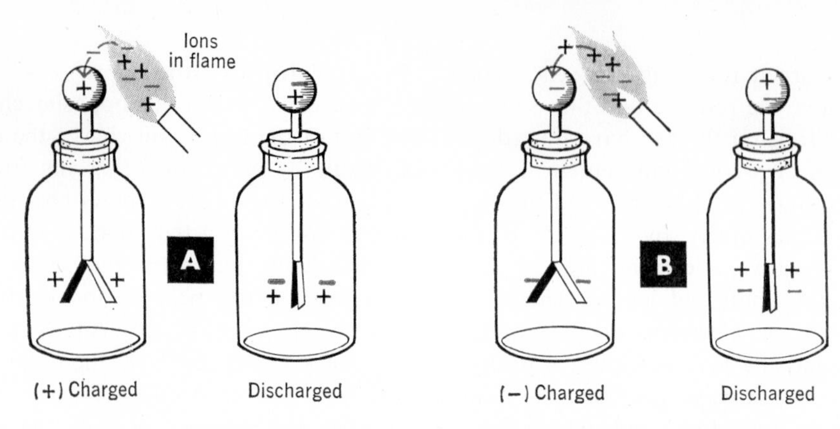

26-10 A flame near the electroscope causes it to lose its charge, whether negative or positive. This fact provides evidence for the existence of charged atoms (ions) in the flame.

on the leaves. Therefore, **charging by induction produces a charge on the leaves opposite to that on the charging rod.**

The electroscope is often used to identify electric charge. If a positively charged electroscope is brought close to a negative charge, the leaves will fall closer together. If, however, it is brought near a positive charge, the leaves are driven farther apart. Can you explain why this happens?

10. Ions

Early in this century scientists began investigating a peculiar fact about the electroscope. They noticed that after an electroscope had been charged and left standing, the leaves tended to collapse slowly, and the electroscope became neutral by itself. This meant there was a leakage somewhere. But where? They were forced to conclude that the charge was escaping into the air. It was known that air will conduct an electric current if it contains charged particles which the scientists named **ions.**

Ions are atoms or molecules of gas which for some reason, that we need not go into at present, have either gained or lost electrons. Hence they are charged. Positive ions, in a gas such as air, will be attracted to a negatively charged electroscope and neutralize its charge. Negative ions in the air will be attracted by a positively charged electroscope and will neutralize it. These ions often cause the charge on an electroscope to leak away with the result that the leaves collapse.

You yourself can test for the presence of ions in a gas by bringing a lighted match close to a charged electroscope. Whether the electroscope is charged positively or negatively does not matter; the leaves of the electroscope will collapse (Fig. 26-10). The molecules of gas in the flame are in violent motion due to their high temperature. Electrons from atoms in some of the molecules become separated and attach themselves to other molecules. Thus some molecules become negatively charged (with an excess of electrons); others become positive (with a deficiency of electrons). These charged molecules are ions. If the electroscope has a positive charge, the knob attracts the negative ions (Fig. 26-10A); if negative, the positive ions (Fig. 26-10B). In either case the charge on the electroscope is neutralized and the leaves come together.

11. Radioactivity and ionization

Radium and other radioactive materials are very effective agents in ionizing the air. The nuclei of such radioactive atoms release high-speed atomic particles (Chapter 45). These particles travel at speeds of tens of thousands of miles a second and knock electrons out of atoms in the air, thus creating an enormous number of ions.

Some scientists believed that most of the ionization of the air was caused by the presence of radium and other radioactive materials in the earth. It was to test this hypothesis that an electroscope was taken up in a balloon, as mentioned on page 333. If the hypothesis is correct, there should be slower leakage from a charged electroscope high in the air than from one on the ground. Actually, however, it was discovered that the leakage was more rapid! But the time spent on the experiment had not been wasted. There was very good reason to think that the ionization of air was caused by something coming from outside the earth, rather than from the radioactive materials in the ground. This eventually led to the discovery of **cosmic rays.** In Chapter 43 we learn what these rays are and why they produce ionization in the air.

Storing electrical energy

You remember from Chapter 25 that electrons and water have the same property of being able to flow through suitable conductors. We know that water can be stored in large tanks, and that when we want to use it we simply have to open a valve. Let us see how we might store electrical energy.

12. Condensers

Electrical energy can be stored in a device called a **condenser** (capacitor). Figure 26-11 shows us how to make one. Take two metal plates and put them close together, face to face, but separated by some insulating material (or **dielectric**, as it is often called). Air is an excellent dielectric, but glass, waxed paper, plastics, and thin sheets of mica are found to be superior for many purposes.

Now attach one of the metal plates to a source of electric charge, as shown in Fig. 26-11A. Let us assume that the source is the negative terminal of some source of high voltage. The other plate is grounded, perhaps by connecting it to a nearby water pipe or simply by touching it with the hand. Electrons flow into plate R from the source of the negative charge. As a result electrons on plate S are repelled and flow into the ground, leaving plate S with a positive charge. When the two wires are removed, we have a charged condenser. We can carry the condenser about with us; the

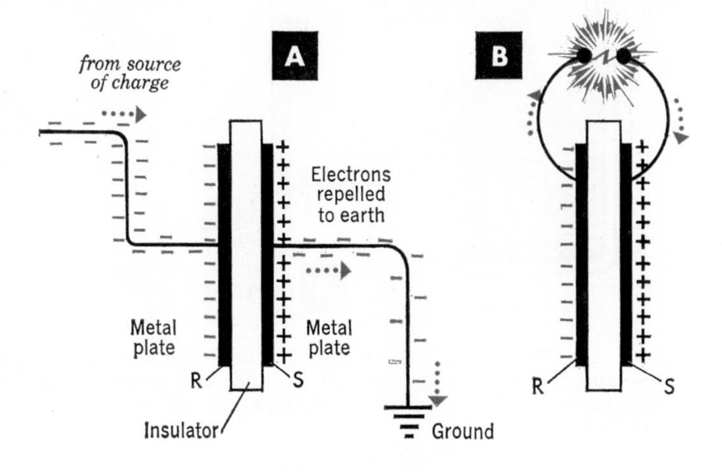

26-11 A condenser is simply two conductors separated by an insulator. A charge on one plate attracts and holds an opposite charge on the other plate. If wires from both plates are touched together, the charges will be neutralized by electrons flowing through the wires.

two plates still retain their opposite charges, and the charges will not leak off easily because the positive charge on one plate attracts the nearby negative charge on the other side of the insulator. If we connect a wire to each plate and bring the ends of the wires close together, a spark will leap across the gap, resulting in the neutralization of each plate (Fig. 26-11B).

One form of condenser which was used several hundred years ago and is still in use today is the **Leyden jar** (Fig. 26-12A). Here a glass jar acts as the insulator. Two pieces of metal foil, one on the inside of the jar, the other on the outside, serve as the two metal plates.

At this point it is well to consider just what is meant by the phrase "storage **capacity** of a condenser." In charging a condenser, work is done moving electrons from one plate (the outside foil in Fig. 26-12A) and moving an equal number to the other plate (the inside foil in Fig. 26-12A). Negative charge is simply moved from one plate to the other, and the total charge on the condenser remains the same. Hence the capacity of a condenser is not its ability to store charge. Its capacity is its ability to store electrical *energy*. If the terminals of a charged condenser are connected together (Fig. 26-11B), the stored energy is converted into the energy of heat, light, and sound in the resulting spark.

Condensers are important in many electrical devices—radios, for example. One way to increase the capacity of the condenser (ability to store electrical energy) is to increase the metal surface used. In the fixed condensers of radio sets, this is frequently accomplished by coiling up two long, thin sheets of waxed paper between thin sheets of metal foil (Fig. 26-12B). The capacity of the fixed condenser is also increased by making the insulator as thin as possible. This brings the plates closer together so that the attraction of their unlike charges is greater and there is greater storage capacity. The capacity also depends on the kind of insulating material used for the dielectric.

Another type of condenser is used in a radio set to enable you to tune in different stations. This condenser (Fig.

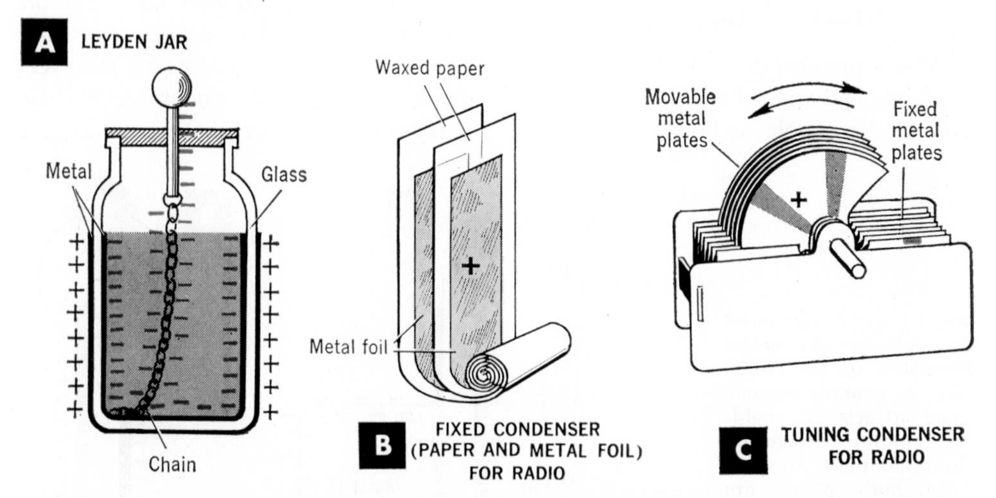

26-12 Three forms of condensers. A shows the Leyden jar, where the jar is the insulator separating the conducting metal foil on inside and outside. In B, sheets of thin metal foil are separated by insulating waxed paper. In C the adjustable metal plates are separated by the insulating air.

26-12C), called a **variable condenser** (air is the insulator), allows you to change the capacity at will. A set of movable metal plates rotates between a set of fixed plates. When more of the surfaces of the plates are brought closer together, the capacity increases. We shall learn more about condensers and their uses in Chapter 35.

13. Electrostatic machines

Another important use of condensers is in high-voltage generators used in atom-smashing experiments. Scientists in the eighteenth century wanted a better method of producing electricity than the slow, inconvenient one of rubbing objects by hand. So they invented the electrostatic machine. One machine produces small charges on metal strips mounted on oppositely rotating glass discs. The electric charges are transferred to condensers which gradually build up a large amount of electric energy.

All electrostatic machines take minus charges (electrons) away from one metal terminal and pile them up on the other. The eighteenth-century machines could build up enough charge to make a spark several inches long. Modern machines, operating on the same principle, can create sparks more than 20 feet long. Figure 26-13 shows an electrostatic generator. This huge machine is used in atomic research.

Westinghouse Electric Corp.

26-13 **This looks like the launching of a space ship. But actually it is an atom-smashing machine known as a Van de Graaff generator. It operates by building up a 4-million-volt charge at the top, which then hurls electrically charged atomic particles at targets at the bottom of the structure.**

14. Lightning

Lightning is, in a sense, nothing more nor less than the spark from a stored-up charge in a huge natural condenser. Such sparks can cause serious destruction when they strike.

In Chapter 20 we saw how summer thunderstorms develop as the result of convection currents caused by the heating of the earth. In this process there is violent motion of cold air, warm air, and water droplets. In a manner not yet completely understood, the water drops trying to fall downward against the uprush of air became charged. Their motion acts like a giant electrostatic machine, charging the droplets that make up the clouds. Some parts of clouds are charged positively, others negatively. Lightning is the spark between two bodies of unlike charge, sometimes from cloud to ground, or from ground to cloud, and at other times from one cloud to another (Fig. 26-14).

Let us assume that a negatively charged cloud is passing over the ground.

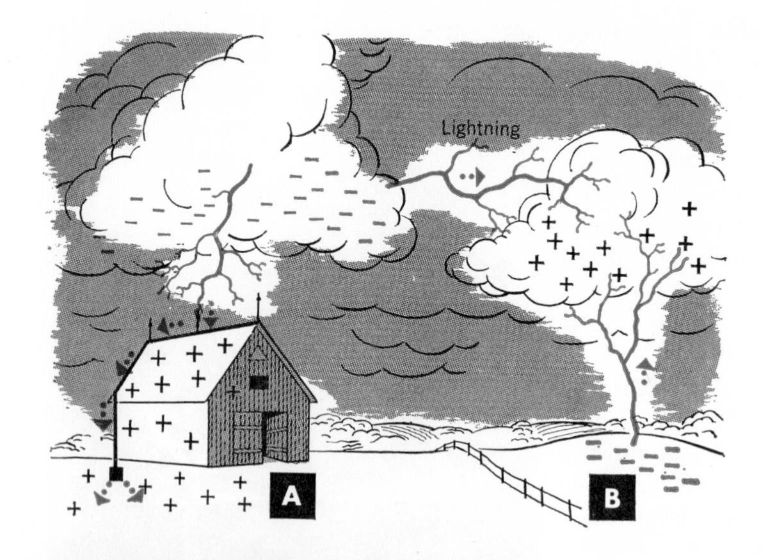

26-14 A simplified view of lightning. We may think of it as a giant spark caused by motion of electric charges from cloud to ground (or reverse), or from cloud to cloud. The lightning rods on the barn conduct the electric current harmlessly to the ground.

Westinghouse Electric Corp.

26-15 The man in the car is quite safe from this 3-million-volt, man-made lightning flash. Repulsion of the electrons keeps the charge on the outside surface of the car. Note how the flash jumps across the insulating front rubber tire to reach the ground.

This negative charge in the cloud repels electrons in the ground and thus gives the ground an opposite, positive charge (A in Fig. 26-14). We now have a condition like that of a condenser: two conductors (the cloud and the ground) are separated by insulating air. As the charge in the negative cloud builds up, it finally becomes great enough to strike through the insulation to the ground. This is the giant spark we call lightning.

The passage of the lightning heats the air along its path and makes it expand suddenly. This expansion sends out a giant compression wave that we call **thunder.**

15. Protection from lightning

Lightning descending from a cloud to the earth tends to strike the nearest conductor. A tall building, a house, a man standing in an open field are all excellent targets. The best way to safeguard a building is to provide a safe, easy path for the electrons to reach the ground should lightning strike. A metal **lightning rod** on the top of the building, connected to pipes or wires that lead into the ground, serves this purpose. Tall skyscrapers are protected by their steel frames. The Empire State Building, the tallest building in the world, has been struck by lightning at the rate of once a minute during a thunderstorm without harm to the building or the people in it.

The most effective lightning rods have many sharp points. Electrons on an object repel each other, tending to concentrate at the sharp points. The strong repulsion at the points tends to make the electrons leak off into the air. Positive charges act in a similar way, except that the electrons leak into the object most rapidly at the sharp points. The loss of charge at the points of the lightning rod tends to neutralize the charge in the cloud and ground and thus helps prevent the lightning from striking the house.

Figure 26-15 shows that a car is a rather safe place to be in for protection from lightning. The metal body conducts the charge around (and not through) the man in the car to the ground.

Going Ahead

THE VOCABULARY OF PHYSICS

Define each of the following:

static electricity	proton
dielectric	charge by induction
attraction	variable condenser
atom	ions
repulsion	cosmic rays
neutron	condenser
electrostatic machines	Leyden jar
positive charge	capacity
electron	lightning
negative charge	thunder
electroscope	lightning rod
charge by contact	

THE PRINCIPLES WE USE

Explain why

1. One object rubbed on another may produce an electric charge, one object becoming negative and the other equally positive.
2. A metal rod held in the hand shows no charge when rubbed on other objects.
3. A charged Bakelite rod attracts bits of sawdust.
4. The bits of sawdust jump off a charged rod after a while.
5. Charging by contact produces the same charge on an electroscope.
6. Charging by induction with a negative charge produces a positive charge on an electroscope.

Explain how

7. Electricity was discovered.
8. To charge an electroscope with the same charge that is on the rod.
9. To charge an electroscope by induction.
10. The study of leakage of charge from an electroscope began the search that led to the discovery of cosmic rays.

EXPERIMENTS WITH ELECTRICITY

Describe an experiment which

1. Indicates that radium from the earth is not the main cause of leakage of charge from an electroscope.
2. Shows that when two insulators are rubbed together and become charged, one object becomes charged positively and the other negatively.
3. Enables you to find out whether a certain object is charged positively or negatively. (Give two methods.)

APPLYING PRINCIPLES OF ELECTRICITY

1. A doctor has just misplaced some valuable radium and suspects that it has been carried away in the refuse. How could he use an electroscope to help him find it?
2. As the paper sheets come out of a printing press or mimeograph machine, they may become unmanageable because they repel each other. What causes this? Can you suggest a way of overcoming this defect?
3. Why are gasoline trucks equipped with a metal chain that drags on the ground?
4. Why are flexible metal rods placed in the path of cars approaching a toll booth?

5. Describe the construction of a condenser and the method of charging it. Give two uses for condensers.
6. Mention three ways in which the capacity of a condenser may be increased.
7. What is the cause of lightning and thunder? How can houses be protected from destruction by lightning?
8. A charged rod is brought near a negative electroscope, and the leaves come together. What is the charge on the rod? How do you know? What explanation would you give to account for the fact that the same leaves go farther apart when another charged rod is brought near?
9. Two aluminum-coated pith balls, A and B, are suspended from insulating silk threads. The balls touch each other. In each of the following cases tell what you would observe (if anything) and explain why it occurs.
   (a) A negative rod is brought near A.
   (b) A negative rod is touched to A and removed.
   (c) A positive rod is brought near A while B is touched with the finger. The finger is removed and then the rod is removed.

---

10. Two negatively charged pith balls 4 cm apart repel each other with a force of 9 dynes. What will be the force of repulsion if (a) one of the balls has its charge doubled, (b) the other ball then has its charge tripled, (c) the distance between them is then changed to 2 cm?

PROJECTS: TRICKS WITH STATIC ELECTRICITY

1. On a dry day in a warm, dimly lighted room you can light a fluorescent lamp tube simply by rubbing it briskly on your sleeve. The static electricity produced is enough to create pulses of current through the tube, provided the humidity is low enough to prevent the charge from leaking off through the air.
2. If a strongly charged glass or Bakelite rod is held close beside a smoothly flowing thin stream of tap water, the stream is spectacularly bent. Two or three rods can be adjusted so as to force the water to fall in a zigzag course.
3. Balloons can be made to do tricks when they have been charged by rubbing them briskly with a woolen cloth. They will adhere to vertical walls and repel one another strongly when supported on long threads.
4. If your school has a toy Van de Graaff high voltage generator you may be able to produce the effect illustrated in Fig. 26-2. Three things are important for success. The hair should be dry and fairly free of oil. The experiment requires a dry day. The experimenter should be very well insulated from ground. And finally, your instructor should be present to see that the conditions are right and all is safe. Stand on two inverted battery jars and be careful not to touch any nearby objects except the source of voltage. Can you give a reason for these instructions?

CHAPTER

27

Electrons, Protons, and Neutrons

In the middle of the nineteenth century a German glassblower named Heinrich Geissler (gīs'lẽr) performed a number of interesting experiments. He fashioned various shapes of glass tubes, filled each with a different gas at very low pressure, and sent high-voltage sparks of electricity through them. A surprising thing happened. The gas in the tube glowed! And each variety of gas gave light of a different color (Fig. 27-1A).

Geissler would be amazed if he were to walk down a lighted street today and see his tubes across store fronts and theaters as neon signs and within the stores as fluorescent lamps. He would be even more amazed to see X-ray tubes, television tubes, and radio tubes. But he would be completely baffled to find that his curious tubes had provided one of the keys to what goes on inside atoms. His work, combined with the work of many other men, laid the basic groundwork for our understanding of atomic energy.

Matter is electrical

The British physicist Sir William Crookes decided to see what would happen if he removed as much of the gas as possible from Geissler's glowing tubes. At very low pressures, with most of the gas removed, Crookes found that most of the glow disappeared. But new effects were noticed. Figure 27-1B shows one experiment with a form of the Crookes tube. When the plate on the left was connected to the negative terminal of a source of high voltage, the shadow of the cross was thrown onto the end of the tube. But there was no such shadow when the connections were reversed. This seemed to indicate that the "something" which formed the shadow came out of

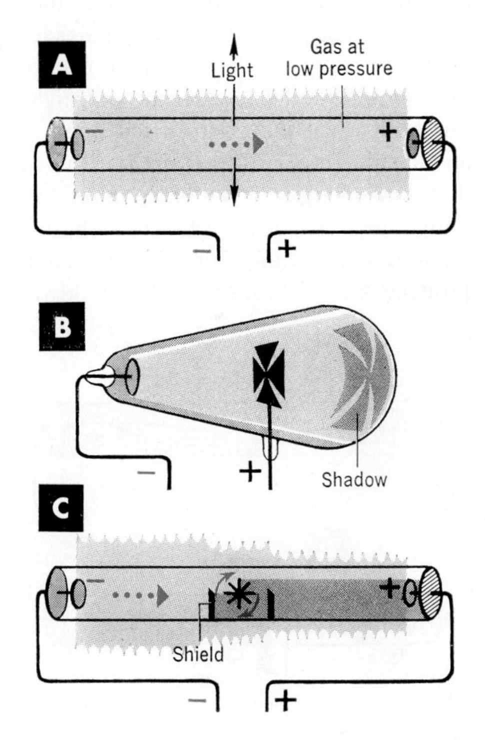

27-1 Cathode rays are produced when a spark passes through a low-pressure tube. In A we note that the passage of these rays causes the gas in the tube to glow. B and C show that cathode rays are a stream of particles from the negative terminal.

the negative terminal but not out of the positive terminal.

In addition, a wheel placed in the tube as shown in Fig. 27-1C rotated in a direction that showed something must have been coming out of the negative terminal.

1. Electrons

Were these shadows caused by light rays of some kind or by the passage of a stream of particles? A narrow slit (S in Fig. 27-2A) was placed in the tube. A screen coated with special chemicals which glowed when struck by the beam was placed along the entire length of the tube to reveal the path of any beam coming through the slit. Now a beam could be observed passing through the slit in a straight line between the two terminals.

If the beam consisted of particles, then according to the Laws of Motion they should be deflected from a straight-line path when they were acted upon by an outside force. It was already known that like charges repel each other and unlike ones attract. By placing a charged plate near the beam, then, Crookes could produce a force whose action would determine whether the rays were electrically charged particles. When positively and negatively charged plates were placed above and below the beam, it was found that the beam was attracted by the positive plate and repelled by the negative one. This ruled out the possibility that the beam consisted of light rays, since it had already been learned that light is neither attracted nor repelled by charged plates.

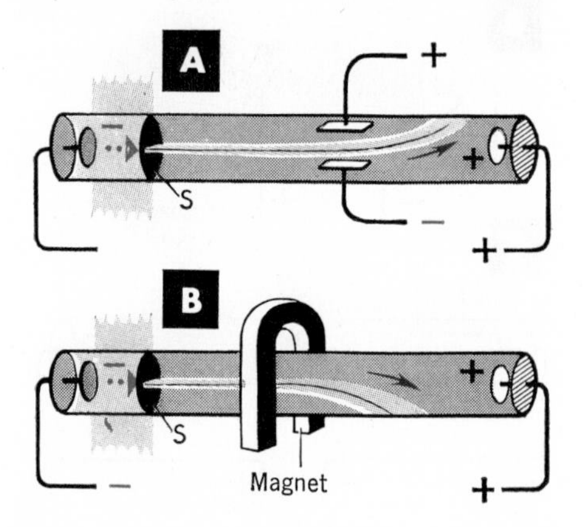

27-2 Cathode rays consist of a stream of negatively charged particles, as shown by their attraction to a positive plate and repulsion from a negative plate. A magnet can also cause the beam to swerve. We know today that the beam consists of electrons.

Similar deflections occurred when magnets were used, as shown in Fig. 27-2B.

So the beam probably consisted of charged particles of some kind. Since these particles were attracted to a positive plate, it was assumed that they must be negative. Since the negative terminal is called the cathode, these streams of particles were named **cathode** (kăth′ōd) **rays.** Additional evidence that the beam consisted of negative particles was obtained by placing an insulated object in the tube in the path of the beam. It was found that the object became negatively charged when it was struck by the beam.

Scientists now began an intensive study of cathode rays. Around the beginning of the twentieth century an English scientist, Sir Joseph John Thomson, performed a brilliant series of experiments similar to those shown in Fig. 27-2, from which he determined the amount of negative charge carried per gram of particles. These particles were given the name **electrons.**

2. Protons

Thomson then turned his attention to an interesting observation made by another physicist. When a hole was drilled in a cathode (negative terminal) placed in the middle of the tube, rays streamed through it, as shown in Fig. 27-3. But these rays traveled in a direction opposite to that of the cathode rays. And, also unlike the cathode rays, they were attracted by a negative plate and repelled by a positive plate. In addition, an insulated object placed in the path of the rays became positively charged. So the particles which made up these rays must have been positive.

But there was an important difference between the action of these positive rays and that of the negative electrons which made up the cathode rays. Cathode rays behaved in the same manner no matter what the gas in the tube might be, but the positive rays behaved differently if different gases were used. Thomson concluded from his experiments that these positive particles were much heavier in some gases than in others. His measurements also showed that *in any gas* these positive particles were thousands of times heavier than electrons.

Thomson's final explanation of the action of the positive rays in the tube was this: He decided that the atoms of any gas he put in the tube consisted of both positive and negative particles, and that the positive particles, which were much heavier than the negative particles, accounted for the positive rays. The positive particles in the atoms of different gases differed in mass and had different amounts of charge, but the number of negative particles in each atom was such that their total negative charge balanced the charge of the positive particle. Thus the whole atom, in its "normal" state, was neutral. From a great variety of experiments Thomson concluded that this picture of the atom was applicable to all kinds of materials.

Thomson followed up his discovery that the positive particles in different gases varied in mass. He found that the lightest positive particles were found in hydrogen gas. These positive particles of the hydrogen atom are now called **protons.**

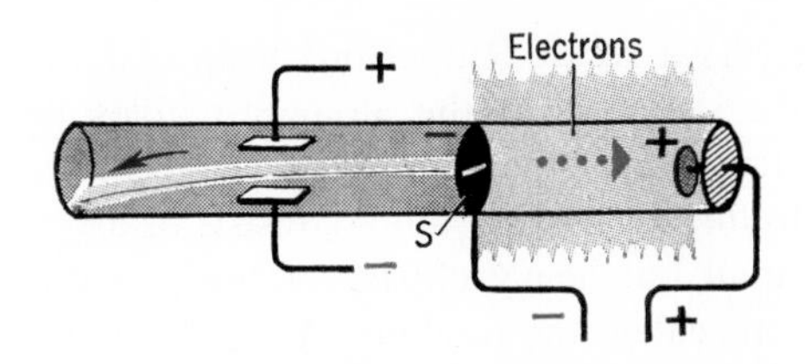

27-3 As the electrons stream from negative to positive terminals, they bombard atoms of gas in the tube, knock off some electrons (ionize them), and make them positively charged. These positively charged atoms travel toward the negative plate and can be made to pass through a hole in the plate to form rays. Their nature can then be studied by experiments in which electrically charged plates and magnets force the beam to swerve.

It was also found that the total masses of the positive particles in atoms of other elements seemed to be whole number multiples of the mass of the proton. Thus an atom of helium, the next lightest gas after hydrogen, contained positive particles whose mass was about 4 times that of a proton. The positive particles of carbon were 12 times the mass of a proton. It looked as though the proton was in some way a basic positive particle in the structure of any atom.

3. Space inside the atom

In 1911 another Englishman, Ernest Rutherford, conducted an experiment in which he shot positively charged helium atoms at high speed through thin sheets of material. He found that most of these "bullets" passed through the sheets without swerving. This must mean that there is a great deal of empty space between the protons and electrons which make up the atoms of the material. However, an occasional particle was deflected, and some were even bounced right back to the source. This seemed to indicate that there was some kind of small nucleus in each atom which did not take up much room but which was capable of bouncing high-speed particles back when struck by them.

As a result Rutherford proposed that the structure of the atom resembles the structure of the solar system. According to this idea, every atom consists of a very small but heavy cluster of positive protons called the **nucleus.** Around this nucleus an equal number of light negative electrons revolve like planets around the sun.

It gives us some idea of the amount of empty space in an atom if we compare the nucleus to a fly in the center of a vast auditorium. The electrons are whirling around this central fly at the outside limits of the auditorium. As another example, if all the matter in the earth could be squeezed together without leaving any empty space inside the atoms, the earth would make up a package no larger than a football!

4. The changing picture of the atom

Rutherford's first picture of the atom as a miniature solar system has since been somewhat modified. In 1913 Niels Bohr (bōr), a Danish physicist, sought to explain the colors emitted by hydrogen gas when a spark is sent through it (see Chapter 43). His explanation was that electrons normally travel in certain fixed orbits. Unlike the planets in the heavens, however, they can jump from one orbit to another. When they do, they either give out light energy or absorb it.

Recent studies also indicate that the electrons move very fast around the nucleus. We cannot say precisely where an electron is at any instant, but must think of it as forming a cloud or region where it might be found. The diameter of the atom is the diameter of the outer electron orbits that make up this cloud.

5. Neutrons

Our picture of the nucleus has also been changed as the result of a discovery made in 1932 by Sir James Chadwick, a physicist working in Rutherford's laboratory. Chadwick found a third fundamental particle to add to the electron and the proton. Although it is approximately the same weight as the proton, it contains no electric charge—it is neutral. Hence its name, the **neutron.** The neutrons are packed into the nucleus together with the protons. We shall learn more about neutrons in Chapters 45 and 46 when we study the nucleus of the atom in greater detail.

A number of other particles have also been discovered in the nucleus of the atom, and one of the most important jobs of physicists today is to study these new particles and to explain them. But neutrons, protons, and electrons are the most important ones for our work in electricity.

6. Matter is electrical

You notice that this theory of atomic structure reduces all the matter on earth to three fundamental particles, electrons, protons, and neutrons, and that it considers matter to be electrical in nature. We shall learn in the last chapter of still more fundamental particles. The fact that these fundamental particles can be arranged in a great many different ways accounts for the many different kinds of atoms known to physicists.

We also see how this theory extends the idea of an "electrical fluid," widely accepted in Franklin's time. Franklin thought that electricity was a special fluid which accounted for electrical effects. We know today that even normal, uncharged matter is electrical in nature. The existence of an electrical charge is due simply to the gain or loss of electrons or (in special circumstances) protons. In most of the examples we shall study, the charge is due to a gain or loss of electrons in the outer orbits of the atom.

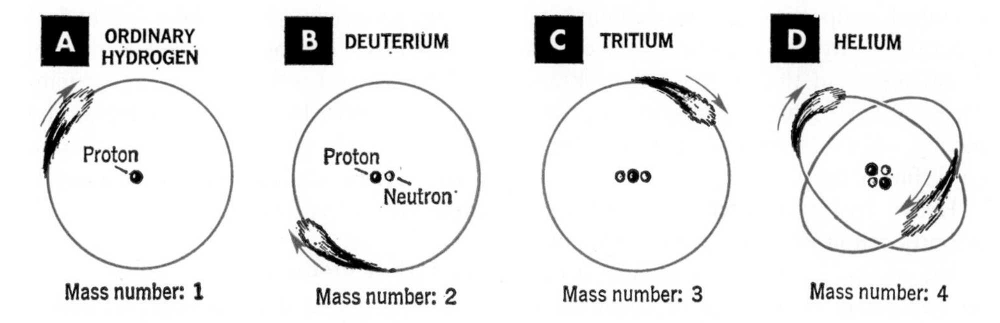

27-4 The four simplest atoms—of mass number 1, 2, 3, and 4. Three of these (A, B, and C) are forms of hydrogen because they all have one proton and one revolving electron. D shows the most common atom of helium; it has two protons and two electrons.

Various atoms

As a result of many experiments of the kind we have mentioned, physicists have built up a picture of the different kinds of atoms and the part that electrical particles play in their structure. Let us look at some of these.

7. Hydrogen

The simplest of all atoms is the hydrogen atom. It consists of a single electron revolving around a proton (Fig. 27-4A). The negative charge of the electron exactly counteracts the positive charge of the proton so that the atom is electrically neutral.

We shall use two different terms that will be of great help in describing atomic structure. One is mass number; the other is atomic number. **Mass number** refers to the *total number* of particles (both protons and neutrons) in the nucleus of the atom. Thus the mass number of helium is 4; its nucleus contains 2 protons and 2 neutrons. The mass number of lithium, whose nucleus contains 3 protons and 4 neutrons, is 7. Mass number is nearly the same as atomic weight.

The other term is atomic number. **Atomic number** refers to the *number of protons* in the nucleus. Thus the atomic number of helium is 2 and of lithium 3.

You notice that mass number does not include electrons. They are so light that they have little effect on the mass of the atom. Since the nucleus of a simple hydrogen atom consists of a single proton without any neutrons, we say that its mass number is 1. And its atomic number, which concerns only protons anyway, is also 1.

It is the number of protons (i.e. the atomic number), and nothing else, that determines what element the atom is. Two atoms of the same element *must* have the same number of protons but they may differ in the number of neutrons. Atoms of the same element which differ from each other in this way are called **isotopes.**

8. Isotopes of hydrogen

There are several isotopes of hydrogen. One, called **deuterium,** was discovered by the American scientist Harold Urey in 1932. The nucleus of this atom contains 1 neutron in addition to the proton (Fig. 27-4B); hence its mass number is 2. Since the nucleus still contains only 1 proton, its atomic number is 1, which means that deuterium must be a form of the element hydrogen.

This heavy hydrogen atom also contains 1 electron, just like an ordinary atom of light hydrogen. It has been discovered that the ordinary physical and chemical properties of an element are determined by the outermost electrons revolving about the nucleus of an atom. These two types of hydrogen atoms, each having only 1 electron, therefore have very similar properties.

Heavy hydrogen occurs in nature in the proportion of about 1 atom to 5,000 of the ordinary kind of hydrogen. It can be separated into its pure form from ordinary hydrogen. Like ordinary hydrogen, it will combine with oxygen to form water, but this water is about 10 per cent heavier than ordinary water, boils at 214.6°F (instead of 212°F), and freezes at 38.8°F (instead of 32°F). This **heavy water** looks and tastes like ordinary water, dissolves the same materials, and generally has much the same chemical properties.

A third kind of hydrogen is called **tritium** (trĭt′ĭ·*ŭ*m). The nucleus of a tritium atom contains 2 neutrons in addition to the 1 proton (Fig. 27-4C); thus its mass number is 3. Its atomic number is still 1, because it still has only 1 proton. Its ordinary physical and chemical properties are like those of the other forms of hydrogen. However, the nucleus is not stable and breaks apart, liberating energy; in other words, tritium is radioactive (see Chapter 45).

An element as we find it in nature is usually a mixture of isotopes. The *average* mass of all the isotopes in the mixture compared with oxygen as 16.000 is called the **atomic weight** of the element. It is usually not a whole number.

9. Helium

An atom of helium has a nucleus consisting of 2 protons and 2 neutrons (Fig. 27-4D). Thus its mass number is 4; and its atomic number (number of protons) is 2. But instead of 1 electron like hydrogen, it has 2. This increase in the number of electrons gives helium entirely different properties from those of hydrogen. Unlike hydrogen, it will not burn. It liquefies and solidifies at different temperatures than hydrogen, and looks and acts differently when in its liquid or solid form. There are isotopes of helium as there are of hydrogen.

10. Other elements

We have said that the different chemical properties of an element are determined by the outermost electrons in its atoms. Since the total negative charge of the electrons in an atom normally balances the positive charge of the protons, we can see that two atoms having the same number of protons will have the same number of electrons as well.

Atoms having the same number of protons (in other words, the same atomic number) all have the same properties and together constitute an **element.** Over the years over 100 different elements have been discovered. A list of them is in the table on page 681.

Molecules

We have said that there are over 100 different known elements. And yet there are millions of known substances. This fact is accounted for by the enormous number of ways that different atoms can combine to form molecules.

11. How atoms form molecules

Most of the substances in the world are **compounds**; that is, they are made of two or more elements chemically combined. A molecule of the compound water, for example, is made up of 2 atoms of the element hydrogen and 1 atom of the element oxygen (Fig. 27-5). We write it as a chemical formula, H_2O. We can form water easily by mixing hydrogen and oxygen in a dry container and igniting the two gases. The gases explode and a great deal of energy is given off in the form of heat. Water vapor and droplets of water are formed. If we take water and pass energy in the form of an electric current through it, the water gradually disappears and we find that the hydrogen and oxygen reappear.

If we ignite the metallic element sodium (chemical symbol: Na) in the presence of the gaseous element chlorine (Cl), they combine rather violently. When the reaction has ended, the sodium and chlorine are gone (if the quantities are chosen properly), and in their place we find some white powder. When we taste the powder and test it in other ways, it turns out to be ordinary table salt, or sodium chloride, NaCl.

The appearance and properties of a compound are very different from the appearance and properties of the separate elements which make it up. Sodium, for example, is a violently active, silvery metal which may actually cause an explosion when thrown into water, and chlorine gas is poisonous. But the two combine to produce harmless table salt.

27-5 Two atoms of hydrogen combine with one atom of oxygen to form a water molecule (H_2O).

12. Borrowing and lending electrons

The scientific explanation of how atoms combine into molecules involves the electronic structure of the atom. According to this theory, some atoms tend to "borrow" electrons, others to "lend" them. When an atom borrows electrons (which have a negative charge), it becomes negatively charged. Chlorine and oxygen and **nonmetals in general tend to borrow electrons.** On the other hand, an atom which lends electrons becomes charged positively, because it will have an excess of protons once some electrons are gone. **All metals tend to be lenders of electrons.** Thus many molecules are held together by the force of attraction between opposite charges—positively charged metals (or positively charged groups of atoms) and negatively charged nonmetals (or negatively charged groups of atoms).

Some atoms can also *share* electrons with other atoms, and this method of combination accounts for the formation of many molecules found in nature. But for our purposes here we shall discuss only the molecules formed by the lending and borrowing of electrons. If you have studied chemistry, you will no doubt have learned more about the electron-sharing process.

Let us consider the formation of salt. We bring an electron lender (sodium) near an electron borrower (chlorine) and increase their energy with the help of heat from a flame. Each lending atom of sodium gives up an electron to a borrowing atom of chlorine (Fig. 27-6). This means that the sodium atoms are now positively charged while the chlorine atoms are negative. Charged atoms, as you remember from page 336, are called ions. The positive ion of sodium now attracts the negative ion of chlorine with the result that they are drawn together into what we may call a molecule of ordinary table salt (Fig. 27-6). Actually chemists have found that the sodium and the chlorine ions are arranged

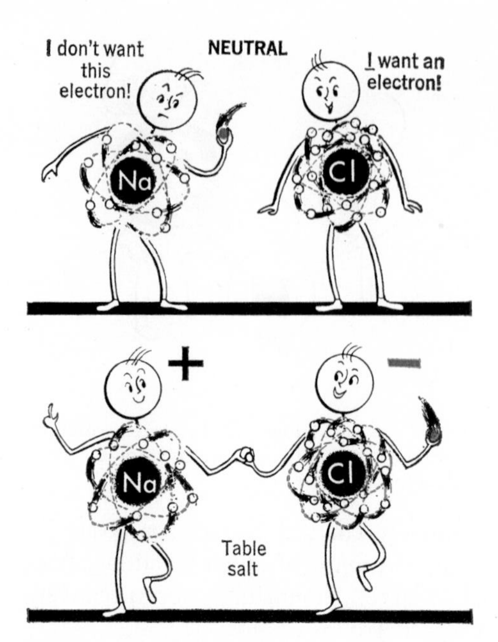

27-6 Ordinary table salt (sodium chloride) is formed when an atom of sodium gives up an electron to an atom of chlorine. Opposite charges then keep them bound together.

alternately in equal numbers in a regular latticework which we observe as a crystal. No one sodium atom belongs to a particular chlorine atom, though for the reasons we have explained they are present in equal numbers and the formula is correctly written NaCl.

Both sodium and chlorine are said to be **active elements** since they lend and borrow electrons readily. A sodium atom will always tend to give 1 electron to a chlorine atom, which is always prepared to take 1 electron. And a chlorine atom will always tend to accept an electron from a sodium atom which has one to give. Atoms of other elements may give up one or several electrons to atoms of elements that are prepared to accept them. Hence there exists a variety of compounds like sodium chloride, zinc chloride, sodium iodide, sodium fluoride, aluminum chloride, and so on. All of these compounds are called **salts**, compounds of two kinds of elements; one of which is a metal that lends electrons, and the other a nonmetal that borrows electrons.

We picture many of the chemical changes in nature as occurring in a similar fashion. As the atoms of two different substances are brought close together, an exchange of electrons may take place. If that occurs, some of the atoms are turned into charged ions, those of one substance becoming positive, the others negative. The attraction of these charged atoms for each other brings them together to form molecules of a new substance. Of course, this is a highly simplified picture of what happens, and there are many kinds of combinations that are possible—sometimes involving thousands of atoms. But basically all chemical changes are really the result of exchanges or sharing of electrons among atoms.

13. Why metals are good electrical conductors

As we have seen, metals are good lenders of electrons. All metal atoms have one or more electrons which they lose easily to other atoms. It is because of these relatively "loose" electrons that metals are such excellent conductors of electricity. If one end of a metal wire is attached to the positive terminal of a source of voltage, such as a battery, and the other to the negative terminal, the loose electrons in the atoms of the wire are repelled from the negative terminal and attracted toward the positive terminal. The flow of electrons along the wire is the **electric current** (Fig. 27-7). Of course, as many electrons flow into the wire at the negative terminal of the battery as flow out of the wire at the positive terminal. Thus the number of electrons is kept the same in the wire.

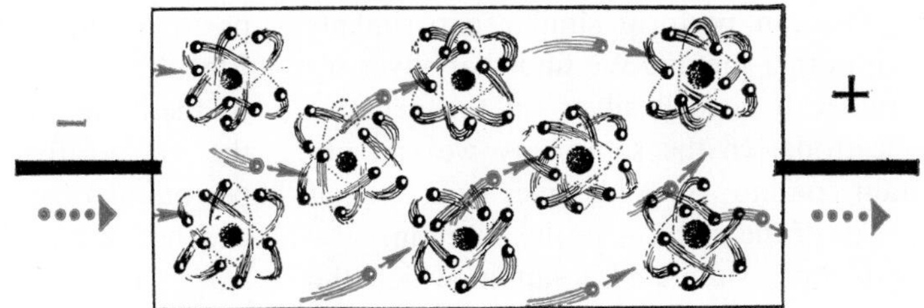

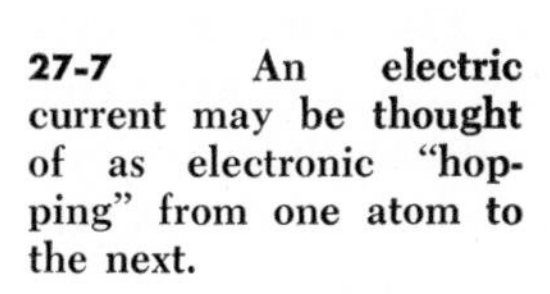
27-7 An electric current may be thought of as electronic "hopping" from one atom to the next.

Electrolysis

A great deal of information about the nature of atoms and molecules has been obtained from the process called **electrolysis** (ē·lĕk′trŏl′ĭ·sĭs). In this process chemical changes are produced by electric currents. Copper plating is a practical application of electrolysis. So also is the manufacture of aluminum.

14. Copper plating

Connect a battery to two carbon rods and place them in a solution of copper sulfate, as shown in Fig. 27-8. In a few minutes a reddish-brown coating of copper will form on the rod connected to the negative terminal.

Now reverse the connections so that the rod with the copper coating is connected to the positive terminal. The copper begins to leave the rod, and a coating of copper forms on the other rod. Apparently the copper will deposit itself only on a negative rod, called an **electrode,** and will be removed from a positive electrode.

15. Why do metals plate out?

Where did the copper come from? And why should it always be deposited on the negative electrode?

The source of the metallic copper must be the solution of copper sulfate. However, the copper in the solution is part of a compound, copper sulfate. Apparently copper atoms in the copper sulfate are attracted by the negative electrode. From this fact it would seem that the copper atoms in the copper sulfate have a positive charge. This seems reasonable when we recall from the previous section that metal atoms (such as copper atoms) lend electrons to the nonmetal part of a molecule (such as the sulfate part) and therefore tend to be positive.

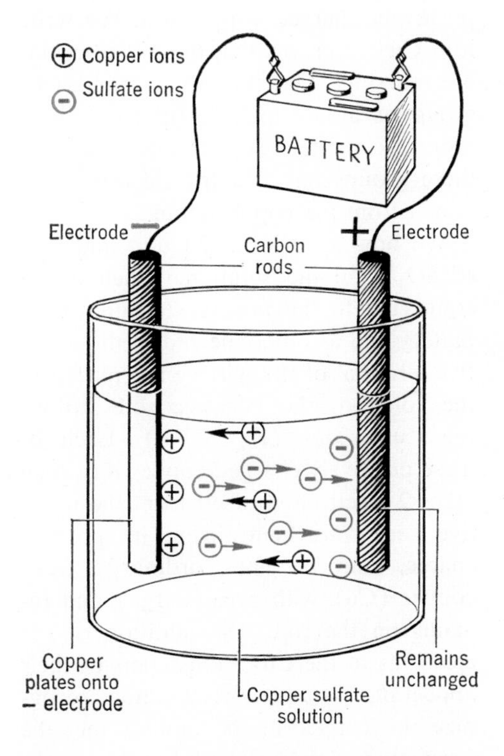

27-8 Copper can easily be plated out of a solution of a copper salt onto a negative terminal. A battery supplies the charge that pulls positively charged copper ions out of the solution and deposits them on a negative terminal.

We can perform similar experiments with other metals. We find that silver nitrate will deposit silver on the negative electrode. In the same way we can obtain coatings of gold, zinc, iron, etc., if a salt of the metal is in the solution.

Copper sulfate is a salt in which the copper atom in each molecule has lent 2 electrons to the sulfate group of atoms. This sulfate group consists of 1 atom of sulfur and 4 of oxygen; its chemical formula is SO_4. Together, the copper atom and the sulfate group make up a molecule of copper sulfate, $CuSO_4$.

It has been found that when copper sulfate dissolves in water, many of its atoms separate. A copper atom which separates from the sulfate group, leaving its two outer electrons behind, is now a positively charged copper ion. We write it as Cu^{++} to show that these 2 electrons are missing. On the other hand, the group of atoms making up the sulfate group is written SO_4^{--} since each of these groups has 2 extra electrons, obtained from the copper atom.

We now have ions of Cu^{++} and ions of SO_4^{--} in the solution. When a rod connected to the positive terminal of a battery and a rod connected to the negative terminal of the battery are placed in the solution, the negative rod attracts the Cu^{++} ions (Fig. 27-8). Each of these positively charged copper ions then takes 2 electrons to itself from the negative rod, thus neutralizing its positive charge, and becomes ordinary "free" copper (Cu^0, with zero charge) and remains on the rod. The addition of the electrons to the Cu^{++} ions changes their appearance completely. We now recognize the copper in its familiar metallic form, that is, copper with no charge.

While the copper is being deposited on the negative rod, something else is happening at the positive rod. In a practical copper-plating arrangement, the positive rod is made of copper metal. Electrons are being taken out of this rod by the battery (Fig. 27-8) and thus keep the rod positive. As a result, some copper atoms of the rod lose their electrons and become positive ions, Cu^{++}. As the SO_4^{--} ions in the solution are attracted to the positive rod, they pull these Cu^{++} ions into solution, thus forming more copper sulfate. If the positive electrode is made of copper, it serves to keep the supply of copper ions in the solution from becoming exhausted.

In all electrical metal plating, the object to be plated is made the negative electrode, the positive electrode is constructed of the plating metal, and the solution contains a salt that produces the ions of the plating metal.

16. Electrochemical equivalent

Since each copper atom that is deposited gains two electrons from the negative electrode, the number of atoms, and therefore the mass, of copper that plates out depends upon the amount of charge that flows into the negative electrode.

The charge of one electron is too small to be used conveniently as a unit of charge. The common unit is the charge of 6.28 billion billion electrons (6.28×10^{18})! This unit of charge is called the **coulomb.** (The amount of charge that flows through a 100-watt light bulb in one second is roughly 1 coulomb.)

Experiments have shown that the number of grams, m, of copper deposited is proportional to the amount of charge, Q, that flows through the cell. This fact was discovered by Faraday and may be written

$$m = ZQ$$

where Z is the mass of material deposited per unit of electrical charge. Z is called the **electrochemical equivalent.**

The electrochemical equivalent is a constant for the element used, but it is a different number for each element.

If m is expressed in grams and Q in coulombs, the electrochemical equivalent, Z, is in grams per coulomb. Table 11 gives the electrochemical equivalents of some common elements.

SAMPLE PROBLEM

How much copper can be deposited by a charge of 5,000 coulombs flowing through a copper-plating cell? Substituting in Faraday's equation:

$$m = ZQ$$
$$= 0.000329 \text{ gm/coulomb} \times 5{,}000 \text{ coulombs}$$
$$= 1.65 \text{ gm}$$

STUDENT PROBLEM

How much silver would be deposited by 2,000 coulombs of charge? (ANS. 2.24 gm)

17. Conducting solutions

In copper plating, the solution of copper sulfate acts as a conductor of electric current. But this flow of current is somewhat different from that in solid metals, where electrons pass from one atom to another. The current in the solution consists of positive copper ions moving through the liquid toward the negative electrode while the negative sulfate ions are moving in the opposite direction toward the positive electrode. This is how salt solutions conduct an electric current.

Substances like copper sulfate which conduct electricity when dissolved in water are called **electrolytes** (ē·lĕk′trō·līts). All salts and acids are electrolytes. So is another group of substances called **bases.** The molecules of all electrolytes are made up of one group of atoms which lends electrons and another group which borrows them.

TABLE 11
ELECTROCHEMICAL EQUIVALENTS (grams per coulomb)

Element	grams per coulomb
Hydrogen	0.00001045
Oxygen	0.0000829
Nickel	0.0003041
Copper	0.0003294
Zinc	0.0003387
Silver	0.001118

A great many materials are not electrolytes. For example, sugar, alcohol, and ether are not electrolytes. These and a host of other materials have a molecular structure which does not permit them to break apart easily (ionize) in water. Hence they are poor conductors of electricity in solution. Figure 27-9 shows one way of determining whether or not a given material is an electrolyte. An electric wire to which a lamp is connected is broken, and each end of the open wire is connected to a carbon rod placed in a solution of the material being tested. Then the wire is attached to a source of current. If the lamp lights, it shows that the solution carries current and that the

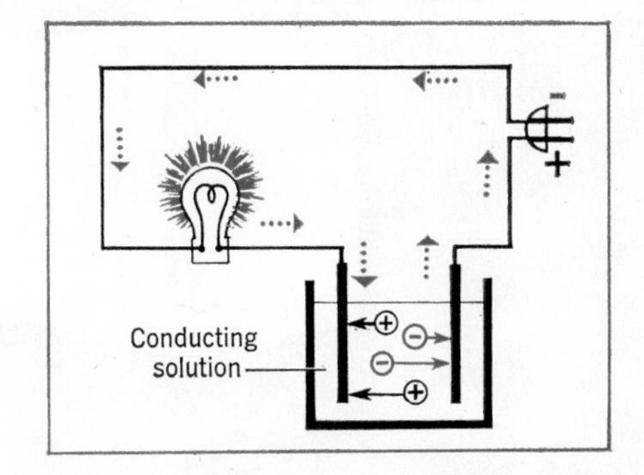

27-9 This experiment will reveal whether a solution conducts electricity (is an electrolyte). The lamp lights up if current flows in the solution. How is the electrical conduction in the solution different from that in metal wires?

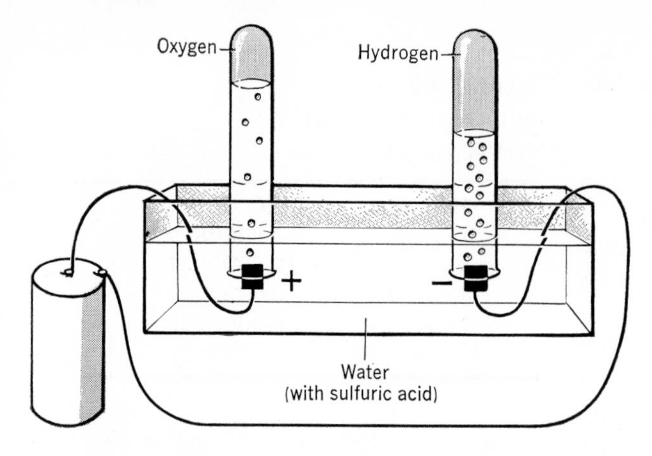

27-10 Water can easily be broken up into its separate elements, hydrogen and oxygen, by means of an electric current. What fact in this experiment provides evidence that water is H_2O?

material is a good electrolyte. If the lamp does not light, then the material dissolved in the water is not an electrolyte, or is a weak one.

18. Uses of electrolysis

You are probably already familiar with the separation of water into hydrogen and oxygen by **electrolysis** (Fig. 27-10).

Even some metals are produced by electrolysis. Of these, aluminum is the most important. Magnesium, a very light metal, is obtained in large quantities from its salts in sea water by means of electrolysis. Both of these light metals are much more suitable than iron for many purposes. As the cost of producing them is being reduced, they are supplementing and even replacing iron for many uses. At present the large amount of electrical current required for their manufacture makes them relatively expensive. Chemicals such as sodium, lye, chlorine, and pure copper are also produced by electrolysis (Fig. 27-11).

The Anaconda Company

27-11 In this copper refinery process impure copper plates are used as the positive terminals in this roomful of electrolytic cells. A current through the cells causes copper atoms to ionize and to be drawn through the solution to be deposited finally as pure (electrolytically refined) copper on the negative terminals.

Electroplating is a kind of electrolysis in which one metal is placed over another. Tin, gold, silver, nickel, chromium, and copper are some of the metals commonly used in this process.

Electroplating is employed where exact impressions of surfaces are required. It can be used in making phonograph records, for instance, and it is commonly used in the production of the plates from which books are printed. The type is set and a wax or plastic impression made. This impression is coated with graphite so that it will act as a conductor. It is then used as an electrode, and copper is plated onto it. Now the copper-coated impression, backed by additional metal to strengthen it, can be used to print the page.

In the next chapter we shall consider a process which is just the reverse of electrolysis. In electrolysis we put in electrical energy to produce chemical changes. But we can also use chemical changes to produce electrical energy. This is the process that occurs in electric batteries. We shall see that the process of electrolysis is closely connected with the action of batteries.

Going Ahead

THE VOCABULARY OF PHYSICS

Define each of the following:

cathode rays	element
electron	compound
proton	active element
nucleus	salt
mass number	electric current
neutron	electrolysis
atomic weight	electrode
atomic number	coulomb
deuterium	electrolyte
heavy water	electrochemical equivalent
tritium	
isotope	

THE PRINCIPLES WE USE

1. Describe our modern picture of the atom and its parts.
2. Describe the structure of ordinary hydrogen, deuterium, tritium, and helium.
3. Describe our picture of the molecule of water; of table salt.
4. Describe how you would go about plating a spoon with silver. Explain silver plating.
5. What kinds of materials will conduct electricity when in solution? Name three that conduct current and three that do not.
6. Why is electrolysis important?

Explain why

7. Ordinary hydrogen and deuterium have almost the same chemical properties, even though they differ in atomic structure.
8. Metals are good conductors of electricity.
9. Nonmetals are poor conductors of electricity.
10. Compounds of metals and nonmetals form rather easily.
11. Metals always plate out of solution onto a negative electrode.
12. A solution of table salt conducts electric current.

EXPERIMENTS WITH ELECTRICITY

Describe an experiment

1. To show copper plating.
2. To show cathode rays.
3. To indicate that cathode rays consist of particles rather than of some kind of light ray.
4. To show positive rays.
5. To indicate that matter consists mainly of empty space.

APPLYING PRINCIPLES OF ELECTRICITY

1. Mention five practical uses for electrolysis.
2. What part does copper plating play in printing books?
3. Physicists have tried all sorts of experiments in which they bombard atoms

with high-speed particles. Can you figure out why it would be difficult to hit the nucleus of an atom with a proton? Why should the proton have high speed? Why is it difficult to hit a nucleus with an electron? Why is a low-speed neutron more likely to hit a nucleus than is an electron or a proton?

4. You are given the following facts: a zinc atom tends to lose 2 electrons, a chlorine atom tends to gain 1, an oxygen atom tends to gain 2, a silver atom tends to lose 1, an aluminum atom tends to lose 3, a sulfur atom tends to gain 2.

   Can you predict five combinations of these atoms to form molecules? In each case state how many atoms of one combine with how many atoms of the other.

SOLVING PROBLEMS IN ELECTRICITY

1. How many grams of silver will be plated on an electrode in a silver nitrate cell by passing a charge of 4,000 coulombs through the cell?
2. What is the electrochemical equivalent of a metal that requires 2,000 coulombs to deposit 0.608 gm of the metal? What is the metal?
3. How much charge is needed to plate 3.00 kg of nickel on an iron electrode?

---

4. A copper-plating cell and a silver-plating cell are connected in series to a source of electricity. 2.35 gm of silver are plated out. How much copper was plated?
5. If an ampere is defined as a current, or rate of flow of charge, of 1 coulomb/sec, how long will it take to deposit 2.00 gm of silver if a current of 4.00 amp flows through the cell?
6. A charge of 10,000 coulombs is used to decompose water. What volume of hydrogen and oxygen at 0°C and 76 cm of mercury pressure will result?

PROJECTS

1. Identifying Electrolytes. Connect two dry cells in series with a flashlight bulb and a pair of metal strips dipped in water as shown in Fig. 27-9. To identify a compound as an electrolyte, dissolve some in the water and observe the bulb. When salt is sprinkled into the water, the bulb lights, indicating salt is an electrolyte. Sugar has no effect at all and is not an electrolyte. Try vinegar, baking soda, glycerine, rubbing alcohol, and fruit juices.
2. A Pure Metal by Electrolysis. Connect a dry cell to two copper wires whose ends dip into a solution of copper sulfate. The wire connected to the negative terminal should be scraped clean of insulation; to the other wire attach a brass screw.

   The flow of current through the solution transfers pure copper atoms to the bare wire where it builds up in a spongy mass. The copper is removed from the brass screw, leaving the zinc to dissolve or fall to the bottom of the jar.

CHAPTER

28

Electricity from Chemicals

One day in the year 1780 a pair of frog's legs had been prepared for study by the Italian scientist Luigi Galvani (gäl·vä'nē). The frog's legs were lying close to an electrostatic machine when someone happened to touch a nerve in the leg with a metal scalpel. A spark jumped from the machine through the scalpel into the frog's leg. And the leg twitched! Galvani decided to find out why.

The voltaic cell

Galvani discovered that the frog's leg could be made to twitch if the tissue of the leg was touched with the free ends of two wires made of copper and iron joined together at their other ends. He thought that this strange occurrence might be due to some kind of electricity in the leg of the frog.

But another Italian experimenter, Count Alessandro Volta, challenged this interpretation. He found that the copper and iron were producing electric currents in the leg. It was this current which caused the leg to twitch. Moreover, the fact that it was a frog's leg turned out to be unimportant. What was important was the fact that there were certain salts and acids dissolved in the tissues of the leg which produced an electric current when they came in contact with the two dissimilar metals. In Chapter 27, we learned that salts and acids and bases, because they can carry electric current, are called electrolytes. Volta was able to obtain current when the frog's leg was dispensed with altogether and a solution of an electrolyte (acid, base, or salt) was used instead.

1. Making a cell

You can repeat Volta's experiment yourself, using a solution of an electrolyte and two wires or plates made of different metals. When these materials are arranged as shown in Fig. 28-1, an electric current sufficient to ring a bell or light a small lamp is produced. This simple source of electrical energy is known as the **voltaic cell.**

We can use different metals in the voltaic cell and compare their voltage-producing abilities. For this purpose it is best to use a voltmeter (this instrument will be studied in detail in Chapter 32), which gives us readings in volts. A **volt** is a unit of electromotive force which measures the ability of a voltaic cell to push a current of electrons through a wire. For our experiment it is convenient to use a voltmeter whose pointer can move either to the right or to the left; that is, which indicates zero at the center of the scale.

Let us use a dilute solution of sulfuric acid, H_2SO_4, in water as our electrolyte. When we place a plate of zinc and one of iron in the solution and connect them

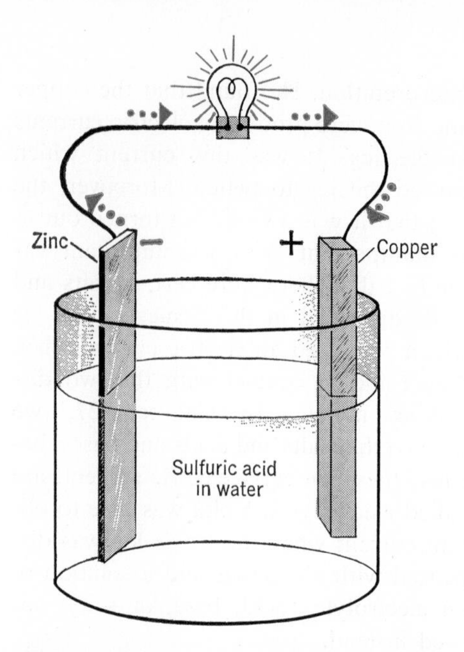

28-1 Batteries are based on the simple experiment diagramed above. Just dip two different metals into an electrolyte solution (such as acid in water), connect the metals, and you get current. The combination of zinc and copper gives enough current to light a small lamp or ring a bell.

to a voltmeter, we see that they produce less than 0.3 volt. The pointer of the voltmeter moves in a direction which indicates that the zinc is the negative terminal. (This direction can be determined in advance by connecting a dry cell to the voltmeter and noting the direction of motion of the pointer and the connections of the negative and positive terminals of the cell to the voltmeter.)

When we use zinc and copper, we get 1.1 volts. Again zinc is the negative terminal. A zinc-carbon combination produces 1.5 volts with zinc again the negative terminal. When we use iron with copper, iron takes the place of zinc as the negative terminal. The voltage we get from this combination is less than from the zinc-copper combination.

Why should two different metals placed in an electrolyte produce current? What determines the different charges (plus or minus) that the metals acquire? And why should the voltages change with different combinations of metals? Let us investigate.

2. Active metals

Put a clean iron nail into a solution of copper sulfate. In a few seconds you find that copper has been deposited on the nail. A chemical test would show that the solution had acquired some iron. But if a copper rod is placed in an iron sulfate solution, none of the iron from the solution is deposited on the copper.

We explain this difference by saying that iron is a more active metal than copper, tending to replace copper in a solution. Zinc is still more active than iron, so that in a solution it tends to replace the iron.

By testing the activity of various metals in this manner, it is possible to find out whether one metal is more active than another. We can then list them in order of activity in what is known as the **electrochemical series** (Fig. 28-2).

Metals high on this list tend to replace those lower down. For example, the metals in the very active category—lithium, potassium, calcium, and sodium—are so active that when put in water they will actually replace some of the hydrogen in the water. The heat developed in this process is sometimes great enough to ignite the hydrogen and cause an explosion.

The metals at the bottom of the list, however, are so inactive that they do not readily replace other metals and do not easily undergo chemical change. It is this inactive quality of silver, gold, and platinum that makes them suitable for tooth fillings, jewelry settings, and money. The farther apart any two metals are on this list, the higher the voltage they will develop when used as electrodes in a voltaic cell. Moreover, the metal that is higher on the list will become the negative electrode of the cell. Hence any two dissimilar metals placed in an electrolyte can form a voltaic cell.

3. Electricity on tap

One of the simple voltaic cells we described in this chapter used the metals zinc and copper in sulfuric acid, H_2SO_4. We explain the action of this cell as follows: Because zinc is more active than hydrogen, zinc atoms dissolve, becoming positive zinc ions, and leave electrons behind on the zinc electrode, thus making it negative. Since hydrogen is more active than copper, positive hydrogen ions take electrons from the copper, leaving the copper positive.

As soon as the zinc electrode becomes slightly negative, its charge prevents any more zinc from dissolving. Similarly, positive charge that builds up on the copper repels hydrogen ions.

However, if the negative and positive electrodes are connected by a wire, electrons flow through the wire (Fig. 28-3), neutralizing both electrodes. The chemical action then continues, removing electrons from the copper and adding them to the zinc, thus keeping a steady flow of current through the wire. If a device such as a small lamp is connected between the electrodes, the flow of electrons from the zinc to the copper may be great enough to cause the lamp to light.

The hydrogen ions which take electrons from the copper become neutral hydrogen atoms and come out of the solution as bubbles of hydrogen at the copper. This formation of hydrogen bubbles is called **polarization.** Unless these bubbles of gas are gotten rid of, they will

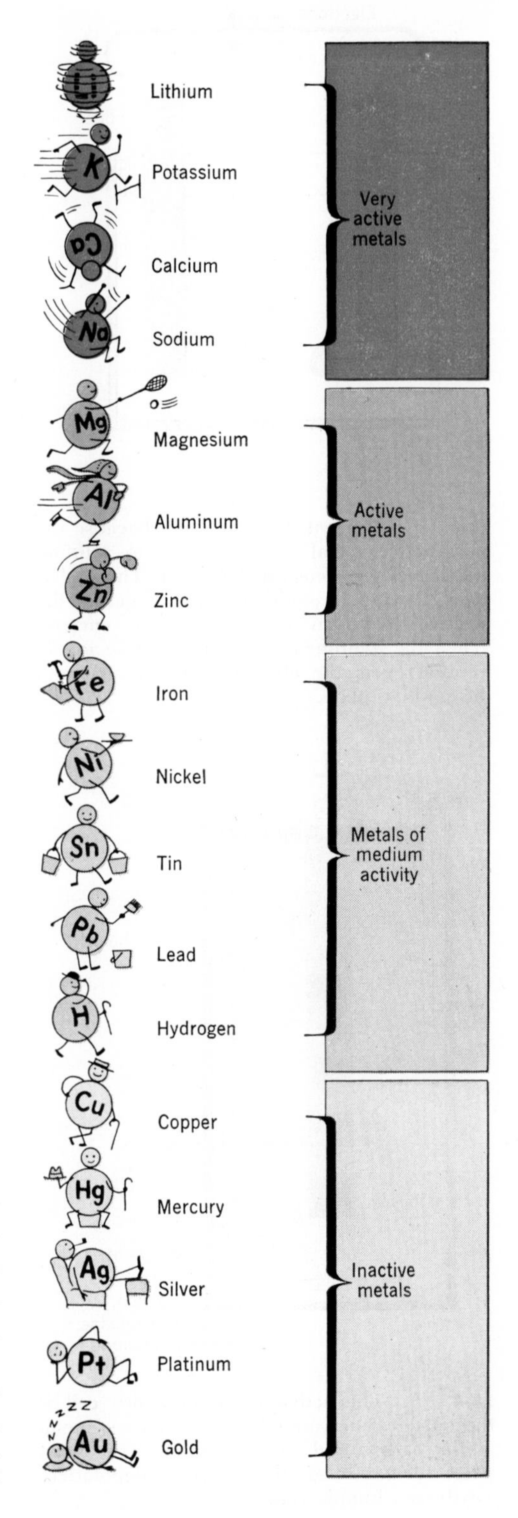

28-2 **In the electrochemical series metals are arranged in the order of their activity. Active metals are at the top, inactive ones at the bottom.**

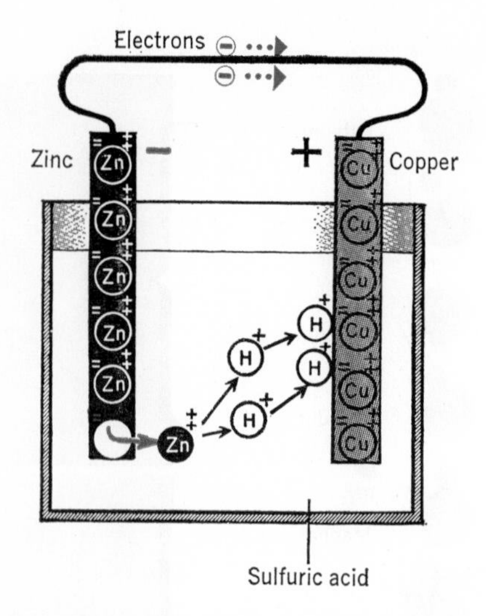

28-3 Current flows in a cell because the most active metal (zinc) goes into solution and leaves electrons on the metal. These electrons, flowing through the wire to get to the positive hydrogen ions, cause the current. When the hydrogen (or other positive ion in solution) gets the electrons, it deposits onto the positive plate.

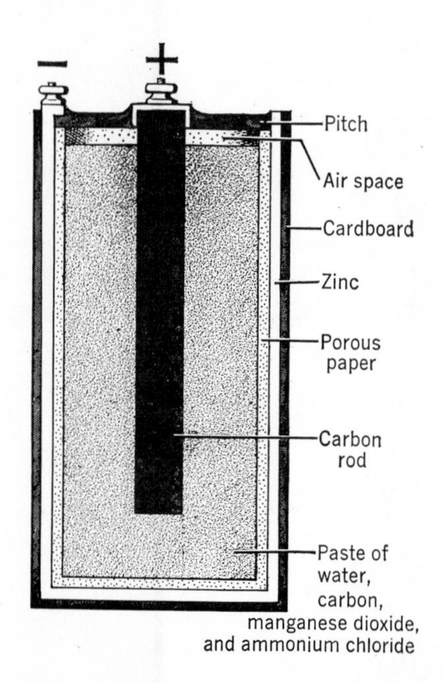

28-4 The active metal in a dry cell is the outside zinc container. The inactive metal is the carbon rod in the center. The moist paste contains electrolyte and depolarizer (hydrogen bubble remover).

form an insulating barrier around the copper and stop the action of the cell. Any chemical capable of furnishing oxygen to oxidize the hydrogen into harmless water will serve as a good depolarizer. Manganese dioxide is one such chemical (Fig. 28-4).

The cell eventually runs down because the zinc plate gradually dissolves, hydrogen is driven out of the solution, and the solution changes from sulfuric acid, H_2SO_4, to zinc sulfate, $ZnSO_4$.

4. The dry cell

The **dry cell** has been misnamed. It should really be called a "moist cell." Every cell must have a solution of an electrolyte. But in the dry cell this solution is in the form of a moist paste to avoid spilling when the cell is moved.

Figure 28-4 shows the construction of a dry cell. The outer container is made of zinc (covered with cardboard) and serves as the negative electrode. The positive electrode, in the center of the cell, consists of a rod of carbon. The electrolyte is a salt called sal ammoniac (chemical name: ammonium chloride) dissolved in water. It is mixed with grains of carbon and a depolarizer (manganese dioxide) to form the paste. A layer of blotting paper separates the zinc from the rest of the cell but still permits the electrolyte solution to soak through in order to carry on the action of the cell. The top of the cell is sealed with pitch to prevent the water in the electrolyte solution from evaporating. A new dry cell produces 1.5 volts.

Because polarization is not entirely eliminated, the dry cell is not satisfactory for continued use. Its best use is in devices where the current needed is occasional, such as in flashlights, doorbells, and portable radios. Its major advantage is that it has little weight and can be easily carried.

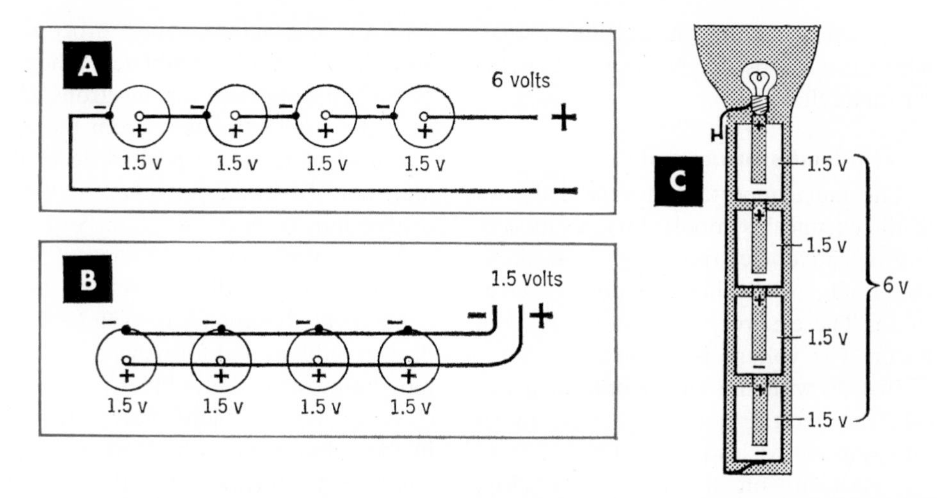

28-5 Piling the flashlight cells into the barrel of the flashlight connects them in series to give more voltage (C). The electrical diagram is shown in A. The parallel connection in B does not increase the voltage and is not used in flashlights, but it is used when large amounts of current at low voltage are needed.

5. Constructing a battery

The 1.5 volts produced by a single dry cell are insufficient for many purposes. But making the cells larger will not increase the voltage above 1.5 because the voltage developed is determined only by the materials of which the electrodes are made. The only advantages of a larger cell are that it will last longer before running down, and its internal resistance will be lower.

In order to achieve higher voltage, several cells are connected so as to increase the push behind the electrons, in other words, to increase the voltage. If we put two dry cells in series, connecting the plus of one cell to the minus of the other, and measure the voltage with a voltmeter, we find that they produce 3 volts. Three cells produce 4.5 volts; four cells, 6 volts (Fig. 28-5A). So we see that each cell in series adds its voltage to that of the others. This method of increasing voltage is used in flashlights (Fig. 28-5C). A **battery** is a group of cells connected together to give higher voltage.

If we connect cells in parallel, as shown in Fig. 28-5B, we find that the voltage is the same as for one cell, no matter how many cells we use. This combination will last longer than one cell, however, and so it is useful where frequent replacement is to be avoided.

The storage battery

In the cells we have been describing, chemical energy is converted into electrical energy—exactly the reverse of the process of electrolysis described in Chapter 27. There we put in electrical energy and got back a pure metal which had chemical energy.

The storage battery makes use of this reversibility of energy. We put substances in it that possess potential energy when the battery is charged. As the battery runs down these substances change to new ones having less energy, the loss in potential energy being converted into electrical energy carried by an electric current. As we shall see, we can restore these substances to their original chemical condition again by putting in electrical energy from an outside source, which it is impossible to do with a worn-out dry cell.

6. The lead storage cell

The most important storage cell uses lead (chemical symbol: Pb) as the active negative electrode and lead dioxide, PbO_2, as the inactive positive electrode. The electrolyte is sulfuric acid in water. The voltage is 2 volts.

Just as with the zinc atoms in a dry cell, the neutral lead atoms tend to go into solution as lead ions (Pb^{++}), leaving electrons on the lead plate (Fig. 28-6A). The action cannot proceed unless the circuit is complete. When the circuit is complete, the electrons travel over the conducting wire from the negative lead plate to the positive lead dioxide, there becoming attached to the hydrogen ions from the solution. So far this action is very similar to that of a dry cell. But the lead dioxide acts as its own depolarizer. It supplies oxygen to the hydrogen bubbles and forms water, thus removing most of the bubbles.

When the Pb^{++} ions meet SO_4^{--} ions in the solution near the lead plate, they combine to form lead sulfate, $PbSO_4$.

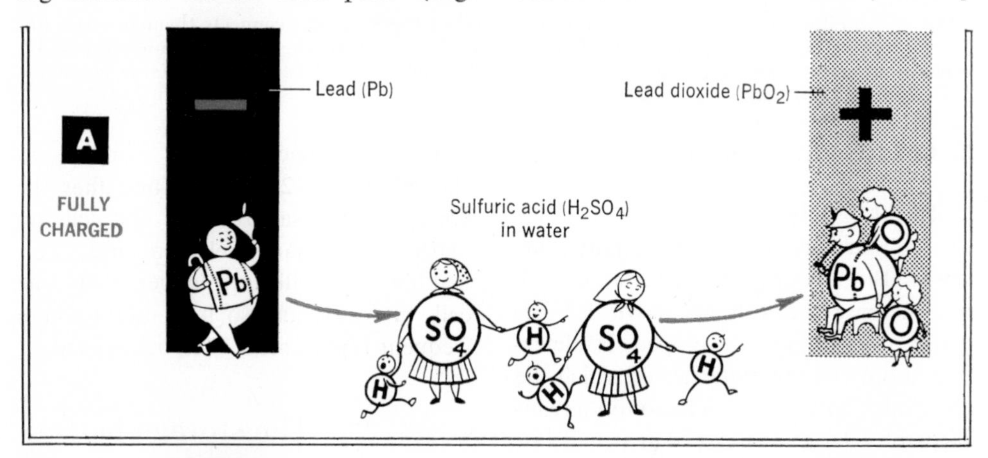

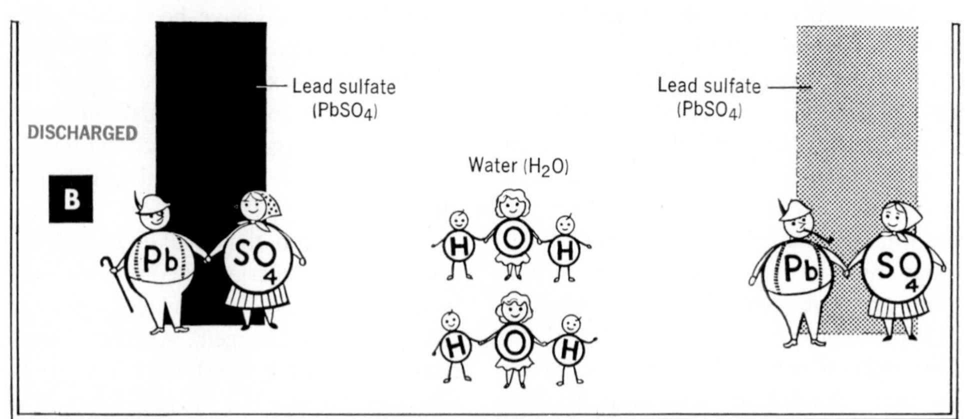

28-6 When a cell discharges, the lead ions (Pb^{++}) combine with sulfate ions ($SO_4^{=}$), and each oxygen atom combines with two hydrogen atoms. In the meantime the transfer of electrons that makes this chemical exchange possible gives us electric current. To recharge the battery we apply electrical pressure (voltage) to separate these groups and return them to their original state.

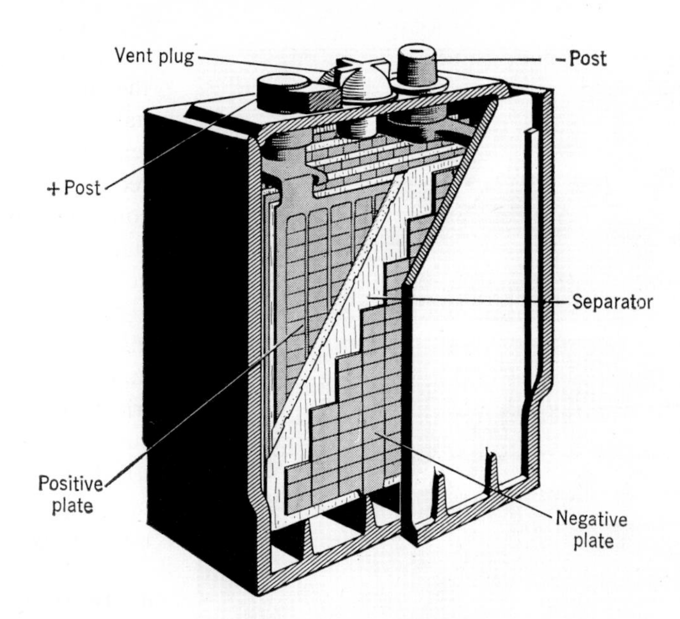

28-7 Large amounts of current from a battery are possible because the metal is packed into many plates close together in each cell. These plates are placed in alternating rows like interlocking fingers of two hands. The positive and negative plates are kept from touching and short circuiting by means of thin insulating separation sheets. The electrolyte is not shown.

Since the lead sulfate does not dissolve easily, it coats the lead plate instead of remaining in solution. At the same time the chemical action at the positive plate, involving hydrogen, oxygen, lead, and sulfate ions, forms water and more lead sulfate. Thus as the cell is used, both plates, originally unlike, become coated with the same material, lead sulfate, and the cell eventually stops working. In this condition it needs recharging.

Figure 28-7 shows the structure of a single lead storage cell. A six-volt battery consists of three such cells connected in series.

7. Recharging the cell

To recharge the lead storage cell, we simply connect the terminals of the cell to a source of direct current with a slightly higher voltage, such as an electric generator or a charger (Fig. 28-8). The negative terminal of the outside source is connected to the negative lead plate of the battery, thus charging it strongly negative. This strong negative charge we have added to the plate attracts positive lead ions away from the lead sulfate to form neutral lead once

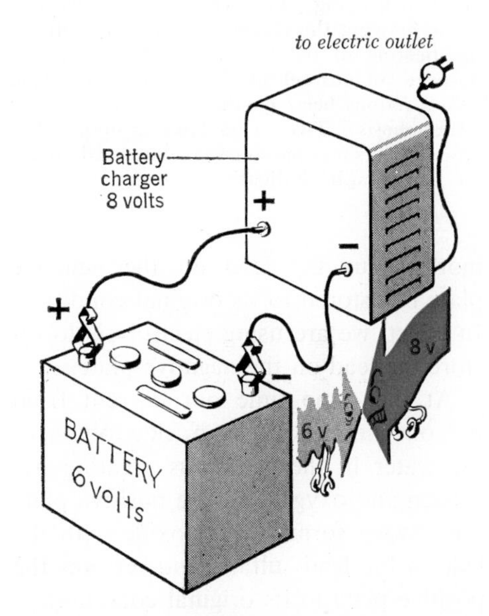

28-8 A battery is charged by reversing the current in it. We do this by connecting the negative of a slightly stronger (higher voltage) current supply to the negative of the battery, while positive is connected to positive. The process of electrolysis creates chemical changes in the battery that restore its materials to the original charged state. In an automobile a DC generator is used to keep the battery charged.

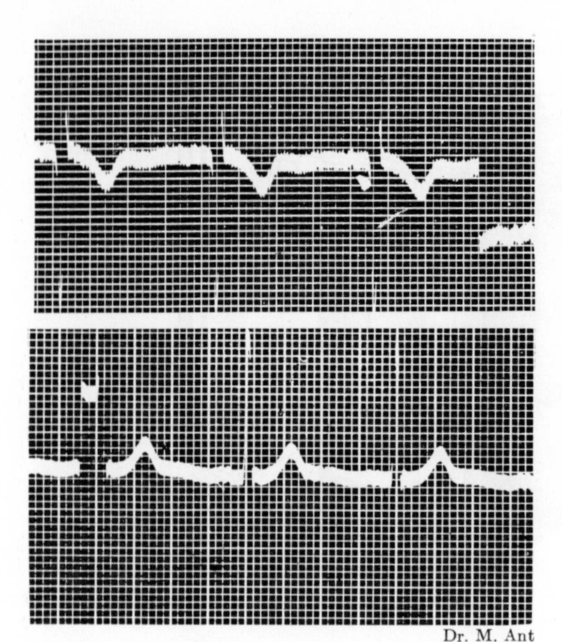

Dr. M. Ant

28-9 **Electrochemical actions play an important part in keeping you alive. Here are two cardiograms showing the electrical changes produced by the beating of the heart. The sudden downward surges of current in the upper photograph indicate a serious heart condition known as "coronary thrombosis." Two years later a similar cardiogram of the same patient shows upward surges (lower photograph) indicating recovery.**

more. Thus the lead on the negative plate is restored to its original condition. In effect, we are using electrolysis to restore the lead on the negative plate.

At the same time our current from the outside produces an electrolysis of the water in the cell, thus creating hydrogen and oxygen. At the positive plate the oxygen forms lead dioxide with the lead in the lead sulfate, and restores the positive plate to its original condition.

The sulfate ions, together with the hydrogen ions formed by electrolysis, combine to form sulfuric acid in the electrolyte once more. The only material we have to supply in this recharging process is water, which gives us the hydrogen and oxygen necessary for the chemical action.

To summarize, we might say that all the materials we put into the cell originally are changed to lead sulfate and water when the cell is discharged. During recharging the lead sulfate and water are converted back to the original materials: lead, lead dioxide, and sulfuric acid (Fig. 28-6).

While the cell is in use it is discharging, and chemical energy of the original materials is being converted into electrical energy in order to light lamps or run motors. During recharging, electrical energy applied from the generator or charger is converted back into chemical energy.

8. Testing the cell

You can see that sulfuric acid is used up in forming the lead sulfate while the cell is discharging. Since sulfuric acid is denser than water, this removal of the acid from the solution reduces the density of the solution. The specific gravity of the solution in a charged battery should be about 1.27. When the specific gravity is down to about 1.15, the battery should be recharged. We determine the specific gravity of the solution by means of a hydrometer.

After repeated recharging, the lead sulfate formed in the battery gradually falls off the plates so that the battery eventually wears out and must be replaced.

9. The automobile storage battery

Most storage batteries used in automobiles have six lead cells arranged in series so as to produce 12 volts (6×2 volts). Each cell is in a separate compartment and contains its own solution. In each cell there are many alternate plates of lead and lead dioxide. The plates are separated by thin sheets of wood or Fiberglas to prevent them from touching and producing a short circuit.

A large plate area is used so that the battery will be able to supply large amounts of current when needed (Fig. 28-7).

The automobile does not need a battery while it is running. When the motor is running, its motor operates a generator (described in Chapter 34) which supplies enough current for all electrical needs of the automobile, including the charging of the battery. But the battery is necessary to start the engine. Furthermore, when the car is parked, the battery can supply current to the lights, radio, heater, and so on.

10. Electrochemical actions

You recall that it was Galvani's observation of a twitching frog's leg that gave rise to our modern knowledge of how electricity can be produced from chemicals. Today we know that the electricity which can make a frog's leg twitch is also responsible for many of the life processes in our own bodies. Moreover, the "messages," or impulses, which travel along the nerves of the human body are electrochemical in nature. Electrical impulses also accompany the beating of the heart. These impulses can be measured and a record of them set down in an **electrocardiogram;** in this way certain diseases of the heart can be detected (Fig. 28-9).

Similar currents called brain waves have been found in the brain. Changes in the pattern of these currents have been found to indicate certain types of mental illness. Similarly, some infections in the body can be detected by changes in electrical current due to chemical changes in body organs. This whole study of electrical currents in the body is still rather new, but is proving useful in understanding the basic processes in living cells.

The dentist uses the facts of electrochemistry when he fills your teeth. He knows that a silver filling placed near a gold inlay, with the slightly acid saliva of the mouth acting as an electrolyte, can set up slight electric currents which are felt as an acid taste, and sometimes even as tiny shocks in the teeth.

Electrochemical processes are responsible for the corrosion of many metals. For example, galvanized iron is iron with a thin coating of zinc to keep it from rusting. Should a hole develop in this coating, so that both iron and zinc are exposed, any water containing mineral salts can set up a small voltaic cell at this spot. The zinc then dissolves away rapidly and the iron begins to rust. You can see why corrosion is a particularly serious problem on ships operating in salt water.

Going Ahead

THE VOCABULARY OF PHYSICS

Define each of the following:

voltaic cell	dry cell
volt	polarization
electrochemical series	electrocardiogram
	battery

THE PRINCIPLES WE USE

1. What materials are needed to make a cell, and how should they be arranged?
2. What is the electrochemical series? What useful information can be obtained from it?
3. Mention three examples of electrochemical actions other than in batteries.
4. Describe the chemical and electrical actions that occur in a zinc-copper cell.

Explain why

5. The more active metal in a cell becomes the negative electrode.
6. Very active metals are not satisfactory for cells.
7. It is incorrect to say that a battery stores electricity.

APPLYING PRINCIPLES OF ELECTRICITY

1. Draw a diagram of a dry cell and label the main parts. Explain the purpose of each of the parts. Label the plus and minus terminals.
2. How are large voltages obtained using cells of small voltage? Draw a diagram showing an arrangement of dry cells that will provide 12 volts for a certain radio.
3. A certain bell requires $1\frac{1}{2}$ volts to operate. The single cell used to operate it is troublesome to change and does not last long enough. Draw an arrangement of 3 cells that will be more satisfactory.
4. Draw a lead storage cell and describe the changes that take place as it is used.
5. How is a storage cell recharged? What changes take place as the cell is recharged?
6. Why are iron screws never used on ships to fasten copper fittings exposed to salt water?
7. How is a storage battery tested? Explain why the test works.
8. If electrolysis can be used to restore the original chemicals in a lead storage cell, why won't the same thing work for a cell made of zinc, copper, and sulfuric acid? Figure out what will happen in electrolysis when these materials are used. You may then see why this cell cannot be recharged.
9. Figure out a reason for the fact that two cells in parallel produce the same voltage as one.

PROJECTS

1. A Simple Cell You Can Make at Home. You can make a simple cell of your own and show that it produces an electric current in the following way. For your cell soak a one-inch square of cardboard in salt water, and while it is still moist press against it a copper penny and an iron washer. You can't feel the voltage, of course, but you can prove that your cell works, as follows. Wind about 50 turns of thin insulated wire in a compact coil around a pocket compass. Adjust the compass so its needle is lined up beneath the coil. Press one exposed end of the wire against the penny and the other against the washer, holding the entire "sandwich" firmly together between your fingers.

   Immediately the compass needle will be deflected. We shall learn in Chapter 30 that this is evidence of the flow of an electric current. Try reversing the connections to your cell, and your compass will swing the other way, showing that you have reversed the direction of current flow.
2. Making a Storage Cell. With sandpaper clean two strips of lead and stand them in a small jar of dilute sulfuric acid not touching one another. This is your storage cell. To charge it connect the plates to two or three dry cells connected in series. After about 5 minutes your cell should be sufficiently charged to light a small flashlight bulb connected to the lead strips.
3. Taking Apart a Dry Cell. Find a dead dry cell and break it apart. Can you find all the parts identified in Fig. 28-4? You will discover, if the cell is dead, that the paste filling is nearly dry. But if you have not destroyed the cell too thoroughly by breaking it open, you can revive it by placing it in a jar of ammonium chloride solution (or even salt water or vinegar).

CHAPTER

29

Magnets

For centuries scientists suspected that there might be a connection between magnetism and electricity. But it was not until 1820 that the Danish scientist Hans Christian Oersted (ûr'stĕd) discovered an important connection between them. While performing a classroom demonstration, Oersted made the observation that a wire showed magnetic properties when a current of electricity flowed through it (Fig. 29-1). This discovery of a connection between magnetism and electricity eventually made possible such useful devices as the electromagnet, electric bell, electric motor, generator, telegraph, telephone, radio, and television.

But before studying these devices, we should know certain fundamental facts about magnetism, facts which were already well established by Oersted's time.

Facts about magnetism

Spread out a handful of small nails and place the flat side of a bar magnet on top of them. When you lift the magnet, the nails come with it. You notice that they are concentrated at the ends of the magnet (Fig. 29-2). These points of concentration are called **magnetic poles.** By experiment you will find that all magnets have at least two poles. In a bar magnet these poles are usually (but not always) located at the ends of the magnet. Although the two poles look alike and seem to act alike, there is an important difference between them, as we shall see from the following experiments.

1. North and south poles

Suspend a bar magnet from a string so that it rotates freely, making sure that there are no iron or steel objects or other magnets near by (Fig. 29-3A). When the swinging magnet comes to rest, you find that one end always points approximately north and the other end south. You can turn the bar through various angles but it always returns to this position. You find that all bar magnets when freely suspended behave in the same way.

Mark with an "N" the pole of the magnet which points north and with an "S" the pole pointing south. Do the same with another magnet and bring the N poles of the two magnets close together. You will find that they repel each other. So do the S poles (Fig. 29-3B). But the N and S poles attract each other (Fig. 29-3C). You can repeat this experiment with many different magnets; you always find that **like poles repel and unlike poles attract.**

As you perform these experiments you will find that the stronger the magnets you use, the greater will be the force exerted between them. You will also find that the farther the poles of the magnets are from each other, the weaker the force between them will be. These ideas are stated more precisely in **Coulomb's Law for Magnetic Poles** which says:

The force exerted by two magnetic poles on each other is directly proportional to the product of the pole strengths and inversely proportional to the square of the distance between them.

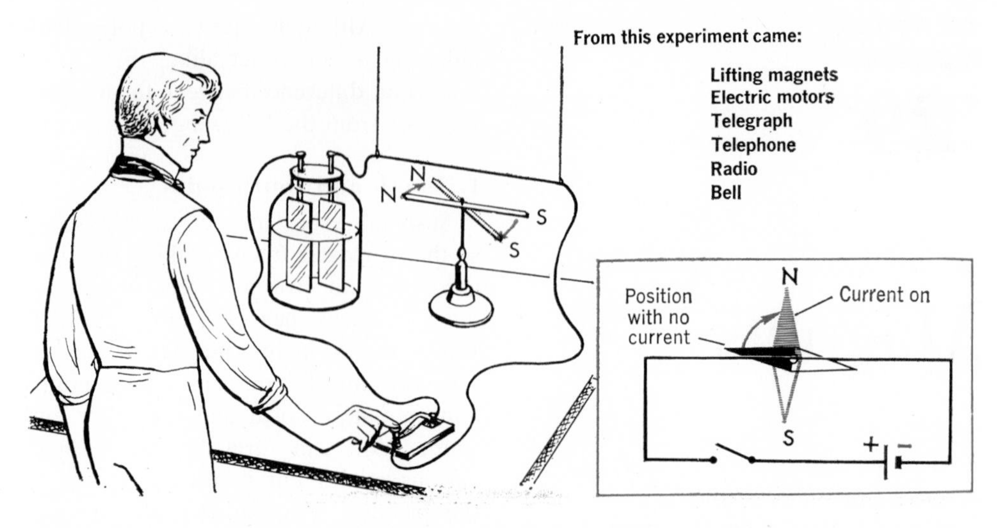

29-1 Important discoveries came from Oersted's simple experiment in which a magnet on a pivot moved when current flowed in a nearby wire.

This law may be written in the following way:

$$F = \frac{km_1m^2}{r^2}$$

F is the force, m_1 and m_2 are the pole strengths, r is the distance between them, and k is a constant. This is another example of an "inverse square law." For other examples see pages 146 and 330.

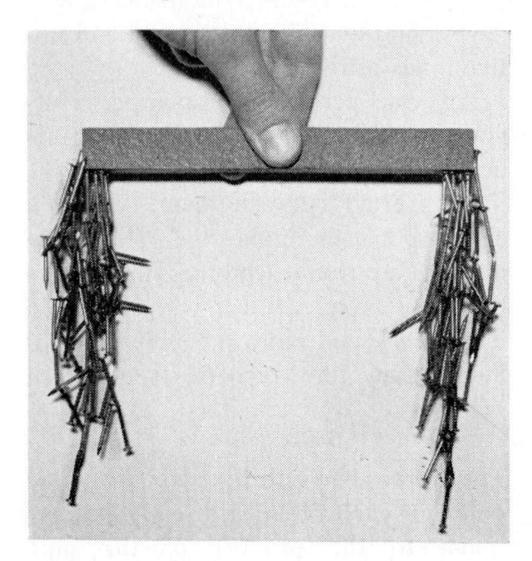

29-2 The poles of a magnet are simply the places where the magnetic effect is greatest. Dipping a bar magnet into nails shows the poles to be at the ends.

2. Making a magnet

Stroke an iron nail with the N pole of a strong magnet. You will find that you have magnetized the nail so that it has become a magnet. When you bring the magnetized nail close to another magnet, such as a compass needle (a small pivoted magnet), you find that the nail, like the compass, has two poles. The like poles of the nail and compass repel each other; the unlike poles attract. Moreover, you find that the end of the nail last touched by the stroking magnet acquires a magnetic pole opposite to that of the end of the magnet you used in stroking it. For example, the end of the nail that has been stroked last with the N pole of the magnet becomes an S pole (Fig. 29-3D).

It is not necessary for an iron object actually to touch the magnet to become magnetized. If the magnet is strong enough, it will cause the iron to be magnetized even at a distance. Magnetizing at a distance is called **magnetic induction.** Figure 29-4 shows how the magnets of loud-speakers are magnetized as they pass under strong electromagnets.

3. Magnetizing with electricity

Take a large unmagnetized nail and wrap some insulated wire around it, as shown in Fig. 29-5. Connect the ends of this wire to a battery. You will find that the flow of current magnetizes the nail, providing it with N and S poles. Since the magnetism here is caused by an electric current, we call the combination of nail and coil an **electromagnet.**

When you shut off the current, almost all of the magnetism disappears from the nail. Remove the nail and let the current flow through the coil alone. You find that the coil produces the same poles that the nail had. But the coil by itself (called a **solenoid**) is a weaker magnet than the combination of coil and nail. Figure 29-4 shows how such a solenoid is used in a loud-speaker. The solenoid (called the **voice coil**) vibrates as a result of being attracted and repelled by a stationary permanent magnet. The paper cone on which the coil is mounted is set into vibration with the coil, and the vibrating cone then creates sound waves in the air.

Atomic magnets

How do we account for the similarity between the magnetism of a permanent magnet and that of an electric current in a coiled wire? The French scientist Ampère (ăm′pēr) provided a simple explanation. Since a current traveling in circles around a coil can produce poles in an iron nail, Ampère suggested that the magnetism of the iron was also caused by tiny circular currents in the iron. However, this hypothesis remained unproved for many years.

When the existence of electrons revolving around the nucleus of an atom was discovered, scientists speculated that each atom might be a tiny electromagnet complete with N and S poles as a result of the spinning of the electrons as they whirled around the nucleus. Today there

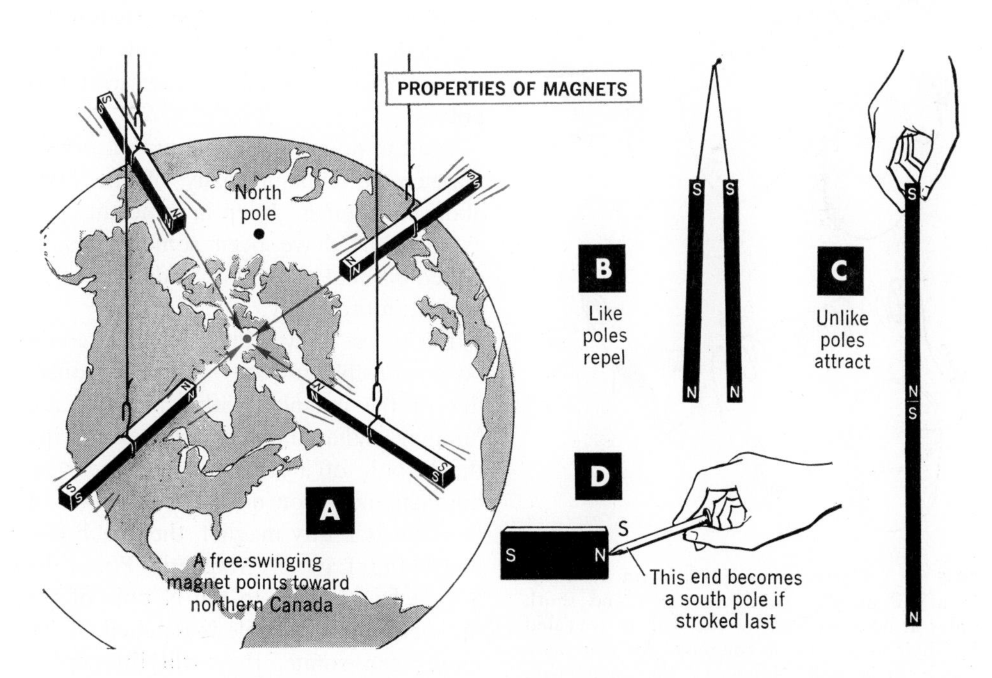

29-3 What four facts about magnetism are shown in A, B, C, and D?

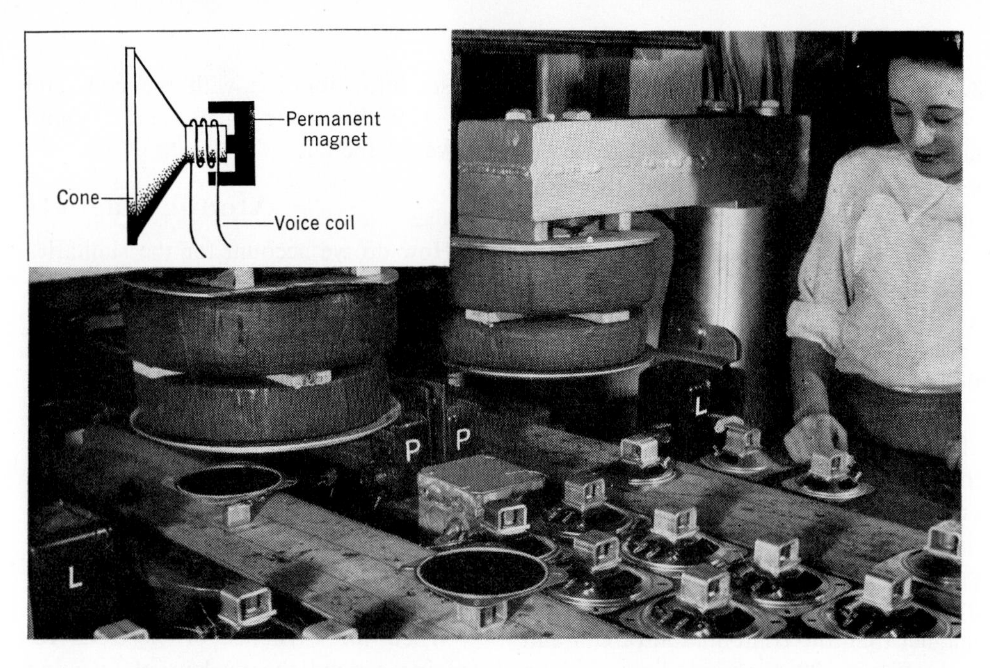

29-4 The loud-speakers passing along on conveyor belts contain Alnico (an alloy of aluminum, nickel, cobalt, and iron) which is magnetized by the strong electromagnets above them. Current for the electromagnet is turned on as the loud-speaker blocks the beam of light from the lamp L to the photoelectric cell P. Inset diagram shows the Alnico permanent magnet, the voice coil, and the loud-speaker cone.

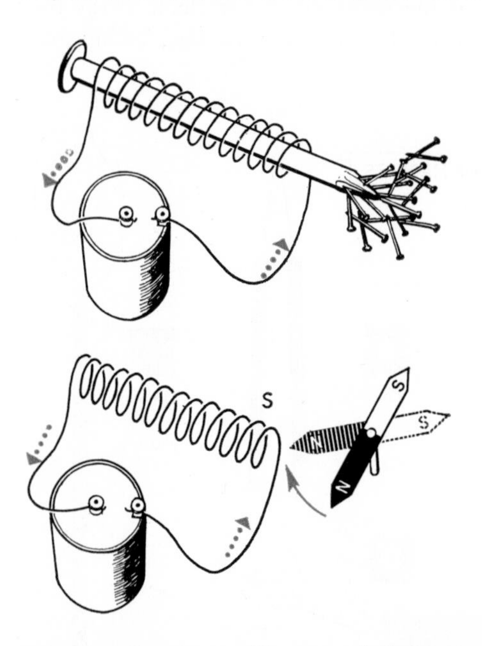

29-5 Current traveling around the turns in a coil of wire produces north and south poles at opposite ends of the coil, as revealed by their action on a compass. An iron core (such as a nail) increases the magnetism enough to pick up small brads or tacks.

is convincing evidence that it is indeed the "spin" of the electrons (similar to the rotation of the earth) that creates the magnetic effect. The net result is that every atom is a small magnet with two poles.

We can now use this picture of atomic magnets to explain many of the facts about magnetism. Keep in mind that our explanations have been somewhat simplified so that you may understand the process more easily.

Let us see how atomic magnets could be responsible for the observed properties of the magnets with which we are familiar. What happens when we bring the N pole of a strong magnet near an unmagnetized iron nail? If each atom of the iron is a tiny magnet, then each one obeys the Law of Magnetic Poles. Its S pole is attracted to the N pole of the magnet, and its N pole is repelled. If the atoms can rotate, they will line up as shown in Fig. 29-6. Suppose we bring the S pole of this magnetized nail near a compass. The south ends of the atoms in the nail are closer to the compass than the north ends and therefore have a greater effect. The total effect of these trillions upon trillions of atomic S poles is to attract the N pole of the compass.

4. What is a magnet?

We see that the difference between a piece of material that is magnetized and one that is not is that the atoms of the magnetized materials are lined up. It is this lining up of atoms that makes a magnet.

Actually, the lining up does not occur atom by atom. The atoms in the magnetized iron we have been considering are grouped together in microscopic grains called **domains.** Within each domain the millions of magnetized atoms are always lined up, whether the material is magnetized or not. It is the domains which line up when attracted by nearby magnets. The top limit of the magnetic strength of any piece of material has been reached when all its domains have been lined up. In this condition the magnet is said to be **saturated.**

There is no attraction between two pieces of unmagnetized iron because the domains in both are disarranged; the attraction of unlike poles is balanced by the repulsion of poles that are alike. On the other hand, if one of the pieces of iron is magnetized, it attracts the unmagnetized iron by lining up the domains in the iron and producing poles (magnetic induction). When the magnet is taken away, some of the domains in the iron remain lined up for a while, so that the iron retains some of its magnetism.

As you have seen, when a magnet is dipped into small nails the attraction is greatest at the ends of the magnet. This is apparently because the S poles of the domains in the magnet are exposed at

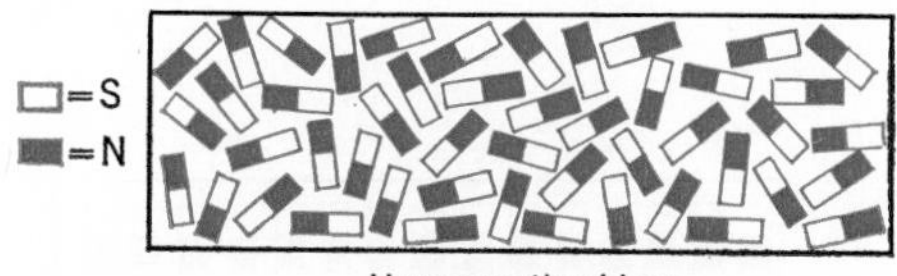

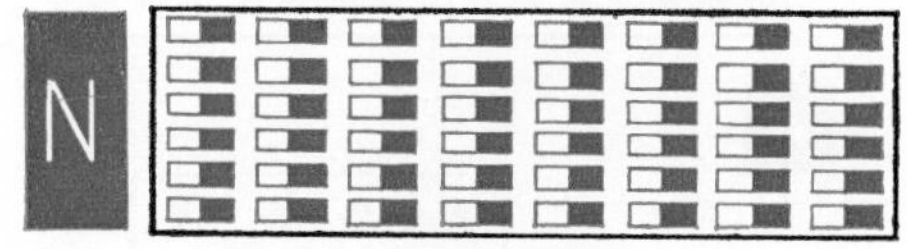

29-6 **Order versus disorder. The disorderly arrangement of domains in a piece of ordinary iron (top) creates such magnetic chaos that all forces cancel out and no magnetic force is noticed. But when the domains are lined up in orderly fashion (below), they all pull together and the magnetic force is observable at each end.**

one end, the N poles at the other. The center of the magnet does not exert as much attraction because the N and S poles of the domains in this region are adjacent to one another, so that their opposite effects cancel out.

Because every domain in a magnet has two opposite poles, the magnet itself must have at least two poles. In the magnets we have been considering, these poles are at opposite ends. Figures 29-7A and B show other possible arrangements.

5. Cutting a magnet in half

If it is true that the domains in a magnet are lined up and that each domain is a small magnet in itself, we should be able to break a magnetized bar in half without destroying its magnetism. We would simply have two magnetized bars instead of one, each with its own poles.

You can test this for yourself. Straighten out a paper clip and magnetize it by stroking it with a magnet. Once magnetized, the wire has an N and an S pole at opposite ends, as you can see by bringing the ends near a compass. When you cut the wire in two, each piece affects the compass in the same way that the larger piece did.

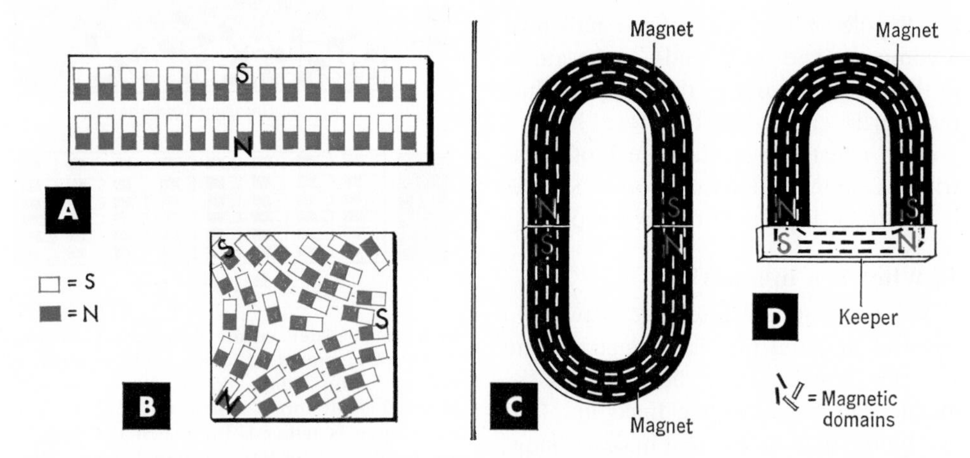

29-7 Magnetic poles form when domains line up. A shows a lined-up arrangement which causes poles to form at the sides of the magnet. B shows a magnet with two south poles and one north pole. C and D show two ways of storing U-shaped magnets to keep the domains lined up and preserve the magnetism.

6. Magnetic and nonmagnetic materials

So far, as you have noticed, we have used iron in our discussion of magnetism. We said that magnetism is caused by the action of the rotating electrons in each atom of the iron. Since the atoms of all materials contain rotating electrons, we would assume that all materials could be magnetized. But in practice a magnet has no noticeable attraction for certain materials, such as wood or glass. How do we explain this difference?

Certain kinds of atoms may be very weak magnets because some electrons in each atom may be spinning one way and some the other way, thus tending to form opposite poles that cancel each other's effect. In addition, most molecules are very complex and their atoms contain electrons spinning in many different directions, thus canceling out the magnetic effect of any one atom. Furthermore, it may be difficult for some kinds of molecules or domains to line up when magnetic forces are applied. As a result of the operation of these factors, some materials may show little magnetic effect and others may show a great deal. An interesting application of the use of a magnetic material is shown in Fig. 29-4.

7. What common materials are magnetic?

Try lifting common objects with a magnet to see if they show any magnetic attraction. You will find that those objects which are made of iron or steel, such as nails, needles, or tin cans (iron coated with tin), are **magnetic**, that is, are attracted by the magnet. Most metal coins, and objects made of wood, paper, glass, rubber, and cloth, show no visible reaction whatever and are therefore called **nonmagnetic.**

It has been found that of all the elements, iron is the most easily magnetized. Other elements can be magnetized, but it is harder to line up their atoms. After iron, the most easily magnetized element is nickel; the metal cobalt is third.

A material which is hard to magnetize will retain its magnetism longer than one which is magnetized easily. Certain types of steel, much harder to magnetize than iron, are used for **permanent magnets.** The domains in these materials are harder to line up, but once lined up, they tend to stay put. On the other hand, pure iron ("soft iron") is good for temporary magnets (electromagnets). The domains in soft iron line up easily. But once the magnetizing force is removed, the domains quickly become disarranged, due to molecular motion, and we find that the magnetism almost completely disappears.

8. Magnetic alloys

We can greatly change the magnetic properties of a metal by making an **alloy;** that is, by mixing the metal with other metals in the molten state. For example, although nickel is the most easily magnetized material after iron, an ordinary five-cent piece shows no magnetic effect because the nickel in the coin has been alloyed with other materials, mostly copper. However, a Canadian nickel, made of a different alloy, is attracted by magnets. A special alloy called **Permalloy** (pûrm′*ă*·loi′) is often used in magnetic devices because it gains and loses magnetism much more readily than iron. Steel, which is an alloy of iron mixed with slight amounts of carbon and other materials, varies greatly in its magnetic properties depending on its composition. **Alnico** (ăl′nĭ·kō), an alloy of aluminum, nickel, cobalt, and iron, makes the strongest permanent magnets known. The material is hard to magnetize but will retain its magnetism almost indefinitely (Fig. 29-4).

9. How to weaken a magnet

If we can magnetize a substance by lining up its domains, we would expect to weaken the magnetism by doing anything that tends to disarrange the domains or atoms in them. One way to weaken a magnet is to heat it. The increased molecular or atomic motion produced by heat gets the atoms and domains out of line. You can test this by heating a magnetized steel wire in a Bunsen burner. After heating, the wire no longer acts like a magnet, as shown when it is brought near a pile of iron filings. Its atoms and domains have been disarranged. Jarring a magnet tends to disarrange the domains and thus has the same effect. You can show this by magnetizing an ordinary nail and then banging it on the table or floor a few times. After being jarred, it shows little attraction for iron.

10. Horseshoe magnets

The most familiar shape for ordinary magnets is the horseshoe (or U shape). In the horseshoe shape, the N and S poles are brought close together, thus increasing the pull of the magnet. When a piece of iron is brought near the magnet (as shown in Fig. 29-7D), the horseshoe shape enables the magnet to line up the domains of iron more easily.

11. Storing magnets

Magnets should be stored in a place where no outside magnetic force will shift the line-up of the domains. If several magnets are stored together, they should be arranged so that the N pole of one touches the S pole of another (Fig. 29-7C). In this way each magnet keeps the domains of the one next to it lined up in the proper direction. In storing horseshoe magnets, a piece of soft iron, called a **keeper,** is usually placed across the two poles. The keeper becomes magnetized with an S pole next to the N pole of the magnet, and thus the domains of the magnet stay lined up (Fig. 29-7D).

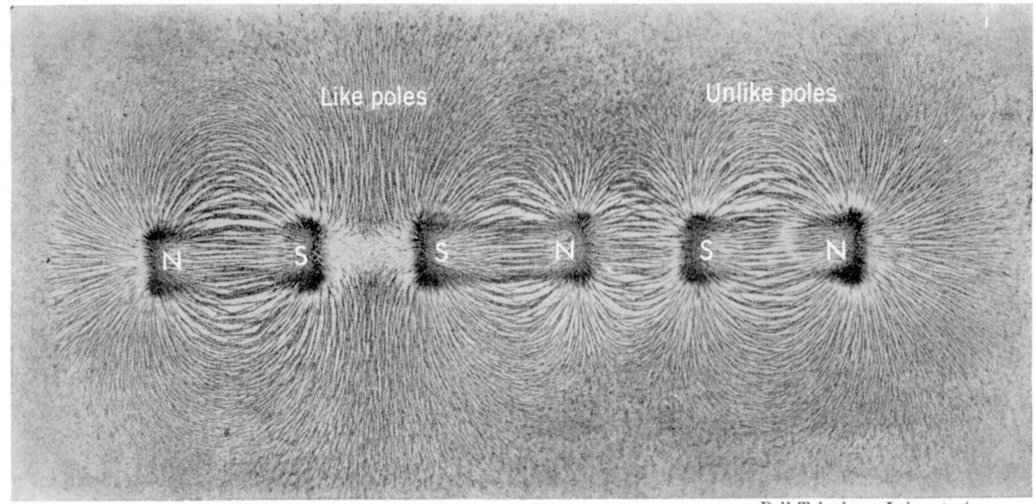

Bell Telephone Laboratories

29-8 Iron filings have fallen from a height onto a sheet of paper placed over magnets. The resulting pattern reveals the lines of force. Note how lines of force from nearby like poles seem to repel each other, while lines of force from unlike poles seem to attract. Follow a line of force from one pole and you will always end up at an opposite pole.

Magnetic fields

We have noticed before that many things in the world are not directly apparent to our senses. Magnetism is one of them. We cannot see, hear, feel, touch, taste, or smell it directly. We can only observe its effect on other things. To talk about such invisible things at all, we have to construct pictures, schemes, and symbols in our minds. The picture we use to describe the magnetic region around a magnet is called a **magnetic field.** We picture this magnetic field as made up of **lines of force.**

12. Lines of force

You can see what we mean by lines of force by performing a simple experiment. Place a sheet of paper over a bar magnet, making sure that the paper fits as flat and as close to the magnet as possible. Now sprinkle iron filings on the paper from a height of about 10 inches, tapping the paper gently as you do so. The filings form a pattern of lines. These filings fall in place along the magnet's lines of force.

Repeat this experiment using the arrangement shown in Fig. 29-8. As you see, the lines of force from like poles look as though they repel each other; those from unlike poles appear to attract. And the lines of force are always most crowded together near the poles of the magnet.

13. Path of a compass

Let us trace one of these lines of force with a small compass. We place the compass near the N pole of the magnet and start pushing it in the direction its own N pole is pointing (Fig. 29-9). We stop every half inch to take our bearings. Each time we stop, we notice that the direction in which the compass is pointing has changed slightly from our last reading. If we allow the compass to "follow its nose" in this way, it eventually reaches the S pole of the magnet, as shown in Fig. 29-9. By taking different starting points near the N pole, we get different paths. These paths the compass takes are the same lines of force we got by using the iron filings. So we see that a line of force is really the path a compass would take as it "followed its nose" from the N to the S pole of the magnet. This means that a line of force emerges from an N pole, travels around to the S pole, and returns through the magnet to the N pole again.

Let us see why our iron filings fall according to this pattern. As each filing falls toward the magnet, it becomes magnetized by **induction.** Since it is free to rotate in mid-air, it tends to line up on the paper in exactly the way a small compass would line up. By tapping the paper gently with a finger, we assist this lining-up process. If we drop enough filings, they will line up so that the S pole of one touches the N pole of the next and we have a continuous chain of filings along the line of force, stretching from one pole to the other.

You will notice from Fig. 29-8 that the lines of force concentrate at the poles of the magnet. The greater the concentration of lines, the greater the strength of the field. We shall see in Chapter 34 that this picture of lines of force is very useful in explaining the operation of the generator and the transformer.

14. Permeability

When an iron block is placed close to a magnet, we find that the lines of force curve out of their normal path and appear to concentrate in the iron (Fig. 29-10). A block of wood or copper has no such effect. This ability of iron and other magnetic materials to concentrate lines of force is called **permeability.** There is a great difference in the permeability of various materials. That of iron is high, while the permeability of air, for example, is very low. Certain kinds of iron can concentrate more than a thousand times as many lines of force per square inch as air can.

Or we can look at the matter in another way. When we place a block of iron close to a magnet, the iron becomes magnetized by induction. It then acts like any other magnet, and the lines of force from the N pole of the magnetizing magnet gather together where they enter the S pole of the iron block. They then pass through the block and emerge into the air at its N pole, continuing back through the air, and returning to the S pole of the magnetizing magnet. From here they go through the magnet itself back to its N pole, thus forming a complete circuit. The iron, being highly permeable, concentrates the lines of force, while the air, with a low permeability, does not.

We shall see in Chapter 34 how these facts about permeability are important

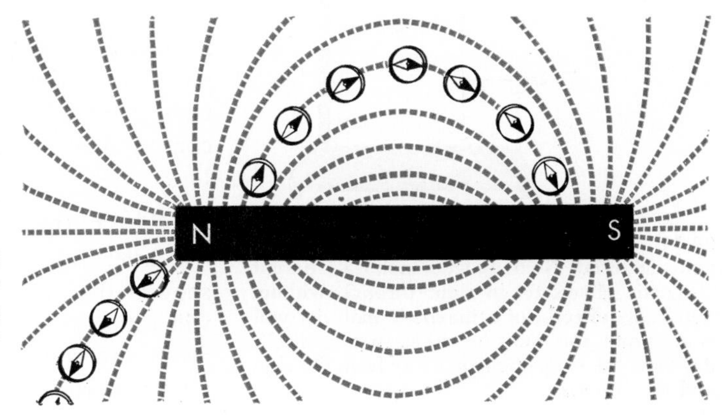

29-9 A line of force can be traced by placing a compass at the north pole and "following its nose."

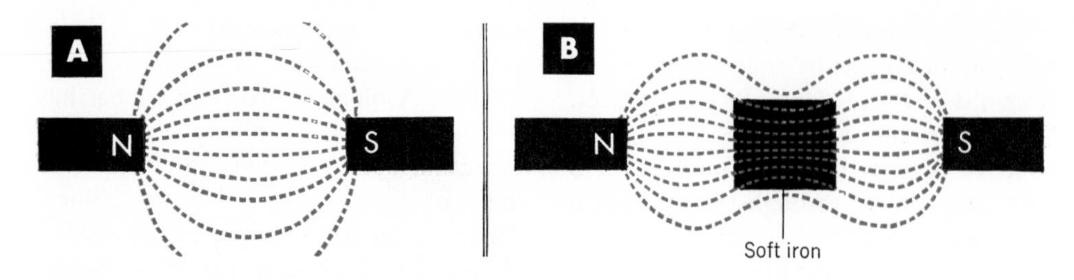

29-10 Lines of traffic progress more easily on a superhighway. Likewise magnetic lines of force tend to follow the easier path provided by the presence of soft iron in the field.

when it comes to designing devices that make use of magnetic fields, such as electric motors. The shape of the motor should be such that the lines of force have to pass through as little air as possible. This is achieved by filling up the space between the magnet poles with iron instead of air. This allows for the greatest concentration of lines of force and consequently the maximum magnetic strength. The transformer, the telegraph sounder, the bell, and the lifting magnet all use the same principle. You used it also if you performed the experiment shown in Fig. 29-5.

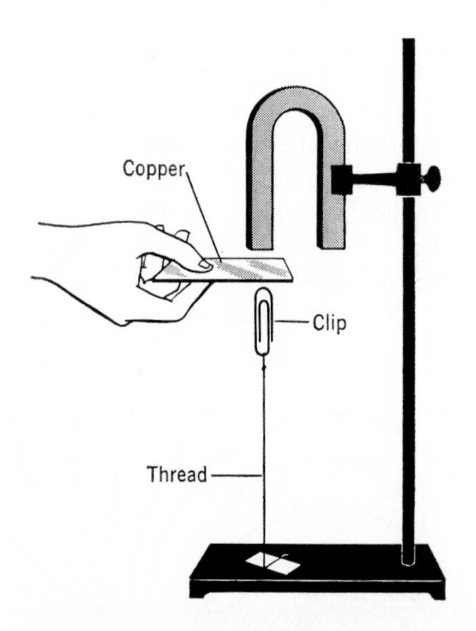

29-11 Here's a simple test for a magnetic material. Nonmagnetic materials permit magnetic effects to go right through without noticeable effect. But a magnetic material will concentrate the lines of force within itself, diverting the magnetic field away from the clip, and the clip falls.

15. Magnetism penetrates materials

Fasten a paper clip (made of iron) to the end of a thread, and then suspend the clip in mid-air below a strong magnet, as shown in Fig. 29-11. Insert sheets of different materials between the magnet and the clip. You will find that nonmagnetic materials, such as paper, wood, glass, copper, lead, cloth, and even the body, have no effect on the magnetic attraction (Fig. 29-12). It passes right through them from the magnet to the clip. But the situation is changed if you hold a large sheet of a magnetic material, such as iron, between the magnet and the clip. Now the magnetic lines of force supporting the clip are diverted through the iron sheet, and the clip falls. From this experiment we see that magnetic lines of force penetrate nonmagnetic materials but are deflected and concentrated by magnetic materials.

16. The magnetism of the earth

Earlier in this chapter (page 367) we showed that a freely suspended magnet will rotate so as to point approximately north and south. We call the pole that points north, that is, that seeks the north, a north pole—really, a north-seeking pole. The pole that seeks the south is called the south pole (south-seeking pole).

If we were to follow the line of force indicated by the north-seeking pole of a compass, we would eventually arrive at the North Magnetic Pole. This is located at a spot north of Hudson Bay. This magnetic pole is hundred of miles away from the geographical North Pole. Considering the earth as a giant magnet, the magnetic pole is the S pole of the magnet. In the Antarctic region of the southern hemisphere, there exists an N pole, the South Magnetic Pole.

Since a compass points to the *magnetic* North Pole and not true north to the geographic North Pole, its readings are in error on most parts of the earth. This error, called **declination,** must be taken into account by navigators (Fig. 29-13).

Just why the earth has magnetic poles is still something of a mystery. Although the central core of the earth is believed to contain a great deal of iron, this iron is in a plastic state. Iron in this state is much too hot for the lining up of atoms or domains.

Nor is it clear why the magnetic poles of the earth lie hundreds of miles away from the geographic North Pole and South Pole, and even slowly change their position.

There is evidence that the planets and stars, including the sun, also have magnetic fields. In fact, the magnetic field of the sun has actually been measured. To explain the earth's magnetism we need further information, some of which may be obtained only with the help of magnetic measuring devices in earth satellites. Meanwhile we can list this unknown reason for the earth's magnetism as one of the many unsolved mysteries of science.

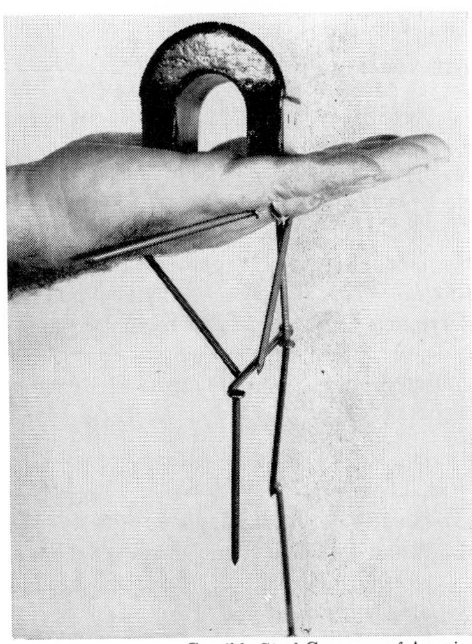

Crucible Steel Company of America

29-12 This powerful horseshoe magnet attracts the nails right through the man's hand, yet the passage of the lines of force causes no feeling in his hand.

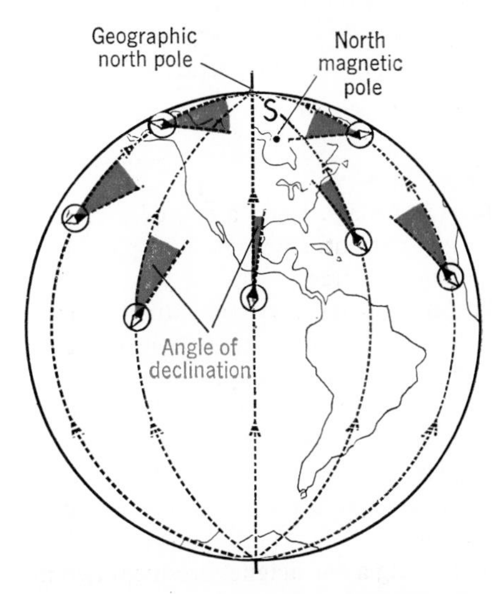

29-13 Can a compass point west, east, south? Yes, a person at the *geographic* north pole will find his compass pointing south toward northern Canada. Navigators can use compasses for accurate work only if they have charts showing the amount of declination (error of the compass) at each location on the earth.

Going Ahead

THE VOCABULARY OF PHYSICS

Define each of the following:

magnetic poles	induction
magnetic induction	nonmagnetic
electromagnet	magnetic field
domains	line of force
permanent magnets	permeability
alloy	geographical poles
Permalloy	declination
Alnico	solenoid
saturated	magnetic
keeper	voice coil

THE PRINCIPLES WE USE

1. State the Law of Magnetic Poles.
2. Name three magnetic elements. Name two magnetic alloys. Name three nonmagnetic materials.
3. Explain the theory of the cause of magnetism.
4. Draw diagrams showing the lines of force for: a bar magnet, a horseshoe magnet, two like poles, two unlike poles, two unlike poles with a piece of unmagnetized iron between them.
5. Describe the magnetic field of the earth.

Explain why

6. A bar magnet and a coil of wire carrying current show similar patterns of lines of force.
7. An iron nail may not attract a nearby iron object even though its molecules have magnetic poles.
8. A piece of unmagnetized iron becomes magnetized when brought near a magnet.
9. A bar magnet cannot have only one pole.
10. Cutting a bar magnet produces two new magnets, each with two poles.
11. Some materials show little magnetic effect as compared with iron.
12. Iron is good for temporary magnets, while certain kinds of steel make much better permanent magnets.
13. Heating tends to weaken a magnet.
14. Jarring tends to weaken a magnet.
15. Magnets are stored with the north pole of one touching the south pole of the next one.
16. It is incorrect to say that compasses always point north.
17. An iron fence post in the United States will be magnetized, and its north pole will be at the bottom.

EXPERIMENTS WITH MAGNETISM

Describe an experiment to

1. Show that a bar magnet has two poles.
2. Discover whether a piece of steel is a magnet.
3. Find out which end of a magnet is the north pole.
4. Make the head of a nail a south pole and then reverse it to a north pole.
5. Show a line of force with a compass.

APPLYING PRINCIPLES OF MAGNETISM

1. How does the navigator of a plane know the exact direction in which the plane is going even though the compass does not point directly north?
2. People working near powerful magnets often find that their watches no longer keep correct time. Explain why. How could a watch be made nonmagnetic?

THOUGHT QUESTIONS

1. Can you figure out a way to make a magnet with two north poles and one south pole?
2. Referring to Fig. 29-9, can you explain why a compass placed equidistant from the ends of a bar magnet will point in a direction parallel to the bar?
3. One end of a bar of metal attracts the N pole of a compass needle. What conclusions can you draw concerning the metal bar?
4. The N and S poles of two bar magnets are 10 inches apart. They attract each other with a force of 5 ounces. What will the new force of attraction be (a) if both magnets are doubled in strength, and (b) if these magnets are then placed 5 inches apart?

CHAPTER

30

Electro-magnets

Many times a day you use an electromagnet to make a telephone call or send a telegram, to ring a doorbell or start a car.

Because of the electromagnet, we find it easy to convert electrical energy into mechanical energy, or motion, and so get work done.

Magnetic effect of currents

We have already learned about Oersted's significant discovery that current flowing in a wire creates a magnetic field around it (Chapter 29). Let us now study this effect in greater detail.

Set up a straight vertical wire through which an electric current flows (Fig. 30-1). If you place a compass nearby to find the poles in the wire, you discover that there are none. The compass simply "follows its nose" round and round the wire without entering the wire or coming to any point of concentration that could be called a pole. If you reverse the direction of current in the wire by reversing the terminals of the battery, the effect is similar except that the compass "follows its nose" in the opposite direction. We see from this experiment that the lines of force around a wire carrying current are circular.

1. Direction of the lines of force

From repeated experiments with wires carrying current, a handy rule has been devised to tell us what direction the lines of force will take. Here it is. Draw an arrow along the wire showing the direction in which the electrons are flowing (from the negative to the positive terminal of the cell). Point the thumb of your *left* hand in the direction of the arrow. Your other fingers will curve around the wire in the direction of the lines of force (Fig. 30-1). In other words, the curved fingers of your left hand show the direction that the north pole of a compass would follow. This is called **the left-hand rule for lines of force around a single wire.**

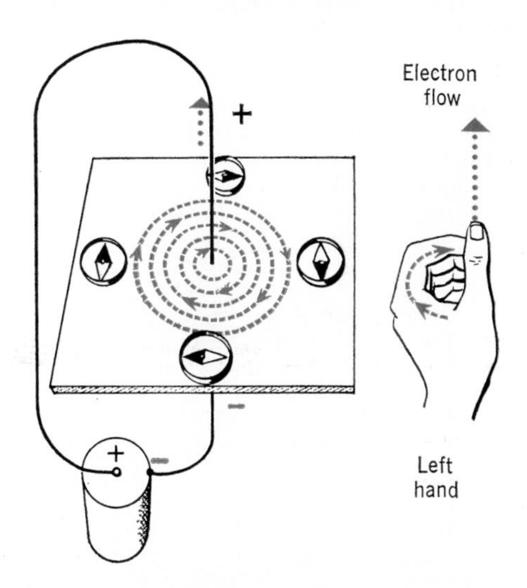

30-1 A single wire with current flowing has circular lines of force. The direction of lines of force (direction north pole of a compass would follow) can be found by pointing the thumb of the left hand in the direction of electron flow in the wire. The curved fingers show which way the circular lines of force curve.

New York Central System

30-2 This electromagnet depends for its strength on a large electric current flowing through the coil inside it, on the use of many turns of wire on the coil, and on the use of an iron core and cover.

2. Making a strong electromagnet

Let us coil up our wire into a solenoid, as shown in Fig. 30-3A (insulated wire must be used). Now the circular lines of force in the adjacent coils reinforce each other. The effect is to produce lines of force which come out of one end of the coil, travel around the outside, and go into the other end like the lines of force of a bar magnet. Now the coil has a north and a south pole, whereas the straight wire did not.

Another left-hand rule can be used to tell us the kind of poles formed by the solenoid. Draw arrows around the coil showing the direction of the flow of electrons. Curve the fingers of your *left* hand around the coil in the same direction as the arrows. The thumb will then point toward the end of the coil that is a *north pole*. We shall refer to this rule as **the left-hand rule for solenoids and electromagnets** (Fig. 30-3B).

If we reverse the current we find that the poles also reverse.

If we place a piece of iron in the center of this solenoid, the domains in the iron are lined up by the lines of force of the coil so that the magnetism of the iron

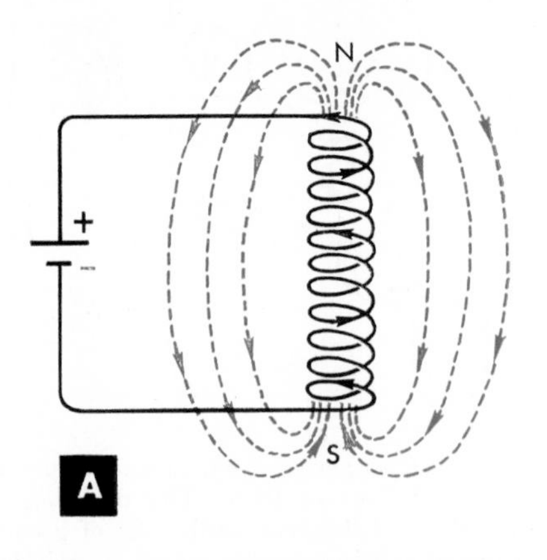

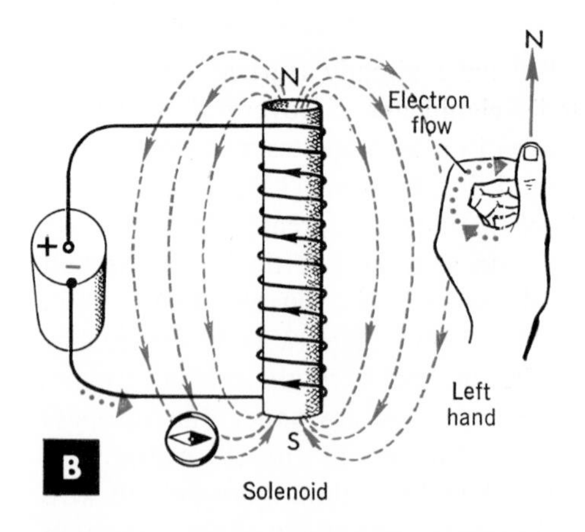

30-3 If the wire is coiled, a north and south pole are produced. If the fingers of the left hand are curved in the same direction that the electrons flow around the coil, the thumb will point to the north pole of the solenoid.

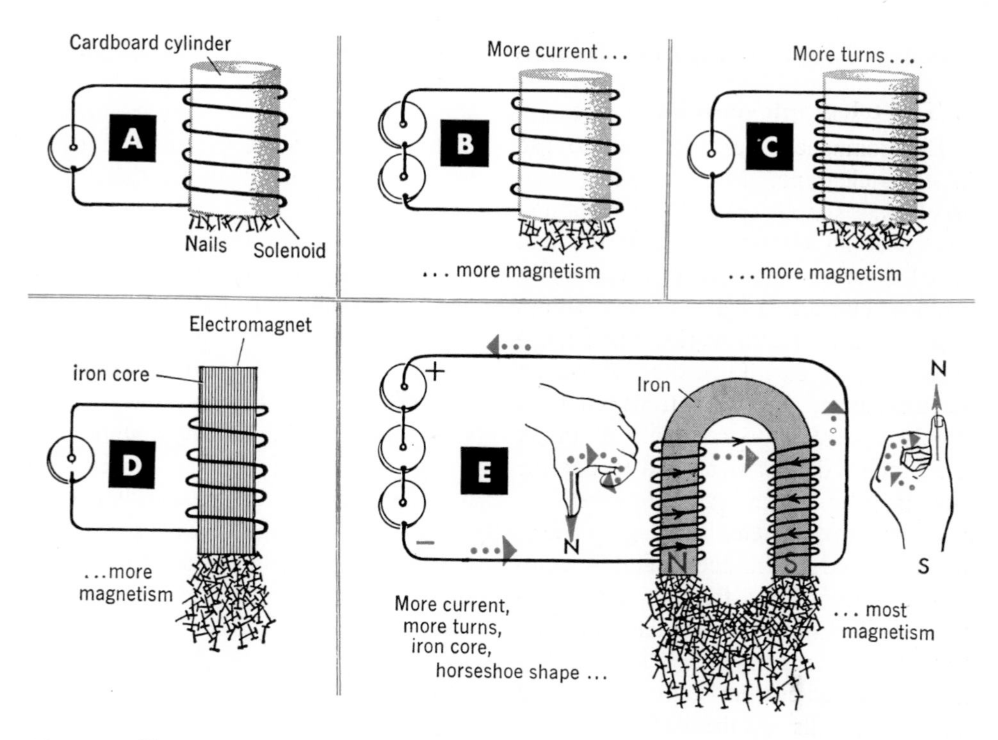

30-4 Four ways of increasing magnetism. Note how winding the coils in opposite directions in E produces opposite poles at the ends.

is now added to that of the coil. As shown in Fig. 30-4D, this iron **core** increases the strength of the magnetism many times. We now have an *electromagnet*.

There are still other ways in which we can increase the strength of the electromagnet. We can increase the current (Fig. 30-4B). We can add more turns to the coils of wire (Fig. 30-4C). As a matter of fact, the strength of two similarly constructed electromagnets can be compared by comparing their *ampere-turns*. Ampere-turns, as the name implies, are found by multiplying the number of amperes by the number of turns. For example, a 500-turn electromagnet operating on 2 amperes would be about as strong as a 1000-turn electromagnet operating on 1 ampere. The strength of each electromagnet would be 1000 ampere-turns.

We can also increase the strength of the magnet by bending it into a horseshoe shape, thus concentrating the lines of force into a smaller space. Figure 30-4E shows how each of these changes increases the weight-lifting ability of a horseshoe electromagnet in which dry cells are used as the source of current.

We can get an even stronger magnet by surrounding the coil with iron. Because of the high permeability of iron, the lines of force are kept within the iron, almost none of them escaping. Most of these lines of force are concentrated at the bottom of the electromagnet, where they are most effective for the job of weight-lifting. Interesting uses of electromagnets are shown in Figs. 30-2, 30-5, and 29-4. See if you can name a few more applications of electromagnets.

Electromagnets in use

3. The telegraph sounder

Electromagnets were first put to use in the telegraph. Morse's early telegraph was simplicity itself. It consisted chiefly of an electromagnet at the end of a long wire connected to a battery to provide current. Figure 30-6A shows how such a telegraph works. The sounder is an iron bar, I, pivoted above an electromagnet, E. The transmitter is simply a key, K, to complete or break the circuit. All that is needed to operate this one-way telegraph is to open and close the key. When the telegraph key, K, is pressed down, the circuit is completed; current flows through the sounder, the electromagnet attracts the iron bar, and a click is heard. When the key is released, the current stops flowing and the spring pulls up the bar against the stop, making another click. This combination of two quick clicks constitutes a **dot.** To signal a **dash,** the sender holds the key down for a longer time.

One of the wires in such a system can be eliminated by grounding both ends of the circuit, as shown in Fig. 30-6A.

If messages are to be sent to and from a number of stations, several keys and sounders are set up in series, as shown in Fig. 30-6B. Since the circuit must be broken in only one place (at the sender's key), some way must be found for the current to pass through each station when the key in that station is not being used. For this purpose each telegraph key is provided with a **line switch,** L in Fig. 30-6, in parallel with the key. This line switch must be closed when messages are not being sent by a station.

All line switches are kept closed when no messages are being sent. Since the current is on all the time in such a system, anyone wanting to send a message simply opens up the line switch at his station and breaks the circuit. He then sends his message by dots and dashes, making and breaking the circuit at will. The electromagnets in every station along the way receive current, and the sounders in each station operate simultaneously.

4. Relays

When a message is to be sent a great distance, the resistance of the long wire

U.S. Steel Corp.

30-5 The electromagnet part of a 3-million-pound cyclotron. It can produce atomic particles with an energy of 400 million electron volts!

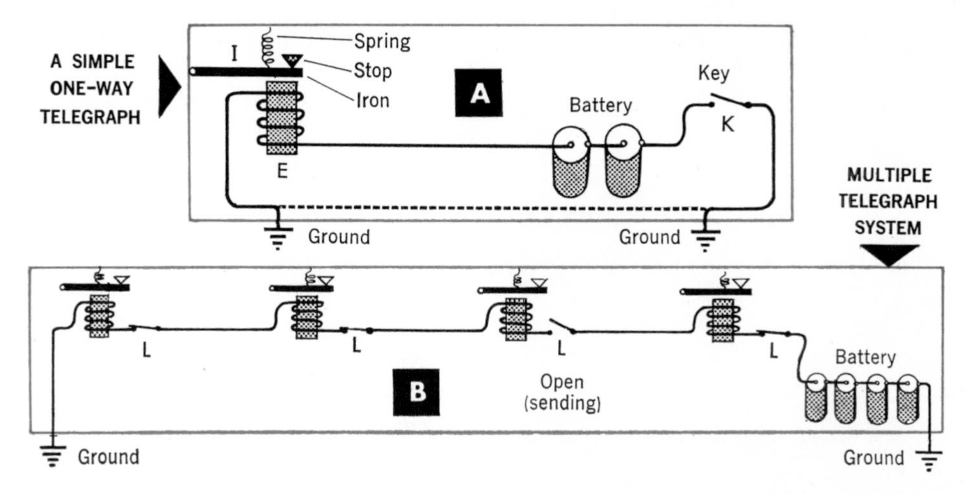

30-6 A simple one-way telegraph system (A) can be made from a key, electromagnet, iron bar, spring, and battery. If a number of key-sounder units are connected in series (B), it is possible to communicate with a number of stations. How can you tell which unit is sending messages in B?

will be large, and the current may be too weak to pull down the bar of the sounder against the tension of the spring. A device called a **relay** becomes necessary. Resistance will be discussed in Chapter 31.

A relay is constructed in almost the same way as the sounder (Fig. 30-7). When current flows in the electromagnet, E, of the relay, an iron bar, Y, is attracted by the magnet. This electromagnet is relatively strong even with a weak current, because of the very many turns of wire. The bar touches a contact point, P, thus acting as a switch to close a local circuit connected with a stronger source of current from a battery, B. This second circuit operates a sounder or any other electrical device, D. The movement of the bar toward the magnet stretches a sensitive spring, S, which pulls the bar back when no current is flowing through the main circuit. No current is then flowing through either circuit.

A relay can be used whenever a current is too feeble to operate a device. For example, suppose we wish to open a kitchen door (Fig. 30-8). When the waiter blocks a beam of light entering a photoelectric cell (Chapter 43) a tiny current is interrupted. This current would be too weak to operate most devices. But by means of a sensitive relay, the interruption of this small current causes the relay to turn on a much larger current, sufficient to operate the motor that opens the door (Fig. 30-8).

A relay can be used to shut off current as well as to turn it on. This is done by placing the contact point at A on the

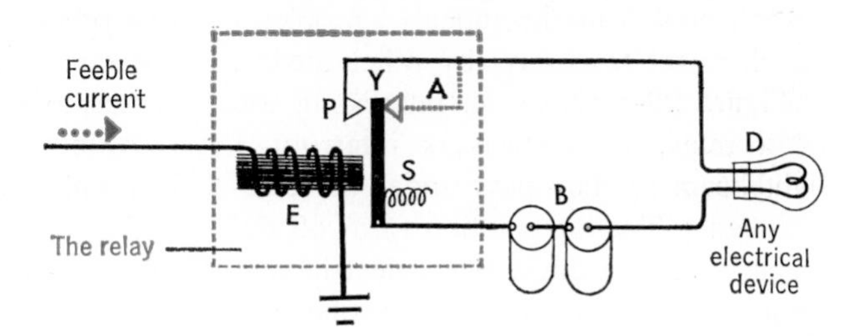

30-7 Like the runner in a relay race, the electrical relay passes on the signal from one electrical circuit to another. A feeble signal entering the electromagnet E pulls an iron bar (Y) acting as a switch to turn on the device (D).

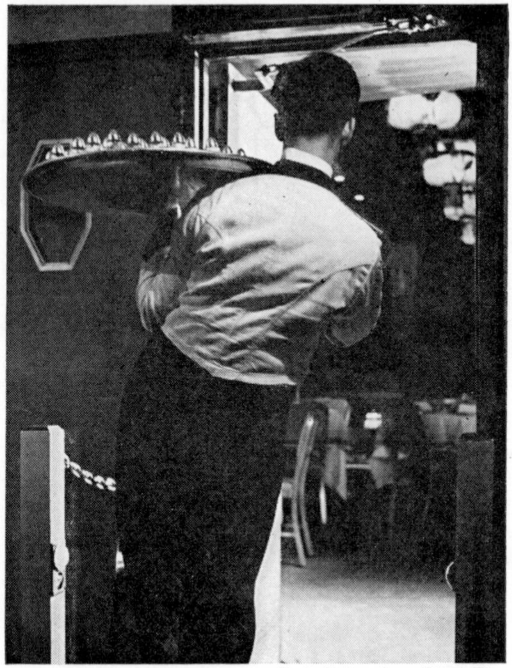

V. de Palma from Black Star

30-8 One use for relays. Here a photoelectric cell is in use to operate a relay that turns on and off a door-opening mechanism.

other side of the bar (Fig. 30-7), instead of at P.

The relay is important to many modern control devices where large currents must be handled by feeble signals. Automatic elevators likewise use relays in this way. When you step into the elevator and press the button for a particular floor, you turn on a small current that operates a relay; this relay in turn controls the much larger current needed to operate the elevator motor. Thousands of relays are also used in some giant computers (or "mechanical brains") such as the one shown in Fig. 25-11, which occupies several rooms and solves complicated mathematical problems which might take days by other means.

Figure 29-4 shows an interesting use of a relay. Loud-speakers interrupt a light beam as they pass under the electromagnet. The interruption turns off the feeble current from a photoelectric cell, causing a relay to turn on a strong current in the electromagnet. As a result, the Alnico magnets in the loud-speakers are magnetized automatically. This operation is a simple example of what is called **automation**, the replacement of human guidance by automatically operated machinery.

5. The electric bell

With a slight change in the circuit, the electromagnet can be made to move an iron hammer back and forth when the current is on, thus "ringing" a bell. Figure 30-9 shows how this is done. When you press the push button, P, current flows into the electromagnet through the binding post, A. Instead of returning from the electromagnet to the battery, the current is first sent to a "contact point," C, which touches a springy brass tongue, T, attached to the iron hammer, H. To eliminate one wire, the body of the bell itself is often used as a "ground" to get the current from D to F and back to the battery B.

When the push button is pressed, the circuit is completed and current flows. The electromagnet now attracts the iron hammer which strikes the bell. But this motion of the hammer pulls the brass tongue away from the contact point so that the circuit is broken at C (Fig. 30-9). Then the springiness of the tongue makes the hammer rebound to its original position; here it again touches the contact point C to complete the circuit once again. Now the process will begin all over again as long as someone has his finger on the push button. When this pressure on the button is removed, the circuit is broken and the hammer ceases to vibrate.

A buzzer differs from the electric bell only in that there are no hammer and no bell. The buzz is caused by the tongue vibrating against the contact point.

The electric motor

One of the most important uses of the magnetic effect of electricity is in the electric motor. Compare the quiet, clean, efficient electric motor with a gasoline engine. Imagine what it would be like to operate an electric fan or a vacuum cleaner with a gasoline engine! The noise, the smell, and the difficulty of getting the engine started would be serious drawbacks indeed. But with an electric motor, all you have to do is to turn a switch and you get prompt, efficient service.

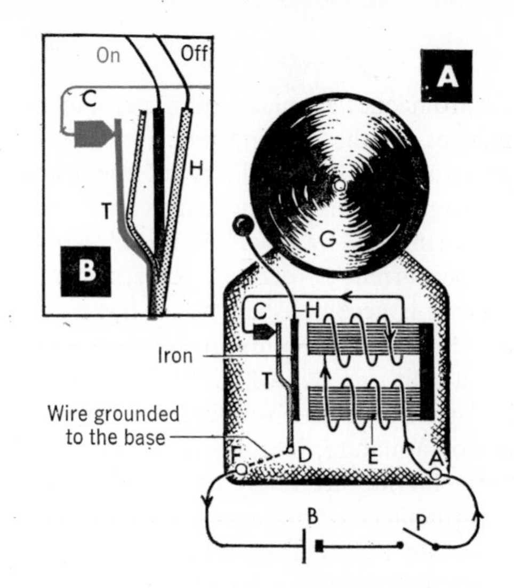

30-9 The electric bell. The use of a contact point (C) through which current passes to the vibrating tongue (T) makes the current go on and off. As a result, the attraction of the electromagnet is also on and off, thus causing the hammer to vibrate back and forth to ring the bell.

6. Rotating electromagnets

To make an electric motor practicable, a way had to be found to make a magnet spin continuously. It is easy enough to make a pivoted permanent magnet vibrate back and forth for a short while; the attraction of a stationary magnet will do that. But when a pole of the rotating magnet gets as close as possible to an opposite pole of the fixed magnet and then starts to move away on the other side, the attraction pulls it back (Fig. 30-10A). The back-and-forth movement of the swinging magnet gradually decreases until the magnet comes to rest.

The problem of making a magnet rotate continuously was solved by using an

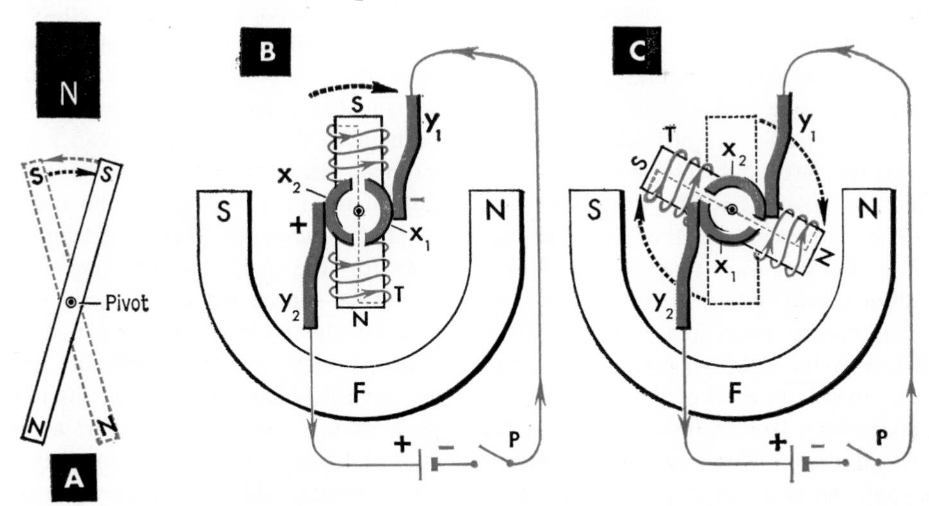

30-10 A motor cannot be made from a rotating permanent magnet near a stationary one (A) because it simply swings back and forth. But with the use of an electromagnet, a commutator (split ring, X_1 and X_2), and brushes (Y_1 and Y_2), current in the rotating electromagnet is made to reverse at just the right position to keep the magnet spinning around.

electromagnet in place of a permanent magnet for the rotating part (called the **armature**). The advantage of the electromagnet is that its polarity can be reversed merely by reversing the direction of the current it receives. Reversing the current is accomplished by an ingenious device called a **commutator.** This is simply a split ring, X in Fig. 30-10B, which is fastened to the rotating electromagnet or armature, NS. The two halves of the split ring are really the terminals of the spinning electromagnet. Two brushes, Y, conduct current to and from the electromagnet through contact with the opposite halves of the split ring. A large **field** (or stationary) **magnet**, F, surrounds the armature, causing it to rotate.

We turn on the motor by pressing the push button, P. Current then flows into the armature, NS, causing it to become magnetized. We can locate the north and south poles of the armature by following the direction of the electron flow. We see that the electrons flow from the negative terminal of the battery to a brush, Y_1, then to one of the halves of the commutator, X_1, thence around the coils of the electromagnet, NS, and back to the other half of the commutator, X_2, and from there out through the other brush, Y_2. The circuit is thus complete. By using the left-hand rule, we locate the poles. The lower end is a north pole.

The attraction of the field magnet causes the armature to rotate in the direction shown by the arrow. As soon as the armature poles reach the position closest to the field magnet, the brushes may touch both ends of the split ring for a short time. But an instant later the inertia of the moving armature carries it beyond the point of closest approach to the position shown in Fig. 30-10C. Now the brush, Y_1, is touching split ring X_2 instead of X_1, and the current in the electromagnet is reversed. Because of this reversal of current, the N pole of the armature changes to an S pole. Where there was attraction before, between armature and field poles, there is now repulsion. The net result is that the rotation of the armature continues as before, in the direction shown by the arrows.

As you see, the purpose of the commutator is to reverse the current in the armature at just the right instant. The purpose of the brushes is to conduct the current to and from the armature without twisting the wires. The reversal of the current takes place every half turn, since the splits in the commutator are located a half turn apart. In this way the armature is alternately attracted and repelled by the field magnet so as to spin continuously in the same direction at all times.

7. Improving the electric motor

The efficiency of this simple electric motor can be improved in a number of ways. Thus, if the iron armature core is made cylindrical to fill up the space between the poles, the magnetism is greatly strengthened (Fig. 30-11A). Then, if we substitute an electromagnet for the permanent field magnet, we can have a stronger field magnet. Thus the attraction of the field magnet for the rotating magnet is increased. We can get our current for the field electromagnet from the same source we use for the armature. If the coils of the armature and field magnet are connected in series, we have a **series motor** (Fig. 30-11B). Such motors are found to exert a great deal of force from the moment they begin to rotate. This is why they are used in trolley cars, elevators, and trains, where a large pulling force is needed from the start.

A motor in which the armature and field magnet coils are connected in parallel is called a **shunt motor** (Fig. 30-11C). These motors are found useful in such devices as fans and vacuum cleaners, where the speed of operation should be kept fairly steady and where it is not important to have a large force at the start.

We can also improve our electric motor if we use several coils in the armature instead of one (Fig. 30-11D). The single-coil armature has several disadvantages. The forces of attraction and repulsion change considerably as the position of the armature changes, which means that the motor is not always exerting the greatest possible force. In addition, the single-coil armature is not always self-starting, because it may stop in a position where its poles are closest to the poles of the field magnet. In this position the attraction of the field magnet causes no motion of the coil when the current is turned on again.

All these defects are remedied if we wind a number of separate coils around the armature, connecting each one to a separate pair of **segments** on the commutator (Fig. 30-11D). Each coil is thus connected to the circuit for a short time at its point of greatest pull. Hence rotation is caused by a steady series of one-after-the-other attractions instead of by a changing pull. This is a bit like the action of an eight-cylinder gasoline engine as compared to an engine of one cylinder. Figure 30-11D shows coil A while it is being attracted. In another moment, coil B will move over to this position and begin operation, followed in turn by coils C and D.

Giant motors to run our modern industrial civilization have been developed, based on the principles we have just explained (Fig. 30-12).

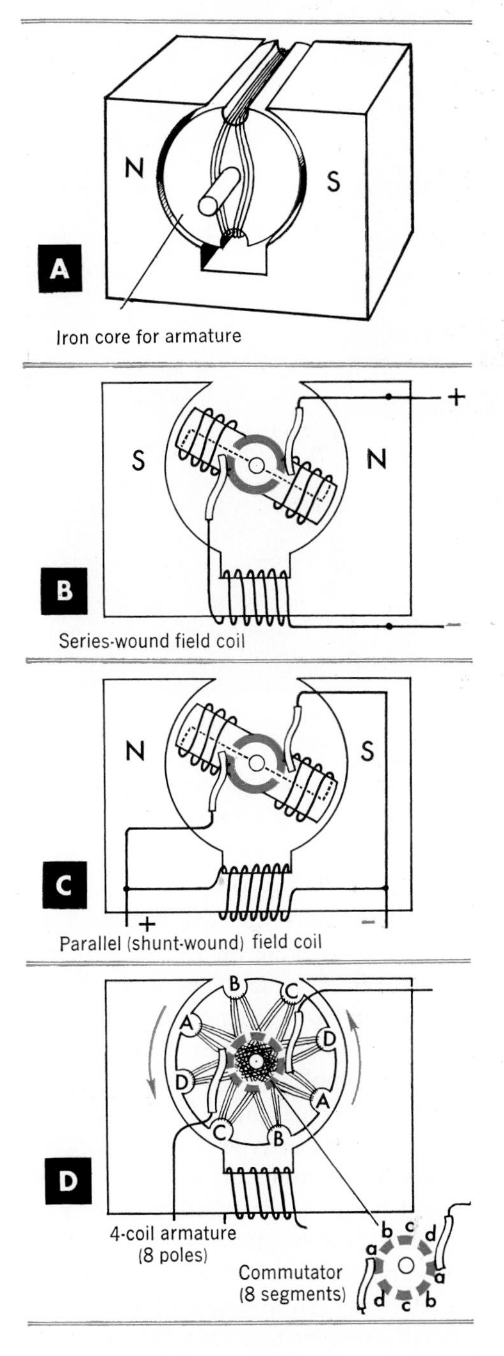

30-11 Electric motors can be improved by replacing air spaces with iron (A), using an electromagnet or field magnet (B or C), and winding several coils on one armature (D).

General Electric Co.

30-12 This gigantic 7,000-horsepower electric motor for a steel mill can do the physical work of about 50,000 men. Note how small the man appears in comparison to this gigantic motor.

8. AC and DC motors

All the motors we have discussed up to this time operate on DC. DC, or **direct current**, is the type produced by batteries and DC generators. Direct current flows in one direction as long as the circuit is complete.

But most homes and factories are supplied with another type of current, called **alternating current**, or AC. Alternating current is a back-and-forth type of current; the current surges one way, then stops and reverses. To obtain the greatest efficiency from alternating current, specially designed motors are generally used. But many small motors similar in construction to the one described in the previous section (Fig. 30-11D) are designed to operate with reasonable efficiency on both AC and DC. A motor of this type is called a **universal motor.**

A universal motor uses an electromagnet as the field magnet, just like the DC motor in Figs. 30-11B and C. You might think that the reversal of current would reverse the poles of the electromagnet in the armature and thus reverse the direction of motion. This would be true if the field magnet had permanent poles. But if the field magnet is an electromagnet, supplied by the same source of current as the armature, then its poles also reverse at the same time that those of the armature reverse. This double reversal keeps the direction of rotation always the same (Fig. 30-13). Alternating current will be the subject of Chapter 35.

We have seen in this chapter how Oersted's simple discovery that a wire carrying electric current has a magnetic field led to many useful and important inventions. Telegraphs, relays, lifting magnets, bells, and electric motors, are by no means all of the widespread applications of the magnetic effects of electric current. In Chapters 34 and 36 we shall see how the principle of the electromagnet is applied to generators, transformers, radio, and television. No doubt the future will see even greater development of devices using magnetic fields produced by electric currents.

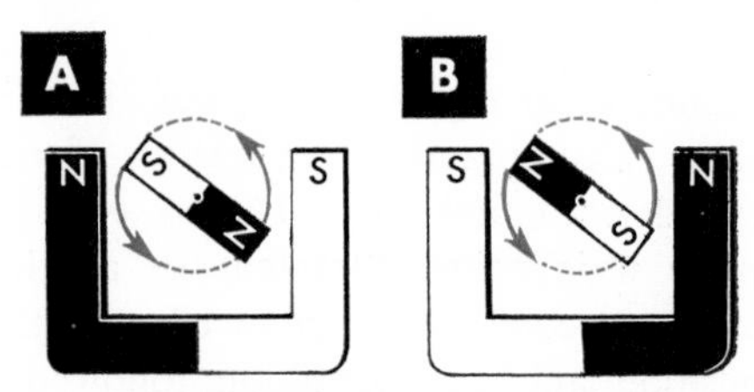

30-13 A double reversal of poles in an electric motor keeps the armature spinning in the same direction. But this occurs in a motor only if the field coil is also an electromagnet. Such a motor can operate on either AC or DC.

Going Ahead

THE VOCABULARY OF PHYSICS

Define each of the following:

left-hand rule for a single wire
core
left-hand rule for electromagnets
automation
dot
dash
line switch
relay
commutator
armature
field magnet
series motor
shunt motor
commutator segments
direct current (DC)
alternating current (AC)
universal motor

THE PRINCIPLES WE USE

Explain why

1. A line switch is necessary in addition to the sending key of a telegraph set.
2. The current in the armature must reverse if an electric motor is to work.
3. An iron core increases the strength of an electromagnet.
4. Ampere-turns are a measure of the strength of an electromagnet.
5. Bending a magnet into the shape of a horseshoe increases its strength.
6. Relays are needed in the operation of an elevator.
7. Several coils are usually wound on the armature of a motor instead of just one coil.
8. A series motor is used to operate an elevator.
9. A shunt motor is used to operate an electric fan.

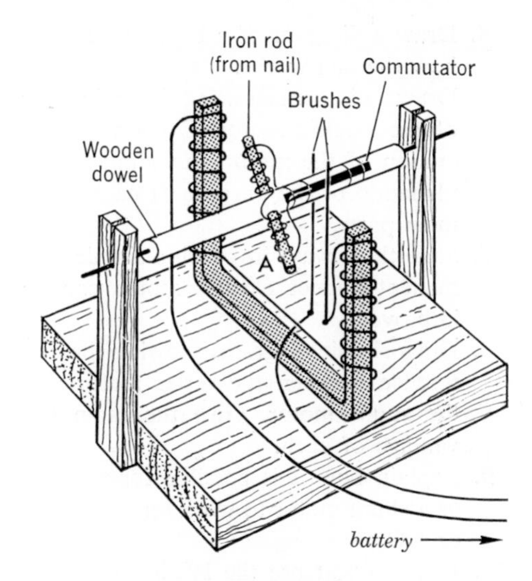

30-14 Can you make this electric motor? Try it.

APPLYING PRINCIPLES OF ELECTROMAGNETISM

1. State four ways in which an electromagnet can be strengthened.
2. Draw a diagram of a two-way telegraph system showing the details of windings of the electromagnet. Explain the operation of the system. Label the north and south poles of the magnets.
3. How is it possible to eliminate the return wire in a telegraph system?
4. Draw a diagram to show how a relay enables a photoelectric cell to turn on a 110-volt motor. Explain its operation. State two uses for relays.

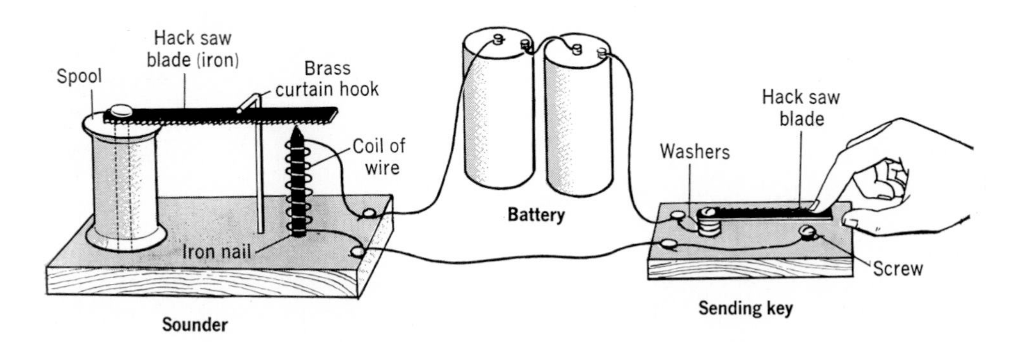

30-15 Make this simple telegraph set from a hack-saw blade, nails, wood, and wire.

5. Draw a diagram of a bell, and explain how the bell works.
6. Draw a diagram showing the main parts of the electric motor. Explain the operation of the motor, including an explanation of the purpose of each of its main parts.
7. State three ways in which a simple permanent-magnet electric motor can be improved.
8. Draw simple diagrams showing the difference between the series motor and the shunt motor. Give the main advantage of each.
9. Explain how it is possible for a series or a shunt motor to work on either AC or DC.
10. How do you use the left-hand rule to find the direction of current flowing in a wire?
11. As you know, alternating current reverses itself at intervals. Why isn't it possible to use AC instead of a commutator to reverse the poles of the armature in the simple permanent-magnet motor shown in Fig. 30-10?
12. What change would you make in the wiring of a vibrating bell to change it into a gong that rings once when pressed?
13. Will a permanent-magnet motor work on AC? Explain.
14. Will reversing the connections at the battery reverse a series or shunt motor? Explain.
15. How would you go about reversing (a) a permanent-magnet motor, (b) a series motor, (c) a shunt motor? Explain each case, using a diagram if helpful.
16. Would an electromagnet work on AC? Explain.
17. Imagine that electrons inside a television picture tube are moving straight towards you. Are the magnetic lines of force around them clockwise or counterclockwise?
18. Alternating current at 110 volts is commonly carried to household appliances through a double wire (Fig. 25-5). What can you say about the magnetic field around such a double wire when the appliance is turned on?
19. You are given two sealed wooden boxes of identical size and shape and weight. One contains a bar magnet; the other contains an electromagnet and a battery. Can you tell which is which? If so, how?
20. A certain electric motor is 60% efficient. What do you think has become of the other 40% of the input energy?

General Electric Co.

31-6 This generator, driven by the steam turbine connected to it on the same shaft, delivers 80,000 kilowatts. This power is enough to supply the needs of a small city. At 5 cents a kilowatt-hour how much would it cost to run this generator all day at full capacity?

horsepower ($\frac{1,000}{746}$). But 1 horsepower is 550 foot-pounds per second, so 1,000 watts equal 1.34×550 or 737 foot-pounds per second. The work done in 1 hour by 1,000 watts (1 kwh) is $737 \times 60 \times 60$ or 2,660,000 foot-pounds.

The cost of electricity in most cities in the United States ranges from about 2 to 7 cents per kilowatt-hour. Generally a sliding rate is applied by which those who use large quantities of electricity can obtain it at lower rates.

SAMPLE PROBLEM

Let us see how we would determine the cost of operating a 200-watt lamp 4 hours a day for 30 days at a rate of 4 cents per kilowatt-hour.

$$200 \text{ w} = \frac{200}{1,000} \text{ kw} = 0.2 \text{ kw}$$

The lamp is operated for a total of 4 hours × 30, or 120 hours. The energy used (work done) is:

$$\begin{aligned} \text{Work} &= \text{Power} \times \text{Time} \\ &= 0.2 \text{ kw} \times 120 \text{ hr} \\ &= 24 \text{ kwh} \end{aligned}$$

Since each kilowatt-hour costs 4 cents, the total cost is 24×4 cents, or 96 cents.

This could be shortened into one operation, as follows:

$$\begin{aligned} \text{Cost} &= \frac{\text{Watts} \times \text{Hours} \times \text{Cost of 1 kwh}}{1,000} \\ &= \frac{200 \times 120 \times 4¢}{1,000} \\ &= 96¢ \end{aligned}$$

STUDENT PROBLEM

At 6¢ per kwh, how much will it cost to run a 50-watt lamp for 9 hours? (ANS. 2.7¢)

14. Electrical energy: man's servant

By means of electricity we can obtain the amount of work done by one horse in one hour for an average of about 4 cents. This gives some indication of the reason for the enormous development of electricity during the past century. By simply pressing a button today, we can summon the service of electric motors to lift elevators, move cars and trains, pump water, and do a host of other things cheaply and conveniently. Since a man can exert about $\frac{1}{10}$ of a horsepower over a period of time, the physical labor of a man for an hour can be duplicated with electrical energy for about $\frac{2}{5}$ of a cent.

We can see why the use of labor-saving machinery and the substitution of motors for physical labor have already made possible a completely new type of civilization. If all physical labor such as lifting, pushing, and pulling were replaced by automatic machinery, operated by electric motors, human beings would be free for other types of work in which brain and skill would count more than brawn.

Going Ahead

THE VOCABULARY OF PHYSICS

Define each of the following:

coulomb	kilowatt
ampere	resistance
electromotive force	ohm
potential difference	watt
ammeter	joule
voltmeter	kilowatt-hour
volt	watt-second

THE PRINCIPLES WE USE

1. State Ohm's Law. Write it as a formula.
2. If the resistance of a circuit is increased, and the E.M.F. kept constant, what happens to the current?

SOLVING PROBLEMS OF ELECTRIC POWER

1. What is the resistance of a circuit in which 20 v cause 0.5 amp to flow?
2. How many dry cells (1.5 v each) are needed to cause 0.5 amp to flow in a circuit of 30 ohms?
3. Fill in the unknowns in the table below:

Amperes	*Volts*	*Ohms*
(a) 2	20	?
(b) 4	?	6
(c) ?	200	10
(d) ?	90	450
(e) 6	?	3

4. How much current will flow when 4 dry cells are in series with a resistance of 45 ohms?
5. How much voltage is needed to obtain 0.05 amp in a 640-ohm circuit?
6. How much resistance is in a circuit if 90 v cause a current of 3.5 amp to flow?

---

7. What is the cost of using a 400-w heater 60 hr a month at a rate of $3\frac{1}{2}$¢ per kwh?
8. What is the cost of operating a 50-ohm heater on 120 v for 200 hr at 4¢ per kwh?
9. A lamp operates on 220 v and uses 0.87 amp. (a) What is the resistance of the lamp? (b) What is the power used by the lamp? (c) How much will it cost to operate this lamp for 86 hr at 3.2¢ per kwh?
10. An electric motor uses 495 w of power when operated on 110 v. (a) How much current flows in the motor? (b) What is its resistance? (c) What will it cost to operate it for 45 hr at 2.9¢ per kwh?
11. A 50-w lamp has a resistance of 350 ohms. (a) What is the voltage of the lamp? (b) How much current flows?
12. An electric meter has a current flow of 0.023 amp when used with 35 v. (a) What is the resistance of the meter? (b) How much current will flow in this meter when used with 58 v?
13. How many dry cells in series are needed to cause approximately 0.4 amp to flow in a circuit of 88 ohms?

14. 6 dry cells are used in series to operate a large bell of 10.5 ohms resistance. How much current will flow in the bell?
15. For how many hours can a 100-w lamp be used at a cost of 1¢ in a region where electricity sells for 3¢ per kwh?
16. An electric iron is rated for 600 w, 120 v. (a) What current flows through the iron? (b) What is the resistance of the iron? (c) How much energy does it use in 10 hr? (d) How much will it cost to operate it for 10 hr at 5¢ per kwh?
17. The motor of an electric refrigerator operates on 110 v and uses 2 amp. To keep the refrigerator cool, the motor operates one fourth of the time. What is the cost of operating the refrigerator for 30 days at 5¢ per kwh?

---

18. What is the horsepower of a 60%-efficient electric motor that operates on 3 amp at 120 v?
19. How efficient is a 4-hp electric motor that operates on 20 amp at 200 v?
20. What is the horsepower of a 70%-efficient electric motor that uses 5 amp of current at 220 v?
21. What is the efficiency of a 2.5-hp motor that uses 6.8 amp of current at 550 v?
22. A pump forces water into the bottom of a tank measuring 10 ft wide, 20 ft long, and 5 ft deep from a lake 50 ft below the bottom of the tank. An electric motor operating on 120 v and using 4 amp is used to operate the pump. If the entire system of motor, pump, and pipes is 20% efficient, how long will it take to fill the tank?
23. A 100-v, 5-amp lamp is operated on 150 v. What resistance in series with the lamp will keep the current at the proper amount?
24. A 100-w bulb is operated for 150 hr, a 60-w bulb for 100 hr, a 1,000-w stove for 90 hr, a 250-w refrigerator for 150 hr. What is the total cost of operation if electrical energy costs 4¢ per kwh?
25. How many ft-lb of work can be done in 10 min by a 220-v motor drawing 12 amp if the motor is 80% efficient?
26. How many gal of water can be pumped into a 50-ft-high tank in 1 hr by the motor of the previous problem? (1 gal of water weighs 8.3 lb).

CHAPTER

32

Electrical Circuits

Try the following experiment. Measure with an ammeter the current flowing in a 1-foot length of a high-resistance wire such as nichrome, using a battery as the source of voltage (Fig. 32-1A). Connect this wire in series with another piece of the same wire of equal length and measure the current flowing through both of them (Fig. 32-1B). You find that the current has been reduced to one half. Add a third wire of the same length in series, and the current is reduced to one third.

Series circuits

1. Resistance increases with length of wire

From the foregoing experiment, we see that connecting resistances in series increases the resistance of the entire circuit, which reduces the current. We use this fact in calculating the resistance of wires. Spools of wire are usually marked in ohms per foot of length. For example, if a spool of high-resistance wire is marked 3 ohms per foot, the resistance of 2 feet of the wire will be 6 ohms, 5 feet 15 ohms, 10 feet 30 ohms, etc. So we see that resistance is proportional to the total length of wire.

2. Resistances in series

Suppose that we connect two unequal resistances in series. What will be the resistance of the combination?

Let us suppose that one resistance is 5 ohms and the other 3 ohms. When we connect them in series we are simply adding their resistances, as if they were a single piece of wire of resistance 8 ohms. The resistance of the two together in series is 5 + 3 or 8 ohms.

To find the total resistance, R, of several devices in series, we simply *add the separate resistances,* r_1, r_2, r_3, etc. That is,

$$R = r_1 + r_2 + r_3 + \cdots$$

STUDENT PROBLEM

Find the total resistance of a 10-ohm resistance, a 75-ohm resistance, and a 3-ohm resistance all connected in series. (ANS. 88 ohms)

3. Does current ever get used up?

It is a common error to refer to current as having been "used up." If current is flowing through the 5-ohm and 3-ohm resistances in series, it might seem that the current flowing through the 5-ohm resistance would be partly used up and leave less for the 3-ohm resistance. Let us consider what happens in the case of flowing water.

When a gallon of water flows into the kitchen sink, the water does not get used up but simply flows into the outlet pipe. At the same time another gallon of water flows into the house through the main water pipe to take the place of the gallon that has flowed down the drain. In the same way, if a certain number of

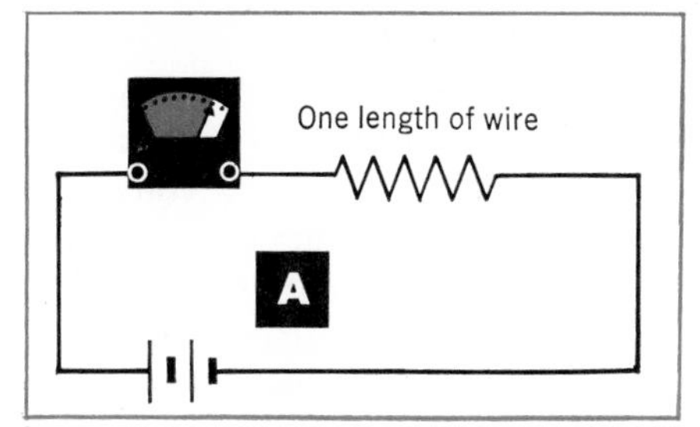

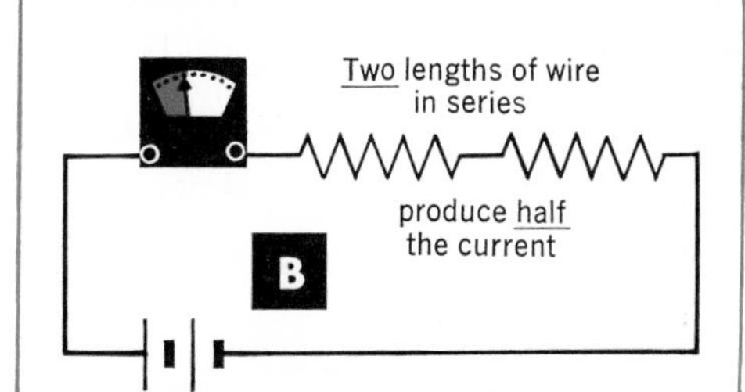

32-1 The longer the wire, the greater the resistance. Doubling the length of wire produces double the resistance. Resistance is proportional to length of wire.

electrons flow out of a device, an equal number come in at the inlet.

In other words, the flowing electric charge which constitutes a current never does get used up.

Electric charge can neither be created nor destroyed.

This is called the **Law of Conservation of Charge** and was first stated by Benjamin Franklin.

We can prove this by experiment if we connect a number of ammeters in a series circuit, as shown in Fig. 32-2. The ammeters read the same no matter where we place them in the circuit, indicating that the current is the same in all parts of a series circuit. That is:

$$I = i_1 = i_2 = i_3 = \cdots$$

4. Voltage in a series circuit

What is the potential difference (voltage) across each of the resistances in a series circuit like that of Fig. 32-3?

In this diagram we have written in all our given data and represented by letters the unknown quantities we wish to find. You, too, should always draw a diagram in this manner when you solve circuit problems.

The potential difference across each resistance is found by using Ohm's Law, that is, by multiplying the value of the resistance by the current through it ($e_1 = i_1r_1$ and $e_2 = i_2r_2$). We are given the value of each resistance, r_1 and r_2, but in order to find the voltage across each we must first find the current through it. Using Ohm's Law again, and remembering that the total resistance of a series circuit is the sum of the separate resistances:

$$I = \frac{E}{R}$$

$$I = \frac{120}{200 + 40} = 0.50 \text{ amp}$$

Since the current is the same everywhere in a series circuit, the total cur-

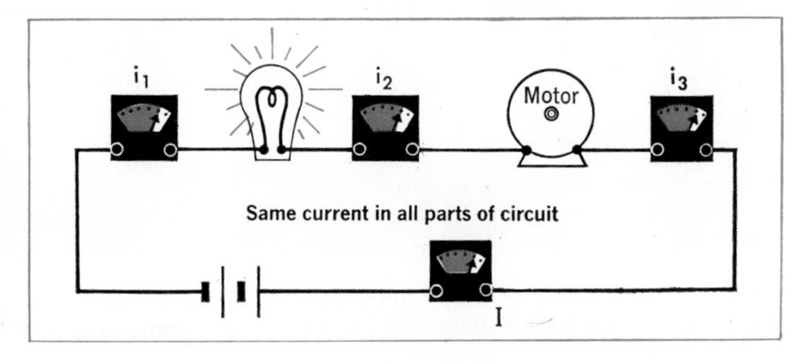

32-2 Does a motor or lamp "use up" current? This diagram of an experiment shows that current does not get used up. The ammeters show that current is the same in all parts of this circuit.

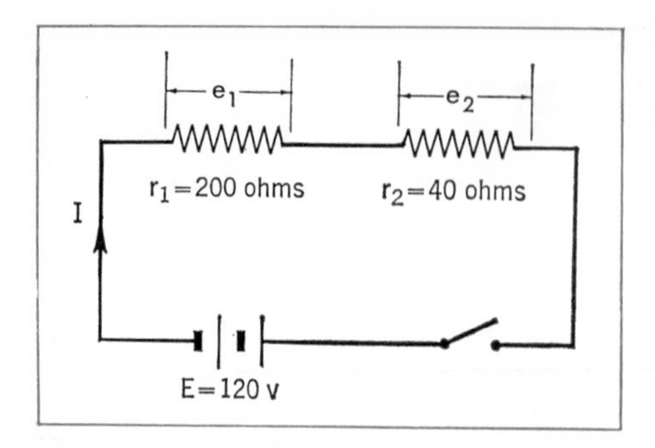

32-3 In a series circuit, the sum of the potential differences around the circuit equals the applied voltage ($E = e_1 + e_2$).

rent, 0.50 ampere, flows through each resistance. We have all the information we need to find the potential difference across each of the resistances separately.

For the 200-ohm resistance,

$$e_1 = Ir_1$$
$$e_1 = 0.50 \times 200 = 100 \text{ v}$$

For the 40-ohm resistance,

$$e_2 = Ir_2$$
$$e_2 = 0.5 \times 40 = 20 \text{ v}$$

The sum of the potential differences ($e_1 + e_2$) is 120 volts, the applied voltage from the battery. It is, in fact, a general rule that **the sum of the potential differences in a series circuit equals the total voltage applied,** or

$$E = e_1 + e_2 + e_3 + \cdots$$

This rule continues to apply to even the most complex circuit.

5. Voltage drop

The potential difference across a resistance is often referred to as the **voltage drop.** There must be a voltage drop between any two points along a conductor carrying a current, no matter how small its resistance. Even a power line has some resistance, and therefore there must be a voltage drop, *E,* given by $E = IR$, between any two points along the wire. This formula shows that the larger the current flowing through a power line (or any other resistance), the greater will be the voltage drop (or "line drop") in the line itself, and hence the less will be the potential difference available at the end of it.

Have you ever noticed the lights dim when someone in your house turns on a device drawing a good deal of electric current? This occurs because the added current being drawn from the power line increases the voltage drop in the power line, and hence less voltage is available to each lamp in your house.

6. Calculation of voltage drop in a line

In order to reduce voltage drop in a power line, conductors of very low resistance are used to bring electricity from the powerhouse to the consumer. Let us assume that the powerhouse generates electricity at 120 volts, that the resistance of the several miles of cable leading to

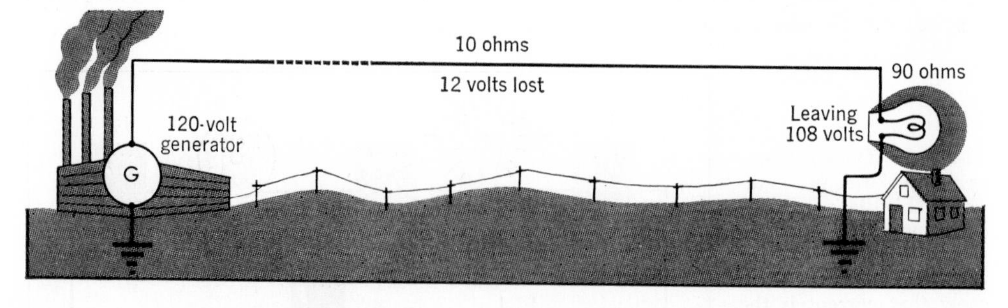

32-4 Resistances of power lines must be low or the voltage drop along the line will be too great, and too little voltage will appear at the consumer's end of the line.

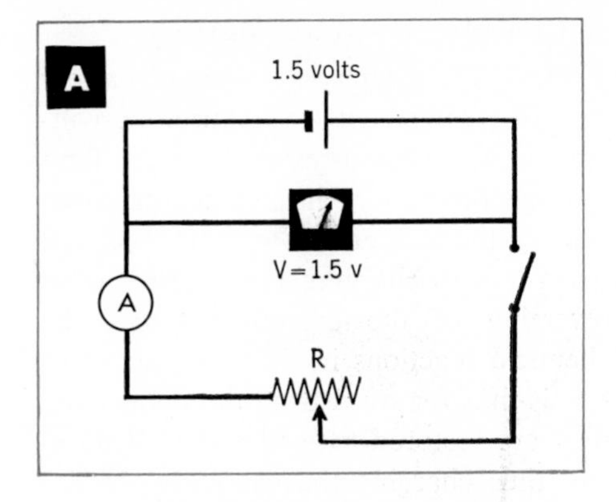

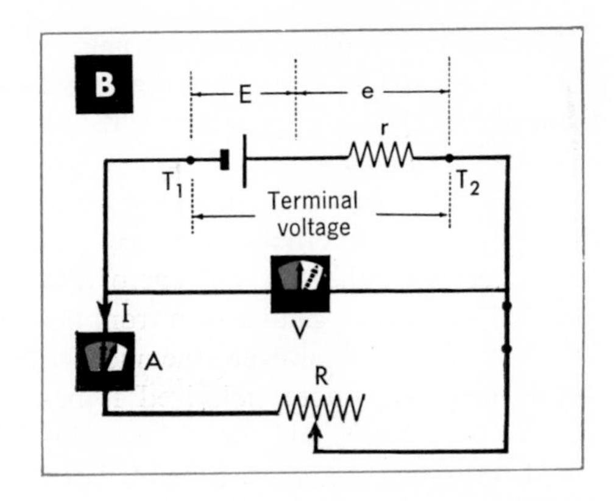

32-5 When no current is drawn from a cell (A), its terminal voltage is also its E.M.F. But as soon as current is drawn from it (B), the terminal voltage falls.

the consumer's home is 10 ohms, and that a 90-ohm lamp designed for use with 120 volts is connected into the circuit in the home (Fig. 32-4). You note that the 10-ohm cable is in series with the lamp. This cable will use up voltage in proportion to its resistance. The total resistance of the lamp and cable is 100 ohms (10 + 90), of which the 10-ohm cable accounts for $\frac{10}{100}$, or $\frac{1}{10}$ of the resistance. Thus the cable requires $\frac{1}{10}$ of the voltage; that is, $\frac{1}{10}$ of 120 volts, or 12 volts. The lamp itself gets only $\frac{9}{10}$ of the voltage, or 108 volts, and therefore does not operate at maximum brilliance. If the cable had a greater resistance, it would require an even larger proportion of the voltage, and the lamp would get even less voltage. This shows us why conducting cables of power lines must be of low resistance if losses due to the voltage drop in power lines are to be kept down.

7. Internal resistance and terminal voltage

Suppose we connect up the circuit shown in Fig. 32-5A. The cell has an E.M.F. of 1.5 volts and the voltmeter reads 1.5 volts with the switch open. When the switch is closed, however, the reading of the voltmeter drops a little. If the resistance, *R,* of the external circuit is decreased, the voltmeter shows a still lower reading. Why is this? There must be a voltage drop inside *the cell* that increases as the current flowing through the cell increases.

Let us redraw the circuit of Fig. 32-5A so that it looks like the circuit in Fig. 32-5B. Here T_1 and T_2 represent the terminals of the cell; and between them is the **terminal voltage,** the potential difference between the terminals of the cell. Between these terminals there is a source of E.M.F., *E,* and a resistance, *r,* which is the **internal resistance** of the cell to a flow of current.

In order to find the current flowing in this circuit we consider the internal resistance, *r,* and the external resistance, *R,* to be in series. We then have, by Ohm's Law:

$$I = \frac{E}{R + r}$$

The terminal voltage of the cell must be the E.M.F., *E,* minus the voltage drop across the internal resistance in the cell ($e = Ir$):

$$\text{Terminal voltage} = E - Ir$$

Most new cells have a low internal resistance, a fraction of an ohm; the larger the cell, the less the resistance. If the external resistance is more than a few ohms (usually true for apparatus using dry cells), the internal resistance has little effect on either the current or the terminal voltage, as can be seen from the above equations. However, the more a cell is used, the more polarized it becomes (page 359), and the more its internal resistance increases. Finally, when the cell is no longer able to provide sufficient terminal voltage and current to the external circuit, it must be discarded (if it is a dry cell) or recharged (if it is a storage cell).

SAMPLE PROBLEM

A cell whose E.M.F. is 2.0 volts and whose internal resistance is 0.05 ohm is connected to a 0.8-ohm resistance. What is the current through the resistance and what is the terminal voltage of the cell?

Applying the equation to find the current, we have:

$$I = \frac{E}{R + r}$$

$$I = \frac{2.0}{0.8 + 0.05} = 2.35 \text{ amp}$$

The terminal voltage is then found from the equation:

$$\begin{aligned}\text{Terminal voltage} &= E - Ir \\ &= 2.0 - 2.35 \times 0.05 \\ &= 1.88 \text{ v}\end{aligned}$$

STUDENT PROBLEM

Find the current and the terminal voltage of a cell whose E.M.F. is 1.5 volts and whose internal resistance is 0.1 ohm, if it is connected to a 1.5-ohm resistance. What percentage of error in the current would result from neglecting the internal resistance? (ANS. 0.938 amp, 1.41 v, 6.7%)

8. Charging a storage battery

In order to charge a storage battery we must connect it to a source of voltage that will cause current to flow through the battery in a direction opposite to the current flow from the cell when it is discharging (Fig. 32-6). This reversing of the current reverses the chemical reactions in the cell. Lead sulfate is thus removed from the plates, and they are restored to lead and lead dioxide, thus charging the cell (Fig. 28-6). In order that the current may be reversed, the charging voltage must be greater than the E.M.F. of the battery and connected so as to oppose this E.M.F. How large a charging current should be used?

If too large a charging voltage is used, the current will be so large that it will produce too much heat in the internal resistance and ruin the battery.

SAMPLE PROBLEM

Suppose a 6-volt battery with an internal resistance of 0.06 ohm is to be charged at a rate of 15 amperes. How large a charging voltage is required?

Looking at Fig. 32-6 you will notice that the charging voltage applied to the battery is the terminal voltage. Since the current through the circuit is oppo-

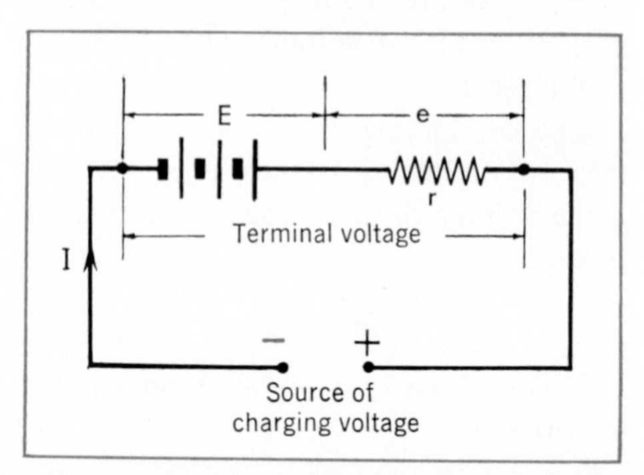

32-6 When a storage battery is being charged, the terminal voltage is equal to the E.M.F. of the storage battery, E, plus the voltage drop, e, in the internal resistance, r.

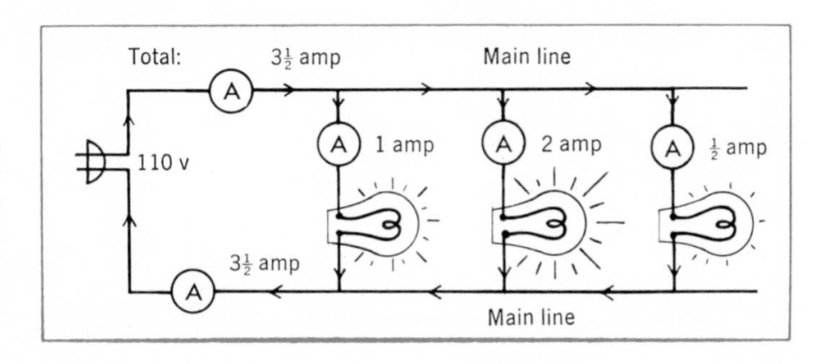

32-7 How does the diagram of the experiment shown here indicate that the current in the main line of a parallel arrangement is the sum of the separate currents?

site to the flow when the battery is discharging, the voltage drop across the internal resistance is now opposite in polarity. It must therefore be *added* to the E.M.F. of the cell in order to find the terminal voltage. We can express this in the formula:

$$\text{Terminal voltage} = E + Ir$$

Substituting the given charging current in the equation, the charging voltage, which is the same as the terminal voltage, is:

$$\text{Terminal voltage} = 6 + 15 \times 0.06 = 6.9 \text{ v}$$

While a battery is being charged, its internal resistance decreases. This increases the charging current for a given charging voltage. The charging voltage must therefore be reduced after a while so that the current will not become too great and overheat the battery.

STUDENT PROBLEM

What voltage is needed to charge a 2.00-v, 0.02-ohm cell at a rate of 10 amp? (ANS. 2.20 v)

Parallel circuits

In Chapter 25 we learned that most household appliances are connected in parallel. Let us see how we would calculate the current in this situation.

9. Currents in parallel

Try the experiment shown in Fig. 32-7. Connect three lamps in parallel with ammeters to measure the current flowing in them. First measure the current when each lamp is turned on separately. Then turn the lamps on two or three at a time. You note that **the current in the main line is always the sum of the separate currents.** When you turn a lamp on or off, it has no effect on the current through any of the other lamps.

Each lamp is independent of the others because it is connected by a separate path to the two wires of the main line. This is not so in a series circuit. There the current flowing in one lamp must flow through all the others in the circuit and is, therefore, affected by them.

Since current cannot be used up, all the current flowing into each of the separate branches of the parallel circuit must come out of one main-line wire and return to the other main-line wire through the main-line ammeters. Thus the main-line ammeters record the total current flowing in the separate branches. Since each lamp receives the same current whether operating alone or combined with others in a parallel circuit, we always treat each lamp separately in calculating the current it will receive. Then we get the total current by adding up the separate currents. Thus

$$I = i_1 + i_2 + i_3 + \cdots$$

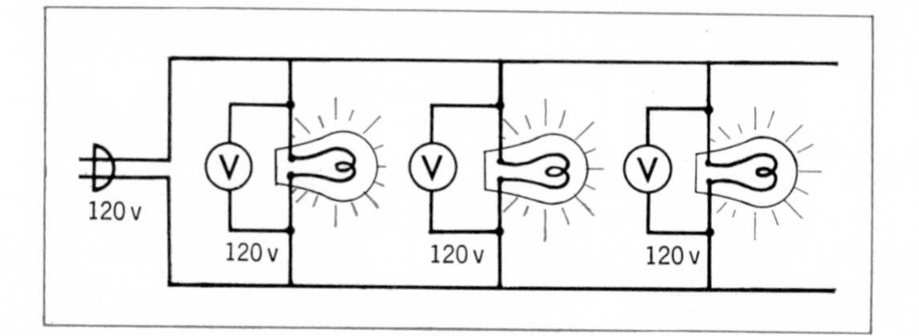

32-8 These voltmeters show that each device in parallel gets the same voltage as every other device, and also the same as the original source of voltage.

10. Voltages in parallel

Measure the voltage applied to each of the three lamps in parallel while the current is flowing, and then measure the voltage when each of them is shut off (Fig. 32-8). The voltage is the same whether or not current flows through the lamps. Thus in a parallel circuit each device gets the full voltage of the main line:

$$E = e_1 = e_2 = e_3 = \cdots$$

11. Resistances in parallel

We saw that in a series circuit we could obtain the resistance of the entire circuit by simply adding up the separate resistances. In a parallel circuit we calculate the total resistance in the following way. If the individual resistances are r_1, r_2, r_3, . . . their combined resistance, *R*, may be found by the formula:

$$\frac{1}{R} = \frac{1}{r_1} + \frac{1}{r_2} + \frac{1}{r_3} + \cdots$$

If there are more than three resistances in parallel, we simply add a term for each additional resistance.

SAMPLE PROBLEM

Suppose 20 ohms and 30 ohms are connected in parallel. What is their combined resistance?

From the formula

$$\frac{1}{R} = \frac{1}{20} + \frac{1}{30}$$

Multiplying through by 60R,

$$60 = 3R + 2R$$
$$R = 12 \text{ ohms}$$

TABLE **12**

COMPARING SERIES AND PARALLEL CIRCUITS

	Series	Parallel
Current	The current, I, is the same in all parts of the circuit. It can be calculated by treating all the resistances as one large resistance.	Each branch is independent of the others. The total current, I, is the sum of separate currents in each branch. $I = i_1 + i_2 + \ldots$. Calculate the current in each device as though it were alone.
Resistance	Add up separate resistances to obtain the total resistance. $R = r_1 + r_2 + \cdots$	The combined resistance is less than any one resistance in the circuit. $\frac{1}{R} = \frac{1}{r_1} + \frac{1}{r_2} + \cdots$
Potential difference	The sum of the potential differences is the total E.M.F. applied.	The potential difference across each resistance is the same and is the E.M.F. of the battery.

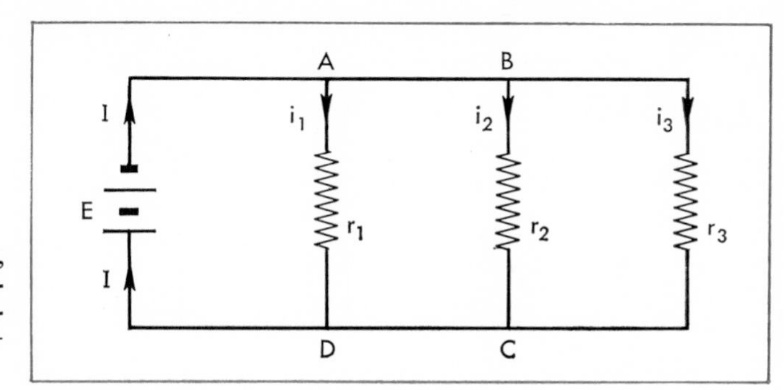

32-9 Total current, I, in this parallel circuit divides into three smaller currents, i_1, i_2, i_3. $I = i_1 + i_2 + i_3$.

We see that 12 ohms is the equivalent resistance (or combined resistance) of 20 and 30 ohms in parallel. Thus a single 12-ohm resistance would permit the same flow of current as the total current through a 20- and a 30-ohm resistance connected in parallel.

Notice that the equivalent resistance of 12 ohms is *less* than either of the two separate resistances. How do we explain this fact? Adding more resistances in parallel provides more paths for the current and, therefore, permits more current to flow. Hence connecting resistances in parallel reduces the combined resistance. This is just the opposite of what occurs in a series circuit. There connecting more resistances in the circuit *increases* the total resistance of the circuit.

STUDENT PROBLEM

What is the combined resistance of 4 ohms and 6 ohms connected in parallel? (ANS. 2.4 ohms)

12. Derivation of the formula for parallel resistances

Suppose the resistances r_1, r_2, and r_3 are connected in parallel to a source of voltage, E, as shown in Fig. 32-9. The potential difference across each resistance in this parallel circuit is the same, and is equal to E. The current through each resistance, however, will be different, depending upon the value of the resistance. Let us apply Ohm's Law to each resistance separately. We then have the following equations:

$$(1) \quad i_1 = \frac{E}{r_1}, \qquad i_2 = \frac{E}{r_2}, \qquad i_3 = \frac{E}{r_3}$$

But all the resistances connected together act like one resistance, R, connected to the source of voltage, E, from which is drawn a current, I. Ohm's Law applies to the effective resistance, R, of the combination:

$$(2) \qquad I = \frac{E}{R}$$

The total current, I, in a parallel circuit is found by adding up the separate currents (Section 9) and so we write the equation

$$(3) \qquad I = i_1 + i_2 + i_3$$

Substituting in equation (3) the values of I, i_1, i_2, and i_3 obtained from equations (1) and (2), we have:

$$\frac{E}{R} = \frac{E}{r_1} + \frac{E}{r_2} + \frac{E}{r_3}$$

Dividing both sides of the equation by E:

$$\frac{1}{R} = \frac{1}{r_1} + \frac{1}{r_2} + \frac{1}{r_3}$$

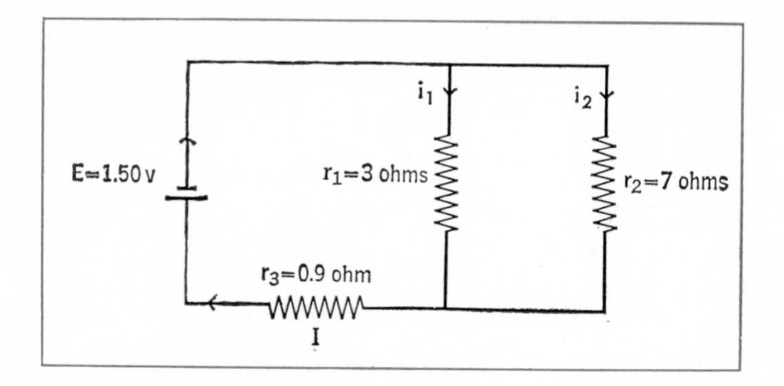

32-10 A series-parallel circuit. The resistance of the parallel combination r_1 and r_2 is called R_p.

13. Series-parallel circuits

Suppose we have two resistances, r_1 and r_2, connected in parallel, and this combination in turn connected in series with a third resistance, r_3, as shown in Fig. 32-10. Let us use what we have learned to find the quantities I, e_1, e_2, i_1, and i_2, as shown in the diagram.

We start by finding the combined reresistance, R_p, of the parallel-connected resistances:

$$\frac{1}{R_p} = \frac{1}{r_1} + \frac{1}{r_2} = \frac{1}{3} + \frac{1}{7}$$
$$R_p = 2.1 \text{ ohms}$$

We can now find the total resistance, R, in the circuit by adding this effective resistance of the parallel combination to the 0.9-ohm resistance connected in series with it:

$$R = 2.1 + 0.9 = 3.0 \text{ ohms}$$

The total current, I, is now found by using Ohm's Law:

$$I = \frac{E}{R}$$
$$I = \frac{1.5}{3.0} = 0.50 \text{ amp}$$

The potential difference, e_3, across the 0.9-ohm resistance, r_3, can be found by applying Ohm's Law to this resistance alone:

$$e_3 = Ir_3 = 0.50 \times 0.90 = 0.45 \text{ v}$$

We can also find e_1 or e_2, the potential difference across the parallel combination, by the same method, since this combination acts like a single resistance of 2.10 ohms:

$$e_1 = e_2 = IR_p$$
$$e_1 = e_2 = 0.50 \times 2.10 = 1.05\text{v}$$

Notice that $e_3 + (e_1 \text{ or } e_2) = 0.45 + 1.05 = 1.50$ volts, the total applied voltage, which confirms the fact that the sum of the voltage drops around a series circuit is equal to the applied voltage.

It now remains to find i_1 and i_2, the currents in the two branches of the parallel circuit. We can do this by applying Ohm's Law to each of the two branches separately, since we know their resistances and have just found the potential difference across them (1.05 v).

$$i_1 = \frac{e_1}{r_1}$$
$$i_1 = \frac{1.05}{3} = 0.35 \text{ amp}$$

(through the 3-ohm resistance)

and

$$i_2 = \frac{e_2}{r_2}$$
$$i_2 = \frac{1.05}{7} = 0.15 \text{ amp}$$

(through the 7-ohm resistance)

Notice that $i_1 + i_2 = 0.35 + 0.15 = 0.50$ amp, the total current I flowing into the parallel combination.

STUDENT PROBLEM

In the problem just discussed, let $r_1 = 30$ ohms, $r_2 = 60$ ohms, and $r_3 = 40$ ohms. $E = 120$ v. Find all the information called for in the problem. (ANS. $e_1 = 40$ v, $e_2 = 40$ v, $e_3 = 80$ v, $i_1 = 1.33$ amp, $i_2 = 0.67$ amp, $I = 2$ amp)

Resistance of wires

We have seen that a long wire has greater resistance than a short one of the same material and thickness. The resistance is proportional to the length. But what happens to the resistance if we use a thicker wire?

14. Resistance of thick wires

From the experiment shown in Fig. 32-11A, we can see the effect that the thickness of a wire has on its resistance. Two identical lamps in separate circuits are each connected with resistance wires in series. The resistance wires in each circuit are of equal length, and are made of the same material, but are of different thickness. The lamp connected in series with the thin wire gives a dimmer light than the other lamp, showing that less current is flowing. This must mean that the thick wire has less resistance than the thin one.

We can understand why this is so if we come back to our familiar comparison between wires carrying electric current and pipes carrying water. The wider the pipe, the greater the flow of water for the same pressure pushing it.

Experiment shows that a wire which has twice the diameter of another will have $\frac{1}{4}$ the resistance. If the diameter is three times as great, the resistance of the wire will be only $\frac{1}{9}$ as great. Figure 32-11B shows why this is so. Wire Y, which is twice as thick as X, has four times the cross-sectional area, and thus it is equivalent to four parallel wires the size of X. Therefore it has four times the ability to carry the current, or $\frac{1}{4}$ the resistance of X. If wire Y were three times as thick as X, it would be equivalent to 3×3 or 9 wires the size of X, and Y would have $\frac{1}{9}$ the resistance of X. So we see that:

The resistance of a wire is inversely proportional to its cross-sectional area.

15. Resistance of different materials

The kind of material in the wire is another factor, besides length and thickness, that affects its resistance. Figure 32-12 shows two similar lamps connected in separate circuits in series with wires of the same length and thickness

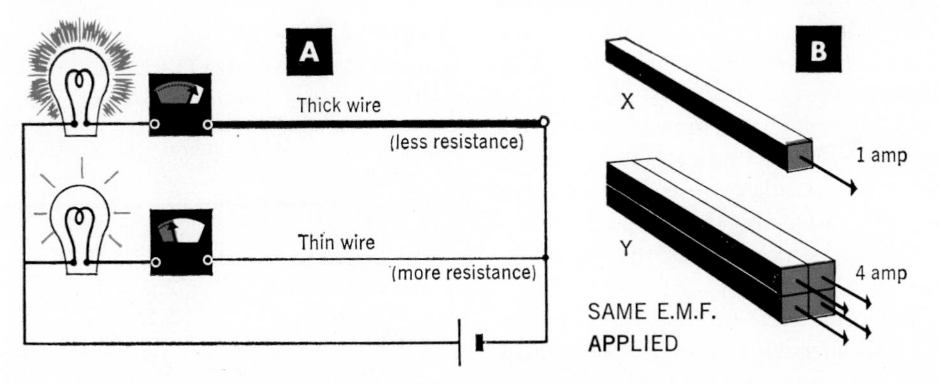

32-11 Thick wires (A), like thick pipes, permit more current to flow and thus have less resistance. If the cross-section area of a wire is made 4 times as great, the wire becomes equivalent to 4 single wires and thus has one-fourth the resistance (B).

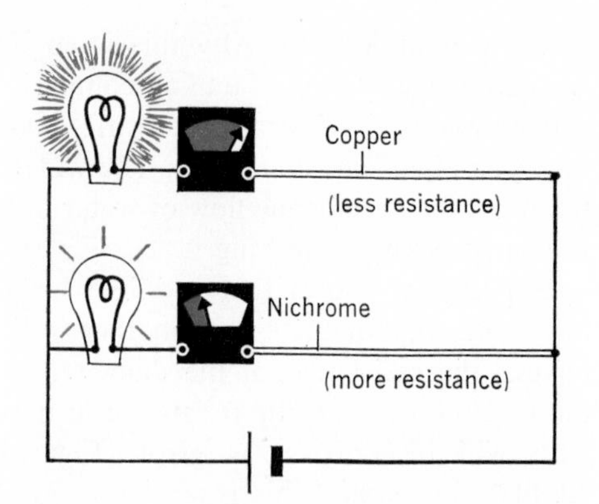

32-12 Does copper or nichrome have less resistance? Selecting copper and nichrome wires of equal length and diameter, we find that more current flows in the lamp and meter with the copper wire in series than with the nichrome. Therefore copper has less resistance.

but of different materials (copper and nichrome). The lamp in series with the nichrome is dim, while the one in series with the copper is bright. Since both branches of the circuit have the same voltage applied to them, we conclude that for the same dimensions nichrome has a higher resistance than copper.

Notice how we "control" these experiments to determine the effect of various factors. If we want to find the effect that the length of wire has on resistance, we compare two wires which are alike in all respects except length, and we apply the same voltage to each. To note the effect of thickness, we select wires alike in all respects except thickness. By varying only one factor at a time, it is possible to find the effect that each factor has on resistance. In other words, the one factor which is varied may then be postulated as the factor which causes the effect observed.

If the temperature of a wire changes, its resistance also changes. The resistance generally increases with temperature. This effect is not noticeable when the temperature change is small. But when a large change in temperature occurs, as in Fig. 32-13, then the change in resistance is very noticeable. It is for this reason that Ohm's-Law experiments with lighted lamps are often unsatisfactory. The hot filament of a lamp has many times the resistance of the unlit lamp. Can you figure out how to use this principle to design an electric thermometer?

16. Comparing the resistance of materials

Table 13 gives the resistance of wires 1 centimeter long and 1 square centimeter in cross-sectional area. The resistance of such a wire of standard size is called the **resistivity** (or **specific resistance**) of the material.

We see from this table that silver is the best conductor of all. But because silver is expensive, it is used only in special cases where low resistance is very important. Copper is almost as good a conductor as silver and has the advantage of being much cheaper; hence it is the material most frequently used for conducting electricity. Where weight and cost are both important considerations, aluminum is used, since the weight of an

TABLE **13**

RESISTIVITY OF MATERIALS

The resistivity (specific resistance) is given in ohms per centimeter of length for a cross-sectional area of 1 square centimeter, at a temperature of 20°C.

Silver	1.6×10^{-6}
Copper	1.7×10^{-6}
Aluminum	2.8×10^{-6}
Tungsten	5.5×10^{-6}
Iron	10×10^{-6}
German silver	33×10^{-6}
Nichrome	100×10^{-6}
Glass	2×10^{13}

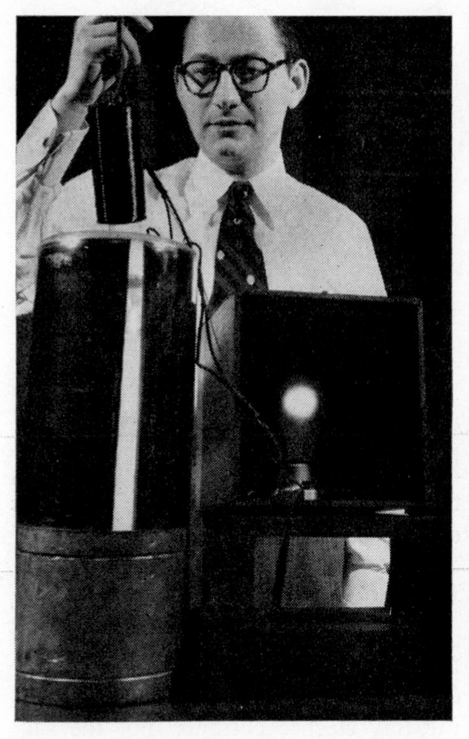

Memorial Hospital, New York

32-13 When the coil in series with the lamp (left) was dipped into liquid nitrogen at a temperature of −319° F (right), the lamp became brighter. The resistance of the wire in the coil decreased at the lower temperature.

aluminum wire is only about $\frac{1}{3}$ that of a copper wire of the same size. In transmission lines across the country, where the wires suspended from tower to tower must be as light as possible, aluminum is generally used. Nichrome is a high-resistance wire often used in electric toasters and heaters. Tungsten (wolfram) is used for the filaments of electric light bulbs. It would be very instructive for you to examine the various electric household appliances in your home to discover where high-resistance wire is most useful.

17. Computing wire resistance

Using Table 13 we can find the resistance, *R,* of wires of any length, *L,* and cross-sectional area, *A,* by the following formula.

$$R = \frac{L \times \text{Resistivity}}{A}$$

If *L* is in centimeters and *A* in square centimeters, then *R* will be in ohms.

SAMPLE PROBLEM

Find the resistance of 3 meters of copper wire whose cross-sectional area is 0.0001 square centimeter.

Substituting,

$$R = \frac{300 \text{ cm} \times 0.0000017 \text{ ohm/cm/cm}^2}{0.0001 \text{ cm}^2}$$

$$= \frac{3 \times 10^2 \times 1.7 \times 10^{-6}}{10^{-4}}$$

$$= 5.1 \text{ ohms}$$

STUDENT PROBLEM

Find the resistance of a tungsten wire 10 m long and 0.05 cm^2 in cross-sectional area. (ANS. 0.11 ohm)

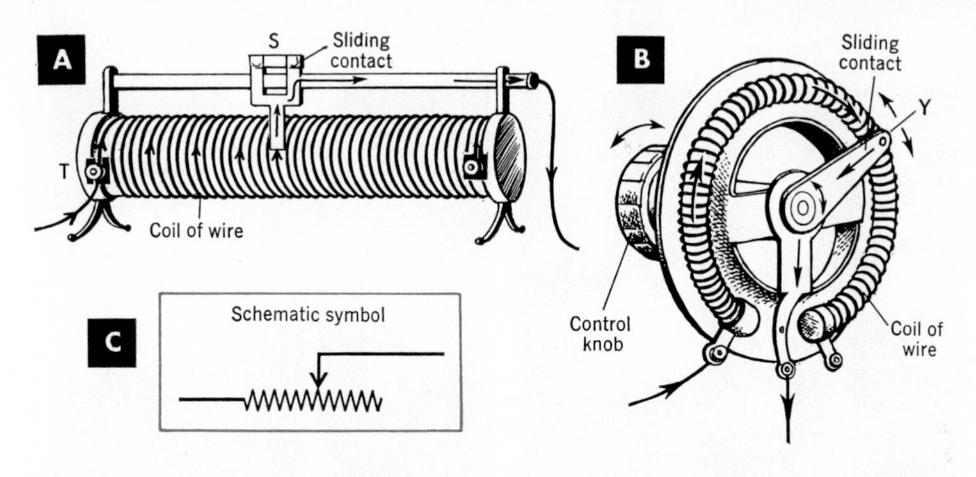

32-14 Currents can be changed in circuits by using variable resistances called rheostats.

18. Rheostats

For many electrical purposes it is necessary to insert a variable resistance in a circuit to enable one to change the current. Figure 32-14 shows one type of variable resistance, or **rheostat.** Coils of insulated high-resistance wire are wound around an insulated cylinder. A metallic slide rests on an exposed part of the wire, making contact with it. The resistance then depends upon the length of wire between the end terminal T and the metal slide S. Rheostats are used as controls on radios, motors, and many other types of electrical apparatus.

Electric meters

Up to this point we have obtained a great deal of information by using the ammeter and the voltmeter, but we have not yet explained how the instruments themselves work. How do they make their measurements? How are they constructed? How do we connect them in a circuit?

Electrical energy is easily changed into other forms of energy. We convert it into mechanical energy by means of electromagnets. In electroplating we change it into chemical energy. We can use electrical energy to produce heat and light. To measure electrical quantities, the motion produced by the magnetic effect is most commonly used.

19. The galvanometer

The most important measuring instrument is the **galvanometer** (găl′vȧ·nŏm′-ê·tẽr) (Fig. 32-15). It is used to measure tiny electric currents. For the sake of simplicity it may be thought of as a DC motor which can rotate only part of a turn because it lacks a commutator. It has a very low resistance.

The current to be measured passes through a coil which is wound around a soft-iron armature, A, pivoted between the poles of a permanent magnet. Whenever current flows, poles are produced in the moving coil at N_1 and S_1. As the coil rotates it stretches a hair spring. An increase in current produces a stronger attraction between these poles and the field poles at N_2 and S_2 and causes a greater rotation of the coil. The coil stops turning when the backward pull of the spring balances the magnetic attraction. A pointer attached to the coil measures the rotation of the coil. AC cannot be used because the armature would no sooner start to rotate in one direction than the reversal of the current would start it rotating in the opposite direction. Hence it would remain stationary.

20. Connecting the ammeter

In all of the experiments in which we have used an ammeter, its connection in the circuit was always in series. This was necessary because all the current to be measured had to pass through the ammeter (Fig. 32-16A). If we attempted to use a galvanometer instead of an ammeter in order to measure current, the galvanometer would in all probability be ruined.

There are two reasons why we cannot use the galvanometer directly in series. First, it is a sensitive instrument and is so constructed that a tiny current is sufficient to move the pointer to the end of the scale. Let us assume that 0.01 ampere can move the galvanometer pointer the full scale, that is, to the end of the dial. If the current we are measuring is more than this amount, as it usually is, it is too great for the galvanometer to withstand, and the instrument, of course, is ruined.

Second, the galvanometer has a resistance of its own. Hence when we connect a galvanometer into a circuit its resistance reduces the very current it was intended to measure. As a result our measurements will be incorrect.

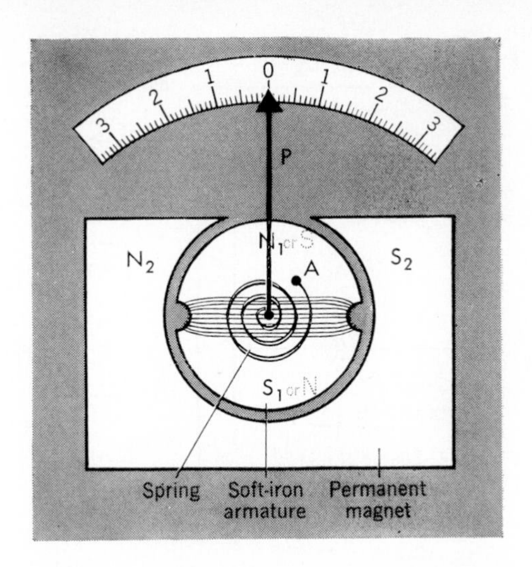

32-15 The galvanometer, which is used to detect current, is simply an electromagnet rotating between the poles of a permanent magnet. The greater the current the greater the attraction and rotation of the electromagnet against the pull of the spring.

21. Converting a galvanometer into an ammeter

How can we convert a galvanometer into an ammeter? Suppose we want the pointer of the meter to reach the end of the scale when 1 ampere flows through the instrument. Suppose, too, that only 0.01 ampere flowing in the coil will make the pointer move full scale. Then we must have some way of by-passing 0.99

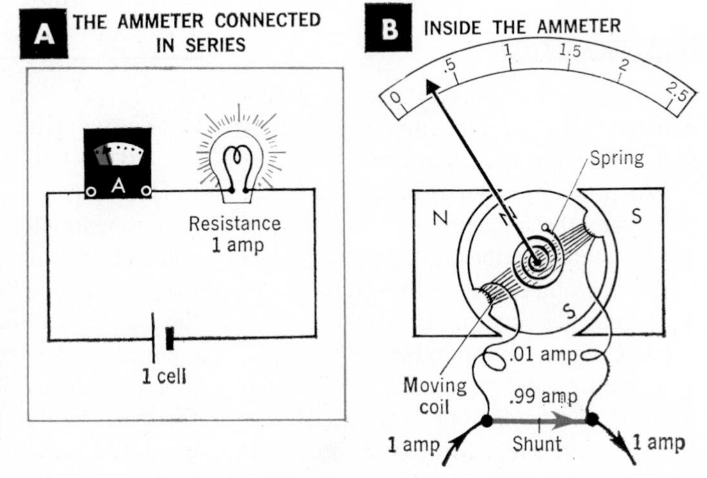

32-16 Ammeters are always connected in series in a circuit (A). But inside the ammeter (B), a low-resistance shunt, through which most of the current flows, is connected in parallel with the coil.

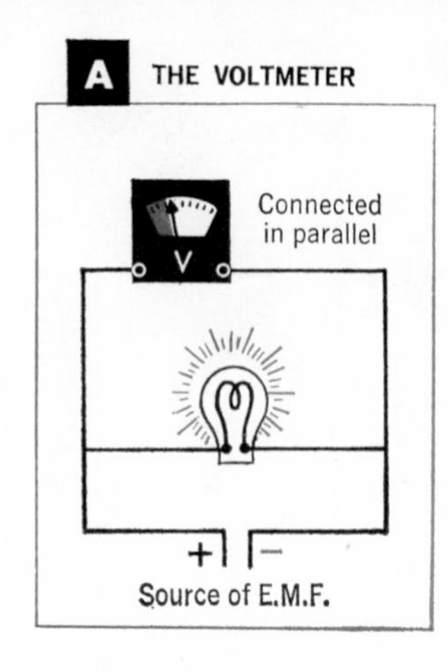

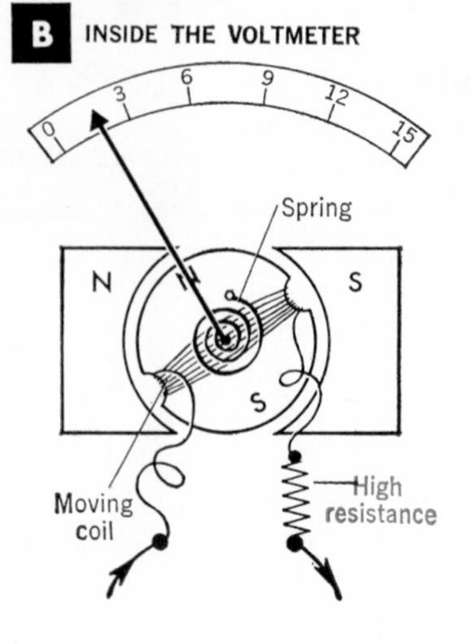

32-17 Unlike the ammeter, a voltmeter is connected in parallel with a device (A). Inside the voltmeter (B) there is a high resistance in series with the moving coil.

ampere away from the coil, permitting only 0.01 ampere to pass through the coil. This can be done by providing a low-resistance **shunt** or by-pass, as shown in Fig. 32-16. Hence **an ammeter is simply a galvanometer with a low-resistance shunt in parallel with its coil.**

The shunt has such a small resistance compared to the rest of the circuit that the ammeter now has little effect in reducing the current in the circuit.

Suppose that the galvanometer coil has a resistance of 10 ohms and that only 0.01 ampere flowing through it will make the pointer move full scale. Knowing these two facts about a galvanometer enables us to design an ammeter of any desired range greater than 0.01 ampere.

SAMPLE PROBLEM

How can we convert such a galvanometer into an ammeter whose full-scale deflection is 10 amperes?

When the ammeter reads 10 amperes, there must be a current of 0.01 ampere flowing through the coil, whose resistance is 10 ohms. Thus, the voltage drop, E, across the coil is $E = 0.01 \times 10 = 0.1$ v. Connected in parallel with the coil is a shunt of low resistance, R. Through it flows the remaining current, $10.0 - 0.010 = 9.99$ amp. Moreover, since it is in parallel with the coil, the voltage drop, IR, across it will be the same as the voltage drop of 0.1 volt across the coil, or

$$9.99\ R = 0.10\ \text{v}$$

$$R = \frac{0.10}{9.99} = 0.01\ \text{ohm}$$

Thus, a shunt of 0.01-ohm resistance will convert the galvanometer into an ammeter whose full-scale deflection is 10 amperes. Notice that our method is summed up in the expression

$$I_{coil}R_{coil} = I_{shunt}R_{shunt}$$

STUDENT PROBLEM

What shunt resistance is required to convert the same galvanometer into an ammeter whose full-scale deflection is 1 amp? (ANS. 0.101 ohm)

22. Converting a galvanometer into a voltmeter

How can we convert a galvanometer into a voltmeter? We must be able to connect this instrument directly across the source of E.M.F. if it is to measure the potential difference or voltage. If we connect a low-resistance galvanometer directly across a battery or generator, the current through it will be far too great and will burn out the galvanometer. Suppose we are to measure 100 volts, and the resistance of the galvanometer is 20 ohms. If we place the galvanometer directly across the 100-volt source without any protection, the current flowing through it will be:

$$I = \frac{E}{R} = \frac{100\text{ v}}{20\text{ ohms}} = 5\text{ amp}$$

If this particular galvanometer coil can take a maximum of only 0.01 ampere, the large current of 5 amperes would ruin it. Therefore, we need a high resistance in series with it to limit the current to 0.01 ampere. **A voltmeter is simply a galvanometer having a high resistance in series with it** (Fig. 32-17).

Voltmeters are often designed with several different resistances (each having a separate terminal) so as to give various ranges of voltage measurement.

The differences between a voltmeter and an ammeter are summarized in Table 14 below.

SAMPLE PROBLEM

How much resistance in series with the galvanometer coil above will reduce the current to 0.01 ampere (and hence give full-scale deflection) when 100 volts are applied? We find that the total resistance needed to keep the current to 0.01 ampere when 100 volts are applied is

$$R = \frac{E}{I} = \frac{100\text{ v}}{0.01\text{ amp}} = 10{,}000\text{ ohms}$$

The galvanometer coil already has a resistance of 20 ohms. So if we put a resistance of 9,980 ohms in series with the 20-ohm resistance of the coil itself, we shall have a total resistance of 10,000 ohms. The current in the voltmeter is now limited to 0.01 ampere with 100 volts, and the pointer will swing the full scale. In this way we have constructed a voltmeter whose range is 100 volts.

STUDENT PROBLEM

What series resistance is required to convert the same galvanometer into a voltmeter whose full-scale deflection is 25 v? (ANS. 2,480 ohms)

23. Electrical measurements

To measure the resistance of an electrical device we connect an ammeter in series and a voltmeter in parallel with the device (Fig. 32-18) to measure the current and voltage. We can then calculate the resistance by using Ohm's Law. If the voltage across a device is 116 volts and the current through it is 2.3 amperes, the resistance would be

$$R = \frac{E}{I} = \frac{116\text{ v}}{2.3\text{ amp}} = 50.5\text{ ohms}$$

We can use the same arrangement to measure the power of electrical devices. We know that Power = E.M.F. × Current (P = EI) (page 396). If we ar-

TABLE **14**

COMPARISON OF AMMETER AND VOLTMETER

	Ammeter	Voltmeter
Method of connection in the circuit	in series with device	in parallel with device
Resistance used to convert from the galvanometer	low resistance	high resistance
How this resistance is connected inside the instrument	low resistance connected in parallel with moving coil	high resistance connected in series with moving coil

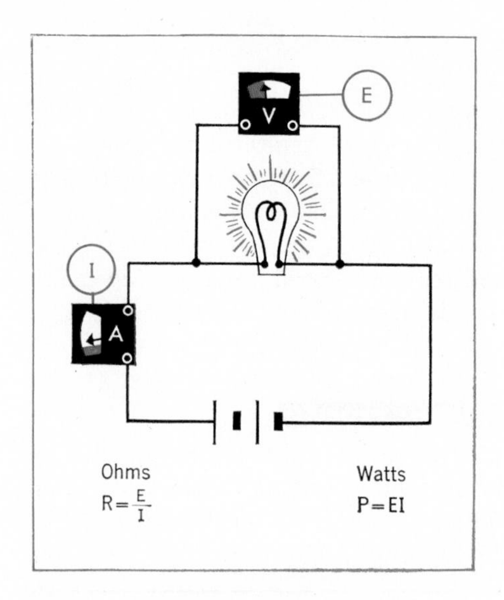

32-18 An ammeter and a voltmeter connected in a circuit as shown here can be used to measure resistance or power.

range our ammeter and voltmeter so as to measure the amperes and volts in a given electrical device, then we can easily obtain the power in watts by multiplying the two readings.

The electric meter in a home must measure not watts of power but kilowatt-hours of electrical *energy* used. It is called a **watt-hour meter.** Time is the important additional factor in this measurement. It makes a big difference in cost if we use a lamp for 1 hour instead of 100 hours. Electrical energy is measured by a small electric motor whose revolutions turn several pointers. The total number of revolutions of the motor is proportional to its speed times the time it has been running. But the motor is built in such a way that its speed is proportional to the power being drawn from the power line. Hence the total number of revolutions is proportional to power times time —or electrical energy. This quantity (usually kilowatt-hours) can be read off the several dials on the meter (Fig. 32-19). An aluminum disk attached to the motor rotates between the poles of a magnet and acts as a brake to stop the motor instantly when the current is turned off.

24. The Wheatstone bridge

The physicist, Wheatstone, devised a very accurate way of measuring resistance. His circuit is called a **Wheatstone bridge** and is shown in Fig. 32-20. Here R_x is the resistance to be determined, R_N is a resistance of known value, and R_A is a resistance which may be varied but whose value is accurately known at all times. R_B is also a resistance of known value.

If R_A is changed, the current, i_g, through the galvanometer will also change, and its direction will be reversed by some values of R_A. However, a value of R_A can always be found that makes $i_g = 0$. The bridge is then said to be "balanced" and we can work out a simple equation for R_x, the unknown resistance, under these conditions.

If the bridge is balanced so that $i_g = 0$, then there can be no potential difference between points D and B;

General Electric Co.

32-19 A modern watt-hour meter. How would you go about reading the number of kilowatt-hours of electricity used?

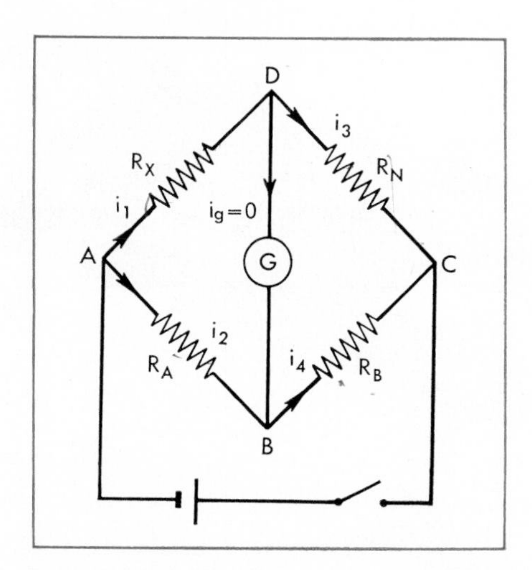

32-20 The circuit of a Wheatstone bridge, used for measuring an unknown resistance, R_X.

therefore the voltage drop across R_A equals the voltage drop across R_x. Hence, by Ohm's Law:

(1) $$i_1R_X = i_2R_A$$

Similarly, the voltage drop across R_N, which is i_3R_N, equals the drop across R_B, which is i_4R_B. Since $i_g = 0$, it must follow that $i_3 = i_1$ and $i_4 = i_2$. Therefore:

(2) $$i_1R_N = i_2R_B$$

Dividing equation (1) by equation (2):

$$\frac{i_1R_X}{i_1R_N} = \frac{i_2R_A}{i_2R_B}$$

$$\frac{R_X}{R_N} = \frac{R_A}{R_B}$$

$$R_X = \frac{R_A}{R_B} R_N$$

So we see that if R_A, R_B, and R_N are known accurately, R_x can be calculated accurately.

A Wheatstone bridge is sometimes made with R_A, R_B, and R_N all enclosed in a box. The different values of resistance are set by turning knobs (Fig. 32-21). A simpler and cheaper form of the circuit is known as a slide-wire bridge (Fig. 32-22). In this version, R_A and R_B are different sections of one long piece of nichrome wire stretched taut along a meter stick. A sliding contact on the wire connected to the galvanometer allows different lengths of wire to be used for R_A and R_B. If the cross-section of the wire is constant, R_A is proportional to the length L_A and R_B is proportional to L_B. Therefore, when the bridge is balanced:

$$R_X = \frac{L_A}{L_B} R_N$$

So it is not necessary to know the values of R_A and R_B in ohms if we can measure L_A and L_B accurately with the aid of the meter stick.

Most ammeters and voltmeters made from galvanometers are not accurate to better than about 2 per cent, and since resistance in the ammeter-voltmeter method is determined by two meter

Leeds and Northrup Co.

32-21 This box contains the three resistances needed to operate a Wheatstone bridge. What is the advantage of enclosing them in a box?

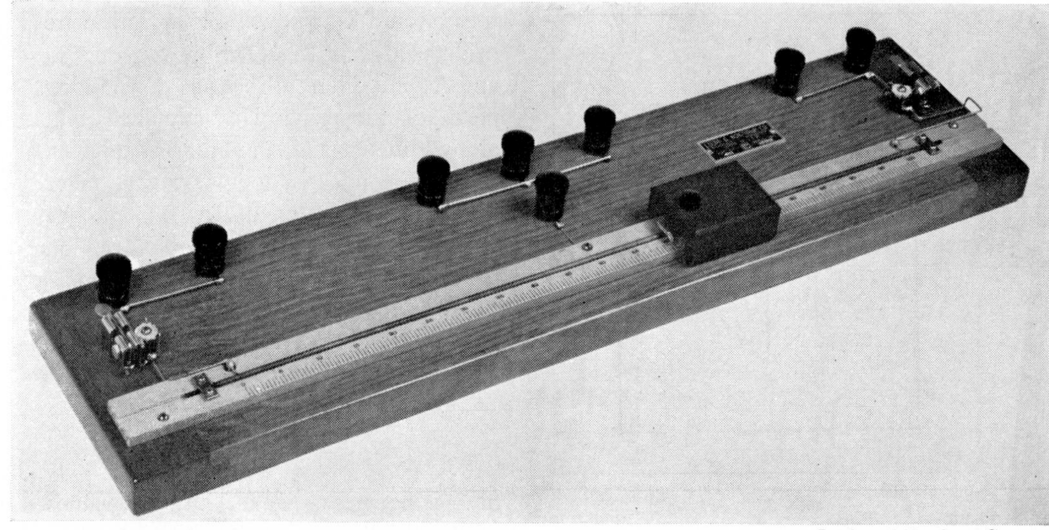

Leeds and Northrup Co.

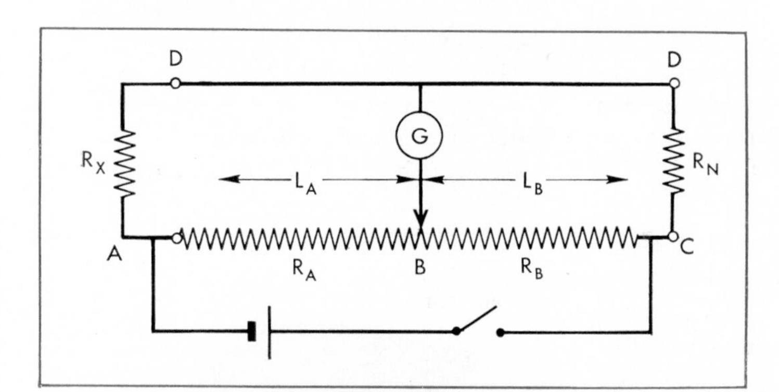

32-22 AC is a length of nichrome wire stretched taut along a meter stick. A sliding contact on the wire connected to the galvanometer allows various fractions of the wire to be used for R_A and R_B. The photograph shows a student's slide-wire Wheatstone bridge.

readings (E/I), the error can be as high as 4 per cent. In the Wheatstone bridge, however, all we need is a sensitive galvanometer that will respond to minute currents. It does not have to be accurate, since we balance the bridge by adjusting the current to zero. The accuracy of the Wheatstone bridge depends only on the accuracy of the values of R_A/R_B and R_N. The result is that the bridge method can easily give an accuracy within better than 1 per cent. For still more precise measurements other "balanced" circuits similar to the Wheatstone bridge can be used. Many electrical quantities such as inductance and capacitance (Chapter 35) are measured by comparable circuits.

Going Ahead

THE VOCABULARY OF PHYSICS

Define each of the following:

resistivity	specific resistance
rheostat	Law of Conservation of Charge
galvanometer	terminal voltage
ammeter	internal resistance
shunt	Wheatstone bridge
voltmeter	watt-hour meter
voltage drop	

THE PRINCIPLES WE USE

1. Two lamps exactly alike are connected in a circuit so that current flows in both. In one experiment they are connected in series and in another experiment they are connected in parallel.

Compare the current flowing in each case (parallel and series) with the current if only one lamp were used. Compare the resistance of both lamps in series and in parallel with that of one lamp. Compare the voltage applied to each lamp in series and in parallel with that applied when each of them is used alone.

2. How are total resistance and current in a series circuit calculated if the individual resistances are known?
3. How is the total current calculated in the main line of a parallel circuit if the current in each branch is known?
4. Describe two methods of calculating voltage drop in a series circuit in which current and resistances are given.
5. State four factors that affect the resistance of wires. Tell how the resistance is changed if the factors are changed.
6. Draw a diagram showing how an ammeter and a voltmeter are connected in a circuit to make measurements of the current through and the potential difference across a lamp.

Explain why

7. Current in the main line is greater if several devices are in parallel than if they are in series.
8. All devices in a series circuit have the same current.
9. Current cannot be "used up."
10. The total resistance of a parallel circuit decreases as more devices are placed in the circuit.
11. A thick wire has less resistance than a thin wire of the same material and length.
12. Aluminum is used for wires in transmission lines.
13. Silver is sometimes used for electric wires even though it is expensive.
14. An ammeter should have a very low resistance.
15. A voltmeter should have a high resistance.
16. A voltmeter should be connected in parallel in a circuit, but an ammeter should be connected in series.

APPLYING PRINCIPLES OF ELECTRICITY

1. Name three good conductors of electricity and tell under what conditions each is used. Why is each used?
2. Describe a common type of rheostat and give two uses for it.
3. Describe how a galvanometer operates.
4. What change in a galvanometer must be made in order to convert it into an ammeter? into a voltmeter?
5. Describe how you would measure the resistance of a lamp. How would you measure the power of the lamp?
6. Why is it important to have low-resistance wire for main lines that carry current to consumers?
7. How would you use a Wheatstone bridge circuit to measure resistance?
8. A pupil made a mistake in the laboratory and connected an ammeter in parallel with a lamp. What do you think happened? What would happen if a voltmeter were placed in series with the lamp?
9. Precisely how would a galvanometer be ruined if it were connected across a source of voltage?
10. If n equal resistances of r ohms each are connected in parallel, the total resistance $R = \frac{r}{n}$. Prove it.
11. Find out how an ohm-meter works.

SOLVING PROBLEMS IN ELECTRICAL MEASUREMENT

1. Two devices of 6 ohms and 10 ohms resistance are connected to an 80-v line. Calculate the current that would flow in the line if the devices were in series; calculate the current that would flow if they were in parallel.
2. Repeat problem 1 using 3 resistances of 2 ohms, 8 ohms, and 5 ohms connected to a 40 v line.
3. Three devices of 15 ohms, 20 ohms, and 30 ohms are in parallel in a 120-v circuit. (a) Find the total current. (b) What single resistance (in place of all three devices) will conduct the same current?

4. A certain wire has a resistance of 80 ohms. What would be the resistances of a wire 4 times as long? of a wire 4 times as thick?
5. Which has greater resistance, a silver wire 10 m long and 1 mm^2 in cross-sectional area, or an aluminum wire 30 m long and 2 mm^2 in cross-sectional area?
6. Which has greater resistance, an iron wire 9 ft long and 0.30 in^2 in cross-sectional area, or a nichrome wire 6 ft long and 0.50 in^2 in cross-sectional area?

---

7. Calculate the potential difference across each of the devices in Problem 1 when in series.
8. Calculate the potential difference across each of the devices in Problem 2 when in series.
9. Calculate the current through, and the potential difference across, each of the devices in Problem 3 if they are in series.
10. Three resistances of 40, 60, and 100 ohms are connected in series in a 120-v circuit. (a) How much current flows through each? (b) What is the potential difference across each? (c) What current would flow through each if these resistances were in parallel?
11. A carbon-arc lamp with a resistance of 3.5 ohms is designed to operate with 12.5 amp of current. If it is used on a 120-v circuit, how much resistance must be placed in series for the lamp to operate properly? What is the potential difference across the lamp?
12. Two resistances of 10 ohms each are connected in parallel. The parallel combination is then placed in series with another 10-ohm resistance and connected to a 90-v battery. How much current flows in the main circuit?
13. If the battery in Problem 12 has an internal resistance of 2 ohms, how much current will flow in the main circuit?
14. A 30-ohm and a 60-ohm bulb are connected in series to a 120-v source. Find (a) the total resistance, (b) the total current, (c) the current through each bulb, (d) the voltage across each bulb, (e) the power developed by each bulb, (f) the total power developed.
15. Repeat Problem 14 with the bulbs connected in parallel.
16. A 30-w bulb and 60-w bulb are each connected to a separate 120-v outlet in the laboratory. Find (a) the voltage across each bulb, (b) the current in each bulb, (c) the total current flowing in the main wires (assume no other appliances are connected), (d) the resistance of each bulb. (e) Which bulb glows brighter? Why?
17. The 30- and 60-w bulbs of Problem 16 are connected in series to the same outlet. Find (a) the total resistance, (b) the total current, (c) the current flowing in each bulb, (d) the voltage across each bulb, (e) the power developed by each bulb. (f) Which bulb glows brighter? Why?
18. A 25-ohm and a 60-ohm resistance are connected in parallel, and the combination is connected in series with a 50-ohm resistance and a 12-v battery. Find (a) the current flowing through each resistance, (b) the voltage drop across each resistance.
19. Three appliances are connected in parallel across a 120-v circuit. If the resistance of the first is 10 ohms and the second 15 ohms, and the total current drawn from the line is 20.6 amp, what is the resistance of the third appliance?
20. What is the total resistance of four 150-ohm lamps connected in parallel? How much current will flow through each of them when they are all connected in parallel across a source of 120 v?
21. Eight identical Christmas-tree bulbs are connected in series across a 120-v source. If 3 amp flow through the combination, find (a) the resistance of each bulb, (b) the voltage drop across each bulb, (c) the power of each bulb, (d) the total power.
22. Find the resistance of (a) 5 m of nichrome wire 0.01 cm^2 in cross-section, (b) 200 m of aluminum wire whose

cross-section is 0.3 cm^2, (c) 500 m of copper wire whose *diameter* is 0.01 cm.

23. A rheostat is to have a maximum resistance of 150 ohms. How many centimeters of German silver wire 0.002 cm^2 in cross-section will be needed?

---

24. The voltage across a 10.5-ohm heater plugged into the living room outlet is 115.5 v. The voltage at the fuse box of the house is 118 v. Find (a) the current through the heater, (b) the resistance of the wires from the fuse box to the heater, (c) the rate at which energy is being wasted in the wires that supply the heater, and (d) the full cost of operating the heater for 100 hr at 4¢ per kwh.

25. A galvanometer has a resistance of 25 ohms. It is deflected full scale by a current of 0.005 amp. Find the value of the shunt necessary to convert the galvanometer into an ammeter with a range of (a) 0.1 amp, (b) 0.01 amp, (c) 1.5 amp, (d) 10 amp.

26. Given the galvanometer in the preceding problem, find the series resistance necessary to convert it into a voltmeter whose full-scale deflection is (a) 0.5 v, (b) 3 v, (c) 150 v, (d) 500 v.

27. In the circuit shown in the figure below, find (a) the current through r_1, (b) the current through r_2, (c) the voltage across r_3, (d) the voltage across r_4.

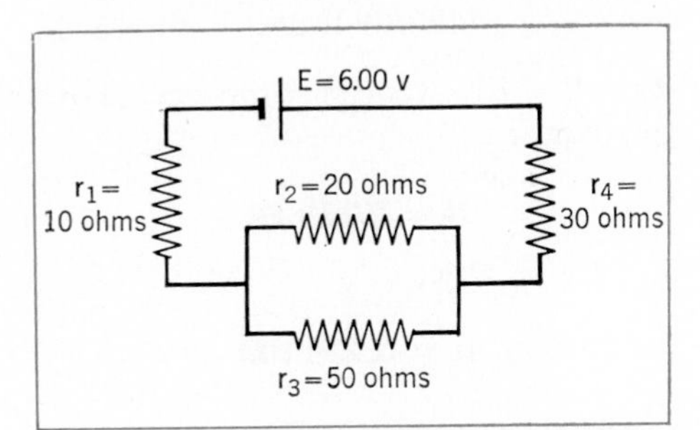

28. A dry cell with an E.M.F. of 1.5 volts and an internal resistance of 0.2 ohm is connected to a 1.3-ohm resistance. What is the current flowing and what is the terminal voltage?

29. Solve Problem 28 when three 1.5-v, 0.2-ohm cells are connected in series to a 1.3-ohm external resistance.

30. Solve Problem 28 assuming three cells are used and connected in parallel.

31. A high-resistance voltmeter connected to a cell reads 1.52 v. When a low-resistance ammeter is connected directly to the terminals of the cell in place of the voltmeter, it reads 30 amp. What is the internal resistance of the cell?

32. A cell is connected in series with a switch, an unknown resistance, *R*, and an ammeter. A voltmeter is connected to the terminals of the cell. With the switch open, the voltmeter reads 1.52 v. With the switch closed, the voltmeter reads 1.48 v and the ammeter reads 10.9 amp. What is the resistance of the cell? What is the resistance of the external resistance, *R*?

33. How much voltage must be applied to a 12-v automobile battery whose internal resistance is 0.05 ohm to charge it at a rate of 10 amp?

34. What is the charging current through the battery in Problem 33 if the charging voltage is 12.5 v?

35. In Problem 33, how much lead is re-formed on the negative plate in 20 min? (The electrochemical equivalent of lead is 0.00107 gm/coulomb).

36. How much energy is wasted per minute in charging the cell in Problem 33? What form does the energy take?

37. What series resistance must be used in charging a 6-v, 0.04-ohm battery from a 12-v source if the charging current is to be 15 amp?

CHAPTER

33

Heat and Light from Electricity

You have surely had the experience of trying to unscrew a bulb from its socket in a lamp, only to find it so hot you had to wait awhile. Obviously, both the light and the heat came from the electric current.

The change of electrical energy into heat energy is the simplest energy change that electricity undergoes. All we have to do is to send a current through a resistance. As we increase the current, the wire gets hotter until it finally glows. If the current is great enough, the wire may glow white-hot; this is what happens as current flows through the filament of a lamp.

Heaters

We can think of the heat produced by a current as the result of "electrical friction" in the wire. This friction is caused by the resistance that the wire offers to the passage of electrons. But this is only a rough picture of what happens. Actually the molecules of the wire are given heat energy when they are struck by the electrons flowing along the wire. This energy originated in the source of voltage.

1. How much heat?

How much heat does a wire produce when electricity flows in it? It has been found that when current flows in a wire, almost all of the electrical energy lost is converted into heat. Therefore, the heat energy produced in a wire will be equal to the electrical energy we put into it.

We saw in Chapter 21 that: Energy (or Work) = Power × Time. Electrical power is usually expressed in watts. If we measure the power in watts and the time in seconds, then we find the energy in watt-seconds (also known as joules):

$$\text{Energy} = \text{Power} \times \text{Time} = Pt$$

$$\text{Energy in watt-seconds} = \text{Watts} \times \text{Seconds}$$

If we wish to find out how many calories of heat energy, *H*, will be produced in a wire over a certain period of time, we must also know the electrical equivalent of heat energy. It has been found by experiment that 1 watt-second = 0.238 calorie. Hence

$$\text{Calories of heat energy, } H = 0.238\ P\ (\text{watts}) \times t\ (\text{seconds})$$

Since P = EI, we can write the above equation as

$$H = 0.238\ EIt$$

or, since E = IR,

$$H = 0.238\ I^2Rt$$

SAMPLE PROBLEM

How many calories does a 25-watt bulb give off to a room in a minute?

$$\begin{aligned} H &= 0.238\ Pt \\ &= 0.238 \times 25 \times 60 \\ &= 357\ \text{cal} \end{aligned}$$

SAMPLE PROBLEM

Suppose we wish to predict the time required for a 300-watt heater to heat 1,000 grams of water from 20°C to boiling (100°C). The problem is worked out at the bottom of the page, using a formula which was derived in Chapter 18.

According to the calculation, the 300-watt heater will require 1,120 seconds or 18.6 minutes to heat the water to the boiling point.

STUDENT PROBLEM

How long will it take to heat 1,100 gm of water from 20°C to 60°C using a 550-w electric heater? (ANS. 336 sec or 5.6 min)

2. Practical heating devices

Most electric heaters are simply coils of resistance wire through which current passes, causing the wire to become red-hot. If you examine a toaster, broiler, or heater operated by electricity, you will always find a long wire, usually made of nichrome. The wire is wound around a material that does not conduct electricity and that can withstand high temperatures without injury. For greater compactness the wire may be coiled. Reflectors are often provided to concentrate the heat radiation.

In electric irons the hot wire is imbedded in an insulated base; the heat is conducted through the metal of the iron to the clothing on the ironing board. Electric soldering irons are made in a similar fashion. Electric heating pads and blankets are also constructed in this way, the insulation used being flexible asbestos.

It should be remembered that although the heating coils are made of high resistance *materials,* nevertheless such a short *length* is used that its resistance is still *low*.

Most heating devices take much more current than the average lamp. The current in a 120-volt broiler is about 6 to 10 amperes as compared with less than 1 ampere in an ordinary lamp. The power used by the average heater or broiler may be 1,000 watts, or even higher.

The electric lamp

An **incandescent electric lamp** does not seem to have much resemblance to a heater, but the two devices are essentially the same. A lamp is nothing but a white-hot wire inside a glass bulb, whereas a heater is a wire that is only red-hot (Fig. 33-1).

3. Tungsten filaments

Lamps using glowing wires were made as early as 1845, but they did not work well because all known wires burned or melted before they got white-hot. Edison wanted to find a wire that would not burn or melt at high temperatures. It was easy enough to avoid burning. He simply surrounded the wire by a glass bulb from which the air had been pumped out. Now the wire could not

$$\text{Heat in calories} = \text{Specific heat} \times \text{Mass} \times \text{Change in temperature} = 0.238 \text{ Watts} \times \text{Seconds}$$

$$\Delta H = S \times m \times \Delta t = 0.238\,Pt$$

$$= 1 \times 1{,}000 \times 80 = 0.238 \times 300 \times t$$

$$t = \frac{1 \times 1{,}000 \times 80}{0.238 \times 300}$$

$$= 1{,}120 \text{ sec} = 18.6 \text{ min.}$$

U. S. Steel Corp.

33-1 This fiery steel bar is so hot that it glows. The filament of a lamp likewise glows because it is so hot. The lamp's filament is heated by the passage of electric current.

burn because there was no oxygen in the bulb.

But the problem of melting was harder to solve. Edison eventually found that certain carbonized threads gave good results. The carbon-filament lamp which he produced as a result of thousands of experiments with different kinds of carbonized filaments could operate at a temperature of about 1,900°C. Today, instead of carbon filaments, we use tungsten wires, which usually operate at 2,800°C. Tungsten is a metal (sometimes called wolfram) with one of the highest melting points known. Because of their higher operating temperature, tungsten-filament bulbs give almost 6 times as much light as carbon-filament bulbs for the same amount of electrical energy. It was not until scientists learned how to draw the metal tungsten into fine wires that it was possible to use it in making lamp filaments.

The filaments in the earlier bulbs would evaporate too rapidly, and therefore have too short a life, if operated at a temperature of more than 2,200°C. It may surprise you to learn that a hot metal wire can evaporate. You will recall that heat is the result of motion of molecules. At the high temperature of over 2,000°C the molecular motion is so great that many molecules simply jump out of the wire into the surrounding space. You have noticed the blackening of an old bulb. This is due to a coating of metal that has evaporated from the filament. A way was found to reduce evaporation, so that the bulb could operate at a higher temperature and with its efficiency raised. How was this done?

4. Gas-filled tubes

Suppose that we fill the bulb with an inactive gas like argon, which does not react chemically with tungsten. The gas atoms impede the evaporation of the tungsten wire. In this way the bulb can operate at higher temperatures for a longer period of time. Because of this improvement, modern tungsten bulbs give about two and a half times as much light for the same power as tungsten bulbs of the older vacuum type (Fig. 33-2).

Some photoflood bulbs used by photographers operate at higher temperatures than ordinary bulbs and thus give more light. But such bulbs are not suitable for most purposes because the filaments evaporate rapidly and the bulbs last only a few hours. The ordinary house bulb is designed to last about 1,000 hours.

Incandescent bulbs consume their rated power and give their best illumination when they are operated at the rated voltage. For example, if you examine a particular bulb, you might find on it the information 100 w—120 v. This means

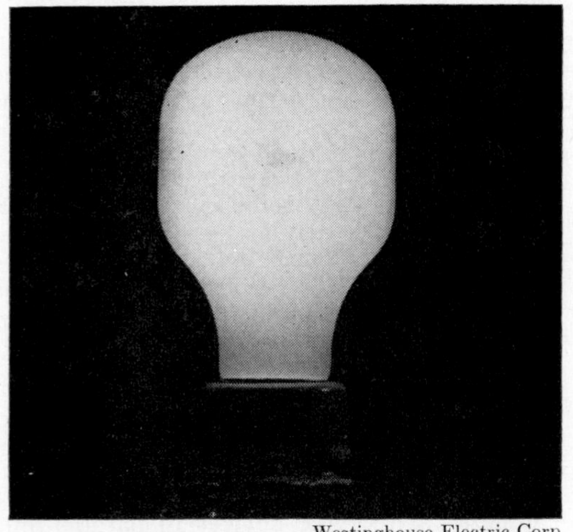
Westinghouse Electric Corp.

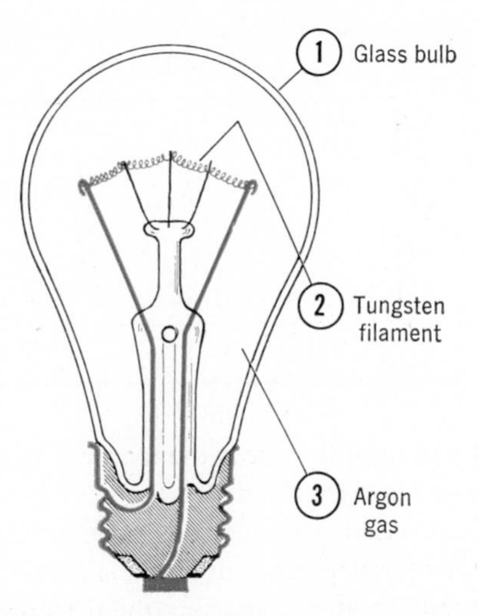

33-2 Modern lamps range in size from the giant 5,000-watt lamp used in airports to the pygmy lamp used by surgeons to light up the interior of the body. Before these modern lamps could be made, scientists had to overcome three major problems. The solutions to these problems are shown in the drawing on the right: (1) burning of the filament, (2) melting of the filament, and (3) evaporation of the filament. The new Eye-Saving white bulb (left) has an interior coated with millions of glass-deflecting particles.

that it consumes 100 watts when operated at 120 volts. At a lower voltage the current and power would be less; temperature would be lower, and the bulb would give less light. If the bulb were operated at a higher voltage than 120, although more light would be produced, the filament would evaporate too fast, and the bulb would not last.

5. Other sources of light

When current is sent through a gas at low pressure, the gas glows with its own characteristic color (see page 583). Mercury vapor in a tube at low pressure gives off a blue-white light, sodium vapor a yellow light, and neon gas a red light.

The light from **fluorescent lamps** is produced in a different manner. The 120 volts used in most house circuits is not great enough to cause the electrons to jump across the space between the terminals of a fluorescent lamp at the start of its operation. Therefore, the circuit is arranged as shown in Fig. 33-3. At first, when switch S is closed, current does not pass through the mercury vapor in the tube but through the two filament wires, A and B, which are connected in series through wires C and D. Filaments A and B heat up when current flows through

33-3 Fluorescent lamps light by the passage of electricity through low-pressure mercury vapor. The lamp must first be started by completing the side circuit CSD to heat the filaments A and B. Then this circuit is broken and the current flows through the mercury vapor from A to B.

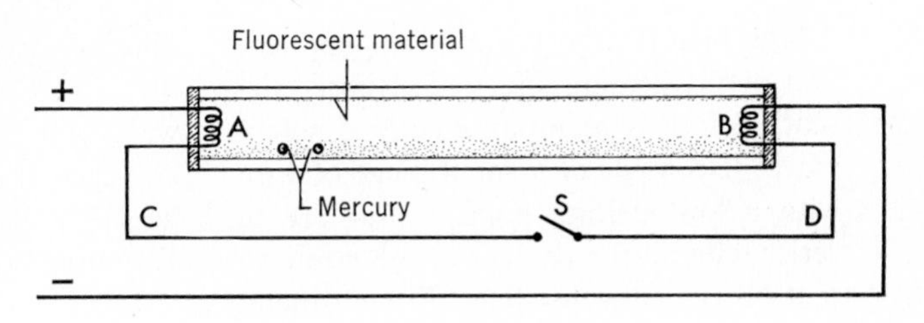

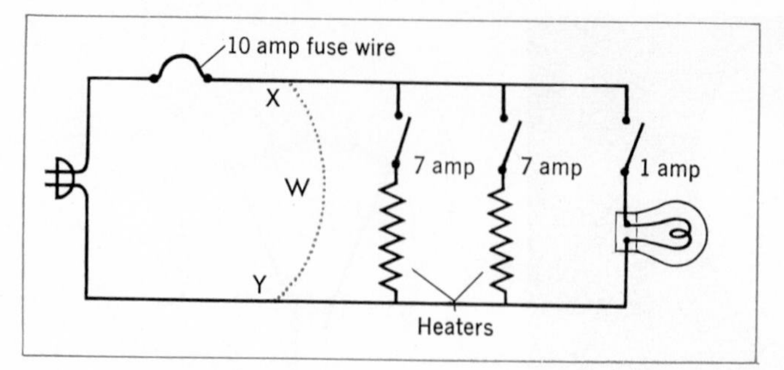

33-4 A 10-ampere fuse will melt and break the circuit when all three devices are turned on. This occurs because the 15-ampere current $(7+7+1)$ overloads the circuit. If a short-circuit connection XY is made with wire W, the very small resistance causes a very large current that melts the fuse wire in a flash and breaks the circuit.

them until they heat the mercury droplets and vaporize them. Now when the switch is released, the electrons flowing through the mercury vapor in the tube collide with mercury atoms, causing them to give off light.

A large part of the light given off by the mercury atoms is of the invisible ultraviolet kind (see Chapter 43). But there are chemicals which are able to emit visible light when they are struck by ultraviolet light. Such **fluorescent** chemicals coat the inside of the fluorescent lamp to make visible the light from the mercury vapor in the lamp.

In another variety of lamp, the carbon arc, the current is permitted to pass through carbon vapor between two carbon rods in series with a limiting resistance. The vapor between the carbon rods gets very hot and produces an intense source of light highly suitable for spotlights and movie projectors. A discussion of the operating costs of various common sources of light is found in Chapter 41.

Fuses

The **fuse** is a very important application of the heating effect of electric current. A fuse is really an electrical safety valve. It consists of a small piece of wire with a low melting point. Fuses are in series with all the devices in the home so that all current must pass through them.

6. Overloading the wires

Try this experiment. Connect a 10-ampere fuse or a piece of fuse wire in series with three devices in parallel and connect to the main line (Fig. 33-4). Insert a 100-watt lamp (using about 1 amp of current) in one socket, and heaters using about 7 amperes each in the other two sockets. Turn on the lamp first and the two heaters one after the other. Very soon after the second heater is turned on, the fuse wire melts, producing a bright spark. All of the devices stop operating at once. The circuit has been broken by the melting of the fuse wire. This happened because we were using a 10-ampere fuse (which melts if more than 10 amperes flow through it), and the current of the three devices added up to 15 amperes. When we turned on the second heater, the circuit became overloaded.

When a house is being built, the electrician selects electric wire of large enough diameter to conduct safely, without overheating, the amount of current the house will probably use. He usually divides the circuit in a house into a number of branches in parallel, each going to different groups of rooms, and each having a separate fuse. If the wire in the walls will safely carry no more than 20 amperes of current without overheating, he might install a fuse to melt at about 15 amperes.

Some new houses are protected not by

fuses but by a kind of switch called a *circuit breaker*. When the current drawn by the appliances in the house becomes dangerously large, the current is also great enough to operate either an electromagnet or a bimetallic strip whose action interrupts the current.

7. Fuse ratings

A physics teacher once had the experience of asking for 15-ampere fuses in a hardware store only to have the salesman give him 30-ampere fuses, remarking as he did so, "They're twice as good as the 15's and no more expensive." This may sound very reasonable, but after our discussion of fuses you can see that the salesman was all wrong. If the circuit in a house is designed to operate safely at only 15 amperes, a 30-ampere fuse permits more than the safe amount of current to flow, the wires may overheat, and a fire may result. When in doubt about what rating of fuse to use, it is better to play safe and select a lower rating rather than a higher one.

If a fuse melts ("blows"), try to find the cause and correct it before installing a new fuse. When you do replace the fuse, be sure you use another fuse of the same rating. If the new fuse also melts at once, you know the circuit is still overloaded; this may be because you are using too many appliances at one time. Or there may be a dangerous short circuit somewhere in the line, and you should remove the device causing the short circuit, or call an electrician immediately to get it fixed.

Electricity from heat

Most energy changes can be reversed. Just as electricity can produce heat, so heat can produce electricity. The method by which this is done is simple enough. Figure 33-5A shows that you merely have to heat two wires of different materials at their point of contact. As the temperature rises, the current increases. A junction of two different metals used in this way is called a **thermocouple.**

8. Electrical thermometers

The amount of electrical power produced by the thermocouple is too small to be practical for large-scale use, but it can be used to make a highly sensitive and convenient form of thermometer. Suppose, for example, that an airplane

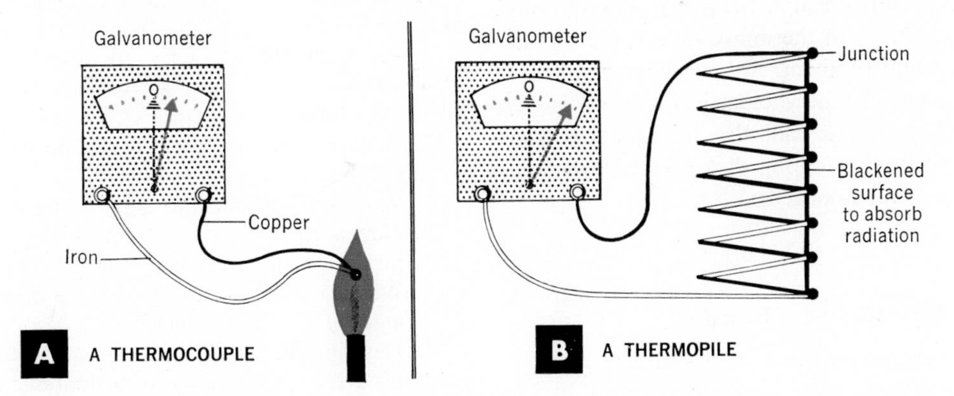

33-5 Producing electricity from heat is simplicity itself. Just heat the junction of two wires of different materials as in A. To increase the current, we make a thermopile by placing a number of junctions in series as in B. When used with a telescope, the thermopile has a blackened surface to absorb radiation from distant planets, enabling us to determine the temperature of the planet.

pilot wants to measure temperature inside the hot cylinder of his plane. A thin electrically insulated junction can be imbedded in the metal of the cylinder. This is connected to a galvanometer pointer on the dashboard calibrated in degrees of temperature.

The E.M.F. of the thermocouple can be magnified by constructing a series of junctions, as shown in Fig. 33-5B. The E.M.F. of each junction adds to that of the others. This arrangement is called a **thermopile.** It is sensitive to extremely small temperature changes. For example, when used with a telescope, which concentrates radiation, a thermopile can detect the heat of a candle fifty miles away. The heat rays from the candle are focused on the thermopile, and a galvanometer measures the increase in temperature. In this way astronomers have measured temperatures on the planets.

9. Obtaining electrical energy

The lamps and heaters that we have studied in this chapter would be of little value if we did not have some way of obtaining electrical energy cheaply.

How is electrical energy produced in powerhouses? What causes the electrons to rush through the wires to our homes to perform the many tasks which we demand of them? In our next chapter we shall see how electrical energy is generated. We shall also see how it is transmitted over long distances to consumers several hundred miles away.

Going Ahead

THE VOCABULARY OF PHYSICS

Define each of the following in one or more sentences:

fuse	thermopile
fluorescent lamp	incandescent lamp
thermocouple	circuit breaker

THE PRINCIPLES WE USE

1. How is heat produced from electricity?
2. What effect does increased current through a resistance have on the heat produced?
3. How much more heat will be produced if the current through a resistance is doubled? if it is tripled?

Explain why

4. Electricity produces more heat in an iron wire than in a copper wire of the same dimensions placed in series with it.
5. A 100-watt bulb is dimmer than a 50-watt bulb connected in series with it.
6. The temperature of the filament of a 100-watt lamp may be 2,800°C while the temperature of a 1,000-watt heater may be only 1,200°C.
7. A 20-ampere fuse used where a 10-ampere fuse is required can be dangerous.

APPLYING PRINCIPLES OF ELECTRICITY

1. Why is nichrome wire selected for heaters?
2. Mention five devices that make use of heat produced by electric current.
3. What advantages does electrical heating have as compared with other sources of heat?
4. What is one serious disadvantage of electrical heating?
5. Why do designers of incandescent lamps try to raise the temperature of the filament?
6. Why is the filament surrounded by a glass bulb?
7. Why is tungsten used for lamp filaments?
8. Why is argon put into lamp bulbs?
9. How are photoflood bulbs made to give more light than ordinary bulbs?
10. Why should a 120-volt bulb be operated only at that voltage?
11. Describe the operation of an arc lamp.
12. Describe the operation of a fluorescent lamp. Diagram the electric circuit.
13. Describe how a fuse is used to protect electrical circuits.
14. A fuse melts and breaks the circuit in your house. How could you tell whether

it was caused by an overload or by a short circuit?

15. Why is it foolish to jam a penny into a fuse socket as a temporary measure?
16. Describe three ways in which short circuits might occur at home.
17. Describe a practical use for the thermocouple.
18. Why should fuse wire have more resistance per inch than the copper wire it protects?
19. Why must the wires that lead current into the lamp bulb through the glass be selected of material with the same coefficient of expansion as glass?

SOLVING PROBLEMS OF ELECTRICITY

1. How many calories of heat will be produced by 2 amp of current flowing in a 20-ohm wire for 20 sec?
2. How much heat is produced in a wire carrying 5 amp in a 120-v circuit for 5 min?
3. A 600-w, 120-v hot plate heats 864 gm of water from 10°C to 30°C. Find (a) the number of calories of heat absorbed by the water (assume no losses), (b) how long it takes for the water to reach 30°C.
4. A 60-w, 120-v immersion heater is placed in a liter of water for two minutes. How many calories of heat does the water gain? (Assume no losses.)
5. If the water in Problem 4 starts out at 15°C, what will be its temperature at the end of the two minutes? (Assume no losses.)
6. How long would it take a 400-w electric heater to make 200 gm of water at 20°C begin to boil?
7. How many calories of heat will be produced by 8 amp of current flowing in a 60-ohm wire for 5 min?
8. An electric heater is being used to boil 9 kg of water in a tank. The heater has a resistance of 20 ohms and uses 6 amp of current. If no heat is lost to the surroundings, how long will it take for the water to reach the boiling point, starting from 20°C?
9. A heater, to be used on a 120-v circuit, is to produce 5,000 calories of heat per minute. (a) What current should flow in it? (b) What resistance should it have? (c) How much would it cost to operate this heater for 400 hr a month at a rate of 4¢ per kwh?

---

10. How long would it take for a 50-w heater to boil off 300 gm of water completely, starting at 20°C? Assume that no heat is lost to the surroundings.
11. (a) How much heat is developed in 1 min in a 40-w lamp operating on a 120-v circuit? (b) How much heat is developed in 1 min in a 60-w lamp? (c) Suppose the two lamps are placed in series. How much heat will develop in each lamp on the 120-v circuit? (d) How do the sum of the answers to (c) compare with the sum of the answers to (a) and (b)? Explain.
12. What should be the resistance of a nichrome-wire heater coil that is to produce 10,000 cal of heat an hour when used on 120 v?
13. A piece of nichrome wire with a resistance of 2 ohms per ft is to be made into a heater coil that can produce 2,500 cal of heat per min on a 120-v circuit. How long a piece should be used?
14. How long would it take a 300-w electric heater to raise 1 liter of ethyl alcohol from 20°C to its boiling point (78°C)?

CHAPTER

34

Generating Electrical Energy

Heavy snow, ice storms, or high winds sharply remind people that our modern world depends heavily upon electricity. When power lines are down, homes are darkened and left without modern conveniences and business life slows to a crawl.

Behind the radio, the telephone, the electric light, and electric transportation are the powerhouses containing giant **generators** which whirl at steady speed day and night, summer and winter, being stopped only for an occasional repair job.

How do these generators work? How do they produce electric energy?

Generators

A decade after Hans Christian Oersted had shown that electric current in a wire could produce magnetic effects, the English scientist Michael Faraday performed experiments to see if the reverse were true.

1. Faraday's experiment

To repeat one of the important experiments Faraday performed, connect a coil of insulated wire to a galvanometer as shown in Fig. 34-1. Then thrust a magnet back and forth inside the coil. The pointer of the galvanometer moves; current has been produced by a moving magnet. Current produced in this way is called **induced current.** The modern generator that produces most of our electric energy today is based on simple experiments like this one performed by Faraday in 1831.

Faraday's experiments disclosed several interesting facts. When he pushed the magnet into the coil, the pointer of the galvanometer swung one way. When he pulled it out of the coil, the pointer swung the other way, showing that the flow of current had been reversed. But

Roger Horowitz

34-1 Modern giant generators came from simple experiments like this one, first performed by Faraday in 1831. Can you describe this experiment?

when the magnet was motionless, the flow of current stopped. If he moved the coil instead of the magnet, current was also produced. Faraday found that he could increase the flow of current in three ways. He could move the magnet or the coil *faster*. He could make the magnet *stronger*. Or he could use *more turns of wire* in the coil.

It is not necessary to have many turns of wire in order to get current. Even a single loop of wire moved near a magnet will produce a tiny current, hardly noticeable in most galvanometers. The effect of more turns of wire is to increase the current.

Faraday noted that if he moved a wire near a horseshoe magnet in the direction of the arrow AB in Fig. 34-2, no current was produced at all. However, if he moved the wire in the direction AC, he obtained a maximum of current. When he moved the wire in other directions, the amount of current varied between the two extremes.

Faraday combined most of these facts into one short statement:

The E.M.F. (electromotive force or voltage) induced is proportional to the number of lines of force cut per second.

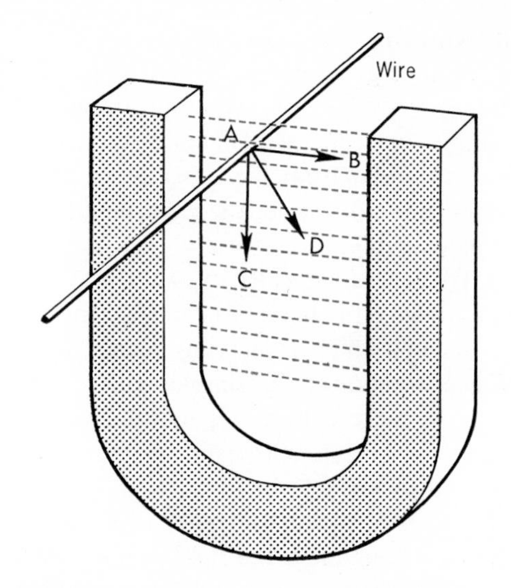

34-2 According to Faraday, E.M.F. is produced by lines of force being cut when a wire moves near a magnet (or vice versa). Which direction of motion of the wire AB, AC, AD will produce the most E.M.F.? Which direction will produce no E.M.F. at all?

2. Cutting lines of force

Let us examine this statement to see if it really does cover the facts revealed by Faraday's experiments. Faraday visualized the lines of force as being "cut" whenever the wire moved across them. For example, the lines of force of the horseshoe magnet shown in Fig. 34-2 run from the north pole to the south. A wire moving in the direction of AC cuts across the lines of force, producing an E.M.F. But when the wire moves parallel to the lines of force, in the direction AB, then no lines of force are cut and no E.M.F. is produced. When the wire moves at an angle to the lines of force, in the direction AD, it cuts fewer lines of force per second than when it moves at the same speed in the direction AC, and we therefore find that less E.M.F. is produced.

Likewise if the magnet moves faster, it cuts more lines of force per second and so produces more E.M.F. If we use a stronger magnet, there are more lines of force to cut (because they are more concentrated) and so the production of E.M.F. is greater. The use of a coil instead of a single wire increases E.M.F. because each turn produces its own E.M.F., and since the turns of a wire are in series, the E.M.F. increases in proportion to the number of turns. Thus we see that Faraday's simple rule covers most of the facts he observed.

If 100 million (10^8) lines of force are cut per second, the E.M.F. induced will be 1 volt.

3. Making a generator

A huge generator in a modern power-house seems a far cry from Faraday's simple experiments. To understand how one followed from the other let us consider one of Faraday's experiments in greater detail.

To generate electricity we start with a wire which is moving near a magnet or a magnet which is moving near a wire. Since rotation is a simple, convenient way to cause continuous motion, we arrange a rectangular coil of wire, the **armature** (EFGH in Fig. 34-3), so that it rotates between the poles of a strong horseshoe magnet, the **field magnet.** We get current out of the coil by connecting its ends to two **slip rings,** R and S, that rotate with the coil. Stationary **brushes,** T and U, against which the slip rings slide, then conduct this current to the outside without twisting the wires.

Let us follow the part EF of the coil in Fig. 34-3 as it moves around in a circle. We can follow the coil more clearly in Fig. 34-4A, where the wire EF in Fig. 34-3 is represented as a point because it is perpendicular to the paper. Let us assume that the wire moves through the angle JZK, taking one second to do so. The magnetic lines of force are pictured as straight lines drawn from the north pole of the magnet toward the south.

In position J, the wire is moving parallel to the lines of force. The vector arrow at J shows the velocity at this point. We see that no lines of force are cut, and hence no E.M.F. is produced. In position K (after the first second), the velocity arrow shows that the wire is cutting lines of force at the average rate of 4 per second, as you can see by counting them. In position L, the wire is cutting a perpendicular path across the lines of force; it thus cuts them at the maximum rate, that is, 9 per second. So in this position we have the maximum E.M.F. In position M, the E.M.F. is less than at L, since the vector arrow shows that only

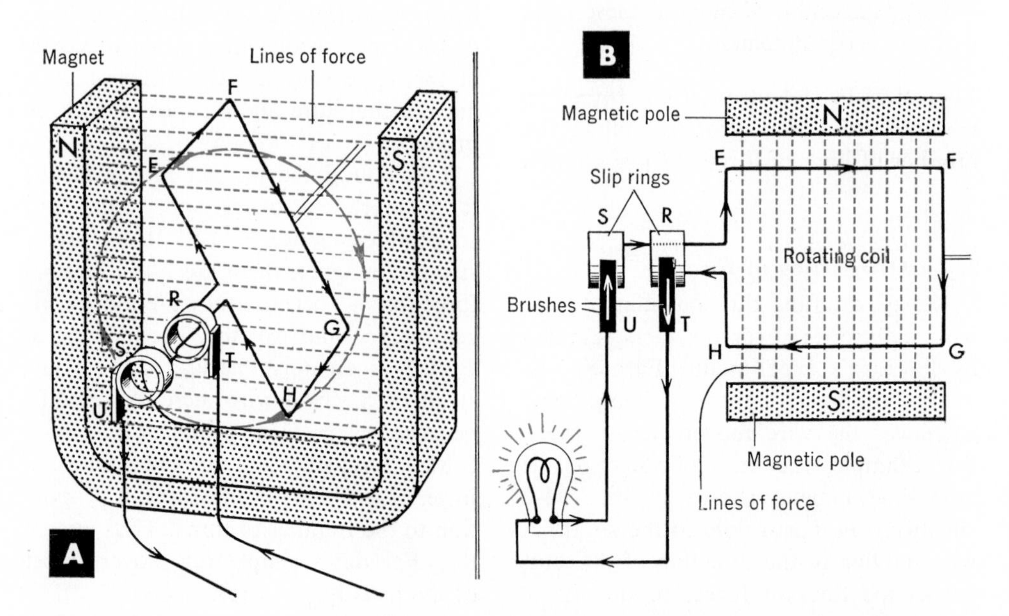

34-3 Side view (A) and top view (B) of a simple generator. The coil rotating between the poles of a magnet cuts lines of force to produce current that flows through slip rings S and R and brushes U and T to a lamp or other device.

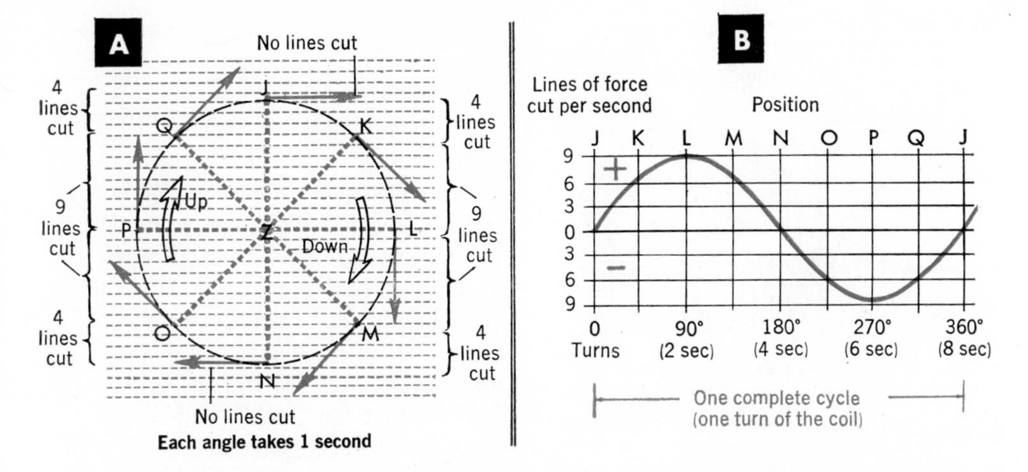

34-4 In A imagine a wire at right angles to the page at J and moving in a circle from J to K to L to M, N, O, P, Q, and back to J. Can you see why the current increases and decreases according to the graph shown in B?

4 lines of force are being cut per second. In position N, at the bottom of the diagram, the wire is again moving parallel to the lines of force, and the production of E.M.F. is reduced to zero.

Now the wire starts on its way up the other side. This reversal of the direction of motion reverses the direction of E.M.F. in the wire. Except for this reversal in the direction of E.M.F., the lines of force are cut in exactly the same way as before. Thus at position N no lines of force are being cut at all, 4 are being cut per second at position O, 9 at position P, 4 at position Q, and again none at all at position J. So we are back at the point where we started. In our experiment we have completed one **cycle.**

4. Alternating current

We can picture the facts stated in the previous section in the form of a graph, as shown in Fig. 34-4B. This wave-shaped graph is known as a **sine curve;** it plays an important role in many branches of scientific study and mathematics. It gives us a picture of voltage and current changes in a circuit to which a generator is connected. In the situation we have been studying, the sine curve shows the changes of an alternating current (AC) with time.

Thus far in our discussion we have considered the action of only one part of the coil, the part marked EF in Fig. 34-3. Let us see what happens in the other parts of the coil. You notice that when part EF is moving upward, part GH is moving downward. So the E.M.F. generated in GH always acts in an opposite direction to that generated in EF (Fig. 34-3B). The final result is that the E.M.F.'s in the two parts of the coil help each other in pushing current around the loop. If EF and GH both produced E.M.F.'s which cause current to flow in the same direction, the two E.M.F.'s would oppose each other and no current would flow at all.

What about portions EH and FG of the loop shown in Fig. 34-3? As you see, they move in a plane parallel to the lines of force, so that they never cut across the lines of force. Hence these portions of the wire do not contribute to the production of E.M.F., and we need not consider them in our analysis.

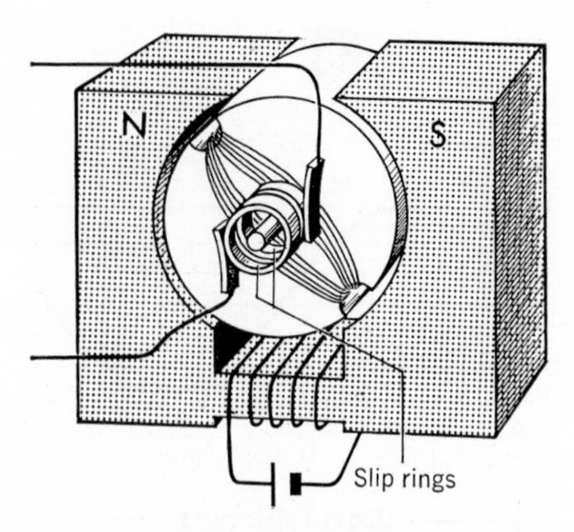

34-5 **Getting closer to a real generator. What three improvements can you see in this AC generator as compared with the one shown in Fig. 34-3? What is the battery for?**

You notice that the current reverses twice during one turn of the coil, that is, during one cycle. It is because of this reversal that we refer to the current in the coil, and in the circuit to which it is connected, as **alternating current.** The electrons surge back and forth exactly in time to the turning of the coil. If we turn the coil faster, so that it makes 10 turns per second, then the electrons also make 10 round trips per second. We then say that this particular alternating current makes 10 cycles per second. If the coil rotates 20 times per second, then a 20-cycle-per-second current is produced. Current in most homes is 60-cycle alternating current. Actually, the giant generators in powerhouses cannot rotate as fast as 60 times per second. Instead, the rotating armature contains many coils on its circumference. If 60 of these coils pass a stationary magnetic pole in one second, then 60-cycle AC is produced in the coil.

Most electric clocks use alternating current. The clock motor is known as a synchronous motor because it rotates in time to the alternations of current. If an AC generator in a powerhouse is kept rotating at a steady speed, then all electric clocks receiving current from this generator will run at a constant speed, and will keep correct time.

5. Improving the generator

We can get much more current from the generator shown in Fig. 34-3 if we make a few simple changes in its design. We can increase its E.M.F. in ways that we have already mentioned. If we make the field magnet stronger we will have more lines of force to cut. Or, the more turns we have in the coil the greater the E.M.F. And the E.M.F. will increase if we rotate the coil faster. All of these improvements make use of Faraday's discovery that the E.M.F. is proportional to the number of lines of force that are cut per second.

We may further improve the generator if we wind the coil of the armature on a cylinder of soft iron (Fig. 34-5). In this way, because of the high permeability of soft iron (see page 375), we increase the number of lines of force passing between the poles of the field magnet. If the field magnet is a permanent magnet (Fig. 34-3A), the generator is called a **magneto** (măg·nē′tō).

In large generators a stronger field magnet is provided by using an electromagnet to produce the lines of force (Fig. 34-5). The generator must have a field magnet with poles that do not change. Hence a separate source of direct current (DC) is needed, since the AC produced by the generator would cause the poles of the field magnet to change in time to the alternations of the current. The amount of electric energy needed to operate ("excite") this field magnet is relatively small.

In many generators, instead of the armature rotating in a stationary magnetic field the magnetic field is made to rotate inside the coil of a stationary armature (Fig. 34-6). This type of generator has the advantage of having no armature

General Electric Co.

34-6 These four sections fit together to form part of a giant generator 30 feet in diameter and weighing 43,000 pounds.

brushes, since the armature is stationary.

Moreover, by using many poles on the same rotor shaft, and a similar number of coils on the stationary part, 60-cycle AC can be produced with a low speed of rotation (Fig. 34-6).

6. Making a DC generator

Generators are useful in automobiles. They produce the current needed for the ignition system, the lights, the horn, and also for charging the battery. But an AC generator cannot be used because charging requires that the minus terminal of the battery always be connected to the minus terminal of the charger (page 363). We must therefore have some way of producing DC for this purpose. The job is done by substituting a **commutator** for the slip rings. This is an ingenious device that we have already studied in connection with the electric motor (see Chapter 29).

As you can see from Fig. 34-7A, the ends of the coil, instead of touching the same brush at all times, now reverse connections every half turn. If we place the brushes at the proper position, this reversal takes place at the instant that the current in the coil reverses. This double reversal keeps the current flowing in the same direction at all times in the *outside circuit*. Thus the *outside circuit* has direct current, that is, current that does not reverse in direction, even though the current in the armature itself is always alternating.

The direct current we get in this way is not steady but pulsating (Fig. 34-7B). The current flow starts and stops, starts and stops, but always keeps moving in the same direction. Thus the current produced by this particular generator is pulsating DC.

Now that direct current is coming out of the brushes we can dispense with the permanent field magnet of Fig. 34-7A and substitute an electromagnet. It is now possible to use the direct current coming out of the generator to supply current for the magnet. When the generator is started from rest we depend on the residual magnetism of the field magnet to induce the initial current in the armature.

There are two different ways to con-

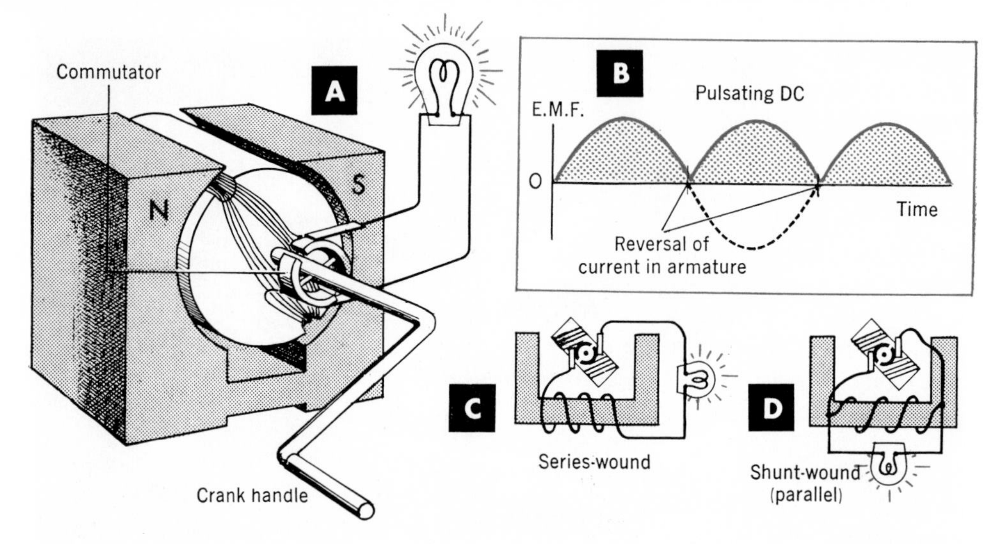

34-7 A simple change from the separate slip rings of the AC generator to a single split ring, or commutator, converts it into a DC generator (A). The commutator reverses the outgoing current every half turn to keep it flowing in the same direction (B). The generated DC current can now be used to magnetize the field coils (C and D).

nect the field-magnet winding to the rotating armature coils. They can be connected either in series or in parallel (shunt) as shown in Figs. 34-7C and 34-7D (see Chapter 25).

DC generators and motors

If you compare the DC generator shown in Fig. 34-7C with the DC motor in Fig. 30-10 you see that they are nearly identical. In fact, a motor can be run as a generator. You can test this for yourself if you connect the wires from a DC motor to a galvanometer and spin the armature by hand. The motor is now a generator and produces DC. Since the motor is turning at slow speed the current will be very weak.

7. Reversibility of energy

Here we have a fine example of the reversibility of energy changes. Using the DC motor, we put in electrical energy and get out mechanical energy. When we put mechanical energy into the same motor (by spinning the armature), we get out electrical energy.

This effect of the reversibility of energy gives electric trains an advantage in hilly country. When a train is going up a hill, the locomotive uses electrical energy (Fig. 34-8) and the train gains potential energy due to the increase in height. On the way down the other side

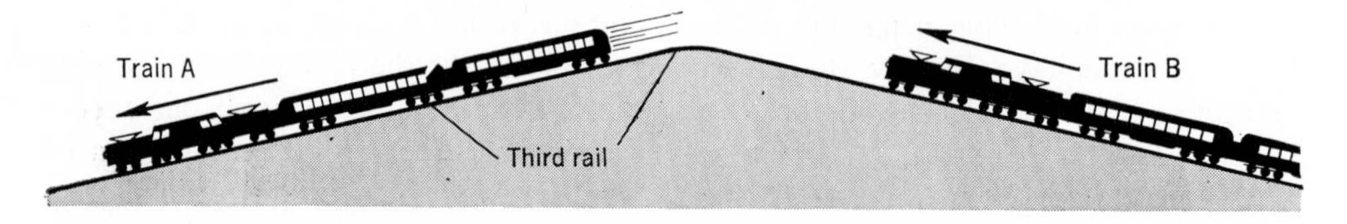

34-8 Down helps up. On the way downhill the DC motors in train A are turned by the wheels to generate E.M.F. which pushes current back into the rails, helping train B climb the hill.

of the hill this potential energy is converted into kinetic energy of motion. Now the wheels of the train turn the armature of the motor, generating electrical energy which is fed back into the third rail or overhead wire. The E.M.F. produced in this way helps the central generator maintain the voltage in the line and at the same time acts to brake the train; this electrical energy can be used by another train going up a hill.

8. Back E.M.F.

Whenever a coil of wire cuts a magnetic field, an E.M.F. is induced in it, whether it be in a generator or in the motor of an electric locomotive coasting downhill, in an electric motor in a machine shop, or in your refrigerator. Every electric motor also acts simultaneously as an electric generator, because the coils of its armature are cutting through a magnetic field. This induced voltage has a direction *opposite* to that of the voltage applied to the motor, and so it has the effect of reducing the current flowing in the armature. Figure 34-9 is a schematic diagram of the armature of a motor in which r is the resistance of the armature and e the **back E.M.F.** induced in the armature by the generator action. It is represented by a cell in series with the armature resistance. This back E.M.F. is always less than the applied voltage, E, and since its polarity is opposite to that of E, the net voltage forcing current through the armature is $E - e$. Applying Ohm's Law to the armature, we have these equations for the current through the armature:

$$I = \frac{E - e}{r}$$

$$Ir = E - e$$

$$e = E - Ir$$

This shows that **the back E.M.F. is equal to the applied voltage minus the voltage drop (Ir) in the resistance of the armature.**

SAMPLE PROBLEM

What is the back E.M.F. in a motor whose armature has a resistance of 0.2 ohm and which draws 10 amperes when operated from a 120-volt source? Applying the above equation, we have:

$$e = E - Ir$$

$$e = 120 - 10 \times 0.2 = 118 \text{ v}$$

STUDENT PROBLEM

What is the back E.M.F. in a motor whose armature has a resistance of 0.3 ohm and which draws 12 amp from a 110-v source? (ANS. 106.4 v)

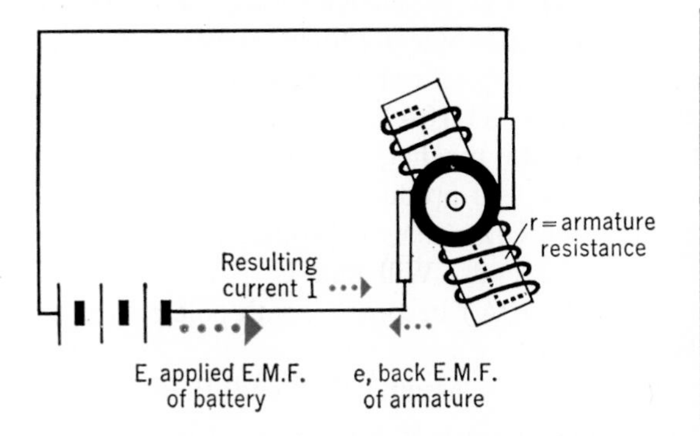

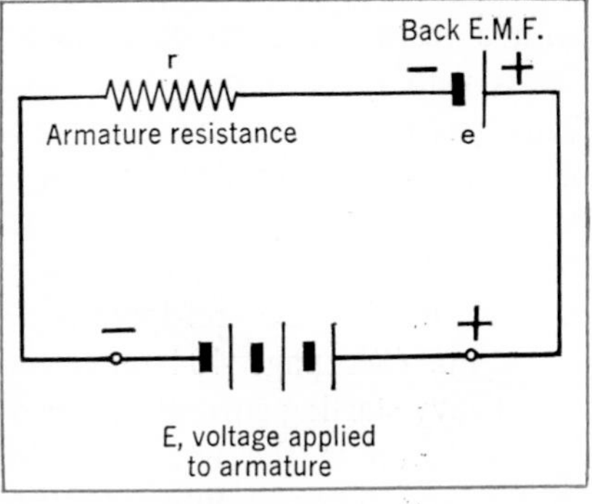

34-9 **The back E.M.F. in the armatures of a motor opposes the applied voltage. The battery determines the value of E. What determines the value of e?**

9. Starting current in a motor

You learned on page 433 that the faster a coil cuts a magnetic field, the greater is the induced E.M.F. How does this apply to the back E.M.F. in a motor? If the motor is not running, there will be no motion of the armature and no back E.M.F. This is the situation at the instant of closing the switch to start the motor. If there is no back E.M.F. to oppose the applied voltage, the current through the armature will be very large. In the motor whose back E.M.F. was found to be 118 volts when the armature rotates at full speed, we would find that while the armature is at rest and there is no back E.M.F. being induced in it ($e = 0$), there will be a **starting current** of 600 amperes. This is calculated as follows:

$$I = \frac{E - e}{r}$$

$$I = \frac{120 - 0}{0.2} = 600 \text{ amp}$$

This very large current flows only for an instant, since as the motor speeds up, the back E.M.F. increases, causing the current in the armature to decrease rapidly. But in the few seconds until this happens, the armature may burn out. Hence large DC motors with low armature resistance are arranged to operate with a resistance such as a rheostat in series with the armature when starting. This resistance may be cut out either manually or automatically as the motor speeds up. Perhaps you have noticed how the lights in your kitchen dim a bit at the instant the refrigerator motor starts up. (See page 404 on line drop.) The heavy starting current causes a large voltage drop in the line supplying the motor and lights, resulting in reduced voltage applied to the lights, thus causing dimming.

STUDENT PROBLEM

What will be the starting current in the motor in the student problem at the end of Section 8? (ANS. 367 amp)

---

10. Conservation of energy

Try this experiment with a small hand generator (or magneto) that can operate a small incandescent lamp. Crank the generator with the lamp disconnected and no current flowing. Notice how easily the handle turns. Then suddenly connect the lamp. At the instant the current starts flowing, the generator becomes harder to turn!

This does not seem so strange if you consider the Law of Conservation of Energy. When the circuit is completed, electrical energy is produced as the armature coil rotates in the magnetic field. This electrical energy is turned into heat and light energy in the lamp. The energy for the lamp is supplied by the extra work you must do to keep turning the armature. If you cease your efforts or open the circuit, the production of electrical energy will stop.

11. Lenz's Law

You can look at it in another way. Picture a coil of wire placed near a magnet. Let us imagine that the magnet is on frictionless "roller skates" and that we give it a slight push toward the coil (Fig. 34-10A). Because of the motion of the magnet, its lines of force cut the coil and produce current. The current in the coil makes it acquire poles at opposite ends. Will end E of the coil (Fig. 34-10A) become a south pole or a north pole?

Let's assume that end E of the coil becomes a south pole as a result of the flow of the induced current. This would attract the north pole of the magnet on

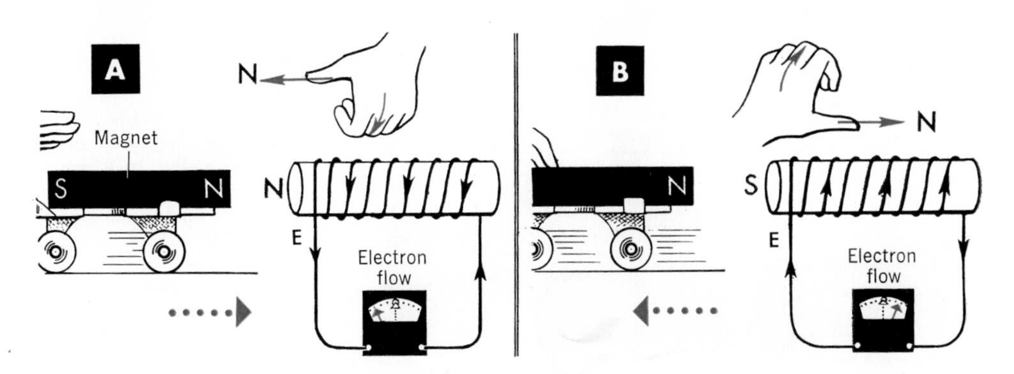

34-10 A magnet on a roller skate helps explain Lenz's Law. Can you use the Law of Conservation of Energy to explain the direction of current flow in the coils in A and B?

the roller skates, so that the magnet would roll faster toward the coil. This extra speed would cause lines of force to be cut more rapidly. Hence more current would be induced in the coil, giving it stronger poles, thus making the attraction between magnet and coil still greater. Thus, assuming end E becomes a south pole, if we were to give the magnet only a slight push toward the coil it would gain speed and kinetic energy. At the same time the coil would gain electrical energy as the induced current flowed through it.

So, on the assumption that E becomes a south pole, we see that we would be getting both electrical and mechanical energy from nowhere, since we would not be doing any work to produce it. We would have constructed a kind of "perpetual motion machine."

This would contradict the Law of Conservation of Energy, which has been found by experiment to hold true at all times. So E cannot be a south pole.

Could it become a north pole? Let's see. If E is a north pole, then it repels the approaching north pole of the magnet, making the magnet slow down and stop. The magnet must now be pushed to make it keep moving (that is, work must be done) to induce current in the coil (Fig. 34-10A). Thus we see that we get the extra electrical energy in the coil at the expense of the extra work done to move the magnet toward the coil. The faster we push the magnet, and the more current and energy we produce, the greater becomes the strength of the repelling north pole at E, and the greater the force needed to push the magnet. So the Law of Conservation of Energy makes it obvious to us that E must become a north pole.

These facts are expressed in **Lenz's Law**, which states:

Whenever a current is induced in a wire the field produced by the current is always in such a direction as to oppose the motion causing it.

12. Finding the direction of induced current

In Fig. 34-10A we saw that when the north pole of a magnet is pushed toward a coil of wire, current flows in the coil in such a way that a repelling north pole is produced at the end of the coil near the magnet. If we pull the north pole of the magnet away, the current in the coil will reverse (Fig. 34-10B). Now we have a south pole at the end of the coil near the magnet; this attracts the north pole of the magnet and opposes its motion.

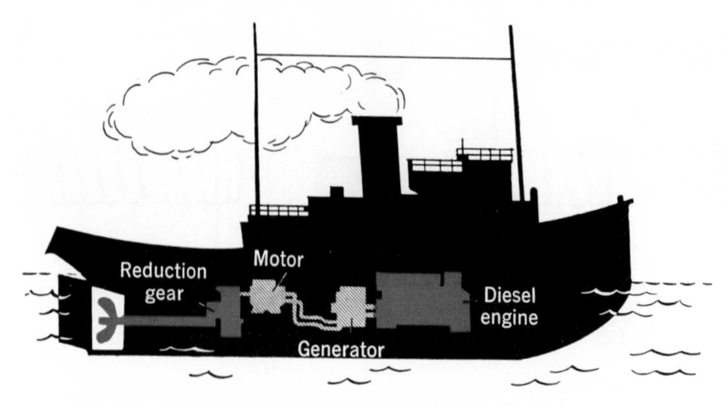

34-11 This diesel-electric tugboat drives its propeller in a roundabout way. Since the diesel engine is powerful but difficult to operate at low speeds, it is made to turn a generator to produce the current for a more easily controlled electric motor.

We can always find the direction of the current induced in a coil by using Lenz's Law. Thus in Fig. 34-10 the north pole is being pulled away from the coil. Lines of force will be cut by the coil and current induced. The direction of the current will be such as to oppose the motion of the magnet, which is being pulled away. So we conclude that the end of the coil near the north pole of the magnet must become a south pole in order to attract the magnet back. Using our left-hand rule (page 380), we then figure out the direction of electron flow as pictured in Fig. 34-10B.

13. Energy for the generator

Does a generator create charge? We have seen that current is a flow of electrons, and that electrons exist in the wire all the time. It is only when the electrons flow in the wire that we become aware of electricity and its effects. Therefore the generator cannot be said to "create charge." It merely makes the electrons move. Nor does a generator create energy. The source of the electrical energy is the mechanical energy applied to the rotating armature of the generator. The generator must be turned by some outside source of mechanical energy such as water power from a dam. Where cities are located far from water power the energy for the generator usually comes from burning coal, which operates a steam turbine which turns the generator. On farms, a wind-operated propeller is often used to turn the generator to produce electrical energy. This electrical energy can then be stored in batteries as chemical energy to be used when the wind is not blowing.

The electric company must take Lenz's Law into account. As consumers use more energy, the generator tends to slow down for the same reason a hand generator becomes harder to turn when current flows. So at those times of the day when demand is high, more coal must be burned and greater steam pressures applied to keep the generators turning at the same speed.

Figure 34-11 shows an interesting use of generators in boats. It is much easier to control the speed of a boat operated by an electric motor than one operated by other types of engines. Therefore, many boats use diesel engines to turn a generator to produce electric current for operating an easily controlled electric motor. The gain in convenience more than makes up for the cost, particularly for ferryboats and other boats needing quick changes of speed. The energy in these boats comes from the oil burned in the diesel engine. The familiar diesel-electric locomotive is another application.

Stepping up voltage

Electrical energy produced at the powerhouse must be transmitted over wires to the consumer. In Chapter 32 we learned of the problems that arise when these transmission wires are very long. The longer the wire the greater its resistance. And the higher the resistance of the line, the greater the line drop (page 404) and the loss of electrical energy in the line itself. If the line is too long, it doesn't pay to transmit the electrical energy at all; so much energy will be wasted as heat in the line that there will be little left for the consumer at the end of the line.

This waste of energy has been cut down by using high-voltage lines, so that today it is practical to transmit electrical energy several hundred miles. Again it was Faraday whose discoveries paved the way for the invention of the **transformer,** which made high-voltage power lines possible.

14. Getting currents from other currents

While Faraday was experimenting to obtain electric current from magnets, he made an interesting discovery. He took an iron ring and wound a coil of insulated wire (A in Fig. 34-12) around it. Then he connected the ends of the coil to a cell, B, through switch S. This coil of wire he called the **primary** (first) **coil.** He then wound another coil (C in Fig. 34-12), on the same iron ring. This second coil he called the **secondary coil.** Faraday then connected the secondary coil to a galvanometer, G. When he turned on the current in the primary coil A, he expected that its magnetic effect might produce current in the secondary coil C.

To his surprise a flick of the galvanometer needle indicated only a momentary current in the secondary at the very instant the current was turned on in the primary. Thereafter the needle went back to zero, showing that there was no current in the secondary even though the switch was closed and there was a steady current in the primary. But at the moment switch S was opened and current ceased flowing in the primary, a sudden flick of the galvanometer needle in the *opposite* direction again showed a momentary current in the secondary. Thus Faraday found that the primary coil A could induce a voltage in the secondary coil C *only at the moment when current in the primary was changing.* A steady current in the primary had absolutely no effect on the secondary coil.

Later experiments showed that the iron ring was not absolutely necessary for this effect, although it did magnify the current considerably. The same thing happened if a primary coil was placed on a table and a secondary coil suspended above it with empty space between. The secondary coil showed current only at the instant that the current in the first coil was shut off, turned on, or *changed* in some way.

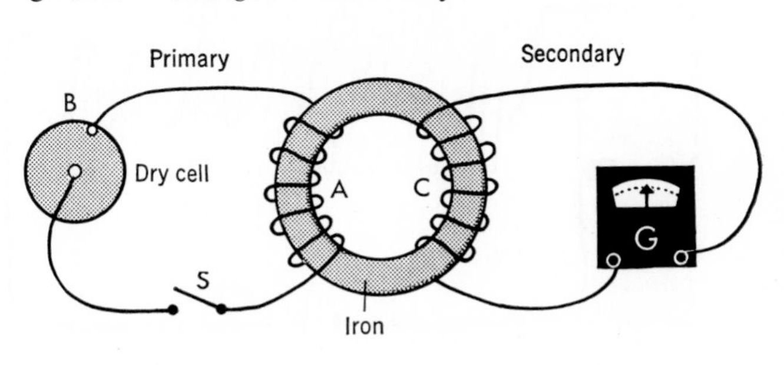

34-12 This experiment, first performed by Faraday, led to our modern high-voltage transmission lines. When the switch, S, is closed, energy is transferred from the primary circuit to the separate secondary circuit by moving magnetic lines of force. As the lines of force cut coil C, they induce a voltage which causes a pulse of current through the galvanometer, G.

15. Moving lines of force

Faraday based his explanation of the above experiment on the idea that the lines of force come out of the primary coil and spread outward as soon as the current is turned on. The current in the primary does not reach its maximum immediately. It takes time for the current in the wire to increase from zero to its final steady amount, even if that time is only a fraction of a second. As the current in the primary increases, the magnetic field around it also increases in strength. As long as the current in the primary continues to *increase,* the lines of force in the growing magnetic field spread outward away from the coil.

If there is a wire (conductor) near by, the moving lines of force from the primary cut across it and induce a voltage in this secondary, just as the cutting of stationary lines of force by a moving wire produces current in a generator. But as soon as the current in the primary coil reaches its maximum value and becomes steady, the magnetic field of the primary becomes stationary and no voltage is induced.

When the current in the primary is shut off, it takes a brief moment for the lines of force in its magnetic field to collapse back into the primary coil from which they came. During this moment that they are moving inward, these collapsing lines of force cut across nearby wires and induce voltage. This return movement of the lines of force is opposite to their direction of motion when the current was being turned on. Hence the induced current in the secondary is also in the opposite direction, as shown by the pointer of the galvanometer. Thus the secondary always produces AC.

16. The induction coil

It did not take long to put Faraday's experiment with primary and secondary coils to practical use in the **induction coil** —an arrangement of a primary and a secondary coil such that high voltage is induced by moving lines of force (Fig. 34-13). The E.M.F. induced in the secondary was increased by using a secondary coil of thousands of turns and bringing it close to the primary—in fact, by surrounding the primary with the secondary. In this way more of the moving lines of force from the primary were able to cut the secondary.

Instead of turning the battery current in the primary on and off by hand, a vibrator was used to "make and break" the circuit rapidly and automatically (Fig. 30-9).

This vibrator is really a "buzzer," arranged in the same way as the ordinary

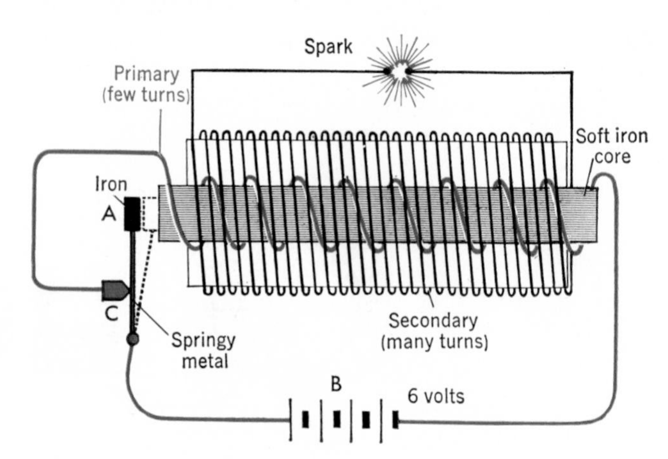

34-13 This spark coil (one form of induction coil) can produce a 10,000-volt spark from a mere 6 volts. The vibrator at A (similar to the electric bell) makes and breaks the circuit in the primary coil to induce a high-voltage current in the many turns of the secondary coil.

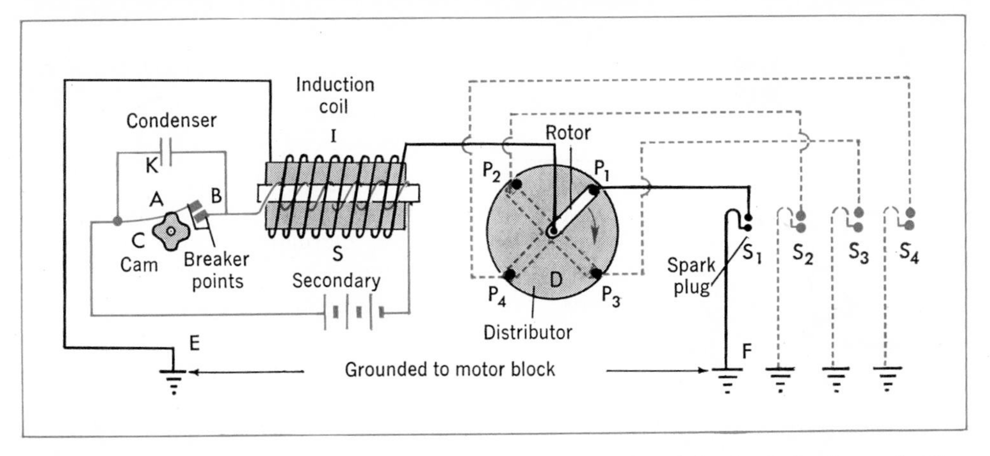

34-14 Your car couldn't run without the sparks produced by its induction coil. The breaker points, B, geared to the engine break the current from the battery to induce a high voltage in the induction coil, I. The current is conducted to the proper spark plug, S_1 to S_4, by means of the distributor, D.

bell, whose armature also vibrates back and forth (Fig. 34-13). In the figure, the current passes from battery B to magnetize the primary coil of the induction coil, then crosses over from contact point C to the springy metal strip and flows back to the battery. As soon as current begins to flow in the electromagnet (the primary), the attraction of the magnet pulls the iron strip A closer and the circuit is broken at C. Now the current ceases and the magnetism disappears. The metal bar springs back to touch C and turn the current on again. This make-and-break process is repeated over and over so as to create an intermittent (on-and-off) current which causes the magnetic lines of force of the primary coil to sweep out and in, out and in. These moving lines of force continually induce an E.M.F. in the wires of the secondary coil.

When the number of turns in the secondary is increased to many thousands, the E.M.F. developed in it may be 100,000 volts or more. Sparks several inches long can be made to jump between the terminals of the secondary. And the high voltage in this type of induction coil comes from a battery of 6 volts or less!

17. The ignition system of a car

The ignition system of an automobile engine makes use of the spark from such an induction coil to ignite the gasoline vapor in the cylinders. The spark need not be continuous, since it is needed only at certain moments. Thus instead of using the vibrator constantly to make and break the circuit in the primary, a single "break" at the right instant is all that is needed.

Figure 34-14 shows the ignition circuit of an automobile. A cam, C, is arranged to rotate with the engine. The cam pushes up the arm, A, which separates the **breaker points** at B. This separation of the breaker points breaks the primary circuit and shuts off the current. The lines of force in the collapsing magnetic field of the primary, P, cut the many turns of the secondary, S, and create a sudden high voltage. At that very instant the revolving rotor of the **distributor,** D, touches one of the **points,** P_1, that leads to a spark plug, S_1. Both the cam, C, and the rotor are attached to the same shaft and turn at the same speed, so that the breaker points are pulled apart to make a spark whenever the ro-

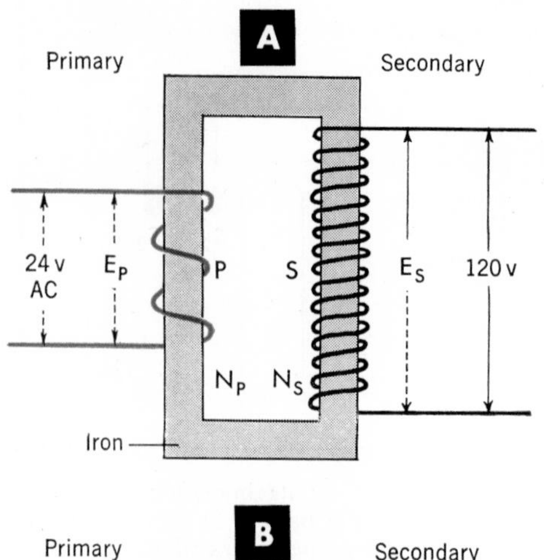

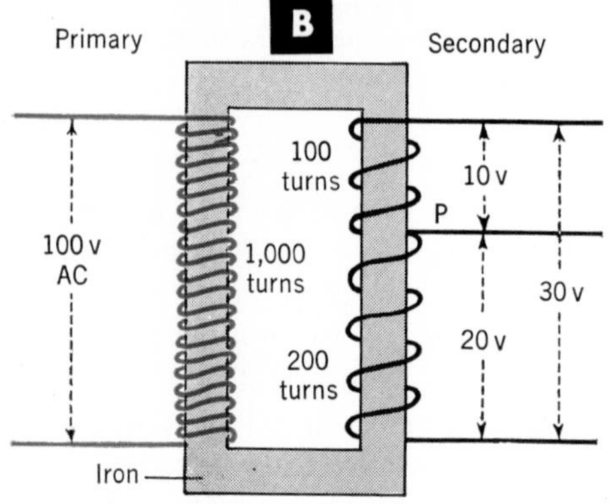

34-15 Two common forms of transformers are shown in A and B. Note in B how different voltages are obtained by connecting a wire to the middle of the secondary coil. How can you determine which of these transformers is step-up and which is step-down?

tor touches a contact that leads to a spark plug.

The secondary circuit is completed by grounding to the motor block at E and F. A condenser, K, is connected across the breaker points to reduce the tendency of sparks to jump across the gap between the points as they separate. The electrons can flow into the condenser instead of jumping in a spark across the high-resistance air. By reducing these sparks, the condenser preserves the breaker points, which otherwise would become pitted by the sparks jumping across the gap.

The transformer

As you know, E.M.F. is induced in the secondary coil only when the current in the primary is changing. If we eliminate the vibrator and battery and replace them with alternating current in the primary circuit, then the magnetic field will still be continually changing. Thus a continually changing voltage and current will still be induced in the secondary.

18. Changing the voltage

If there are more turns in the secondary than in the primary, we have a high-voltage **step-up transformer.** If there are fewer turns, the voltage is lowered and we have a **step-down transformer.** A soft-iron or silicon-steel core serves to concentrate the lines of force and reduce energy losses.

When we use a properly constructed core and the number of turns of wire in the secondary is the same as in the primary, we find by experiment that the voltage of the secondary is the same as that of the primary. If the secondary has twice as many turns, it produces twice the voltage; if it has half the number of turns, just half the voltage is produced. In other words, we say that the voltages of the secondary and primary coils are proportional to the number of turns in each.

This may be expressed by a formula. In Fig. 34-15A, E_p is the voltage applied to the primary and E_s is the voltage induced in the secondary. N_p is the number of turns in the primary coil and N_s is the number of turns in the secondary coil. Since the voltages of the secondary and primary coils are proportional to the number of turns in each, we can state that:

$$\frac{E_s}{E_p} = \frac{N_s}{N_p}$$

SAMPLE PROBLEM

Suppose that a transformer has 200 primary turns and 1,000 secondary turns. If 120 volts are applied to the primary, what voltage will be induced in the secondary? Here $N_p = 200$, $N_s = 1{,}000$, and $E_p = 120$ volts. Substituting:

$$\frac{E_s}{E_p} = \frac{N_s}{N_p}$$

$$\frac{E_s}{120} = \frac{1{,}000}{200}$$

$$E_s = \frac{1{,}000 \times 120}{200}$$

$$= 600 \text{ v}$$

[NOTE. Such a transformer is called a 5:1 step-up transformer (1,000 turns/200 turns = 5:1), and hence the secondary voltage will be 5 times that of the primary: 5 × 120 v = 600 v.]

STUDENT PROBLEM

A transformer has 400 primary turns and 6,000 secondary turns. If 120 v are applied to the primary, what voltage will be induced in the secondary? (ANS. 1,800 v)

■ ■ ■ ■ ■ ■ ■

We make use of the turns ratio in a transformer to obtain any voltage we please in the secondary. Figure 34-15B shows how three different voltages can be obtained by connecting a wire (a "tap") near the middle of the secondary coil. The step-down (voltage-reducing) transformers in toy trains have a sliding wire on the secondary which makes it possible to select different numbers of turns and thus provide different voltages. At lower voltages the current in the train motor is less, and the train slows down. Remember that a transformer always produces AC, never DC, and that the transformer must be fed with a changing current like AC, never DC.

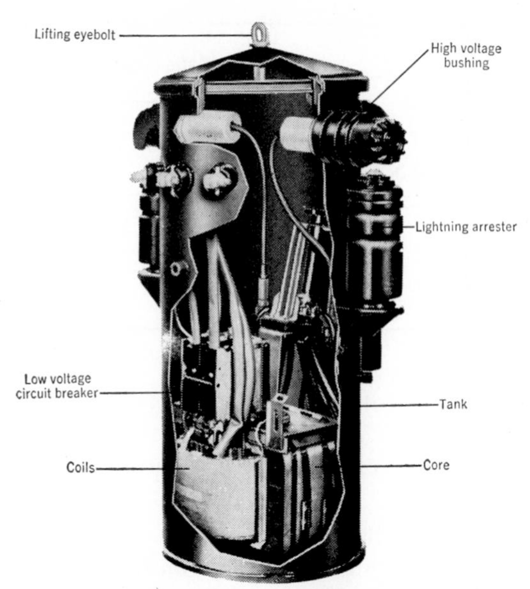

Westinghouse Electric Corp.

34-16 The familiar black box at the top of power line poles looks like this inside. The actual transformer with its coils and iron core is at the bottom of the case. Above it is the equipment to switch it on and off and to protect it against lightning. Why are such large insulators necessary?

19. Something for nothing?

If we can put 10 volts of AC into the primary of a transformer and step it up to 100 volts, it might seem that we are getting something for nothing. But you remember that we found in our study of machines that we gain force only at the expense of speed and distance. Just so with transformers. We find that we have to pay for a gain in electromotive force (voltage) by a loss in the amount of current. Thus if we step up the voltage 10 times, the maximum current that can flow in the secondary is less than that in the primary by exactly the same ratio; in other words, it is only $\frac{1}{10}$ as great.

But doesn't this contradict Ohm's Law, which says that current is proportional to voltage? Shouldn't more voltage produce more current?

We must watch our conditions. So far we have discussed Ohm's Law only un-

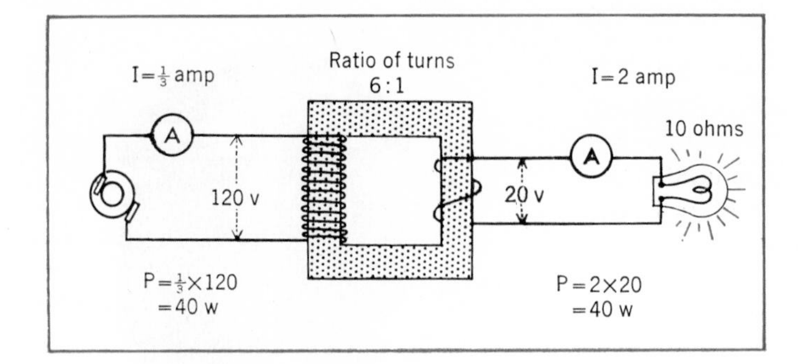

34-17 Voltage may go up or down. Current may go down or up. But the power input and power output are always the same (neglecting losses). In what way does this transformer resemble a pulley system?

der conditions dealing with DC and with a single complete circuit. Will the law apply to AC? Will it apply when there are two separate coils of wire in the transformer?

You note that in the transformer there are two entirely separate electrical circuits (Fig. 34-15A). The connection between these two circuits is not electrical but magnetic. We should not expect to be able to use Ohm's Law to calculate current in the secondary from the voltage in a different electrical circuit—the primary—since Ohm's Law applies only to voltage and current in a single circuit.

But there is one law that all devices obey, so far as we can judge from innumerable experiments. This is the Law of Conservation of Energy. Energy and power output cannot be greater than energy and power input. Let us see how this law works out for the transformer.

20. Power output equals power input

Figure 34-17 shows a step-down transformer which has 6 times as many turns in the primary as in the secondary. The voltage on the primary or input side is 120. This operates a 10-ohm lamp in the secondary.

Since the secondary has $\frac{1}{6}$ as many turns as the primary, it receives $\frac{1}{6}$ as many volts, or 20 volts. How much current can this 20-volt E.M.F. produce in the 10-ohm lamp in the secondary? Using Ohm's Law we get:

$$I = \frac{E}{R} = \frac{20\text{ v}}{10\text{ ohms}} = 2\text{ amp}$$

(current in secondary)

Now let us find the power output of the secondary. We get this by using our formula:

$$P = EI$$
$$P = 20\text{ v} \times 2\text{ amp} = 40\text{ w}$$

(output from secondary)

Since the power input in the primary must supply this 40 watts of output, at least 40 watts of power are needed in the primary. How much current will this require?

$$P = EI$$
$$40\text{ w} = 120\text{ v} \times I$$
$$I = \frac{40}{120} = \frac{1}{3}\text{ amp}$$

(current in primary)

Thus the voltage is stepped down from 120 until it is only $\frac{1}{6}$ as great, or 20 volts. At the same time the current is stepped up from $\frac{1}{3}$ ampere to a current of 2 amperes, 6 times as much. The power remains the same, neglecting losses. If we take losses into account the power output will be slightly less than the input. But transformers exist which are more than 98 per cent efficient. So for practical purposes we usually assume that we have 100 per cent efficiency, and that power output is equal to power input.

For a transformer that is 100 per cent efficient, then, $E_pI_p = E_sI_s$. But for actual transformers E_pI_p is more than E_sI_s, and the efficiency is

$$\text{Efficiency} = \frac{\text{Power output}}{\text{Power input}}$$
$$= \frac{E_sI_s}{E_pI_p} \times 100 \text{ (expressed as a \%)}$$

STUDENT PROBLEM

What is the power output of an 85%-efficient transformer that draws 8 amp from a 120-v line? (ANS. 816 w)

■ ■ ■ ■ ■ ■ ■

Inefficiency in a transformer is caused mainly by heat losses due not only to current flowing in the coils, but also to unwanted current induced in the core of the transformer. Currents induced in the core are called **eddy currents.** The flow of eddy currents is impeded and the efficiency of the transformer thereby increased by constructing the core of flat sheets of soft iron, called **laminations,** insulated from each other with shellac. Figure 34-18 shows a small transformer, of the sort used in radios, in which the laminations are visible.

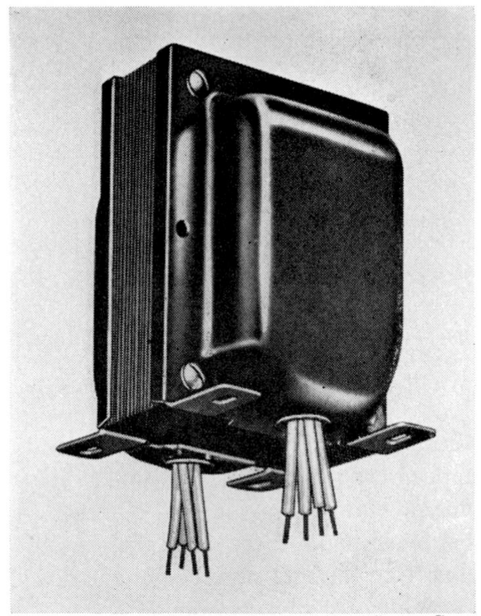

Chicago Standard Transformer Corp.

34-18 **Look inside your radio, and you will probably find a transformer like this one for supplying high and low alternating voltages. Notice the laminations of the core visible along the edge of the case.**

21. Cutting costs with transformers

We said earlier that the most important use of transformers is in the transmission of electrical energy from the powerhouse to the consumer. What is the job they do here?

There are always heating losses in electrical wires. They are called **I^2R losses** since the number of watts of power wasted in heating the power line is I^2R. (Here I is the current through the power line, and R is the resistance of the line; see Chapter 32.) We can cut down these losses in two ways. We can reduce either the current, I, or the resistance, R. It is easy to figure out how to reduce resistance: we use a better conducting material and make the wires thicker. But heavy wires of silver running for long distances would be too expensive.

Can the current be reduced to save power? We can give the consumer the amount of power he requires and still reduce the current in the transmission line by using transformers. At the powerhouse a transformer steps up voltage so that the same amount of power can be transmitted with much *less current.* For example, if the voltage is stepped up 100 times, the same power, $E \times I$, can be transmitted with $\frac{1}{100}$ as much current. The heating loss in the transmitting wires, I^2R, will then be only $(\frac{1}{100})^2$ or $\frac{1}{10,000}$ as great: that is, the loss is proportional to the square of the current. So when current is to be sent any great distance, AC

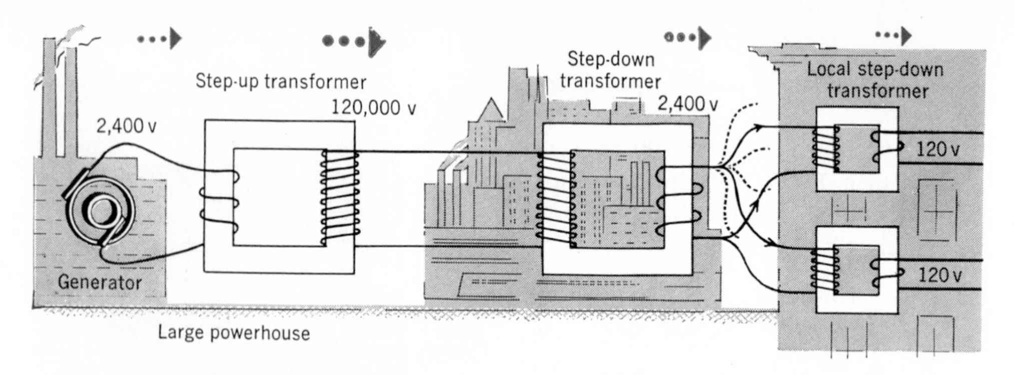

34-19 Up and down goes the voltage from generator to consumer. Why is high voltage used from powerhouse to local area? Why is it stepped down to 120 volts for the consumer? Why is AC more useful for transmission of current than DC?

generators are used; the voltage is stepped up, resulting in a similar stepping *down* of the amount of current. At the other end of the transmission line other transformers step the voltage down for delivery to the consumer.

22. Transmitting electrical energy

The transformer makes it possible to transmit electrical energy several hundred miles from a power station. This could not be done with DC since we have no easy and efficient way of increasing the voltage and at the same time decreasing the current so as to reduce heating losses (I^2R losses) in the power line.

Figure 34-19 shows how electrical energy is transmitted from a powerhouse to a city. At the powerhouse the voltage may be stepped up to 120,000 volts. When it reaches the city, it is stepped down again to 2,400 volts. It is then distributed to local transformers where it is further stepped down to 120 volts. This figure, 120 volts, is chosen for use in homes and factories because it is reasonably safe.

The upper limit of voltage is set by the effectiveness of the insulators supporting the high-voltage wires away from nearby objects. The limit today is in the neighborhood of 1,000,000 volts. Figure 34-20 shows transformers which step up 69,000 volts to 500,000 volts.

Going Ahead

THE VOCABULARY OF PHYSICS

Define each of the following:

generator	breaker points
induced current	brushes
cycle	step-up transformer
sine curve	step-down transformer
magneto	I^2R losses
transformer	starting current (of a motor)
primary coil	armature
secondary coil	commutator
eddy currents	distributor
laminations	slip ring
back E.M.F.	field magnet
induction coil	

THE PRINCIPLES WE USE

1. State Lenz's Law and give an example of its operation.
2. What is the connection between Lenz's Law and the Law of Conservation of Energy?
3. How can current be produced from magnets?
4. Explain why a DC motor can be used as a DC generator and vice versa.
5. How can we calculate the voltage in the secondary of a transformer?
6. In what ways is a transformer like a mechanical machine?

Explain why

7. It is possible that no current will flow in a wire moving in a magnetic field.
8. The faster a wire moves, the greater the E.M.F. induced.

9. A coil of wire will have more E.M.F. induced in it than a single wire.
10. A strong magnet causes more E.M.F. than a weaker one.
11. It is harder to turn a hand generator when it is used to light a bright lamp than when the circuit is open.
12. The rotating coil of a generator produces alternating current.
13. The current of DC generators may be used for magnetizing field coils, but the current of AC generators may not.
14. More coal must be burned in a powerhouse as a storm approaches, even though the generator is kept rotating at a constant speed.
15. Transformers will not operate on direct current.
16. The coils of a transformer are wound on an iron core.
17. When voltage is gained in a transformer energy is not being created.
18. The iron core of a transformer is made of laminations.

EXPERIMENTS WITH INDUCED CURRENTS

Describe an experiment to
1. Show that current is induced in a complete circuit only when lines of force are cut by the wire.
2. Show three ways of increasing the production of E.M.F. when lines of force are cut.
3. Illustrate Lenz's Law.
4. Show that a changing current in one coil can induce a changing current in a nearby coil.

APPLYING PRINCIPLES OF INDUCTION

1. Name the four main parts of an AC generator and describe the purpose of each.
2. State a practical use for alternating current for which direct current could not be used.
3. State three ways of increasing the E.M.F. of a generator.
4. State a practical use of electricity for which AC is not suitable.
5. How can DC be produced from the AC induced in the armature of a generator?

Westinghouse Electric Corp.

34-20 **With these giant four-story high, half-million-volt transformers, engineers are striving for even higher voltages. Why?**

6. What kind of DC is produced by a DC generator?
7. Mention four important sources of energy for generators.
8. Describe a situation in which the reversibility of energy changes in a DC motor is put to practical use.
9. Draw a diagram of an induction coil that converts 3 v DC into high-voltage AC. Explain how it works.
10. Draw a diagram of the ignition system of a car and explain how it works.
11. What determines the upper limit of the voltage that can be carried by a power transmission line? Why?
12. Why is it possible to transmit AC over long distances more cheaply than DC?
13. State two types of losses that occur in electrical transmission lines.
14. Describe an electrical transmission system from a powerhouse to a distant city.

1. Find the missing quantities in each line of the following table:

	PRIMARY			SECONDARY		
	Voltage	*Current*	*Turns*	*Voltage*	*Current*	*Turns*
(a)	100 v	6 amp	800	?	?	2,400
(b)	60 v	10 amp	120	?	?	6,000
(c)	20 v	?	?	180 v	20 amp	450
(d)	300 v	24 amp	1,200	?	?	200
(e)	250 v	8 amp	3,000	?	?	600
(f)	120 v	?	?	5 v	5 amp	60

---

2. Find the power input and output in each case in Problem 1, assuming no losses.

3. Find the power output in (a), (b), and (c) in Problem 1, assuming that each transformer is 90% efficient.

4. A power plant is generating power at 2,400 v and 500 amp, and transmitting it over a power line whose resistance is 3 ohms. (a) If the voltage is not stepped up at the power station, how much power (in kw) is lost in the power line? (b) If the voltage is first stepped up to 24,000 v, how great will be the power loss in the power line? (c) For both (a) and (b), find the number of kw delivered to the consumer at the end of the power line. (d) For both (a) and (b), find the line drop and voltage at the consumer's end of the line.

5. How much current flows through the armature of a motor whose resistance is 0.55 ohm when 120 v are applied to the motor and the back E.M.F. is 117 v?

6. At the moment of starting, a 120-v motor draws 20 amp. Almost at once the current it draws falls to 1 amp. Find (a) the resistance of the motor, (b) the back E.M.F. when it is running.

7. A generator supplies 150 amp at 600 v to a power line leading to a factory. The line has a resistance of 2 ohms. Find (a) the power generated, (b) the voltage at the factory, (c) the power delivered to the factory.

---

8. Suppose the transformer shown in Fig. 34-17 were 90% efficient. What would the current in the primary be? (Remember that the losses take place in the power and in the current but not in the voltage.)

9. A television set is operated from a 120-v 60-cycle AC house circuit. (a) The filaments of the vacuum tubes operate on 6 v. What kind of transformer is required? What is the ratio of the windings? (b) The plate of a certain tube in the set operates at 600 v. What kind of transformer is required? What is the ratio of the windings? (c) The filament of one of the tubes draws 2.00 amp. If the filament transformer is 80% efficient, what current does the primary of the transformer draw from the house circuit to supply this tube?

10. Fifty lamps, each requiring 0.5 amp at 110 v, are lit by means of a generator that is 75% efficient. What is the horsepower of the gasoline engine used to run the generator?

CHAPTER

35

Alternating Current

Radios, television sets, electric clocks, telephones, transformers, and the electric motors in most familiar appliances operate on alternating current, and in the preceding chapter you learned that alternating current is important to the long-distance transmission of electric power. Hence it is important to understand the nature of alternating current and how it is put to work.

Voltage, current, and resistance

1. Effective voltage and effective current

The simplest type of electric generator is the alternating-current generator (page 434).

The voltage change as an armature makes one complete revolution or **cycle** through 360° can be plotted on a graph (Fig. 35-1). The voltage is 0 at 0°, reaches a peak when the armature has turned through 90°, and then at 180° returns to 0. During the second half of the cycle the voltage reverses direction, following the same type of curve, and again returns to 0 at the end of a cycle, or 360°. The plotted curve is of course a sine curve. Voltage varying in this way is known as **sinusoidal voltage.**

The varying current which flows as a result of the sinusoidal voltage also may be plotted on a graph, and normally will also be sinusoidal. In most AC lines in the United States the **frequency** of the alternating current is 60 cycles per second.

How do we measure the magnitude of the voltage generated by an alternating-current generator? Since the voltage is always changing, we must decide on some average value. We could measure the value of the **peak** (or maximum) **voltage** generated (E_p in Fig. 35-1). But the voltage has this peak value for only an instant in each cycle; most of the time its value is much less than the maximum. Many devices operating on alternating current act as if the applied voltage were 0.707 of the peak value, E_p. Usually it is more convenient to use this value, 0.707 E_p, which is called the **effective value** of the voltage. Similarly we often refer to 0.707 times the peak current, I_p, as the effective value of the current.

$$E_{eff} = 0.707\ E_p$$

Similarly

$$I_{eff} = 0.707\ I_p$$

Why is the effective value used? Because it has been found that if 1 effective ampere of alternating current flows through a resistance, heat energy is produced at the same rate as if 1 ampere of DC flowed through the same resistance. Whenever alternating voltage and current are expressed, it is understood that effective values are meant, and it is these values that AC meters indicate.

Thus when we speak of 110-volt, 60-cycle current, the 110 refers to the effec-

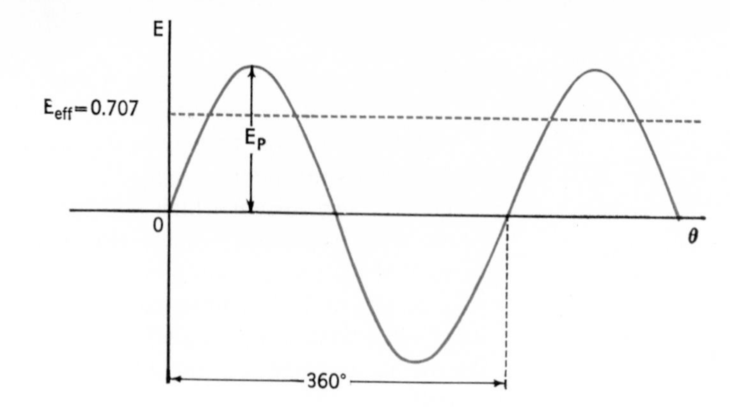

35-1 A sinusoidal alternating voltage showing the peak value, E_p, and the effective or RMS value, E_{eff}. θ (theta) is the angle through which the armature rotates.

tive value, E_{eff}. And when we speak of 0.5 ampere flowing through a lamp operating on AC we refer to the effective value of current, I_{eff}.

It can be shown mathematically that the effective value of an alternating current is simply the square root of the average (or *mean*) of the squares of the values of current throughout one cycle. Hence the effective value of the current is often referred to as the **root-mean-square (RMS) value.**

SAMPLE PROBLEM

Suppose we wish to calculate the peak voltage produced by a 110-volt AC generator. (The 110-volt rating of the generator is understood to mean 110 volts effective.) Applying our formula:

$$E_{eff} = 0.707\ E_p$$
$$110 = 0.707\ E_p$$
$$E_p = \frac{110}{0.707} = 155\ \text{v}$$

STUDENT PROBLEM

What is the peak voltage on a 550-v AC line? (ANS. 778 v)

2. Impedance

An AC circuit that contains only resistance behaves exactly like a DC circuit. The resistance in the AC circuit is usually called the **impedance**—something that impedes the flow of current. As we shall see in the following paragraphs, condensers and coils also impede the flow of alternating current, and so their effect must be included in the impedance of the circuit. The symbol used for impedance is Z, and its value is measured in ohms just like resistance, R. Thus Ohm's Law for an alternating-current circuit is

$$I = \frac{E}{Z}$$

SAMPLE PROBLEM

What are the effective current and the peak current in a circuit which has an impedance of 20 ohms and to which is applied 110 volts AC? Using Ohm's Law to find the effective current

$$I_{eff} = \frac{E_{eff}}{Z}$$
$$I_{eff} = \frac{110}{20} = 5.5\ \text{amp}$$

The peak current would then be found as follows.

$$I_{eff} = 0.707\ I_p$$
$$I_p = \frac{5.5}{0.707} = 7.78\ \text{amp}$$

STUDENT PROBLEM

What is the peak current flowing through a 110-v light bulb whose impedance is 120 ohms? (ANS. 1.29 amp)

3. E.M.F. and current as vectors

It is convenient, as we shall see later, to represent alternating currents and voltages as vectors rather than as sine curves. How is this possible? Examine Fig. 35-2A and B carefully. A shows the graph of an alternating voltage as a sine curve, and B shows a **rotating vector,** whose length is equal to E_p, the peak voltage represented in A. As the voltage in A is generated, starting from the left, the vector B rotates counterclockwise through the angle *theta* (θ). At $\theta = 45°$ the perpendicular, *e,* from E_p to the horizontal is equal to the magnitude of the instantaneous voltage generated at 45° as shown in A. As E_p rotates further, the perpendicular, *e,* increases to the value of E_p at 90°, then decreases to 0 at 180°, increases in a negative direction to $-E_p$ at 270°, and reaches 0 again at 360°. In the course of one full cycle the length of *e* increases and decreases with the sine of angle θ just as the magnitude of the sinusoidal voltage in A increases and decreases with the sine of θ. If the frequency of the voltage being generated is 60 cycles, then we can think of the rotating vector E_p as making 60 revolutions every second in a counterclockwise direction.

4. Adding alternating voltages

Suppose we have two alternating-current generators connected in series and supplying current to a circuit (Fig. 35-3A). Generator A supplies a peak of 30 volts and generator B supplies a peak of 20 volts, but the two generators are not "in step," or in **phase,** as we usually say. To be in phase, the two generators must reach their peak voltages at the same time. Generator A reaches its peak voltage one fourth of a cycle or 90° ahead of generator B (Fig. 35-3B). In other words, we say that the voltage from A *leads* the voltage from B by a **phase angle of 90°** (one fourth of a cycle), or the voltage from B *lags* that from A by 90°.

What if we wish to find the total voltage supplied to the circuit by the two generators? Obviously we cannot add them together by simple addition unless they are in step with one another (in phase) nor can we subtract by ordinary subtraction unless they are just 180° out of phase. When the two voltages are 90° out of phase, it is possible to add the two curves, A and B, on the graph in Fig. 35-3. To do this it would be necessary to add the values of E_p for every point on the graphs of A and B and plot the

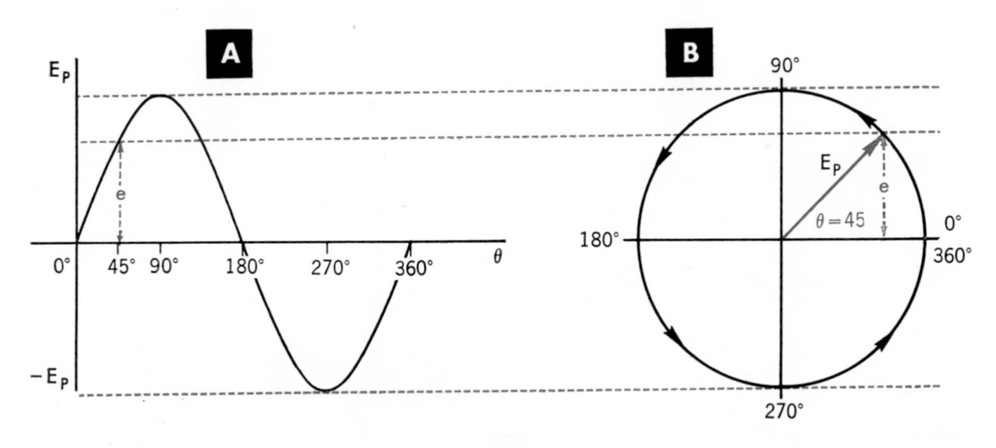

35-2 A rotating vector, B, can be used to represent an alternating current, A.

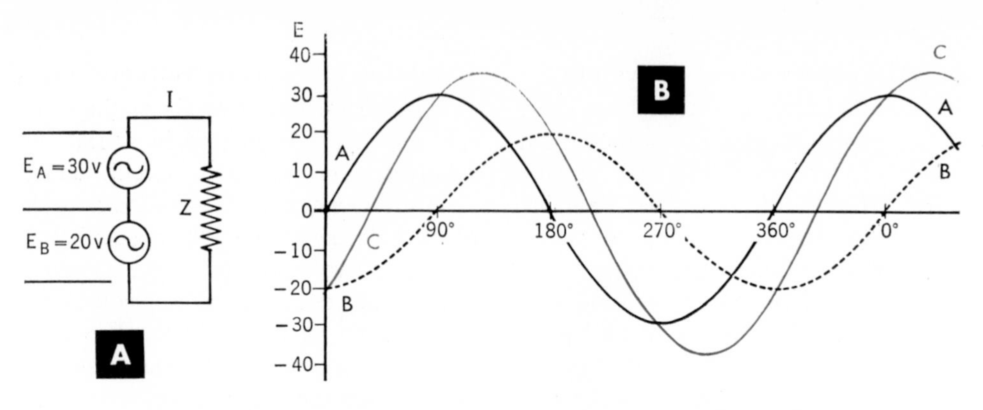

35-3 Two out-of-phase generators, E_A and E_B, produce a resultant voltage (curve C) that is less than the arithmetical sum of the two out-of-phase voltages, A and B.

sums to get curve C, which is the sum or resultant of the two voltages. This is, however, a very laborious process, and it is much easier to represent the two voltages as rotating vectors (Fig. 35-4) and get their resultant, C, by using a parallelogram method such as the one we used to add forces (Chapter 8). Since the two generators are 90° out of phase, the angle between their voltage vectors is a right angle. Notice that the resultant voltage, C, differs in phase (has a different direction) from both A and B.

SAMPLE PROBLEM

Calculate the resultant voltage in the above circuit. The vector diagram in Fig. 35-4 shows us that

$$C^2 = A^2 + B^2$$
$$C = \sqrt{30^2 + 20^2} = 10\sqrt{13} = 36.0 \text{ v}$$

STUDENT PROBLEM

What is the resultant voltage supplied by two generators, differing in phase by 90°, whose voltages are 30 v and 40 v? (ANS. 50 v)

Inductance and reactance

5. Inductance in an AC circuit

Let us examine how a coil of wire impedes the flow of AC. Suppose we have a coil connected to an AC generator. Let us assume that the coil has negligible resistance. You would expect that, because of the very low resistance, the current would be very large. Actually it is not at all large if the coil has a soft iron core or a large number of turns of wire. From what you learned about Lenz's Law you will see why.

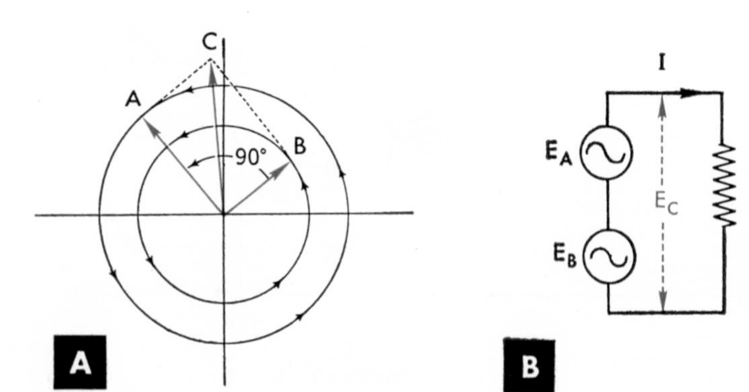

35-4 The resultant of two AC voltages 90° out of phase (E_A and E_B) is most easily found by finding the resultant (E_C) of the two rotating vectors.

The alternating current flowing in the coil is always changing and therefore creates a changing magnetic field which continually cuts the coil. This means that a voltage must be induced in the coil and, by Lenz's Law, this voltage, called the **voltage of self-induction,** must oppose the generator voltage.

The ability of a coil to produce such a voltage of self-induction is called the **inductance** of the coil, and is greatest, as we have just seen, when the coil has a soft iron core or many turns of wire.

The greater the inductance of a coil, the greater will be the voltage of self-induction when the current is changing at a given rate. The unit of inductance is called the **henry.**

A coil has an inductance of 1 henry if an E.M.F. of 1 volt is induced in it when the current through the coil is changing at a rate of 1 ampere per second.

The effect of the opposing voltage of self-induction in an AC circuit is to impede the flow of current. This opposition to the flow of current is called **reactance.** The reactance of a coil depends not only on its inductance (number of turns and the amount of iron in the core), but also on the frequency of the alternating current. As you know, the higher the frequency, the greater the rate at which the current will be changing, and hence the greater will be the opposing voltage of self-induction.

Reactance, being a form of impedance, is measured in ohms. If the frequency of the AC is f and the inductance of the coil in henries is L, then the reactance, X_L, is given by the formula

$$X_L = 2\pi fL$$

Note that the reactance depends not only on the inductance but also on the rate (frequency) at which the current changes.

Sylvania Electric Products Inc.

35-5 This coil contains many turns of wire wound on an iron core. It therefore has a high inductance. Inductance is a property of the coil that does not depend on the frequency of the applied voltage.

SAMPLE PROBLEM

Suppose a 10-henry coil is connected to a 110-volt, 60-cycle line. How much current will flow through the coil?

$$X_L = 2\pi fL$$
$$X_L = 6.28 \times 60 \times 10$$
$$= 3{,}770 \text{ ohms}$$

Now, using Ohm's Law and noting that the only impedance in the circuit is X_L,

$$I = \frac{E}{Z}$$
$$I = \frac{110}{3{,}770} = 0.0292 \text{ amp}$$

STUDENT PROBLEM

What current would flow in the coil in the above sample problem if the voltage supplied was 550 v at 25 cycles? (ANS. 0.350 amp)

---

6. Power in a reactive circuit

There is one important difference between the effect of reactance on a circuit and the effect of resistance. Whenever a

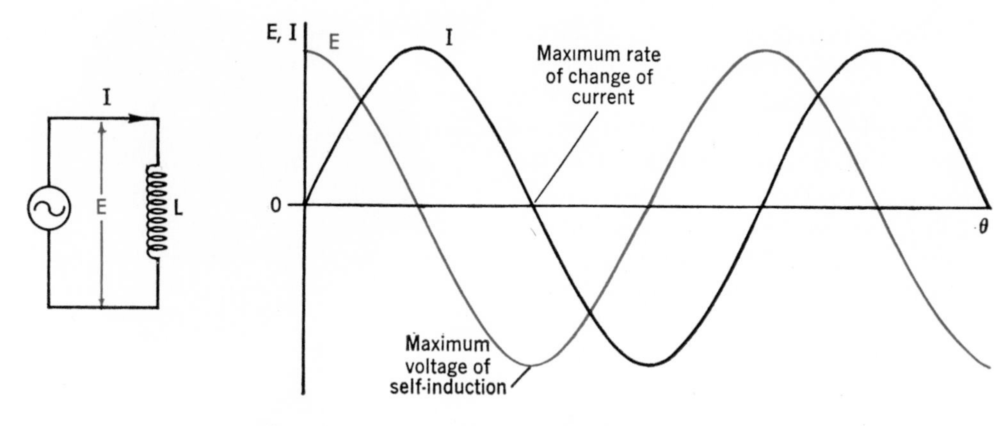

35-6 The voltage in an inductive circuit *leads* the current by 90°.

current flows through a resistance, power is consumed. However, when current flows through a reactance only, no power is consumed at all. Why is this? Remember that the maximum voltage is being induced in a coil when the magnetic field (and consequently the current) is changing most rapidly. The current is changing most rapidly when it is going through zero (changing direction). Hence the voltage is a maximum at the instant the current is zero, and is thus a quarter of a cycle out of step with it. That is, in a circuit containing only reactance the voltage is 90° out of phase with the current (Fig. 35-6), and we say that the voltage *leads* the current by 90°.

At every point on the graph in Fig. 35-6 we could multiply the voltage by the current to get the instantaneous power. Then we could add all these separate values of the power together for one cycle. We would find that the total amount of negative power (where E is positive and I is negative or E is negative and I is positive) would just equal the total amount of positive power. The resultant power would therefore be zero. In other words, no power would be consumed in the circuit. During one half of a cycle power is put into building up the magnetic field about the coil (a magnetic field represents a storage of energy). During the other half of the cycle this same power is given back to the circuit (as the magnetic field collapses). Of course if there is resistance in the circuit (and every coil has some resistance) then power (equal to I^2R) will be consumed by this resistance in producing heat, but the reactance, even though it impedes the flow of current, consumes no power.

7. Resistance and inductance in a circuit

Figure 35-7A shows a circuit containing both resistance, R, and inductance, L. Let us first find the impedance, Z, of the combination of L and R when a voltage, E, is applied to the circuit, and we know R and L (the inductance of the coil). We can then find the current through the combination from $I = E/Z$. We note that E_L, the voltage across the coil, *leads the current* through it by 90°, whereas E_R, the voltage across the resistance, *is in phase with the current*. Hence when we add these voltages, we must add them as vectors, and their resultant will equal the applied voltage E, as shown in Fig. 35-7B. The impedance of the circuit will then be $Z = E/I$, but we do not know I. However, we do know that the applied voltage, E, is the vector sum of the unknown values of $E_L + E_R$. If we apply Ohm's Law to each of these two voltages we have

$$X_L = \frac{E_L}{I}$$

$$R = \frac{E_R}{I}$$

and their vector sum is

$$Z = \frac{E}{I}$$

Since in each case we are dividing by the same quantity, I, we can redraw our vector diagram as an *impedance diagram* (Fig. 35-7C) to find Z from X_L and R. The diagram gives

$$Z = \sqrt{R^2 + X_L^2}$$

Now we can find I from Ohm's Law

$$I = \frac{E}{Z} = \frac{E}{\sqrt{R^2 + X_L^2}}$$

In the impedance diagram, the angle θ is called the *phase angle.* It tells how much the applied voltage, E, leads the current, I, in the circuit. Using the table of trigonometric functions (page 682), we can find θ:

$$\tan \theta = \frac{X_L}{R}$$

110-volt, 60-cycle AC is applied to a coil whose inductance is 0.2 henry and whose resistance is 100 ohms. What is its impedance? What current will flow through it and what will be the phase angle between the current and the applied voltage?

First we must calculate the reactance of the coil:

$$X_L = 2\pi fL$$
$$X_L = 6.28 \times 60 \times 0.2 = 75.3 \text{ ohms}$$

Knowing the resistance, we can now find the impedance:

$$\begin{aligned} Z &= \sqrt{R^2 + X_L^2} \\ &= \sqrt{100^2 + 75.3^2} \\ &= \sqrt{10^4 + 0.567 \times 10^4} \\ &= \sqrt{1.567 \times 10^4} \\ &= 100\sqrt{1.567} = 125 \text{ ohms} \end{aligned}$$

This gives for the current

$$I = \frac{E}{Z}$$

$$I = \frac{110}{125} = 0.880 \text{ amp}$$

The tangent of the phase angle is

$$\tan \theta = \frac{X_L}{R} = \frac{75.3}{100} = 0.753$$

and in the table of trigonometric functions (page 682) we locate 0.753 and find that $\theta = 37°$.

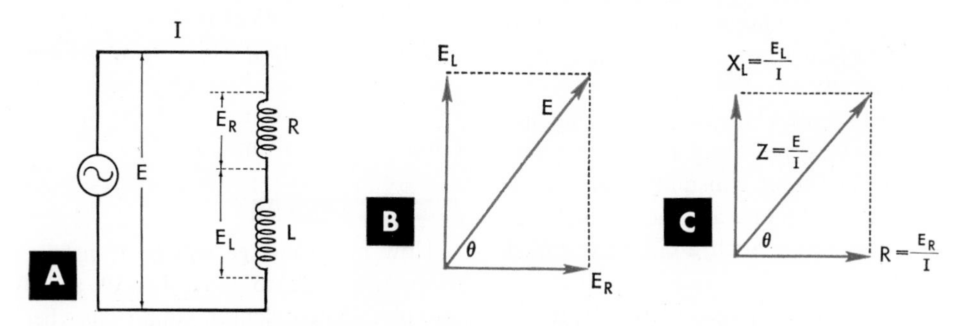

35-7 **(A) A circuit containing inductance and resistance, (B) the corresponding voltage vector diagram, (C) the impedance diagram.**

STUDENT PROBLEM

A 0.1-henry coil whose resistance is 10 ohms is connected to a 110-v, 60-cycle source. What is its impedance? What current flows? What is the phase angle? (ANS. $Z = 38.9$ ohms, $I = 2.83$ amp, $\theta = 74°$).

---

8. Uses of inductance in AC circuits

Inductances (or **inductors** as they are sometimes called) have many uses. Whenever, for example, we wish to reduce the current in an AC circuit without consuming power, an inductance can be used. This is done in theaters to dim the house lights. If a rheostat were used it would have to be very large to dissipate the large amount of heat produced. A coil having a very low resistance is placed in series with the lights. It has a negligible effect until an iron core is slowly pushed into the coil. Its inductance then rises rapidly, increasing the reactance. This increase in reactance decreases the current through the lights and makes them dim.

A small iron-core coil, known as a **ballast coil,** is placed in series with fluorescent lights (page 427) to limit the current when the mercury vapor starts conducting. This is necessary because the resistance of the conducting mercury vapor is so low that on 110 volts the current would be so large that it would burn the light out if it were not for the ballast coil in series.

Inductors are also used with condensers to change pulsating or varying DC to a steady DC of constant voltage such as is needed to operate radio and electronic apparatus. Such a circuit is called a **filter.**

Transformers (page 446) are really inductances coupled together so that the rapidly changing current in the primary coil induces voltages in the secondary coil. It is the reactance of the primary that makes the current drawn by the transformer small when no power is being drawn from the secondary.

9. The capacity of a condenser

A condenser, like a coil, has reactance in an AC circuit. In order to describe and measure its reactance, we must first learn how the capacity or **capacitance** of a condenser is measured.

When a battery drives electrons off one condenser plate and onto the other, the charge, Q, that is transferred from one condenser plate to the other is proportional to the capacity, C, of the condenser and is also proportional to the electromotive force, E, that is driving the charge from one plate to the other. Thus

$$Q = CE$$

This is similar to saying that the number of things, Q, that you can pack into a suitcase depends upon the capacity, C, of the suitcase and the pressure, E, you exert in cramming them in.

If Q is measured in coulombs and E is measured in volts, then the capacity, C, is measured in **farads.** A condenser has a capacity of 1 farad if 1 volt can succeed in transferring 1 coulomb of charge (6.3×10^{18} electrons) from one plate to the other. The farad is too large a unit to be practical for most purposes, and a smaller unit, called a **microfarad,** is commonly used instead. A microfarad is equal to one millionth of a farad ("micro" means "millionth").

SAMPLE PROBLEM

How much charge will be transferred from one plate to the other of a 100-microfarad condenser that has been placed across 500 volts DC? How many electrons is this?

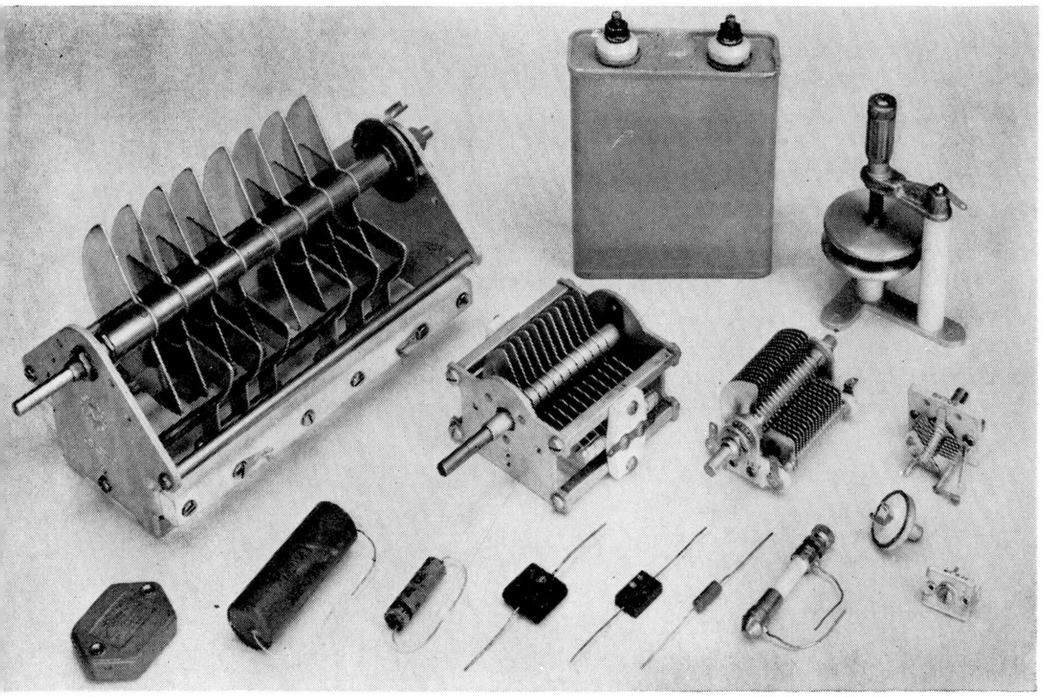

The American Radio Relay League

35-8 Condensers can be made in a variety of forms but all of them work in the same way. Some of those pictured here will look familiar if you have ever examined a radio or other electronic device.

Here $C = 100 \times 10^{-6}$ farad. Applying the formula:

$$Q = CE$$
$$Q = 100 \times 10^{-6} \times 500$$
$$= 0.05 \text{ coulomb}$$

This is $0.05 \times 6.3 \times 10^{18}$, or 3.1×10^{17}, electrons.

STUDENT PROBLEM

Find the charge on a 200-microfarad radio condenser when it is connected across 350 v. (ANS. 0.07 coulomb)

10. Capacitance in an AC circuit

When a condenser (page 337) is connected to a source of AC, it too acts like a resistance. Although no electrons actually move *through* the condenser, they surge in and out as the voltage across the condenser rises and falls through each cycle. Thus a movement of charge occurs throughout the circuit just as if the condenser were a conductor. The reactance, X_C, of a condenser whose capacity in farads is C is

$$X_C = \frac{1}{2\pi fC} \text{ ohms}$$

Like the voltage of self-induction, the voltage across the condenser is 90° out of phase with the current, but instead of leading the current it **lags** the current by 90°.

SAMPLE PROBLEM

What current will flow in a circuit consisting of a 20-microfarad condenser connected to a 110-v, 60-cycle line? Remembering that a microfarad is 10^{-6} farad, we find the reactance of the condenser is

$$X_C = \frac{1}{2\pi fC}$$

$$X_C = \frac{1}{6.28 \times 60 \times 20 \times 10^{-6}}$$

$$= 133 \text{ ohms}$$

The current that flows will be

$$I = \frac{E}{Z}$$

Since Z is all in the form of a **capacitive reactance**, X_C,

$$I = \frac{E}{X_C}$$

$$I = \frac{110}{133} = 0.827 \text{ amp}$$

STUDENT PROBLEM

What is the reactance of a 16-microfarad condenser at 60 cycles? (ANS. 166 ohms)

11. Resistance and capacitance in a circuit

In circuits containing inductance and resistance we found that the applied voltage *leads* the current. The reverse is true in circuits with capacitance and resistance. Here the voltage across *lags* behind the current. Figure 35-9 shows the circuit and the resulting vector diagrams.

To find the impedance of such a circuit we use the formula:

$$Z = \sqrt{R^2 + X_C^2}$$

and the phase angle is found from

$$\tan\theta = \frac{X_C}{R}$$

with the help of trigonometric tables (page 682). Note that the phase angle here is *negative,* since the vector for X_C is below the line of R, which is in phase with the current. This is because the voltage across the condenser *lags* the current.

SAMPLE PROBLEM

What are the impedance and phase angle in a circuit consisting of a 10-microfarad condenser in series with a 300-ohm resistance when supplied with 60-cycle AC?

$$X_C = \frac{1}{2\pi fC}$$

$$= \frac{1}{6.28 \times 60 \times 10 \times 10^{-6}}$$

$$= 267 \text{ ohms}$$

$$Z = \sqrt{R^2 + X_C^2} = \sqrt{300^2 + 267^2}$$

$$= \sqrt{9 \times 10^4 + 7.13 \times 10^4}$$

$$= 100\sqrt{16.1} = 401 \text{ ohms}$$

$$\tan\theta = \frac{X_C}{R}$$

$$\tan\theta = \frac{-401}{300} = -1.34$$

$$\theta = -53°20'$$

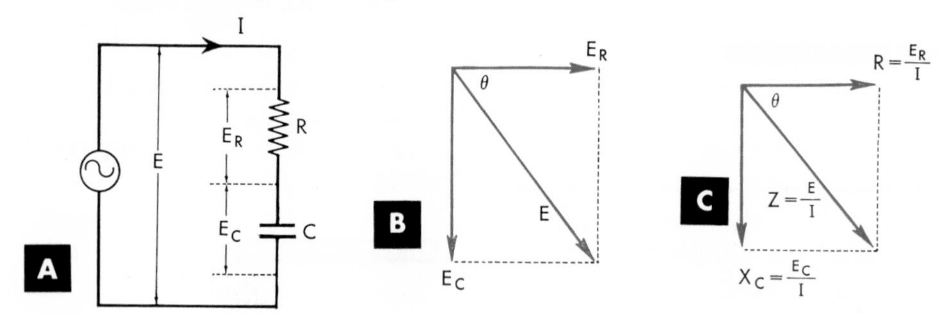

35-9 (A) A circuit containing capacitance and resistance, (B) the corresponding voltage diagram, and (C) the impedance diagram.

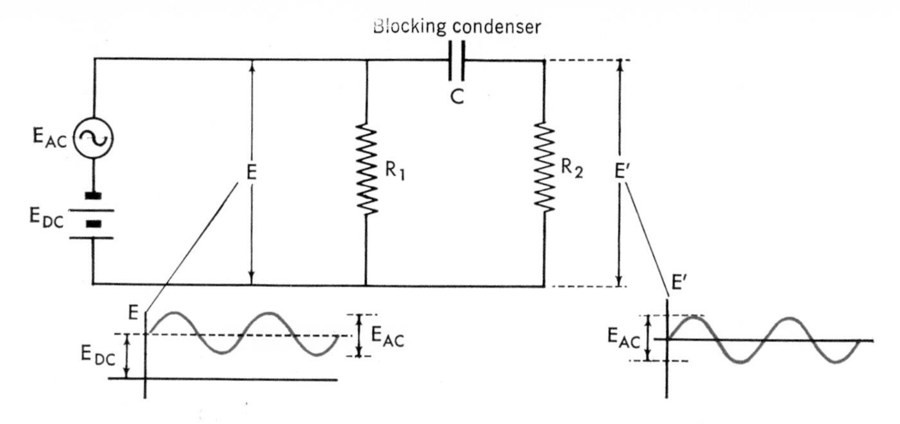

35-10 The combined AC and DC voltage, E, is separated into two parts by a blocking condenser, C, so that only the AC voltage appears across R_2.

STUDENT PROBLEM

What is the impedance of a 20-microfarad condenser at 60 cycles in series with a 100-ohm resistor? (ANS. 167 ohms)

12. Uses of condensers in AC circuits

One of the most common uses of condensers in an AC circuit is to separate an AC voltage from a DC voltage when both occur together, as they do in many circuits. Condensers used in this way are called **blocking condensers** or **by-pass condensers.** Figure 35-10 shows a circuit that uses a blocking condenser. The frequency, *f*, of direct current is zero. Since $X_C = 1/2\pi fC$, the value of X_C approaches infinity as *f* approaches zero. This means that DC current does not get through to R_2, and so no DC voltage appears across R_2. But if the capacitance *C* is sufficiently large, the condenser will offer a very low impedance to the AC voltage, and so nearly all of it will appear across R_2.

13. Resistance, inductance, and capacitance in a circuit

Figure 35-11A shows a circuit containing all three of the circuit elements that contribute to impedance. Let us find the impedance, the current, and the phase angle of this circuit at 60 cycles. X_L and X_C are determined as follows:

$$X_L = 2\pi fL = 6.28 \times 60 \times 1 = 377 \text{ ohms}$$

$$X_C = \frac{1}{2\pi fC} = \frac{1}{6.28 \times 60 \times 26.5 \times 10^{-6}}$$
$$= 100 \text{ ohms}$$

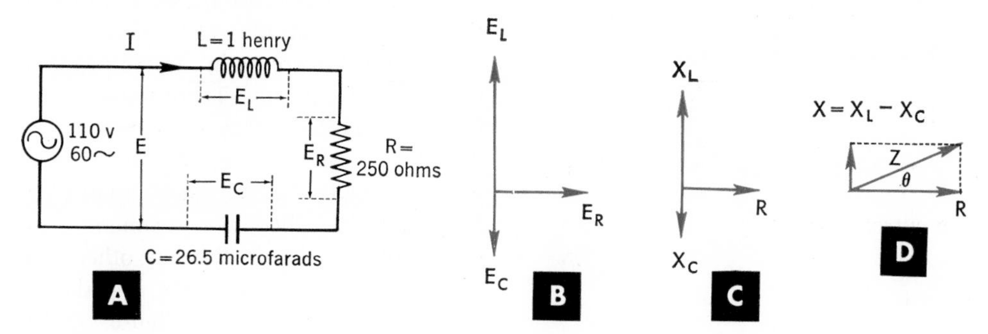

35-11 A circuit containing inductance, capacitance, and resistance with the resulting vector diagrams.

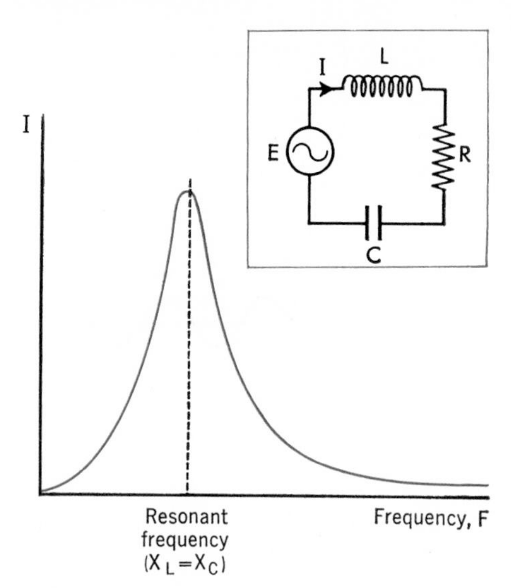

35-12 **The current in a resonant circuit is a maximum when the frequency is the resonant frequency ($X_L = X_C$). At the resonant frequency, Z = R and the current is equal to E/R. The resonant frequency is equal to** $\frac{1}{2\pi\sqrt{LC}}$.

The voltage across the coil is 180° out of phase with the voltage across the condenser. E_L therefore *leads* the current by 90°, and E_C *lags* the current by 90° (as shown in the vector diagram Fig. 35-9B). Hence the reactances X_C and X_L are also 180° out of phase with each other (Fig. 35-11C). Thus we take their arithmetical *difference* (Fig. 35-11D) to get their combined reactance. From Fig. 35-11D we see that the total impedance of the circuit is

$$Z = \sqrt{R^2 + (X_L - X_C)^2}$$

$$Z = \sqrt{250^2 + (377 - 100)^2}$$

$$= \sqrt{250^2 + 277^2} = 373 \text{ ohms}$$

The current will be

$$I = \frac{E}{Z}$$

$$I = \frac{110}{373} = 0.295 \text{ amp}$$

The phase angle is

$$\tan\theta = \frac{X_L - X_C}{R}$$

$$\tan\theta = \frac{277}{250} = 1.11$$

$$\theta = 48°$$

STUDENT PROBLEM

Find the impedance, phase angle, and current in the above sample problem if the frequency of the alternating current is 31 cycles per second. (ANS. Z = 250 ohms, $\theta = 0°$, I = 0.44 amp)

14. Resonance in a series circuit

Resonance exists in a series AC circuit when the opposing effects of the reactance X_L and the reactance X_C counteract each other. Let us examine the circuit in the previous section more carefully. First, consider the expression for the impedance

$$Z = \sqrt{R^2 + (X_L - X_C)^2}$$

If X_L were equal to X_C, then the above expression would reduce to

$$Z = \sqrt{R^2} = R$$

All we need to do to make $X_L = X_C$ is to use AC of the correct frequency, for as we increase the frequency (starting with $f = 0$) we see that as f increases $X_L = 2\pi fL$ must also *increase* proportionally while at the same time $X_C = 1/2\pi fC$ must *decrease* in inverse proportion to the frequency. There must therefore be some frequency for which $X_L = X_C$. At this frequency, the impedance will be equal to the resistance alone. This frequency is called the **resonant frequency** of the circuit. For any other frequency, higher or lower, the impedance will be greater than *R*. Figure 35-12 shows how the current in a circuit varies with frequency.

Notice that the current rises as the frequency rises with its accompanying fall in impedance, reaching a maximum value at the resonant frequency where $Z = R$. Notice too that current falls off at higher frequencies as the impedance again increases. This curve could be plotted experimentally by connecting to the circuit an ammeter and a generator of variable frequency and plotting the current against the frequency.

When $X_L = X_C$, the condition for resonance, we have

$$2\pi fL = \frac{1}{2\pi fC}$$

and

$$f = \frac{1}{2\pi\sqrt{LC}}$$

Thus if we know L and C we can calculate the frequency at which the circuit will be resonant and have its lowest impedance.

You will also note that the phase angle

$$\theta = \frac{X_L - X_C}{R} = 0$$

because

$$X_L - X_C = 0$$

SAMPLE PROBLEM

Let us calculate the resonant frequency of the circuit in Fig. 35-12.

$$f = \frac{1}{2\pi\sqrt{LC}}$$

$$f = \frac{1}{6.28\sqrt{1 \times 26.5 \times 10^{-6}}}$$

$$= 31.0 \text{ cycles/sec}$$

STUDENT PROBLEM

What is the resonant frequency of an 80-millihenry coil and a 2-microfarad condenser? (ANS. 398 cycles/sec)

▪ ▪ ▪ ▪ ▪ ▪ ▪ ▪

In radio circuits, as you will see in the next chapter, a coil and condenser are used together to "tune" a radio set. Either the inductance of the coil or the capacitance of the condenser in the tuning circuit is made variable so that different values of L or C can be selected by the tuning dial. In this way, X_L is made equal to X_C for the frequency of the desired broadcasting station, and for this station alone the radio circuit has a low impedance.

15. Power factor

In Section 6 of this chapter it was shown that no power is consumed by a coil because the voltage and current are 90° out of phase. The same is true of a condenser. However, if there is resistance in the circuit, power is consumed because the phase angle is not 90°. The power cannot be calculated by simply multiplying the effective voltage, E, by the effective current, I. We must take into account the phase angle, θ, between the voltage and the current. If the phase angle is 0°, then the power is indeed EI. But if the phase angle is 90°, then (as we saw in section 6 of this chapter and from Fig. 35-6) the power consumed is zero. For any phase angle, θ, the power is given by

$$P = EI \cos\theta$$

where $\cos\theta$ is called the **power factor.** The power factor reaches a maximum value of 1 when $\theta = 0$, as is true when we have resonance.

SAMPLE PROBLEM

How much power is consumed in a 110-volt circuit if a current of 20 amperes flows and the voltage leads the current by 60°? Substituting in the above equation we have

$$P = EI \cos\theta$$

$$P = 110 \times 20 \times \cos 60°$$

$$= 110 \times 20 \times 0.5 = 1{,}100 \text{ w}$$

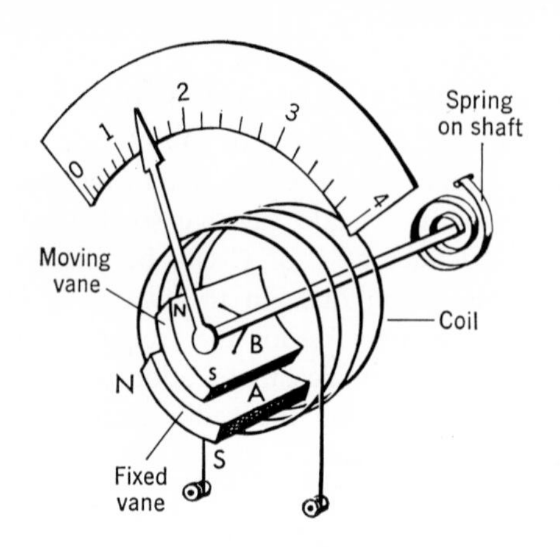

35-13 A moving-vane AC meter. The repulsion of the fixed and moving vanes causes the needle to move.

Notice that if we tried to measure the power consumed in the above circuit with an ammeter and a voltmeter we would get 2,200 watts—an error of 100 per cent! However, phase angle θ can be calculated from $\frac{X_L - X_C}{R}$, which is equal to tan θ (see Section 14). Cos θ can then be looked up in the trigonometric tables (page 682).

STUDENT PROBLEM

What would be the power consumed in the above problem if the phase angle were 88°? (ANS. 76.7 w)

AC meters and motors

The direct-current galvanometer described on page 414 will not measure alternating currents. This is because, as you remember, the direction in which the needle moves across the scale depends upon the direction in which the current flows through the meter. If such a meter is connected in a 60-cycle AC circuit, the current reverses direction so rapidly that the inertia of the moving coil and needle causes them to remain stationary or vibrate slightly no matter how strong the current.

16. AC meters

The cheapest type of AC meter is the **moving-vane** type. It contains a fixed coil surrounding two pieces (vanes) of soft iron, A and B (Fig. 35-13). A is fixed but B is mounted on a movable axis fixed to the pointer. The current through the coil magnetizes both pieces of iron in the same direction regardless of the direction of current flow. That is, the two "top" ends of A and B may *both* be north poles as shown in the figure and then an instant later when the current through the coil reverses they will *both* be south poles. Since like poles repel, there will always be a force of repulsion between the vanes which pushes the needle over the scale, always in the same direction. The greater the current, the greater will be the repulsion and the farther the needle will move against the resisting force of a hairspring. Moving-vane meters are not very sensitive or accurate, but they are cheap and rugged and so find a great deal of use.

The **dynamometer** type of AC meter is more accurate but more expensive. It is made like a DC galvanometer, but instead of a permanent field magnet it has a field coil (Fig. 35-14). The current to be measured flows through the field coil

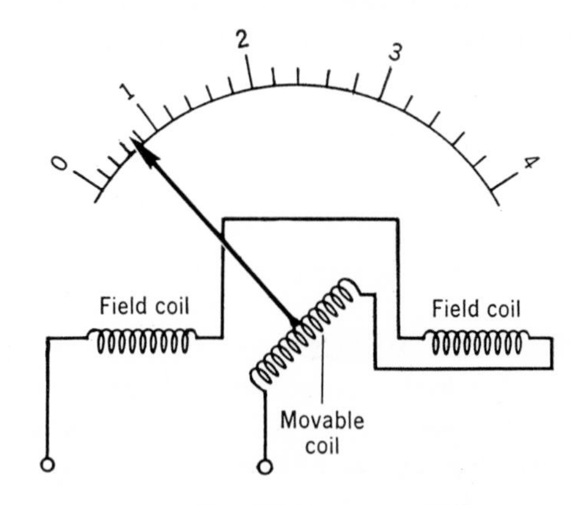

35-14 A dynamometer AC meter is built like a DC galvanometer except that, instead of a permanent magnet, it has a pair of field coils.

35-15 (A) A rectifier-type AC meter. The rectifier changes the AC to DC which is then read on a DC meter. (B) A thermocouple AC meter. The AC current flowing in the heating coil raises the temperature of the copper-iron junction and causes a DC current to flow through the meter.

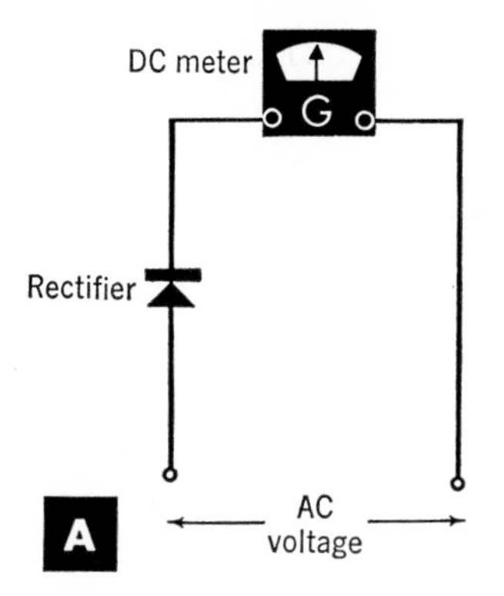

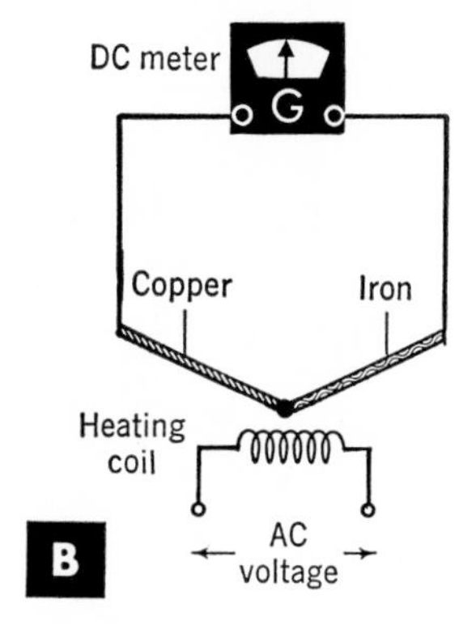

and the movable coil in series. When the current reverses through the movable coil it also reverses through the field coil, and so the force on the needle always remains in the same direction. These meters are not as sensitive as DC permanent-magnet meters, because it is difficult to develop a strong field using coils unless a large current is drawn.

Another type of AC meter is the **rectifier** type (Fig. 35-15A). It consists of DC meter in series with a rectifier. **A rectifier is a device that allows current to flow only in one direction. In other words, it converts AC to DC.** It acts in much the same way as the valve in a pump which allows water to flow through a pipe in only one direction. In a rectifier AC meter, the rectifier is made of copper oxide, selenium, or germanium, through which current can flow in only one direction. A DC meter may then be used in series with the rectifier to measure the current.

Still another type of AC meter is the **thermocouple meter** (Fig. 35-15B). This meter depends upon the **thermoelectric effect.** If wires of two different metals such as copper and iron are joined together and heated at the junction, a DC current will flow around the circuit formed by the iron and copper and connecting wires. The higher the temperature at the junction, the greater the current. An AC thermocouple meter contains a tiny heating coil through which the AC flows. The flow of current raises the temperature of the iron-copper junction, causing DC to flow through the galvanometer, whose scale is calibrated to read the current flowing in the AC circuit.

Another type of AC meter is a **vacuum-tube voltmeter** (VTVM) in which AC is amplified by vacuum tubes, rectified, and then applied to a DC meter. Such meters are complex and expensive but are very sensitive.

AC meters generally read effective (RMS) values.

17. AC motors

Some small motors using AC are made in the same way as DC motors and operate on either AC or DC (page 388). Such motors are not very efficient, however. Most large AC motors are of the **induction** type. That is, they have no brushes or commutators but depend upon the principle of the transformer to induce current from the stationary field into the rotating armature. By properly applying voltage of different phase to one field winding after another, the magnetic field is made to increase in one pair of windings (Fig. 35-16) while it is

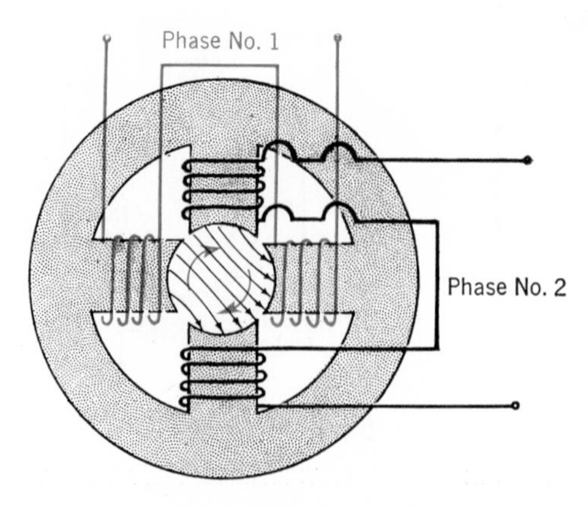

35-16 A two-phase rotating field. If Phase 1 and Phase 2 differ by 90°, the magnetic field in the center will rotate at the same rate as the frequency of the alternating current.

decreasing in the pair of field windings at right angles. The effect is to cause the resulting magnetic field to rotate, carrying the armature around with it. The speed of rotation depends upon the frequency of the alternating current. Such induction motors are also called **synchronous motors**; that is, their rotation is synchronized with the alternations of current. Electric clock motors, phonograph motors, and large motors such as the one in Fig. 30-12 are examples.

In some synchronous motors, like the one in Fig. 35-16, the separate phases are provided from a special power line. The common **three-phase motor** is supplied by three wires, each of the three combinations of two wires carrying currents differing in phase by 120° from the currents in the other two combinations (Fig. 35-17). In most small motors running on the ordinary 110-volt supply, however, the phase of part of the current supplied to the motor is shifted in the mechanism of the motor itself. A **shaded-pole motor** is of this type.

18. Converting AC to DC

It is often necessary to change AC to DC, since some types of electrical apparatus work only on DC (vacuum-tube amplifiers, electroplating equipment, battery chargers, etc.). This is usually done with selenium or silicon rectifiers of large power capacity, or by means of vacuum tubes called **diodes**, which are described in the next chapter. It can also be accomplished by using AC to run an AC motor which drives a DC generator. This is called **a motor-generator set.**

In automobiles, boats, and airplanes, the only source of voltage is low-voltage DC, usually 6, 12, or 24 volts. Radio apparatus requires much higher DC voltages (from 100 to several thousand volts). The simplest way to get this high DC voltage is to use the low-voltage DC

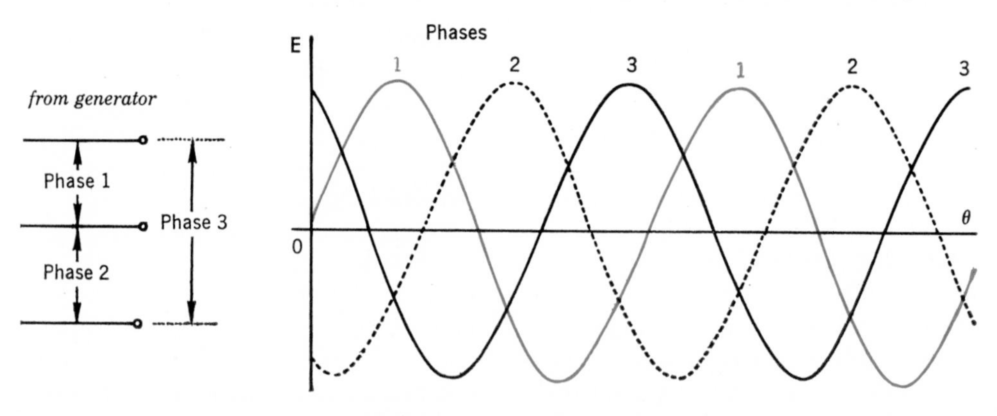

35-17 A three-phase AC supply. The voltage between any pair of wires is the same but differs in phase by 120° from the voltage obtained from the other two pairs.

to operate a motor which drives a high-voltage DC generator. When motor and generator are combined in a single unit having a low-voltage field winding in common, this device is called a **dynamotor.**

Automobile radios commonly use a **vibrator supply** to change the DC of the battery to AC. It operates like a combination electric buzzer and relay. The rapidly vibrating armature opens and closes contacts, which rapidly reverse the current direction, converting DC to AC. This AC is then stepped up by a transformer to the required high-voltage AC which is then rectified into DC.

Going Ahead

THE VOCABULARY OF PHYSICS

Define each of the following:

RMS voltage
impedance
phase angle
cycle
dynamotor
inductance
frequency
thermocouple meter
capacitive reactance
lag
blocking (by-pass) condenser
resonant frequency
rectifier
3-phase motor
dynamometer
induction motor
vibrator
reactance
power factor
peak voltage
rotating vector
phase
henry
ballast coil
moving-vane meter
synchronous motor
thermoelectric effect
shaded-pole motor
capacitance
microfarad
vacuum-tube voltmeter
farad
filter
sinusoidal voltage
effective value
voltage of self-induction

THE PRINCIPLES WE USE

1. What is the impedance of a circuit at its resonant frequency equal to if it consists of R, X_L, and X_C in series?
2. Name four kinds of alternating-current meters.
3. What are the three important circuit elements in an AC circuit?
4. What are the two kinds of reactance?

Explain why

5. A resonant circuit has its lowest impedance at the resonant frequency.
6. A permanent-magnet field cannot be used in a dynamometer-type AC meter.
7. An inductance is used to dim theater lights.
8. A DC galvanometer cannot be used directly to measure AC.
9. Power in a reactive circuit cannot be calculated by taking the product of voltage and current.
10. DC may burn out a transformer.

PRACTICAL APPLICATIONS

1. How can you increase the resonant frequency of an AC circuit?
2. In what two ways could you increase the AC current through a variable condenser without changing the magnitude of the applied voltage?
3. How would you find the sum of the voltages of two out-of-phase generators connected in series?
4. How would you calculate the peak current flowing through a 110-v, 100-w light bulb?

PROBLEMS

1. What is the peak voltage obtained from a 60-cycle, 110-v line?
2. An AC circuit draws 20 amp on a 110-v line. What is its impedance?
3. A circuit with a 20-ohm impedance is connected to a 220-v RMS line. What is the peak current that flows?
4. A peak voltage of 160 v causes an RMS current of 10 amp. What is the impedance of the circuit?

---

5. What is the reactance at 60 cycles of (a) a 5-henry coil, (b) a 0.1-microfarad condenser?

6. What are the impedance and phase angle at 400 cycles of a 0.5-microfarad condenser in series with an 800-ohm resistance?
7. What is the power factor of the condenser and resistance in Problem 6?
8. A 0.25-henry coil is connected in series with the condenser and resistance of Problem 6. At 400 cycles, what will be (a) the impedance, (b) the phase angle?
9. How much current would flow in the circuit of Problem 8 if it were supplied with 1,000 v?
10. What is the resonant frequency of the circuit in Problem 8?
11. What size condenser must be used with a 500-microhenry coil to be resonant at 10 megacycles?
12. An electric hot-water heater contains a 12-ohm resistance coil. The heater is placed in 10 liters of water and 8 amperes of alternating current flow through it for 15 minutes. Find the temperature of the water at the end of this time, assuming that all the heat produced is used to heat the water.
13. When an inductance of negligible resistance is connected across 125-volt 25-cycle AC, a current of 4 amperes flows through it. What current would flow through it if it were connected across a source of 110-volt 60-cycle AC?
14. A coil has an inductance of 0.25 henry. What is its reactance when connected across a 1000-cycle source of voltage?
15. A source of 60-cycle 110-volt AC is connected across a 40 mfd condenser. Find (a) the reactance of the condenser, and (b) the current in the circuit.
16. (a) Find the inductance of a coil that has a reactance of 200 ohms when used on a 25-cycle source of voltage. (b) Find the capacitance of a condenser that has a reactance of 200 ohms when used on a 25-cycle source of voltage.
17. A coil with an inductance of 0.10 henry and a resistance of 8 ohms is connected across the 100-volt 60-cycle line. Find (a) the reactance of the coil, (b) the impedance of the coil, (c) the current through the coil, (d) the phase angle between the current and the voltage, (e) the power factor, and (f) the power produced in the coil.
18. Across a 60-cycle 110-volt source there are connected in series a 75-ohm resistor, a coil whose reactance is 30 ohms and resistance is 8 ohms, and a condenser whose reactance is 90 ohms. Find the voltage drop across the resistor, the coil, and the condenser.

CHAPTER

36

Electronics and Communication

Did you know that fighting in the War of 1812 continued for several weeks after the peace treaty had been signed in Ghent? There was no way of getting the news of the treaty from Belgium to the armies in America quickly. Today this would not happen. Radio, telephone, or telegraph would bring the news almost instantly to all corners of the earth. It is hard for us to realize what it was like to live only a century and a half ago when the telegraph was in its infancy and telephones and radio were unknown.

The telephone

The first big advance in communication came when Samuel F. B. Morse developed practical long-distance telegraphy in 1844. Telegraph poles and wires began to spread over the country. Several decades later the transatlantic cable was laid and telegraphic messages were being sent to and from Europe.

Alexander Graham Bell's invention of the telephone came in 1876. The next year Thomas Edison added his carbon-granule transmitter to the telephone. Since then telephone wires have criss-crossed the country and spread out under the streets of our cities.

1. Changing sound into electricity

The idea behind the telephone is simple enough. The transmitter (or **microphone**) shown in Fig. 36-1 consists of a flexible **diaphragm** (dī′ă·frăm), *D,* which vibrates in time with the compressions and rarefactions of the sound waves that strike it (see Chapter 22 for a review of the nature of sound). When a sound wave with a frequency of 100 vibrations per second reaches the transmitter it forces the diaphragm to vibrate at the same frequency.

Behind the diaphragm is a small box (*C* in Fig. 36-1) which contains loose carbon grains. Current from the battery passes through these grains. When a compression of the sound wave reaches the microphone the flexible diaphragm is pushed inward. The carbon grains are squeezed so that they make better contact with each other, and the resistance of the grains to the passage of current is lowered (Fig. 36-2A). Hence the current in the circuit increases whenever a compression causes the diaphragm to move inward.

When a rarefaction of the sound wave reaches the microphone, the diaphragm moves outward. This decreases the pressure on the grains so that they tend to separate, there are fewer points of contact, and resistance to the flow of current is increased (Fig. 36-2B). Thus current decreases whenever a rarefaction reaches the transmitter and causes the diaphragm to spring outward. If a diaphragm vibrates with 100 vps, in re-

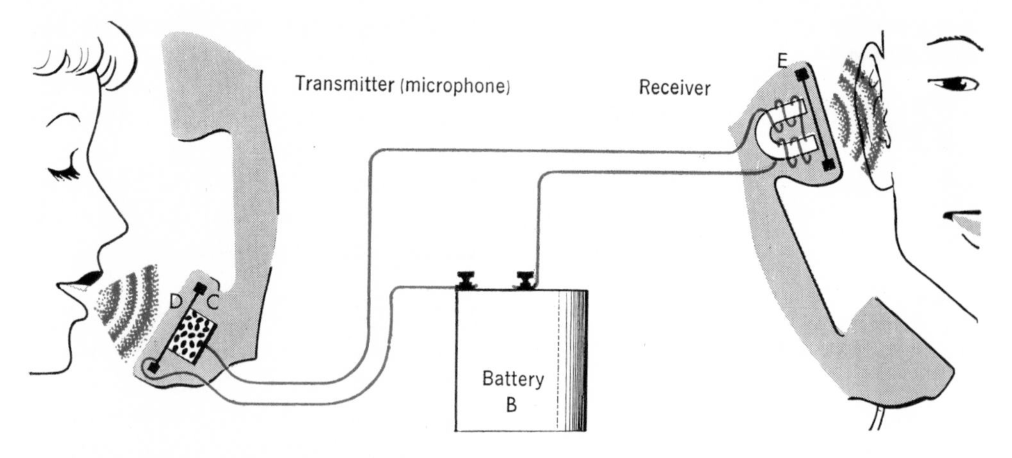

36-1 The telephone is amazingly simple. See if you can name the parts of the microphone and receiver and explain how they work.

36-2 In A, compression from a sound wave squeezes the carbon particles, improves their electrical contact, and increases the current. By means of the increased and decreased pressures of the sound waves, increased and decreased currents are allowed to flow in the transmitter.

sponse to a sound wave of 100 vps, then the current increases and decreases 100 times a second. In this way the vibrations of sound are converted into similar changes in the electric current. These electrical changes have the same timing as the vibrations. And this changing current can be sent through a wire to the listener far away.

You see that it is actually not sound that travels along the telephone wire but a changing electric current. This type of changing current in the simple telephone circuit is called **pulsating direct current.** It is DC because it always travels in the same direction.

2. The telephone receiver

This pulsating electrical current is converted back into sound by an electromagnet (E in Fig. 36-1) in the receiver. As the current increases and decreases, so does the strength of the electromagnet. A springy iron disk near E is alternately pulled close to the electromagnet and released. In this way the pulsations of electric current are converted into the motion of the iron disk, and the vibrations of the disk set the air at the receiver vibrating at the same frequency as the

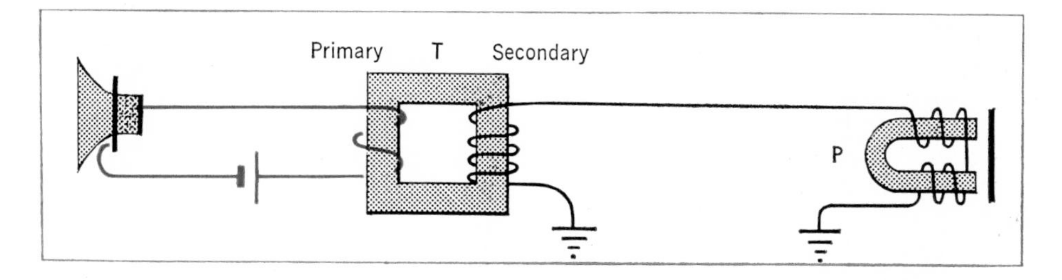

36-3 A step-up transformer in a telephone circuit helps in two ways. First, it steps up the voltage of the changes in the current. Second, it makes unnecessary the transmission of the battery current over long distances. Only the changes in current are transmitted over the wire.

original sound. Thus a sound wave is sent out from the receiver which has the same number of vibrations per second as the original sound.

Figure 36-3 shows an improved telephone circuit.

The electron tube

The first electron tube, or **vacuum tube,** was made by Thomas A. Edison in 1883, five years before radio waves had even been discovered. Edison had become annoyed with the blackening of the inside of the glass bulb of his newly invented incandescent lamp. In an effort to find out the cause of this undesirable effect, he inserted a metal plate in the bulb and connected a galvanometer between the plate and the filament (Fig. 36-5A). To his surprise he found that the galvanometer showed that a small current flowed when the filament was hot, and that the hotter the filament, the greater was the current. Edison concluded that the filament was emitting electricity as a consequence of its being hot. This emission of electrons by in-

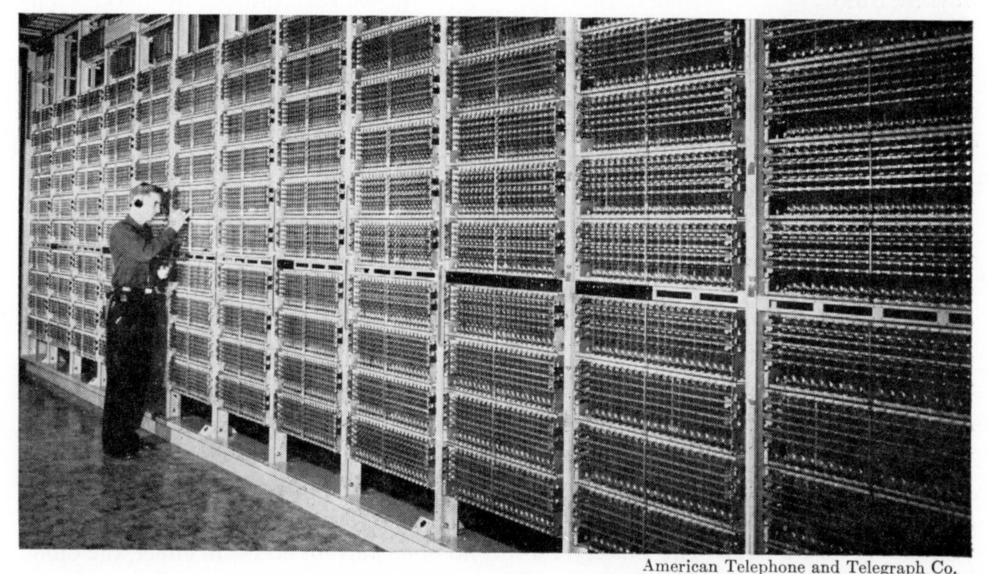

American Telephone and Telegraph Co.

36-4 This picture shows only a small part of a large automatic telephone exchange. Here calls are switched by means of electromagnets which in turn are operated by the dial on the caller's telephone. Can you see why dust must be kept out of this room?

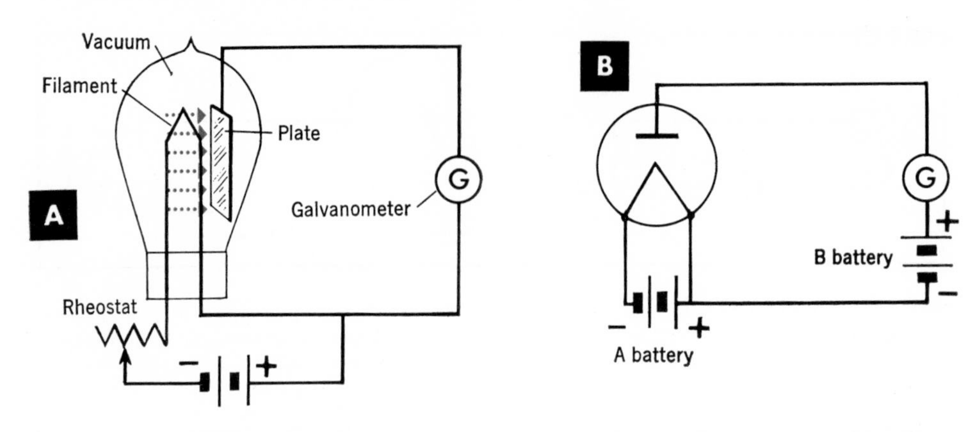

36-5 (A) Edison showed that electrons flow from the hot filament of a light bulb to a metal plate inside the bulb. (B) A battery will increase the flow of current if it makes the plate positive.

candescent solids is called **thermionic emission** ("thermionic" meaning "heat").

We can picture the emission of electrons by a hot solid in a vacuum as being similar to the evaporation of molecules from a hot surface. The higher the temperature, the greater the rate of evaporation of electrons into the space inside the glass bulb.

3. The diode

It was not until 1904 that Edison's light bulb with its two electrodes (filament and plate) was put to work. An English scientist, J. A. Fleming, realized that the electrons emitted by the hot filament (also called the cathode) would conduct a current (a flow of electrons) from the filament across the vacuum to the plate (or anode), but not in the reverse direction. Fleming used this two-element tube, or **diode** ("di-" meaning "two"), as the electronic valve to change AC to DC, a process which is called **rectification.**

Let us see how such a tube works. Suppose we set up the diode as Edison first did (Fig. 36-5A). The battery furnishing current for the **filament circuit** is called the **A battery.** You will note that we have inserted a rheostat (variable resistance) in the filament circuit to enable us to control the amount of current that lights the filament and hence to control the filament temperature. The circuit consisting of the plate, galvanometer, and the filament of the tube is called the **plate circuit.** As the resistance of the rheostat is slowly reduced to permit the filament temperature to rise, the current in the plate circuit (which the galvanometer registers) slowly increases. The amount of current, however, is usually small—measured in milliamperes (a milliampere is a thousandth of an ampere). If we pull electrons from the filament by placing a battery in the plate circuit, connected so that it gives a positive charge to the plate (Fig. 36-5B), we find that the plate current is greater. The higher the voltage of this plate-circuit battery (called a **B battery**), the greater the plate current. However, if the polarity of the battery is reversed, the galvanometer reads zero, since a negatively charged plate repels the electrons streaming from the filament and turns them back before they reach the plate, so that no plate current flows. In effect, the diode acts as an electrical conductor when its plate is positive, and as an insulator when its plate is negative.

4. The diode as a rectifier

Now let us do as Fleming did and replace the B battery with a source of AC. We find that the DC galvanometer indicates a flow of current in one direction (you will remember that a DC meter will not give any reading at all when only AC flows through it). Thus we see that AC can be converted to DC by a diode, since it conducts current in only one direction. Since the diode is a conductor during only a half cycle, and a nonconductor during the other half cycle, a graph of the rectified current would look like that shown in Fig. 36-6. You can see from the graph why this process is called **half-wave rectification.**

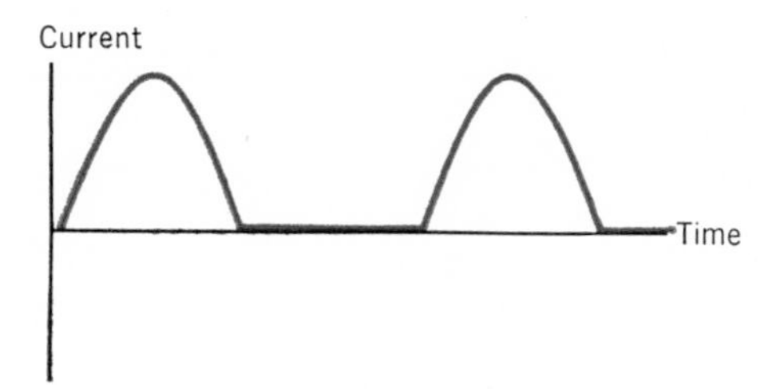

36-6 A graph of how current changes with time in a half-wave rectifying circuit. The diode allows current to flow only during half of each cycle.

We can replace the galvanometer with any device that requires DC for its operation. For example, the charging of a storage battery requires DC. Figure 36-7 shows how the battery to be charged would be inserted in the plate circuit. Since during the charging process electrons must be forced into the negative terminal of the battery (page 363), the negative terminal must be connected to the plate. In a battery charger used in garages (Fig. 28-8) the voltages for operating the filament and plate circuits are obtained from two separate secondaries of a step-down transformer.

5. Smoothing out pulsating DC—filters

The graph in Fig. 36-6 shows that a half-wave diode rectifier does not produce steady DC; it gives a pulsating direct current. Pulsating DC, however, is not satisfactory for all DC needs. For example, in certain circuits of a radio the use of pulsating DC would produce a hum in the speaker. Steady DC of constant voltage must be used. A battery will provide such DC, but the pulsating DC output of a diode can be used if it is changed into a steady DC. This can be accomplished by the use of a **filter.** A simple filter consists of a condenser and an iron core inductance arranged as shown in Fig. 36-8.

The pulsating DC voltage from the rectifier can be thought of as being the sum of a steady DC component (the average value of the pulsating DC) and an AC component, as shown in Fig. 36-8. What is the effect of the inductance, *L,* and the condenser, *C,* on each of these components? The reactance of the inductance ($X_L = 2\pi fL$) is very large for AC but practically zero for DC, whereas the reactance of the condenser ($X_C =$

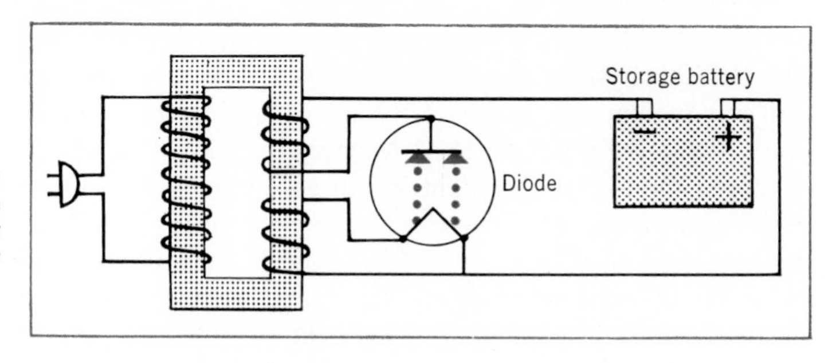

36-7 A diode is used with a step-down transformer to give direct current to charge a low voltage storage battery.

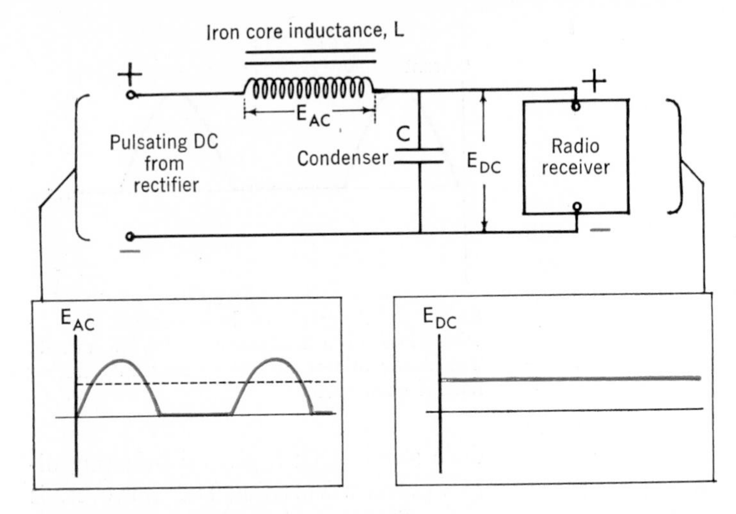

36-8 A filter changes pulsating direct current, E_{AC}, into steady direct current, E_{DC}, by separating the alternating components of voltage from the average DC component of voltage.

$1/2\pi fC$) is very large for DC but nearly zero for AC. The result is that nearly all the voltage of the AC component appears across the coil and very little AC voltage appears across the condenser. On the other hand, practically all the DC voltage appears across the condenser. Thus we have separated the AC and DC parts of the pulsating DC, and only steady DC is supplied to the radio receiver in Fig. 36-8.

Radio

The most amazing development in communication came with the invention of radio. This made it possible to eliminate wires between the transmitter and the receiver.

6. The invention of radio

As so often happens with scientific discoveries, the experiments which led to the discovery of radio were at first carried out with another purpose in view. The British physicist James Clerk Maxwell was seeking an answer to the question, "What is light?" In 1873 Maxwell advanced the theory that light is caused by an **electromagnetic wave** set up by the vibrating motion of electrical particles. He predicted that there might be electromagnetic waves similar to light waves and moving at the same speed, but of longer wave lengths. Maxwell was unable to tell how to detect these "radio waves" (as we call them today). But only fifteen years after Maxwell stated his theory, the German scientist Heinrich Hertz produced radio waves from a simple discharge of a spark, and received them using a single loop of wire as both his antenna and receiving set.

7. Electromagnetic waves

You remember Faraday's experiment with the transformer (page 443). Here a primary coil was made to induce E.M.F. in a neighboring coil by sending out lines of force that cut across the secondary coil. If we remove the iron core from Faraday's two coils, we have a device similar to a radio transmitter and receiver.

Figure 36-10 shows how we can send a "message" across a short distance with this simplified "radio" set. The lower coil is connected to an alternating current. Another coil, several inches above it, is connected to a lamp. AC in the primary sends out moving lines of force which induce enough current in the secondary to light the lamp. We could transmit messages by inserting a key or switch in the primary. In this way we can make the lamp connected with the secondary go on and off. This effect corresponds to a simple kind of dot-dash wireless set. Figure 36-9 shows a similar experiment using an actual radio transmitter, in which current induced in the lamps is sufficient to cause them to light up at a distance of several feet.

A radio wave is called an **electromagnetic wave** because it consists of electric and magnetic fields which travel through space. We can get an idea of radio waves if we imagine the lines of force from the primary moving outward. The radio wave, however, consists not only of magnetic lines of force but also of electric lines of force created by electric charges. And instead of going out and back, as happens with the transformer, the lines of force in a radio wave go out and do not return. They continue to travel outward with the speed of light waves: 186,000 miles per second—fast enough to go around the earth during the time it takes to blink your eyes!

You might imagine that the 60-cycle AC in the wires of your home would send out radio waves because the lines of force around them are changing. However, the properties of such electrical circuits are not very efficient for this purpose. At the moment of turning a switch on or off in your home a high-frequency spark may be created in the switch setting up, momentarily, a very rapidly alternating current. The conditions are suitable at the moment for producing electromagnetic waves. This is why when you turn a light on or off in your living room in the evening you may

Westinghouse Electric Corp.

36-9 **Magic wands? Not at all. A nearby short-wave radio transmitter induces enough current in the lamps to cause them to light up.**

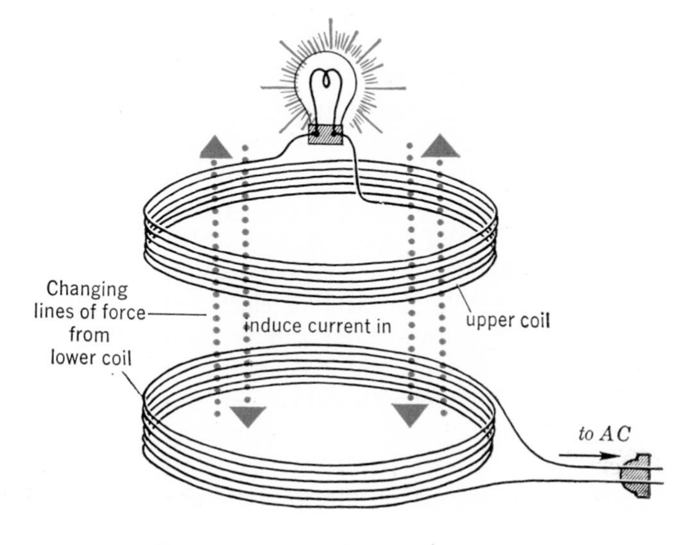

36-10 **Transmission of energy through a distance without wires. The lamp in the upper coil can be made to light from current induced in it by lines of force moving up and down from the lower coil (primary). In effect, this is a very elementary kind of radio transmitter and receiver.**

The American Radio Relay League

36-11 Amateur radio links this boy directly with other amateur operators all over the world. In radio transmitters and receivers induced high frequency voltages are put to constant use.

hear a click in the radio set near by, if the set is on. In the same way buzzers, bells, electric motors, electric shavers, and lightning produce high-frequency sparks which generate electromagnetic waves which may be picked up by a radio set as static. Automobile radios must therefore be designed to eliminate the interference from sparks from the ignition system.

8. Radio transmitters

The first radio transmitters employed by Guglielmo Marconi produced radio waves by means of sparks. Later, powerful high-frequency (referred to as **radio-frequency** or **RF**) generators (capable of about 20,000 alternations per second) were used. With the help of electronic tubes, today's transmitters can produce alternating currents with frequencies as high as billions of alternations per second. Each station generates alternating current at its own particular radio frequency by means of the vacuum-tube oscillator described in Section 13. Thus a station that is tuned at 790 on the dial has a frequency of 790 kilocycles (kc) per second. Another station, located at 1,450 on the dial, is producing alternating current of 1,450 kilocycles—that is, 1,450,000 cycles per second!

The rapidly alternating current produced at the transmitter travels up and down the transmitting antenna and sends out electromagnetic waves of the same frequency as the current. These waves, sweeping across distant metal objects—wires or antennas—induce in them alternating currents similar to those in the broadcasting station's antenna. We can think of the transmitting antenna as the primary, and the receiving antenna as the secondary, in a kind of gigantic but very inefficient transformer.

It is simple enough for the receiving antenna to pick up the feeble signals. The difficulty comes in magnifying these weak currents and converting them into sound. Also we have to be able to pick out the desired radio station by selecting its particular frequency from among all the different frequencies present in the antenna. Let us see how these difficulties are overcome.

9. Changing the carrier wave

Like a telephone transmitter, the microphone at the radio station converts sound waves into changing electric currents having the same frequency as the sound waves producing them. This frequency range is known as **audio frequency (AF).** In Fig. 36-12 these changing currents are represented as the **voice current** (or **modulation current**). At the same time special vacuum-tube circuits produce a radio-frequency alternating current. This current is known as the **carrier current** and is pictured in Fig. 36-12B. The voice current is mixed with the carrier current and changes the intensity of the carrier current. The result is a **modulated carrier current,** shown in Fig. 36-12C. This method of mixing vibrations is called **amplitude modulation (AM).** In AM we change the amplitude, or intensity, of the carrier current without changing its frequency.

Once we have our modulated carrier current we amplify it by special electrical circuits. This amplified current is then sent to the transmitting antenna.

The high-frequency AC carrier current creates electromagnetic waves which sweep out in all directions from the transmitting antenna.

10. Converting the electrical signal into sound

How do we get the original sound from the changing current induced in our antenna (L in Fig. 36-12D)? To simplify matters let us suppose that the antenna receives waves from only one radio station.

If we send the small induced current in the antenna directly into a telephone receiver (R in Fig. 36-12D), in the hope of reproducing the original sound, we find that we don't get any sound at all. The RF alternations of the current, about a million per second, are too rapid for

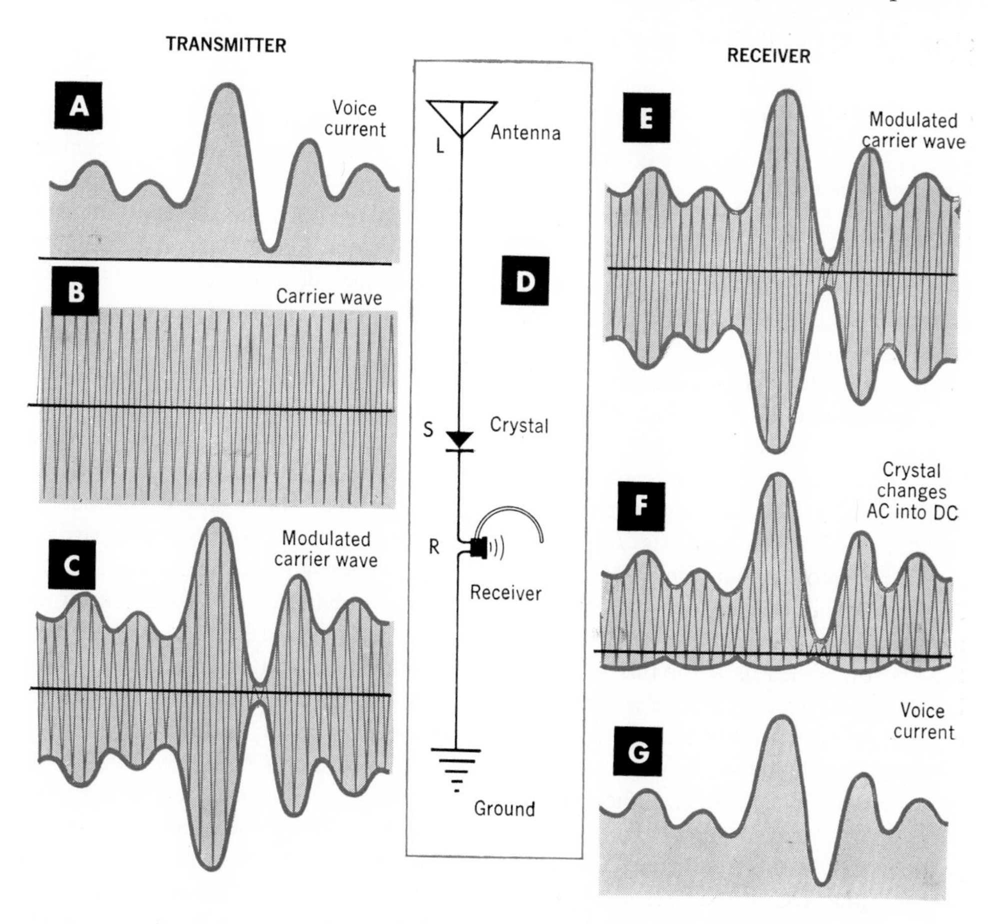

36-12 Here we see the basic changes that take place in the voice current (A) and carrier current (B) in all radio transmission. C shows the mixed or modulated carrier current that is transmitted as a radio wave. This wave is received (E), converted into current, and unscrambled by the radio receiver (F) to reproduce the original sound (G).

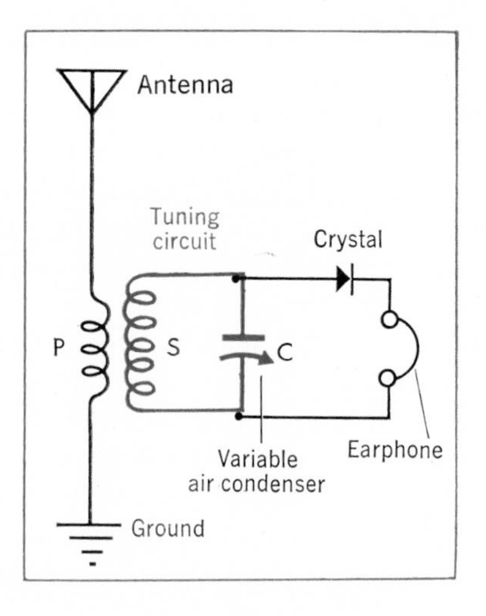

36-13 The particular frequency of a certain radio station is selected from all those in the antenna by means of the tuning circuit consisting of a transformer, P–S, and a variable air condenser, C. This circuit resonates to a particular radio frequency in much the same way that a tuning fork can be made to resonate to an air column of the proper sound frequency.

the iron disk in the receiver; the disk simply cannot vibrate that fast. But if we place a small **crystal detector,** usually made of the elements germanium or silicon, in the circuit at S (Fig. 36-12D), then we can hear the original sound. This crystal has the property of permitting current to flow just in one direction, so that the crystal converts AC (Fig. 36-12E) into a pulsating DC (Fig. 36-12F). This conversion, you will remember, is called rectification. A diode tube is sometimes used for this purpose, but it is bulkier and, unlike a rectifying crystal, requires power to heat an electron-emitting filament. Now we can get sound from the earphone. Although this earphone cannot respond to the extremely fast radio-frequency pulses of the pulsating DC, it does respond to the more gradual audio-frequency peaks of current, and hence the original sounds are reproduced in the earphone. This process of rectifying the high-frequency signal in the antenna, leaving the pulsating audio-frequency DC, is called **demodulation** or **detection.**

So you see that if we are close enough to a radio station we can receive the signal from the station and turn it into sound if we have the following equipment: an antenna in which the radio wave induces a current; a crystal to convert this alternating current into pulsating direct current; and an earphone to produce sound from these pulsations. The operation of the earphone is described on page 472 and the operation of the speaker on page 482 (Fig. 29-4).

11. Tuning in on a station

Now suppose we are faced with the problem of selecting one station out of the many which are inducing current in our antenna. We call this process **tuning.** If several stations are transmitting signals, all these signals induce currents in the antenna of our receiver. At first glance it might seem an impossible task to pick out one signal from such a scramble and reject the rest. But it can be done. Let us see how.

Each radio station has a carrier wave of a particular frequency which is assigned to that station alone. So the job of tuning amounts to selecting a particular carrier-wave frequency.

The receiving circuit is arranged as shown in Fig. 36-13. Instead of feeding the signal from the antenna directly into the crystal detector and then to the earphones, we pass it through the primary of a transformer (P in Fig. 36-13). In most radios this transformer consists of two coils, primary and secondary.

The secondary coil picks up the high-frequency AC signal by induction. Now we place a variable condenser, consisting of two sets of metal plates separated by air (see page 337 and Fig. 26-12), in the circuit at C. This condenser and

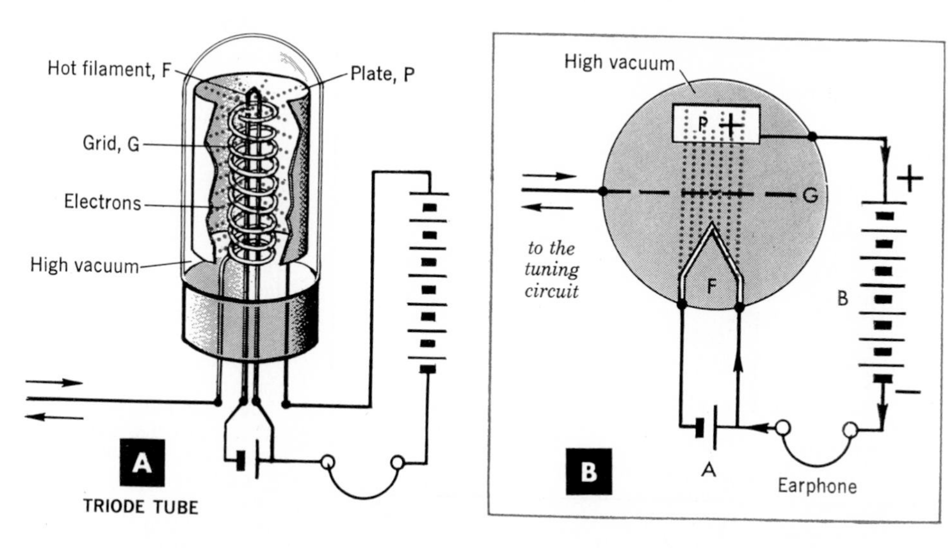

36-14 The vacuum tube is the heart of all electronic devices such as radio, television, and radar. A typical tube is shown in cutaway in A. B is a symbolic diagram of the tube and a typical circuit for its use. A hot filament, F, sends out electrons which are attracted to a positive plate, P. Slight changes in the electric charge on the grid, G, produce much greater changes in the current to the plate and so are magnified.

coil S form the tuning circuit. We change the capacity of the condenser by rotating one set of its plates. In this way the tuning circuit is able to pick out (resonate with) any frequency one desires. For example, if one desires to tune in on a station broadcasting on 780 kc, he adjusts the capacity, *C*, of the variable tuning condenser so that its reactance, X_C, will equal the inductive reactance, X_L, of the coil when an alternating current of 780,000 cycles is flowing. That is, one adjusts the reactance of the tuning circuit so that the circuit is **resonant** to 780,000 cycles (see Chapter 35). The frequency that is in resonance with the tuning circuit experiences minimum impedance and flows with greater ease than current of any other frequency and is therefore selected by the circuit. This is very much like the way a resonating tube of air vibrates in time to a sound of the same frequency (Chapter 23).

This large current flowing in the resonant circuit of the secondary coil and variable condenser produces a large voltage across the condenser, which causes a current to pass through the crystal detector and the earphone to produce sound. But the signal is usually so weak that it must be amplified after detection. This is where the vacuum tube comes in, and it is also where the complications in radio work begin. Since we cannot go too deeply into the subject of radio in this book, we shall limit our discussion to a simple one-tube amplifier.

12. The vacuum-tube amplifier

Just how does the vacuum-tube amplifier work? In a simple three-element vacuum tube (called a **triode;** tri′ōd), four wires enter the glass wall of the tube, which is represented by the circle in Fig. 36-14B. There is a high vacuum in the tube. One wire goes to a metal **grid,** G; another goes to a metal **plate,** P; the other two wires go to a **filament,** F. These filament wires are connected to a battery, A. The current from this **A battery** makes the filament hot so that it will emit electrons. These electrons are pulled

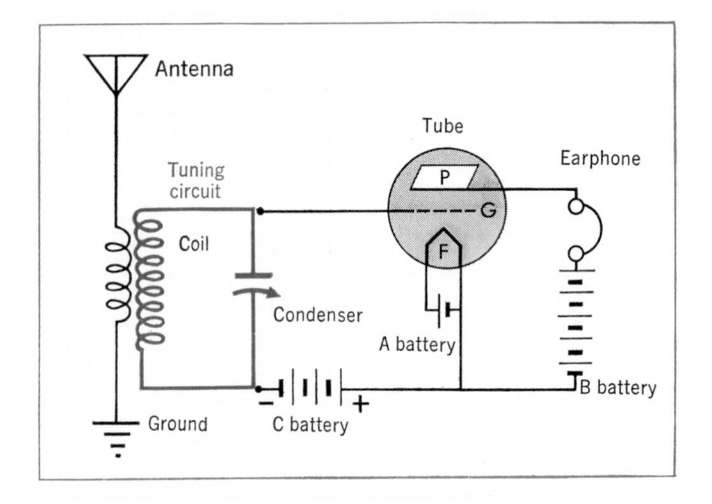

36-15 A simplified one-tube set. The AC high-frequency signal selected by the tuning circuit is fed to the grid of the radio tube G and causes an amplified pulsating DC current in the plate circuit, P–B. The earphone responds to pulsating DC to produce sound. All batteries could be replaced by a power supply.

across the space of the tube by a positive charge put on the plate by the **B battery**, B.

The grid, consisting of a screen or coil of wire, is supplied with the varying voltage from the tuning circuit (Fig. 36-15). The grid is charged more negative whenever this varying current tends to cause electrons to flow into it, and it is more positive whenever the electrons tend to flow out of it. When the grid is at its maximum negative potential, many electrons from the filament are repelled and the current to the plate is a minimum. But when the charge on the grid becomes more positive, electrons are attracted, and the flow of current to the plate is increased. Thus the changes in the electrical charge on the grid produce similar changes in the flow of electrons across the tube. But the grid is placed close to the filament so that a *slight* change in weak grid voltage produces a *much larger* change in the strong current traveling to the plate. The plate current that results therefore has changes in it that are greatly magnified (*amplified*) as compared to the original signal. This plate current operates the earphones.

We can amplify the signal from the plate circuit (the output of the tube) still more by sending it through a step-up transformer, from which it can be fed in turn into the grid (the input) of a second tube which amplifies it further. Additional step-up transformers and tubes make for still greater amplification. Because the surface of a loud-speaker is larger than that of the earphone, it can set a greater volume of air into motion and consequently it produces louder sounds (Fig. 29-4). But, of course, greater energy is necessary to move the larger surface and volume of air. This additional energy is provided by the current from the B battery or some other source of DC.

The foregoing method of amplification is also used in phonographs, sound movies, telephone circuits, and hundreds of other electronic devices. A triode can even be used as a relay if the weak charge on the grid is used to turn the plate current on and off. Much use is made of triodes in electronic computers (or "brains").

Of course this description of how tubes operate in a radio set has been greatly simplified. In actual radio sets many improvements are made by inserting condensers, resistances, and coils in various ways. Should you decide to study practical radio circuits you will find them more complex than the highly simplified description we have given.

13. The vacuum-tube oscillator

The triode is the heart of a vacuum-tube oscillator. You will remember that the plate circuit is called the **output** of the triode and the grid circuit the **input** (page 482). Now the DC from the B battery can be made to go and stop (oscillate) at a frequency of thousands of cycles per second, producing high-frequency pulsating DC, if the output of the plate is arranged to control the input to the grid. Any tiny fluctuation in the DC output of the tube causes coil A (Fig. 36-16) to induce an AC in coil B by ordinary transformer action. The AC in coil B is fed to the grid of the triode, causing further and more violent change in the DC output of the plate. The changing DC of the plate circuit once more goes through coil A which, in turn again, induces a larger AC in coil B, and the process (called **regeneration** or **feedback**) continues back and forth—output feeding input, and input controlling output. The frequency of the oscillations produced is controlled by a condenser and inductance which form a circuit whose resonant frequency can be changed at will by adjusting the variable condenser (Fig. 36-16). The frequency of the oscillation is given by $f = \frac{1}{2\pi\sqrt{LC}}$ (page 465).

This principle of feedback used in the triode to produce high-frequency (RF) currents is used in broadcasting transmitters for producing the carrier wave.

14. Frequency modulation

The currents induced in an antenna which cause static are added to the signal in the antenna and change the **amplitude**

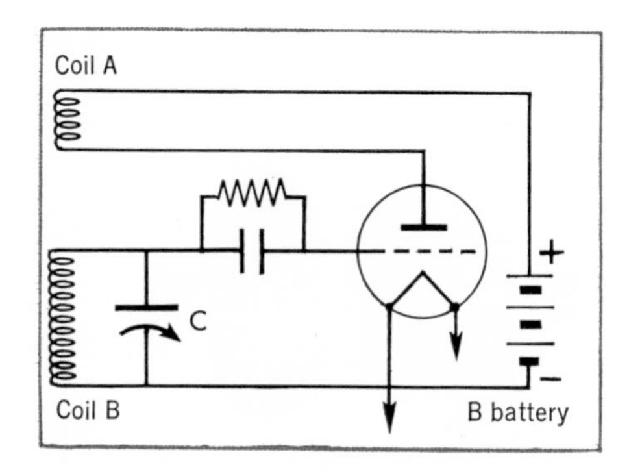

36-16 A simple vacuum tube oscillator circuit. Pulsating DC in coil A (plate circuit) induces alternating current in coil B (grid circuit) which has the effect of changing the plate current. Hence weak oscillations of plate current are strengthened.

of the signal. To reduce static a system of **frequency modulation (FM)** has been developed to replace the amplitude modulation we have just described.

In frequency modulation, as the name suggests, the voice vibration does not change (modulate) the amplitude of the carrier wave (represented by the height of the wave). Instead it changes (modulates) the *frequency* of the carrier wave (Fig. 36-17). The unwanted outside signals from static still cause changes in the amplitude of frequency-modulated waves, but changes in amplitude play no part in producing the sound in FM receivers. Hence static has little effect on the final sound that is produced in FM radios.

In practice, FM broadcast signals are sent out on waves of higher frequency than normal AM radio broadcast waves. Unfortunately, this limits the distance at which the signals can be received; they can be transmitted only to the horizon, which often may be less than 100 miles, depending on the height of the transmitting and receiving antennas. This is because these high-frequency waves are not reflected around the curve of the earth's surface by the **ionosphere** (an ionized layer of air 50 to 200 miles up).

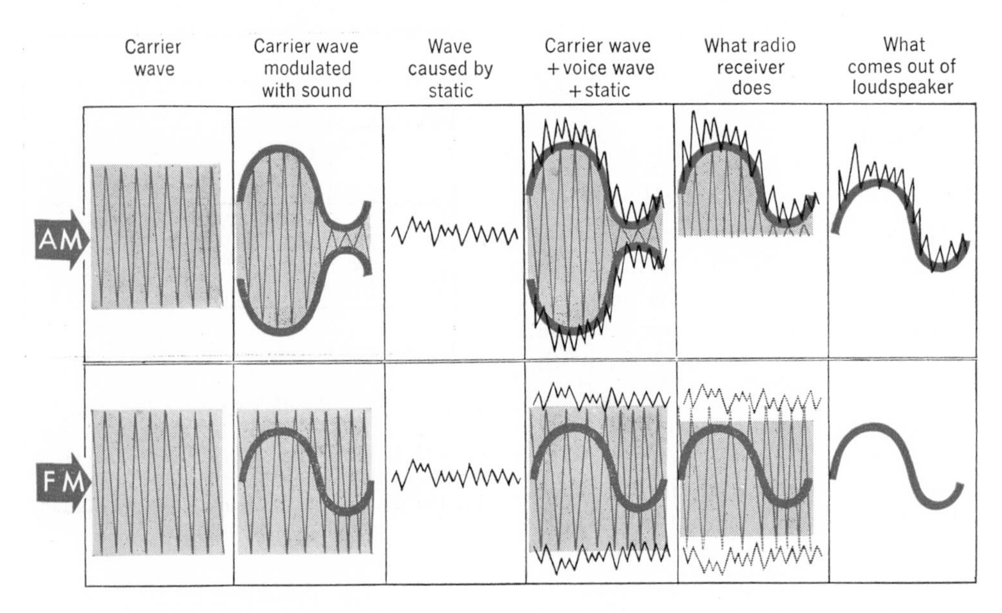

36-17 FM (frequency modulated) radio transmission has much less static than AM (amplitude modulated) because the voice vibrations change (modulate) only the frequency of the carrier wave. In AM transmission the voice wave changes the amplitude of the carrier wave and so does the static.

Motorola, Inc.

36-18 This recent model of a radio transmitter enables the fire fighter to receive and give directions. There are even smaller models of these "walkie-talkie" radio devices.

15. The future of radio

The rapid development of both FM radio broadcasting and of the new medium of television has hardly reached its peak. In the next few years we will witness expansion of radio broadcasts to include telephoning from cars (already a reality) and even telephoning from person to person walking in the street. Small radio sets ("walkie-talkies," Fig. 36-18) have already been developed that make direct conversation possible between people far from a telephone. New methods of printing electrical circuits with metallic "ink" on sheets of plastic reduce wiring space in receiving sets and thus make lower costs possible. Transistors as small as peas are replacing the bulky vacuum tube for amplification in many applications (Fig. 36-19) including radio transmitters operating in earth satellites. So one day you may carry your radio-telephone around with you as a "wrist watch" or in your pocket.

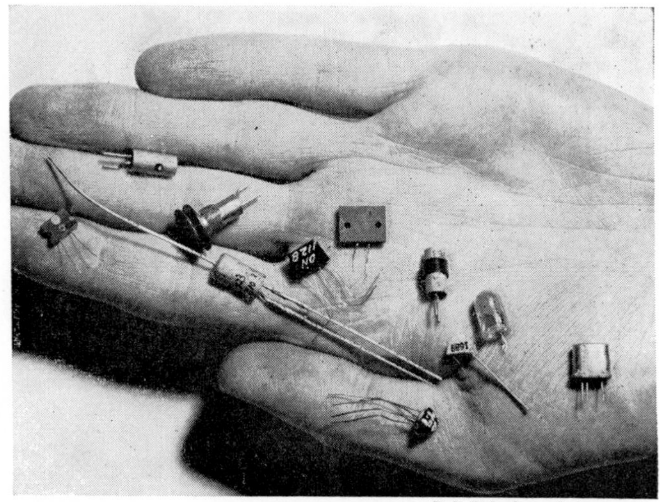

36-19 These tiny models of the transistor are revolutionizing radio. They consist of junctions of special metals similar to the crystal of a crystal set, and can be made to perform many of the functions of radio tubes. Their advantages are that they are much smaller than radio tubes, use much less power, and have almost unlimited life.

Bell Telephone Laboratories

Pictures through wires

Suppose you wanted to send a picture by telephone. You could criss-cross the picture with a network of tiny squares like those of graph paper. The person at the other end of the line would have a similar page of squares, each one numbered to correspond to the squares covering your picture. Then you could tell him which of the squares on the picture were black, dark gray, gray, or white. If he filled in his squares carefully, according to your directions, he would end up with a rough copy of the picture. If you find it difficult to believe that pictures can be sent in this way, try this experiment. Examine a newspaper photograph with a strong hand magnifying glass—or better still, a microscope. Figure 36-20 shows a magazine picture, enlarged many times so as to reveal that it is composed of individual black and white spots. The trick in getting good reproduction is to make the spots as small as possible. In this way the individual black and white spots do not stand out as separate from each other but blend together to form the picture.

Although it is possible to send such a picture by telephoning instructions, it certainly would be time-consuming and therefore costly and impractical. There is a better way to send the "instructions" automatically. This method makes use of the photoelectric effect in a photoelectric cell or "electric eye."

16. The electric eye

To transmit a picture we need a device at the sending end which can convert the light and dark areas of a picture into electric current. This job is performed by a photoelectric cell, or "electric eye."

It is known that light can "kick out" electrons from atoms of certain materials. This is called the **photoelectric effect.** Metals like sodium, potassium, selenium, and cesium are particularly useful for this purpose.

Figure 36-21 shows how a light-sensitive metal of this kind can be used in an electric eye. The piece of metal, M, is placed in an evacuated glass tube. When the light falls on the metal, some electrons are kicked out of the atoms of metal. Some of these electrons are captured by the nearby plate P. If the plate is made positive in charge by connecting it to a battery, its attraction for negative

36-20 Magazine or newspaper photos and television pictures resemble each other in that both are made up of separate white and black spots. Here, at the left, is a photograph as it would appear in a magazine and, at the right, a magnified portion of the same photo.

electrons will pull almost all of them across the space to the plate. The more light that strikes the sensitive metal, the greater the number of electrons that are emitted, and the greater the current flowing to the plate.

In order to transmit a picture automatically, the picture need not actually be ruled into squares, although it is convenient to *imagine* that such squares exist. The picture is wrapped around a cylinder and revolved in front of a photoelectric cell. Light from each successive tiny "square" on the picture is sent through a lens and focused, one "square" at a time, in order, on the cell. The light impulses are thus converted into electrical impulses which can be transmitted by telephone wire. After each revolution the cylinder advances sideways a little for

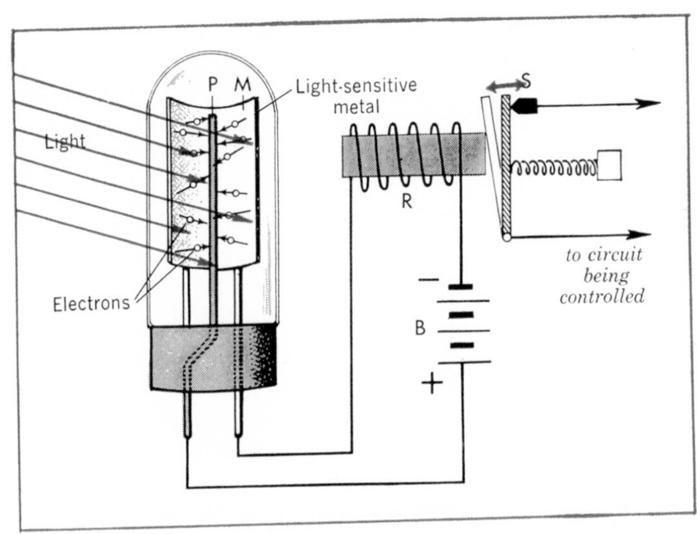

36-21 The electric eye converts light energy into electrical energy by means of a light-sensitive metal. Light striking the metal, M, kicks out electrons that flow across the tube to a positive metal plate. This current caused by the light can be used to operate a relay, R, and turn a switch, S, on or off. How could this arrangement be used as a burglar alarm?

the next row of "squares" to be transmitted. At the receiving end there is a light bulb whose brightness is controlled (electronically) by the incoming pulsating current. The brightness of the bulb changes in step with the brightness of the "squares" as they are focused on the photoelectric cell at the transmitter. The light from this flickering bulb is focused at a spot on film which is mounted on a revolving cylinder the same size as the transmitting cylinder. The two cylinders turn in step so that a series of light and dark spots will appear on the film when developed in the same order that the "squares" of the original picture were focused on the photoelectric cell. With this system it is possible to send the electrical picture impulses by radio as well as by wire.

17. Other uses of photoelectric cells

The electric eye has many other uses. It can be used in connection with a street-lighting system so that the lights are turned on automatically as dusk approaches. It can be used to open garage doors by reacting to the lights from headlights. It is used in public buildings to open doors when a person passing through breaks a beam of light. In the same way traffic or parts moving along a factory assembly line can be counted automatically by the interruptions of a light beam. Figure 29-4 shows how such a photoelectric cell is used to turn on the current for magnetizing loudspeaker magnets. This same electric eye is essential to projecting movies with sound. Here the cell converts the light and dark streaks of the sound track (Fig. 24-6) into changing electric currents; these changing currents are then amplified to produce sounds in a speaker. We may hear these sounds as music or speech.

U.S. Navy

36-22 A rather efficient type of photoelectric cell has been developed that can convert sunlight into electrical energy sufficient to operate a telephone circuit or the radio transmitter of an earth satellite. A group of these cells is called a solar battery; it is behind the oblong window in this picture of an earth satellite.

Television

In television, moving pictures are transmitted from one place to another. We cannot use the simple rotating drum method described in Section 16 for this purpose. It is not fast enough. If we are to transmit moving pictures, we must scan the picture many times a second. Special camera tubes have been developed to do this. The ones in actual use today are very complex, but they all work on the same basic principle. They all employ a rapidly moving beam of electrons to give high scanning rates. The type of camera

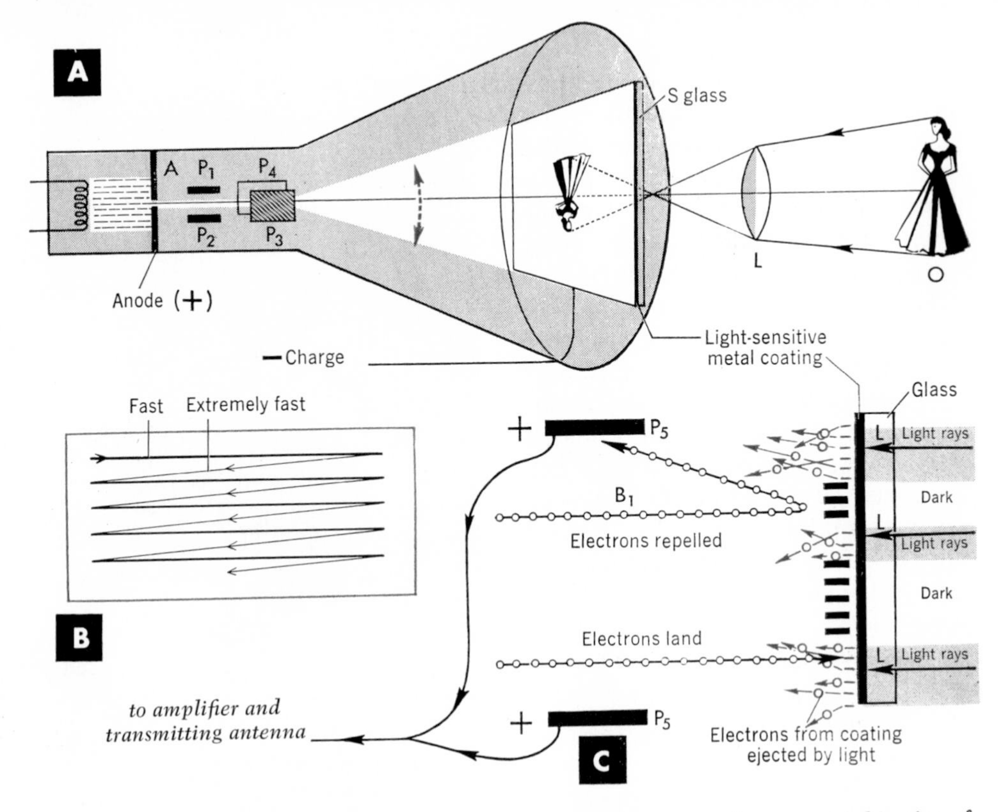

36-23 (A) The camera tube used in television broadcasting stations is a combination of electron gun and camera. (B) The electron beam scans the image on the face of the tube. It is repelled by strongly charged (dark) areas of the image. (C) As the beam scans the image, a changing electric current is produced that is picked up by plate P_5, a metal ring around the edge of the tube face. It is then amplified, mixed with a carrier wave, and broadcast.

tube we are about to describe illustrates the fundamental principles used in all modern camera tubes.

There are three main parts to a camera tube: an electron gun, a kind of photoelectric cell, and a lens to form an image. How are these parts put together to make the television camera work?

18. The electron beam

A beam of electrons is given off from a hot filament (at left in Fig. 36-23). These electrons are speeded up by the attraction of a positive charge on the anode (positive plate) at A. Some of the electrons shoot through a hole in the anode to form a narrow **electron beam.** This beam then passes between two pairs of deflecting plates, P_1, P_2, and P_3, P_4. To see how the deflecting plates work assume that the upper plate, P_1, is charged positive. It then attracts the negative electrons in the beam and they swerve upward. At the same time the lower plate, P_2, is negative and repels the electrons—also in an upward direction. As the charge is made stronger on both plates, the swerving of the beam increases. Suppose the charge on the upper plate, P_1, is negative instead of positive. In that case the plate will repel the beam of electrons downward. If at the same time the charge on the lower plate, P_2, is positive, it will also pull the electron beam downward. By varying the voltage on the two plates, the electron beam can be made to move up and down at any desired frequency.

After the beam passes between the deflecting plates P_1 and P_2, it travels between the other pair of deflecting plates P_3 and P_4. However, this set of plates is at right angles to the first set. An alternating voltage on these plates will cause the electron beam to swerve from side to side instead of up and down.

By a clever arrangement of properly timed alternating voltages, the electron beam is made to move back and forth across the screen of the tube, S, starting at the top and working down toward the bottom as shown in Fig. 36-23B. As soon as the beam reaches the bottom of the tube, it starts all over again at the top. Thus the beam scans the screen of the tube in a series of separate horizontal lines, one below the other. This scanning motion can also be achieved by using properly constructed electromagnets instead of the charged plates.

19. The camera-tube photoelectric cell

Now let us see how a picture is formed on the target screen, S. From outside the tube a lens casts an image of the object being televised, O, on a special thin metal coating on the inner surface of the glass, S. The glass plate serves as a wall of the vacuum tube, but permits light to pass through to form an image on the metal coating inside.

The thin metal layer is made of a material with two special properties. In the first place it must be photosensitive—that is, electrons must be ejected from it when light strikes. It must also be of rather high resistance for a metal, so that electrons do not readily travel sideward across it.

At the start of operation of the tube the thin metal plate is charged negatively (Fig. 36-23C). Now, as the light comes through the glass at a certain spot, electrons are ejected. Since the metal is very thin, many of them are ejected on the opposite side into the vacuum tube (Fig. 36-23C). The loss of electrons from a spot makes it less negative than a neighboring spot. A nearby spot illuminated by less light from the image will have fewer electrons ejected and so will not lose as much negative charge. Because of the high resistance of the metal to the passage of electrons sideward across the plate, these differing charges remain on the metal for a short time. In other words, the light image is transformed into an electrical image in which light areas are represented by less negative charge and dark areas by greater negative charge. The electrons that are ejected are now pulled out of the way by a positive charge near the end of the tube.

20. Forming the electrical image

Now let us return to our electron beam. It is scanning the end of the tube back and forth in horizontal lines, and, as it does so, it shifts from the top of the image to the bottom. As we mentioned before, the photosensitive metal plate is negatively charged. The amount of charge is so chosen that the approaching electron beam, B_1 in Fig. 36-23C, is slowed down by the charge, stopped, and reversed just before it strikes the metal. The reversing electron beam is then attracted by a positive plate, P_5. But, as you have seen, the light image has produced varying negative charges on the metal plate; those areas struck by more light have become less negative. Suppose the electron beam is approaching one of these less negative spots on the metal (a light area, L). It is not repelled as strongly as it would be if it approached a dark spot whose negative charge was reduced only slightly. As a result, a number of electrons reach that spot and stay there.

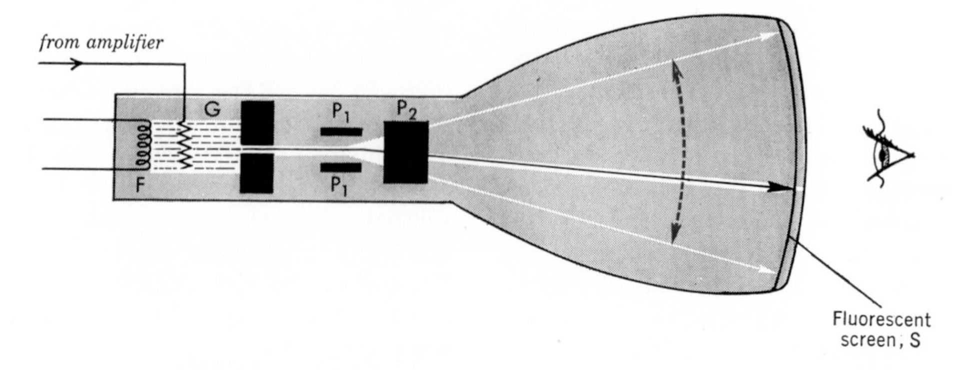

36-24 The picture tube in your receiving set is an electron gun with special deflecting plates (or coils). As changing charges on P_1 and P_2 cause the beam to scan the screen back and forth, up and down, the intensity of the beam is changed by varying charges on the grid, G. These charges are those caused by the incoming television signal. The changing intensity of the electron beam causes lighter and darker areas on the fluorescent screen, S, which produce the image that we see.

On the other hand, suppose the beam is approaching a spot that is illuminated by less light, and whose negative charge is therefore reduced only slightly. In that case almost all the electrons are repelled back to the plate P_5. Thus, the electron beam returning from dark spots on the metal plate is more intense than the beams returning from the light areas.

As the beam scans across the metal, rapidly touching light and dark spots in succession, it creates a changing current on P_5, with more electron current representing dark areas and less current representing light areas. This changing current on the positive plate P_5 can then be conducted out of the tube and amplified by vacuum-tube amplifiers.

Notice that the succession of electric impulses, stronger and weaker in proportion to the bright and dark spots that are scanned, is similar to the much slower method of transmitting a "still" picture by means of a rotating drum and a photocell. This was discussed earlier in the chapter (p. 485).

Of course, the electrical impulses from the camera tube must be broadcast to all the television sets in the area. So the changing signals of the electrical picture are now mixed with a very high-frequency carrier current and go out as a modulated carrier wave from the transmitting antenna in much the same manner that the voice vibration currents are mixed with carrier waves and transmitted in radio broadcasting. This carrier wave is amplitude modulated (AM) and is referred to as the **video signal**; the sound part of a television broadcast (referred to as the **audio signal**) is an FM signal.

21. Forming the picture in the picture tube

When the video signals reach the antenna of a television receiver, they are transformed into electrical currents in the antenna. The currents are detected and amplified by vacuum-tube arrangements of the kind described in the section on radio. They are then fed into a picture tube, which performs the job of converting the changing electrical impulses into a picture (Fig. 36-24). The front of this tube is the fluorescent screen S, which carries the final picture that we see. As in the camera tube, an electron beam is made to scan this screen exactly in time with the original scanning in the television studio. Now the incoming signal is fed into a grid of the tube, G, which controls the number of electrons in the beam. A strong negative charge on the grid repels electrons in the beam and permits fewer to get by. A weak negative charge in the grid will let more electrons go by. Thus the electron beam of the picture tube varies in strength as it scans the screen of the tube. The circuits are so arranged that a weak electron beam corresponds to a dark spot on the object being televised and a strong beam to a bright spot.

The screen on which the electron beam falls is made of a fluorescent material which glows when struck by electrons. If the beam is intense, the screen glows brightly at that spot. If the beam is weak, the screen glows faintly at that spot. But these bright and dark spots now correspond exactly to the original dark and light spots of the scanned image in the studio. Therefore we get an identical image on the screen of the picture tube.

Earlier in the chapter we saw that when a picture is duplicated spot by spot with our graphing method, it gives only a rough duplication if the spots are large. But if the spots are very small, then the eye cannot detect the separate spots and they blend together to form a continuous image. That is why the number of lines in the scanning process must be very great—the greater the better (Fig. 36-25). Television engineers have decided upon 525 lines as a practical number to use for the television picture. As scanning starts, the electron beam whips across the screen, returns to a point just below the starting point, whips across the screen again, and so on, until the 525th line is reached at the bottom of the picture. The beam is then brought back to the start and the scanning process is repeated. In modern TV sets these scanning lines are interlaced to reduce flicker.

The scanning process is repeated 30 times in one second—a speed rapid enough so that the eye cannot detect the separate pictures.

RCA

36-25 As the number of scanning lines increases, the clarity of the television image improves. Compare the rough 60-line television picture (top) with the 120-line picture (center) and the 240-line picture (bottom). Regular television images have 525 lines.

We see then that the television image is formed in the same basic manner as in our spot-for-spot graphing method described earlier in the chapter (page 485). But the big difference is that, with the television system, the spot-for-spot duplication of the image occurs automatically, and with such great rapidity that millions of electrical changes are registered each second. The speed of scanning motion of the beam of a large picture tube averages about 6 miles a second. Yet despite this enormously rapid motion, the electron beam can register changes from all the individual spots on the screen. It is the great mobility of electrons (because they have practically no inertia) that makes electron tubes so useful in radio and television.

22. Color television

In **color television,** mirrors and colored filters are used to split up the light of the scene being televised into three separate colored images, one red, one green, and one blue. Any color can be produced by making combinations of these three colors in the proper proportions. Each of the three images is scanned by its own electron gun and is broadcast separately but simultaneously in a way similar to that used in black and white broadcasting. The transmitting station broadcasts a spot from the red image followed by a spot from the green image, and then by a spot from the blue image. These form a group. Over 500 such triple-colored groups are scanned per line. At the receiver, the picture tube (Fig. 36-26) has three

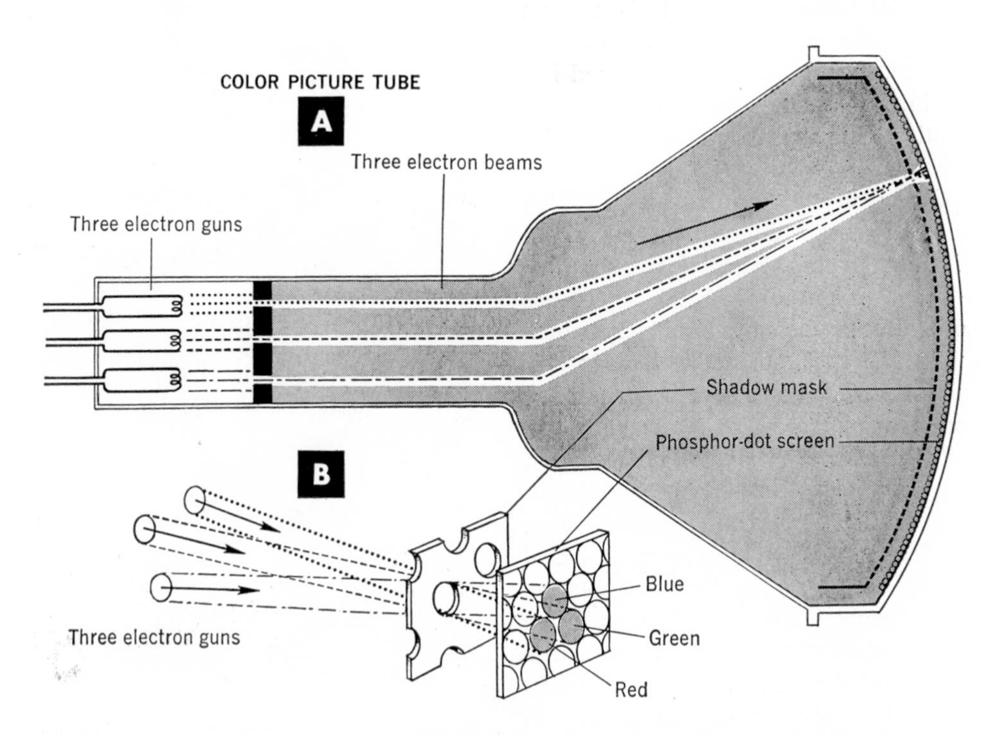

36-26 (A) A common form of color picture tube contains three similar electron guns. (B) Electrons from one gun produce only one color on the screen because of the arrangement of holes in the shadow mask and the arrangement of phosphor dots on the screen. But when the action of all three guns is co-ordinated properly a picture in full color is produced on the screen. Grids, anodes, and other parts have been omitted for simplicity.

U.S. Navy

36-27 Looking like searchlights, these radar antennae are used to observe and control guided missiles fired from this modern cruiser. Note the more conventional radar antenna further to the left.

electron guns, each supplying a beam whose intensity is controlled by one of the three colored image signals. The three beams scan the screen of the tube together. The screen has thousands of groups of **phosphors**—chemicals that glow when struck by a beam of electrons. Each group consists of three spots—one that glows red, one that glows green, and one that glows blue. A mask having one tiny hole for each group of phosphors is mounted inside the tube close to the screen so that the three electron beams intersect at a hole, and each beam is focused onto the spot of colored phosphor it is meant to strike. Hence, each glowing spot on the screen of the picture tube consists of three tiny colored spots which blend to give the eye the impression of a single color. Varying the intensity of each of the spots can produce the impression of any desired color.

23. Radar

A television system transmits a definite image from one place to another. Radar performs a somewhat different "seeing" job. It enables us to detect invisible objects many miles away in any direction. The word **radar** comes from abbreviating *RA*dio *D*etection *A*nd *R*anging.

Radar acts very much like the familiar flashlight we use to find our way in the dark. From the bulb of the flashlight, light waves radiate outward in all directions. Some of these rays strike the parabolic metal reflector behind the lamp, which focuses them into a narrow beam. When the light waves of this beam strike an object, they bounce off; part of the beam is reflected back to our eye to form an image of the object. As we point the flashlight in different directions, one object after another is illuminated.

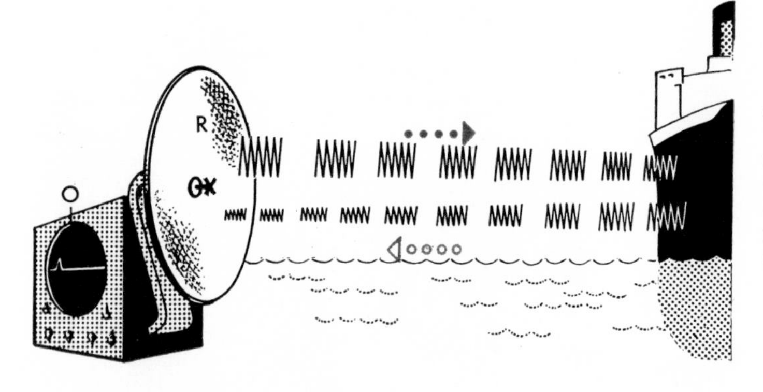

36-28 Radar detects objects from the reflections of rapid pulses of radio waves. Distances can be measured by timing the return using an oscilloscope, O.

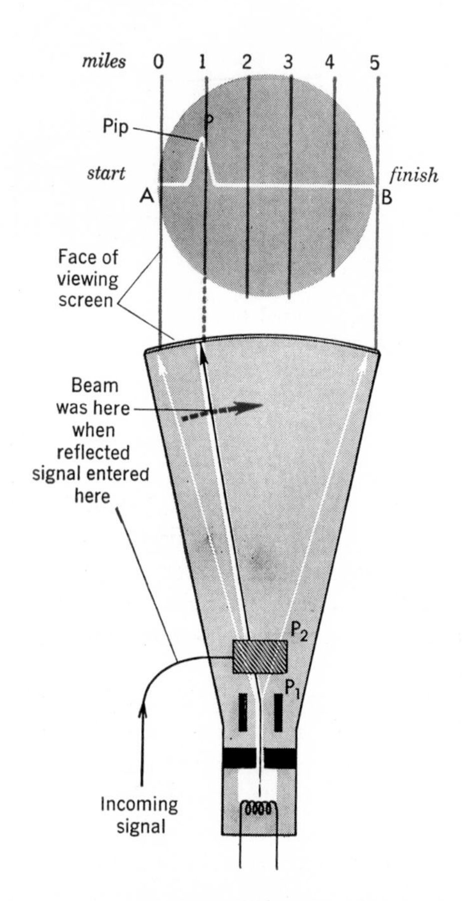

36-29 The "scope" in radar is like a television tube. As the beam rapidly travels across the tube from A to B, an incoming signal caused by the return of a reflected radio wave causes the beam to swerve upward and make a "pip" on the screen. Distance of the pip from A enables the observer to measure the distance of the object.

Radar operates in much the same fashion except that we use a beam of radio waves (called **microwaves**) instead of waves of visible light; special equipment is needed to detect the reflected waves.

Just as with a flashlight, a parabolic reflector (R in Fig. 36-28) is used to focus the radio waves into a beam. Whenever this radio beam strikes an object, part of the beam is reflected back. Some of these reflected waves strike the original reflector, which now acts as a receiving antenna. During the instant it receives a pulse it automatically ceases to broadcast. Once the reflected signal is received it is fed into electronic tube circuits where it is detected and amplified. After detection and amplification it passes into a cathode-ray tube (Fig. 36-29 and O in Fig. 36-28).

The main part of the radar viewer is the cathode-ray tube known as an **oscilloscope** or **scope**—similar to a television picture tube. In one simple receiving arrangement an electron beam moves back and forth so as to make a horizontal line AB on the face of the tube (Fig. 36-29). The electron beam starts moving across the scope from A to B at the same instant that the radio-wave pulse is sent out. When the reflected signal returns it is amplified and impressed on deflecting plate P_2 of the oscilloscope.

This deflecting plate pulls the beam up to register what is called a **pip** (P in Fig. 36-29).

Every object in the path of the radar wave reflects back a signal that shows up on the screen of the scope. The distance of the pip from the starting point of the beam shows how far away the object is. The larger the object the more radio energy it will reflect; and this strength of the reflected wave is indicated by the height of the pip.

When the electron beam reaches the end of the scope, a new pulse is sent out by the transmitter, and the electron beam then jumps back to begin its journey across the scope all over again. This happens many times during a single second, so that we have a succession of separate pips at the same spot on the screen. The pips occur so rapidly that the separate images merge into one continuous image.

The rate at which the pip moves indicates the speed of the object. And all of this radar scanning can take place in absolute darkness or fog.

United Press-International

36-30 This large radar antenna is part of a chain of interconnected stations built to detect all airplanes approaching this country across the Arctic.

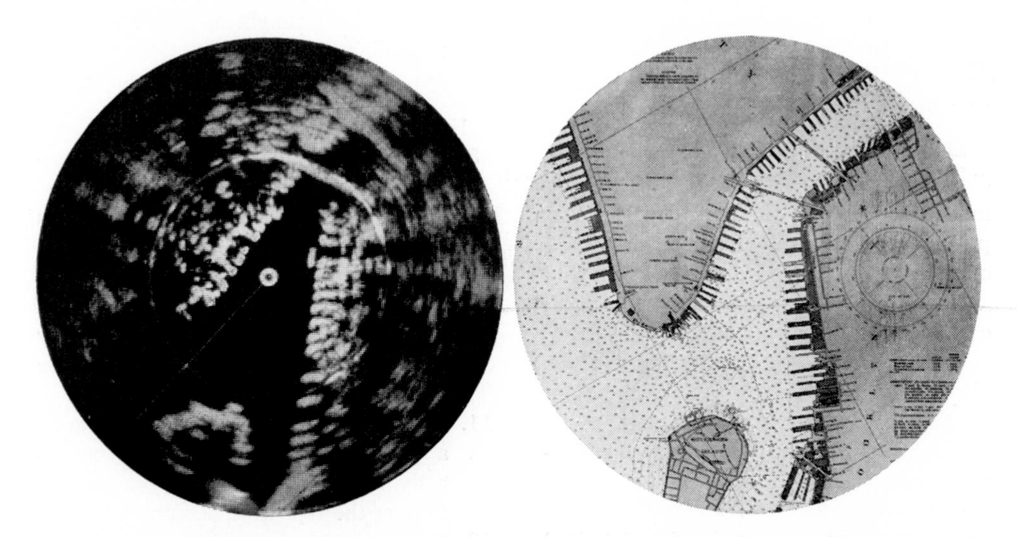

36-31 On the left is a PPI radar view of New York harbor, showing particularly the outline of the East River and the docks along its edges. The spot in the center of the picture shows the point directly beneath the airplane taking the picture. On the right is a chart of the same area for comparison.

If we want to detect objects in every direction around us, we slowly rotate the radar antenna. In one type of radar set, called a PPI (plan position indicator) set, radar beams are sent out while the antenna is moving in a complete circle. Electrical circuits controlling the electron beam on the scope are also changed so that a rough picture of the surroundings is formed, as indicated in Fig. 36-31.

Radar finds its most important use in military defense (Fig. 36-27) and in guiding planes and ships in foggy weather. By means of radar all planes within many miles of an airport can be located and safely navigated by radio instructions from the airport, even though there is a fog.

If the plane itself is equipped with radar the pilot can detect the presence of mountains and other obstacles ahead. By projecting the beam downward he can also measure distance from the ground. This overcomes the chief defect of the altimeter, which only measures altitude as height above sea level.

Going Ahead

THE VOCABULARY OF PHYSICS

Define each of the following:

diaphragm
microphone
pulsating DC
electromagnetic wave
carrier current
rectification
diode
modulated carrier current
amplitude modulation (AM)
crystal detector
demodulation (detection)
tuning
static
thermionic emission
filament circuit
audio frequency (AF)
radar
oscilloscope (scope)
pip
plate circuit
half-wave rectification
filter
voice (modulation) current
radio frequency (RF)
resonance (electrical)
vacuum tube
triode
grid
plate
filament
A battery
B battery
frequency modulation (FM)
regeneration
feedback
amplitude
phosphor
target screen
microwave
electron beam
ionosphere
video signal
audio signal

THE PRINCIPLES WE USE

1. Describe in detail how vibrations of sound are converted by a microphone into pulsations of electric current.
2. Describe how the telephone receiver converts electrical vibrations back into sound.
3. How were radio waves discovered?
4. Explain the main difference between AM and FM broadcasting. Draw a diagram showing the difference.
5. Describe the operation of a photoelectric cell. Use a diagram.
6. In what way does transmission of pictures by television resemble transmission of sound by radio?
7. In what way does radar resemble the use of a searchlight at night? How do they differ?

Explain why

8. The disk of a telephone receiver cannot be made of brass.
9. If a signal from a receiving antenna is passed through an earphone, no sound is heard; but a crystal placed in the circuit enables one to hear sound.
10. A distant lightning flash and a nearby electric shaver (in use) both cause static in a radio set.
11. The timing of the scanning motion in a television picture tube must be exactly the same as that in the camera tube at the television station.
12. We see a continuously moving image on the television screen even though there is actually only a moving spot of light.

13. When the image on a television screen is examined closely, horizontal lines are seen.
14. Radar signals are sent out in brief pulses.
15. The radar antenna must rotate continuously if a radar picture of the entire surrounding area is desired.
16. Describe the scanning path of the electron beam in the television camera tube and tell its purpose.
17. Describe how a picture tube produces an image.
18. Describe the use of the cathode ray tube in radar.
19. Describe how the distance and approximate size of an object are detected by means of radar.

APPLYING PRINCIPLES OF ELECTRONICS

1. Draw a diagram of a simple one-way telephone circuit and explain its operation.
2. In what way does the transmission of energy from one coil of a transformer to another resemble the transmission of messages by radio?
3. How are radio waves created? Mention two simple ways.
4. Look up the radio page in your newspaper and find out what range of frequencies and wave lengths are used for radio broadcasting.
5. Trace the changes that occur in the voice vibration and carrier wave from the transmitter to the earphone of the receiver. Draw diagrams to show these changes.
6. How is it possible for many stations to send signals to your antenna without getting the signals mixed up?
7. Draw a diagram of a crystal radio receiver containing a tuning arrangement.
8. Draw a diagram of a triode; label and describe the parts.
9. Describe how a triode can amplify small current changes.
10. Describe how a diode can convert AC signals into pulsating DC.
11. Draw a diagram of a simple one-tube amplifier and explain its operation.
12. Why does FM broadcasting give less static and interference than AM?
13. How are pictures transmitted by electricity through a wire?
14. Mention three practical uses for photoelectric cells.
15. Explain how a television camera tube converts an image into a series of electrical signals.

SOLVING PROBLEMS IN ELECTRONICS

1. The velocity of radio waves is the same as that of light, 186,000 mi/sec. How can you find the wave length of a wave sent out by a radio station operating on a frequency of 1,500 kc? Calculate it.
2. Design a two-way telephone system.
3. Draw a telephone receiver circuit to show how a bell can be rung from the telephone exchange and shut off when the receiver is lifted off the hook.
4. Some intercommunication systems use a kind of receiver that can also be used as a transmitter (called a sound-powered telephone). How can a receiver work as a transmitter? (HINT. The receiver is an ordinary earphone and the principles of the generator and electromagnet are used.)
5. In some cities police are detecting speeding vehicles by stationing small radar sets on roads. How could radar be used to measure the speed of approaching vehicles?
6. In an early system of television the picture was scanned and received mechanically, using synchronous motors that could be kept rotating at steady speed. The motor turned a disk having a spiral of evenly spaced holes. Explain how such a system might work. What are some serious disadvantages of such a system?
7. As you know, the velocity of a radio wave is 186,000 mi/sec. If an electron beam travels across a radar screen in $\frac{1}{5,000}$ sec, how far away is an object if a pip from it is observed halfway across the screen?

EXTENDING YOUR STUDY

Wide World

36-32 **This powerful new instrument of astronomy is a radio telescope. Making use of many of the principles of physics you have already studied, radio telescopes can detect objects more remote than even the farthest galaxies seen in conventional light-gathering telescopes.**

"The earth we inhabit is a member of the solar system—a minor satellite of the sun. The sun is a star among the many millions which form the stellar system. The stellar system is a swarm of stars isolated in space. It drifts through the universe as a swarm of bees drifts through the summer air. From our position somewhere within the system, we look out through the swarm of stars, past the borders, into the universe beyond.

"The universe is empty, for the most part, but here and there, separated by immense intervals, we find other stellar systems, comparable with our own. They are so remote that, except in the nearest systems, we do not see the individual stars of which they are composed. These huge stellar systems appear as dim patches of light. Long ago they were named "nebulae" or "clouds"—mysterious bodies whose nature was a favorite subject for speculation.

"But now, thanks to great telescopes, we know something of their nature, something of their real size and brightness, and their mere appearance indicates the general order of their distances. They are scattered through space as far as telescopes can penetrate. We see a few that appear large and bright. These are the nearer nebulae. Then we find them smaller and fainter, in constantly increasing numbers, and we know that we are reaching out into space, farther and even farther, until, with the faintest nebulae that can be detected with the greatest telescope, we arrive at the frontiers of the known universe." *

But the telescope is not the last of the instruments with which we explore space. The telescope explores the world of light; what of "telescopes" that are not dependent on light?

Radio Astronomy

During the Second World War, British radar apparatus picked up strong radio signals which were at first mistaken for enemy efforts to jam radar reception. Only long and careful investigation showed finally that the radar signals were being received from the sun! Of course the pulse being sent out by the radar transmitter was not being returned as an echo from a target; the received signal was being emitted by the sun itself. It consisted of erratic bursts of electromagnetic waves in the 100-centimeter range of wave lengths. The conclusion was finally reached that the sun gives out radio "noise" all the time, but occasionally (especially at a time of great sunspot activity) it flares up and for a few hours emits radio energy as much as a million times faster than before. It was one of the flare-ups that caught the attention of the British radar apparatus.

Since the war, this odd discovery has been followed up with the help of much more sensitive equipment. Today there are many large radar-like receivers in existence, called **radio telescopes.** Of course a radio telescope does not transmit radio pulses like a radar transmitter (Fig. 36-32); it simply receives the incoming radio signals from outer space. One of the largest, near Manchester, England, has a steerable parabolic antenna 250 feet in diameter connected to a very sensitive radio receiver. Pictured on the opposite page is a similar but smaller one in Massachusetts.

What do we "see" with a radio telescope? Like light from the darkening twilight sky, weak radio radiation comes from all directions. Indeed, when your television set is tuned to receive waves of 5.0 to 5.6 meters wave length (Channel 2) about 80 per cent of the "snow" appearing on your screen is caused by radio waves entering in all directions from outer space. In addition to this "background radiation" a careful survey reveals radio waves coming in from several thousand separate starlike sources, concentrated in spots much like the light from the stars you can see scattered about in the twilight sky. The strongest source of all (after allowing for the effect of distance) is in the constellation of Cygnus and is emitting a billion billion (10^{18}) times as much radio energy as our flared-up sun.

Of course one of the first things astronomers did was to see what their ordinary light-gathering telescopes could show them of these point sources. Disappointingly, only about 60 of the several thousand radio sources have been identified with objects that can be picked up by light-gathering telescopes. Some of them are shown in Fig. 36-33. The strong source in Cygnus turns out to be two galaxies (each like our Milky Way system of stars) in the process of collision. Other sources, such as the Crab Nebula, are the remnants of stars that exploded some time ago. Yet a third type of radio wave emitter consists of dust clouds in our own galaxy illuminated by a nearby star.

* From Edwin Hubble, *The Realm of the Nebulae*, Yale University Press, 1936, Chapter I. From "Explorations in Space: The Cosmological Program for the Palomar Telescopes," *Proceedings of the American Philosophical Society*, Volume 95, 1951.

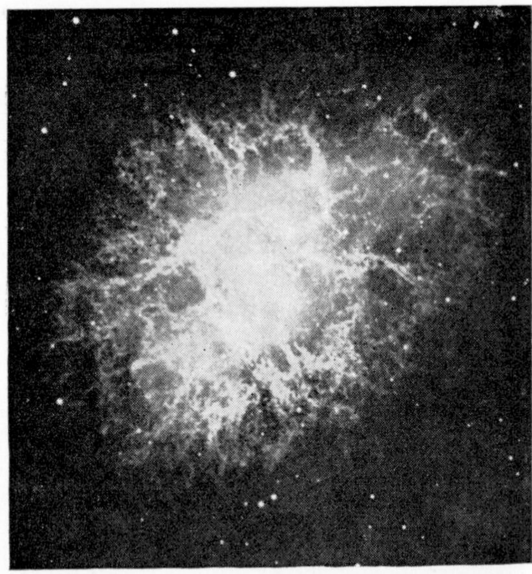

Mt. Wilson & Palomar Observatories

36-33 All three of these objects, visible through large optical telescopes, are known to emit radio waves as well as light. The strong source (top left) in the constellation of Cygnus is a pair of colliding galaxies. The picture at top right shows the remains of a supernova, called the Crab Nebula. Above is pictured a region of dust and stars known as the Horsehead Nebula.

Perhaps the most significant of all is the radiation being emitted from the center of our own galaxy, the Milky Way. Since it is obscured from view by dust clouds, it seemed that we might never be able to observe this important region of space. Now, by means of radio waves that easily penetrate such obstructions, we may at last be able to "see" it and learn what secrets it holds concerning the origin and age of our system of stars in the universe.

Within our solar system it has been found that not only the sun but even our moon and the planet Jupiter are emitting radio waves.

Why do objects emit radio waves? You will recall from page 231 (Stefan's Law) that all objects at a temperature above absolute zero emit electromagnetic radiation. The hotter they are the shorter are the shortest wave lengths they emit. Hence objects that emit visible light must be hotter than objects that emit only the longer wave length radio waves. Thus the moon and Jupiter (whose visible light is *reflected* from the sun, not emitted) are warmer than the surrounding space, and like the several thousand other radio objects in the sky, emit radiation of long wave lengths. The sun, like the other objects in Fig. 36-33, is much hotter so that it emits not only radio wave lengths but the shorter wave lengths of visible light as well (and, indeed, even the invisible and very short ultraviolet and X rays, page 588). The radio telescope therefore reveals to us many objects in the sky that are too cool ever to be revealed to our unaided eyes.

What do these radio waves tell us about the universe around us? One of the most exciting things is that some of these waves apparently come from objects beyond the reach of even the biggest light-gathering telescopes (page 559). The 200-inch telescope on Mt. Palomar reveals galaxies as far as 2 billion light years away, but the radio telescope that picked up the "bright" source in Cygnus could have detected a source as bright as this at 6 billion light years away. Studies of the "red shift" (page 587) suggest that at a distance of about 6 billion light years galaxies must be speeding away from us at about the speed of light itself. Beyond that distance they will become invisible. Hence the radio telescope holds out the hope that we may now be able to "see" to the very edge of the observable universe. Information on the population of galaxies and their speeds at this distance will help answer two of the major questions in modern science: How big is the universe, and how old is it?

Another important question that radio waves are already helping us to settle is the structure of our own Milky Way galaxy. From a million light years away our galaxy probably looks something like our nearest neighboring galaxy in the constellation of Andromeda (Fig. 36-34). Though our galaxy probably contains close to 10 billion stars (you see fewer than 2,000 of them when you look at the Milky Way with your unaided eye), many of them are cut off from our view by intervening clouds of dust and hydrogen gas in the space between the stars. These clouds of dust and hydrogen contain up to about 100 atoms per cubic centimeter, compared with 10^{10} atoms per cubic centimeter found in a first-rate laboratory *vacuum*. In space, where there are no clouds of dust and hydrogen, there may be only 1 atom per cubic centimeter. Now it has been dis-

Lick Observatory

36-34 **This distant galaxy of stars, the Andromeda spiral, is similar in shape and size to our own Milky Way. Like our own galaxy it also emits radio waves.**

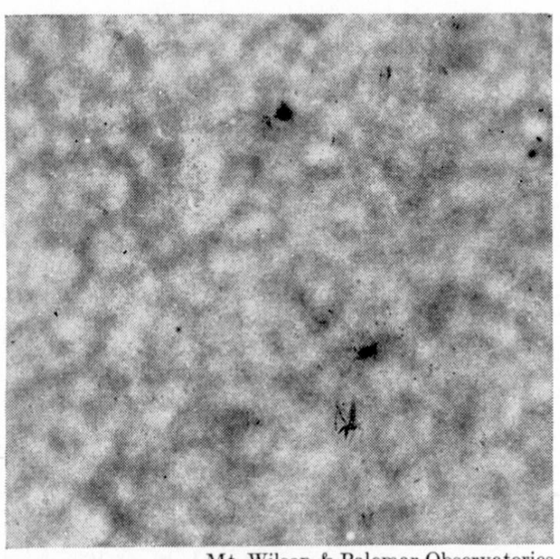

Mt. Wilson & Palomar Observatories

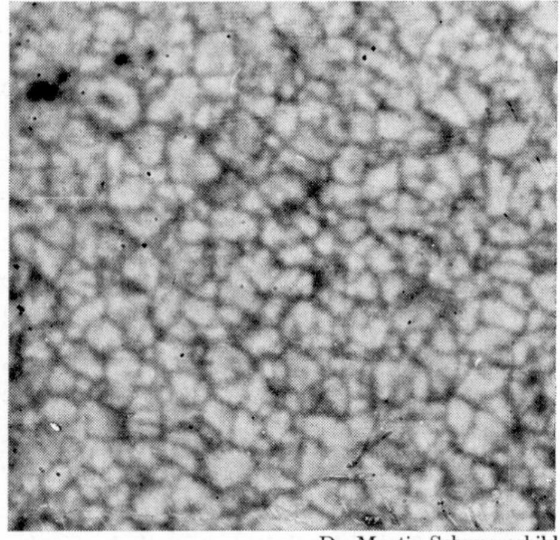

Dr. Martin Schwarzschild

36-35 **This picture gives an idea of how much our atmosphere obscures our view of the sky. Left: a picture of the sun's surface taken through a telescope at ground level. Right: how the sun's surface appears through a telescope that was carried above most of the atmosphere by a balloon. Notice how much more clearly the granular appearance of the sun's surface appears.**

covered that the continuous "background radiation" reaching us from all parts of the sky is emitted largely by the atoms of hydrogen scattered through interstellar space. Unlike the radio "noise" emitted by stars and other objects, the atoms of hydrogen in interstellar space radiate only one wave length—21 centimeters. Although this radiation reaches us from all directions, it is strongest from the direction of the spiral arms of our galaxy, where much of the hydrogen is concentrated. We are thus able to trace the outlines of the spiral arms. Moreover, the radiation from the arms which are moving toward or away from us exhibits the Doppler effect (page 587), and from the change in the 21-centimeter wave length we can find out how rapidly they are moving along our line of sight. Thus radio astronomy is giving us a picture of our Milky Way system of stars impossible to achieve by earlier means. It confirms that our galaxy is a slowly rotating spiral arranged rather like cream swirling in a half-stirred cup of coffee.

The Doppler effect in the 21-centimeter radiation has even been detected from other galaxies, and indicates that they are moving with a speed identical to that found from the visual "red shift" observation made with optical spectroscopes.

It is a pity that our atmosphere is opaque to all radiation except the visible region and the radio region. Outside our atmosphere space is undoubtedly flooded with radiation of other regions of the spectrum, which would reveal still further mysteries of the universe if we could only detect it. The only way to escape the filtering action of the atmosphere is to establish an observatory on an earth satellite or on the airless moon itself. From there we might hope to have an unobstructed view. Meanwhile, radio telescopes, a new extension of our sense of sight, continue to reveal more and more, bringing us closer and closer to an understanding of the fundamental questions that we ask ourselves about the nature of the vast universe in which we live.

UNIT SIX

Light Energy—

The Motion of Electromagnetic Waves

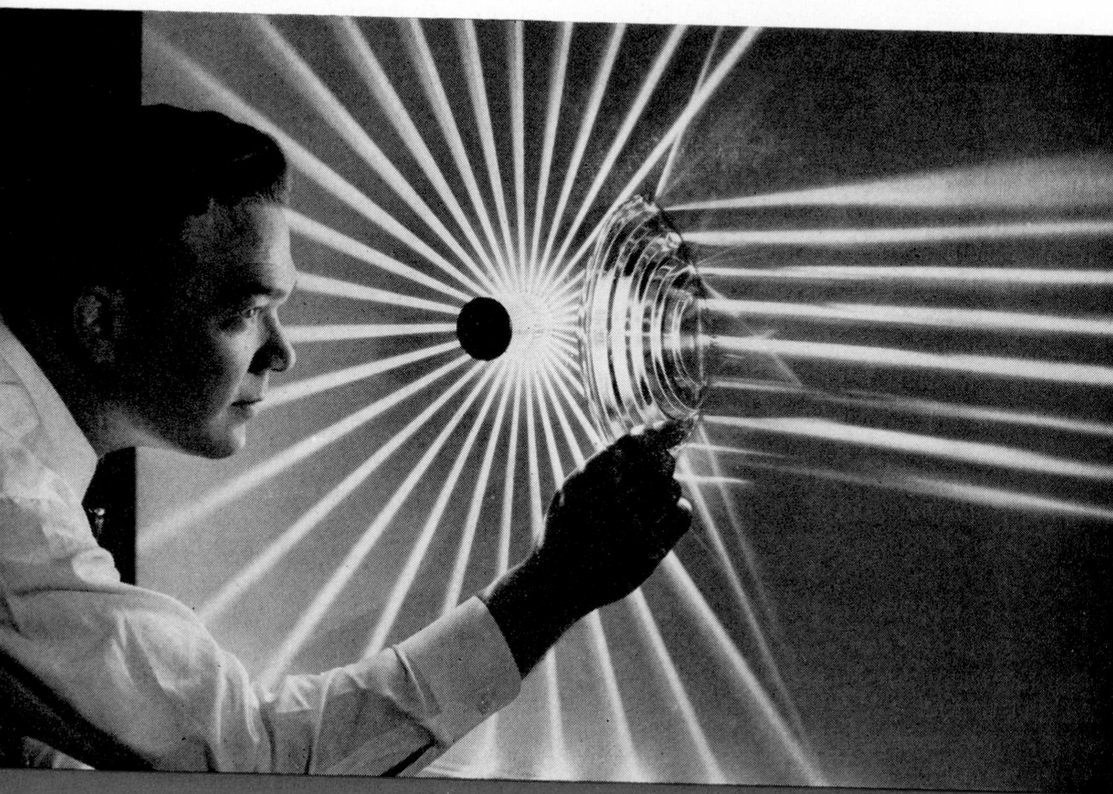

General Electric Co.

What would the world be like without light from the sun? You might answer that there would be eternal darkness. But, even more important, there would be no life at all. For the energy from sunlight is needed for plants to grow—and without plants, there would be no food. Then, too, our planet would rapidly cool off. All water and even the air would freeze.

For hundreds of years physicists have studied the mysterious form of energy called light. These studies have resulted in such useful devices as cameras, telescopes, eyeglasses, microscopes, and projectors, which enormously extend our knowledge of the world around us. Moreover, we know today that the light rays we can see are only a small portion of the kinds of light that actually exist. Today we use invisible X rays, gamma rays, infrared, and other rays.

CHAPTER

37

The Behavior of Light

What is Fig. 37-1A supposed to show? You see nothing but a black rectangle. Actually an intense beam of light is passing across the room from left to right, as shown in the top view of Fig. 37-1A. The room, however, is dust free, so that there is nothing in the path of the beam to reflect light to the eye, and therefore you see only darkness.

Now, a hand appears in the beam (Fig. 37-1B), but the person whose hand it is is invisible, although he is standing just beyond the light beam (top view of Fig. 37-1B). The hand has reflected some of the light to the eye.

As soon as two erasers covered with chalk dust are struck together in the path of the light beam, the beam itself as well as objects in the room become dimly visible (Fig. 37-1C). The objects can be seen because not all of the light now passes straight through the room. Some of it bounces off the particles of chalk dust, hits the different objects around the room, and is reflected off them to the eye.

Some properties of light

Our observations in Fig. 37-1 teach us that three conditions are necessary if we are to see any object that is not luminous (giving off its own light). First, there must be a source of light. Second, the light must strike the object. Third, the light must be reflected from the object to the eye.

1. What happens when light strikes objects

You notice that the path of the light beam in the above experiment is straight. Careful measurements show no curving of the beam. Now let us place a mirror in the path of the beam and then sprinkle the path with dust so that the beam is visible. Figure 37-2A shows the result.

You notice that when the beam bounces off the mirror it keeps its original width. Moreover, no light is scattered out of the beam by the mirror. This reflection from a smooth, regular surface is called **regular reflection.**

Now let us substitute a piece of white paper for the mirror. As shown in Fig. 37-2B, part of the light goes through the paper (is **transmitted**) but is scattered in all directions on the other side. The piece of paper is said to be **translucent.**

Some of the light doesn't pass through the paper, but is reflected from it. We see that this light is not reflected in a straight beam, as it was by the mirror, but is reflected in all directions. Thus the paper can be seen from any direction. This scattered type of reflection from a rough surface is called **diffuse reflection.**

If we put many different kinds of objects in the path of a light beam, we notice that the smoother the object, the more it acts like a mirror. The rougher the object, the more it acts like paper and reflects light diffusely.

We get interesting results if we place a pane of glass in the path of the beam (Fig. 37-2C). A small part of the light undergoes regular reflection from the smooth surface of the glass, but most of the beam goes right through. Since this part of the beam keeps its original direction, we say that the glass is **transparent.**

Figure 37-2D shows what happens if you illuminate a piece of black cardboard. The cardboard is neither translucent like the piece of white paper, nor transparent like the glass. It permits no light to pass through at all. Objects that prevent light from coming through are said to be **opaque** (ō·pāk′). There is also little reflection of light from the black surface, and the surroundings are only slightly illuminated.

Now place white paper and black cardboard side by side and illuminate them. The white paper appears brighter than the black cardboard because it reflects more light. The black cardboard

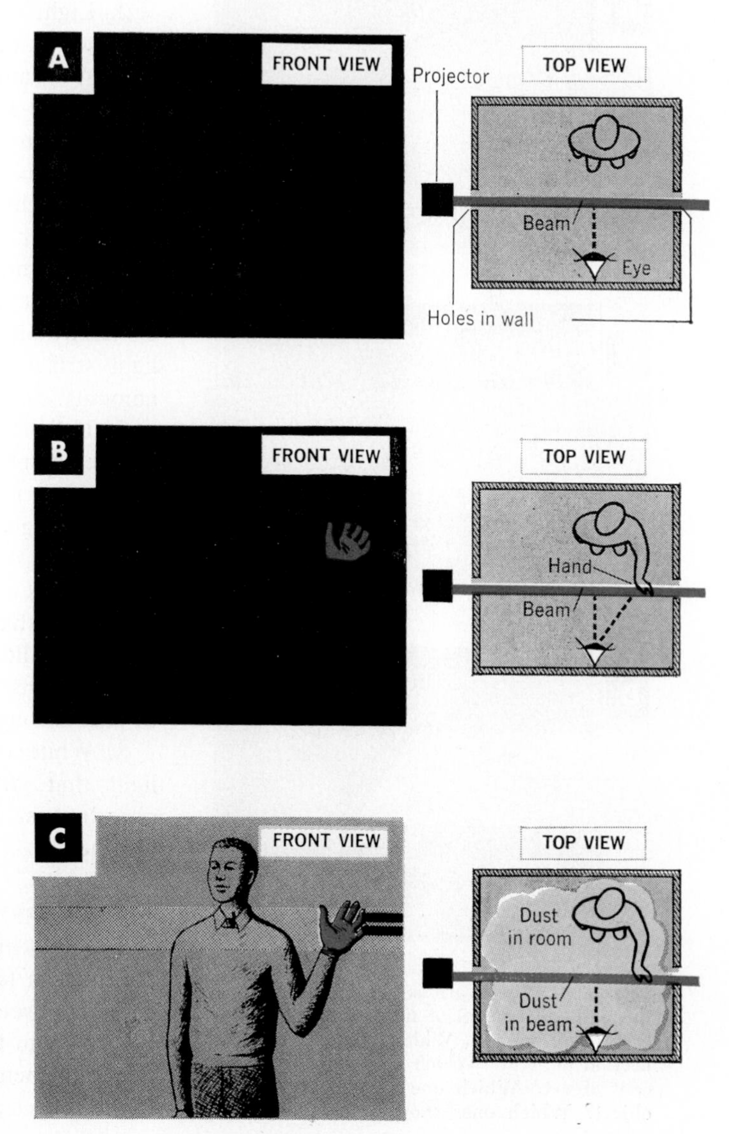

37-1 If you are in the dark about the meaning of the black picture in A, don't worry about it. More light is shed on the subject in B and C as materials in the path of the beam of light reflect light to your eye.

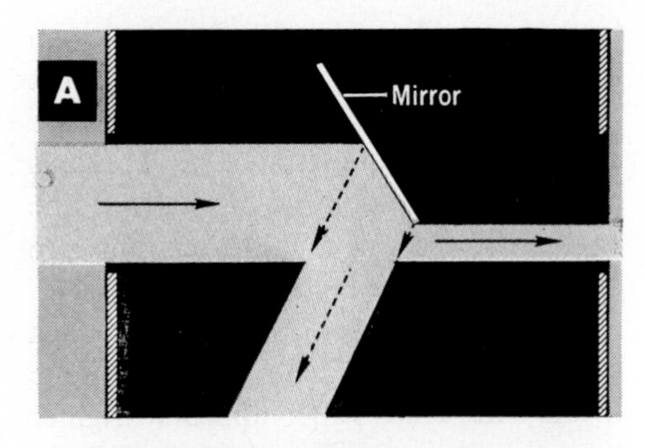

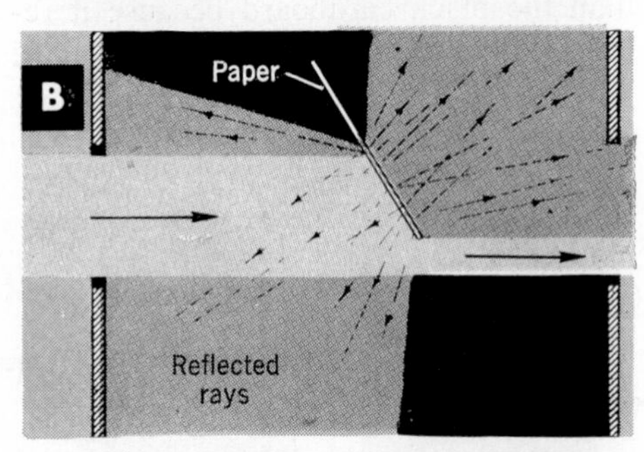

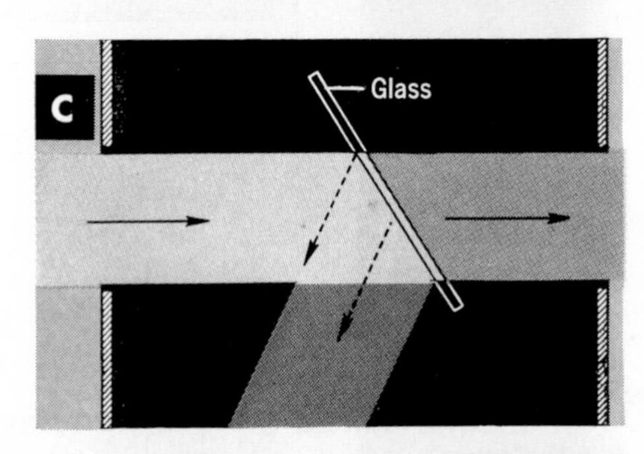

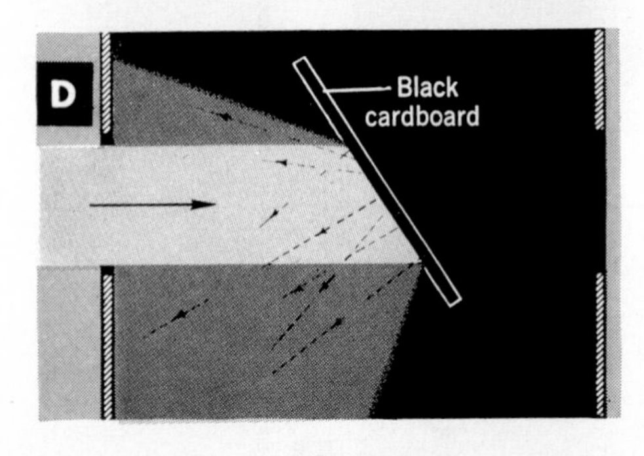

37-2 When various objects are placed in the path of a beam of light, several different things can happen. Which diagrams show reflection of light? Which one shows a translucent object? Which one shows a transparent object? Which ones show opaque objects?

converts into heat a great deal of the light that strikes it. If you touch the cardboard after a few minutes of illumination, it feels warmer than the white paper.

2. Summarizing our observations

Let us review what we have learned:

1. To see an object we must have a source of light; the light must strike the object; and the light must be reflected from the object to the eye.

2. Light travels in straight lines until it strikes an object (Fig. 37-3).

3. A beam of light is visible only when there is dust (or other reflecting particles) in its path (Fig. 37-3).

4. Three things may happen to a beam of light when it strikes an object:
   a. It may be reflected.
   b. It may be transmitted.
   c. It may be absorbed.

Generally all three things happen when light strikes a surface, but in different amounts.

5. If light strikes a smooth object, it is reflected regularly. If it strikes a rough object, it is reflected diffusely.

6. If light scatters upon passing through an object, the object is said to be translucent. If the light beam retains its regular shape after going through, the object is called transparent.

7. Objects which do not permit light to pass through are called opaque.

8. White objects reflect most of the light that strikes them. Black objects absorb most of the light and convert it into heat.

3. Light travels in straight lines

The sixteenth-century forerunner of the modern television set was called the **camera obscura.** It consisted simply of a dark room from which a small hole in the wall opened onto the street outside. Light reflected from people passing by in the street entered the hole. Because light travels in straight lines it formed their images upside down on the opposite wall (Fig. 37-4A).

4. Pinhole camera

To see how the camera obscura works and why it shows images upside down, let us construct one in the form known as a **pinhole camera,** as shown in Fig. 37-4B. When this camera is pointed at bright objects, small inverted images of the objects appear on the waxed-paper screen W. You can make the image larger by pulling out the sliding box which contains the screen (position 2 in Fig. 37-4B).

Let us assume that the object we are viewing is the horse and rider CD. They are illuminated by light coming from all sides. They reflect the light diffusely and scatter it in all directions. Let us first study the head C of the rider. Rays of light reflected from C travel outward in all directions. Some of the light rays hit the front of the pinhole camera. Most of these are reflected or absorbed by the front of the camera. But the rays of light that reach the pinhole pass straight through in a very narrow beam. Inside the box this beam of light from the head C of the rider continues in a straight path until it reaches the screen. Since this beam is traveling downward at an angle from the head of the rider, it strikes the lower part of the screen at C′, where the image of the head is formed.

The rays of light from a foot of the horse (D in Fig. 37-4B) also travel in a straight path through the hole in the box. But the direction of these rays is upward from the foot, so that they strike the upper part of the screen at D′, forming an image there of the foot of the horse.

The image on the screen is upside

Don Knight

37-3 These searchlight beams provide additional evidence that light travels in straight lines. What conclusion can you draw about the air from the fact that the beam is visible? Can you find 3 instances of reflection of light in this photograph?

down. Tracing the rays from the left and right sides of the object will reveal that the image is also reversed from side to side. If the screen is farther away from the pinhole, as shown in position 2 of Fig. 37-4B, then the diverging (spreading apart) rays from the horse spread farther apart before they reach the screen. Thus the image on the screen will be larger than it was before.

The screen we use in the box must be made of waxed paper or some other material that is translucent. If we used a transparent material, such as glass, the beam of light would travel straight through and we would see no image on the screen.

If we substitute a film for the screen in our pinhole camera, we can take photographs with it. But the images formed by such a camera are clear only if the pinhole is very small. Because the tiny pinhole admits only a small amount of light, long exposures are necessary. That is why the pinhole camera is not used more frequently. The action of the pinhole camera illustrates the fact that light travels in straight lines.

5. Shadows

Figure 37-5 shows how light from a small source causes an object to cast a shadow. Because light travels in straight lines we can predict the exact shape and position of the shadow of an object on a screen by drawing straight lines from the source of light to the edges of the object and extending these lines to the screen (Fig. 37-6A).

If the source of light is broad and diffuse, instead of small and sharp, then the edges of the shadows are diffuse (Fig. 37-6B). Each point of light on the surface of a broad source sends out rays of light. Each point is in a different position, and hence the shadow resulting from it will also be in a different position. Consequently, many overlapping shadows are created. One part of the screen may be completely dark because it gets no light at all from any part of the source of light. Such a completely dark part is known as the **umbra** (H in Fig. 37-6B). Other parts of the screen will get light from some parts of the luminous source but not from others. Such regions will be only partly illuminated, and may be considered to be partial shadows. Such an area of partial shadow is called a **penumbra** (E, F, G, in Fig. 37-6).

A single small, intense source of light creates shadows which are very sharp and dark. The sharp contrast between

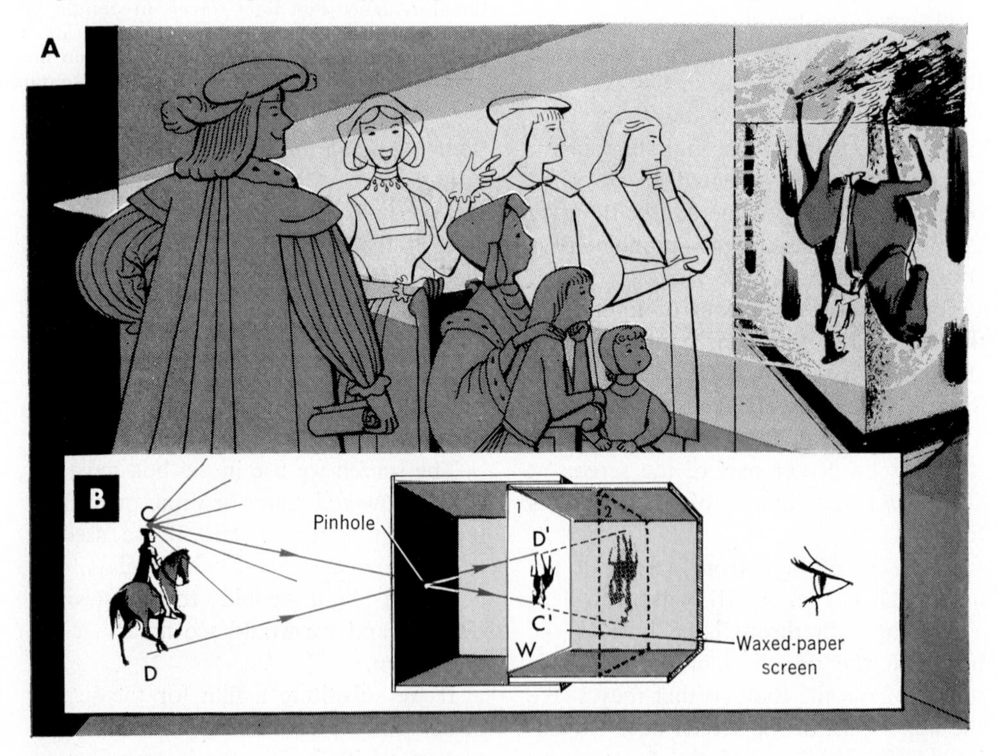

37-4 TV—or rather CO (camera obscura)—in the sixteenth century. Inverted images of objects outside are cast on a wall as light enters from the bright, sunny outdoors through a hole in a dark room. Try this yourself some time.

Ruge from Black Star

37-5 This shadow is sharp because the source of light is small. Why is the beam of light from the source of light invisible? What illusion is being created here?

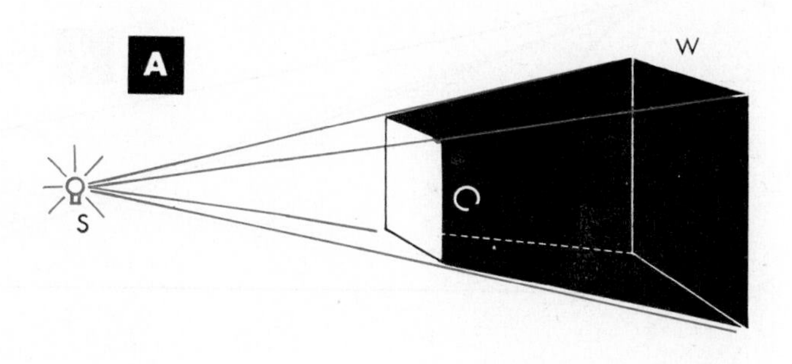

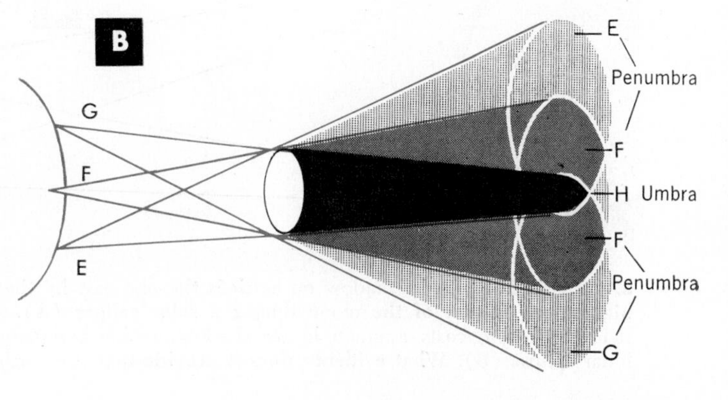

37-6 The exact shape and size of a shadow can be predicted from the fact that light travels in straight lines. In A a small point of light produces a sharp shadow, W, on the wall. But if the source of light is broad (B), the overlapping shadows reduce the dark part (umbra) and increase the half-shadow part (penumbra).

light and dark produced in this way can cause eyestrain. That is why so many lamps are broad diffuse sources of light. Eyestrain is cut down by the use of translucent glass bulbs, globes, or shades on lamps, or diffuse reflection from the ceiling.

6. Eclipses

An eclipse is the result of the giant shadow cast by the earth or the moon. A **solar eclipse** is one in which the sun is blocked from our view by the moon (the shadow of the moon is cast on the earth) (Fig. 37-7A). In a **lunar eclipse** the shadow of the earth is cast on the moon (Fig. 37-7B). The round shape of the earth's shadow on the moon during a lunar eclipse is one way we know that the earth is round.

The study of eclipses has contributed a great deal to the advance of science. For example, in 1919 an expedition was sent to Brazil to take photographs of the stars near the sun during an eclipse. The purpose of the trip was to check on one of the predictions made by Albert Einstein in connection with his Theory of Relativity. According to Einstein's theory, a ray of light passing near the sun should be bent by the sun's strong gravitational pull. If this were so, then a star A in line with the sun (Fig. 37-8) would be seen at A′. Ordinarily a star which is almost in line with the sun cannot be seen at all, because the sun's light is too strong. But during an eclipse, when most of the sun's light is blocked out by the moon, such stars can be observed.

Photographs taken during this eclipse indicated that Einstein's prediction was correct. Stars near the sun did appear in different positions from those they were known to occupy. This discovery

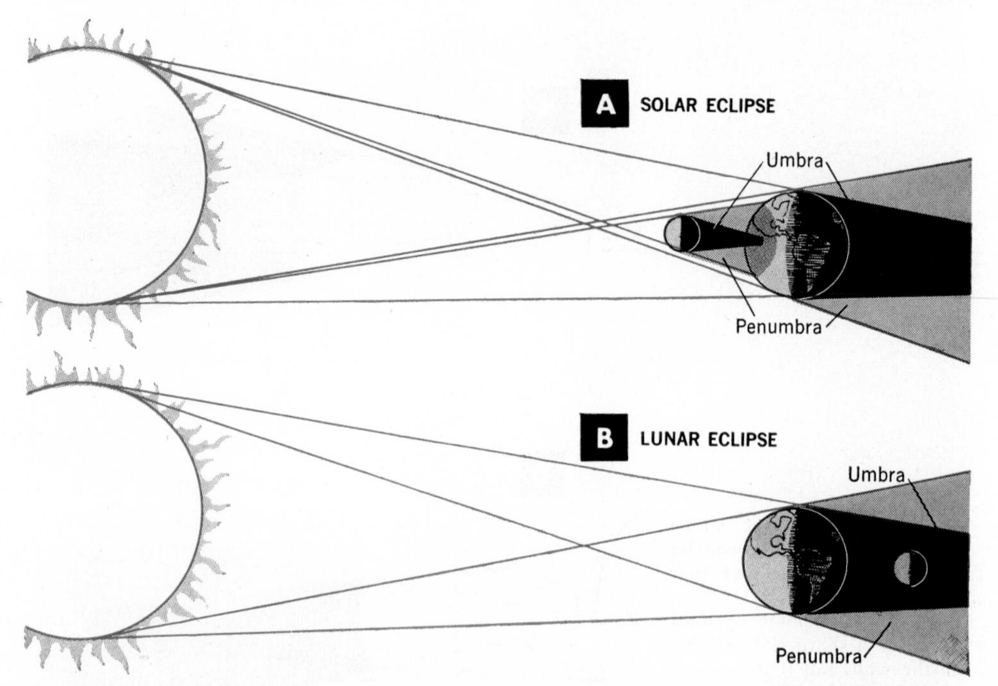

37-7 The biggest shadow on earth is the one cast by the earth causing night. The next biggest one is that of the moon during a solar eclipse (A). The earth, being much larger than the moon, casts a much larger shadow, which completely covers the moon during a lunar eclipse (B). What evidence does B provide that the earth is round?

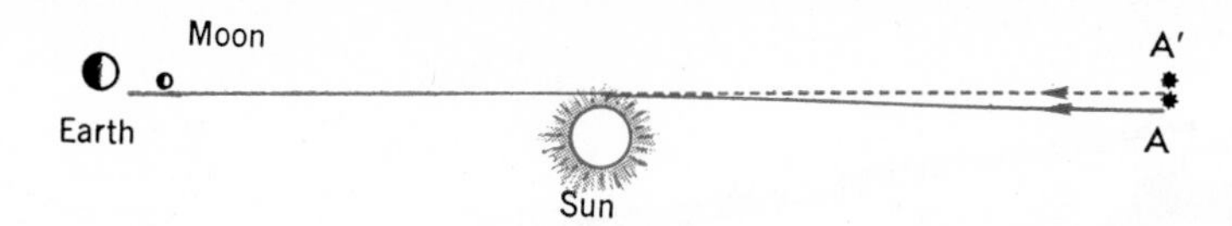

37-8 Einstein's prediction that gravitation of massive bodies would make a beam of light bend is checked during solar eclipses. The bending of the light causes a star, which normally would appear in the direction A, to appear to be in direction A′. This observation can be made only during an eclipse.

was strong indication that the Theory of Relativity was correct, since a prediction based upon it turned out to be correct. It also revealed that light does not always travel in a straight line through a vacuum, but may be bent when passing near a massive object like the sun.

The speed of light

Shortly after the planets were first examined by telescope in the seventeenth century, it was discovered that it takes time for light to travel from one place to another. This discovery came about from the observation that one of the moons of the planet Jupiter seemed to fall behind schedule in its revolution around that planet. The revolution of the moon would appear to slow down until it was about 1,000 seconds behind schedule, then it would appear to speed up gradually until it was back on schedule. These mysterious changes in time of revolution occurred about once a year. Why did they occur?

In 1676 the Danish scientist Römer (rû′mĕr) gave an explanation for this strange occurrence. He suggested that since the earth was farther away from Jupiter at one time of the year than at another, it might take more time for the light from Jupiter's moon to reach us. Careful observations showed that the light arrived 1,000 seconds late only when it had the additional diameter of the earth's orbit to cross. This distance is 186,000,000 miles, since we are 93,000,000 miles from the sun. Dividing 186,000,000 miles by 1,000 seconds, we arrive at 186,000 miles per second as the velocity of light.

7. The fastest speed known

At this enormous speed of 186,000 miles per second, light could travel more than 7 times around the earth in 1 second. One trip around the earth would be completed in about the time it takes to blink an eye! It is because light travels at this enormous speed that we see things happening around us at practically the very instant they occur. For example, it takes only about $\frac{1}{10,000}$ second for the light from a lightning flash 20 miles away to reach our eyes.

But the situation is different when we are dealing with the enormous distances of interplanetary space. Light traveling from the sun takes about 2 hours to reach the planet Pluto. Light from Alpha Centauri, the star nearest to the earth, takes about 4 years to reach us on earth. The light which strikes us from some distant star groups called galaxies actually started out on its journey more than 500,000,000 years ago.

Because these astronomical distances are so great, we need an enormously long unit of length for measuring them conveniently. The distance traveled by light in one year's time, called the **light year,** is the unit of length used by astronomers. Thus Alpha Centauri, the nearest star to the earth, is said to be about 4 light years away, because it takes light from that star 4 years to reach us. Can you figure out how many miles this is?

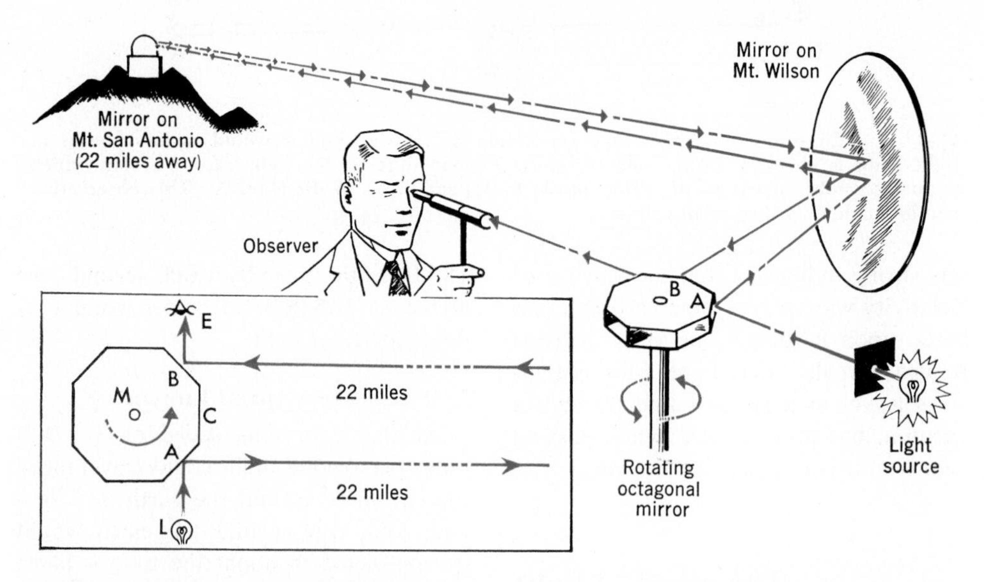

37-9 Michelson's historic, accurate measurement of the speed of light was performed by timing a light beam over a 44-mile measured course. Accurate timing to a tiny fraction of a second was achieved by means of an octagonal mirror rotating at a known rate.

8. Measuring the speed of light

In 1902 an American scientist, Albert Michelson (mī'kĕl·sŭn), measured the speed of light as 186,285 miles per second—with an error probably as small as a few miles per second. Figure 37-9 shows how he did it. Lamp L was used as the source of light. The light from the lamp was reflected from side A of the octagonal mirror M. After traveling a total distance of 44 miles, the light returned to side B of the mirror, when the mirror was stationary. However, when the mirror was rotated at the correct speed, side C moved around to position B in just the proper time to reflect the light, which was then seen by the eye at E. The time it took for side C to get into proper position for reflection to occur was calculated from the known speed of rotation of the mirror. During this time the total distance traveled by the light was 44 miles. The speed of light could easily be calculated from this known distance and the known time it took for side C to replace side B.

9. An example of the scientific method at work

The first careful attempt to measure the speed of light was made by Galileo in the early seventeenth century. He had earlier measured the speed of *sound* by stationing an assistant with a gun several hundred yards away. Galileo then fired a gun at the same instant that he started a clock. The assistant fired his gun at the instant that he heard Galileo's shot. When Galileo heard the sound of his assistant's shot, he observed the time that had elapsed from the instant when he fired the first gun. Dividing the distance the sound had traveled (from him to his assistant and back) by the time elapsed, he found the velocity of sound to be 1,100 feet per second.

Reasoning that the same method, using lights instead of guns, could be used to measure the velocity of light, Galileo performed the same experiment at night with a pair of lanterns. He first uncovered his own lantern. As soon as his assistant saw the light, he uncovered the second lantern, and Galileo stopped his clock on seeing the light. Of course, Galileo found that he was unable to measure the few millionths of a second that it took the light to travel back and forth.

From what we know about the speed of light today, it is easy to see how crude this experiment of Galileo's was, and dismiss it as a waste of time. However, though Galileo was unable to measure the speed of light, his experiment was far from worthless. He had learned from this experiment that light travels at a very high speed indeed, and other scientists, learning of Galileo's work, did not waste their time performing similar experiments, but were able to plan better methods using either a very long distance (Römer) or ingenious methods for accurately measuring very short intervals of time (Michelson).

There are two very important lessons to be learned from Galileo's attempt to measure the speed of light. First, when a scientist is doing research, he is investigating the unknown and is not certain what his results will be (if he knew exactly what the outcome would be, there would be little reason for doing the research!). Indeed, the great majority of research experiments do not turn out quite as predicted. But such research is very valuable, for by observing what actually does happen and then changing accepted theories to explain the unexpected results, man is able to increase his store of scientific knowledge.

A second lesson to be learned from Galileo's experiment is this: A free interchange of the results of scientific work among the world's scientists (through scientific journals and books) enables each one to take advantage of the results turned up in experiments all over the world. Today there are many hundreds of magazines in existence devoted entirely to publishing results of scientific research, both pure and applied. This means less useless repetition and speeds up the rate at which man increases his knowledge of the laws of nature.

What you learn in this book has resulted from the labor of many hundreds of able scientists starting from the dawn of history. There is still much to be discovered by this generation and succeeding generations of people who have curiosity about the unknown and the determination to try to find the answers. Perhaps you will be one of those to push back the unknown in science one step further.

What is light?

We have defined energy as the capacity for doing work. We consider light to be a form of energy, because it can do work. Perhaps this may seem strange to you, but a few examples will suffice to show that it is so.

10. Light as a form of energy

As you probably know, the energy of the sun's light is captured by the green leaves of plants in the process of **photosynthesis** and is stored up as the potential energy of the food produced by the plants. In this way the energy of light plays an important part in supplying us with food and fuel. Light and other forms of radiation from the sun also warm up the earth, supplying heat energy to keep us alive. This heating of the earth by the sun's light rays also causes the giant convection currents in the air called wind, which can be used to push sailboats and operate windmills.

Light can also cause motion directly. Careful measurements have shown that light exerts a very tiny push when it

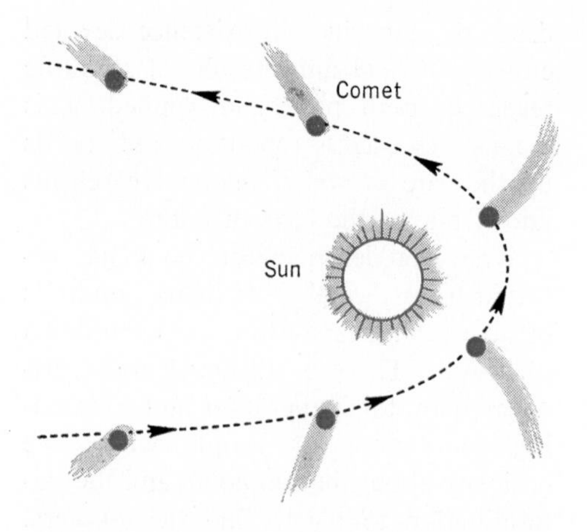

37-10 Evidence that light can exert pressure is provided by observations of comets. The pressure of the sun's light on the thinned-out gases in the comet's tail causes the comet to leave the neighborhood of the sun traveling tail first.

strikes an object. As shown in Fig. 37-10, the tail of a comet, which is made up of dust particles and gas, is pushed outward by the light streaming out from the sun. As a result, the comet leaves the neighborhood of the sun traveling tail first.

In our study of the photoelectric cell (Chapter 36) we have already seen that light can also be converted into electrical energy.

Light also causes the chemical change on photographic films that enables us to take pictures. Finally, it is light which enables us to see, by affecting certain nerve endings in the back of the eye.

All of these examples of light doing work or being converted into other forms of energy show that light is a form of energy.

11. Particles or waves?

Several centuries ago, the Dutch scientist, Christian Huygens (hī′gĕnz), proposed the theory that light is some kind of wave motion. He pictured a light wave as being similar to a water wave, spreading out in ever-widening circles. When a water wave strikes a large smooth object it bounces off, retaining its original shape. A beam of light is reflected from a mirror in much the same way.

From our study of waves in sound and electricity we might expect that there would be waves of high and low frequency. According to the wave theory of light it is these different frequencies that account for the different colors of light.

A short while after Huygens proposed his wave theory, Sir Isaac Newton proposed a different one. Newton said that light consists of tiny particles. According to his theory, these tiny particles of light shoot out from luminous or glowing objects and travel at a great speed in all directions. When they strike another object, some are bounced off (reflected) like balls bouncing off a wall. According to Newton's theory, different colors are due to different kinds of particles.

Both Newton's theory and Huygens' could be used to explain all the facts known about light at the time. But the particle theory had the weight of Newton's prestige, so it was accepted by most scientists until about 1800. In Chapter 44 we shall examine evidence that has accumulated since that time.

Going Ahead

THE VOCABULARY OF PHYSICS

Define each of the following:

regular reflection	umbra
translucent	penumbra
diffuse reflection	solar eclipse
transmission (of light)	lunar eclipse
transparent	light year
opaque	photosynthesis
camera obscura	pinhole camera

THE PRINCIPLES WE USE

1. Tell what happens to a beam of light as each of the following is placed in the path of the beam: chalk dust, a mirror, a piece of white paper, a piece of black cardboard, a piece of glass, a smooth object, a rough object.
2. What three conditions are necessary for us to see an object that is not luminous (giving off light)? Show how these conditions apply to a certain object (a tree, for example).
3. Mention three important facts which show that light is a form of energy.

Explain why

4. The revolution of Jupiter's moon apparently falls behind schedule by about 1,000 seconds every year.
5. A comet goes away from the sun tail first.
6. The image in a pinhole camera becomes larger as the screen is moved away from the pinhole. (Use a diagram.)
7. The image in a pinhole camera is inverted and reversed. (Use a diagram.)
8. A solar eclipse lasts for only a few minutes at any place on earth, while a lunar eclipse may last for several hours.

Describe an experiment to

9. Measure the speed of light.
10. Show regular reflection.
11. Show diffuse reflection.
12. Show that light travels in straight lines.

13. Why are frosted bulbs (or globes over clear bulbs) preferred for illumination in the home?

SOLVING PROBLEMS IN THE BEHAVIOR OF LIGHT

1. A square card 1 inch on each edge is placed 3 inches from an unfrosted lamp. The shadow falls on a screen placed 2 inches behind the card. Draw a diagram whose exact dimensions will help you locate the shadow. What is the size of the shadow?
2. Draw diagrams showing why a tiny bulb (point source) produces a sharp black shadow, while a frosted bulb produces a diffuse shadow. Label the umbra and penumbra.
3. Draw a diagram showing the cause of an eclipse of the sun.
4. Draw a diagram showing the cause of an eclipse of the moon.
5. How many miles are there in a light year? Show your calculation.
6. It takes 4 years for light to reach us from the nearest star. How many miles away is it?
7. When you look into a store window on a bright day you can see a clear reflection of yourself. But at night, when the inside of the store is brightly lit, the reflection is very faint. Explain why.

PROJECT: A PINHOLE CAMERA

Make a pinhole camera from a tin can with a tightly fitting lid. In the bottom of the can cut out a hole at least $\frac{1}{4}$ inch in diameter, and fasten over it a sheet of aluminum foil. Use a fine needle to make the tiniest possible hole in the aluminum. (Why should it be so small?)

In total darkness in a closet, tape a piece of camera film to the lid of the can, being careful that the emulsion side is up. Press the lid on the can, and you are ready to go.

Set the camera where it can remain motionless for 10 minutes to an hour (time determined by trial) viewing a brightly-lit scene. Why is no shutter necessary if you set it up quickly? At the end of the exposure, develop the film.

CHAPTER

38

The Path of Light

When you look into a mirror you see yourself and your surroundings in reverse. How does this happen?

Reflection

In Chapter 37 we experimented with the reflected beam of light from a mirror and noticed that the beam retained its shape. To understand how the mirror works, we must study this mirror reflection more carefully.

1. The laws of reflection

Let us try an experiment. Adjust a projector to give a beam of parallel rays. Now let this beam fall on a tiny hole in a piece of cardboard. We can call the very narrow beam that passes through the hole a **ray** of light. Let this ray of light strike a mirror at different angles, and each time measure the angle at which the ray is reflected from the mirror (Fig. 38-1). We find for each measurement that the angle at which the ray strikes the mirror is the same as that at which the reflected ray leaves the mirror.

Instead of measuring angles between the rays and the mirror surface, as in Fig. 38-1A, it is customary to measure the angle between the ray and the perpendicular to the surface. In Fig. 38-1D, the line I represents a ray called the **incident ray,** which strikes the mirror at point P. From P we draw a line PN, perpendicular to the mirror; this line is called the **normal.** Now we measure the angle which the normal makes with the incident ray. This is angle *i,* called the **angle of incidence.** When we measure the **angle of reflection,** *r,* which the **reflected ray** R makes with the normal PN, we find that it always equals the angle of incidence.

Our experiments lead us to the **First Law of Reflection:**

When a ray of light is reflected, the angle of incidence equals the angle of reflection.

We also find that:

The incident ray, the normal, and the reflected ray all lie in the same plane.

This is the **Second Law of Reflection.**

Figure 38-2 shows a periscope. Periscopes are used in many ways, such as in submarines to scan the surface of the ocean when the submarine is submerged. In the photograph it is being used in an atomic-energy laboratory to see what is going on behind the brick shield without the danger of exposure to atomic radiations.

2. Images in a mirror

If you set a mirror on edge on a piece of graph paper you can discover several important facts about the formation of images in a mirror (Fig. 38-3A). Place the back edge of the mirror on one of the ruled lines of the graph paper. Make sure that the mirror is standing perpendicular to the paper. Now make a dot, S,

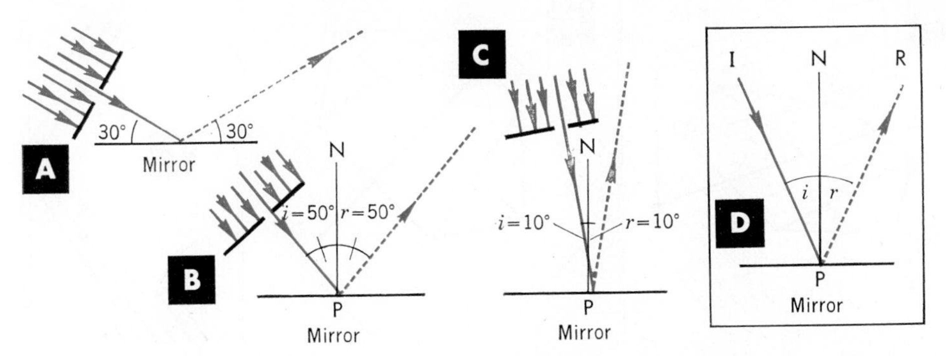

38-1 Light is reflected by smooth surfaces so that the angle at which it is reflected is equal to the angle at which it strikes. Thus, in D, angle *i* equals angle *r*, since PN is perpendicular to the mirror surface.

5 squares in front of the mirror. When you look into the mirror, you notice that the image is as far behind the mirror as the dot (the object) is in front of it. We can see that each point on any object and the corresponding point on the image are equidistant from the mirror, and that the mirror is on the perpendicular bisector of the line between them.

We can use this fact to locate the image of an object such as the triangle in Fig. 38-3B by making a dot behind the mirror for each vertex of the triangle. R′ is the equidistant image of R, S′ of S, and T′ of T. Now we connect the three dots to form the image of the triangle. We note that object and image are equal in size and are equidistant from the

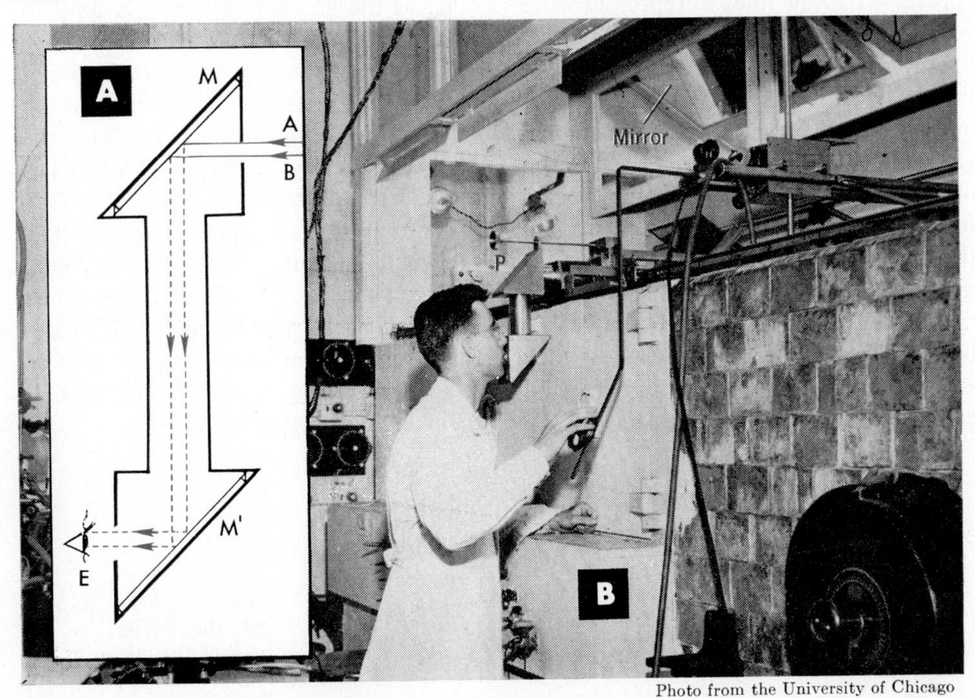

Photo from the University of Chicago

38-2 Danger—radioactivity! How do you handle the stuff behind a brick wall? Mirrors on the ceiling and a periscope containing reflecting surfaces enable this man to see around corners and use his mechanical hands. Why is the image seen right side up?

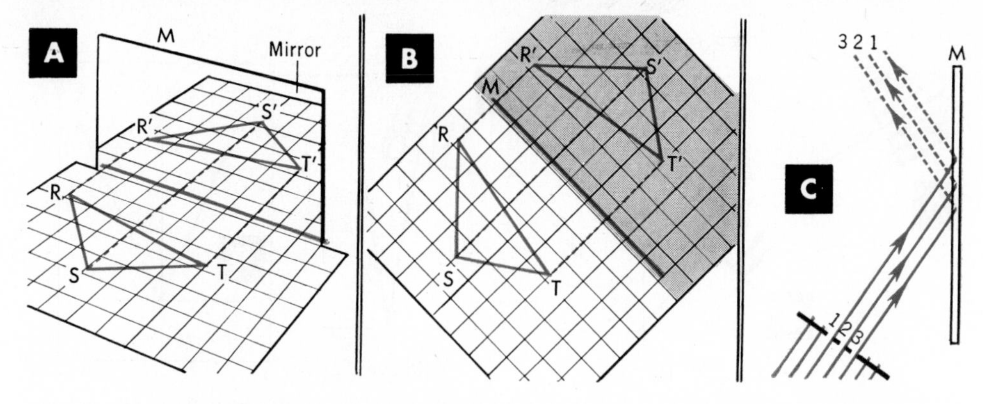

38-3 Can you come up with 3 facts about mirror images from looking at the experiment in A? B shows how you may predict the exact position of the image on paper. In C observe that the mirror reverses the rays of a beam and thus causes a reversal of all images.

mirror, but the position of the image in the mirror is reversed from left to right as compared to the position of the object.

3. Why is the image reversed?

By performing a simple experiment, we can see why the image in the mirror is reversed. Draw lines to represent three parallel rays, (1, 2, and 3) which strike a mirror, as shown in Fig. 38-3C. All three rays strike the mirror at the same angle; thus all are reflected at equal angles, and they come out parallel to each other. But the reflected rays are reversed from left to right. Ray 1, which was on the left as the rays struck the mirror, is now on the right. Ray 3, which was on the right, is now on the left side of the beam. This crossing of light rays as they are reflected from a mirror is what causes the reversal of the images.

4. Why image distance equals object distance

If a ray R_1 is reflected to your eye, as shown in Fig. 38-4, you see it as coming from somewhere back along the line AB. With one eye alone you are quite unable to judge *how far* along AB the image lies—unless, of course, you have a scale such as the graph paper in Fig. 38-3 to help you.

If you now move your eye to C so that it receives ray R_2, you will think that the object lies somewhere back along the line CD. Clearly, if the image of O lies along line AB and also along line CD, it can only be at the point I, where they intersect. All other rays, R_3, R_4, etc., that you might draw will likewise, when extended back, pass through I. Be careful to draw them so that their angle of incidence on the mirror is the same as their angle of reflection.

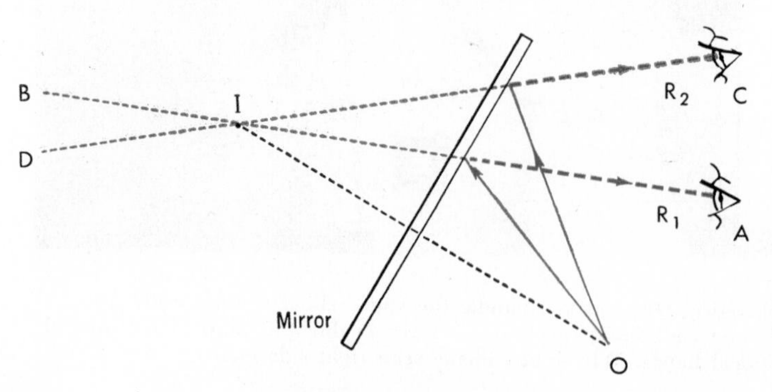

38-4 Why do we see the image as far back of the mirror as the object is in front of it? Rays R_1 and R_2 actually come from O but *seem* to come from I.

I is the *virtual* image of O. The term virtual means that there is no light actually coming from O. It only *seems* to, since the eye is unable to detect the bending of the ray as it bounces off the mirror. If you measure the diagram you will find that O is as far in front of the mirror as I is behind it, and you will find that OI is perpendicular to the mirror.

With the help of geometry you may be able to prove that this will be true for all positions of O.

5. How our eyes judge distance

From Fig. 38-4 you will see that one eye alone cannot reliably judge the distance of an object unless assisted by a scale or by memories of past experience. Two eyes are necessary, whether we are looking at reflections in a mirror or looking directly at the object itself.

The farther an object is from you, the more nearly parallel are the rays IA and IC reaching your two eyes. Nerves in your eye muscles communicate this fact to your brain, and your brain, from past experience since infancy, is able to judge the distance to the object you are looking at. In a similar way range finders, used to guide the focusing of a camera or the firing of artillery or even the distance to the nearer stars, measure the directions of the object from two different viewpoints, A and C, and deduce the distance from the angle between them.

6. Diffuse reflection

You know that a rough surface does not reflect a beam of light in the same way that a mirror does (page 504). The rays from a rough surface are scattered, producing diffuse reflection. Again the Laws of Reflection provide an explanation.

Even though the rays are scattered, the angle of incidence of each ray is equal to its angle of reflection. Figure

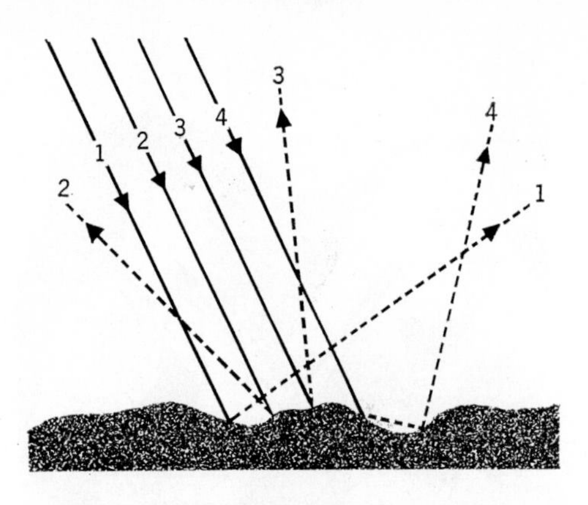

38-5 Every which way go the reflected rays of a beam of light as they strike a rough surface. Actually each ray obeys the Laws of Reflection.

38-5 shows four rays in a beam of light moving parallel to each other. Striking a rough surface, each ray hits a different imperfection at a different angle, and therefore they all bounce off in different directions, producing diffuse reflection.

It is important that light illuminating a house be properly diffused. Painting the walls and ceilings to make the surfaces rough causes diffuse reflection. As a result, light rays bounce unevenly from all surfaces in the room, thus spreading the light to give an even illumination.

The kind of paper on which a book is printed produces a similar effect. If the pages are smooth and glossy, they produce mirror images of the source of illumination. When such images are reflected from the page to the reader's eye, they create a glare which makes reading difficult (Fig. 38-6).

7. Reflectors

Up to this point we have discussed reflection only from flat, or plane, surfaces. Light rays from a point source diverge (spread out) from one another after they have been reflected from such a surface (Fig. 38-7A). But if the reflecting surface is curved into a shape known as a **parabola,** the rays can be reflected so that they come out parallel

38-6 A reading lamp is in the best position when the light comes from behind and above the reader. The eye then sees by diffuse reflection (B). In the lamp position in A, mirror reflections of the lamp seen in the table and book make reading hard on the eyes.

to each other. This parallel beam does not spread apart as it travels, and can therefore give intense illumination to objects at a great distance.

Figure 38-7B shows how a reflector of this type is used in a searchlight. If the light source is properly placed close to the mirror, the beam can be concentrated as much as desired. Such reflectors are used in flashlights and automobile headlights. They are also used to concentrate the light of floodlights and photographers' flash bulbs, and they play an important part in projectors, microscopes, photoenlargers, and other optical devices. Methods of finding the image formed by curved mirrors are discussed in detail in Chapter 39.

Refraction

Look at the pencil in the jar of water in Fig. 38-8. It seems to have been broken just at the water level. Let us see why this illusion occurs.

8. Light rays from air into water

What happens when a single beam of light from a projector is sent through water? In Fig. 38-9A six different rays from a single light source have been drawn striking the water, each at a different angle. Each ray takes a different path through the water.

You notice that all the rays travel in a straight path through the air until they reach the water. But at the surface of the water rays 1, 3, 4, 5, and 6 bend,

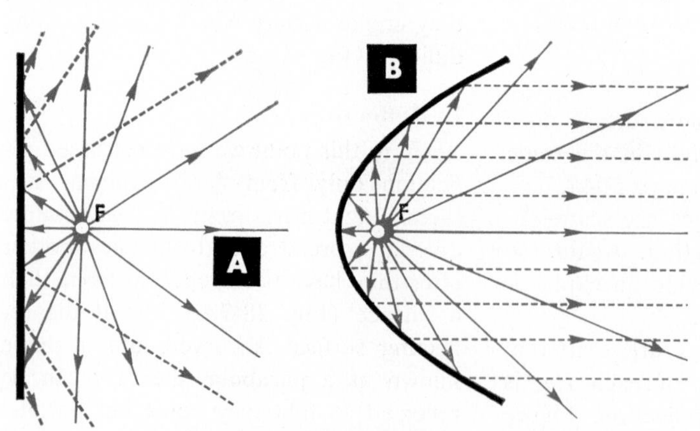

38-7 The special parabolic shape of the reflecting surface in B causes rays of light from a source of light at the focus to come out parallel instead of spreading out as in A. Why is it an advantage for searchlights to make use of this set-up?

and their direction changes. Once a ray is in the water it continues in a straight line, so we see that bending takes place only at the surface between the air and the water. The bending of a light ray as it passes from one transparent material into another is called **refraction.**

Ray 2, which enters the water at a right angle to the surface, is the only ray that does not bend as it enters the water. Any ray perpendicular to the surface of the water follows a straight path from the air through the water.

You can predict how any ray will bend as it enters water from air by drawing a diagram (Fig. 38-9B). The ray from I (the incident ray) will strike the surface at point P. The normal will be N′PN. The dotted line PF will show the straight path the ray would have taken if it had not entered the water. Except

Frank Murray

38-8 Light passing through transparent materials does strange things. Look closely and find three illusions in this picture. Can you explain the illusions?

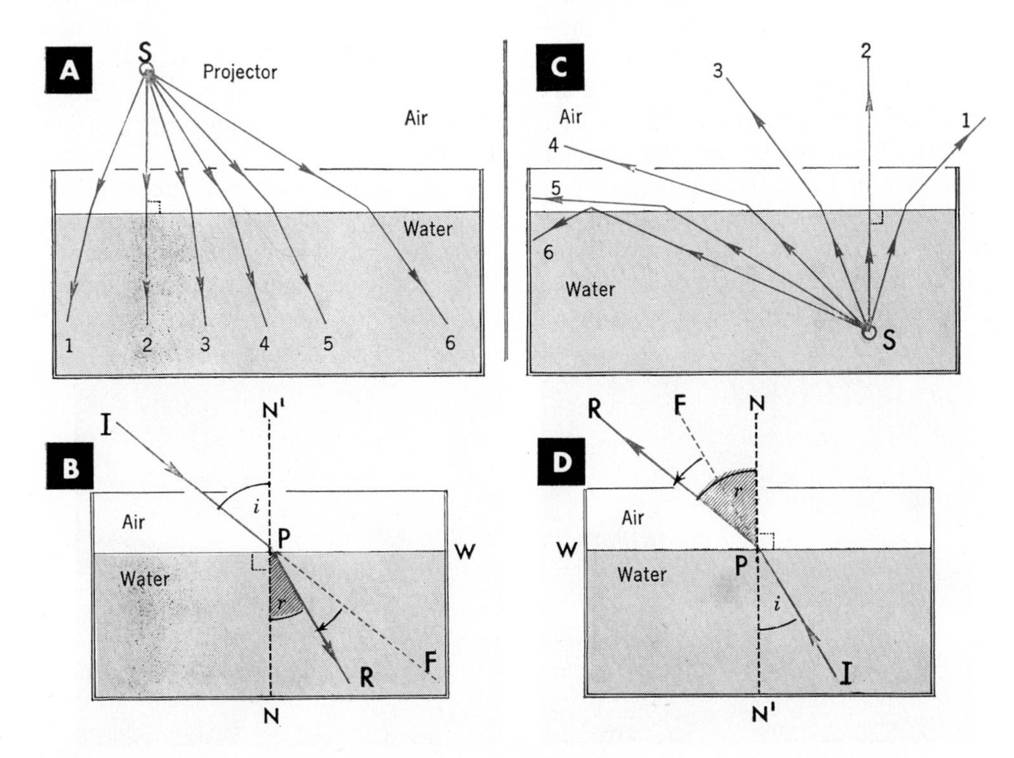

38-9 Light rays passing from one transparent material to another bend according to definite laws. A and B show the kind of bending observed as light passes from air into water, while C and D show how light bends from water into air.

along PN the actual path, PR, of the ray is *closer to the normal* than if the ray had continued straight. This change in direction is caused by refraction.

The angle i that the incident ray I makes with the normal is called the **angle of incidence.** The angle r that the **refracted ray** R makes with the normal is called the **angle of refraction.** We find that when a ray of light enters water from air obliquely, the angle of refraction is always less than the angle of incidence, and they always lie on opposite sides of the normal.

9. From water into air

What happens if the ray is coming out of the water instead of entering it? Figure 38-9C shows six rays coming from a light source *below* the water surface. The rays bend in the reverse manner from those entering the water from the air. In this situation we draw the incident ray IP (Fig. 38-9D) and show the direction it would have taken had it continued in a straight path (PF in Fig. 38-9D). Then we draw the normal NPN′. We find that the ray bends *away from the normal* when passing from water into air; thus the angle of refraction, r, is greater than the angle of incidence, i. This is just the opposite of what takes place when the ray passes from air to water. But again an incident ray which is perpendicular to the surface (ray 2) does not bend but passes out of the water in a straight line.

10. The critical angle

As you can see, ray 5 in Fig. 38-9C comes out almost parallel to the surface of the water. A ray which is bent a bit more than this will not be able to escape from the water, but will be reflected back into it. This happens with ray 6. It is reflected back into the water by the surface of the water, which now acts as a mirror. The last ray to get out, ray 5, has an angle of incidence that we call the **critical angle.** The critical angle is the angle of incidence that has an angle of refraction of 90 degrees. Obviously 90 degrees is the greatest angle of refraction possible. If a ray strikes the surface of the water at an angle greater than the critical angle (about 47° for water), it will not be able to escape from the water but will be *totally reflected* internally.

A 45-degree right-angle glass prism will produce total internal reflection because the critical angle for glass is about 43 degrees when light strikes it as shown in Fig. 38-10.

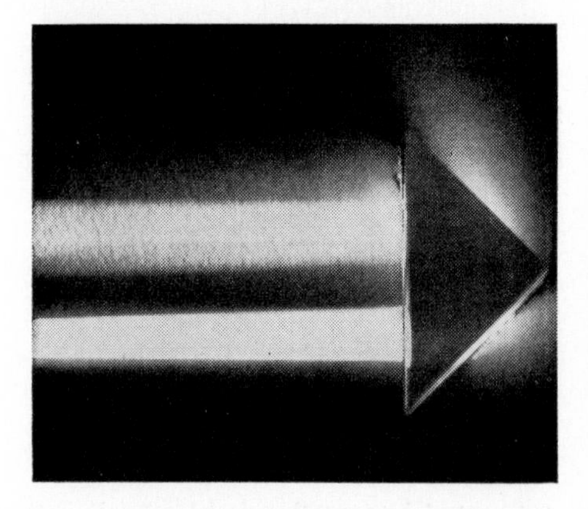

38-10 A triangular prism can serve as a mirror in binoculars and other optical instruments. What advantage do such prisms have over mirrors?

11. What causes refraction?

The wave theory of light enables us to explain the way light bends when it enters a transparent material. This explanation is based on the fact that light waves travel more slowly in matter than in a vacuum.

Let us imagine a beam of light traveling toward a piece of glass as shown in Fig. 38-11. The parallel lines represent the positions of the front of one of the light waves at different moments as it travels forward. One side, A, of the wave strikes the glass first and is slowed down before the other side, B, reaches the surface. Part B is still traveling at the normal speed it has in air. This side-to-side slowing down of the wave causes it to swing around as it enters the glass. Thus the direction of the wave shifts and as a result is bent closer to the normal. This is exactly what we found in our experiment with light rays which pass from air into water.

Once the entire wave is inside the glass all of its parts move together again at the lower speed, and the whole beam now travels in the new direction through the glass.

What happens when a beam goes out of glass into the air? Part A′ of the wave strikes the surface first and gets out into the air (Fig. 38-11). It speeds up as soon as it enters the air because its velocity is changed in the new material. Meanwhile the rest of the wave, which is still in the glass, travels at a slower speed until it enters the air, causing the entire wave to swerve. You can see now why a beam of light entering a material perpendicularly is not bent; the front of the wave is slowed down from side to side at the same instant.

This bending of light is similar to what happens when a line of marchers changes direction in a parade. To hold the line together the man on the inside must walk more slowly, while the man on the outside swings the line around.

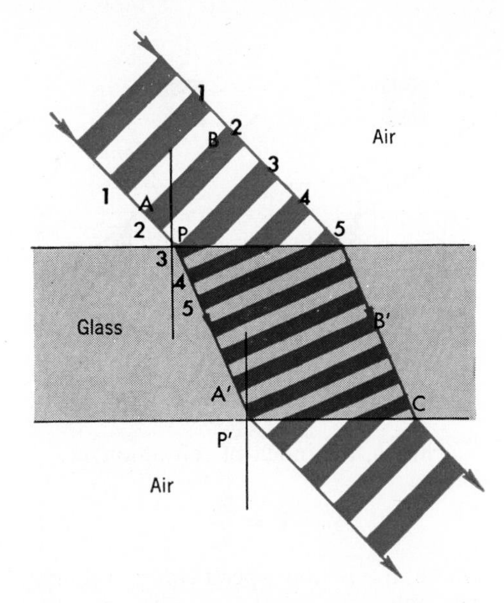

38-11 Why does a light wave bend as it travels from air into glass and out into air again? The slower speed in glass causes the wave to swerve just as a car would swerve if one wheel went off the road into the mud.

12. Refraction in other transparent materials

Every transparent material refracts rays of light as glass does. The only difference is in the amount of refraction. Figure 38-12A and B shows a ray of light entering water and another ray entering glass. Both rays are at exactly the same angle of incidence, *i*. The ray which enters the glass is refracted much more sharply than the one which enters water. Other materials, such as diamond, bend light even more sharply than glass (Fig. 38-12C).

The number we use to describe the amount of bending for each material is called the **index of refraction** of that material.

The index of refraction of a substance is defined as the speed of light in a vacuum divided by the speed of light in the substance.

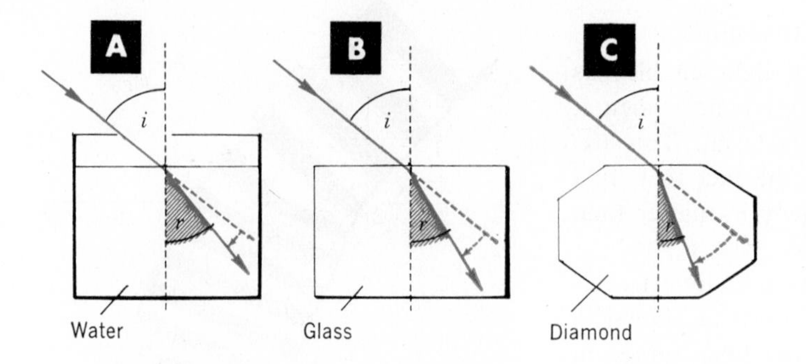

38-12 The rays strike the transparent materials in A, B, and C at the same angle. Diamond (C) bends the light more than glass (B), and glass more than water (A).

Thus k, the index of refraction, is:

$$k = \frac{S_v}{S_s}$$

where S_v is the speed in a vacuum, 186,000 miles per second (3×10^{10} cm/sec), and S_s is the speed of light in the substance. The values of the index of refraction of some common substances are shown in Table 15. Why must k always be greater than 1?

SAMPLE PROBLEM

What is the speed of light in diamond? From Table 15, k is 2.42. Substituting in the above equation:

$$2.42 = \frac{186{,}000}{S_s}$$

$$S_s = \frac{186{,}000}{2.42} = 76{,}800 \text{ mi/sec}$$

STUDENT PROBLEM

What is the speed of light in water? (ANS. 140,000 mi/sec)

TABLE **15**
INDEX OF REFRACTION

Substance	Index
Vacuum	1.0000
Air	1.0003
Water	1.33
Glass (light crown)	1.51
Glass (dense flint)	1.71
Diamond	2.42

13. Snell's Law

It is not easy to measure directly the speed of something that travels as fast as light, as we saw in the preceding chapter. But it is very easy to measure the speed of light in a substance *indirectly* by a method worked out by the Dutch physicist Willebrord Snell in 1621. Studying the bending of light when it passes into a substance, Snell discovered that

Whenever light is bent on entering a substance from air (or vacuum), the sine of the angle of incidence divided by the sine of the angle of refraction is constant for a given substance.

It was eventually proved, after the speed of light had been measured in different substances, that the constant obtained from Snell's Law is actually equal to the index of refraction of the substance. Expressed mathematically, **Snell's Law** is:

$$\frac{\sin i}{\sin r} = k$$

From this you can see that the easiest way to find the speed of light in a substance is to measure the angles of incidence and refraction of a ray passing into the substance and then use the ratio of the sines to calculate k. From this formula you will also see that the higher the index of refraction, the greater the bending that takes place.

SAMPLE PROBLEM

A ray of light enters a substance at an angle of incidence of 30 degrees. Its angle of refraction is found to be 22 degrees. What is the index of refraction of the substance? Substituting in Snell's Law:

$$k = \frac{\sin i}{\sin r}$$

$$k = \frac{\sin 30^\circ}{\sin 22^\circ}$$

Using the table of sines on page 682, we find that sin 30° = 0.500 and sin 22° = 0.375. Substituting:

$$k = \frac{0.500}{0.375} = 1.33$$

Looking at the values of the index of refraction for some common substances in Table 15, we see that the substance is probably water.

STUDENT PROBLEM

A ray of light enters a substance at an angle of 45°. The angle of refraction is 28°. What is the index of refraction of the substance? (ANS. 1.5)

■ ■ ■ ■ ■ ■ ■

Snell's Law can be predicted by both the wave theory of light and Newton's particle theory. The prediction from the wave theory assumes that light travels *slower* in water than in air, whereas the prediction from the particle theory assumes that light travels *faster* in water. Obviously, a crucial test of the two theories lies in an accurate measurement of the speed of light in water. Though the speed of light in air was known in the time of Huygens and Newton, there was no adequate way of measuring short enough time intervals to find the speed of light through the very short distance encountered in a tank of water of reasonable size. It was not until about 200

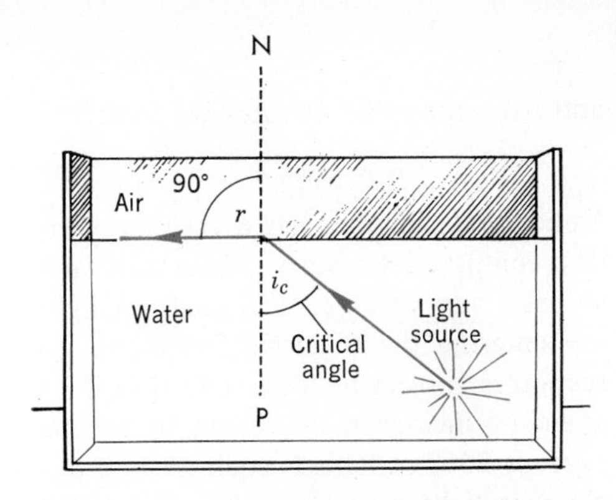

38-13 The critical angle for light passing from water to air is the angle of incidence when the angle of refraction is 90°.

years after Newton's time that Foucault, using an apparatus similar to the one later used by Michelson (page 512) measured the speed of light in water and found it to be *less* than the speed in air. This was the real death blow to Newton's particle theory of light.

We can use Snell's Law to show that there is a definite relation between the critical angle of a substance and its index of refraction. Referring to Fig. 38-13, we see that the critical angle is reached just as the angle of refraction of the ray leaving the substance is 90°. Since the ray is *leaving,* not entering, the substance, the angles of incidence and refraction are interchanged when we apply Snell's Law:

$$k = \frac{\sin r}{\sin i} = \frac{\sin 90^\circ}{\sin i_c}$$

$$\sin i_c = \frac{\sin 90^\circ}{k} = \frac{1}{k}$$

Thus we see that the larger the index of refraction, the smaller sin i_c, and hence the smaller the critical angle, i_c.

SAMPLE PROBLEM

What is the critical angle of diamond? Since the index of refraction of diamond is 2.42, we have:

$$\sin i_c = \frac{1}{2.42} = 0.413$$

and from the table of sines we have:

$$i_c = 23°24'$$

Notice that diamond has a very large index compared to most substances, and so has a very small critical angle. A light ray entering the diamond from any direction is subject to total reflection if its angle of incidence on trying to escape exceeds 23°24′. This is such a small angle that light rays are reflected several times inside the diamond, and come out in many different directions. It is this small critical angle that makes diamonds sparkle and seem to catch the light. Diamonds are cut in geometric shapes that make use of the critical angle to give the maximum internal reflection.

14. Effects of refraction

The property of refraction in transparent materials accounts for a number of curious effects we notice in nature. The pencil in the jar of water shown in Fig. 38-8 is an example. If we take a ray from the pencil and follow it to the eye to find out how it bends, it appears as shown in Fig. 38-14A. The eye, tracing the ray back to the point it *seems* to come from, sees the pencil in the water at P_1, and not where it really is. As a result, the pencil appears to be bent or broken.

Another example is the bending of the sun's rays slightly towards the vertical as they enter the earth's atmosphere from outer space. Because of this fact we see the sun at a point higher in the sky than that which it really occupies (Fig. 38-15). This increases the time from sunrise to sunset, thus making the day slightly longer than it would be if there were no refraction of the sun's light by the air around the earth.

In winter when a window is open above a hot radiator, objects beyond it appear to shimmer. This effect is caused by the refraction of the light as it passes from cold to hot layers of air. Cold air has a higher index of refraction than warm air. As the layers of air move, the direction from which the bent light is coming shifts rapidly back and forth, so that objects appear to change their positions. A similar shimmering effect occurs above a hot beach or road on a summer day.

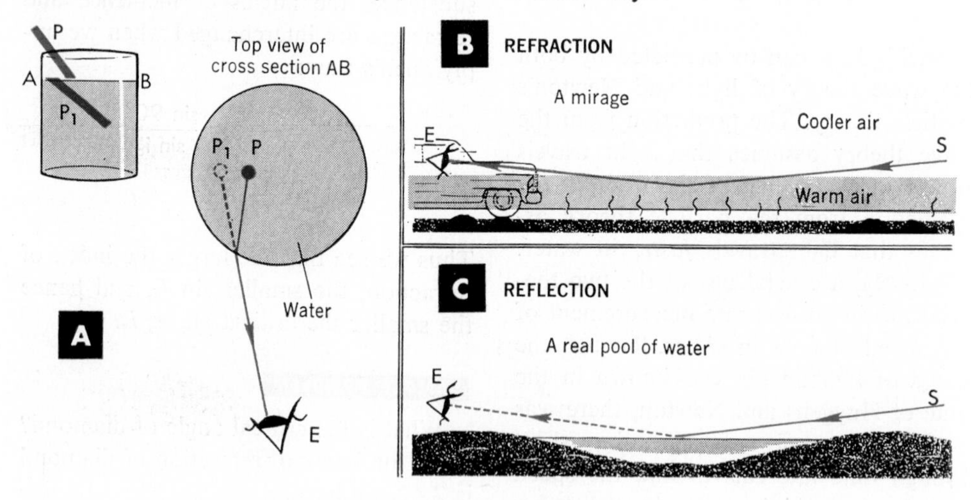

38-14 **Refraction causes many illusions. A explains why the pencil in Fig. 38-8 seems to be in a different place from where it really is. B shows how refraction causes mirages that conjure up false pools of water in the distance.**

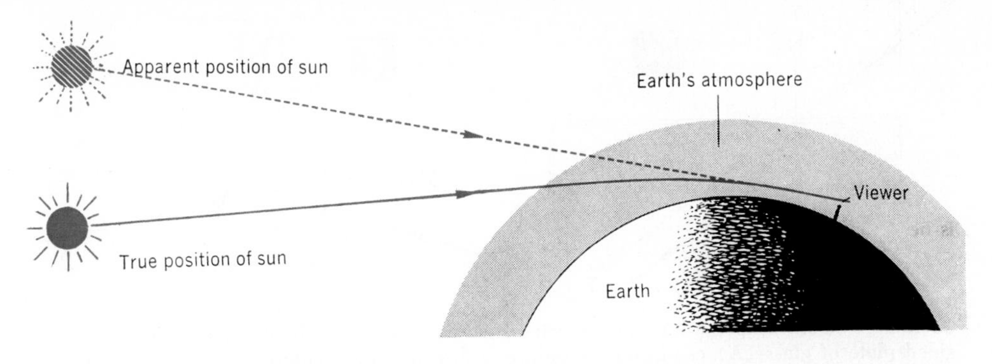

38-15 When we see the sun just rising, it is really still below the horizon. The earth's atmosphere bends the sun's rays.

15. Mirages

The bending of light as it travels through layers of air of different temperatures is what causes **mirages** (mĭ-räzh'ĕz). You have heard how people stranded in a desert without water suddenly see pools and lakes in the distance which disappear on closer approach. When riding in a car on a summer day you may have "seen" a pool of water in the road ahead. But when you reached the spot, there wasn't any water there at all. Let us see why this happens.

Suppose there actually is a pool of water on the ground a certain distance away, as shown in Fig. 38-14C. Light from the sky at S will be reflected from the pool to your eye. This bright reflected image of the sky comes to be connected in your mind with a pool of water.

But now consider a situation where the ground is hot and the air near the ground is warmer than the air higher up. Light from the sky at S in Fig. 38-14B is bent as it comes from the cool air into the warm air near the ground. This refracted light then bends upward into your eye without ever touching the

U.S. Weather Bureau

38-16 Can a mirage be photographed? This is a photograph of a mirage in Death Valley, California. What looks like a large lake is really a mirage. Since the light rays are real and not imaginary they will affect the photographic film.

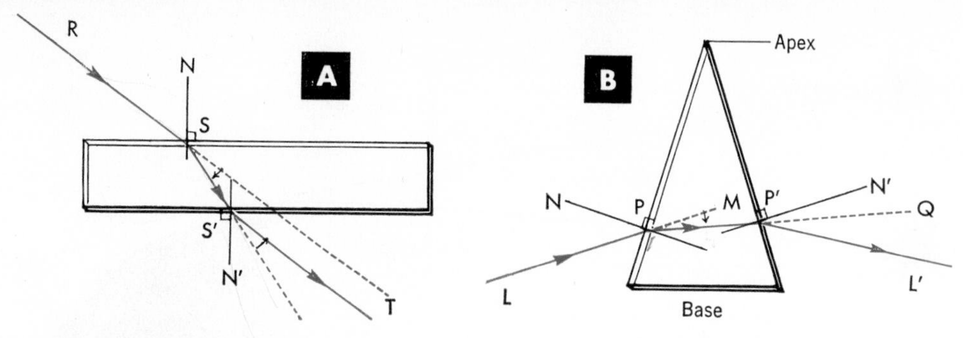

38-17 Following the laws of refraction, we can see why a light ray entering a parallel-sided plate of glass (A) comes out in the same direction as it entered but displaced slightly. We can also explain why a light ray traveling up from the base line of a triangular prism (B) is bent so as to turn toward the base line.

ground at all. The effect of this refraction is the same as if the light had been reflected from a pool of water. Because your eye cannot distinguish between the two, you are fooled into thinking that the refracted light is reflected light from water.

Refraction in lenses

You probably know that the most important practical application of refraction is the lens, which is used in optical devices such as the camera, projector, eyeglasses, microscope, and telescope—and even in the eye. Before studying the manner in which a lens works, let us first see what happens when light passes through glass of different shapes.

16. Parallel-sided glass plates

Figure 38-17A shows how a ray travels through a glass plate having parallel surfaces. You notice that the ray comes out of the glass plate traveling parallel to the path it had when it entered. In effect, the ray has simply been displaced sideways. If the glass is thin, this side-

Frank Murray

38-18 What's the matter with this window? Nothing that better panes of glass wouldn't fix. The front and back of the glass panes are slightly wavy. Acting like triangular prisms, the glass bends the light into different directions and distorts the view of distant objects. In what way does this resemble the shimmering of air over a hot radiator?

ways displacement is so small as to be scarcely noticeable. Since rays coming out of this glass plate are traveling parallel to the line along which they entered, all objects seen through the plate seem perfectly normal if the plate is perfectly flat.

Most window glass is not perfectly flat. Look through a nearby window pane, moving your head from one position to another. You will probably notice that objects viewed through some parts of the window appear distorted (Fig. 38-18). If the glass were perfectly flat like the plate glass used in store windows this would not be so. But where there is a slight imperfection in the surface, light rays change their direction in passing through the glass, so that objects seem to shift their positions if viewed through the imperfect section of the pane.

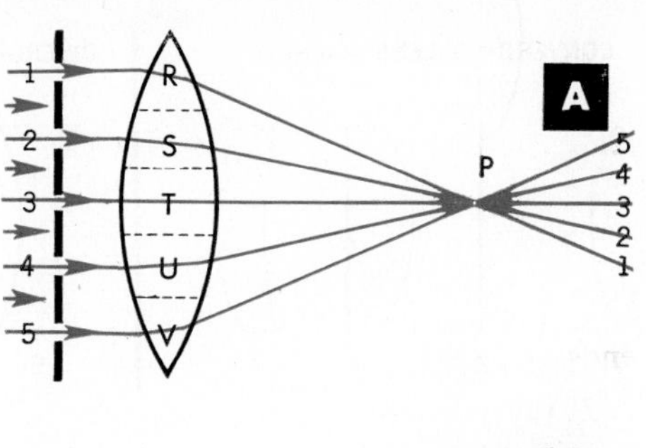

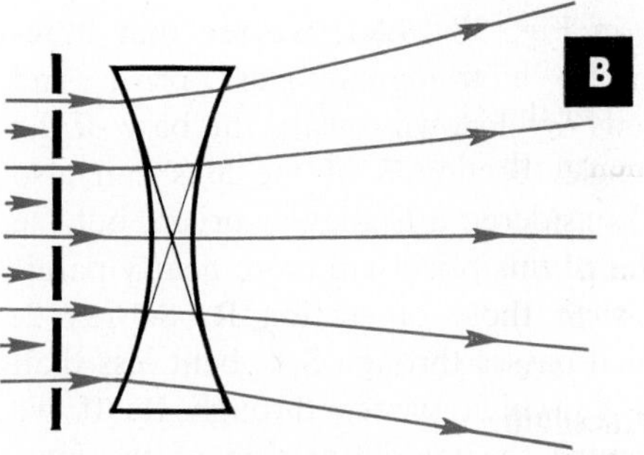

38-19 The bulging surfaces of convex lenses bend rays of light to bring them to a focus (A) while the inward-curving surfaces of concave lenses spread them farther apart (B).

17. Triangular glass plates; prisms

Figure 38-17B shows the observed path of a ray of light traveling through a triangular plate of glass (**prism**). The ray is bent twice, once when it enters the glass and again when it leaves. At both points it is bent closer to the base of the glass triangle.

We could have predicted this from our knowledge of the refraction of light. Light ray L strikes the glass at point P (Fig. 38-17B). At this point we erect the normal N, at right angles to the surface of the glass. Then we draw the line PM showing the direction the ray would have taken if the glass were not there. Because the ray has entered a material of greater refractive index, it bends toward the normal. At P′, where the ray hits the second surface of the glass on the way out, we draw the line P′Q to show the direction the ray would have taken had it not been bent on leaving the glass. Now we draw the normal P′N′. We see that the ray P′L′ bends away from this normal as it passes out of the glass into air. This predicted path is similar to the path actually observed.

18. How a lens bends light

Now let us take a lens in which both surfaces bulge outward as shown in Fig. 38-19A. This is called a double **convex** (outward-bulging) **lens.** We use as our source of light a projector that creates a parallel beam. Let the light pass through a series of slits in a piece of cardboard so as to form the group of rays labeled 1, 2, 3, 4, and 5. You notice that the lens brings these rays together at point P, although they were parallel when they left the projector. This type of lens is called a **converging lens** because it converges, or brings the rays together.

We could have predicted this effect from our study of the way light rays pass through glass prisms. If we examine a small section near the edge of the lens

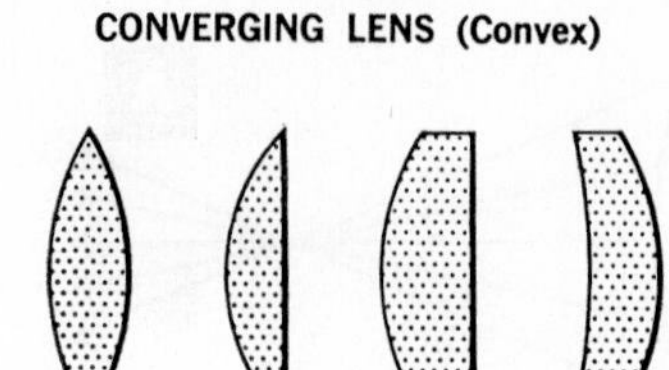

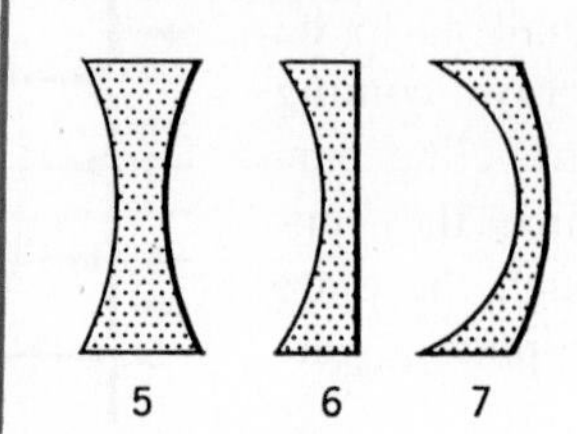

38-20 Converging lenses are thicker in the middle. Diverging lenses are thinner in the middle. Most optical instruments contain combinations of such lenses.

(R in Fig. 38-19A), we see that it resembles a triangular glass prism and bends ray 1 down toward the base of the triangle. Section S of the lens can also be considered a triangular prism, but the sides of this prism are more nearly parallel than those of section R. So ray 2, which passes through S, is bent less than ray 1, which passes through R. If we examine the middle section of the lens, T, we see that it resembles a parallel-sided plate of glass. Ray 3 strikes both surfaces of the lens along the perpendicular and goes straight through without bending.

When we examine sections U and V of the lens, we find that they also resemble triangles of glass, but the triangles are upside down. Thus ray 4, which strikes U, is bent upward toward the base; section V is a triangle with a greater angle than U, and so ray 5 is bent more than ray 4.

The net result of all these actions is to bring the rays together so that they meet at a point. This point P is called the **principal focus.** Its distance from the center of the lens is called the **focal length.**

If the lens is **concave** (that is, curves inward) the situation is just the reverse. Now the rays of light which enter the lens are spread farther apart (Fig. 38-19B). We can think of a concave (diverging) lens as two pieces of triangular glass which are placed tip to tip rather than base to base. Thus light rays are bent toward the bases at the outer edges of the lens and diverge.

You notice that it is the thickness of a lens at the middle as compared with its thickness at the edge which determines whether the lens is of the converging or diverging type. In Fig. 38-20, lenses 1, 2, 3, and 4 are all thicker in the middle than at the edges, and converge rays of light. Lenses 5, 6, and 7 are thinner in the middle than at the edges, and therefore diverge rays of light.

This action of a lens in converging or diverging rays of light is the basis of all optical devices. In the next chapter we shall study it in greater detail.

Going Ahead

THE VOCABULARY OF PHYSICS

Define each of the following:

incident ray
normal
angle of incidence
angle of reflection
reflected ray
principal focus
refraction
refracted ray
angle of refraction
critical angle
index of refraction
mirages
prism
convex (converging) lens
concave (diverging) lens
parabola
focal length
virtual image

THE PRINCIPLES WE USE

1. State the Law of Reflection. Draw a diagram to illustrate it.
2. A ray of light strikes a mirror at an angle of incidence of 23°. Draw and label the incident ray, the normal, the reflected ray, the angle of incidence, and the angle of reflection. Check the accuracy of your drawing by folding

the paper on the line representing the normal and holding it up to the light.

3. Repeat the procedure of the previous question for an angle of incidence of 67°.
4. Explain how images are formed by a flat mirror.
5. Compare an object and its image in a mirror as to (a) distance from the mirror, (b) size, (c) shape, (d) position.
6. (a) Draw a line on your paper to represent a mirror. Then draw an arrow nearby to represent an object. Construct the image by using the facts learned in this chapter. Draw all construction lines carefully with a ruler. (b) Repeat, using a triangle as your object.
7. Draw a diagram to show how a rough surface diffuses a parallel beam of light.
8. Explain how the refraction of light depends on its speed in a material.
9. Draw a diagram showing how light bends as it passes (a) from air into glass, (b) from water into air.
10. State Snell's Law.
11. Explain what is meant by the critical angle. Illustrate your answer using a 45° right prism.
12. Explain total reflection.

Explain why

13. A piece of furniture when highly polished acts like a mirror.
14. A mirror image is really an optical illusion.
15. A ray of light bends as it passes from air into water.
16. A boy trying to spear a fish must aim below the spot where it seems to be. (Use a diagram.)
17. The air seems to shimmer over a hot radiator.
18. You sometimes get an optical illusion of a distant pool of water in a road on a hot day.
19. A lens thicker in the middle than at the edge will converge rays of light.
20. A lens thinner in the middle than at the edge will diverge rays of light.

APPLYING THE LAWS OF REFLECTION

1. Describe the part played by diffuse reflection of light in good home illumination.
2. Why is rough paper more suitable than smooth, glossy paper for reading?
3. Why is it best to have the source of illumination behind you or to one side as you read?
4. How does an automobile headlight make use of reflection to produce good illumination at a distance? Use a diagram.
5. A light ray strikes the surface of water in a thick glass container at an angle and then passes through the water and the flat glass bottom. Draw a diagram showing the path of the ray.
6. How much of your face can you see in a mirror 3 inches square?
7. If you have had geometry, try to prove that the image formed by a flat mirror must be as far back of the mirror as the object is in front of it. (See Fig. 38-3B.) The only assumption you are permitted to make is that the angle of incidence is equal to the angle of reflection. Limit the object to a single point and use only two rays.
8. The corner-shaped aluminum foil attached to the balloon in Fig. 38-21 is designed to reflect radar waves back to the source. We have seen in this chapter that the same principle applies to light waves. Prove by geometry that a ray of light striking two mirrors at right angles to each other will always reflect the ray back in the opposite direction, parallel to the original ray.

SOLVING PROBLEMS OF REFLECTION AND REFRACTION

1. What is the speed of light in flint glass?
2. The speed of light in a substance is 160,000 mi/sec. What is the index of refraction of the substance?
3. What will be the angle of refraction in flint glass of a light beam whose angle of incidence is 30°?
4. The angle of refraction of a light beam passing into water is 30°. What is the angle of incidence?

Westinghouse Electric Corp.

38-21 **This picture provides evidence of similarity between light and radar waves. The reflecting surfaces at right angles reflect the waves right back at the source. Explain the peculiar shape of the tinfoil radar reflector under the weather balloon. Why is it used? Could you use this idea if you wanted to signal a distant observer with light waves?**

5. When light strikes a certain substance at an angle of incidence of 45°, the angle of refraction is found to be 30°. What is the index of refraction of the substance?

6. When light passes into a certain material at an angle of incidence of 22°, the angle of refraction is found to be 18°. What is the speed of light in the substance?

7. A ray of light inside a diamond strikes the surface at an angle of incidence of 10°. What is the angle of refraction in air?

8. Two mirrors are hinged together at an angle of 45°. A ray of light strikes one mirror at an angle of incidence of 30° and then is reflected from the other mirror. What is the angle of reflection from the second mirror?

9. A mirror is rotated at the rate of 2° per sec. At what rate does the angle between the incident and the reflected ray change?

10. How long does it take a ray of light to travel 20 ft through water?

---

11. (a) What is the ratio of the speed of light in water to the speed of light in glass? (b) A ray of light passing from water into glass strikes the glass with an angle of incidence of 45°. What will be the angle of refraction in the glass?

12. From the results of Problem 11, what do you think is meant by the term *relative index of refraction?* What is the relative index of refraction of water to glass?

13. What must be the angle of incidence of a ray of light going from air to water in order that the angle between the reflected and refracted rays is 90°?

14. Using the index of refraction, calculate the critical angle of (a) water, (b) flint glass.

15. Will a ray of light that enters at right angles to the surface of one of the small faces of a 45°–45°–90° prism made of flint glass emerge from the other small face or the hypotenuse? Explain.

16. A ray of light enters the small face of a 20°–70°–90° triangular flint-glass prism perpendicularly. What will be the angle between the entering ray and the emerging ray?

CHAPTER

39

Lens Images

Hold any convex lens, such as an ordinary magnifying glass, near a blank wall, with light coming through a window behind you. What will you see?

As you move the lens close to the wall, you will discover that at a certain distance it casts an image of the window on the wall. There are several things you should notice about this image. First, it is in clear focus only when the lens is a certain definite distance from the wall. Second, the colors in the image are the same as those of the object. Third, the image is much smaller than the object. Finally, the image is upside down and reversed. Let us see why the image formed by a convex lens has these characteristics.

Images

In Chapter 38 we saw that a convex lens converges a beam of parallel rays of light to a point. Let us experiment with a convex lens and examine the image formed by a point source of light as it is moved from one position to another.

In Fig. 39-1A we place our point source, S, on a line, PQ, which is perpendicular to the lens through its center. This line, PQ, through the center of the lens is called the **principal axis.**

1. Inverted images

Now let us use two new sources of light of different colors (Fig. 39-1B). One source, R, is a source of red light and is situated above the principal axis. The other, G, is a source of green light and is below the principal axis. Each of these light sources sends out rays which strike the lens. The red light focuses on the other side of the lens at R′ below the principal axis, the green light at G′ above the principal axis. If we place a screen at R′G′ we get two light spots, one red and one green, but inverted in position.

It makes no difference whether we use two sources of light or a hundred. Some of the light from each source enters the lens to be focused at a definite point on the other side, as we saw in Fig. 39-1B. If we place a screen at the point where the light converges, we get an exact duplicate of the sources of light, point for point and color for color. But the image is upside down because all light which starts from above the principal axis is focused below it, while light starting below the axis is focused above.

It might seem odd that all these different rays of light from spots on the object can cross each other passing through the lens and come out completely separate. But that is a property of waves. One wave can cross another without either being affected in any way.

It will help us to understand the reason for the inversion and reversal of an image if we think of a lens as made up of three sections of glass. As Fig. 39-1C shows, section E at the middle of the lens is very much like a glass plate with parallel sides. A ray of light striking E bends twice in passing through the lens and

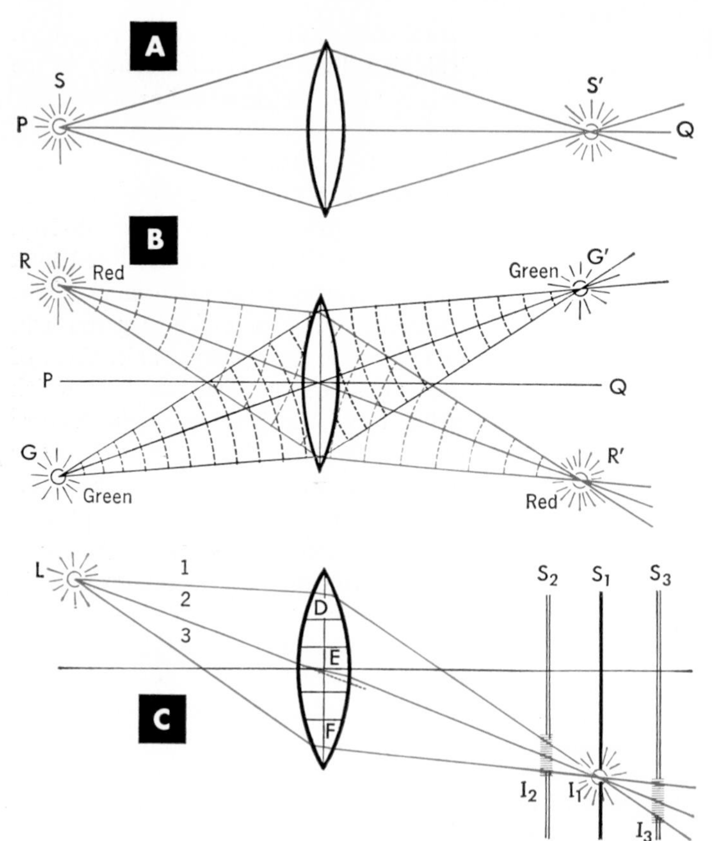

39-1 A convex lens can bring together rays of light to form an image (A). In B we can observe why images formed by convex lenses may be inverted. C shows why the screen that catches the image must be moved back and forth until the image is "in focus."

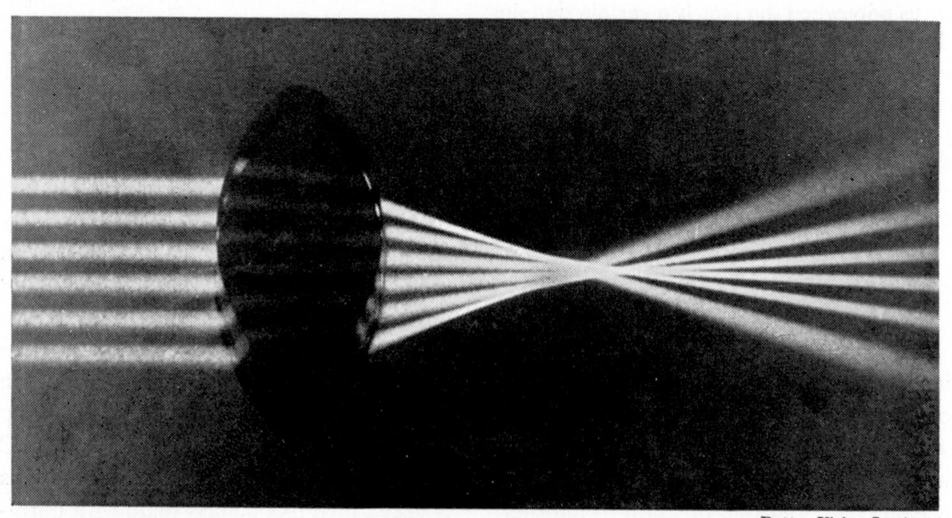

Better Vision Institute

39-2 Rays that strike a convex lens from a source of light are converged to a focus and then pass through each other to spread apart.

comes out traveling parallel to its original direction. Since most lenses are thin, the displacement due to bending is extremely slight. For practical purposes we can say that a ray of light striking the center of the lens will travel through in a straight line. On the other hand, a ray such as 1, striking the triangular upper part of the lens at D, is bent downward toward the base of the triangle. A ray such as 3, which strikes the triangular-shaped bottom of the lens at F, bends upward toward the base. Thus the three rays meet somewhere below the principal axis at point I_1.

2. Focusing the image

As we have seen, an image is in focus only when the screen is placed at a particular distance from the lens. As shown in Fig. 39-1C, if we place the screen at S_2, nearer the lens, the image of every spot of light comes out a blur or circle of light. The same thing happens if we place the screen at S_3, beyond the converging point. Only when the screen is at S_1 do all the rays from each spot on the object converge to a point on the screen. At this position of the screen the image is said to be in focus (Fig. 39-3).

3. Changing the size of the image

Let us move our *lens* back and forth, forming images of objects at various distances. We start with an object which is at an enormous distance: the sun. First we focus the image of the sun onto the screen. Suppose we find that it is in sharp focus when the screen is just 1 foot from the lens. The rays from such a distant object are parallel to each other, and the brilliant image of the sun will be at a point called the **principal focus** of the lens. The focal length of this lens is 1 foot. Next we focus the image on a house 100 feet away, and measure how far from the lens our screen must be placed. We

Gary Earle

39-3 **The camera lens that took this picture converged rays of light from the nearest flowerpot to a focus on the film. Objects farther back are blurred because rays of light from them were not focused on the film by the camera lens. What would have happened to the image of the near flowerpot if the photographer had tried to focus clearly on the distant objects?**

repeat these measurements for other objects, choosing them closer and closer to the lens until our last object is placed right up against the lens itself. Table 16 shows the results we get for a lens of any focal length, F. It gives us the relationship between three things we are changing: the distance of the object from the lens, the distance of the image from the lens, and the size of the image obtained. Instead of measuring distances in feet, we have referred to the distances in terms of focal length, F.

You notice from Table 16 that as we bring the object closer to the lens, the screen on the other side has to be moved farther away if we are to keep the image in focus. As the object is moved toward the lens, the image moves away from the lens. Moreover, the closer the object gets

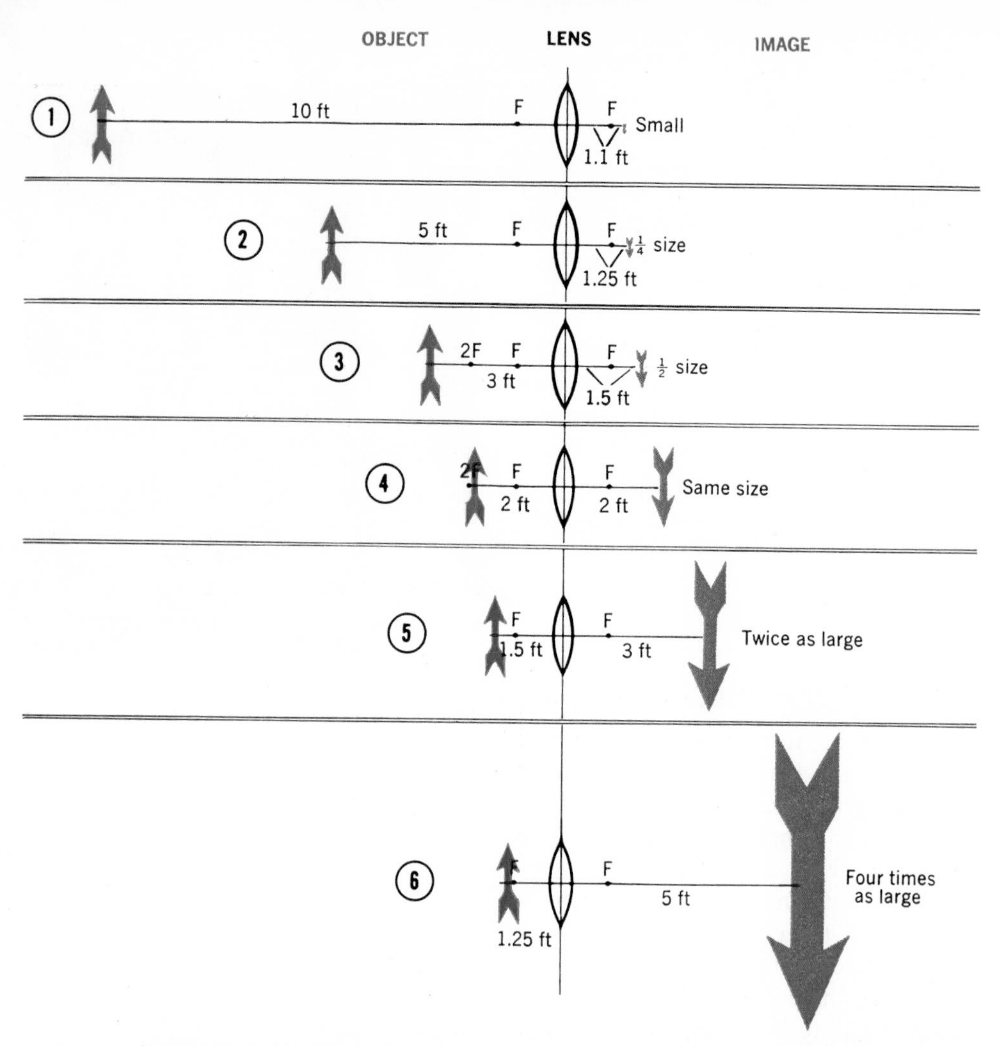

39-4 Image size and distance from lens depend upon the location of the object. As the object comes nearer to the lens, the focused image on the screen moves away from the lens and becomes larger. What do you observe about cases 2 and 6, 3 and 5? What do you notice about case 4?

to the lens, the larger the image becomes. A number of the situations in Table 16 are pictured in Fig. 39-4, in which we have assumed F to be 1 foot.

When we move the object from a great distance away to a distance only 10 feet from the lens, we have to adjust the position of the screen only a small fraction of a foot to keep the image clear, only $\frac{1}{10}$ of a foot, to be exact. By the time we reach a distance of 2 feet (2F), image and object are exactly the same distance from the lens and are both the same size.

But as we move the object closer and closer to the lens beyond the 2F mark, each movement of the object requires a much greater movement of the screen if we are to keep the image focused. And the size of the image likewise increases at a faster rate the closer we bring the object to the lens. When the object is 1.25 feet from the lens, the screen is 5 feet away and the image is 4 times as wide as the actual object. If we now move the object to 1.11 feet from the lens, the screen must be 10 feet away and the im-

A LONG FOCAL LENGTH

F_1

B SHORT FOCAL LENGTH

F_2

39-5 "Strong" and "weak" in lenses refer to the ability of a lens to converge rays of light. The more sharply curved surfaces (greater bulging) of the "stronger" lens cause sharper bending of rays and a shorter focal length.

age will be 9 times as wide as the actual object. But at this point if we move the object only slightly nearer, to a distance of 1.01 feet, the screen must be moved 10 times as far away (to a distance of 100 feet) and the focused image will be 100 times as wide as the original.

4. Focal length

The focal length can be found either by focusing a beam of parallel rays onto a screen or by focusing some object so far away (say beyond 100 ft) that the light rays from a point on it are very nearly parallel. The distance between the screen and the lens is its **focal length.**

We find that the lens we used in the above experiment will not form an image on the screen (**a real image**) if the object is closer to the lens than 1 foot, which is the focal length.

Suppose we use a lens which has a more sharply curved glass surface than the lens of 1-foot focal length used in the experiment above. This lens bends rays more sharply and converges light more readily (Fig. 39-5); it therefore has a shorter focal length. If the focal length of the lens is 6 inches instead of 1 foot (12 in) then we observe the same sequence of image distances and sizes, but they are just half those shown in Fig. 39-4.

Image diagrams

Suppose we wish to predict the position and size of the image before we try to locate it on a screen by actual experiment? How can this be done?

With the information we now have, we can locate the image of any object at any

TABLE **16**
RELATION OF DISTANCE OF OBJECT TO DISTANCE OF IMAGE

Distance of object	Distance of image	Size of image as compared with object
1. very far away	F (one focal length)	exceedingly small
2. far away	slightly beyond F	very small
3. beyond 2F	between F and 2F	small
4. 2F	2F	same size
5. between F and 2F	beyond 2F	larger
6. slightly more than F	far away	very large
7. less than F	no real image obtainable on a screen	larger

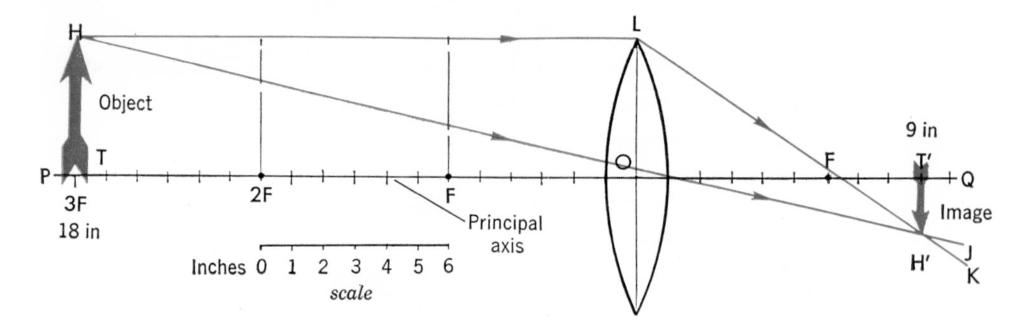

39-6 How can we locate the position and find the size of the image? Ray HO from the head of the arrow passes through the center of the lens O and gives us one location line. Ray HL, parallel to the principal axis, passes through F and gives us the other.

distance from any lens by drawing a diagram to scale. We can simplify this construction by locating the image of only one spot on the object. Suppose our object is the arrow HT shown in Fig. 39-6. Let us try to locate the image, H′, of the head, H, of the arrow. Once we locate this image of the head we can drop a perpendicular line from it to the principal axis PQ. This would locate the image of the tail of the arrow T′.

5. Locating the image with a ray diagram

We can locate the exact position of the image if we have some way of predicting where the rays from H will converge to a focus after passing through the lens. If we can trace the path of any two rays coming from H, their point of intersection will show us where all the other rays from H will meet. Which two rays shall we use?

We have already seen that ray HO, which strikes the center of the lens, can be considered as traveling through it in a straight path (page 535). This gives us the path of one light ray from the head, H, of the arrow (Fig. 39-6). Now we need only one more ray to locate the exact position of the image. The ray parallel to the principal axis is a good one to choose. Let's see why.

If you look back at Fig. 38-19, you will observe that all the rays of light that strike a lens parallel to the principal axis are refracted through the focus. Thus in Fig. 39-6 if we choose the light ray HL that travels parallel to the principal axis from the head of our arrow, we can be sure that it too will be refracted through the principal focus. If the object is larger than the lens, we still draw the ray parallel to the principal axis until it reaches the plane of the lens (OL extended), and from this point beyond the edge of the picture of the lens, we draw the ray through the focus. The image of H is formed at the intersection of the two rays at H′. Dropping perpendicular H′T′ gives us the entire image.

Let's apply this method to an actual problem. Suppose we place an object 18 inches from a lens which has a focal length of 6 inches. How far from the lens must we place our screen to get a clear image? And how will the size of the image compare with the size of the object? We tackle the problem in the following stages:

1. Since our object is 18 inches from the lens, we select a convenient scale which can show this distance on our sheet of paper. Suppose we choose a scale of 1:6; in other words, 1 inch on the paper represents an actual distance of

6 inches. As shown in Fig. 39-6, we mark off the important points F and 2F on each side of the lens. F stands for the principal focus, 1 focal length away from the lens (1 inch from the lens according to our scale). Point 2F will be drawn 2 inches from the lens.

2. Now we draw our object on the paper. Since it is 18 inches from the lens we show it on the paper (using our scale of 1:6) as being 3 inches away. We represent the object by a perpendicular arrow HT.

3. Refer again to Fig. 39-6. As you can see, we draw ray HO, which starts from the head of our object and strikes the center of the lens. This ray passes through the center of the lens in a straight line toward J.

4. We must draw a second ray from H which runs parallel to the principal axis. We know that this parallel ray HL will pass through the principal focus F. So we draw LF, extending it toward K.

5. We find that our two rays from H cross at point H′. This is where we must place our screen to get a focused image; in other words, it is the point where all the rays from H will converge after passing through the lens.

6. To obtain the complete image we drop a perpendicular from H′ to the principal axis. This gives us H′T′ which is the image of the arrow HT. The distance from the lens to H′T′ turns out to be $1\frac{1}{2}$ inches on the paper. According to our scale that means an actual image distance 6 times as great, or 9 inches.

7. We measure the size of the image H′T′, and compare it with the size of the object HT. The image turns out to be one half the size of the object.

6. Information from a ray diagram

Now we have all the information we set out to find. We know that if an object is placed 18 inches from a lens whose focal length is 6 inches, the image will be half the size of the object and will be located 9 inches from the lens. This is precisely the result we obtain by experiment when we measure the image distance and size using an actual lens of 6-inch focal length and an object 18 inches away.

By this method we can draw diagrams to show the images cast by an object at any distance from any lens. Figure 39-7 shows several additional diagrams.

As you see, all these diagrams follow the pattern which we set down in Table 16 (page 537). When the object is far away (Fig. 39-7A), the image is located slightly beyond F and is small. This diagram represents the images formed by a camera (Chapter 40) or the eye. When the object is at 2F (Fig. 39-7B), the image is also at 2F from the lens on the other side, and image and object are equal in size. When the object is between F and 2F (Fig. 39-7C), the image is beyond 2F and is larger in size. This diagram represents the image formed by an enlarger. When the object is very close to F (Fig. 39-7D), the rays come out of the lens traveling nearly parallel to each other. These rays meet at a very great distance to form an image. The projector in a theater illustrates this.

When the object is brought nearer to the lens than 1 focal length (as in Fig. 39-7E), the rays from it spread apart at too great an angle to converge after passing through the lens, and no image can be focused on the screen no matter where the screen is placed.

7. Virtual images— the magnifying glass

Suppose the object is closer to the lens than 1 focal length (Fig. 39-8). In this situation, the rays from the object are not converged by the lens and therefore cannot focus on a screen. Instead, whenever a convex lens is held closer to an object than 1 focal length the lens becomes a magnifying glass (Fig. 39-7E). The image can no longer be formed on a screen; it can only be seen by looking at it *through* the magnifying glass.

Figure 39-8 illustrates what we see through a magnifying glass. Rays L_1 and L_2 from the head of the arrow at H have passed through the lens and been bent. The eye sees rays L_1 and L_2 as if they had come directly from H′. Thus the distance between the head and the tail of the arrow seems much greater than it really is, and the object appears magnified. You also notice that the image in this case is not inverted but appears right side up.

This kind of image is called a **virtual image** in contrast to the real images we have studied so far. A **real image,** you will remember, is one that can be formed

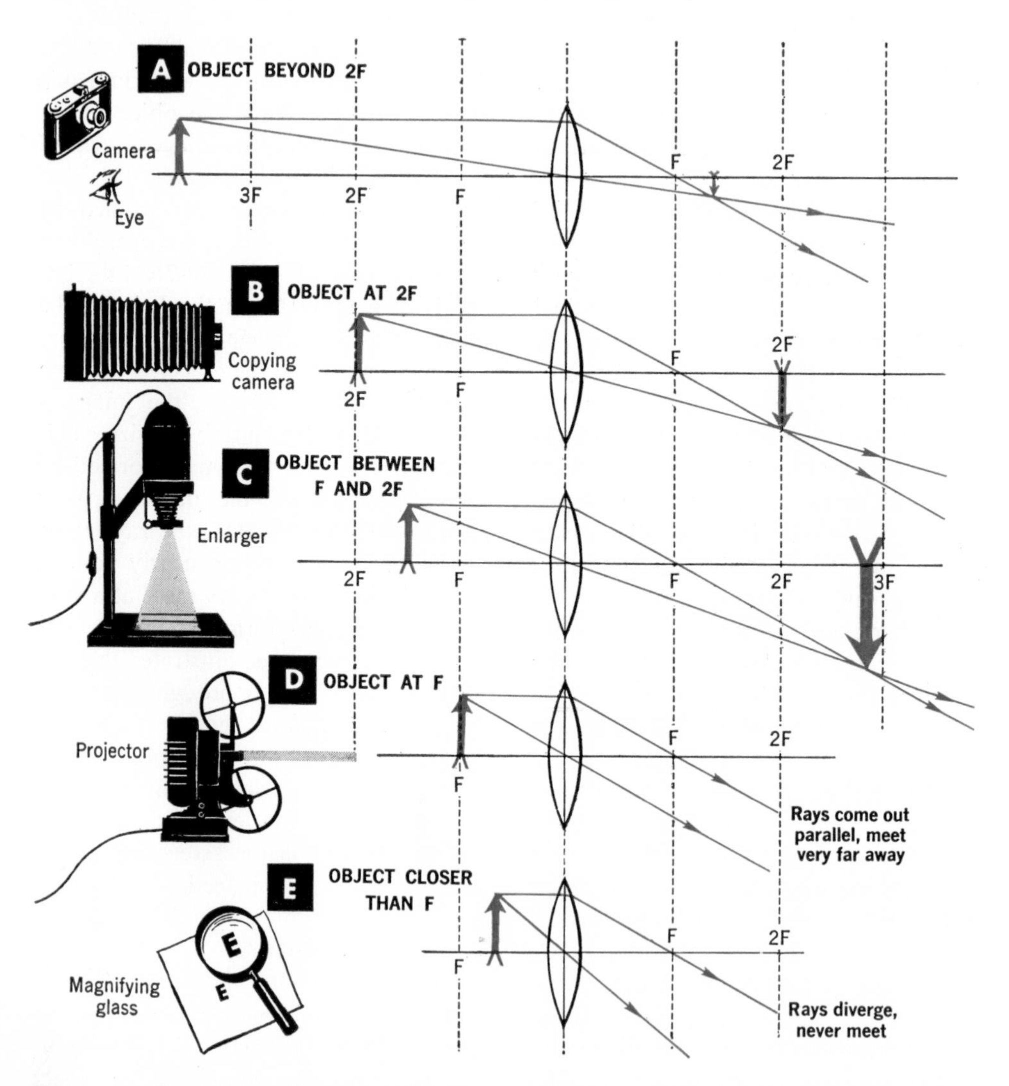

39-7 These image diagrams enable us to explain the observations made in Fig. 39-4. We can apply these lens diagrams to many important optical instruments. What's different about E?

39-8 A lens used as a magnifying glass creates an optical illusion. The lens bends the rays from the object in such a way as to produce an enlarged, erect virtual image. Now you can explain the enlarged image of the ruler in Fig. 38-8.

by the actual intersection of rays on a screen. But a virtual image cannot. It is obvious from Fig. 39-8 that if we place a screen at H′, light falling on it from the object will not have passed through the lens, and so no image can be formed on it. An image seen in a plane mirror is also of this kind: the image is behind the glass of the mirror, although no actual light rays come from where we see the image.

So we see that there are three important differences between the image on a screen formed by a convex lens, and the image formed by the same lens used as a magnifying glass. First, the image on the screen is real while the image formed by the magnifying glass is virtual. Second, the real image is inverted while the virtual image is erect. And third, the real image formed by a convex lens is always on the opposite side of the lens from the object while the virtual image is on the same side. You have to look into the lens at the object to see a virtual image.

Image calculations

It takes time to predict the size and position of an image by drawing a ray diagram, and there is a good possibility of error unless great care is used. There is, however, a formula which can be derived from our ray diagram by means of geometry and algebra. This formula does the job speedily and reduces the chances of making a mistake.

8. The lens formula

In Fig. 39-9 we call the distance from the object to the lens D_o, the distance from the image to the lens D_i, and the focal length F. If we know any two of these quantities we can calculate the third by means of the following formula:

$$\frac{1}{D_o} + \frac{1}{D_i} = \frac{1}{F}$$

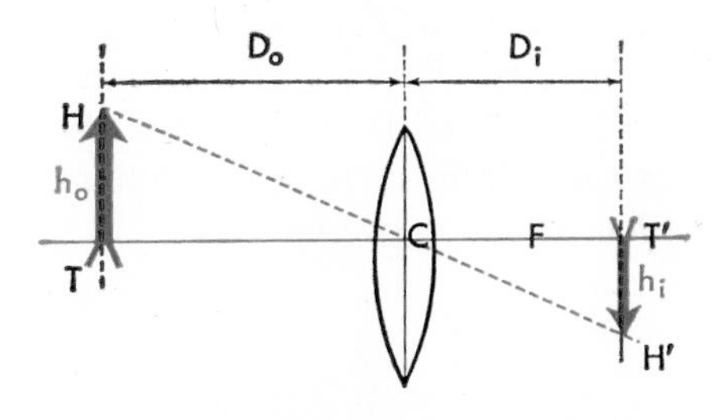

39-9 Can you prove from this diagram that the sizes of image and object are in the same proportion as the distances of image and object? If you want a larger image, should you bring the object closer to or farther from the lens?

Let us use this formula to find the answer to the problem we solved on page 538 with a lens diagram. Our object, you remember, is 18 inches away from a lens whose focal length is 6 inches. We want to find how far from the lens the image will be. In other words, $D_o = 18$ inches, D_i is unknown, and $F = 6$ in. Substituting in our formula we get:

$$\frac{1}{18} + \frac{1}{D_i} = \frac{1}{6}$$

We multiply both sides by $18D_i$, our least common denominator:

$$\frac{18D_i}{18} + \frac{18D_i}{D_i} = \frac{18D_i}{6}$$

$$D_i + 18 = 3D_i$$

$$D_i = 9 \text{ in}$$

So we see that our image distance, D_i, is 9 inches. This is the same answer that we got with much more labor by constructing a ray diagram.

STUDENT PROBLEM

An object is placed 15 inches from a convex lens whose focal length is 10 in. What is the image distance? (ANS. 30 in)

■ ■ ■ ■ ■ ■ ■

How can the lens equation be used to solve problems involving virtual images? Let us look at a problem.

Suppose the object, HT, in Fig. 39-8 is 5 centimeters from the lens, which has a focal length of 10 centimeters. Where does the image appear? Substituting in the lens equation, we have:

$$\frac{1}{D_o} + \frac{1}{D_i} = \frac{1}{F}$$

$$\frac{1}{5} + \frac{1}{D_i} = \frac{1}{10}$$

$$\frac{10D_i}{5} + \frac{10D_i}{D_i} = \frac{10D_i}{10}$$

$$2D_i + 10 = D_i$$

$$D_i = -10 \text{ cm}$$

The minus sign means that the image is on the same side of the lens as the object, and so is a *virtual* image.

STUDENT PROBLEM

An object is placed 8 inches from a convex lens of 12-inch focal length. What is the image distance? (ANS. −24 in)

■ ■ ■ ■ ■ ■ ■

If we are given the position of the virtual image and the object and asked to find the focal length of the lens, we have to be careful about signs. As you learned above, the image distance, D_i, of a virtual image is negative, and so we must substitute a negative value for D_i in the lens equation. Here is a problem.

A virtual image is produced 12 inches from a lens when an object is placed 4 inches from the lens. What is the focal length of the lens? Since the image is virtual and on the same side of the lens as the object, we must substitute *minus* 12 inches for D_i:

$$\frac{1}{D_o} + \frac{1}{D_i} = \frac{1}{F}$$

$$\frac{1}{4} - \frac{1}{12} = \frac{1}{F}$$

$$3F - F = 12$$

$$F = 6 \text{ in}$$

STUDENT PROBLEM

A virtual image is formed 6 inches from a lens when an object is placed 3 inches from the lens. What is the focal length of the lens? (ANS. 6 in)

---

9. Calculating the size of the image

We found in Fig. 39-6 that the image was one half the size of the object. We can also calculate this by formula.

You notice in Fig. 39-9 that the ray which passes through the center of the

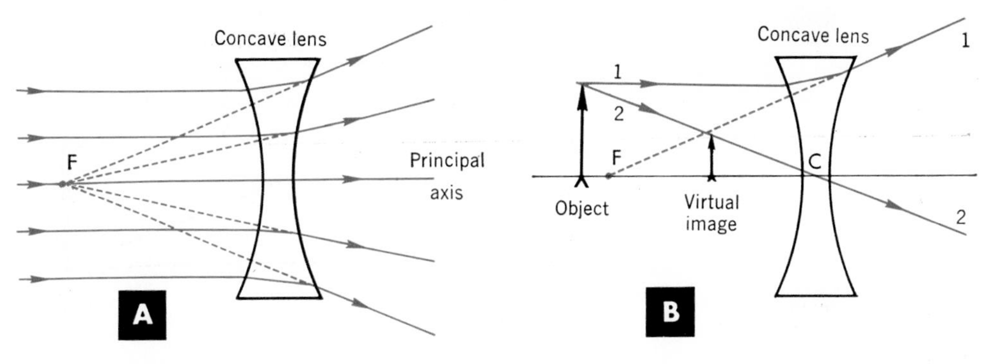

39-10 In A we see that a concave lens diverges rays that pass through it and in B, a ray diagram shows how the image is located. Concave lenses always produce reduced virtual images.

lens creates two similar triangles, HTC and H'T'C. We can make a simple proportion between the parts of these triangles.

$$\frac{\text{Size of image (H'T')}}{\text{Size of object (HT)}} = \frac{\text{Distance of image } (D_i)}{\text{Distance of object } (D_o)}$$

This ratio of size of image, called h_i, to size of object, h_o, is called the **magnification,** M. Hence

$$M = \frac{h_i}{h_o} = \frac{D_i}{D_o}$$

You can see from the above formula that when an image is in focus magnification depends only on the distances of the image and the object from the lens. In the special situation when these distances are equal, then sizes are equal. This occurs at 2F from the lens. If the focused image is 100 times as far from the lens as the object, then it will be magnified 100 times.

SAMPLE PROBLEM

An object 2 centimeters long 5 centimeters from a lens forms a virtual image at a distance of 10 centimeters. What is the magnification and size of the image?

We find the magnification by substituting in the formula

$$M = \frac{D_i}{D_o} = \frac{10}{5} = 2$$

and the size of the image is found from:

$$M = \frac{h_i}{h_o}$$

$$2 = \frac{h_i}{2}$$

$$h_i = 4 \text{ cm}$$

STUDENT PROBLEM

An object 5 inches long placed 3 inches from a lens forms an image at a distance of 7 in. (a) What is the magnification? (b) What is the length of the image? (ANS. (a) 2.33, (b) 11.7 in)

10. Images with a concave lens

Figure 39-10A shows how rays parallel to the principal axis diverge when they pass through a concave lens. They spread out as though they were all diverging from F, the principal focus. The image formed by a concave lens can be located by means of a ray diagram drawn in the same way as was done with a convex lens. Ray 1, parallel to the principal axis (Fig. 39-10B), leaves the lens as

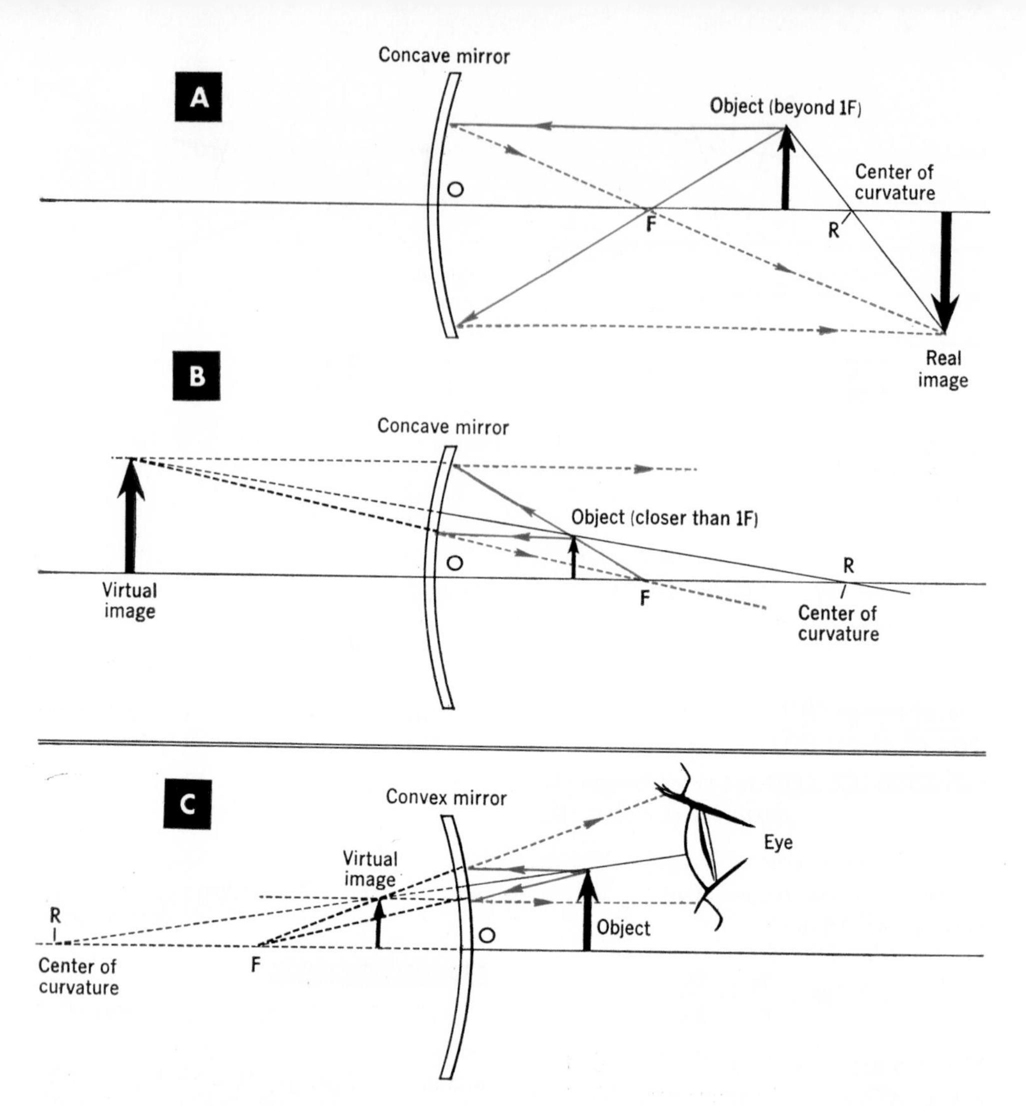

39-11 (A and B) *Concave* mirrors form images in a manner similar to *convex* lenses. (C) A *convex* mirror forms images like those formed by a *concave* lens.

though it were coming from the principal focus. Ray 2 simply passes straight through the center C of the lens. These rays diverge after passing through the lens, so an erect virtual image results. No matter where the object is placed:

A concave lens always gives a reduced erect virtual image.

The lens equation can be used to solve problems involving concave lenses, but, as diagram A in Fig. 39-10 shows, the principal focus is really a virtual focus, since light does not really diverge from F but only *appears* to. Hence:

The focal length of a concave lens is always negative.

SAMPLE PROBLEM

Where is the image formed and what is the magnification if an object is placed 7 inches from a concave lens whose focal length is 3 inches? Remembering that we must substitute a minus value for F, we have:

$$\frac{1}{D_o} + \frac{1}{D_i} = \frac{1}{F}$$

$$\frac{1}{3} + \frac{1}{D} = \frac{1}{-7}$$

$$7D_i + 21 = -3D_i$$

$$21 = -10D_i$$

$$D_i = -2.1 \text{ in}$$

The fact that the image distance is negative means that the image is on the same side of the lens as the object and is virtual. The magnification is:

$$M = \frac{D_i}{D_o} = \frac{2.1}{3} = 0.7$$

The minus sign has no meaning in calculating the magnification and so it is dropped.

STUDENT PROBLEM

An object is placed 4 cm from a concave lens whose focal length is −5 cm. What is (a) the image distance, (b) the magnification? (ANS. (a) −2.22 cm, (b) 0.556)

---

11. Images with curved mirrors

Concave and convex mirrors form images in a manner similar to lenses, but a concave mirror acts like a convex lens and a convex mirror acts like a concave lens. The bending of the rays is the result of the Law of Reflection rather than the Law of Refraction. It can be shown by geometry that the focal length of a mirror is *always* equal to one half the radius of curvature, which is not true for a lens. The lens formula

$$\frac{1}{D_o} + \frac{1}{D_i} = \frac{1}{F}$$

can also be used to solve curved-mirror problems. The focal length of the convex mirror must be considered negative, and the focal length of the concave mirror positive.

Figure 39-11 shows how images are formed in concave mirrors, A and B, and in a convex mirror, C. Can you give a practical application for each case?

Going Ahead

THE VOCABULARY OF PHYSICS

Define each of the following:

principal axis
virtual image
principal focus
real image
magnification

THE PRINCIPLES WE USE

1. You are given a convex lens and asked to find its focal length. Describe what you would do.
2. Draw a diagram showing the difference between a lens of long focal length and one of short focal length.
3. Describe the changes that take place in (a) the size of the image, (b) the distance of the image from the lens, as an object is brought from very far away right up to the lens.
4. A lens forms an image of size equal to that of the object when the object is 14 inches from a lens. (a) What is the focal length of the lens? (b) How far away should the object be to form the closest real image? (c) What is the closest an object can be brought to this lens and still form a real image? (d) Describe the image when the object is 8 inches from the lens.
5. Under what conditions can a lens be used as a magnifying glass?
6. State three differences between the image formed by a magnifying glass and an image formed by a convex lens on a screen.
7. Describe the changes that take place in (a) the size of the image, (b) the distance of the image from the lens, as an object is brought from very far away right up to a concave lens.

PROBLEMS IN LENS AND MIRROR IMAGES

1. Draw ray diagrams to find the images formed by a lens of 4-inch focal length for each of the following object distances: 20 in, 12 in, 8 in, 6 in, $4\frac{1}{2}$ in, 4 in, 3 in. What is the image distance and magnification with each lens? Can you estimate your answers before diagraming the problem?
2. Use the lens formula to make calculations of image distance and magnification for each object distance in Problem 1.
3. Repeat Problem 1 for a lens of focal length 3 cm and object distances of: 12 cm, 6 cm, 4 cm, 3 cm, 2 cm.
4. Use the lens formula to make calculations of image distance and magnification for each object distance in Problem 3.

5–10. Find the unknown quantities for each of the following lenses:

	F(cm)	D_o(cm)	D_i(cm)	M	h_o(cm)	h_i(cm)
5.	5	10	?	?	3	?
6.	4	?	12	?	?	5
7.	−115	15	?	?	4	?
8.	6	6	?	?	?	4
9.	?	4	8	?	6	?
10.	10	?	−12	?	?	6

11. A movie projector has a lens 2 inches in focal length. (a) If the film is $2\frac{1}{32}$ inches from the lens, where must the screen be placed? (b) What is the magnification? (c) How big an image will a 16-mm film have? Check your answers with a ray diagram.
12. A camera lens has a focal length of 8.1 cm. (a) How far should the film be from the lens if the object being photographed is 80 cm from the lens? (b) What is the magnification?
13. In Problem 12, what is the height of the biggest object that can form an image on a film 6 cm high?
14. An object placed 100 inches from a convex lens has its image formed 25 inches from the lens. Where should the object be placed so that its image will be the same size as the object?

15–18. Find the unknown quantities:

		F(cm)	D_o(cm)	D_i(cm)	M	h_o(cm)	h_i(cm)
15.	concave mirror	4	?	−12	?	2	?
16.	convex mirror	?	10	−7	?	1	?
17.	convex lens	10	?	?	4	3	?
18.	concave lens	20	infinity	?	—	—	—

19. How large an image of the moon will be formed by a convex lens whose focal length is 20 ft?
20. An image is seen by reflection from the surface of a 4-in diameter steel ball. (a) Where is the image? (b) Describe it completely.

---

21. When an object is placed 10 inches from a certain lens, an enlarged image is formed on a screen. Which of the following statements are true? Which are false? Which are "maybe's"? (a) The focal length is more than 10 inches. (b) The image is virtual. (c) The image is inverted. (d) To obtain an image of equal size move the lens away from the object. (e) When the object is 20 inches from the lens, the image will be smaller than the object. (f) This lens will serve as a magnifying glass when the object is 8 inches from the lens.
22. Derive the lens formula

$$\frac{1}{D_o} + \frac{1}{D_i} = \frac{1}{F}$$

by geometry. (Use similar triangles HTO, H′T′O, and LFO, H′T′F in Fig. 39-6.)

23. A certain lens forms an image 5 inches away when the object is 8 inches away. Using this information construct the ray diagram. What is the focal length? Check the diagram by calculating the focal length with the lens formula above.

CHAPTER

40

Optical Instruments

You have seen how light rays are bent as they pass from one transparent material into another or are reflected from shiny surfaces. You have learned how lenses and mirrors can refract and reflect light rays to form images visible to the eye.

How many instruments can you name in the familiar world around you whose action depends upon the refraction or reflection of light rays? In this chapter we shall examine some of the most important ones, beginning with the most important and familiar one of all—the human eye.

The eye

Examine your eyes in a mirror (Fig. 40-1A). Notice that in the center of both white eyeballs is a colored disk about ½ inch in diameter. This disk is called the **iris diaphragm.** It is opaque: that is, it does not permit light to pass through. But in the center of the iris is a small black disk called the **pupil.** This disk is actually an opening which permits light to enter the eye. The size of the pupil changes automatically with the intensity of light. It appears black to you because you are really looking into a "dark box" which receives light but does not give it out.

The iris is covered by a protective layer of transparent tissue called the **cornea** (kôr′nē·ȧ).

Behind the iris is a converging (convex) **lens,** made of transparent tissue (Fig. 40-1B). In back of the lens the eyeball is filled with a transparent, jellylike substance through which the light passes after traveling through the lens. At the very back of the eye is the **retina** (rĕt′-ĭ·nȧ). The retina is the screen upon which the image is formed. It consists of millions of tiny nerve cells which send "messages" to the brain when struck by light (Fig. 40-2). It is these messages which tell us what we are seeing.

Thus there are three main parts of the eye as far as obtaining an image is concerned: the iris diaphragm, which controls the opening of the pupil and regulates the amount of light that enters the eye; a converging lens, which produces the image; and the retina, on which this image is focused.

1. How the eye accommodates for distance

In the previous chapter we saw how the screen on which the image is focused must be moved closer to the lens or farther from it as the distance between object and lens changes. But in the human eye the distance between the screen (the retina) and the lens is fixed and cannot be changed. So it would seem that we could get an object clearly in focus only if the object were a certain fixed distance away from us.

However, the lens is flexible and is equipped with muscles which can change its shape, making it thicker or thinner in the middle and hence regulating its focal length. In short, we change the shape of the lens when we look at ob-

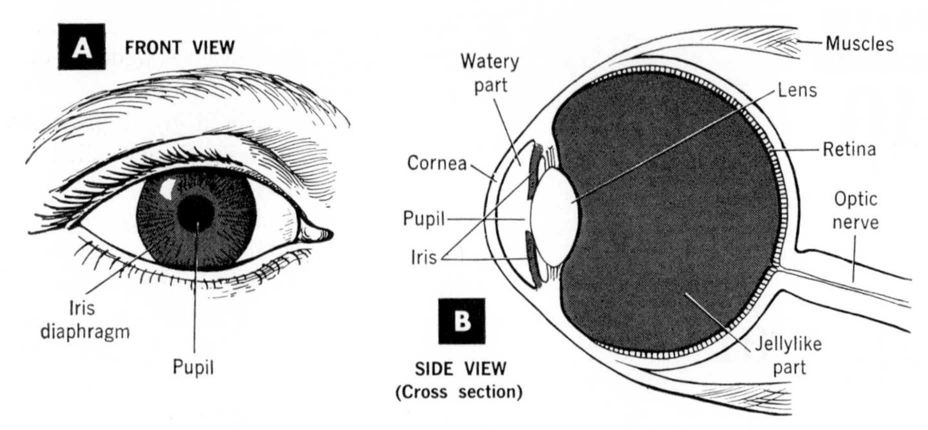

40-1 In many ways the eye and the camera are twins. The eye is really a "dark box" with a lens in front to form an image on the back. The pupil is the opening that lets light through. Does a camera have an iris diaphragm? What is its purpose?

jects at different distances. This ability to change the shape of the lens is called the **power of accommodation.** Figure 40-3A shows how a lens normally focuses light from a distant object: image I of object O is on the retina and is in clear focus. In Fig. 40-3B the object has been brought nearer. The dotted lines show the shape the lens assumes before it has accommodated itself to the change in position of the object, and illustrate how the rays would meet at I′, beyond the retina. The image on the retina would be blurred for this shape of lens. But when the eye muscles shorten the focal length of the lens by thickening it (solid lines in Fig. 40-3B), the rays from the object are bent more. This means that they converge at a shorter distance from the lens, so the image can be focused on the retina. In this simple way the eye adjusts itself for the distance of the object. There is a limit to the closeness with which an object can approach the eye and still be kept in clear focus without strain. In most adults this distance is about 10 inches.

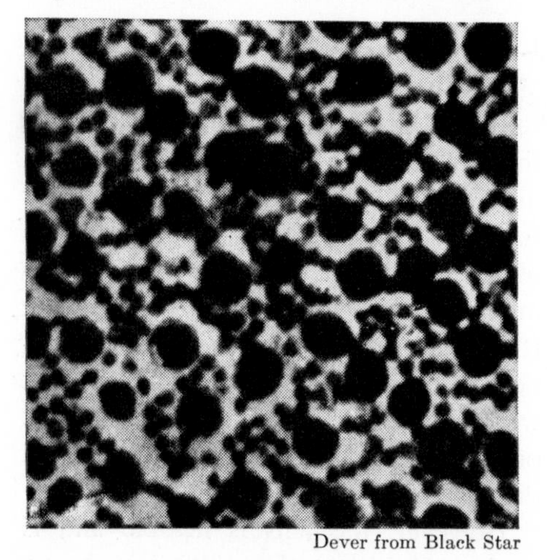

Dever from Black Star

40-2 A microscope view of the retina shows structures called cones (the large round spots), by which we see color. The smaller spots are the rods, which enable us to see in dim light. Each of these cones and rods is connected by a nerve to the brain.

2. Judging distance

How does the eye judge the distance of objects? Actually, the estimation of distance takes place in the brain. The simplest way of judging distance and the only way that is useful at large distances (beyond about 100 feet) is based on experience. Everyone knows the approximate size of an automobile, for example. The farther the automobile is from us, the smaller it appears, because its image on the retina is smaller. This enables us to gauge its distance. People who report "flying saucers" in the sky as being "300

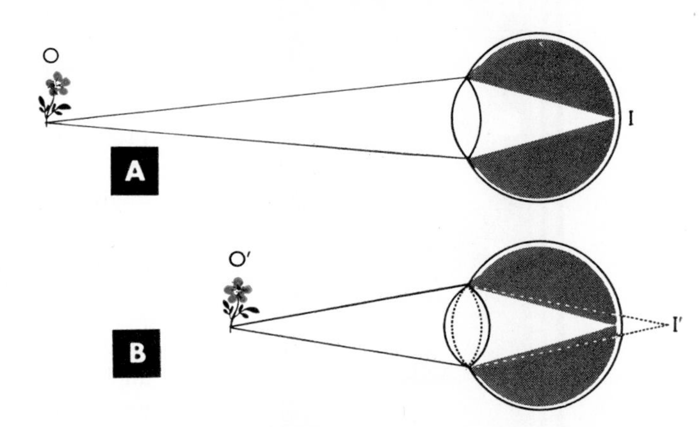

40-3 Our eyes can accommodate for objects at different distances and focus the images clearly, by changing the curvature of the eye lens. When the object is nearer, the lens is made thicker in the middle, converges the rays more sharply, and brings them to a focus on the retina.

feet long, five miles away, and traveling 1,000 miles per hour" are being carried away by their imagination and are not being careful observers. For unless you know the true size of a flying saucer, you cannot judge how far away it is any more than you could for a car. Because we know the true size of many objects we are able to judge their distance with reasonable accuracy.

For objects that are quite close, the fact that we have *two* eyes aids us in judging distance. As we saw in Chapter 38 (page 519), each eye sees an object from a slightly different angle. The closer the object is the more the eyeballs turn inward. The brain's sensation of this enables us to judge distance. Moreover, because each eye is aimed at a different angle, the farther away the object is, the more nearly identical are the two images. The nearer the object, the greater the difference in the two images. These two different views, fusing together in the brain so that we are aware of only one image, also give us an impression of how far away the object is. This **binocular effect** is the basis of stereoscopic cameras. Such cameras have two lenses, spaced a distance apart equal to the normal eye

Stereo by Henry M. Lester from *Stereo Realist Manual*

40-4 Here are two pictures of the same airplane, one as if seen by your right eye, one as if seen by your left eye. The alignment of the plane against the background differs very slightly. Viewed with a device that properly presents one picture to each eye, a sense of depth is produced.

Impco, Inc.

40-5 The illusion of moving pictures is created by flashing on the screen in rapid succession a series of separate images, each a bit different from the one before. At least 16 pictures per second are needed to create this illusion. Regular movies use 24 pictures per second to reduce flickering. Notice the sound track on the left margin of the film.

spacing. They take two pictures at once, each from a slightly different angle (Fig. 40-4). When the developed film is viewed properly, each eye sees a different picture and the illusion of depth is very apparent —giving a "three-dimensional" view.

3. Why don't we see things upside down?

Our study of the lens taught us that an image formed on a screen is always upside down. The lens in the eye is no exception to this rule. The image on the retina is actually upside down! Why, then, don't we see objects that way? Through experience since earliest childhood a person *learns* what is "up" and what is "down," and having learned from experience his brain interprets the position of what is seen.

4. Persistence of vision

When we see a moving picture, we have the illusion that we are viewing a continuous scene. But really we are watching a series of separate still images flashed on the screen in extremely rapid succession, about 24 images in one second (Fig. 40-5). The eye does not notice any break between images because after one image is gone the nerve cells on the retina continue for a short time to send the impulses connected with that image to the brain. This phenomenon is called **persistence of vision.**

Vision persists for about $\frac{1}{16}$ second after the light from a particular image is gone. In a regular movie projector the image is flashed on the screen for about $\frac{1}{30}$ second. During this $\frac{1}{30}$ second the film in the projector is actually stationary. Then, while a shutter covers the lens, a small metal arm slips into the sprocket holes at the side of the film and pulls it down one frame, bringing the next picture into view less than $\frac{1}{16}$ sec-

40-6 All of us have at least one blind spot. Close your left eye and look at the X with your right eye. As you bring the page closer, the circle disappears. Still closer, the circle appears and the square disappears. The image disappears when it happens to fall on the blind spot of the retina, where all the optic nerves meet and leave the eye.

ond later. But because of persistence of vision the eye continues to send impulses to the brain during this short break between images. If there were a longer time between the separate pictures, the eye would see the image flicker. Because the images follow each other so rapidly on the retina of the eye we have the illusion of continuous motion. Television, too, depends on the persistence of vision.

Eye defects

Many people have eyeballs that are too long, too short, or somewhat out of shape, with the result that their vision is poor. In such cases properly chosen glasses can usually restore normal or nearly normal eyesight.

5. Farsighted eyes

In one type of eye defect the eyeball is too short, and the image focuses beyond the retina (Fig. 40-7B). We call such a person **farsighted.** He can see distant objects clearly, but he cannot examine a nearby object at the normal 10-inch distance because the image would come to a focus at a point behind his retina. In order to see a close object clearly he must push it away from him beyond the 10-inch distance, perhaps to 20 or even 30 inches. When he does this, the image moves toward the lens, finally falling on the retina (Fig. 40-7A).

This eye defect is easily remedied by using convex glasses. If a convex lens is placed in front of the eye, as in Fig. 40-7C, the eye lens is assisted in converging the light rays, and the rays focus properly on the retina.

6. Nearsighted eyes

Some people have the opposite difficulty: their eyeballs are too long. They can focus images of nearby objects but not those of objects which are far away. A person with this defect is said to be **nearsighted** (Fig. 40-8). He cannot focus objects on his retina unless they are brought close to his eyes, which is often impossible (Fig. 40-8A).

To remedy nearsightedness the oculist prescribes glasses with concave lenses. The concave lens diverges the rays of light slightly apart before they enter the eye, so that the image of distant objects focuses properly on the retina, as shown in Fig. 40-8C.

7. Astigmatism

Another common defect of vision is called **astigmatism** (ȧ·stĭg′mȧ·tĭz′m). Here the lens of the eye (or the cornea),

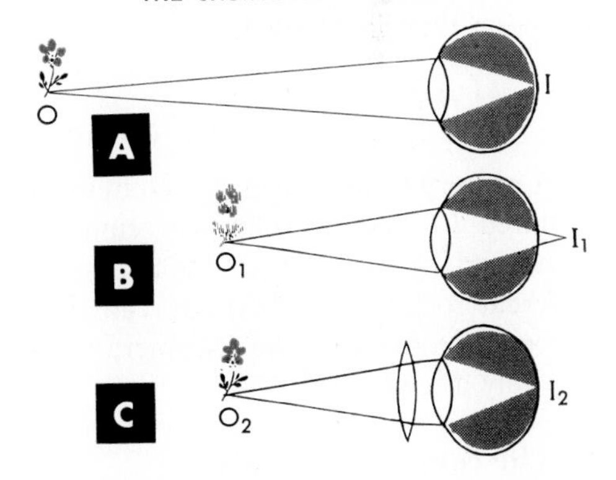

40-7 The farsighted eye can focus distant objects properly (A), but has trouble bringing the rays to a focus if the object comes too near (B). A convex lens in the form of glasses assists the eye in converging light rays to a focus on the retina again (C).

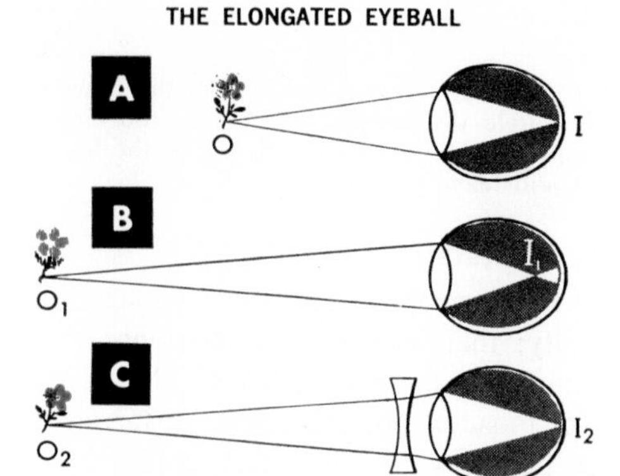

40-8 The nearsighted person finds near objects in proper focus (A). But when looking at distant objects, his long eyeball causes rays to meet before they reach the retina (B). A concave lens spreads apart the rays a bit and the eye lens brings them to a focus once again on the retina (C).

instead of having the proper spherical surface, curves more in one direction than in another, just as a football has a different curve from tip to tip than it has around the middle. A person with astigmatism may see lines running in one direction as blurred even though lines running in another direction are in focus (Fig. 40-9).

For this defect the oculist prescribes a special lens which curves more in one direction than in another.

The camera

A century or more ago, the equipment required for photography was cumbersome and costly. The situation is very different today. Now anybody can take a good picture with a small camera costing only a few dollars.

8. The camera and the eye

The modern camera (Fig. 40-10) is remarkably like the human eye. Just as with the eye, the most important part of the camera is the convex lens. Like the eye, the camera has an iris diaphragm which controls the amount of light which enters the camera. But in place of the retina of the eye which catches the image, there is a sheet of photographic film.

While the eye is open most of the time during our waking day, the camera is open only for that instant when a picture is being taken. So the camera is provided with a device called a **shutter** which keeps light out completely when the camera is not in use. Once the shutter is open, the amount of light which enters the camera is determined by the size of the opening of the iris diaphragm (called the lens opening) and the length of time the shutter remains open.

When we are taking a picture the object is usually far from the camera. The image therefore is formed close to the principal focus of the lens (see Chapter 39). This means that the film is placed at a position from the lens slightly greater than the focal length, so as to record the focused image.

40-9 This figure will enable you to find if you have astigmatism. Look at it through one eye at a time. If some of the spokes of this wheel appear to be in focus while others are blurry and out of focus, you may be suffering from this rather common eye defect caused by improper curvature of the surface of your eye.

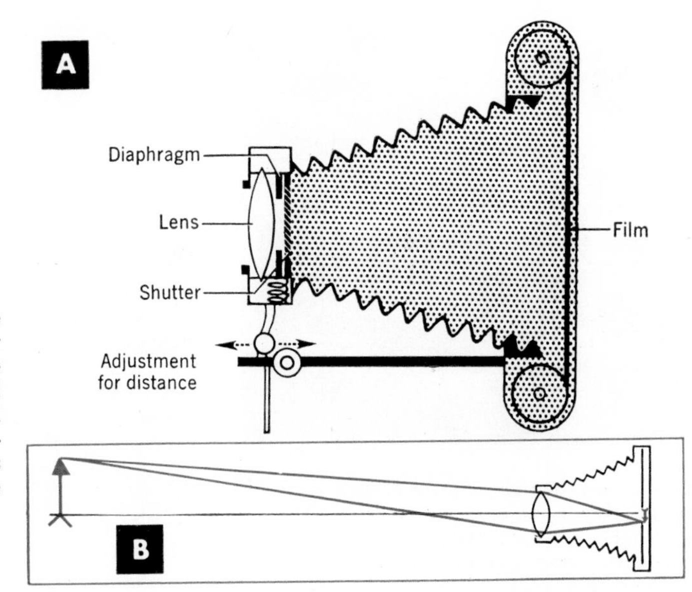

40-10 Click! The shutter springs open for an instant, a focused image appears on the film, and chemical changes record the image. Why is a diaphragm needed? Why is an adjustment for distance needed?

9. Focusing the camera

As with the eye, the camera must also accommodate for distance. The nearer the object is to the lens the farther away the focused image will be. In most cameras a focusing adjustment is provided to change the distance between the lens and the film. This adjustment is marked off according to the distance of the object from the camera. Cheaper cameras have a **fixed focus.** They have no adjustment for distance and so their usefulness is limited.

10. Getting the proper exposure time

A picture taken under conditions of good exposure comes out with all the proper shades of light and dark. Too much light causes **overexposure** of the film. Too little light causes **underexposure** of the film (Fig. 40-11). To get the proper exposure two important adjustments have to be made—the size of the iris diaphragm, or lens opening, and the time that the shutter remains open. In order to make those two adjustments properly, you must take into account the amount of light available at the time you take the picture, as well as the sensitivity to light of the film you are using.

To be sure you get the proper exposure, it is wise to use an **exposure meter.** This tells you exactly how much light is available and how to adjust the shutter speed and the lens opening (usually marked on the camera as the **f number**) in order to get the right exposure. One popular type of exposure meter consists of a photoelectric cell which is connected to a galvanometer (see page 414). You point the exposure meter at the scene being photographed. The light which enters the photoelectric cell produces a tiny electric current which causes the pointer of the galvanometer to move. The pointer indicates the amount of light available for your picture and the lens opening that goes with each shutter speed.

11. Photographing moving objects

When selecting the shutter speed, you must also take into account the motion of the object. Suppose we are taking a side view of a car traveling 50 feet per

40-11 Just the right amount of light must hit the film to make a proper exposure (above right). Too little light causes underexposure (above left), and too much causes overexposure (right). How can you be sure to get the proper exposure?

Eastman Kodak Co.

second (about 34 mi/hr). If the shutter of the camera is open for $\frac{1}{25}$ second, the car moves 2 feet during that time, so its image will be blurred.

The usual way to avoid this blurring effect is to cut down the time during which the shutter is open. The faster the motion we want to photograph, the more rapidly the shutter must operate to prevent noticeable blurring. Some cameras have shutters that open and close in $\frac{1}{1,000}$ second or less. It is even possible to photograph a rifle bullet cracking a bottle by using a lamp which flashes on for as short a time as two millionths of a second (Fig. 40-13). The average picture is taken at a shutter speed which varies from $\frac{1}{25}$ to about $\frac{1}{250}$ second. The faster the shutter speed, the larger the lens opening must be in order to get sufficient light.

12. Spherical and chromatic aberration

All optical instruments that contain lenses are subject to defects produced by the lenses, which cause blurred images. One of these defects is called **spherical aberration.** As you see in Fig. 40-12, a simple lens does not bring all rays from the object into perfect focus. Rays 1 and 5, which pass through the edge of the lens, are bent too much, so that they do not come to a focus at the same point as the other rays, 2, 3, and 4. This defect, which causes blurring and distortion, occurs because of the spherical shape of the lens. This cannot be avoided in cheap lenses because the spherical shape is much easier to grind than other shapes. Spherical aberration can be reduced somewhat by making the lens spherical on the front surface and flat, or only slightly curved, on the back.

We can avoid most of the defect of spherical aberration if we reduce the opening of the lens enough to cut out rays 1 and 5. We do this by simply closing down the iris diaphragm in front of the lens. But this remedy has the disadvantage of allowing less light to enter, and therefore cannot be used where light conditions are poor.

Another common defect is known as **chromatic aberration,** in which the lens acts as a prism and forms rainbow-colored images (see Chapter 42).

13. Improving the lens

In the past hundred years lens-makers have learned to overcome to a great extent these defects of lenses, and to produce images quite free from distortion, blurring, or rainbow-colored edges. To achieve better pictures, they use a combination of lenses. The better cameras use an **anastigmat** (ăn·ăs′tĭg·măt′) **lens.** This lens is made up of several separate parts, each ground to careful specifications, put together as shown in Fig. 40-14. The process of making these lenses is costly and increases the price of a camera considerably.

With this more expensive lens, which is wider and permits more light to enter the camera without distorting the image, you can take pictures where the light is poor. Or you can use this "fast" lens with a fast shutter to take pictures of rapidly moving objects.

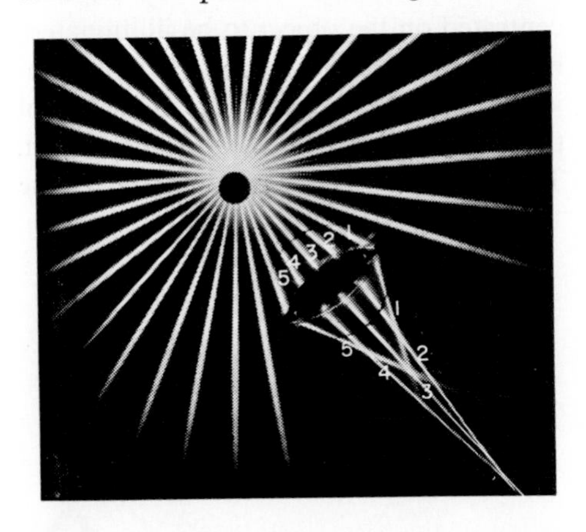

40-12 One serious defect of lenses, spherical aberration, occurs because rays passing through the edges of the lens (1 and 5) focus at a different place from those that pass through near the center of the lens (rays 2, 3, 4). Can you think of a very simple way to prevent spherical aberration?

General Electric Co.

40-13 This photo, taken in two-millionths of a second, catches the glass cracking after a bullet has passed through it. The bullet has cut the tape at the left to set off a momentary flash of light.

Lens-makers face still another difficulty. Every time light strikes the surface of a lens (or any other transparent substance), some of it is reflected. By the time light has passed through the lens combination shown in Fig. 40-14, a large fraction of it will have been lost by reflection, and the resulting image will be faint. This loss can be nearly eliminated by applying a thin film of hard, transparent material to each lens surface. If this film is of precisely the right thickness, the rays of light reflected from the outer surface of the film and from the surface of the film which is in contact with the glass will interfere with each other. The result is that little light is reflected, and most of the light is transmitted, an effect to be explained more fully in Chapter 44.

Nearly all camera lenses and binocular lenses made today are **coated lenses** of this sort.

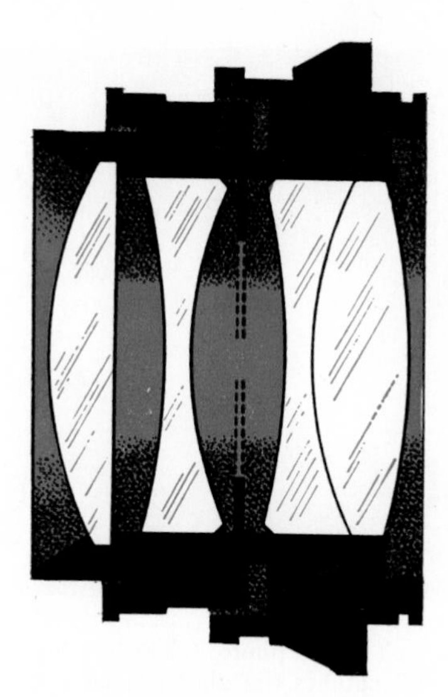

40-14 A good camera lens must be made of a combination of lenses. The anastigmat lens shown here, used in better cameras, is made of four separate lenses.

Enlarging images

Once the film of a camera has captured an image and the negative is developed by chemicals, it is then a simple matter to enlarge the picture on the original negative to any size.

14. The enlarger

From our study of the convex lens we learned that to obtain an enlarged image we must place the object between 2F and the principal focus, F, of the lens. The image is then beyond 2F on the other side of the lens and is larger in size than the object (Fig. 39-7C). To obtain an enlarged image an **enlarger** is used. If we place a negative somewhere between F and 2F from the lens with a light source behind it, we get an enlarged image on a sheet of photographic paper.

15. The projector

The **projector** (Fig. 40-15) is an optical instrument very much like the enlarger. It uses a convex lens to produce on a screen an enlarged image of a lantern slide or a movie film. Here it is important to get the brightest image possible. To accomplish this we use a special projector lamp which provides a strong concentrated source of light. Such a lamp may use as much as 500 to 1,000 watts of power. A concave mirror (M in Fig. 40-15) is placed behind the lamp to reflect the light that would otherwise be lost at the back of the projector. In addition, the rays of light that would ordinarily be lost in directions LA′ and LB′ are bent by the two **condensing lenses,** C_1 and C_2, so that they are concentrated on the object to be illuminated, namely the slide or film, S. The projector lens, P, then forms an image of the lantern slide on the distant screen. The slide must be placed in the frame

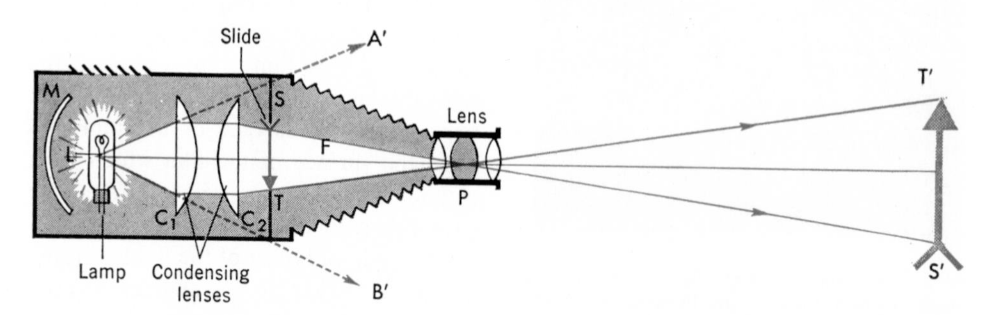

40-15 The projector has two main optical systems. The combination of mirror M, lamp L, and condensing lenses C_1 and C_2 is designed to concentrate as much light on the slide as possible so as to produce a bright image. The combination of lenses in P, called the objective, is designed to produce a clearly focused and undistorted image of the slide ST.

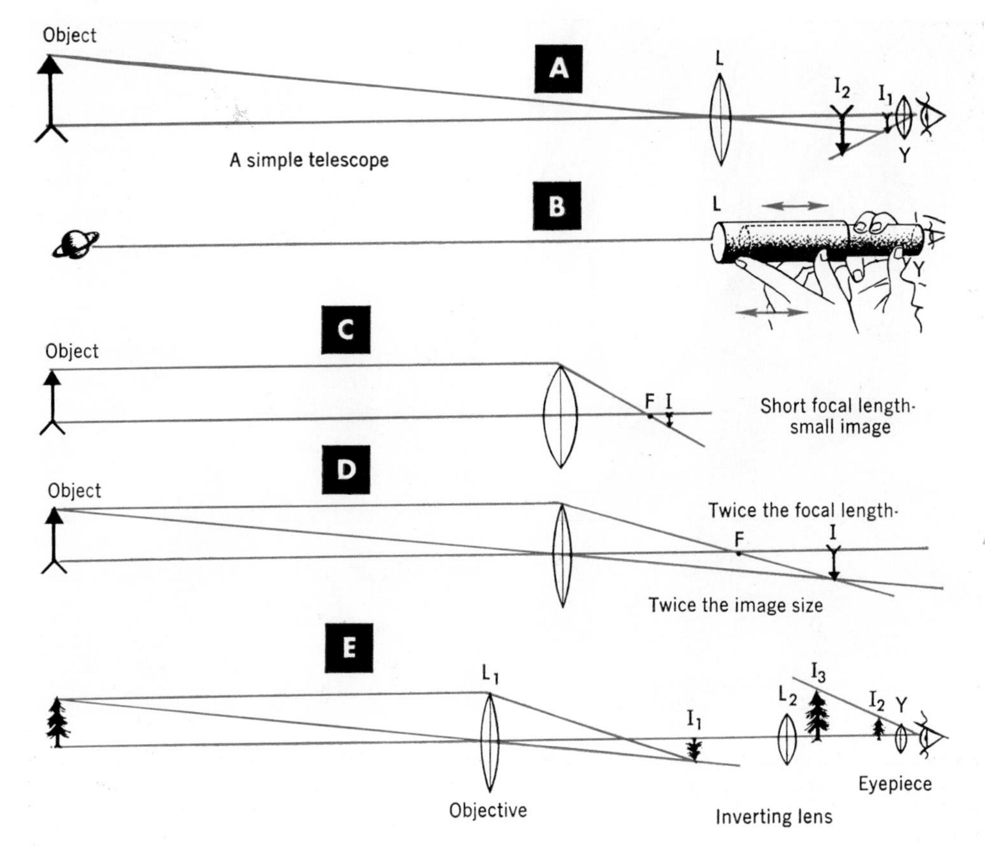

40-16 **Two lenses, an objective L and an eyepiece Y, make up a simple telescope (A and B). To increase magnification, we use a long focal-length objective lens (C and D). To make an erect image, we invert the image with an extra lens L_2 (E).**

in an inverted position so that the image on the screen will be right side up. If you want a larger picture, how would you move the projector? What happens to the brightness of the image as its size is increased?

16. The refracting telescope

Suppose we use a convex lens to form the image of a distant object. The image might be formed on a frosted-glass screen. We could then use a second lens as a magnifying glass to examine the image on the frosted-glass screen. With these two lenses we have made a simple astronomical telescope. The frosted-glass screen may be removed, as it is not really essential.

The front lens of our telescope (L in Fig. 40-16A and B), called the **objective**, forms a small, real, inverted image, I_1, slightly more than 1 focal length away from the lens. The second lens, Y, called the **eyepiece**, magnifies this image in exactly the same way as the magnifying glass described on page 539. For the image to be magnified, you will remember, the object (here the image formed by the objective) must be placed at a distance less than the focal length of the eyepiece. Then the eyepiece forms an enlarged, virtual image of the distant object, I_2, still inverted as it was formed by the objective. Such a telescope is suitable for use in astronomy, where the fact that the image is upside down is im-

Yerkes Observatory

40-17 The refracting telescope at Yerkes Observatory in Wisconsin. Its objective lens is 40 inches in diameter, the largest in use anywhere.

Edmund Scientific Co.

40-18 This homemade reflecting telescope uses a mirror for its objective. The mirror is in the end of the tube at the right. Compare it with the mirror in Fig. 40-19.

material. In actual telescopes the magnifying glass (eyepiece) is placed in a sliding tube so that we can focus the image by moving the eyepiece back and forth (Fig. 40-16B).

A telescope whose objective is a lens is called a **refracting telescope.** As we shall see, it is also possible to use a concave mirror as the objective, and then we have a **reflecting telescope.**

We determine the magnification of the telescope by selecting the focal lengths of our two lenses. Figure 40-16C shows the real image of an object formed by a certain objective lens. Figure 40-16D shows the image of the same object produced by a lens of greater focal length. We notice that the lens of greater focal length produces a larger image. Telescopes of high magnification are made with long tubes because the objective lens is selected to have a large focal length.

The eyepiece also has an effect on magnification. The magnification of an eyepiece increases as its focal length decreases. So we increase the magnification of our telescope by using as the objective a lens of long focal length, together with a magnifying glass, or eyepiece, whose focal length is short.

The magnifying power of a telescope is the focal length of the objective divided by the focal length of the eyepiece.

It has been found that the only way to get a bright image in a telescope and still have a very high magnification is to increase the diameter of the objective lens. Because of this increase in diameter, the lens gathers more light, thus providing a brighter image. The largest telescope lens in the world is that of the Yerkes Observatory in Wisconsin. It is 40 inches in diameter. Telescopes of the mirror type—explained later in this chapter—can be made much larger; the mirror that acts as an objective in the Mount Palomar telescope in California (Fig. 40-19) has a diameter of 200 inches (about 17 feet!).

17. The terrestrial telescope

You remember that the objective lens of a telescope produces an inverted image. The magnifying eyepiece does not invert the image again (Fig. 40-16A), so when we look through a telescope made of two convex lenses we see an inverted image.

But we want the image to be right side up when we are using a spyglass or **terrestrial telescope,** that is, a telescope to look at objects on the earth. One way to accomplish this is to add another lens (L_2 in Fig. 40-16E) between the objective and the eyepiece to invert the image formed by the objective lens. The image is then magnified by the eyepiece. Thus in Fig. 40-16E the objective lens, L_1, forms an image at I_1. Another lens, L_2, is placed so as to form an image of the first image at I_2. The image which was already inverted has now been inverted again and is therefore right side up. It is then magnified by the eyepiece Y.

In some telescopes the image is put right side up by the use of a concave lens eyepiece. This type of telescope was first used by Galileo.

18. The reflecting telescope

The largest telescopes on earth are the reflecting telescopes used in astronomical observatories. All these telescopes make use of the converging ability of concave mirrors. Figure 40-18 shows a small telescope of this type.

In the reflecting telescope, the mirror substitutes for the objective lens and forms an image of a distant object practically at F. If an eyepiece were placed at F, the observer's head or camera would tend to block the light falling on the objective. Therefore, a small mirror (or reflecting prism) is usually placed in the path of the converging rays before F to reflect the rays to the eye outside the telescope tube.

As with the lens, a mirror which has a spherical surface produces spherical aberration, causing a blurred and distorted image. This defect can be more easily remedied in a concave mirror than in a lens of the same size because only one surface of the mirror has to be

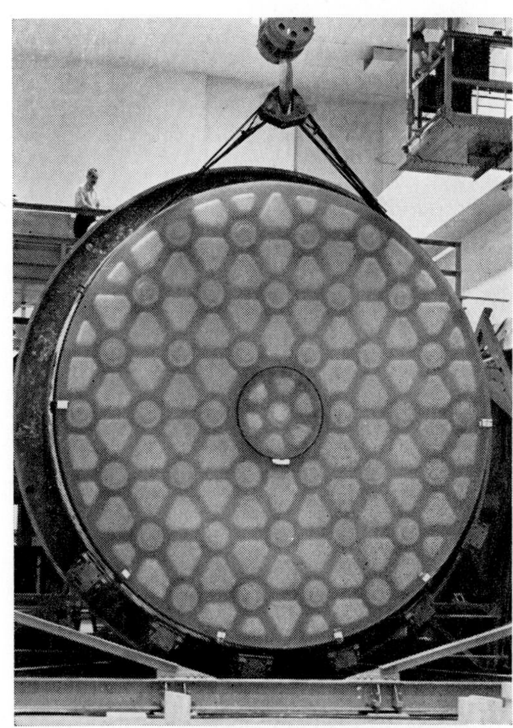

Mount Wilson and Palomar Observatories

40-19 Grinding, polishing, and testing the 18-ton mirror for the world's largest telescope at Mount Palomar was a real engineering job.

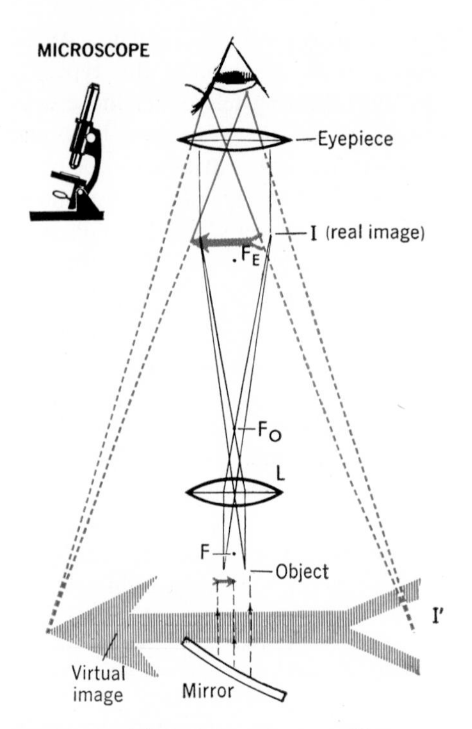

40-20 The microscope resembles an enlarger in that it produces a large image of a small close object. The enlarged image, I, can be photographed with a film or examined with a magnifying eyepiece, E. Note that the image, I′, is inverted.

ground. We give the surface a parabolic shape (Fig. 38-7). With this shape all rays from a distant point along the axis of the telescope are brought into perfect focus. Such a mirror is called a parabolic mirror.

19. Reflecting prisms

In the periscope shown in the atomic laboratory in Fig. 38-2, reflecting prisms are used instead of mirrors. Figure 38-10 shows how such a prism reflects beams of light. The action here depends upon the fact that rays striking a glass surface at an angle of incidence greater than about 43 degrees are totally reflected. This angle of 43 degrees is the critical angle described in Chapter 38.

Binoculars make use of such prisms to shorten the length of the instrument by reflecting the light back and forth, and to erect the image. Prisms have an advantage over mirrors in that they do not have a silvered surface that has to be resilvered every few years.

Also, it should be remembered that a mirror can never reflect 100 per cent of the light that falls on it. A prism, on the other hand, gives almost total reflection.

20. The microscope

In the **compound microscope** shown in Fig. 40-20, we place the object just outside F, the focal length of the lens, L, and hence form a real, enlarged, inverted image far away at I. Then we use a magnifying glass, or eyepiece, to enlarge the image still further. A microscope can be used with the eye, or with a camera to take a picture. A **photomicrograph** taken

A. Devaney, Inc.

40-21 Electron microscopes like this one can magnify more than a hundred thousand times.

with such a camera-microscope combination is shown in Fig. 40-2.

In the microscope the problem of distortion and blurring due to spherical and chromatic aberration is a serious one. To reduce these defects the objective lens may be made up of as many as seven separate lenses. With such an arrangement we can magnify an object many hundreds of times. Using beams of electrons in place of rays of light, it is possible to build an **electron microscope** (Fig. 40-21) for magnifications of up to 100,000 diameters, or 50 times greater than with the best optical microscope.

Going Ahead

THE VOCABULARY OF PHYSICS

Define each of the following:

iris diaphragm
pupil
cornea
lens (of the eye)
retina
accommodation
farsightedness
nearsightedness
astigmatism
persistence of vision
binocular effect
shutter
fixed focus
overexposure
underexposure
exposure meter
spherical aberration
chromatic aberration
anastigmat lens
electron microscope
coated lens
enlarger
projector
condensing lenses
objective
eyepiece
terrestrial telescope
reflecting telescope
compound microscope
photomicrograph
refracting telescope

THE PRINCIPLES WE USE

1. What is the function of each of the following parts of the eye: iris diaphragm, pupil, cornea, lens, retina, nerve cells?
2. How does the human eye accommodate for various distances?
3. Draw diagrams showing how the eye lens focuses light from an object far away and focuses again when the object is brought closer.
4. Draw a diagram showing why objects close to a farsighted person seem blurred. Draw another diagram showing how a lens can be used to correct this defect.
5. Redraw the diagram in Question 4 to apply to a nearsighted person.
6. How could you tell by examining someone's glasses whether he is nearsighted, farsighted, or has astigmatism?
7. State five ways in which the camera is like the eye.
8. State four ways in which the camera differs from the eye.
9. Name four optical instruments which produce enlarged images.

Explain why

10. It is necessary for the eye to accommodate for different distances.
11. It is difficult for most adults to see objects clearly when they are brought closer than 10 inches from the eye.
12. It is difficult for a person to see when he comes out of a dark place into bright light.
13. You do not see objects upside down even though the image is upside down in your eye.
14. A flicker appears whenever a movie projector is slowed down sufficiently.
15. Movies give the illusion of continuous motion even though separate pictures are flashed on the screen.
16. It is necessary to adjust the lens of a camera to take a picture of a close object.
17. It is usually best to snap a picture at the highest possible shutter speed.
18. Telescopes have long tubes.

APPLYING PRINCIPLES OF OPTICAL INSTRUMENTS

1. Explain why speeding up a movie camera when taking pictures produces slow motion when the film is projected at the normal rate.
2. The distance between the lens and retina of the eye of a certain person is $\frac{1}{2}$ inch. How large is the image of a 6-ft man standing 100 ft away?

3. How is the proper exposure for a picture obtained? What four factors must be taken into account?
4. Cheap cameras and expensive cameras can both take clear pictures under the proper conditions. For what purposes are the more expensive cameras more desirable? Why do they cost more?
5. Draw a diagram showing the cause of spherical aberration. How is it remedied (two methods)?
6. Explain the operation of a projector. (Use a diagram.)
7. A certain projector forms an image of a slide on a screen. If the image is too small, how could you obtain a larger one?
8. Explain the operation of a refracting astronomical telescope. (Use a diagram.)
9. How is the magnification of a telescope increased (two methods)?
10. Give two reasons why an objective lens of large diameter is better for a telescope.
11. Why is a concave mirror, instead of a convex lens, used as the objective in large telescopes?
12. How is an erect image produced in a terrestrial telescope?
13. Draw a diagram to show how a microscope produces a magnified image.
14. Explain why a photographer prefers to use a lens of long focal length in his studio camera, but uses a lens of short focal length in his "candid" camera.
15. An engineer wishes to design a home enlarger which will enlarge a 2-in × 3-in negative to a size of 24 in × 36 in. A lens of 4-inch focal length is mounted on a stand that moves vertically upward on a track. The engineer wishes to figure out how high a track he needs. How high above the table will the lens be when it is in position to produce the magnification he wants?

SOLVING PROBLEMS OF OPTICAL INSTRUMENTS

1. Draw a diagram showing how a camera lens with a 4-inch focal length forms an image of an object 24 inches from the lens.
2. Draw a diagram showing the image formed when a film 1 inch high is placed 6 inches from the lens of an enlarger of 4-inch focal length. How large is the final picture? Check your answers by calculation with the lens formula on page 541 and page 543.
3. A certain camera is adjusted for distance by pulling the front of the camera out to the proper distance as marked off on a track. If the lens has a focal length of 3 in, what will be the distance from the lens to the film when the object is 8 ft away?
4. Suppose you happen to have two convex lenses, one of 2-inch focal length and the other of 30-inch focal length, which you wish to use in making a telescope. (a) What will be the magnifying power of the telescope? (b) How long will the telescope be?
5. Imagine you have a lens whose focal length is 12 in. You wish to use it as the lens in a home-made projector you are building. If you plan to have the projection screen 15 ft from the lens, how far behind the lens must you arrange the film holder?
6. To what distance would you have to move the lens in Problem 5 if you decided to set up your projector in a hall 40 ft from the projection screen?
7. (a) Find the magnifying power of the telescope in Fig. 40-18 if the focal length of the mirror is 50 inches and the focal length of the eyepiece is $1\frac{1}{2}$ inches. (b) What would happen to the brightness of the image if the eyepiece were replaced by one whose focal length is $\frac{3}{4}$ inch?

CHAPTER

41

Illumination

Most jobs can be accomplished without eyestrain, and more efficiently, if the worker has enough light to see by. In order to find just how much light is needed for a given job, engineers have measured illumination under different conditions. From these studies they have obtained information which is useful in designing methods of lighting houses, stores, and factories.

In this chapter we shall see how illumination is measured.

Measuring light-producing power

When scientists first began to measure light, the candle was in common use. Hence it was natural for them to take the candle as their standard. Their name for this standard survives to this day: light-producing power is measured in **candlepower,** just as mechanical power is measured in horsepower, another standard for an earlier day.

1. Candlepower

But this standard is not accurate enough for our measurements today. Scientists have established a new and more accurate standard. Platinum placed in a special box is heated to its melting temperature, 1,773°C. The top of this box has a hole exactly 1 centimeter square which allows light to escape from the hot, glowing platinum inside (Fig. 41-1). This light will always be the same provided the platinum is at melting temperature. We define the light-producing power of this box of glowing platinum as 60 candlepower. Candlepower actually is a unit of power; it is a measure of the rate at which light energy is emitted by a source, as well as being a measure of the brightness of the source.

Obviously it is not practical to carry a box of molten platinum around with us. So we use a standard electric lamp whose brightness has been compared with the light from the standard box of molten, glowing platinum. (We shall see later in this chapter how the two lights are compared.) Once we know the candlepower of this electric lamp, we can use it as a secondary standard to measure light in place of the original standard.

2. Changing the amount of illumination

You could make up for the dimness of a reading lamp by putting in a brighter bulb. In this way you could increase the amount of **illumination** on your book. By illumination we mean the amount of light energy falling on a unit area every second. Other things being equal, a lamp of 100 candlepower will give twice as much illumination on your book as one of 50 candlepower.

But the candlepower of the source is not the only thing that affects illumination. You could make up for the dimness of the lamp by bringing your book closer to the bulb. So distance has an important effect on illumination too. What is the connection between the distance of a lamp and its brightness of illumination?

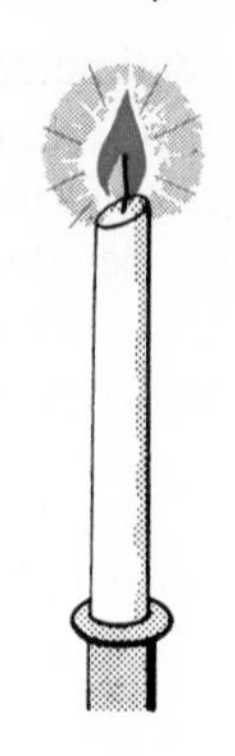

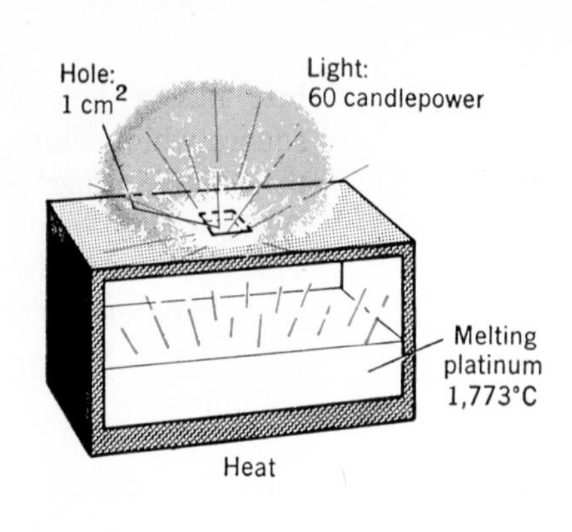

41-1 Old and new standards of light. Can you explain why the old standard candle was replaced by the box of glowing platinum with a hole cut in the top to let out some light? Would it be easy for a scientist to duplicate the 1,773°C temperature—even without a thermometer?

3. How distance affects illumination

By experimenting with a diverging beam of light on a screen we can see how much the illumination decreases with an increase in distance. Select a point source of light such as an arc lamp (without a lens). Now cut a hole exactly 1 inch square in a piece of cardboard (H in Fig. 41-2) and hold the cardboard 1 foot from the arc, L. The light shining through the opening in the cardboard spreads out as it travels away from the arc lamp. Now place a screen, S, 2 feet from the arc and 1 foot from the cardboard. You notice that the beam passing through the 1-inch-square hole in the cardboard makes a square of light on the screen which is 2 inches on each side (Fig. 41-2B). This square on the screen contains 4 square inches (2 in × 2 in). In other words, light passing through a hole 1 inch square, 1 foot away from the arc lamp, spreads out to cover 4 square inches on a screen 2 feet from the lamp. This means that each square inch on the screen gets only $\frac{1}{4}$ of the light that passes through the hole. So the illumination at 2 feet is only $\frac{1}{4}$ what it was at 1 foot.

When we place the screen 3 feet from the arc light, the beam passing through the 1-inch-square hole in the cardboard (1 ft from the light source) makes a square on the screen which is 3 inches on each side (Fig. 41-2C). This square contains 9 square inches; thus each square inch 3 feet away gets only $\frac{1}{9}$ of the light that passed through the hole in the cardboard at 1 foot. We repeat for 4 feet (Fig. 41-2D) and find the light covers 16 square inches on the screen; at 5 feet, 25 square inches; at 6 feet, 36 square inches. Thus we see that:

Illumination decreases as the square of the distance from the source of light.

This relationship of illumination to distance is an example of the use of inverse squares. This kind of law applies to many other effects in nature besides light. Loudness of sound, the force of gravity, and magnetic and electrical forces all diminish with the square of the distance.

4. How to calculate illumination

We can calculate the illumination on a surface if we know the brightness of the source of light and its distance. We have seen that the illumination varies inversely with the square of the distance from the source of light. It is also directly proportional to the intensity (candlepower) of the source. These facts are expressed in the following formula:

$$\text{Illumination} = \frac{\text{Candlepower}}{(\text{Distance})^2}$$

or

$$I = \frac{C}{D^2}$$

Let us apply this formula to a situation in which we have a 1-candlepower source of light illuminating a surface 1 foot away.

$$I = \frac{C}{D^2} = \frac{1}{1^2} = 1?$$

Our answer is 1. But 1 what? The unit chosen for measurement of illumination is the **foot-candle.** One foot-candle is the amount of illumination given by 1 candlepower at a distance of 1 foot.

SAMPLE PROBLEM

How much illumination will we get on a surface 5 feet from a 100-candlepower lamp? Using our formula we find:

$$I = \frac{C}{D^2} = \frac{100}{(5)^2} = \frac{100}{25}$$
$$= 4 \text{ ft-candles}$$

In other words, a 100-candlepower lamp gives the same illumination at 5 feet that we would get from 4 candlepower at a distance of 1 foot.

Using this formula we can predict the illumination a lamp will cast on a surface any given distance away.

STUDENT PROBLEM

What is the illumination from an 80-candlepower lamp at a distance of 4 feet? (ANS. 5 ft-candles)

5. Measuring candlepower

A device for measuring the candlepower of a lamp is called a **photometer** (fô·tŏm'ê·tẽr). One simple form consists of two wax blocks (W_1 and W_2 in Fig. 41-3) with a piece of aluminum foil between them. When this double wax block is placed between the two lamps, side W_1 is illuminated by the standard lamp while the other side W_2 is illuminated by the unknown lamp. The aluminum foil between the two blocks keeps the light of one lamp from illuminating the block on the other side. The double wax block is moved back and forth between the lamps until, at position P, the two blocks appear equally bright. At that point the illumination from each lamp is equal. The distance of each lamp from the block is then measured.

We obtain the exact candlepower of the unknown lamp as on the next page.

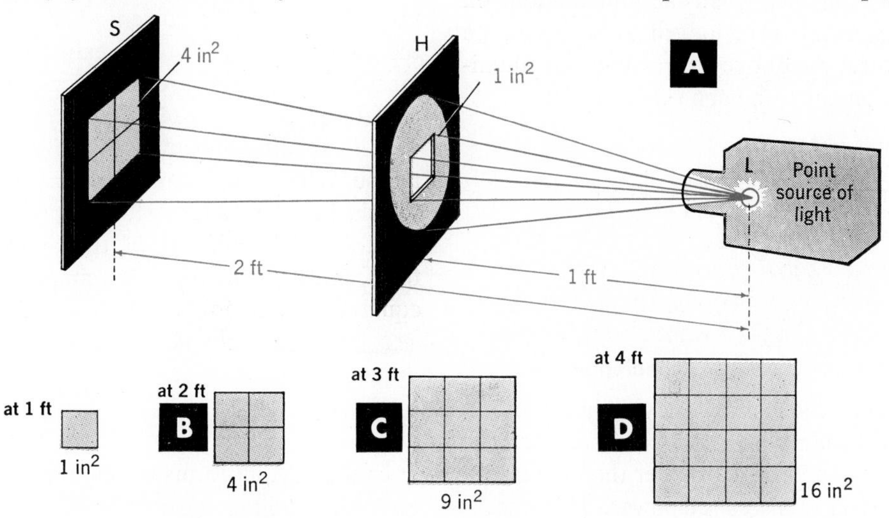

41-2 Illumination decreases as the square of distance (inverse square law) because the light covers larger areas at longer distances. Note how the areas covered at 2, 3, and 4 feet increase to 2^2, 3^2, and 4^2 (or 4, 9, 16) times as much.

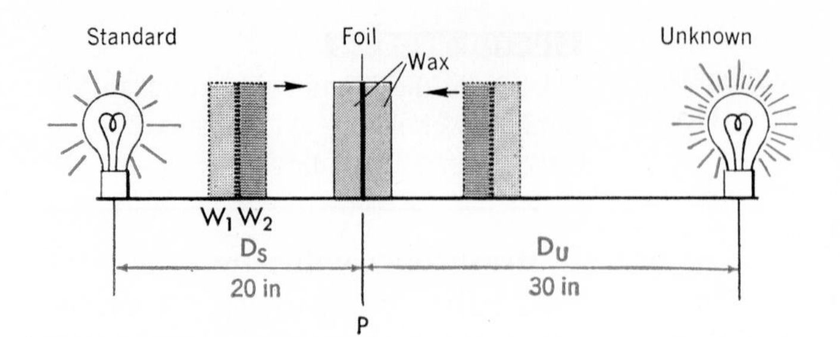

41-3 We can find the candlepower of an unknown lamp by finding the distance at which it produces the same illumination as the standard lamp. Wax blocks separated by aluminum foil scatter the light falling on them and enable us to compare the illumination from each lamp.

SAMPLE PROBLEM

Suppose we are using a 10-candlepower standard lamp. We find that it gives the same illumination at 20 inches from the blocks that the unknown lamp gives at 30 inches.

Illumination I_s from the standard lamp is:

$$I_s = \frac{C_s}{D^2} = \frac{10}{(20)^2}$$

Illumination I_u from the unknown lamp is:

$$I_u = \frac{C_u}{(30)^2}$$

Since these measurements of distance were obtained when the illuminations on each block were judged to be equal, we can set the above expressions for illumination equal to each other. Thus:

$$I_s = I_u$$

$$\frac{10}{(20)^2} = \frac{C_u}{(30)^2}$$

$$\frac{10}{400} = \frac{C_u}{900}$$

$$C_u = \frac{900 \times 10}{400}$$

$$C_u = 22.5 \text{ candlepower}$$

Our calculation shows that the unknown lamp has 22.5 candlepower. The unknown lamp is brighter than the standard one of 10 candlepower. This checks with the fact that the unknown lamp gave the same illumination even though farther from the blocks.

STUDENT PROBLEM

It is found that a 25-candlepower lamp gives the same illumination at 15 inches from a screen as an unknown lamp gives at 45 in. What is the candlepower of the unknown lamp? (ANS. 225 candlepower)

■ ■ ■ ■ ■ ■ ■

Standard lamps are calibrated in a similar way against the 60-candlepower box of glowing platinum.

A photoelectric cell provides a still more convenient way of measuring illumination. In Chapter 40 we showed how it is used in photography to measure illumination. The greater the illumination, the greater the current. The dial of the meter can therefore be calibrated to measure foot-candles of illumination at a surface.

With this instrument it is also easy to measure the candlepower of any lamp by placing the meter 1 foot away from the lamp. The reading of the meter in foot-candles will then be numerically equal to the lamp's candlepower.

The efficiencies of different lamps

In the past all forms of illumination—torches, candles, and kerosene lamps—used the chemical energy of fuels as the energy source. Today the most common source of energy for light is electricity.

But only a small part of the energy put into any lamp is converted into light. The rest is transformed into other forms of energy, mainly heat. By comparing the light energy output and the electrical energy input of a lamp, we can find its **efficiency.**

6. Efficiency

The efficiency of any machine is the ratio between power output and power input. For a lamp, which converts electrical energy into light energy,

$$\text{Efficiency} = \frac{\text{Light power output}}{\text{Electrical power input}}$$

Table 17 shows the results of measurements of efficiency for lamps of different kinds.

You notice that the modern argon-filled tungsten lamp gives 114 candlepower for an input of 100 watts of electric power. Thus we get slightly more than 1 candlepower per watt from the tungsten lamp. The carbon-filament lamp, on the other hand, gives only 23 candlepower for 100 watts, or less than $\frac{1}{4}$ candlepower per watt. Thus the modern tungsten bulb gives about 5 times as much light as the carbon-filament bulb for the same cost in electric energy. But even the tungsten lamp is only 2.3 per cent efficient, wasting in the form of heat almost 98 per cent of the electrical energy which is put into it.

The fluorescent light is almost 3 times as efficient as the tungsten bulb, giving 320 candlepower for 100 watts as compared to the tungsten bulb's 114. This greater efficiency explains the wide use of fluorescent lamps, especially in stores and factories where large amounts of light are needed. Fixtures for fluorescent lamps cost more than those for ordinary bulbs, and the tube itself is more than 5 times as expensive as the ordinary tungsten bulb. But the efficiency of the fluorescent lamp usually makes up for its greater original cost (Fig. 41-4).

TABLE **17**

COMPARISON OF EFFICIENCIES OF 100-WATT ELECTRIC LAMPS

Type of lamp	Approximate candlepower per watt	Efficiency
Carbon-filament lamp	0.23	0.5%
Electric carbon-arc lamp	0.96	2.0%
Ordinary tungsten-filament lamp (argon-filled)	1.14	2.3%
Fluorescent lamp	3.20	6.5%
maximum candle-power possible	51.0	100.0%

Sodium-vapor lamps, which produce a yellow light, are used on many highways today because they are even more efficient than fluorescent lamps for high-power lighting and can therefore reduce the lighting costs considerably.

Engineers are constantly seeking to raise the efficiency of lamps. Figure 41-5

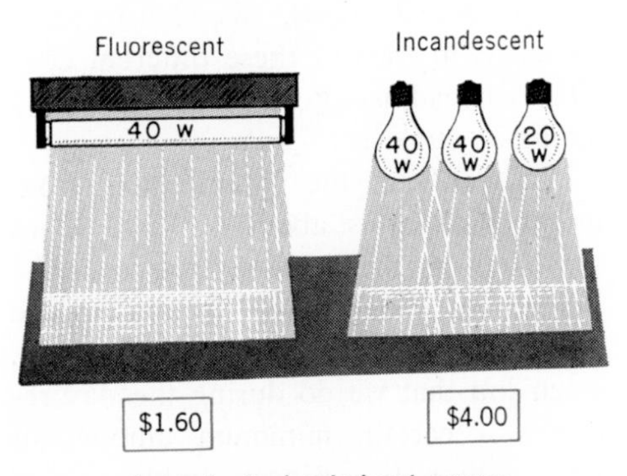

41-4 Can you see why fluorescent lamps are widely used, though the fixture and the bulbs are more expensive than those for incandescent lamps?

Westinghouse Electric Corp.

41-5 This experimental 10,000-watt mercury vapor lamp gives as much light as a hundred incandescent lamps (each 100 watts). Heat radiation is sufficient to light the cigarette in the man's hand. The lamp must be cooled by a stream of water to keep it from melting. Such lamps are more efficient than any used in homes today.

shows an experimental 10,000-watt mercury lamp which has a surface brilliance $\frac{1}{5}$ that of the sun. The heat is so intense it is sufficient to light the cigarette in the man's hand.

7. Fitting the illumination to the job

Table 18 indicates the wide range of illumination which our eyes encounter in various situations. We must adapt ourselves to each of these different conditions by adjusting the diameter of the pupils of our eyes.

As you see, the illumination from bright sunlight is about 400,000 times as strong as the light we get from the full moon. We have to be able to see comfortably between these two extremes. Each job that we do during the day requires a certain minimum amount of illumination. It has been amply proved in factories that poor light can be very costly in terms of accidents, ineffective work, and a general lowering of efficiency and morale.

8. The distance of stars by the inverse-square law

The inverse-square law of light illumination is a universal law. We need not confine its use, as we have, to measuring the illumination on the surface of the earth, at short distances from man-made sources. We can apply the law to the light from a star to measure its distance if we know its candlepower.

Suppose, for example, that a sensitive photoelectric photometer shows the illumination received from a star, I_{star}, to be 2.3×10^{-14} that of the illumination from the sun, I_{sun}. If the star is known by other means to be a source of the same brightness (candlepower) as the sun, we could calculate its distance from the earth by means of the inverse-square law and the known distance from the sun to the earth (9.3×10^7 mi).

Since the two sources are of equal brightness,

$$I_{star} \times D^2_{star} = I_{sun} \times D^2_{sun}$$

or

$$\frac{I_{star}}{I_{sun}} = \frac{D^2_{sun}}{D^2_{star}}$$

Substituting

$$\frac{2.3 \times 10^{-14}}{1} = \frac{(9.3 \times 10^7)^2}{D^2_{star}}$$

$$D^2_{star} = \frac{(9.3 \times 10^7)^2}{2.3 \times 10^{-14}}$$

$$= \frac{86.5 \times 10^{14}}{2.3 \times 10^{-14}}$$

$$= 37.6 \times 10^{28}$$

$$D_{star} = 6.1 \times 10^{14} \text{ mi}$$

(or about 100 light years)

You may well ask how we know that a star whose distance we wish to find has the same candlepower as the sun. As one would expect, stars come in many different sizes and brightnesses. However, the color of a star is related to its candlepower, and from a detailed examination of the colors it emits (see discussion of the spectroscope, page 583), we can see if the star shows the same colors as our sun. If it does, we can have confidence that our distance calculations based on the inverse-square law will be approximately correct.

9. The temperature of the sun

We can also make use of the inverse-square law to measure the temperature of the surface of the sun. We start by measuring the heat produced by the sun's radiation when it falls on a known area on the earth's surface. To do this, we can let the heat be absorbed by a known weight of water and measure its temperature rise in a given time. By the method described in Chapter 18, the number of calories of heat-producing radiation absorbed per minute can be determined by using the heat equation $\Delta H = mS\,\Delta T$. Finally, from the area through which the heat was absorbed, the number of calories received as radiation from the sun per square centimeter per minute can be determined. This value is called the **solar constant.** Its value is 1.94 cal/cm²/min. This is really a measure of the illumination at the surface of the earth, I_{earth}, on a sunny day. It includes not only visible light but also ultraviolet and infrared light, all of which produce heat when absorbed.

Now we can use the solar constant to help us find the intensity of illumination, I_{sun}, *at the sun's surface,* using the inverse-square law as in the preceding section.

$$\frac{I_{earth}}{I_{sun}} = \frac{D^2_r}{D^2_e}$$

Here D_e is the distance from the sun's center to the earth ($= 9.3 \times 10^7$ mi), and D_r is the distance from the sun's center to the sun's surface ($= 0.43 \times 10^6$ mi). Substituting,

$$\frac{1.94}{I_{sun}} = \frac{(0.43 \times 10^6)^2}{(9.3 \times 10^7)^2} = 2.14 \times 10^{-5}$$

$$I_{sun} = \frac{1.94}{2.14 \times 10^{-5}}$$

$$= 9.06 \times 10^4 \text{ cal/cm}^2\text{/min}$$

TABLE 18

ILLUMINATION UNDER VARIOUS CONDITIONS (in foot-candles)

Noonday sunlight	10,000
Hospital operating room	75
Night baseball	30
Close inspection work	30
Reading	10
Street lighting	1/20
Bright moonlight	1/40
Moonless night sky	1/10,000

We now turn to Stefan's Law (page 231) to give us the sun's surface temperature from the calculated value of its surface radiation. Stefan's Law states that the total energy, I, radiated per minute from a square centimeter of a surface is proportional to the fourth power of its absolute temperature, T.

$$I = kT^4$$

The value of k has been determined in the laboratory by measuring the heat energy emitted from surfaces of known temperatures. Its value is 82.2×10^{-12} cal/cm²/min/°K⁴. Therefore $I = 82.2 \times 10^{-12}\, T^4$. Substituting our calculated value for I_{sun},

$$9.06 \times 10^4 = 82.2 \times 10^{-12} T^4$$

$$T^4 = \frac{9.06 \times 10^4}{82.2 \times 10^{-12}}$$

$$= 0.110 \times 10^{16}$$

$$T = 0.575 \times 10^4$$

$$= 5{,}750°\text{K}$$

This result agrees very closely with the temperature computed by other methods.

Convair, General Dynamics Corp.

41-6 This large mirror is used by a western airplane manufacturer to concentrate the sun's radiation onto a small area at the mirror's focus. The very high temperatures produced are useful for studying the properties of materials used in making better airplanes and rockets.

10. Solar power

The solar constant tells us how much *energy* is received on a unit area of the earth's surface *per minute*. In other words, it tells us how much *power* in the form of radiation strikes the surface, assuming the sun is approximately overhead in a clear sky. Let us convert this "illumination" into horsepower. We know from the mechanical equivalent of heat (page 244) that 1 Btu = 778 ft-lb. Since 1 Btu also equals 252 calories, 1 calorie equals $\frac{778}{252} = 3.08$ ft-lb. On a square centimeter of the earth's surface, therefore, the sun must supply $1.94 \times 3.08 = 5.98$ ft-lb/min of power in the form of radiation. On a square meter (approximately 1 square yard) we would have 5.98×10^4 ft-lb/min. In terms of horsepower (1 hp = 33,000 ft-lb/min) this would be

$$\frac{5.98 \times 10^4}{33 \times 10^3} = 1.81 \text{ hp}$$

Thus approximately 12 horsepower of solar energy falls on the top surfaces of an automobile on a sunny day—and is entirely wasted.

The solar energy falling on only one acre of land is sufficient to supply the needs of a small town IF it could only be efficiently turned into a more useful form of energy, and IF it could also be stored for use when the sun is not shining.

The solar energy falling on the entire earth over a period of only two days is greater than the energy stored in all the world's known reserves of coal, oil, and natural gas put together.

A great deal of work has been done in an effort to harness the energy of the sun and put it to practical use, particularly in sunny countries (Fig. 41-6). The major difficulty, not yet overcome, is that conversion of solar energy into any other form of energy is always uneconomical. This means that to obtain useful power from the sun on a clear day a large area is needed to absorb radiation. We are still a long way from using the heat from solar radiation to raise steam to run an economical electric generator. So far the only practical conversion of solar energy into useful electrical energy occurs in highly inefficient (0.5%) but useful photoelectric cells and in solar batteries.

The energy that the sun pours upon the earth is all eventually radiated back into space. But much of it is stored momentarily in the form of heat energy in the surface of the earth and the atmosphere and in the form of latent heat of vaporization of water vapor in the atmosphere. When you realize that the 1.81 horsepower falling on each square meter (average) *cannot be destroyed* (Law of Conservation of Energy) you will appreciate how it is possible for a thunderstorm to have accumulated the energy of a conventional atomic bomb, and for a full-fledged hurricane to produce 500 million million horsepower, compared with 150 million horsepower produced by all the electric generators in the U.S.A. All this energy originated in the sun.

11. Solar batteries

If sunlight is allowed to fall on a photosensitive surface, the energy of the radiation can be changed directly into electrical energy. A large surface of this sort is called a **solar battery** (Fig. 36-22). Though such batteries have an efficiency of only about 10 per cent, they may prove useful for supplying the small amounts of power needed for telephones and even radios in some rural areas. They are, of course, ideal to power tiny radio transmitters in artificial earth satellites. Such a radio transmitter could draw about one watt from every 100 square centimeters of a solar battery.

Going Ahead

THE VOCABULARY OF PHYSICS

Define each of the following:

illumination	photometer
candlepower	efficiency (of lamps)
foot-candle	solar battery
solar constant	

THE PRINCIPLES WE USE

1. State the law of inverse squares.
2. Describe our standard of candlepower.
3. State the formula for calculating illumination.

Explain why

4. The glowing-platinum standard of candlepower is superior to the standard candle.
5. Illumination decreases as the square of the distance.
6. Thunderstorms and hurricanes contain such an enormous amount of energy.

EXPERIMENTS IN ILLUMINATION

Describe an experiment to:

1. Demonstrate the law of inverse squares.
2. Measure the candlepower of a lamp (two methods).
3. Measure the energy of sunlight falling on a unit area.

APPLYING PRINCIPLES OF ILLUMINATION

1. Mention three ways of increasing the illumination on a surface.
2. Compare the carbon-filament bulb, the argon-filled tungsten bulb, and the fluorescent lamp as to their light-producing efficiency.
3. Explain why it pays to use fluorescent lighting in a store instead of ordinary bulbs, even though the fluorescent lamps and fixtures are much more expensive.
4. Why would it pay to use a small tungsten bulb rather than a fluorescent light for illuminating cellar stairs which have to be lighted only occasionally?
5. Suppose you are an engineer planning the lighting system of a factory. What factors would you take into account? What measurements might you make?
6. Why isn't the energy of the sun more widely used as a source of power?
7. In what devices is the energy in sunlight converted directly into electrical energy?

PROBLEMS IN ILLUMINATION

1. How will the illumination a lamp produces at 2 ft compare with that at 4 ft; at 6 ft; at 8 ft; at 10 ft?
2. How many foot-candles of illumination are produced when a 60-cp lamp is placed 5 ft from a surface; 2 ft from the surface?
3. How far away must we place a 40-cp lamp to produce an illumination of 10 ft-candles on a surface?
4. How many 20-cp lamps will be needed to produce an illumination of 8 ft-candles on a surface 5 ft away?

5. You are doing some work requiring close inspection. There is a 25-cp lamp 3 ft above the table. Is the light sufficient for proper vision? (Consult Table 18.) If not, how many candlepower (at that distance) would be just right? What effect would a reflector have on the illumination on the table?
6. Approximately what candlepower could be obtained from (a) a 200-w tungsten lamp, (b) a 200-w fluorescent lamp?
7. The lights in a store are changed from tungsten lamps to fluorescent lamps. Before the change the monthly cost of lighting was $37. What will be the monthly cost after the change?
8. What candlepower is produced by a 100-w lamp whose efficiency is 5%?
9. You are using a photometer to measure the candlepower of a lamp. You find that the photometer appears equally illuminated by both lamps when the standard lamp of 25 cp is 20 inches away and the unknown lamp is 15 inches away. Which is the brighter lamp? What is the candlepower of the unknown lamp?
10. Using the inverse-square law we can calculate that the illumination produced on a wall 20 ft away by a certain 100-cp automobile lamp should be $\frac{1}{4}$ ft-candle. (Check this calculation.) However, a light meter shows that the illumination is actually 3 ft-candles. How do you explain this?
11. A photographer is using two floodlights at a distance of 8 ft from the subject for proper exposure. One lamp goes out and he doesn't have a replacement. How far away should he place the remaining lamp to maintain the same illumination?
12. Which give more illumination, three equal lamps at 3 ft or two equal lamps at 2 ft?
13. In a certain city 200-watt bulbs are used for street lighting. How far apart should the street lights be so that all sections of the street receive enough illumination? (Consult Table 18.)
14. The intensities of light from two stars of the same candlepower are observed to be in a ratio of 25 to 1. How do the distances of the two stars compare?
15. A certain 110-v, 120-cp lamp is 3% efficient. How much current does it draw?
16. How many calories of heat are produced per minute by a 100-cp lamp whose efficiency is 4%?
17. A lamp placed 2 ft from a book gives an illumination of 20 ft-candles. How far must it be moved to reduce the illumination to 15 ft-candles?
18. A 20-cp lamp and a 30-cp lamp are 4 ft apart. Where must a screen be placed to be equally illuminated by the two lamps?

---

19. How far away is a star which gives 3.5×10^{-14} the light of the sun if it is known to be a source of the same brightness as the sun?
20. A man can do sustained work at the rate of about $\frac{1}{6}$ hp. On how many square feet of ground must an overhead sun shine to deliver the same amount of power?

CHAPTER

42

Color

Have you ever noticed the beautiful rainbow colors that are produced when a beam of sunlight passes through the edge of an aquarium?

In the seventeenth century, Sir Isaac Newton also wondered about this rainbow effect. The series of experiments he undertook gave us important knowledge about what color is.

Analyzing color

Newton placed a triangular prism in the path of a beam of sunlight, as shown in Fig. 42-1 and in Plate I. As we know from Chapter 38, the beam of light was refracted and bent toward the third face of the prism. But instead of coming out as a narrow white beam of parallel rays similar to the one that went in, the beam coming out was colored, and the rays in the beam were no longer parallel. The colored light continued to spread apart to form a rainbow of colors, or **spectrum,** on a screen (S_2 in Fig. 42-1). This effect, called **dispersion,** is what causes chromatic aberration in lenses, as you learned in Chapter 40. As shown in Plate I, following page 580, the order of colors in this spectrum was red, orange, yellow, green, blue, indigo, and violet.

1. The spectrum

Where did these colors come from? Were they in the beam of white light when it entered the prism? To find the answers to these questions Newton performed additional experiments with this beam of colored light. He let the spectrum fall upon a hole in the screen so that only the red light came through (Fig. 42-1). Then he placed another prism, P_2, on the other side of the screen so that this red light had to pass through it. The red beam did not disperse into other colors after passing through the second prism. Nor did beams of green, blue, or any other color in the spectrum disperse further into other colors when analyzed in this way. Newton decided that these colors must be the fundamental ones in the spectrum, since they could not be broken up any further. But the original white light, which could be broken up by passage through a prism, he considered to be a mixture of all the fundamental colors in the spectrum. Newton's idea that white light is really a mixture of the spectrum colors has been confirmed many times. Today we accept his theory as fact.

2. What is color?

Many experiments since Newton's time have revealed other important facts about the nature of color. We now have good reason to believe that the difference between one color and another is simply a matter of **wave length** or **frequency.** A violet light wave has a higher frequency than a red light wave (Fig. 42-2). Or, to put it another way, we can say the wave length (distance from crest to crest) of violet light is shorter than that of red light. (At this point it would be wise to review the meanings of "wave length" and "frequency" as they were applied to the study of sound in Chapters 22 and 23.)

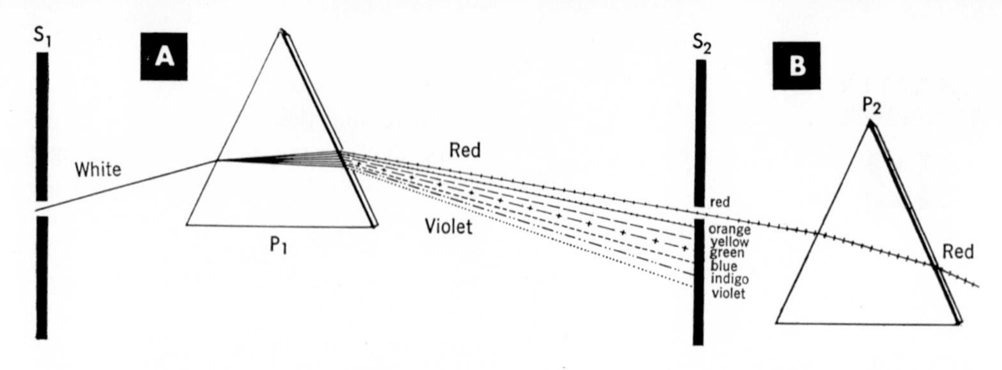

42-1 Newton's famous experiment with color showed that a beam of white light is in fact made up of all the spectrum colors. Attempts to break up the individual colors in the spectrum into more colors by using an additional prism, P_2, came to nought.

Measurement and calculation show that the frequency (rate of vibration) of red light is about 375 trillion (3.75×10^{14}) vibrations per second. Its wave length is about $\frac{3}{100,000}$ (or 3×10^{-5}) inch. This is the lowest frequency (and longest wave length) of any of the visible colors of the spectrum. At the other end of the spectrum, blue and violet light have higher frequencies (and shorter wave lengths) than red. The wave length of violet light, for example, is about 1.5×10^{-5} inch; its frequency is about 750 trillion (7.50×10^{14}) vibrations per second (Fig. 42-2). Other colors have still other frequencies and wave lengths ranging between these two extremes, red and violet.

3. What causes the spectrum?

You remember from Chapter 38 that when light strikes the surface of a transparent material, it slows down and bends as it passes through the material. But suppose the different colors in a beam of white light travel at different speeds in glass? What happens then?

If the velocities of violet and red waves are different from each other, then we would expect the amount of bending to be different for each. The color that is slowed down most by the glass should swerve the most. If we look at the order of colors in Fig. 42-1 and in Plate I, we note that violet is bent more than red in passing through the prism. Therefore, if difference in velocity is the cause of dispersion, then the violet light has been slowed down more than the red. Careful measurements show that the speed of violet light in materials such as glass and water is actually less than that of red light.

4. Filters

When we place a piece of red glass in the path of white light between a prism and a screen, as shown in Fig. 42-3A,

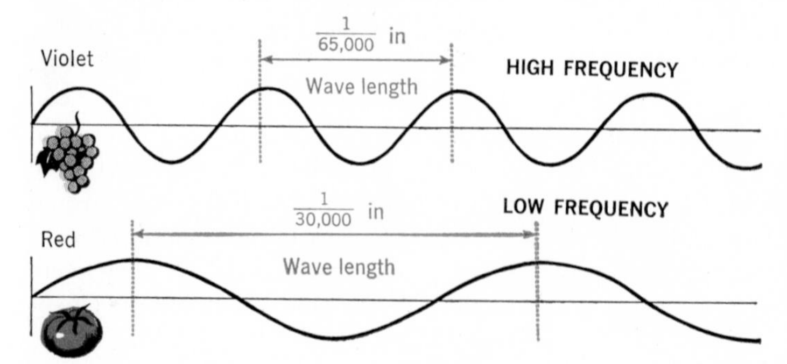

42-2 The key to color is the length of the wave. Red light waves have about double the wave length of violet light.

we note an interesting fact. The only color in the spectrum which shows up brightly on the screen is red. Some of the other colors may appear—but very faintly. Thus we see red glass as red because it permits mainly red light to pass through; almost all the other rays are absorbed and their energy converted into heat. Refer to Plate II.

Now let us try blue glass (Fig. 42-3B). It permits blue to come through to the screen; again almost all the other rays are absorbed. Green glass does the same thing for green. If we checked very carefully we would find that, in all these different experiments, the other colors of the spectrum usually do get through to a certain extent. But the basic color of a transparent glass is determined by the fact that it lets through much more light of one color than of others. Pieces of colored material, called **filters**, are often used in photography and in scientific work to absorb all but a few chosen wave lengths from a beam.

What about a color like purple, which is not observed as a fundamental color in the spectrum? When we use a purple glass filter in our experiment we notice that both red and blue light come through in considerable amounts, and that the middle of the spectrum is gone (Fig. 42-3C). Thus we would conclude that purple is a combination of two fundamental colors, red and blue.

When we try an amber (yellowish-white) filter we find that it allows most of the red, orange, yellow, and green to get through but absorbs the blue, indigo, and violet. We might therefore consider

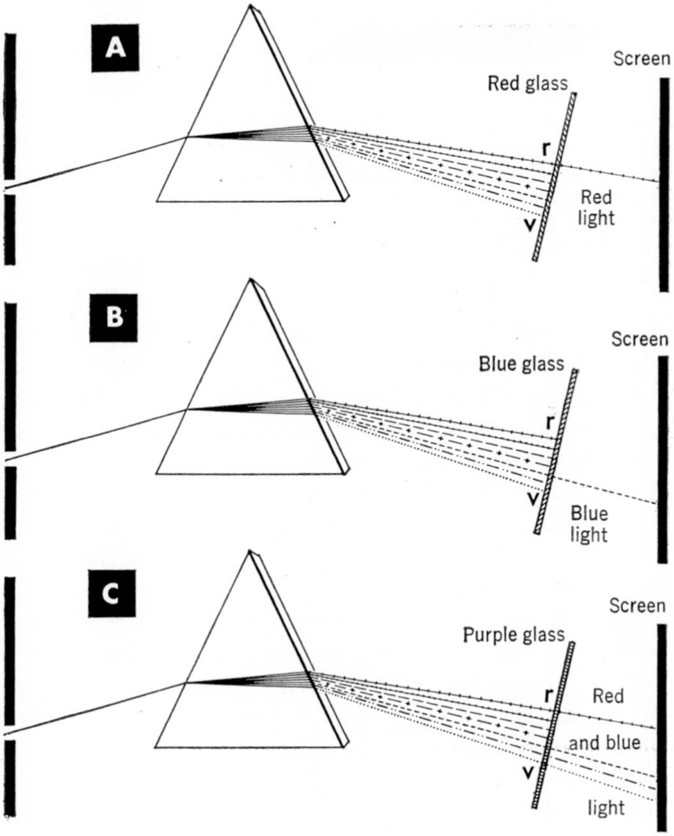

42-3 Why is red glass red? Because it lets through mainly red light and absorbs the others (A). Blue glass lets through mainly blue light (B). But purple glass (C) lets through a mixture of red and blue-violet light, while absorbing the colors in the middle of the spectrum.

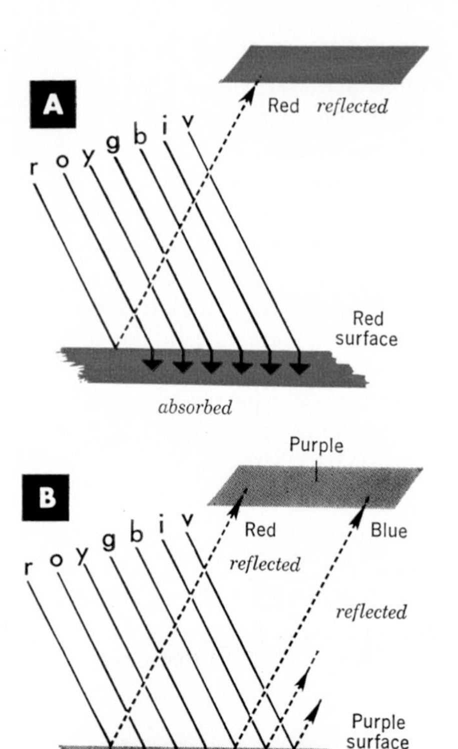

42-4 What makes a red rose red? It is the ability of the rose to reflect mainly red light while absorbing all other colors (A). Similarly, a purple object reflects red and blue-violet light while absorbing the middle colors in the spectrum (B).

the yellowish amber-colored light that comes through as white light minus the blue end of the spectrum.

You might call this process of passing light through a filter **color subtraction.** It is subtraction because the glass takes away (or absorbs) different colors from the spectrum. Refer to Plate VIII.

5. Mixing colored light

Suppose we reverse the process and add light of one color shining on a screen to light of another color projected on top of it. This is exactly what happens when colored lights are used on a stage (as shown in Plate IV). Suppose we turn on red and blue lights at the same time. They combine to form a purple light on the stage. This confirms our earlier analysis that purple is a combination of red light and blue light.

In this same process of stage lighting, when we add blue light to amber we get white. You remember that when we analyzed amber we found that it was composed of all the colors of the spectrum except the blue end. Thus when we add blue to amber we have most of the colors of the spectrum and the light appears white.

We can mix light of different colors in all sorts of proportions. Say we add a small proportion of red light to a light that is mainly white. The result is a pale red or pink. By adding some green to the white we would get a pale green.

We can also get various tints and shades of white by cutting down the amount of any of the fundamental colors that compose white. Thus, if we subtract some of the blue light, we get a white that has an amber or yellowish tint. We can get an infinite variety of shades by mixing basic spectrum colors in different proportions.

6. Colors by reflection

Now let us experiment with colored opaque objects that reflect light. If we disperse white light with a prism and allow the colors to fall upon a red object, we notice that the places where the yellow, green, blue, indigo, and violet formerly appeared in the spectrum now appear as darker spots on the red object. But the red light of the spectrum falling on the red object appears bright red. Refer to Plate V.

Apparently what happens is that the red object can reflect only red light and absorbs (subtracts) all the other colors (Fig. 42-4A). If blue light strikes the red object, it cannot be reflected. Instead it is absorbed and its energy is converted into heat. Since little or no blue light is reflected, the blue portion of the spectrum appears black, which is the *absence* of light. The same is true when the object is placed in any of the other colors. Green light is absorbed. Violet light is absorbed. Only the red light is reflected from the red object, a process called **selective reflection.**

In a similar way, a blue-colored object reflects blue but absorbs all other colors. A purple object reflects red and blue and absorbs the other colors (Fig. 42-4B).

We can check these facts by a further experiment. Project white light onto a piece of red cardboard. Nearby objects appear reddish in hue because of the reflection of the red light from the red cardboard. All the other colors but red in the white beam are absorbed by the red cardboard. If we use green cardboard only the green light in the white is reflected, and nearby objects appear greenish by illumination from the reflected green light. Thus we see that the color of an object depends upon the color of the light it reflects.

Actually, few colored objects reflect just one color. For example, a green leaf reflects mainly green but also small amounts of other spectrum colors. A red tomato (Plate V) reflects mainly red but also other spectrum colors to a slight extent. A lemon absorbs blue from white light and appears yellow (Plate V).

7. Illuminating colored objects with colored light

Thus far we have thrown only a white beam of light on colored objects. White light, you will remember, contains all the colors an object could possibly select for reflection. If we now try substituting light of various colors for the white beam, we get some startling color illu-

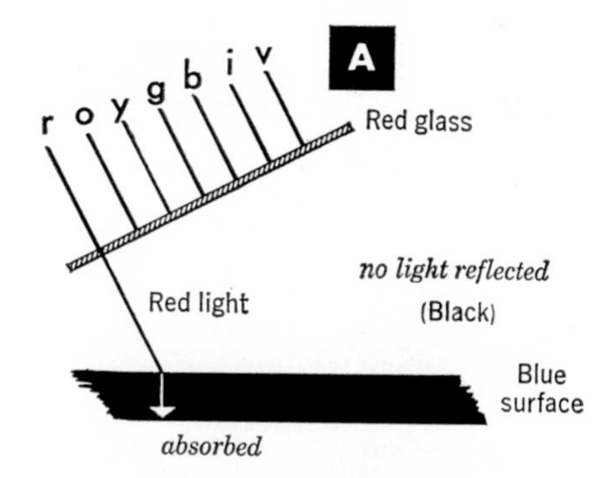

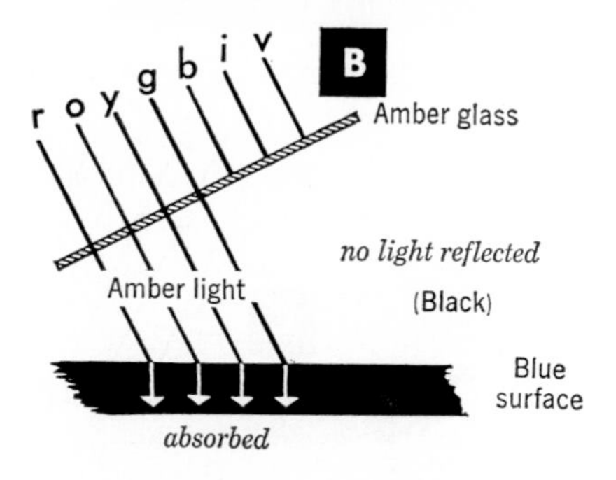

42-5 **How can you make a blue object look black? You need only use a light that has no blue in it. A red light (A) and an amber light (B) both cause the normally blue object to look black because blue light is missing. Why must this experiment be performed in the dark?**

sions. Imagine an actor on a stage wearing a blue jacket. If we turn a beam of red light on him the jacket appears black (Fig. 42-5A). This happens because the blue jacket reflects only blue light and absorbs other colors; therefore, the red light is absorbed, and since no light at all is reflected by the blue jacket it appears black. Black is no color at all. In popular literature ultra-violet light is sometimes inaccurately referred to as "black light."

Now let us try an amber light. When an amber light, which has no blue in it, strikes an object which can reflect only blue, no light is reflected at all, so the jacket still appears black. Suppose we try a purple (red plus blue) light on an amber or yellowish object on the stage.

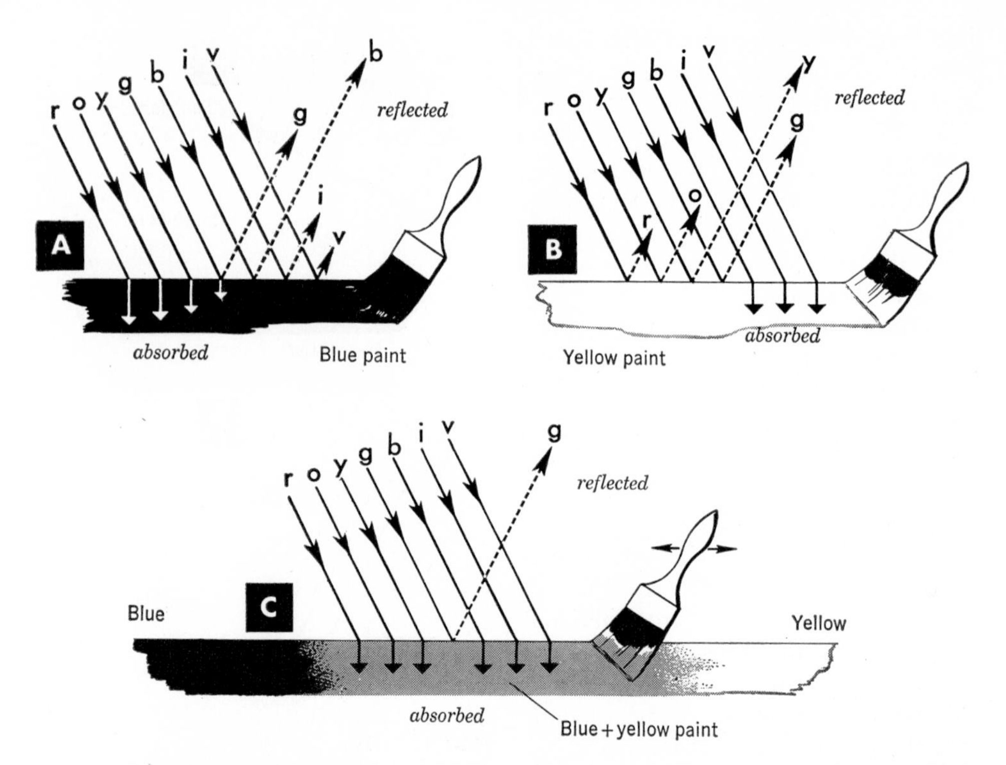

42-6 Why does the blending of blue and yellow paint produce green as the brush in C mixes the blue paint with the yellow? The answer is based on the fact that green is the only color that they reflect in common. All the other colors are absorbed by one paint or the other.

To our surprise, the object appears red. Amber objects reflect both yellow and red light but not blue light. Since only the red light in the purple is reflected, the amber object now appears red.

8. Color illusions

Many common errors in color judgment occur in this way. You may purchase a black suit in a store at night. But when you look at the suit at home during the day you find that it is really dark blue. The store where you bought the suit happened to be illuminated by tungsten bulbs, whose light is deficient in blue. Because there was little blue light in the store, the blue suit couldn't reflect any. At the same time, because the suit was dark blue, it absorbed all colors other than blue. Hence the suit, as you saw it in the store, reflected little light of any kind and appeared black. But in ordinary daylight, which contains plenty of blue light, the suit reflects some of the blue light and a blue color is seen. If white fluorescent lamps had been used in the store, this mistake would not have happened, since they give off a large amount of blue light. Thus we see that:

The color of an object depends on the color of the light that is illuminating it as well as the color which it is able to reflect.

9. Mixing paints

You may wonder how this process of mixing beams of light so as to produce a certain color relates to painting, in which colors are produced by mixing pigments. The two processes are alike in one respect. Both the painter at his easel and the electrician in charge of lighting a stage achieve certain colors by mixing other colors.

But there are important differences between the two processes. Suppose we mixed an amber (yellowish-white) spot of light with a blue spot of light on a stage. We would get white. But if we mix the same colored paints we get a dark green paint! Why? A blue paint is rarely pure blue and probably reflects a certain amount of green in addition to the blue (Fig. 42-6A). A yellow paint reflects a certain amount of green in addition to the yellow and red (Fig. 42-6B). When these two paints are mixed, the yellow paint tends to absorb the blue light which strikes it and the blue paint tends to absorb the yellow and red (Fig. 42-6C). Since *both* paints reflect some green, this color is the only one reflected by the mixture. Therefore, a green color results from mixing the blue and yellow paints. This fact is pictured in Plate VIII.

It is usually difficult to predict the exact color that will be produced by mixing paints. Painstaking measurements with special equipment have to be made to identify each of the different colors which the paints reflect. It is usually easier to try it out by actually mixing them.

How we see color

Though we are still not sure how our eyes detect color, there is good reason to believe that the explanation lies in the action of several different kinds of nerve cells in the retina.

10. Detecting colors

Let us suppose that one particular kind of nerve cell in the retina responds mainly to red light, and to a lesser extent to orange and yellow, which are near red in the spectrum (Fig. 42-7). Let us call this kind of nerve cell R. Suppose that there is another type of nerve cell, G, which responds mainly to green but also to some extent to yellow and blue, which are near green in the spectrum. Finally, let us suppose there is a third type of cell, B, which responds chiefly to the blue, indigo, and violet end of the spectrum, but also somewhat to green, which is near blue in the spectrum.

When white light, which contains all the colors in the spectrum, hits these cells in the retina of the eye, R, G, and B cells will all be affected. This gives us the sensation of all the colors together—which we know as white.

An interesting experiment which provides evidence for this theory of color vision can be performed using Plate VII.

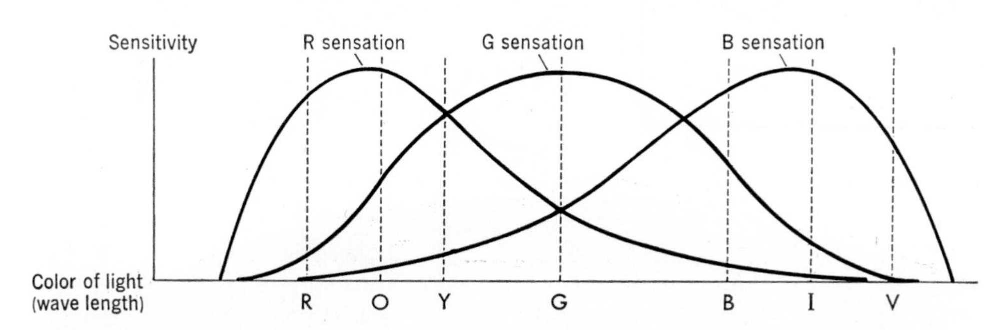

42-7 According to one modern theory of color vision, each color-sensitive cell in the retina is sensitive mainly to one of the three colors, red, green, and blue. The graph lines illustrate the sensitivity of a typical cell to each of the three colors. Can you use this theory to explain the illusion in which a red and a green light combine to give the appearance of yellow?

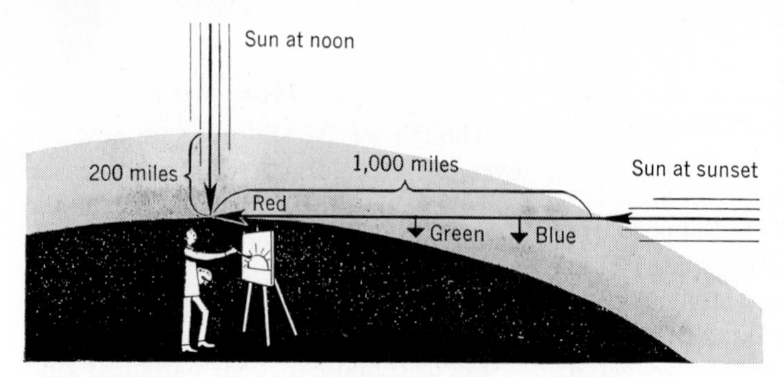

42-8 Why is the sky blue? Why is the sun red at sunset? The answer is based on the fact that red light penetrates the air better than blue light. The red rays from the sunset get through the greater distance of air, but the blue rays scatter.

Place the color page in the text under a strong light. Stare at the dot in the center of the cow for about half a minute without blinking. Then stare at the dark spot on the white surface below until you see an after image of the cow and the magenta background in different colors. Can you explain why this occurs? (HINT. It has to do with nerve fatigue.)

A small percentage of people suffer from **color blindness**; that is, they cannot distinguish between certain colors. A color-blind person, for instance, might not be able to distinguish between the red and green of a traffic light.

A person whom we call color-blind simply sees colors differently from the rest of us. This happens if the R, G, and B cells on the retina of his eye react differently from the way ours do. In certain cases one or more of these different types of cells may not be operating at all.

11. Color in nature

When the sun is low on the horizon, at sundown or sunrise, it appears yellowish or even red. But at midday, when it is directly overhead on a cloudless day, it appears blazing white. This difference in color is due to the fact that the red rays from the sun penetrate the dust and water vapor in the air much more easily than do the blue rays. As shown in Fig. 42-8, the light from the sun must pass through a great many more miles of the

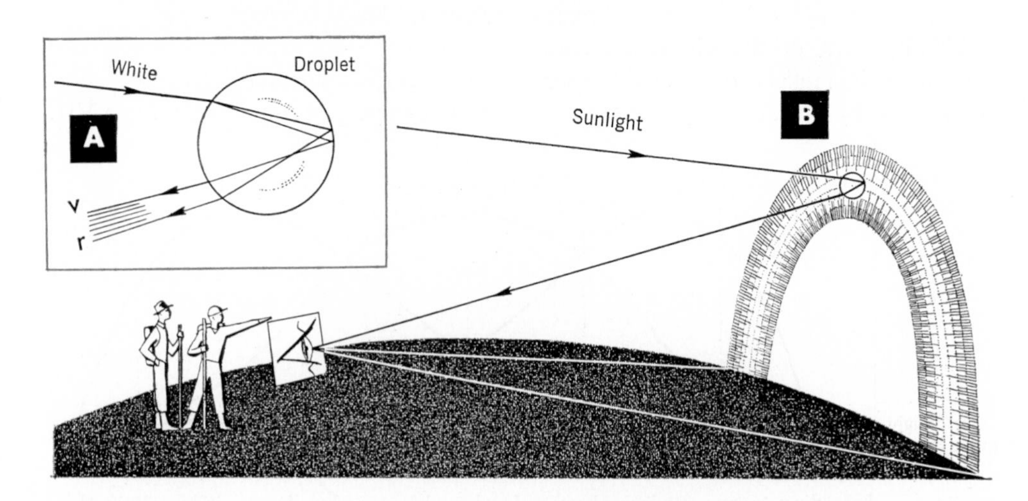

42-9 There is no gold in this rainbow, just the spectrum colors red, orange, yellow, green, blue, indigo, and violet. Rainbows are caused by rain droplets acting as prisms to disperse the white light from the sun. The combined effect of colors from millions of droplets can be seen only at certain angles, producing the rainbow that we see.

Plates I and II. FILTERS AND THE SPECTRUM. In Plate I (right) a prism spreads out into a spectrum, the many colors composing white light. This effect accounts for rainbows and colored sunsets.

In Plate II white light from a projector (below, at left) is split up (as well as reflected) by a prism. But the beam of colored light traveling toward the screen is interrupted by a piece of colored glass next to the prism. At the top the red glass transmits only red light, which appears on the screen in the same place it would in a complete spectrum. In the middle a green filter is seen to transmit only green light, presumably absorbing all other colors. At the bottom the blue filter is seen transmitting only blue light and absorbing all the other colors in the beam.

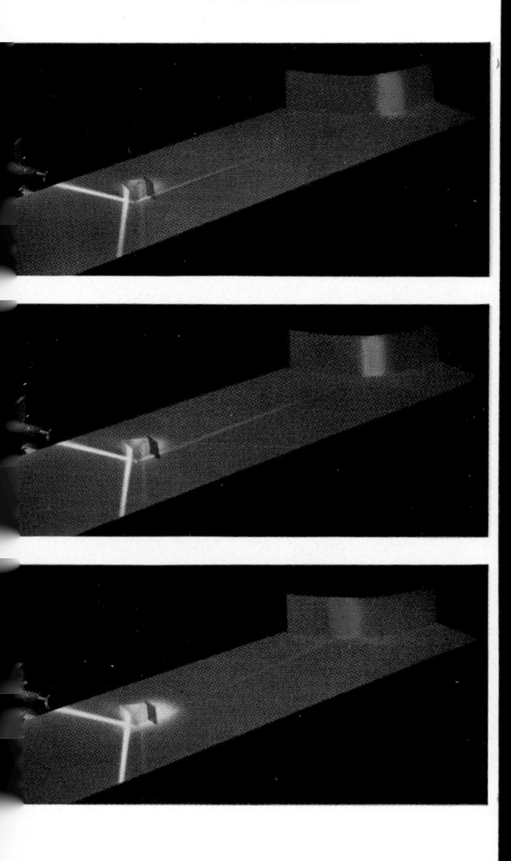

Above, reproduced with permission from the Kodak Data Book, *Color as Seen and Photographed;* at right, Herbert E. Goldberg from *Life* Magazine

Plate III. THREE TYPES OF SPECTRA. All incandescent solids emit a continuous spectrum as in 1, above.

Hot gases (which include vapors from solid material) emit bright-line spectra—a second type—as shown in 2, sodium; 3, mercury; and 4, hydrogen.

If an incandescent solid, emitting a continuous spectrum, is overlain by cooler gases, the cooler gases absorb their characteristic lines from the continuous spectrum. These "absorption lines" appear black, as in the absorption spectrum of the sun in 5. This is a third type of spectrum.

Above, adapted from S. R. Williams, *Foundations of College Physics*, Ginn and Co.; below, reproduced with permission from the Kodak Data Book, *Color as Seen and Photographed*

Plate IV. COLOR ADDITION. Below, by adding red, blue, and green in the proper proportions, all the colors of the visible spectrum can be matched. Red, green, and blue are known as additive primary colors. Notice that the mixture of any two of them, when added to the third primary color, produces white light.

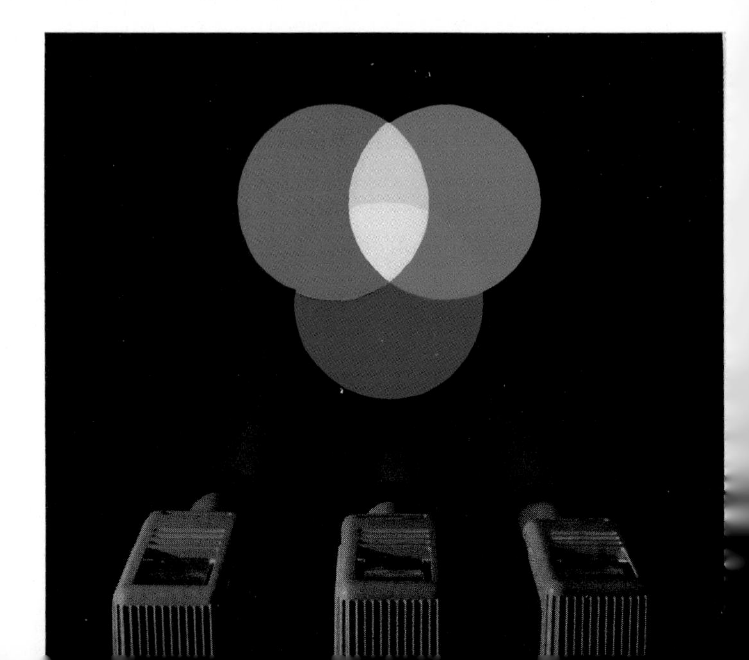

Plate V. REFLECTION OF COLOR. The color of an object in ordinary white light is composed of all the wave lengths the object reflects. Top right, a lemon is seen against a precise color-measuring instrument. The colors it reflects are charted at the left. Notice that it reflects all colors well except blue. By consulting the color-mixing pattern on the opposite page you will see that yellow light plus blue light produces white. Hence white light from which the lemon has absorbed the blue must appear yellow. Why does the tomato in the second picture appear red?

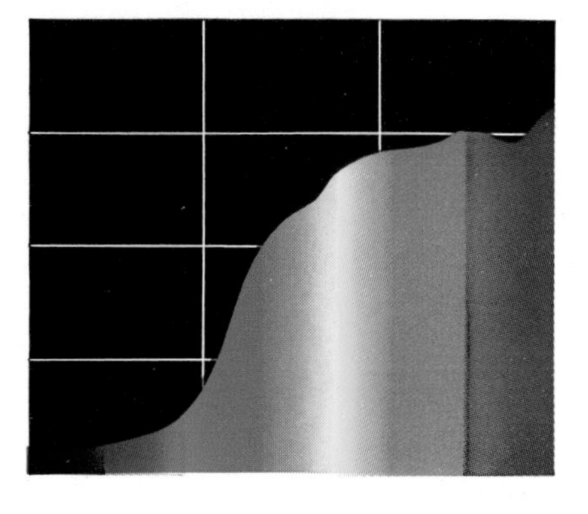

Dmitri Kessel, © Time, Inc.

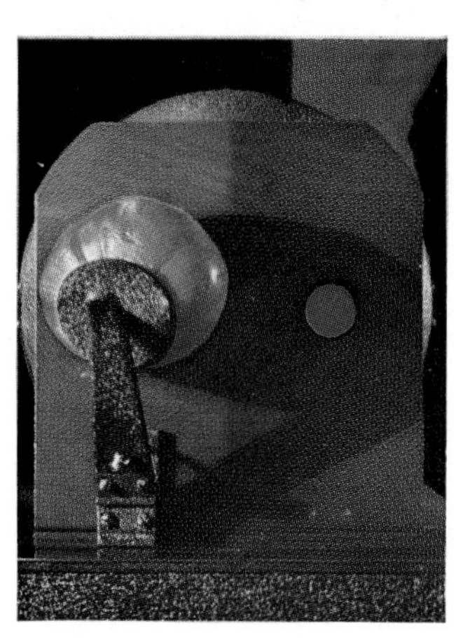

Bausch & Lomb Optical Co.

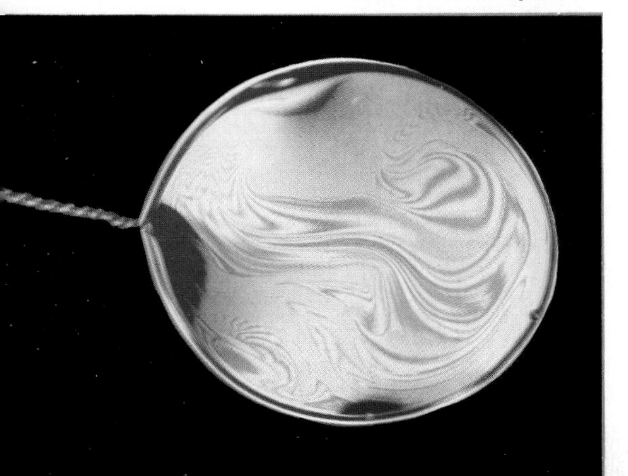

Plate VI. INTERFERENCE. Like oil films on water, this soap film produces its rainbow colors by interference. White light, consisting of all visible wave lengths, falls upon the film from above and is reflected from both the upper surface and the lower surface of the film. The two reflected beams returning toward your eye coincide, and each wave length (color) undergoes interference *except* those for which the two reflected beams are in phase. The surviving wave lengths determine the color of the film at each point.

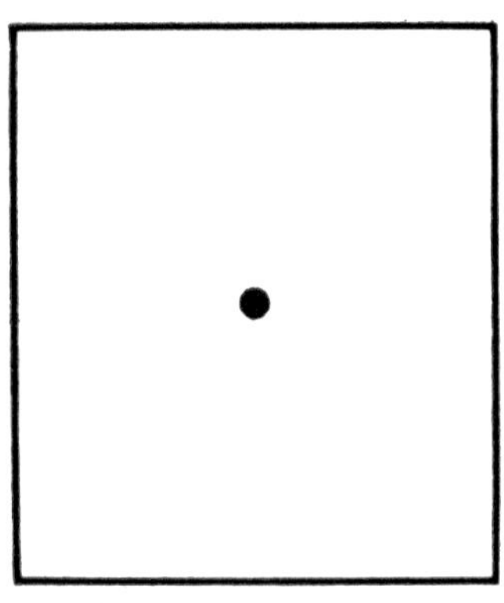

Plate VII. PERSISTENCE OF VISION. Stare at the spot on the cow for 30 seconds without moving your eyes. Then look at the spot on the blank space below. You will see a purple cow in a green field!

You know that white results when purple is mixed with green (Plate V). Hence purple stimulates all your color receptors except those sensitive to green. Your receptors of purple grow tired as you stare at the grass, but your receptors of green do not. When you glance at the white area *all* receptors are fatigued except the receptors of green. Thus the grass looks green.

Why does the cow look purple?

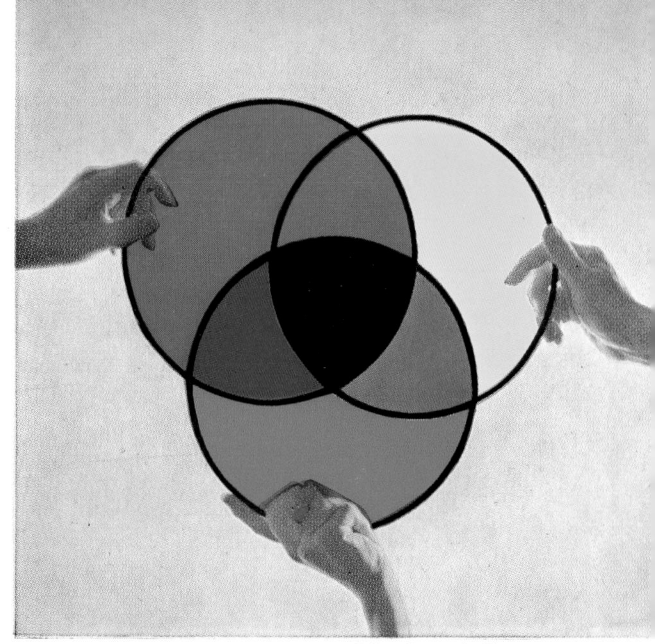

Fritz Goro, © Time, Inc.

Plate VIII. COLOR SUBTRACTION. Yellow filters, like yellow paint, subtract blue from white light, leaving red and green. Blue-green filters, like blue-green paint, subtract red from white light, leaving blue and green. Where both overlap, all colors but green are subtracted. What can you say about the colors subtracted by magenta paint and filters?

Edmund Bert Gerard from *Life* Magazine

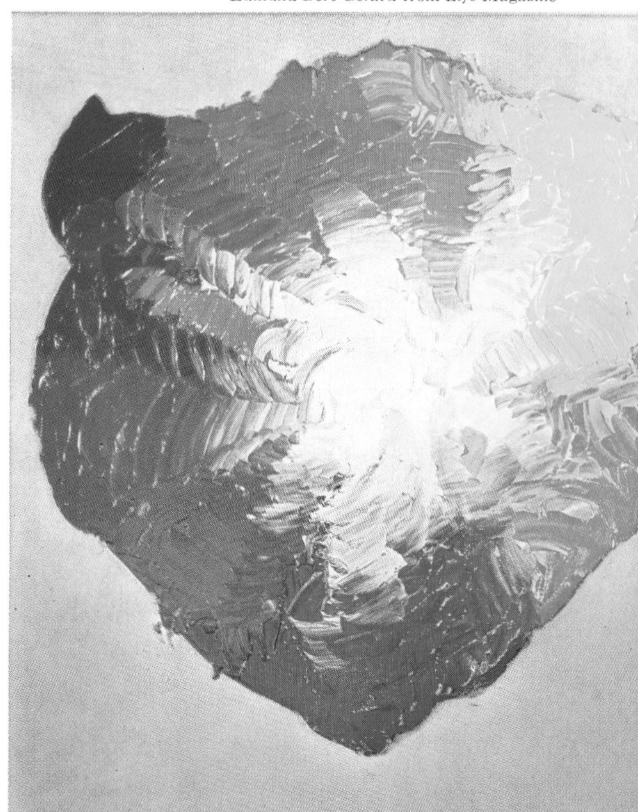

earth's atmosphere at sunrise or sunset than at noon when the sun is overhead. The longer red rays pass through this thick blanket of air rather easily, but many of the shorter blue rays are scattered and do not get through. Thus the sun appears red and the scattered blue rays make the sky appear blue.

The rainbow is another interesting occurrence which can be explained by our knowledge of color. You may see a rainbow when the sun is behind you and there are water droplets in the air in front of you. When light from the sun enters the droplets (A in Fig. 42-9), it is refracted and dispersed into colors just as in a prism. Reflection inside the droplet sends the light back to you. Although the air may be full of water droplets, you get the rainbow spectrum only from those droplets which form an angle of about 42° with you and the sun (Fig. 42-9B). When you change your position, new droplets come into the proper position, so that the angle is maintained and the rainbow appears always in the same direction from you. Figure 42-9B shows how the light from the sun is dispersed by water droplets to form a rainbow.

There is much more to the subject of color than we have had room for in this chapter. In the next chapter we shall learn that there are some "colors" which are actually invisible, and that these are of great importance to modern civilization.

Going Ahead

THE VOCABULARY OF PHYSICS

Define each of the following:

spectrum	filters
wave length	selective reflection
frequency	black
color subtraction	dispersion
color blindness	

THE PRINCIPLES WE USE

1. Compare red light and violet light as to wave length and frequency.
2. Draw a diagram showing how a ray of white light disperses as it passes through a triangular glass prism. Label the colors produced.

Explain why

3. A spectrum is formed when a ray of white light passes through a glass prism.
4. White objects viewed through a piece of red glass appear red.
5. Some objects appear red when viewed through a purple glass, others appear blue, and still others appear purple.
6. A red object appears red in white light but black when green light shines on it.
7. A yellow object appears red in red light and green in green light.
8. The same yellow object appears black when viewed in blue light.
9. Clothing that appears dark blue in daylight appears black in a house illuminated at night by tungsten bulbs.
10. Blue light and yellow light combine to produce white, but blue paint and yellow paint form green paint when mixed.

APPLYING PRINCIPLES OF COLOR

1. A stage is equipped with red, blue, green, and amber (yellowish) light. How would you make each of the following colors: white, blue-green, purple, pink, white with a greenish tint?
2. The owner of a clothing store was strongly advised by an illumination engineer to replace his incandescent lamps with fluorescent lamps. Explain why.
3. It is said that seeing is believing. Do you think this statement is true? Give two examples from your study of color to back up your opinion. Give an example from your own experience.
4. A red light shines on an object and it appears red. Which of the following colors might appear if the object is illuminated with white light: white, green, red, yellow, blue, purple? Explain your answers.

5. What color might you expect to see if you looked through each of the following combinations of colored cellophane wrappers or colored glass plates (one placed over the other): red and blue, red and purple, yellow and blue? Try them out. Try other combinations.
6. What color might you get (a) if purple, blue-green, and yellow lights were used to illuminate a screen at one time, (b) if purple and yellow were used, (c) if yellow and blue-green were used?

PROJECTS

1. Why the Sky Is Blue and Sunsets Pink. Into a glassful of water stir a few drops of milk. Darken the room, and direct a flashlight beam through the milky water towards your eyes.

   When you look directly into the beam through the milk, the beam will look yellow. This is because the particles of milk have scattered the shortest wave lengths (green, blue, and violet) in all directions, removing these colors from the direct beam, precisely as the great depth of the atmosphere does to the sunlight at sunset (Fig. 42-8). The light that gets through consists mostly of the longer wave lengths—yellow here, pink in a true sunset.

   The scattered blue light is easily seen by looking into the side of the glass at right angles to the beam. Watch the deepening of the colors as you add more and more milk to the water.
2. Color by Reflection. No object, however brightly colored, can reflect colors not present in the available light. An impressive home demonstration of this is done in a darkened room with a brightly colored object such as a color photograph. As a light source use a few pieces of borax (sodium borate) in an upturned jar cover into which some rubbing alcohol has been poured.

   Ignite the alcohol, and as the borax is heated it colors the flame bright yellow due to the sodium it contains. In the yellow sodium light you cannot recognize any colors but yellow. Examine colored objects such as colored illustrations and neckties, trying to judge their true colors. Then prepare to be surprised as you switch on the light!

   But does even an incandescent lamp really reveal the "true" colors?

FURTHER READING: MORE ABOUT COLOR

1. Look up the article on color in *Life* magazine of July 3, 1944.
2. All kinds of interesting effects in nature are described in M. Minnaert's *Light and Colour in the Open Air,* Dover Publications, Inc., New York, 1954. 362 pp.

CHAPTER

43

Enlarging the Spectrum

As you read these words you are being hit from all sides by electromagnetic waves you cannot see. Similar to light rays, they are striking you from all nearby radio stations, from all warm objects in your vicinity, from radioactive rock in the earth beneath you, and even from outer space. Indeed, you are giving out invisible rays yourself! Through instruments that can "see" these rays we observe a world very different from our familiar one. In this chapter we shall learn that these rays can give us a great deal of information about the world around us and can be put to many practical uses.

The spectrum

One of the most important instruments for studying light is the **spectroscope.** Let us begin by seeing how it breaks up light into a band of colors called a **spectrum** (plural: **spectra**).

1. Different kinds of spectra

The main part of the spectroscope is a glass prism (P in Fig. 43-1). This is arranged with lenses so that the color spectrum produced by the prism can be examined with an eyepiece or photographed on a film (at F in Fig. 43-1).

As shown in Plate III-1, when white light from an incandescent *solid,* such as the glowing tungsten filament of an electric light bulb, passes through a prism, it gives a **continuous spectrum** which shades gradually from one color into the next. There are no boundary lines between the colors.

If we use a neon lamp as a light source, the spectroscope reveals a spectrum in which the colors show up as bright lines separated by large dark spaces. The brightest of the bright color lines are in the red part of the spectrum, which is why red is the predominant color of neon lights. A spectrum of this type is called a **bright-line spectrum** and is characteristic of incandescent *gases.*

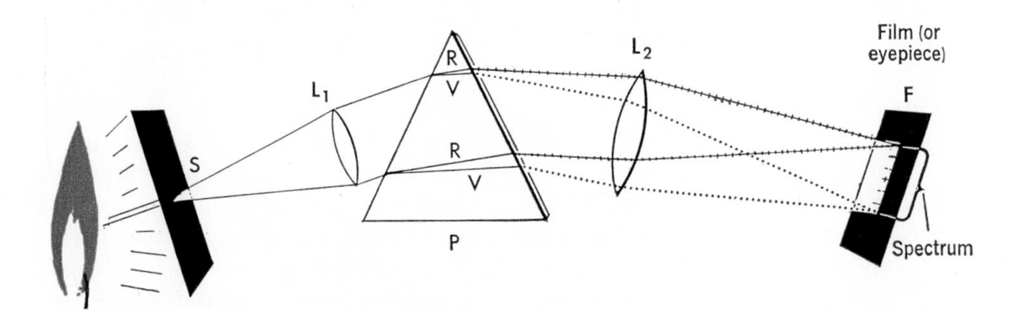

43-1 The spectroscope, basic tool of both the modern criminal investigator and the scientist, is often simply a lens and prism system which makes a bright, clear spectrum of the light coming through a slit.

A mercury-vapor lamp produces a bright-line spectrum in which the blue and green lines are strongest, so the predominant color in this lamp is blue-green. But its spectrum also contains lines of yellow and violet (Plate III-3). Light from a sodium-vapor lamp has just two very bright lines very close together in the yellow region of the spectrum (Plate III-2). When sodium or mercury compounds are placed in a hot flame or in an electric arc, their characteristic color lines will appear in their spectra.

2. Explaining spectrum lines

We saw in Chapter 27 how atoms of the elements differ in the number of protons inside the nucleus and in the number of electrons revolving outside it. Research during the early years of this century has shown that the light of the spectrum lines does not originate in the nucleus of the atom but is produced by events taking place in the surrounding electron layers.

It appears that the solitary electron of a hydrogen atom can revolve only in an orbit whose radius, r, is equal to $0.53 \times 10^{-8}N^2$ centimeter, where N is a whole number (1, 2, 3, 4, etc.). Thus, when $N = 1$, the electron is at a distance of 0.53×10^{-8} centimeter from the nucleus. When $N = 2$, the electron revolves at a distance of $0.53 \times 10^{-8}N^2 = 0.53 \times 10^{-8} \times 2^2 = 2.12 \times 10^{-8}$ centimeters. When $N = 3$, it revolves at $0.53 \times 10^{-8} \times 3^2 = 47.7 \times 10^{-8}$ centimeters, etc. It appears in fact that the electron can revolve only in certain "shells," rather as if its possible orbits had been drawn on the surfaces of the layers of an onion. While more recent work has shown this to be a rather simplified picture, it has also confirmed beyond any reasonable doubt that *only certain orbits are possible*. No electron, for example, can revolve in a circular orbit 0.10×10^{-8} centimeter or 0.70×10^{-8} centimeter from the nucleus.

A similar situation applies to the electrons of heavier atoms. Only orbits at certain distances or levels are possible for each electron.

If you have studied chemistry, you probably have learned that the electrons in the outermost layer, called the valence electrons, are responsible for most of the chemical properties of the atom. These outermost electrons are also responsible for the emission of the visible bright-line spectrum.

To make atoms emit a bright-line spectrum, they must be given energy by strongly heating them or accelerating them in a strong electric field. An atom stores this absorbed energy in a curious way—one of its outer electrons is lifted to a higher-energy orbit *farther out from the nucleus.*

You can see that energy is required to lift an electron into such a "higher" orbit if you have read the description in Chapter 11 of how earth satellites must be given energy to raise them into stable orbits above the earth. In the case of an earth satellite, work is done against the force of gravity and is stored as gravitational potential energy until the satellite falls back to earth. When an electron is boosted to a higher orbit, work is being done against the electrical force of attraction of the positively charged nucleus and is stored as electrical potential energy until the electron falls back "down" to its original orbit.

When the electron falls back to its original stable orbit a few millionths of a second later, it gives up its energy in the form of light. The energy of the emitted light is precisely the energy lost by the falling electron. The energy of the light emitted determines its color. Thus each colored line in a bright-line spectrum corresponds to the loss of energy of an electron in its fall from a higher orbit to a lower orbit.

We have here a most powerful tool for studying the electron structure of atoms. By working backward from the color (and hence the energy) of each line in a bright-line spectrum, it has been found possible to arrive at a picture of the outer electron structure of the atoms of all the elements. By extending the technique to the study of spectrum lines in the invisible X-ray region it has been found possible to work out the location of *all* the electrons, even in atoms as heavy as uranium and lead.

Since the electron orbit structure differs for atoms of each different element, we would expect to find a different bright-line spectrum for each.

3. Fingerprinting the elements

Not only neon, mercury, and sodium, but every other element heated to incandescence in a gaseous state has spectrum lines which identify it. It is therefore possible to take "fingerprints" of all the elements. Suppose we want to find out whether or not sodium is present in a certain material. We place the material in an electric arc, take a photograph of the bright-line spectrum produced, and then compare the photograph with the known spectrum of sodium. If the sodium lines appear in this spectrum, we can be sure that sodium is present. Estimates of the amount of sodium present can be made from the intensity of the lines. This kind of spectrum analysis is often used by scientists when they wish to identify the elements in a material.

When the sun's light is analyzed, the spectrum produced is found to differ from the spectrum obtained from an ordinary lamp. In the spectrum of sunlight numerous dark lines appear against a continuous spectrum (Plate III-5). These dark lines are seen because light having these colors is absorbed as it passes from the enormously hot interior of the sun through the cooler material outside. The cooler vapors of various elements near the surface of the sun absorb light having the same colors they would normally emit. These dark lines are called **Fraunhofer lines,** and the spectrum formed is called an **absorption spectrum.** Absorption spectra can be used to identify the materials that compose the outer atmosphere of the sun as well as that of other stars. When we see dark lines in the yellow region of the sun's spectrum, in exactly the place where the sodium lines occur in a bright-line spectrum, we can be sure that the sun contains sodium.

There are certain strong lines in the sun's spectrum that at one time could not be identified with any of the elements known on earth. Scientists had a set of "fingerprints" but nothing to attach them to. It was not until 1895 that the element responsible for these mysterious lines was identified on earth. It was found as a gas associated with the radioactive element, uranium. This element scientists named "helium," from the Greek word for the sun, "helios."

4. Wonders of the spectroscope

By means of the spectroscope, the astronomer can analyze the light from individual stars. He has found it possible to classify the stars according to spectrum type. From the spectrum of a star he can make important deductions about what is taking place on the surface of the star, what elements it is made of, how hot, how bright, how far away it is, and even what its atmospheric pressure is.

By spectrum analysis of light the astronomer can even tell whether a star is moving toward the earth or away from it. When a star is moving toward the earth, the light waves reaching us from that star are "jammed together" by its approaching motion (side A of Fig. 43-2). Thus the wave length of light from an approaching star is less than it would be if the star were standing still, or moving away. Now we know that wave length determines the color of light, and that wave length is shortest at the violet end of the spectrum. So if a star is approaching us, the lines in its spectrum shift toward violet. If its motion is away from the earth, the spectrum lines shift toward the longer-wave-length red light (side B of Fig. 43-2). The amount of shift will indicate the speed at which the star is traveling. This change in wave length (color) which results from the source of a wave moving either toward us or away from us is known as the **Doppler effect.**

The same thing happens with sound waves. When an automobile is approaching us, the pitch (frequency) of sound produced by its horn sounds higher to us than when the car is moving away. The sound waves are jammed closer together. When the automobile is traveling away from us, the sound waves from the horn are spread out, their wave length is greater, and the pitch of the sound appears to be lower to the stationary observer, although the driver can detect no such change. Hence, as the car passes us and the motion suddenly changes from approaching to receding, the pitch of the note we hear from the horn drops off quite sharply.

5. Detecting double stars

A spectroscope can reveal the existence of double stars revolving around each other, even though the telescope sees only one star in this spot. The revolution of the two stars about each other causes one star to recede while the other is approaching. As a result the lines of the spectrum split in two and then come together again. The time of revolution can be measured by the time it takes the spectrum lines to separate and come together. The speed of motion of each star can also be measured by the amount of separation of the spectrum lines. Once we have these measurements of time and velocity, we can calculate the distance between the two stars. Finally, by using Newton's Laws of Motion, astronomers can actually estimate the mass of each star!

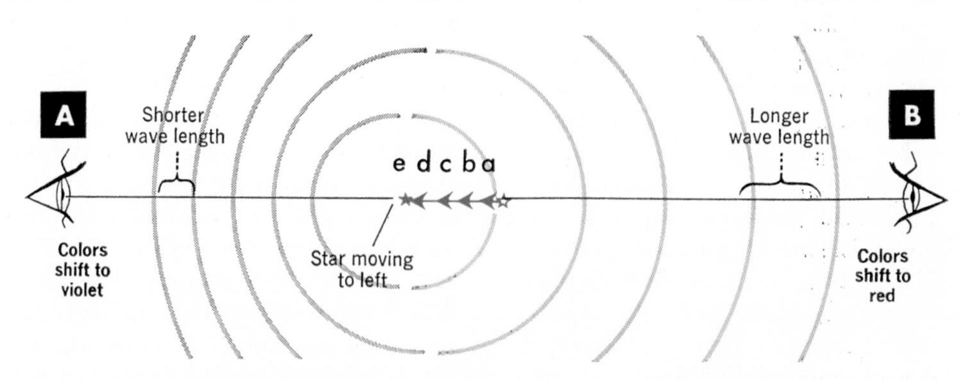

43-2 A source of light (such as a distant star at e) will produce jammed-up waves (shorter wave length) if it approaches the observer (at A) and stretched-out waves (longer wave length) if it travels away from the observer (at B). This effect, known as the Doppler effect, shows up as a shift of the spectrum toward the blue end if the star is approaching (A) and a shift to the red end if the star is receding (B).

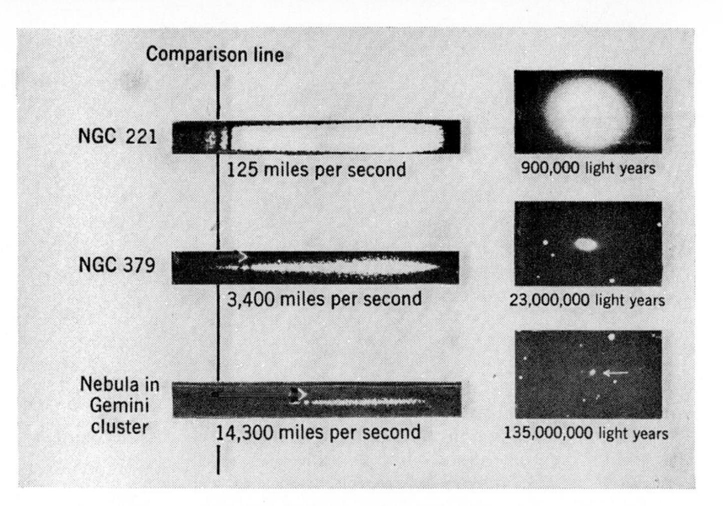

43-3 Theories of the origin of the universe are based on this spectrum evidence. The unusual shift toward the red of spectrum lines from distant galaxies indicates enormous speed away from us. Perhaps these distant speeding star groups all came—along with our sun—from some terrific cosmic explosion billions of years ago.

6. Galaxies and the Doppler effect

Very distant groups of stars, called galaxies, all show a marked shift of their spectrum lines toward the red (Fig. 43-3). This seems to indicate that they are traveling away from the earth at enormous speed. This observation presents modern science with one of its greatest unsolved mysteries. If the "red shift" is due to the Doppler effect, it means that the universe is expanding, perhaps as the result of a cosmic explosion 6 to 13 billion years ago.

Thus the spectroscope has proved useful not only in exploring the tiny, invisible world of atoms, and molecules, electrons and protons; it has been equally valuable in exploring the vast world of outer space.

Extending the spectrum

So far we have considered only those colors in the spectrum which are visible to the eye. But the spectrum actually goes beyond this visible range. Invisible electromagnetic waves play important roles in our world.

7. Radio waves

In Chapter 36 the discovery of radio waves by Maxwell and Hertz was shown to have come from a study of the question "What is light?" Radio waves are the longest waves (lowest frequency) in our electromagnetic spectrum.

Stars emit radiation not only in the part of the spectrum ranging from the infrared to cosmic rays, but also in the radio-wave range. Astronomers are investigating these radio waves with receivers equipped with special antennas. This equipment is called a **radio telescope** (Fig. 36-32).

8. Infrared rays

The discovery of infrared rays came about as a result of a series of experiments begun by Sir William Herschel in 1800. Herschel wanted to find out if different colors in the spectrum produced different amounts of heat. By placing a sensitive thermometer at different places in a spectrum, he found that the colors did produce heat, with the red region producing the most. But when the thermometer was moved into the dark region just beyond red in the spectrum (IR in Fig. 43-4) it registered a still greater amount of heat. Apparently rays of some kind were present in this dark region, although they were invisible to the eye. Herschel had discovered what came to be known as **infrared rays.**

Modern science has learned that infrared rays, whose wave lengths are

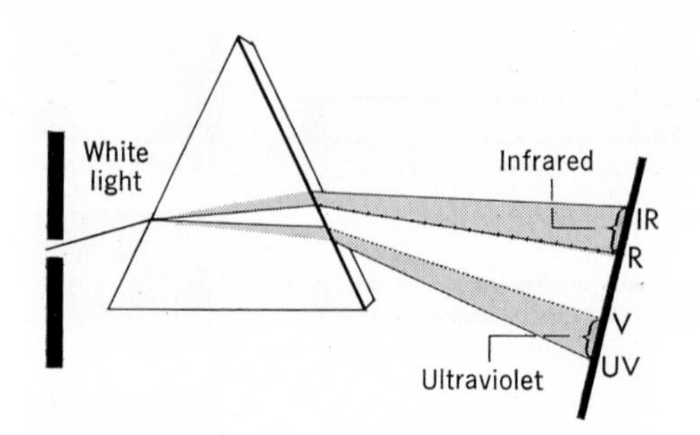

43-4 Invisible kinds of light were first noted from heating effects observed in the dark infrared, IR, region beyond the red. Later, similar heating effects, but less intense, were noted in the ultraviolet region, UV, beyond the violet. The amazing thing is that the light waves we see are but a tiny fraction of those that exist all around us!

longer than those of ordinary red, are the main heating rays of the sun. If they were not present in the sun's radiation, the earth would freeze over. These infrared rays are also given out by heating lamps, toasters, broilers, and other heating devices (Fig. 43-5).

Recently, infrared rays have had an important use in long distance photography. On page 580 it was mentioned that the long red rays from the sun penetrate dust and haze better than the shorter violet and blue waves. But the even longer infrared rays are still more effective in penetrating dust-laden air. A special infrared film sensitive to these invisible rays can take a picture of mountains a hundred miles away in spite of the intervening haze of the atmosphere. Invisible infrared rays can be used to illuminate a scene which the eye sees as completely dark, and a camera containing infrared-sensitive film can take a picture of this scene (Fig. 43-6).

9. Heating effects of different waves

The properties of electromagnetic waves gradually change with the frequency. One kind merges into the next. For example, infrared waves have good heating properties and will not penetrate solids easily. The neighboring radio waves have little heating effect on most solids but good penetrating effects. If a doctor wishes to give an internal heating treatment (**diathermy**), he can find waves of a frequency in the boundary region where radio and infrared overlap. Here there are waves which will penetrate the body and cause internal heating.

10. Ultraviolet rays

Invisible ultraviolet light is also important. **Ultraviolet rays** are the invisible rays beyond the other end of the visible spectrum, next to the violet, which also cause heat. It is the ultraviolet rays in sunlight that cause sunburn. They also stimulate the production of Vitamin D in the skin. Since this vitamin is necessary for the proper growth of bones, it is important that young children receive plenty of sunlight.

Westinghouse Electric Corp.

43-5 This is a simple model of a modern infrared broiler. The steak suspended above the frying pan is broiled by invisible infrared rays!

Eastman Kodak Co.

43-6 **Believe it or not, pictures can be taken in absolute darkness. Here is one. These people are handling film which would be ruined by light, and so must work in the dark. But an invisible infrared light and special film took this picture in the dark.**

Certain frequencies of ultraviolet light have a destructive effect on living tissue. Overexposure to the sun or to a sun lamp is dangerous. In fact, hospitals use lamps producing ultraviolet rays to kill bacteria (Fig. 43-7). These **germicidal** (jûr·mí·sīd'ăl) **lamps** are now also used in homes and public places to kill bacteria in the air. Special ultraviolet sterilizers are similarly employed to kill the bacteria in food while it is being processed.

11. Fluorescence

Ultraviolet light (sometimes referred to as black light) can cause certain materials to **fluoresce.** When the ultraviolet rays strike the fluorescent material (called a **phosphor**), they are absorbed by the individual atoms. These atoms now have additional energy. They may emit some of this energy in the form of visible light waves with longer wave length and lower energy than the original ultraviolet waves.

Modern fluorescent lamps make use of the same principle. For example, the light from mercury vapor is rich in invisible ultraviolet rays. These invisible rays do not produce any useful illumination. But when the inside of the lamp tube containing mercury vapor is coated with the proper fluorescent materials, some of the energy of the ultraviolet light is converted into visible light. We have seen (Chapter 41) that such fluorescent lamps are much more efficient than incandescent lamps.

Fingerprints often show up more clearly when sprinkled with fluorescent powder and illuminated with ultraviolet light. Two samples of paint or two samples of ink of the very same color in ordinary light may fluoresce in very different colors when they are illuminated with ultraviolet light if their chemical compositions are not identical. Hence ultraviolet light is very useful for checking suspected forgeries of stamps, old paintings, or money, for no matter how good a copy the forgery may appear to be under ordinary light it will often look very different indeed when it is placed under ultraviolet.

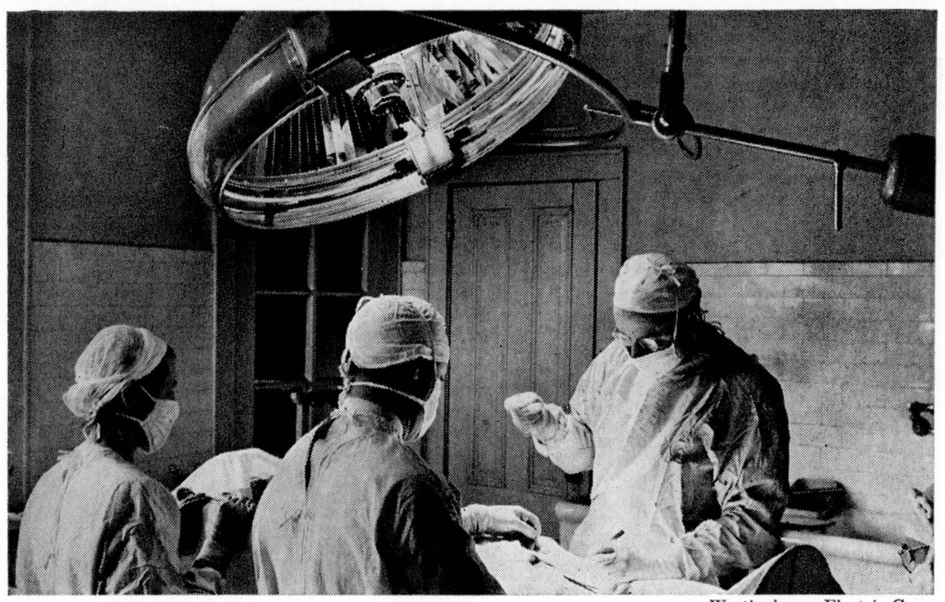

Westinghouse Electric Corp.

43-7 Ultraviolet light helps to kill unwanted germs in the air during an operation. The death-to-germs ultraviolet light comes from the incandescent mercury vapor in the thin quartz tube around the inside of the lower rim of the overhead lamp. Ordinary glass is opaque to ultraviolet.

12. The discovery of the X ray

While radio waves were first being investigated, new invisible short-wave-length rays were discovered at the other end of the extended spectrum, beyond the ultraviolet.

Westinghouse Electric Corp.

43-8 One cannot see these big laundry marks when the blanket comes back, yet they will still be there. They are visible only under ultraviolet lamps. Why are laundries using such ultraviolet marking systems? How could a detective make use of these markings?

The German scientist, Wilhelm Roentgen (rûnt′gĕn), had been experimenting with electric discharges in evacuated tubes. You remember that in Chapter 27 we described the cathode rays that are formed when electrons travel from the negative to the positive electrode in a tube containing a vacuum. While Roentgen was operating a cathode-ray tube, he noticed that some minerals lying near the tube developed a peculiar glow (they fluoresced). When he put a piece of cardboard between the tube and the minerals, the glow still continued. Apparently the radiation from the tube that was causing the minerals to glow could pass through solid cardboard. These penetrating rays that Roentgen had discovered by chance came to be called **X rays.**

13. The use of X rays

It was found that X rays were emitted from the cathode-ray tube when the voltage was great enough to cause electrons from the negative terminal (C in Fig. 43-9) to strike the positive terminal, T,

43-9 X rays are produced when electrons rush across the space of a vacuum tube and crash into a metal target. The higher the voltage, the higher the frequency of the X rays and the greater their penetrating power.

at sufficiently high speeds (measured in thousands of miles per second). The higher the voltage and the faster the speed of the electrons, the higher the energy and frequency and penetrating power of the X rays produced. The X-ray machines used by doctors and dentists require about 65,000 volts to obtain a shadow picture of teeth or bones.

X rays have become an important part of modern medicine (Fig. 43-10). They enable a surgeon to photograph bone fractures, or a lung specialist to detect tuberculosis. And with some diseases, such as cancer, X rays are used to kill malignant cells and thus play an important part in stopping the spread of the disease—sometimes curing it altogether.

In industry, X rays are used to find hidden flaws in rails or machine parts, or to analyze materials. And by bombarding various materials with X rays, pictures are obtained which reveal patterns that give new insight into the arrangement of molecules, increasing our knowledge of the structure of matter.

14. Gamma rays from radium

In Chapter 45 we shall describe in detail how extremely short-wave radiation, called **gamma radiation**, emitted from radioactive materials, was discovered.

Gamma rays from radioactive materials can be used to kill some kinds of cancer cells (Fig. 43-11). But they also have a great capacity for doing harm, since gamma radiation can destroy healthy cells and seriously disrupt normal body functions. In fact, excessive exposure to such radiation can lead to

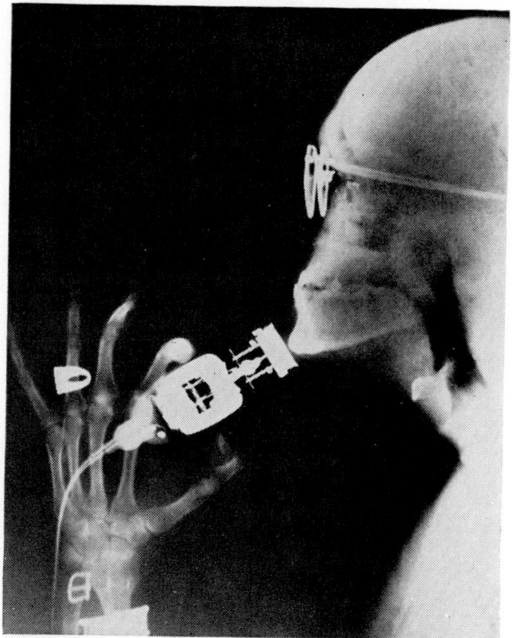

Westinghouse Electric Corp.

43-10 X rays are an important tool of modern industry and medicine because they reveal the interior structure of opaque objects.

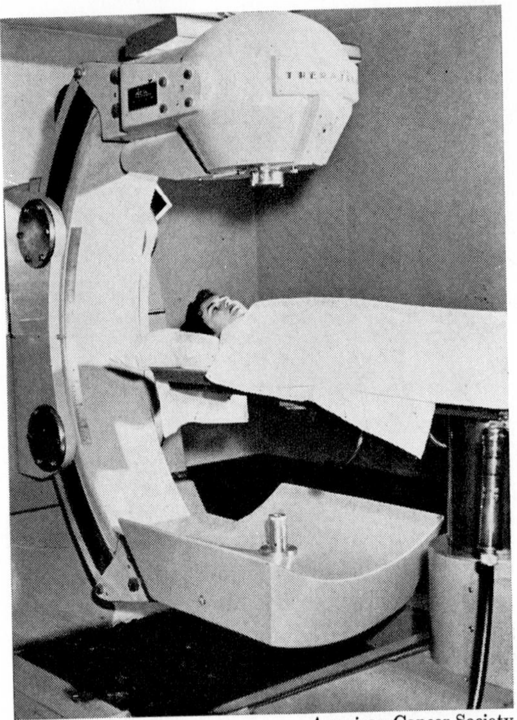

American Cancer Society

43-11 Gamma rays may save this woman's life. The gamma rays from radioactive material in this machine can destroy cancer tissue before it has a chance to spread and cause death. X rays are also used in cancer treatment.

severe illness and even death. This is one of the dangerous effects produced by the explosion of an atomic or a hydrogen bomb. The debris from an exploded bomb radiates an enormous quantity of gamma rays which can kill people a considerable distance away. A new type of warfare is now possible in which the gamma and other rays from atomic and hydrogen bombs can pollute a large area, making it unsafe for human occupation.

Modern atom-smashing machines can produce large quantities of gamma rays as well as high-speed atomic particles of high energy. Studies of the way in which these rays are produced by atoms are giving us a great deal of new information about the atom and the way in which it is held together.

15. Cosmic rays

The most energetic and penetrating rays yet discovered are the cosmic rays. The **primary** (or original) **cosmic rays** which come toward the earth from outer space seem to be mainly protons traveling at enormous speeds. The energy of these particles is measured in units called electron-volts. An **electron-volt** is the energy attained by an electron as it moves through a potential difference of 1 volt. It is equivalent to 1.60×10^{-12} ergs. (See page 396.) The primary cosmic rays (high-speed protons) have energies equivalent to many billions and even trillions of electron-volts. In other words, some of these cosmic-ray particles have energies equivalent to that of an electron subjected to a potential difference of trillions of volts!

When the enormously energetic primary cosmic-ray particles strike atoms in the air, the atoms are virtually blasted apart into their constituent protons, neutrons, and electrons. Dozens of mysterious high-speed particles called **mesons** may also shoot out of one of these smashed atoms (Fig. 43-12). Along with the particles come electromagnetic rays even more penetrating and powerful than the gamma rays from radioactivity. These sometimes have sufficient energy to penetrate several feet of lead. The particles and electromagnetic rays that result from collisions of the primary cosmic rays with atoms in the air are known as **secondary cosmic rays.**

A great deal of research is now going on in cosmic rays. Earth satellites and high-altitude balloons are being sent up carrying cloud chambers and Geiger counters (see Chapter 45) which detect cosmic-ray collisions. At high altitudes these instruments reveal many energetic primary rays that cause the shower of secondary cosmic rays down below at the surface of the earth. Down at the lower levels of the atmosphere we detect mainly the secondary cosmic rays, which rain down upon us every minute of the day and night, but not in quantities great enough to affect us noticeably.

Cosmic rays hold many mysteries which scientists are seeking to unravel. They may hold the secret of new types of atomic energy more powerful than any we have been able to release so far. It is possible that these cosmic rays, which seem to have little effect on living things, may have profound *long-range* effects, causing mutations (changes in inheritance) or perhaps even premature aging. The evidence for this latter speculation is that mice placed near powerfully radiating atomic reactors seem to age more rapidly than usual.

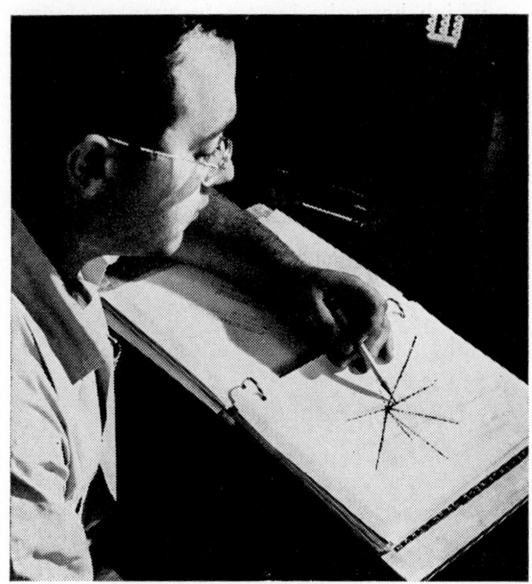

Brookhaven National Laboratory

43-12 This physicist is studying an enlargement of a photograph taken in a balloon 90,000 feet up in the air. The photograph reveals the tracks of an atom smashed to bits by a cosmic ray. From such studies scientists learn something of the nature of atomic structure.

16. A picture of the extended spectrum

Figure 43-13 pictures a giant spectrum containing all the known electromagnetic waves. We can compare it to the keyboard of an enormous piano. But where the keyboard of an actual piano contains about 7 octaves (see Chapter 24), this one contains almost 70. And out of these 70 octaves only one, near the middle of the keyboard, represents the light waves that enable us to see.

Waves of low frequency are on the left-hand side of the keyboard, those of high frequency at the right. On the left, beyond the low-frequency waves of visible red, we have 10 octaves of the invisible infrared waves, whose frequency is lower than that of red. Beyond the infrared come 20 octaves of radio waves,

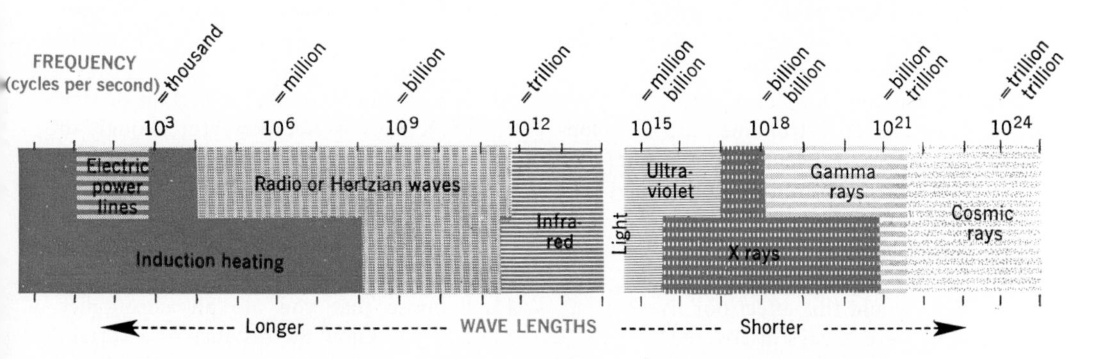

43-13 The light waves that we can see (center) are but a small portion of the electromagnetic wave spectrum. In the higher frequency (short wave length) regions, we find ultraviolet rays, X rays, gamma rays, and cosmic rays. In the region of low frequency (long wave length), we find infrared and radio waves.

which have the lowest frequency of any electromagnetic waves.

On the right side of the keyboard, beyond the high-frequency violet, we have about 5 octaves of invisible ultraviolet whose frequency is still higher. Beyond those come about 11 octaves of X rays. Beyond the X rays, higher in frequency and of more recent discovery, come the gamma rays obtained from radioactive elements, and beyond them, at the far end of the keyboard, the mysterious cosmic rays.

Going Ahead

THE VOCABULARY OF PHYSICS

Define each of the following:

spectroscope	diathermy
continuous spectrum	absorption spectrum
bright-line spectrum	gamma rays
Doppler effect	primary cosmic rays
infrared rays	electron-volt
ultraviolet rays	secondary cosmic rays
germicidal lamp	radio telescope
fluorescent	phosphor
Fraunhofer lines	meson
X rays	

THE PRINCIPLES WE USE

1. Describe three different kinds of spectra and tell how each is formed.
2. With the aid of a labeled diagram, describe the spectroscope.
3. What is the Doppler effect?
4. How can an astronomer use the Doppler effect to detect a double star?
5. How were infrared and ultraviolet rays first discovered?
6. Suppose the sun were suddenly to stop sending infrared and ultraviolet rays. How would this affect our lives?
7. How were X rays discovered?
8. What part does the study of cosmic rays play in modern science?
9. How do atoms give out a bright-line spectrum?

Explain why

10. The cosmic rays that reach us at sea level are not the same as the ones that enter the earth's atmosphere.
11. All lines in the spectrum of a star shift toward the violet end if the star is approaching us, but shift toward the red end if it is receding.
12. Electromagnetic waves of wave length somewhere between radio broadcast waves and infrared waves are used to give diathermy treatment.
13. The study of the spectrum lines emitted by an element has enabled us to work out the arrangement of electrons in its atoms.

APPLYING PRINCIPLES OF ELECTROMAGNETIC WAVES

1. Describe how the spectroscope can be used to detect the presence of minute amounts of chemicals.
2. How does the astronomer make use of the spectroscope?
3. Describe three ways in which infrared rays are useful.
4. Describe four ways in which ultraviolet light is useful.
5. Describe two practical uses of fluorescence.
6. Describe the operation of the X-ray tube.
7. Mention three practical applications of X rays.
8. How does a doctor make use of electromagnetic waves to give a diathermy treatment?
9. If a doctor finds that a certain dosage of X rays is not powerful enough to show the bones in a leg, how can he get more penetrating X rays?
10. Can you design an invisible burglar alarm that will work with ultraviolet light? Make a diagram.
11. Suppose that you are an astronomer and you know the distance of a certain star. Another star $\frac{1}{100}$ as bright shows almost exactly the same spectrum. How might it be possible to calculate the distance of the dimmer star?

12. Today astronomers are beginning to study the spectrum of radio waves which seem to be emitted from certain regions of the sky as a form of "cosmic static." A "radio telescope" used for this purpose is shown on page 498. Find out more about this new field in recent periodicals, and by reading pages 498–502.

PROJECTS

1. Seeing Objects by Invisible Light. There are several inexpensive bulbs on the market rich in blue, violet, and ultraviolet light; some of them are used in germicidal lamps. If you can obtain such a source of ultraviolet light, you can make a number of interesting observations.

   Many objects that look alike in ordinary light differ a great deal when viewed in light rich in ultraviolet. Under the ultraviolet light in an otherwise dark room, some soaps and soap powders fluoresce (page 589) while others do not. Paraffin candles fluoresce, but tallow candles do not. "Your own" teeth fluoresce, but an artificial tooth will not. And fluorescent red and green socks and neckties shine brightly.

   All objects that fluoresce have absorbed invisible ultraviolet light and in its place have emitted light of longer (and hence visible) wave lengths.

2. Detecting Infrared Rays. With the supervision of your teacher, in a well-ventilated room, prepare a saturated solution of iodine crystals in carbon tetrachloride (Carbona). Poured into a small spherical flask, the solution is opaque to all visible light but quite transparent to infrared. Moreover, the spherical shape of the flask serves as a lens to concentrate a beam of infrared rays to a focus.

   Mount the flask directly in front of the projection lens of a powerful slide projector (or better yet, an arc lantern). You can locate the point where the infrared rays come to a focus by moving your hand about in front of the flask. Careful—it's hot! If the source is hot enough, your beam of invisible light will ignite black paper and blackened safety matches at the focus. Why black?

3. Observing Spectra. If a spectroscope is available, you should observe the continuous spectrum emitted by an ordinary incandescent lamp. Why doesn't such a lamp show the bright-line spectrum of tungsten?

   To observe bright-line spectra soak a strip of asbestos paper in salt solution, and wind it around the top of the chimney of a Bunsen burner. The resulting flame emits the bright yellow spectrum of sodium. In a similar way, with fresh asbestos each time, you can produce the spectra of barium (using any soluble barium compound), strontium, calcium, potassium, copper, and lithium.

   You may be able to locate inexpensive 110-volt glow tubes containing neon or argon, whose bright-line spectra are very pronounced. And if possible, try your spectroscope out at night on the lights of a city street.

   To observe an absorption spectrum turn your spectroscope on the sun. But be very careful that you NEVER look at the sun through just the telescope, and be sure that the slit of the instrument is very narrow.

CHAPTER

44

Waves and Particles

What is light made of?

Since the seventeenth century many physicists, including Newton, have been sure light is made of particles. Other eminent physicists, including Huygens, have been sure that light is composed of waves.

Why particles? You may appreciate the argument in favor of particles when you try to imagine how waves can travel through a vacuum. All waves you know of in the familiar world—sound and water waves, for instance—require some material medium to carry them. Remove the air, and sound waves cannot exist. Likewise, with no air between the earth and the sun, how can light waves be transmitted? To many early physicists it seemed clear, as it may to you that only *particles* could travel through empty space. This suggests that light must be composed of particles.

Then why waves? You will appreciate the argument in favor of waves when you try to imagine how particles can travel through water or solid glass. To many early physicists it seemed clear, as it may to you, that only *waves* could travel through solid objects (as sound does). Hence light must be composed of waves!

We have here a dilemma.

Interference of waves

To solve the dilemma of whether light consists of particles or is composed of waves, we shall first explore some of the important properties of ordinary waves. Then we shall see if light exhibits any of these properties. If it does, then we must conclude that light is wave-like.

Then, to make sure, we shall explore some of the important properties of particles. If light can be shown to have any of these properties, we must conclude that light is also particle-like.

You will find that our investigation leads to extraordinary discoveries not only about the nature of light but also about particles of matter itself. The first problem we shall explore is what happens when waves meet each other.

Everyone has seen waves on water and noticed the complicated and unusual patterns that occur when water waves meet. To find out what happens when waves meet, we will first study a very simple example—the meeting of waves traveling along a rope.

1. Standing waves on ropes

If we hook one end of a heavy long rope to a wall and shake the other end, we send a single pulse or wave traveling down the rope (Fig. 44-1A). When it reaches the wall, the hook, which cannot move up or down, flips the wave over and sends it back along the rope. The wave has been reflected (Fig. 44-1B).

Suppose we send two waves down the rope, one following the other with a distance, L, between them. Since both travel at the same velocity, the first wave after reflection will meet the second at a point P, a distance L/2 from the wall (Fig. 44-2A). As the waves pass through each other, the reflected wave tries to pull the rope down while the incident wave tries to pull the rope

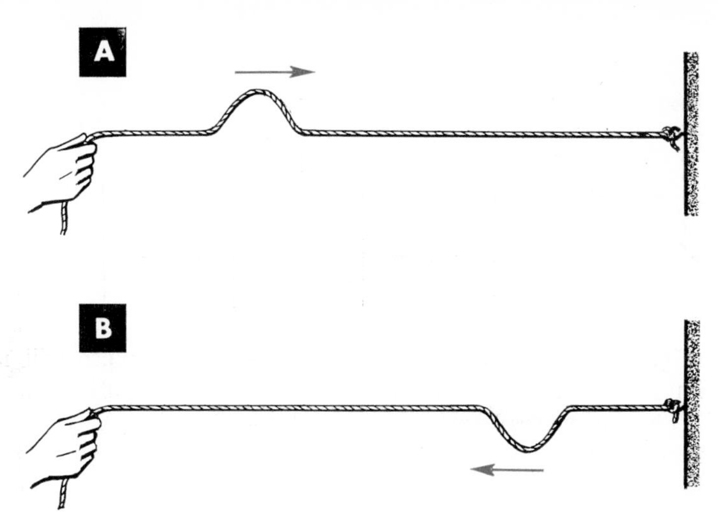

44-1 Reflection of a pulse in a rope. A pulse moving toward the wall (A) flips over and reverses its direction (B) when it gets to the wall.

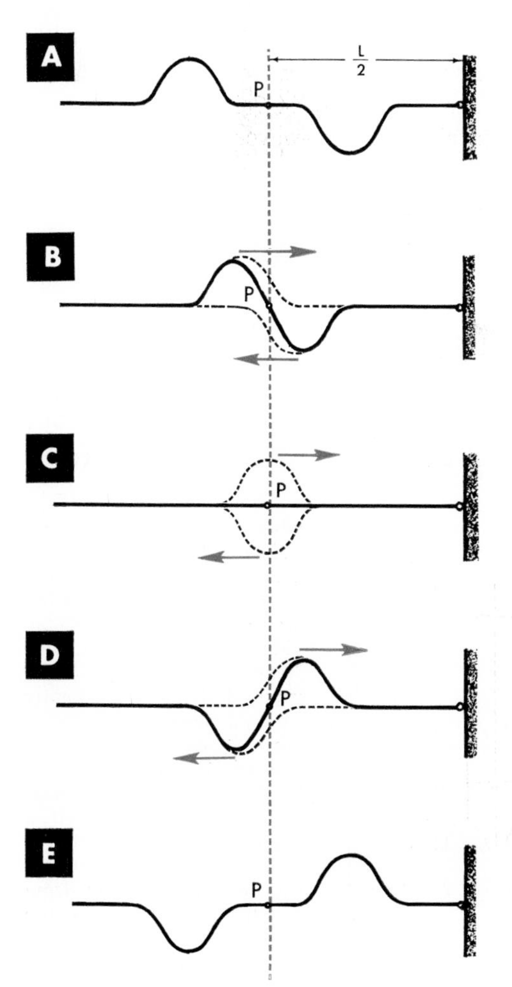

up. The movement of the rope about P will be the resultant of the movement of each wave (Fig. 44-2B, C, D, E). Notice that during the entire collision the point P on the rope, called a **node,** never moves at all. After crossing, the two waves move on, unaffected. This is what happens when all ordinary waves, such as light and water waves, cross.

If we send a series of equally spaced pulses along the rope, the reflected pulses will interfere with the incident pulses (Fig. 44-3) so that P, P′, and P″, half a wave length (L/2) apart, are always nodes.

Now let us shake the rope rapidly up and down, sending a continuous series of waves along the rope (Fig. 44-4A). The reflected waves interfere with the incident waves to give **standing waves.** We see no waves traveling to either right or left. All we see are sections of the rope that move violently up and down (called **maxima**) and points in between (nodes) that do not move at all

44-2 Interference of incident and reflected pulses on a rope. When the pulses pass through each other, point P, half a wave length from the wall, does not move. It is called a node.

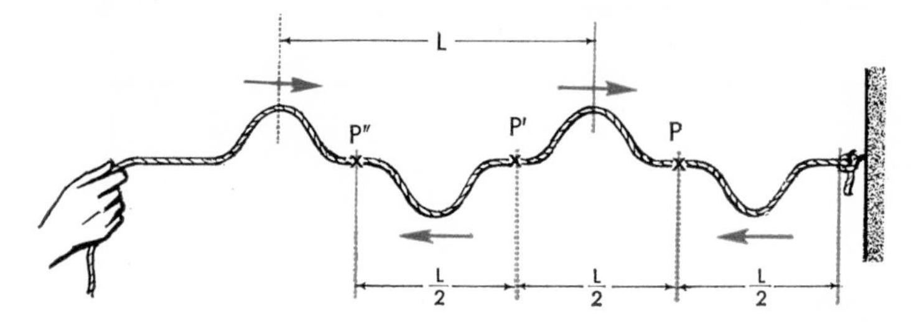

44-3 Reflection of equally spaced waves (pulses) in a rope. Points P, P', and P'' are *nodes.* The rope at these points never moves up or down. These nodes are spaced half a wave length, L/2, apart.

(Fig. 44-4B). The nodes, like the maxima, are half a wave length apart.

Shaking waves along a rope and reflecting them from a wall shows us how **interference** takes place when waves cross; but in a rope, we can produce waves in only one dimension—that is, the waves can travel only along the line of the rope; they cannot travel in any other direction. We shall now do some experiments with waves on the surface of water in order to observe interference in two dimensions.

2. Standing waves on water

Figure 44-5 shows a ripple tank, a very useful apparatus for showing the properties of water waves. If a finger is dipped into the water, a circular wave pulse will move outward and its silhouette can be seen clearly on the sheet of white paper below. If a horizontal ruler is dipped into the tank, a straight wave pulse will move across the tank. This pulse can be reflected from a straight barrier placed in the middle of the tank (Fig. 44-6).

If the straight-wave generator (ruler) is now mounted on the end of a hacksaw blade and made to vibrate up and down in the water, it will send out a series of equally spaced waves that are reflected from the barrier and interfere, giving rise to standing waves just like those in a rope. Figure 44-7 is a photograph of such a standing-wave pattern. The bright lines are the maxima, where the water is moving up and down rapidly. The dark spaces are the nodes, where the water remains undisturbed.

3. Reflection of circular water waves

If we stick a pencil in the ripple tank, we will send out a circular pulse that is reflected from the straight barrier just like light. The reflected wave is circular

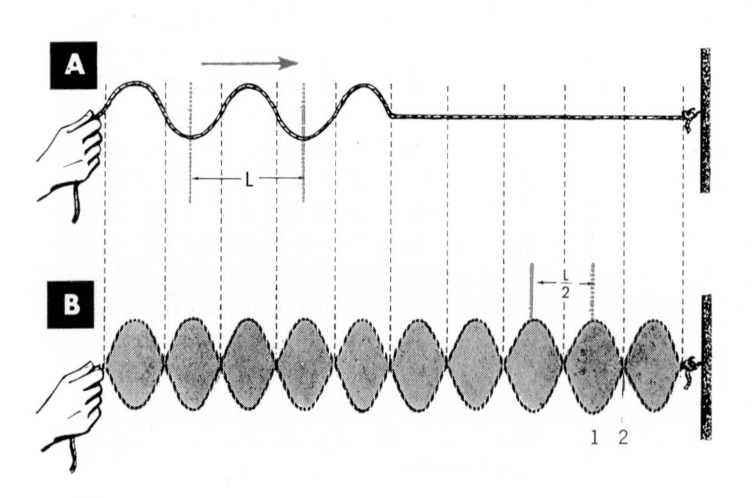

44-4 Standing waves. When a series of equally spaced waves is sent along a rope (A), the reflected waves interfere with the incident waves to produce *standing waves* (B). The rope at 1 moves violently up and down, while at 2 there is no motion at all; 1 is called a *maximum* and 2 is a node. The maxima, like the nodes, are half a wave length, L/2, apart. We do not see the waves traveling toward or away from the wall. We see only the up and down motion near the maxima.

44-5 A ripple tank for showing the properties of water waves. A window frame, A, containing water is mounted on two ring stands. B is a clear glass light bulb and C is a large sheet of paper. A hacksaw blade, D, with a large plastic bead attached to its end, vibrates up and down sending out circular waves that are easily visible on the paper. Straight waves can be generated by replacing the bead by a light horizonal straight edge such as a ruler.

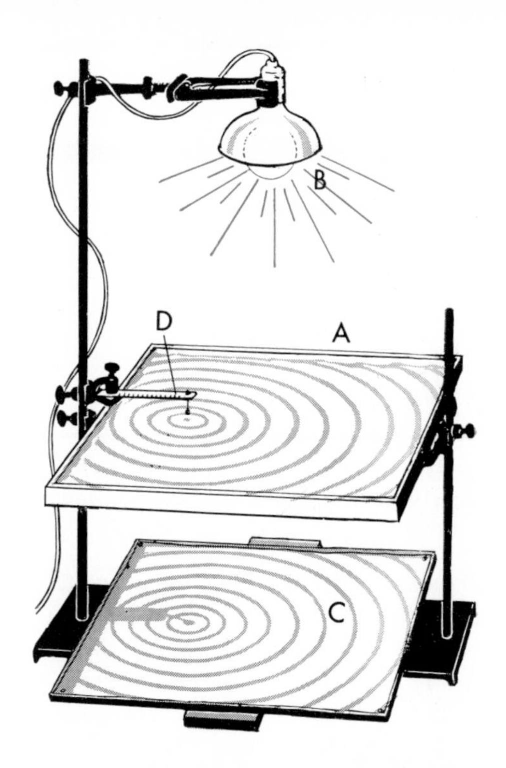

and acts as though it had been generated just as far behind the barrier as the pencil is in front (Fig. 44-8). The point S′ in Fig. 44-8 is a virtual source (a virtual image of S).

Suppose we now vibrate the hacksaw blade with a bead on its end and generate equally spaced circular waves in the tank. They are reflected from the barrier, go off in all directions, and produce interference as shown in Fig. 44-9. The standing waves along the line connecting the source and its virtual image are easy to understand. They give rise to just the pattern we get from straight waves reflected from a barrier; as before, the maxima are half a wave length apart.

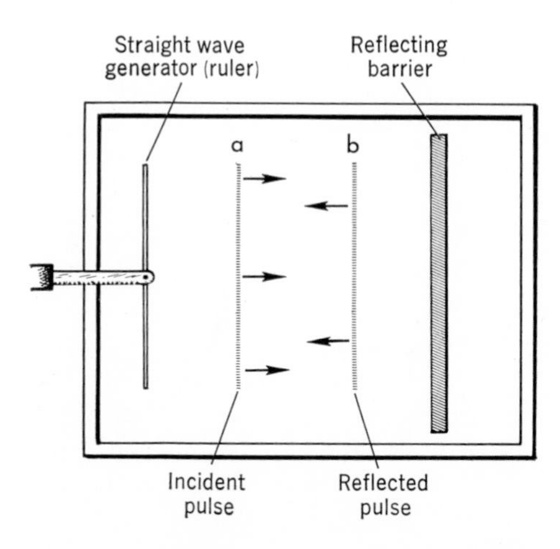

44-6 Reflection of a pulse in a ripple tank. Looking down on the tank we see a pulse, a, generated by dipping a ruler into the water. A previous pulse, b, has been reflected from the straight barrier and is returning toward the generator.

Dr. James Strickland

44-7 When straight waves reflected from a barrier interfere with waves traveling in the opposite direction, the result is a series of motionless "standing waves."

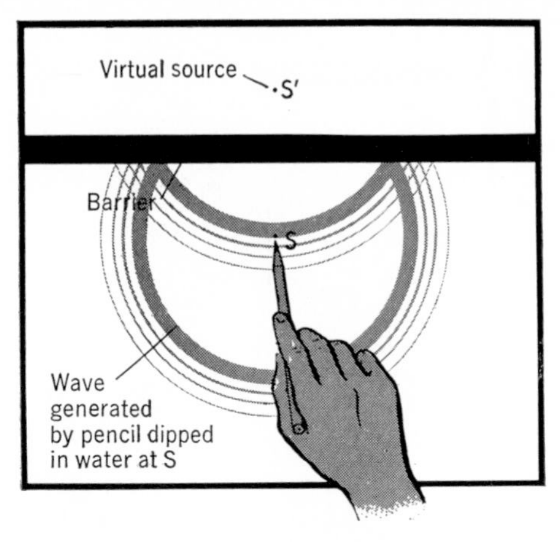

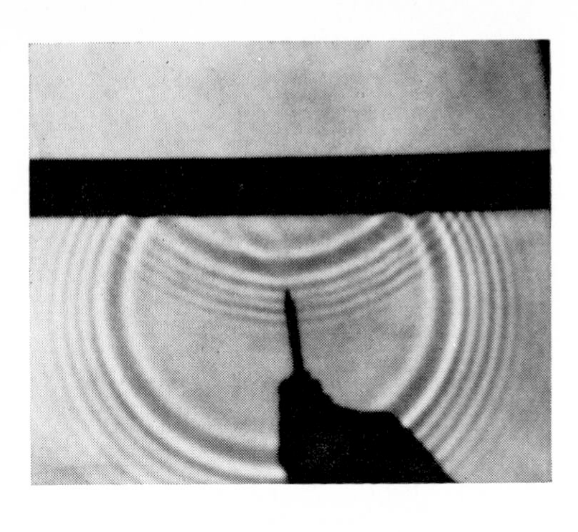

44-8 The reflection of a circular pulse appears to have originated behind the reflecting barrier.

The curving of the interference pattern away from this line is not so easy to explain, but with the aid of the diagram in Fig. 44-10, we can see where the maxima and nodes should occur. In this figure, the curved line AB, called a **nodal line**, marks the position of the node closest to the barrier in the photograph of Fig. 44-9.

We know that for two waves to interfere and produce a node, one must be moving the medium up while the other is moving it down. If the waves are moving in the same direction (rather than in opposite directions as in our earlier interference experiments), interference and node formation will occur when one set of waves is traveling half a wave

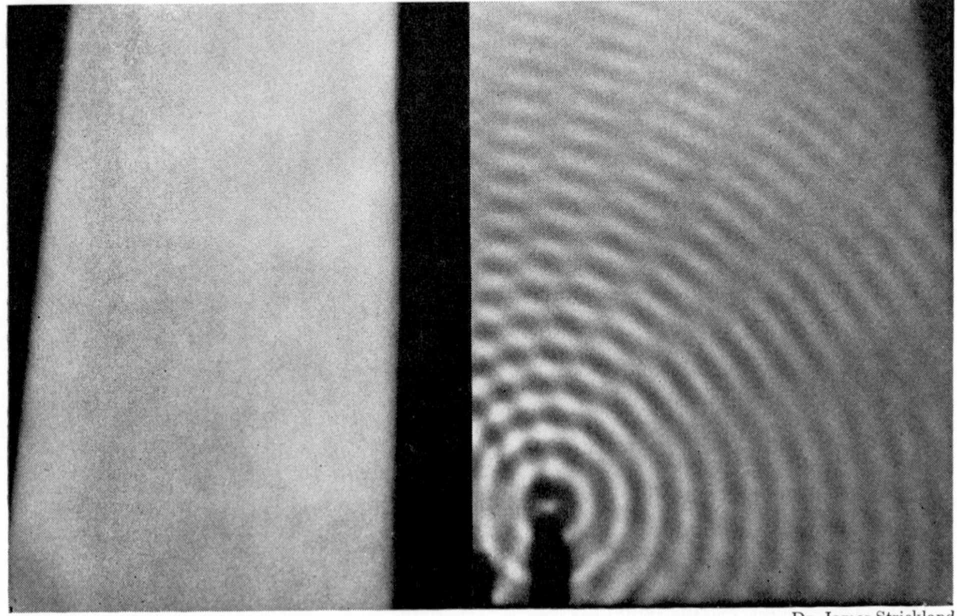

Dr. James Strickland

44-9 When circular waves are reflected from a barrier, they interfere at some points with the unreflected waves. At other points there is reinforcement. In this picture the points where interference occurs are visible as curving lines called nodal lines.

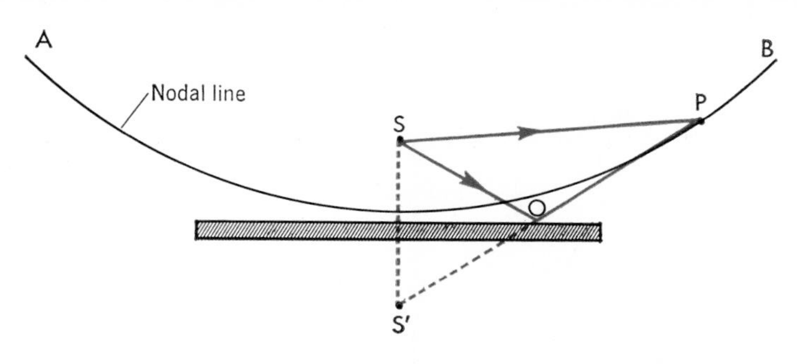

44-10 **Interference by reflection of circular water wave generated at S. A node will occur at P if the path difference SOP minus SP equals one-half a wave length. Curve AB is the locus of all the nodes.**

length behind the other. The crests of one set will then coincide with the troughs of the other and a node will result. Such waves which are out of step are said to be **out of phase.**

P in Fig. 44-10 is any point along the nodal line closest to the barrier. Waves starting from the source, S, arrive at this point by two different paths. They go directly along the line SP and they also follow a longer path along SOP. The difference in path length (SOP–SP) must be equal to half a wave length, since P marks a point on a nodal line where the waves are out of phase. We can say, therefore, that the nodal line marks points where the path difference between the direct waves and the reflected waves is equal to half a wave length.

How do we explain the other nodal lines that we see in Fig. 44-9? These represent other points where crests coming directly from S arrive with troughs from the reflected waves—points where the path difference is $1\frac{1}{2}$ wave lengths, $2\frac{1}{2}$ wave lengths, etc. Every time we shift the path difference by one whole wave length, crests again coincide with troughs and the waves are out of phase.

Between the nodal lines we see maxima, where the water is violently disturbed. These occur where crests coincide with crests—where the waves are "in step" and are said to be **in phase.** This means that the two sets of waves must have a path difference equal to 1, 2, 3, etc. wave lengths.

The interference pattern is completely described by saying that nodes occur wherever the path difference is $(n - \frac{1}{2})L$ (where the waves are out of phase) and maxima occur wherever the path difference is nL (where the waves are in phase). Here n is any whole number (1, 2, 3, etc.) and L is the wave length.

Interference of light rays

You have seen how a single reflecting surface can be used to produce interference of water waves. In 1801 the English scientist Thomas Young showed that under certain conditions two light beams shining on one spot could interfere with each other to produce darkness. Many such interference experiments have been performed since the time of Young. One of the simplest is called the Lloyd's mirror experiment and is surprisingly easy to perform.

4. Lloyd's mirror

In the Lloyd's mirror experiment, a strip of plate glass 30 or 40 centimeters long and 3 or 4 centimeters wide is supported as shown in Fig. 44-11.

A lamp bulb with a single vertical filament is supported just beyond one end of the glass plate and 1 or 2 millimeters in front of the plane of the glass surface. The filament is made precisely parallel to the plate. (An alternative light source that is convenient but open to certain theoretical objections is the following: A thin slit a few tenths of a millimeter wide is put in place of the lamp filament. Behind it is a Bunsen burner with a salt-soaked sleeve of asbestos wrapped

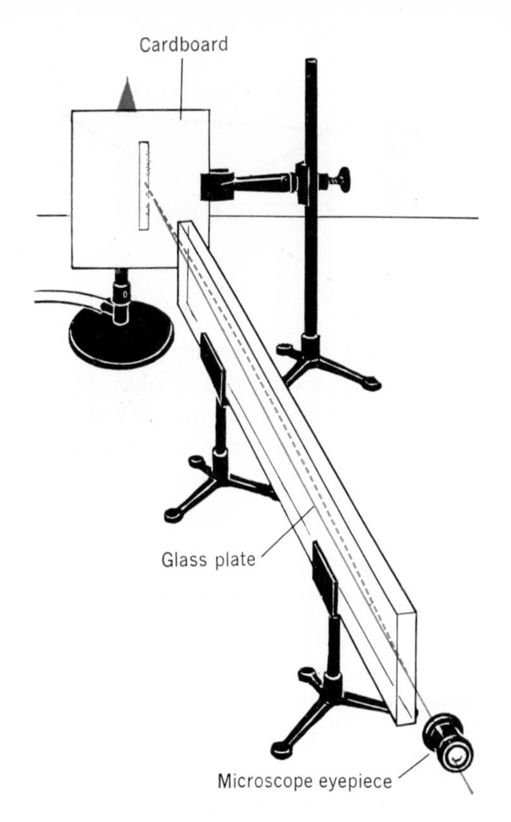

44-11 **Lloyd's mirror is a simple illustration of the interference of rays of light.**

around its top. When the burner is lighted, particles of salt in the flame become heated so that they glow with an intense yellow light.)

If you now place your eye close to the opposite end of the glass plate, in a darkened room, and examine the light source through a microscope eyepiece, or even a single convex lens, you will see at the edge of the luminous area a pattern of vertical yellow lines separated by dark spaces. This pattern is produced by the two beams of light—the beam coming directly to your eye from the light source, and the beam reflected into your eye from the surface of the glass, which acts like a mirror.

If we suppose that light is composed of waves, we can explain this pattern very simply. From the figure you will see that the beam reflected from the mirror has traveled slightly farther than the direct beam—let us say d centimeters farther. If d is a whole number of wave lengths, the two beams reaching your eye will be in phase and will thus reinforce each other. You will see a bright vertical line where they meet. But if d is a whole number of wave lengths *plus a half wave length,* the two beams will be precisely a half wave length out of phase, thus interfering with each other, and where they meet you see the center of one of the dark lines. This phenomenon is called **interference.**

A quantitative study of these lines leads to the conclusion that if the yellow light is composed of waves, their wave length must be close to 5.89×10^{-5} cm.

See if you can explain the pattern of bright and dark lines by supposing that light is composed of particles.

5. Thin films

You can form beautiful interference patterns of light with a simple soap film. Twist a bare wire into a loop about 5 centimeters across. If you dip the loop into soap solution you will observe the familiar colored interference pattern shown in Plate VI. If you support it vertically as shown in Fig. 44-12, the weight of the soap solution causes the liquid to flow downward, making the film thicker at the bottom than at the top. When you look at the film in a darkened room with the yellow light source of the preceding experiment (the Bunsen burner with a salt-soaked asbestos sleeve), you will see a series of horizontal yellow and black bands across the soap film.

This pattern, too, can be easily explained if we suppose that light is composed of waves. Light is reflected to your eyes from both the front and the back surfaces of the soap film. The light from the back surface has traveled slightly farther—through the thickness of the film and back. This extra distance may be such that light will emerge from the film and travel toward your eye *in phase* with the light reflected from the front surface, so that you will see a bright band. For other film thicknesses the two beams may be out of phase, so that a darker region will result. Since the film steadily increases in thickness from top to bottom, there is a continuous change in the phase difference between the two beams from top to bottom, giving alternate bands of brightness and darkness.

If you could somehow adjust the thickness for destructive interference all over the film, you would expect to have a film from which no light would be reflected at all. This is precisely what is done to many of the lenses of cameras, binoculars, microscopes, and other optical instruments. Nonreflecting glass is prepared by coating glass with a hard film of transparent material of just the right thickness. The two beams of light reflected from the front and rear surfaces of the film destructively interfere with each other, and as a result nearly all of the light goes on through the lens into the instrument instead of being uselessly reflected and lost.

Repeat your soap-film experiment, using an ordinary lamp bulb as a light source. When the light source is covered with a red filter, the light and dark bands are farther apart than when it is covered with a blue filter or a green one. This evidence supports the view that red light is composed of longer wave lengths than is green or blue light. By experimenting with filters of various colors and observing the width of the interference bands that result, you may be able to arrange the colors of the visible spectrum in order of decreasing wave lengths. Compare your conclusions with those given in Section 2 (page 573) in Chapter 42.

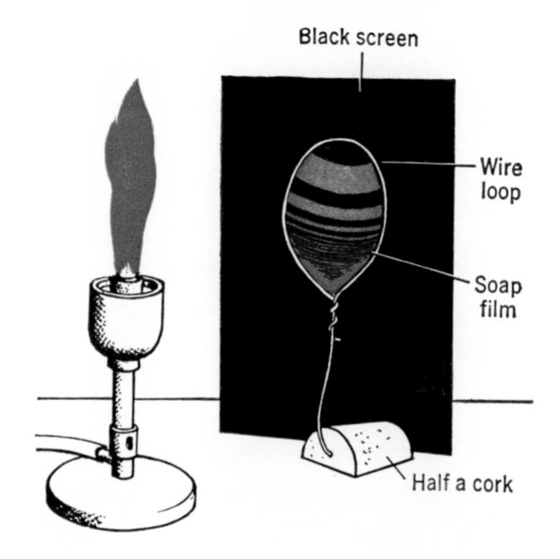

44-12 A vertical soap film is thicker at the bottom than at the top. The variation in thickness gives rise to horizontal bands as the light reflected from front and rear surfaces undergoes interference.

It is clear that glass cannot be made entirely nonreflecting for more than one color of light. Hence as you look at light reflected from the surface of a "coated lens," such as is commonly found in good cameras, you will notice that the lens looks purple. This is because the film thickness cancels out yellow light most successfully; the red and the violet light at the ends of the spectrum are canceled least successfully, and the effect of combining the reflected red and violet light is to produce purple, as you saw in Section 4 (page 575) of Chapter 42.

Again it is difficult to see how a particle theory of light can explain the interference pattern observed in thin films. In interference experiments we have very strong evidence for the wave-like character of light.

Diffraction of waves

Let us now return to the ripple tank and examine another property of waves that is closely related to the properties of light. The ripple tank is arranged as

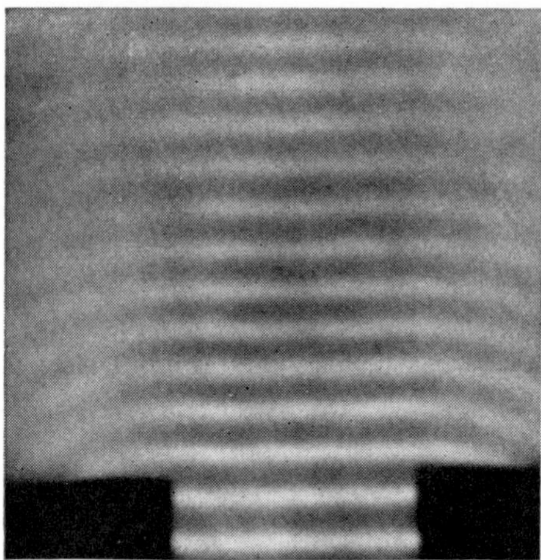

Dr. James Strickland

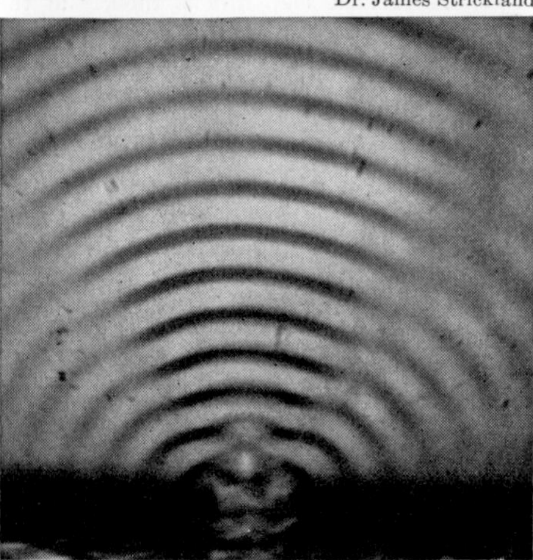

Physical Science Study Committee

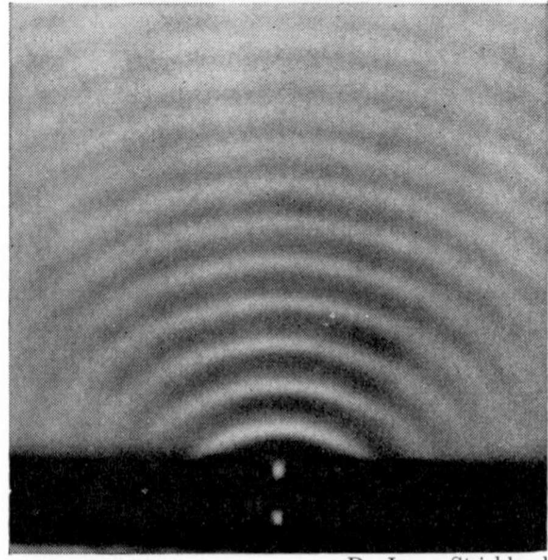

Dr. James Strickland

44-13 The smaller the opening the greater the angle of spreading as the waves pass through it.

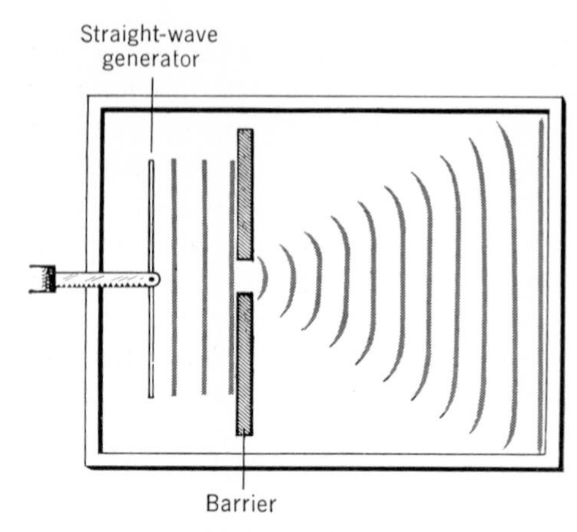

44-14 Diffraction of water waves in a ripple tank. A small opening in the barrier diffracts (spreads out) straight waves.

shown in Fig. 44-14, so that straight waves pass through a narrow opening in a barrier.

6. Diffraction in a ripple tank

The photographs in Fig. 44-13 show that as the opening in the barrier becomes smaller and smaller, the waves spread out more and more, until, when the opening is smaller than the length of the waves, it acts almost like a point source of circular waves. This spreading out is called **diffraction.**

It is important to note that it is not the width of the hole alone that determines the spreading, but *the width compared to the length of the waves.* As the ratio of wave length to opening size (L/a) becomes smaller, the diffraction or spreading becomes greater.

7. Huygens' Principle

Christian V. Huygens (1629–1695) had suggested that every point on an advancing straight wave acts like a point source of circular waves. This is called **Huygens' Principle,** and though not quite correct in detail, it does apply to many situations. A very small opening in the barrier of a ripple tank approxi-

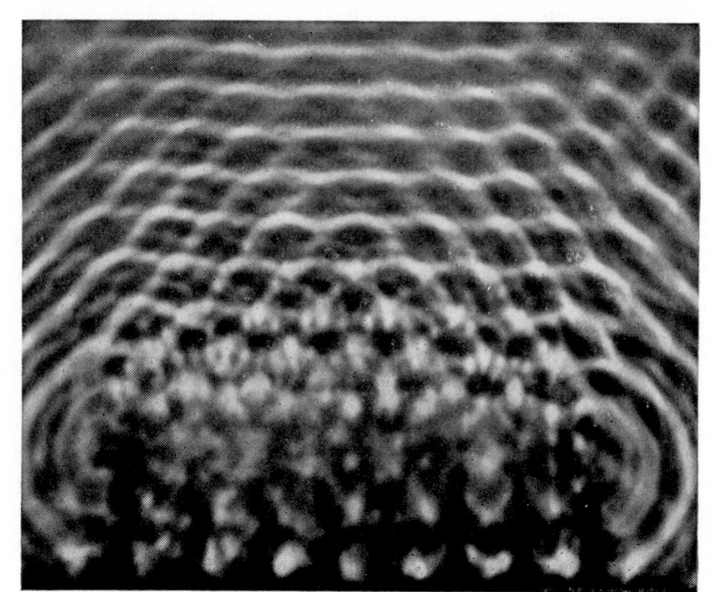

Dr. James Strickland

44-15 This series of eight closely spaced point sources (at bottom of picture) gives rise to straight waves at the top of the picture. All straight waves may be thought of as being generated by a large number of closely spaced point sources.

mates a point on a plane wave, and we have seen in Fig. 44-13C that it acts very much like a point source.

Huygens' Principle raises the question: Will a series of point sources close together in a ripple tank give rise to a plane wave? According to the principle, it will. Figure 44-15 shows waves produced in a ripple tank by a generator made up of many nail points, close together in a line, that move up and down in the water together. As can be seen from the figure, they produce something very much like a plane wave.

8. Diffraction interference patterns

A close look at Fig. 44-13 shows that as diffraction increases, the waves do not spread uniformly. We see what appear to be vague nodes and maxima in the diffraction pattern. This looks like interference, and indeed it is. The parts of the waves from different points along the opening in the barrier arrive with varying path lengths, thus producing nodes and maxima. We can explain this pattern with the help of Fig. 44-16.

Imagine that the opening AB is divided into a very large number of Huygens point sources. We shall analyze the interference pattern from these sources at a distance that is large compared to the width of the opening AB. First, let us suppose that point P in Fig. 44-16 is directly in front of the slit. The path differences between the waves from any of the point sources will be nearly zero, so we should get reinforcement from all of them and hence a maximum in this direction. This is exactly what we observe in the photographs of Fig. 44-13. If we now move the point P to the right of this center line, as shown in the diagram of Fig. 44-16, the path differences between the different point sources increase, and at some point (P in the diagram), the path difference $AC = PA - PB$ is one wave length, L. Waves from point B and from a point in the center of the opening will arrive at P with a path difference of one half wave length ($L/2$) and cancel each other. Similarly, waves from the point just to the left of B and from just to the left of the middle of the opening will also cancel. Moving toward the left, all the point sources in the opening can

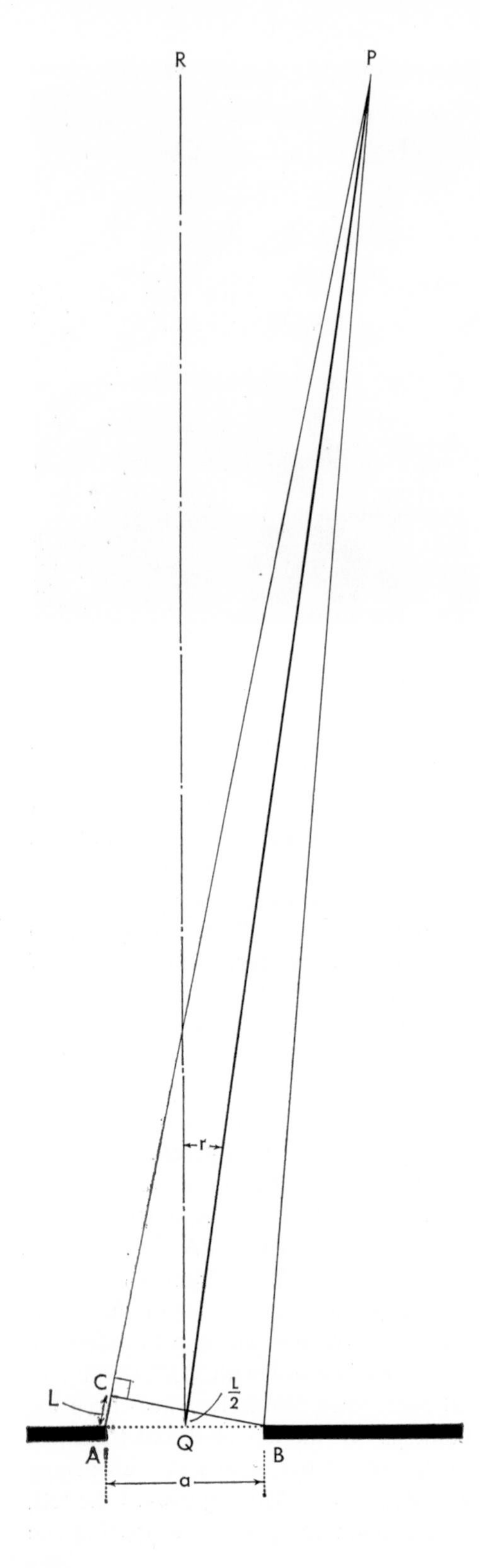

be paired off in this way, each pair producing a node at P.

We can now derive an equation for the angle *r,* which specifies the direction along which we find cancellation, and *a,* the width of the opening. Examine the small triangle ABC in Fig. 44-16. Since P is so far away from AB, the angle ACB is so nearly a right angle that we can, with little error, say that

$$\sin ABC = \frac{L}{a}$$

but, since RQ is perpendicular to AB and PQ is very nearly perpendicular to BC, we can say that ABC = r, and so

$$\sin r = \frac{L}{a}$$

The angle *r,* gives us the position of the first node in terms of *L,* the length of the waves, and *a,* the size of the opening through which the waves pass. If we consider a point P farther to the right, we will increase the path difference and pass alternately through maxima and minima, giving the pattern that we see in the ripple tank in Fig. 44-13.

Diffraction of light rays

If light is really composed of waves, it should exhibit *all* the properties of waves. We have already seen that water waves in a ripple tank spread out, or are diffracted, in all directions when they go through a hole which is not too large compared to the wave length. Do light waves spread out in the same way when they go through a small hole?

We might be inclined to doubt this, since we know that light coming through a keyhole into a darkened room casts a

44-16 If point P is far away from the opening (PB is large compared to AB), then, if the path difference AC is one wave length, L, point P will be a node.

sharp image of the keyhole. No spreading out is observed; the light appears to travel in straight lines like a beam of particles.

From a study of thin-film interference and Lloyd's mirror, we see why keyholes will hardly do. These experiments show that the wave length of light is about 5×10^{-5} centimeter. Hence to get easily observable diffraction effects we must use openings of much smaller size.

9. Diffraction through a single opening

You can easily observe the diffraction of light if you set up a Bunsen burner with a salt-soaked sleeve behind a vertical slit in a dark room (Fig. 44-17A). Stand across the room and hold your hand right in front of your eyes, fingers vertical and *lightly* touching one another. There will be thin slits between your fingers. If you look at the burner through one such slit, you will see a series of vertical yellow bands (Fig. 44-17B). These bands are caused by the diffraction of the light coming through the narrow slit between your fingers. Notice that the pattern is much wider than the width of the slit. The light is actually spreading out after passing through the opening between your fingers.

Now gently squeeze your fingers together, narrowing the slit. As the slit narrows, less light gets through and so the pattern gets fainter, but at the same time it becomes broader (Fig. 44-17C). The narrower the opening, the more the light spreads out as it passes through the slit. You saw the same thing happen in the ripple-tank experiment on page 604.

You know that sound behaves in this way too. When sound waves emerge from your mouth, which is small compared to the wave length of sound, the waves spread out in all directions. Your

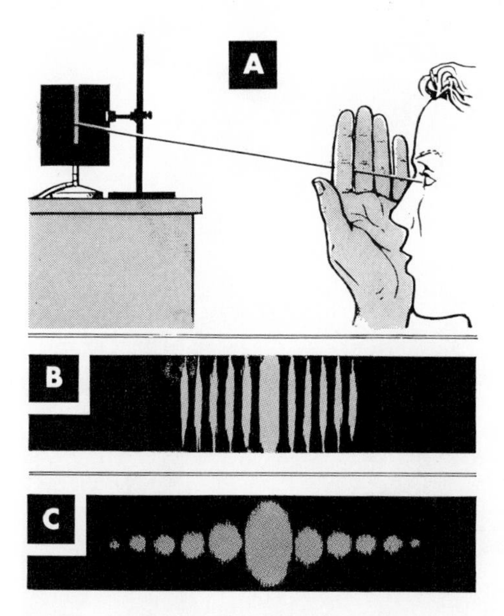

44-17 You can observe the diffraction of light very simply. The burner can even be replaced by a lamp if necessary.

shout can be heard by people off to one side of you as well as by those directly in front. The pattern of sound is spread out as is the light coming through a very narrow slit. If you wish to concentrate the sound of your voice into the forward direction, you must use a megaphone. Coaches and cheerleaders may not know it, but they are reducing the diffraction of sound waves by widening the opening through which the waves pass (Fig. 44-18).

Diffraction is important in the construction of microscopes and telescopes. Light collected from a distant star by a telescope objective should appear as a point on the camera film at the focus. But because of diffraction the point is always spread out into a disk, and the disk is sometimes surrounded by a faint ring. The smaller the telescope objective, the greater the spreading of light passing through it, and the larger the star image, just as the image of the slit broadened when you narrowed the slit between your fingers.

Brooks from Monkmeyer

44-18 Contrary to what you might think, the megaphone does not spread the sound of the cheerleader's voice over an angle. Instead, due to diffraction, it channels the sound of his voice into a narrow beam.

Except for the fact that they are fainter, such broader images might not seem to be a serious drawback. But if the telescope is aimed at two stars close together, the two star images may be so broad that they appear as one, so that we are unaware that we are looking at two separate stars (Fig. 44-19). Here is one of the important reasons for building bigger and bigger telescopes. Larger-size objectives mean less diffraction, which means that points close together, like the stars in Fig. 44-19, can be seen as separate images instead of being blurred together as one. The same reasoning applies to observing fine detail on the surface of, say, the moon. If a rocket ship that has landed on the moon is to be distinguished from a nearby crater, these two objects must be examined with a very large lens or mirror to prevent their diffraction-broadened images from overlapping.

The ability of a telescope or a microscope to show fine detail by reason of its small diffraction is called its **resolving power.** On page 559 you see a picture of the mirror of the 200-inch telescope

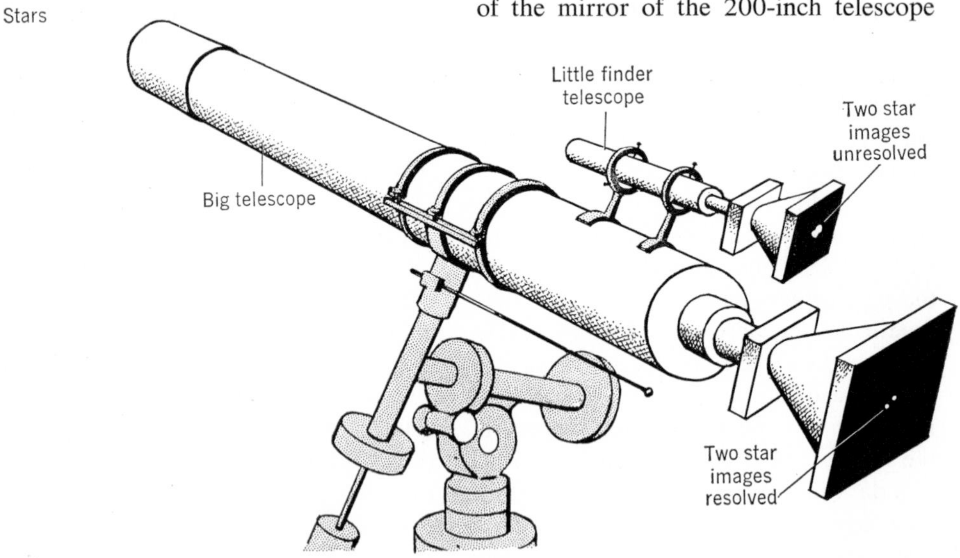

44-19 Large telescopes have greater resolving power than small ones, not because they have higher magnification but because they have larger objectives.

on Mt. Palomar in California, which has the highest resolving power of any light-gathering instrument on earth.

10. The formula for single-opening diffraction

In Section 8 (page 605) of this chapter we saw that the diffraction of water waves is described by the formula, sin r = L/a. The same formula is found to apply to light coming through a thin parallel-sided slit.

If the opening is round, as with a telescope or microscope objective, instead of parallel-sided, the formula must be amended to read

$$\sin r = \frac{1.22\ L}{a}$$

You will notice that this formula is consistent with what you have learned so far about the behavior of light. The smaller the opening *a* becomes, the larger the angle through which the waves spread, whether the opening is a telescope objective or your eye. It does not matter whether the waves pass *through* the opening (as with the pupil of your eye) or whether the waves are *reflected from* the objective (as with a reflecting telescope or a radar antenna); the formula applies to either situation.

We commonly consider that two points of light such as stars are *just* resolved if the center of the image of one lies just on the edge of the other, as shown in Fig. 44-20. For then the combined image will be noticeably elongated, suggesting two points of light very close

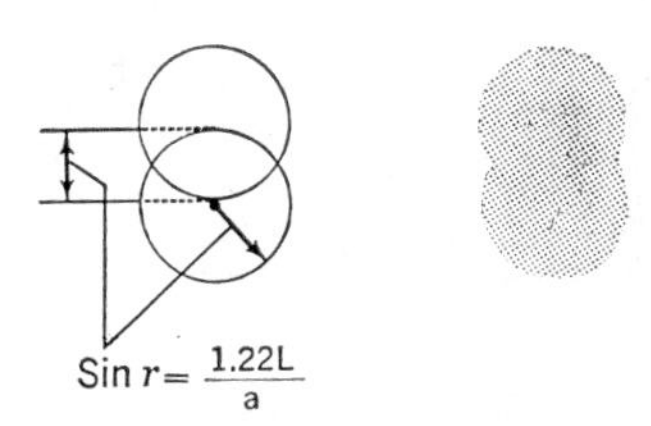

44-20 Two points of light (such as stars) are just resolved if the center of the image of one lies just on the edge of the other.

together. Of course, the two stars are separated in the sky by the same angle as their images, and hence by the angle, *r,* such that sin r = 1.22 L/a (Fig. 44-21).

SAMPLE PROBLEM

Automobile headlights are approximately 120 centimeters apart, and the opening of the pupils of your eyes at night is approximately 0.4 centimeter. Using these figures, find how close a car must be at night for you to see that it is a car with two headlights instead of a motorcycle with one. Take 5.6×10^{-5} centimeter as the wave length of the light you see (yellow).

Taking the distance to the car to be *d* centimeters, since *d* is so large compared to 120 cm we can say that sin r = 120/d. Also since

$$\sin r = \frac{1.22 \times 5.6 \times 10^{-5}}{0.4}$$

our equation becomes

$$\frac{120}{d} = \frac{1.22 \times 5.6 \times 10^{-5}}{0.4}$$

$$d = 7.03 \times 10^{5} \text{ cm}$$

$$= 7.03 \text{ km (about 4.4 miles)}$$

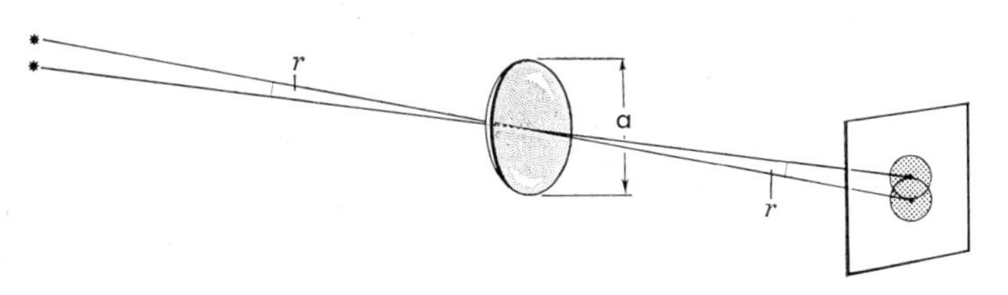

44-21 This diagram shows two stars separated by an angle *r* in the sky, whose images are barely resolved by the telescope objective *a.*

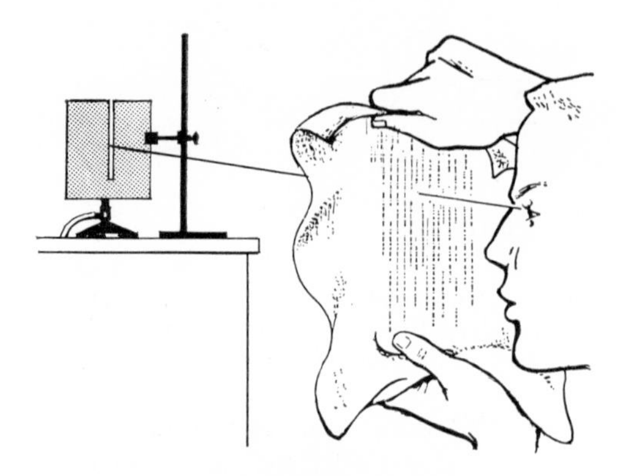

44-22 You can illustrate the action of a diffraction grating by using the simple handkerchief grating shown here.

STUDENT PROBLEM

The moon is 382,000 km away. How close together can two craters be and still be resolved by your unaided eye? Again, use a = 0.4 cm and L = 5.6 × 10^{-5} cm. (ANS. 65 km)

(The answers to these two problems are rather too high due to limitations in the retina of the eye.)

Notice that the resolving power of an instrument becomes better and better (sin *r* grows smaller) as the wave length, *L*, decreases. If we use shorter wave lengths, we should be able to improve the resolving power. In fact, by using ultraviolet light, the resolving power of a microscope can be greatly increased.

Conversely, if the 200-inch telescope were used to receive wave lengths longer than those of visible light (larger *L*), you would expect its resolving power to be poorer (larger angle *r*). The only way to increase resolving power, using a given wave length of light, is to increase *a*, the size of the opening. Thus to examine the sources of the 21-centimeter radio waves reaching us from outer space, the parabolic metal mirrors (antennas) of some radio telescopes (Fig. 36-32) are built several hundred *feet* across.

11. The diffraction grating

You have seen in Section 9 (page 607) that when you narrow an opening through which waves are passing, you increase the angle through which the waves are diffracted. They spread out *more*.

You have also seen (Fig. 44-13) that the broader the opening and hence the larger the number of point sources along a wave front, the smaller the angle of diffraction. The waves spread out *less*.

What will happen then if we allow waves to pass through a *large number of slits* each of which is *very narrow?*

The best way to solve this apparent contradiction is to perform a simple experiment.

Set up a yellow light source and a slit in a darkened room, as in the earlier experiments in this chapter. For the large number of narrow slits use a fine-weave handkerchief. Stand across the room and hold the handkerchief before your eyes with threads parallel to the slit, as shown in Fig. 44-22. You will see a series of vertical lines similar to (but not the same as) those of the single-slit diffraction pattern in Fig. 44-17B. If you now carefully rotate both hands together in such a way that you keep the threads parallel to the slit, you change the slit spacing. Compare the lines with the single-slit pattern. You will see that the effect of using many slits is to make the lines sharper. Because you are using slits very close together, the total pattern is more spread out. The effect is immensely improved if we use a commercial **diffraction grating,** a series of slits, often more than 15,000 per inch, ruled on a film of transparent plastic.

The diffraction grating is a very important and useful tool in modern physics. In order to understand its usefulness we should explain its action in more detail.

In Fig. 44-23 waves are passing through the narrow slits of a diffraction grating (or your handkerchief), which are d centimeters apart, and then striking the distant screen (or retina of your eye), S.

Consider the waves passing through openings T and M. Draw MQ so that QP = MP. Then, as in the derivation of the single-slit formula (page 606)

$$TQ = d \sin r$$

If the waves are to be in phase when they reach the screen at P, then TQ must be a whole number of wave lengths, nL (n = 0, 1, 2, etc.). Hence, as before, nL = TQ and

$$nL = d \sin r$$

This same relationship applies to each adjoining pair of openings all the way across the grating, G, as long as the screen is a long distance away.

For the central image, n = 0. When n = 1, the first bright line will be formed at P on *both* sides of the central image. When n = 2, the second bright line will be formed farther out on both sides of the central image. Even with your handkerchief experiment you should be able to see the central image and these two lines on each side of it. Perhaps you can even see images outside of these, for which n = 3, 4, etc.

Our equation says that when d gets smaller, other quantities remaining the same, sin r and hence r must grow larger. Your handkerchief experiment has already shown that this is so.

If we keep d constant and look at light from a variety of sources (such as a mercury lamp or a neon bulb behind the slit) we see a series of tiny spectra, each similar to those illustrated in Plate III (after page 580); our salt-impregnated Bunsen burner, for example, emits the spectrum of sodium. We may in fact replace the prism of the spectroscope in Fig. 43-1 by such a diffraction grating. *Each* line of the spectrum is repeated several times on the left of the central image and several times on the right, just as in your handkerchief experiment.

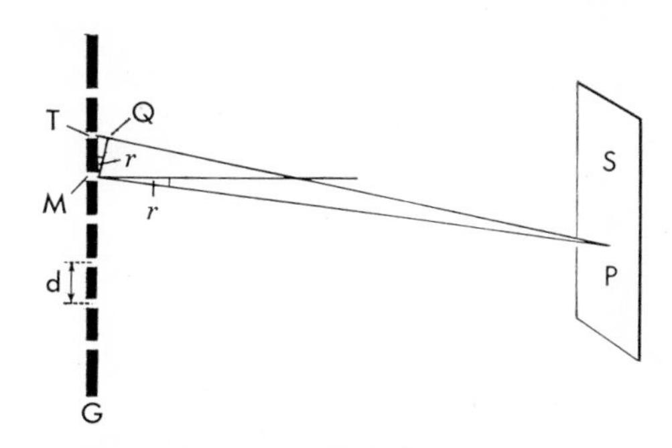

44-23 The simplified geometry of a diffraction grating. TM equals d.

With a diffraction grating the spreading can be made greater than when a prism is used; this is one reason that diffraction gratings are commonly used in spectroscopes. Moreover, it is very simple to find the wave length of the light forming a spectrum line by using the equation derived above.

SAMPLE PROBLEM

Seen through a grating whose lines are 5.00×10^{-4} centimeter apart, the salt-impregnated burner flame shows the nearest lines 6.8° on each side of the central image. Find the wave length of its light.

$$nL = d \sin r$$

Here

$$n = 1$$
$$d = 5.00 \times 10^{-4} \text{ cm}$$
$$\sin 6.8° = 0.118$$

Hence

$$L = \frac{5.00 \times 10^{-4} \times 0.118}{1}$$
$$= 5.89 \times 10^{-5} \text{ cm}$$

Frank Murray

44-24 Sheets of Polaroid material act toward light in a manner similar to that in which the narrow boxes act toward the rope waves in Fig. 44–25. Note that the overlapping sheets held parallel to each other in a vertical direction permit light to pass through. But where the horizontal sheet crosses the two vertical ones, there is complete blackness.

STUDENT PROBLEM

Find the wave length of the red line in the spectrum of hydrogen which appears closest to the central image at 4.7°, using a grating for which $d = 8.00 \times 10^{-4}$ cm. (ANS. 6.56×10^{-5} cm)

Polarization of light rays

Polarization provides a third body of evidence that light is composed of waves instead of particles. Figure 44-24 shows what happens when we look through sheets of special Polaroid material. A strip of Polaroid material is transparent, but if we hold two strips of Polaroid material overlapping at right angles, all light is blocked and the overlapping area appears black. How do we explain this strange action?

In Chapter 37 we pointed out that light is an example of a transverse wave motion in which the vibration of the particles is at right angles to the direction in which the wave travels. We can think of the light wave as resembling the wave sent out along a rope which is tied at one end, A, and moved up and down at the other end, B (Fig. 44-25). Of course, we can also get a wave by moving the rope from side to side in a horizontal direction, or at any other angle for that matter. Now, the Polaroid material acts toward light just as a slotted frame acts toward the wave along the rope. If slot C in Fig. 44-25 is held vertically, it will permit a vertical wave in the rope to pass through, but will not permit a horizontal one.

If two slotted frames are used one behind the other, the wave can pass through if the two slots are parallel, but not if they are at right angles to each other (Fig. 44-25). By turning one of the slots through an angle of 90° we can change the action from stopping the wave to letting the wave through. The action of the Polaroid material strips shown in Fig. 44-24 is comparable to the action of the two slotted frames. Crossing them at right angles caused the two strips to have "slots" across each other (at 90°), and the light was blocked, just as the rope wave was stopped in Fig. 44-25. Light that is vibrating in only one plane is said to be **polarized.**

We see that the ability to polarize light is easily explained by the theory that light is a transverse wave. But how could we explain this action using the theory that light consists of particles moving in a straight line? If light consists of particles, it is difficult to see how the rotation of the Polaroid material could make any difference in letting light through or stopping it, but such is actually the effect.

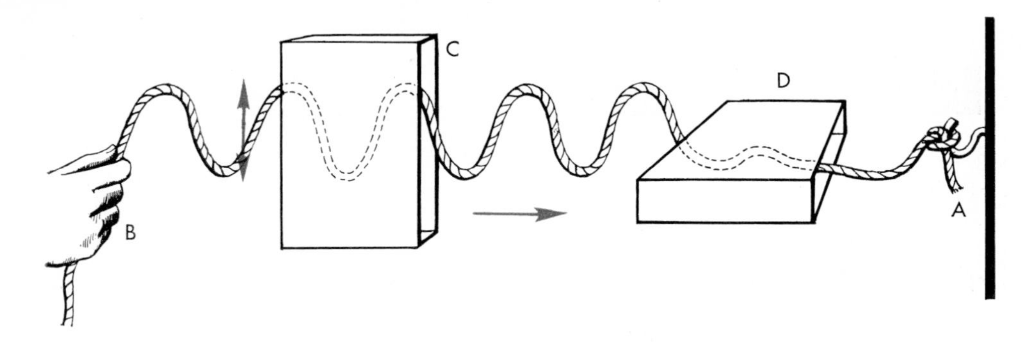

44-25 The rope wave and the slotted frames show how it is possible for two sheets of Polaroid material to let light through in one position, yet stop it if one sheet is turned around. Can you think of a practical use for Polaroid sheets?

12. Uses of polarization

Transparent Polaroid material sheets are coming into increasing use for many purposes. Thus, glasses made of Polaroid material have the effect of removing glare reflected from water or the pavement. It has been found that such reflected light is polarized so that the vibrations from it are mainly parallel to the ground. A pair of lenses made of Polaroid material set into the frames of sunglasses to permit only vertical vibrations to pass through will therefore stop the horizontally vibrating reflections that cause glare. Polarized light also has many interesting applications in photography, in microscope work, in analysis of strains in manufactured parts, and even in astronomy.

Can light be particles?

By this time it may seem clear that light is a form of wave motion. See now if you can use the wave theory of light to explain the following two effects.

13. The photoelectric effect

Streams of electrons pour from the surface of some metals when a strong beam of light shines on it. These are valance electrons (page 584) wrenched loose from atoms by the energy absorbed from the beam of light. This **photoelectric effect** can be made stronger by making the light brighter, as anyone knows who has used a photographer's exposure meter (Fig. 44-26).

But it is more interesting to see what happens when the light is made weaker.

Weston Electrical Instrument Corp.

44-26 The photographer's exposure meter makes use of the photoelectric effect. Light entering a window on the end of the instrument creates a corresponding electric current which is read on the dial at the left.

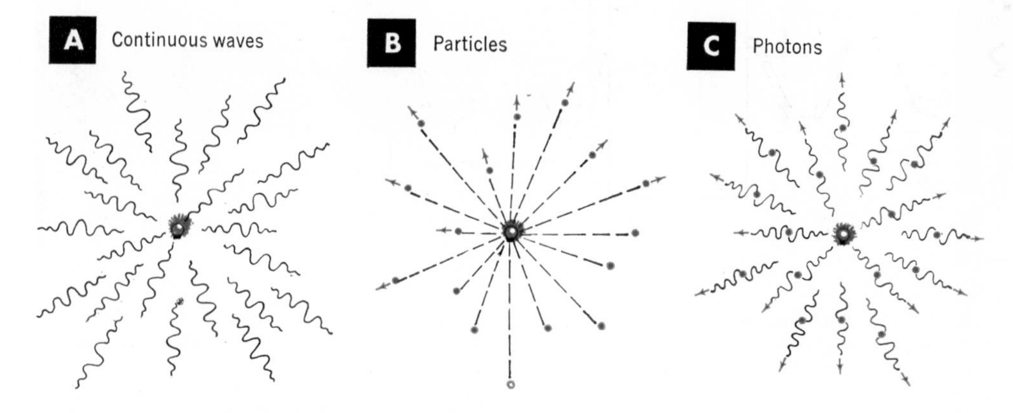

44-27 A modern puzzle. Is light a kind of wave motion (A) or does it consist of particles in motion (B)? Modern theory says both. Today light is viewed as traveling in small packages of wave energy called photons (C).

It is perfectly possible to make the light so weak that only one electron is emitted each second by the light-sensitive surface of an exposure meter. Under this circumstance the very faint light may be illuminating the entire light-sensitive surface, yet the electron emerges from only one atom on the surface. How can one atom absorb energy from light that illuminates an entire surface?

This problem is similar to the problem of explaining how the energy distributed along the length of an average ocean wave could knock down in one blow a telephone pole driven into the shore. As long as the energy of the wave is spread out along its length it can *not* concentrate enough energy on the pole to knock it over in one blow. But if all the energy of the wave could somehow be concentrated at one point then it would easily succeed.

Light waves likewise have sufficient energy to knock electrons off atoms, but *only if all their energy is concentrated at one point,* that is, in a particle. Since we observe that they *do* knock electrons from atoms, we must conclude that light is acting as if it were made of particles! Nor is this the only evidence. There is also the Compton effect.

14. The Compton effect

When X-ray radiation strikes an electron, the electron is given some momentum. The Law of Conservation of Momentum is obeyed precisely, but only if it is assumed that the X ray had momentum and that *all* the momentum took part in the collision.

This can only be true if the momentum of the ray was concentrated in one point—that is, if the ray acted as if it were not a wave but a particle (Fig. 44-27B)! The collision is very similar to the collision between two billiard balls. Try to explain this effect on the basis of wave theory.

Again, this effect can be most easily understood if we abandon our wave theory and accept the particle theory, as A. H. Compton did when he explained it in 1922. The effect has come to be known as the **Compton effect.**

15. Waves or particles?

What are we to think when light behaves as a wave motion in some experiments and as particles in others?

It is we who were at fault in our thinking when we insisted ahead of time that light must be *either* waves *or* particles. We must learn to accept the idea that it apparently has aspects of both. The evidence is indisputable. If such an idea is hard to accept, it merely suggests that man's mind is not the measure of even the observable universe.

Today physicists accept as commonplace the fact that light has a double character (Fig. 44-27C). The evidence you have just been studying is part of the foundation of what is called the **quantum theory,** which is concerned with the nature of radiation and the processes by which atoms emit it and absorb it and even with the structure of atoms themselves.

The situation is even more complex than this. It is worth mentioning in closing that many experiments have been performed which illustrate that *particles of matter act like waves.* Figure 44-28 shows the pattern formed on a photographic film when a stream of electrons passes through a crystal. The orderly arrangement of atoms in the crystal forms, in effect, a diffraction grating, and the pattern that you see is best described by an equation essentially the same as the equation we derived for light waves passing through a diffraction grating.

Such diffraction patterns even enable us to attribute a wave length to electrons. We find that the wave length of an electron decreases as its velocity increases. In electron microscopes, electrons move with sufficient velocity to make their wave lengths shorter than ultraviolet light, and such microscopes therefore give better resolution than any optical microscopes. Indeed, with sufficiently high velocity the effective wave lengths of electrons can be made shorter than the dimensions of an atom. The resolving power of such short wave lengths enables us to examine very fine detail—in this case the very structure of atoms themselves.

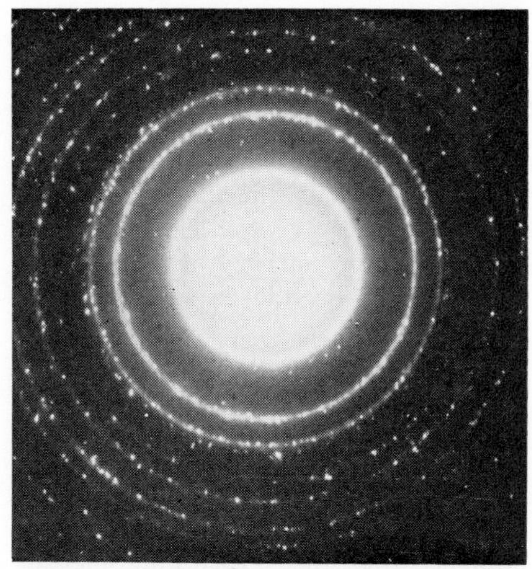

RCA

44-28 When a stream of high-energy electrons passes through very thin crystals, the electrons are diffracted precisely as if they were waves instead of particles. In this picture electrons that have passed through crystals of germanium make this pattern on a film.

In the next unit we shall study information obtained by the use of giant atom smashers. The real value of big atom smashers lies in the fact that the atomic particles they accelerate have such enormous velocities that their wave lengths are exceedingly small—small enough indeed to resolve the structure of the atomic nucleus itself.

In the branch of modern physics called **wave mechanics**, atoms are treated both as waves and as particles.

Going Ahead

THE VOCABULARY OF PHYSICS

Define each of the following:

standing waves	resolving power
node	diffraction grating
out of phase	polarization
in phase	photoelectric effect
diffraction	Compton effect
interference	quantum theory
nodal line	wave mechanics

THE PRINCIPLES WE USE

1. Draw a diagram to illustrate standing waves on a rope.
2. Explain why the rope in your diagram has some points which are motionless.
3. Explain why the maxima of standing waves are half a wave length apart.
4. Explain why an interference pattern appears in the Lloyd's mirror experiment.
5. Why do dark and light interference bands appear across thin vertical soap films?
6. Explain how "coated lenses" work.
7. How does Huygens' principle account for the action of a megaphone?
8. What effect does narrowing the slit have upon the single-slit diffraction pattern? Explain.
9. Why do larger telescope objectives form sharper images of distant objects?
10. What is the advantage of having as many lines per inch as possible on a diffraction grating?
11. Summarize the evidence that light is composed of waves.
12. Summarize the evidence that light is composed of particles.
13. Describe our present view of the nature of light.
14. Why do we believe that particles also have wave properties?

APPLYING THE PRINCIPLES

Explain why

1. In a Western movie a cowboy crouching safely behind a rock can *hear* gunfire directed toward him yet cannot be hit by the bullets.
2. We do not normally see diffracted light when we casually look at things around us.
3. Two sheets of Polaroid material which transmit light when they are in one relative position will stop the light if one of them is then rotated through 90°.
4. Radio-telescope antennas need to be very big.
5. The resolving power of large radio telescopes is generally less than the resolving power of the smaller optical telescopes.
6. You often cannot recognize one of your friends who is far off even though you can clearly see that he is there.
7. The photoelectric effect cannot be explained by means of the wave theory.
8. Giant atom smashers are needed to reveal the structure of the atomic nucleus.

SOLVING PROBLEMS ON LIGHT WAVES

(Assume 5.6×10^{-5} cm for the wave length of visible light in the first two problems.)

1. Two thin vertical posts are 100 cm apart. From about how far away can you resolve them with your unaided eyes when the posts are clearly illuminated?
2. If you use a pair of binoculars whose objectives are 4 cm in diameter, from how far away can you resolve the two posts in Problem 1?
3. Could the radio telescope in Fig. 36-32 resolve two objects separated by the angular diameter of the full moon (30 minutes in diameter) viewed by 21-cm radiation?
4. Find the wave length of a spectrum line for which $n = 1$ and which appears at 5.3° when a grating is used for which $d = 7.30 \times 10^{-4}$ cm.
5. In Problem 4, where would the second order of the same spectrum line appear?

UNIT SEVEN

Atomic Energy—The Motion of Nuclear Particles

Radiation Laboratory, University of California

This long tunnel is the interior of a linear accelerator, a form of "atom smasher" in which relatively heavy fragments of atoms are hurled in a straight line down the central tube extending the length of the tunnel. At the far end, 90 feet away, they collide with atoms of a "target" material. The resulting "splash" of rays and particles from the nuclei of the disrupted target atoms are carefully photographed and measured for clues to the nature of the atomic nucleus.

It may seem odd that the tools of nuclear physics (such as this one) are so very large and powerful, while the nuclei they are designed to study are so tiny that a thousand million of them side by side would form a line still too short to see. But it is just as odd that the energy released from the nuclei of a baseball-sized mass of uranium atoms is enough to destroy a city. Both of these facts are evidence of the enormous energy locked in the nucleus of a tiny atom.

CHAPTER

45

Exploding Atoms

How did we get our first knowledge about atomic energy? One of the earliest indications came from an accidental observation made by the French scientist Henri Becquerel (bĕ'krĕl') in 1896.

Radioactivity

1. Radiation from atoms

While Becquerel was experimenting with some uranium, he put it on top of a package of photographic paper. The photographic paper was in an envelope of heavy black paper to keep out the light. It is not clear what prompted him to develop the paper in the package after a few days. Perhaps he had forgotten whether the paper was exposed or not. At any rate, he developed the paper and found that it had been exposed. The peculiar nature of the exposure caused him to investigate further, and he discovered to his surprise that radiations from the uranium had passed right through the thick black paper.

Experimenting further with these remarkable radiations, Marie and Pierre Curie found that the ore from which uranium was made gave even more intense effects than uranium itself. This suggested that some other material in the ore similar to uranium might be the source of these penetrating radiations. After laborious chemical refining, they managed in 1898 to isolate a tiny bit of radium, a new element, which revealed an astonishing ability to give off enormous quantities of energy continuously.

You can observe similar radiation very simply for yourself.

Take a watch with a luminous dial into a dark room and examine the dial with a strong lens or microscope. The lens reveals an amazing display of fireworks. Actually each of these flashes of light you see represents an atomic explosion. The paint on the dial of the watch has been mixed with a very tiny amount of a **radioactive** material (like radium) whose atoms have the property of breaking down one at a time into other atoms. These atomic explosions make the phosphors in the paint luminous (page 589).

When an atom of radium "explodes," it breaks down into two other atoms—one of the heavy gas called radon, the other of the light gas helium. Before this breakdown, the radium atom weighs 226 times as much as hydrogen, or, we say its **mass number** is 226. The mass number is the number of protons plus the number of neutrons. (Review Chapter 27 for the structure of the atom.) Of the two atoms which result from the breakdown, the radon atom has a mass number of 222, while the helium atom has a mass number of 4 (Fig. 45-1). We write this as follows:

Radium → Radon + Helium
mass 226 mass 222 mass 4

or:

$$_{88}Ra^{226} \rightarrow {}_{86}Rn^{222} + {}_{2}He^{4}$$

The subscript numbers (88, 86, and 2) tell us the number of charged particles (protons) in the nucleus. You will remember from Chapter 27 that this number is called the **atomic number.** In nuclear changes such as shown in our equation above, the sum of the mass numbers before and after the breakdown must remain unchanged ($226 = 222 + 4$), and also the sum of the atomic numbers must remain unchanged ($88 = 86 + 2$).

A century ago such an atomic breakdown into another kind of atom (called **transmutation**) would have been thought impossible. Elements were then thought to be materials which could not be broken down. But with the discovery of radium in 1898 an element was found which actually broke down by itself into other elements. And there was no known way of stopping or speeding up the process.

You will notice that all transmutations originate in the nucleus of the atom. The electrons which surround the nucleus and account for *chemical activity* do not influence nuclear changes.

2. Half-life of radium

When you examine the atomic explosions on the luminous face of a watch, you might assume that all the radium atoms on the dial would be exploded in a short time. But actually only a very tiny fraction of them explode during one second.

It takes 1,600 years for half of any given number of radium atoms to explode. After another 1,600 years the remaining radium is *not* all gone. Only half of it has disintegrated. It would take still another 1,600 years for half of the remainder to disintegrate, and so on. We say that the **half-life** of radium is 1,600 years (Fig. 45-2).

The explosion of each radium atom is

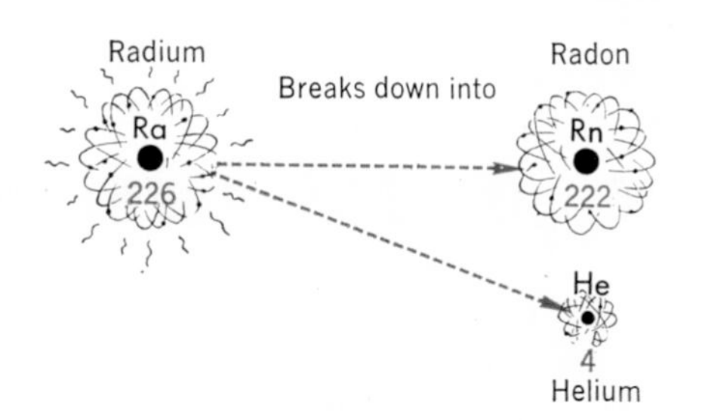

45-1 **Atomic breakdown occurs when a helium atom (mass 4) flies out of the radium atom (mass 226) and leaves a radon atom of mass 222. A large amount of energy is released at the same time.**

so violent that its particles shoot out at speeds of thousands of miles per second. This speed gives each helium nucleus (also called an **alpha particle**) its enormous energy. Along its path each alpha particle rips electrons from thousands upon thousands of any other atoms in the way, turning these atoms into charged atoms or ions. It is this ionization of the atoms in the paint of the watch dial that makes them emit light so that the dial is luminous.

All along its path each rapidly moving atomic particle pushes against molecules, making them move faster and thus creating heat. In a few days a piece of radium can create more heat than the same weight of exploding TNT. And the radium continues to produce heat at this rate for thousands of years! This gives you an idea of the enormous amount of energy that atoms can contain.

3. The family tree of radium

The study of atomic breakdown began with the discovery of radium, but it was soon learned that this breakdown of radium into radon and helium was only one link in a series. Radium itself is the result of other breakdowns that start with the element uranium. And radon atoms, which result from the explosion of the radium atoms, themselves break down further into other elements.

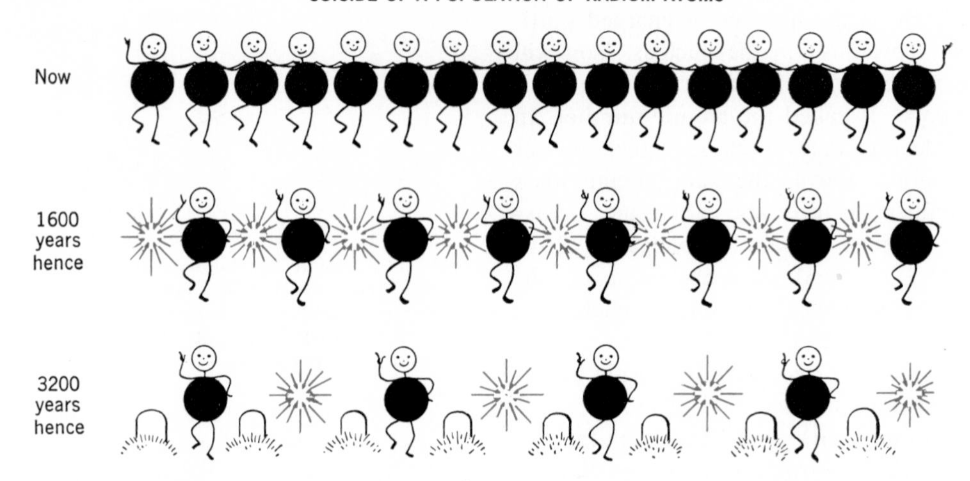

45-2 Radium atoms "pop off" at a definite rate. Half of them explode every 1,600 years. Can you figure out how much of a piece of radium originally weighing one gram would be left after 16,000 years?

These elements in turn break down into others, and so on, until finally, at the end of the series, we have atoms of a variety of lead. These lead atoms in the final step are not capable of further breakdown; with them the exploding process stops.

If we start with uranium as the original atom, we find that radium is the result of the fifth explosion in the series we have been describing (Table 19). After this there are nine more explosions before a stable kind of lead is produced. Note that the half-life of the original

TABLE 19

DESCENDANTS OF URANIUM238: ONE OF THE SERIES OF RADIOACTIVE CHANGES

Element and mass number	Atomic number	Rays emitted	Half-life
Uranium, U^{238}	92	alpha	4,500,000,000 years
Thorium, Th^{234}	90	beta and gamma	24 days
Protoactinium, Pa^{234}	91	beta and gamma	1.2 minutes
Uranium, U^{234}	92	alpha	230,000 years
Thorium, Th^{230}	90	alpha	83,000 years
Radium, Ra^{226}	88	alpha	1,600 years
Radon, Rn^{222}	86	alpha	3.8 days
Polonium, Po^{218}	84	alpha	3.0 minutes
Lead, Pb^{214}	82	beta and gamma	27 minutes
Bismuth, Bi^{214}	83	beta and gamma	20 minutes
Polonium, Po^{214}	84	alpha	0.00015 second
Lead, Pb^{210}	82	beta and gamma	22 years
Bismuth, Bi^{210}	83	beta	5.0 days
Polonium, Po^{210}	84	alpha	140 days
Lead, Pb^{206}	82	none	stable

uranium atoms is 4.5 billion years. Radium, on the other hand, has a much shorter half-life, 1,600 years. The half-life of radon, next in the chain, is only 3.8 days.

Actually the series of atomic breakdowns described in Table 19 is only one of a number of such series. Hundreds of kinds of radioactive atoms are now known, many of them actually produced artificially in atomic reactors or the laboratory.

A piece of pure uranium doesn't remain "pure" very long. The breaking-down process begins at once. Although it would take an enormous number of years for most of the uranium to be converted into lead, atoms of other elements soon appear in it as a result of these series of explosive breakdowns.

All of the elements in this chain, with the exception of the final product, one form of lead, emit one or more of three kinds of rays as they are broken up. These rays, named after the first three letters of the Greek alphabet, are called alpha, beta, and gamma rays (Fig. 45-3). **Alpha rays** are the nuclei of helium atoms traveling at high speed; like the nuclei of any atom they have a positive charge. They are stopped by a few centimeters of air or a piece of paper. **Beta rays** are high-speed electrons; they have a negative charge. They have roughly 100 times the penetrating power of alpha rays. **Gamma rays** are electromagnetic waves like X rays except that some have much higher frequency and have greater energy and penetrating power. They can be detected through thick blocks of iron. They have no charge. (Review Chapter 43.)

Gamma rays are useful in the treatment of cancer, as they kill cancerous cells. But overexposure to gamma rays (or alpha or beta or X rays) can kill healthy cells, produce burns and serious

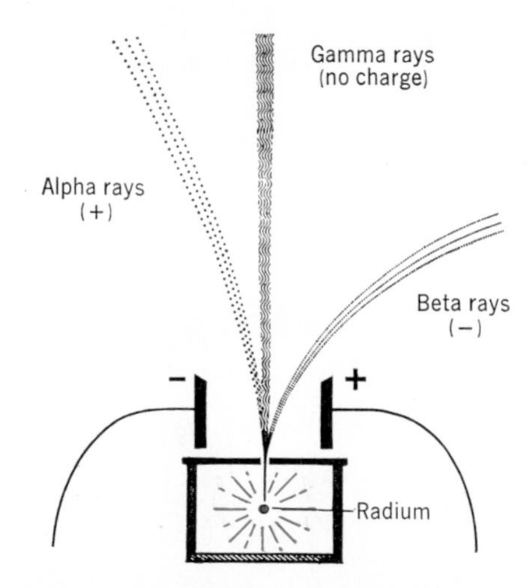

45-3 The three main types of particles and radiation given off by radioactive materials are named alpha, beta, and gamma after the first three letters of the Greek alphabet. They can be separated by using strongly charged plates or magnets. Why should a magnet affect them?

illness, or even cause cancer. For this reason, lead and concrete shielding is used to protect people working with radioactive materials, atom-smashers or X-ray equipment (Fig. 45-4).

4. The cloud chamber

All three types of rays can be detected by several devices. One of them, the **cloud chamber,** was first used for this purpose in 1911 by the English scientist, C. T. R. Wilson.

As shown in Fig. 45-5, the radium or other radioactive material to be studied is placed at R in a closed space containing air or some other gas above a container of water. This air becomes saturated with water vapor. An alpha particle, shooting out of the radium, plows through the atoms of the air, knocking electrons from the outer orbits of some of the atoms. Thus a certain number of the air atoms along the path of the particle become ionized. These ions act as excellent centers for the formation of fog or cloud droplets.

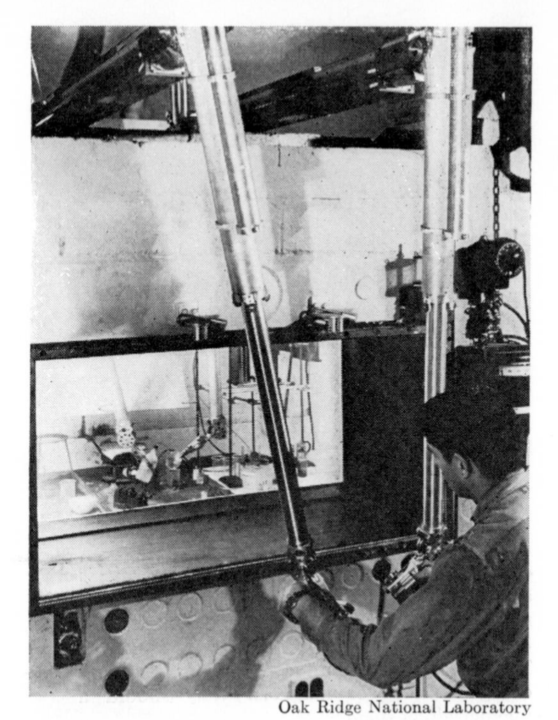
Oak Ridge National Laboratory

45-4 Atomic research is often complicated by dangers from radioactive materials. Here we see a researcher standing behind a special protective window, using mechanical arms to handle his materials.

How can we make these droplets form along the path of ions made by the high-speed particles? As the molecules of water evaporate, the closed air space soon becomes saturated with invisible water vapor. But if the chamber is cooled slightly, some of this water vapor will condense around the ions in the air to form *visible* droplets of water. This cooling is accomplished by slowly squeezing the bulb at the bottom of the chamber and then suddenly releasing it. The sudden expansion of air in the closed chamber cools the air, and the invisible water vapor condenses into tiny visible droplets of water along the path of the high-speed particles. The path of these invisible particles in the cloud chamber is thus revealed in much the same way that the vapor trail from a high-flying airplane reveals its path even though we cannot see the airplane. Cameras placed near the glass top of the container can photograph the path of the rays in the cloud chamber.

5. Recognizing the rays

The greater the energy an atomic particle has, the longer the track it will make in the air chamber before it is stopped. Thus by measuring the length of path, or **range,** of the particle the scientist can estimate its energy. He can decide just what kind of particle it is from the appearance of the path. An alpha particle (helium nucleus) makes a heavy track. The path of the alpha particle is occasionally bent toward the end if the particle comes close to the nucleus of an atom and is deflected by the repulsion of its positive charge. The lighter beta particle (a high-speed electron) produces a light, beaded track which is often made

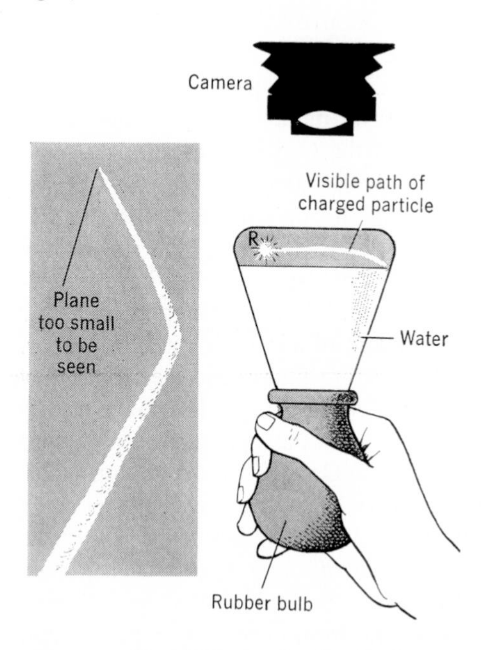

45-5 The Wilson cloud chamber enables you to see the path of radioactive particles in the same way that you detect the vapor trail of an invisible plane high in the sky. Relaxing the squeezed bulb causes a cloud of droplets to form around the electrified particles along the path of the original high-speed particle.

zigzag from many collisions with electrons surrounding the nuclei of atoms.

Since these two particles, the alpha and beta, are charged, they will swerve as shown in Fig. 45-3 when strong magnets or oppositely charged plates are placed near the chamber. You may remember from your study of Oersted's experiment (Chapter 29) that moving electrons (and protons) are surrounded by a magnetic field. Hence magnets are able to move these streams of charged particles in the same way that they move a wire carrying a current (a stream of electrons). The direction in which the particles swerve near a magnetic pole shows which are positive and which are negative. From the amount of swerving, the mass and speed of the particles can actually be calculated. Using such facts the physicist can identify the particles in a cloud-track photo and determine many of their properties (Fig. 45-6).

Rays of the third type, the gamma rays, are not charged and therefore do not swerve when passing close to magnets or charged plates. Their presence is detected by the fact that when they are absorbed by atoms they knock high-speed electrons out of them. The start of the zigzag paths which these electrons take shows where the gamma rays have been absorbed.

Columbia University

45-6 Plenty is happening in this cloud-chamber photo of a beam of mesons (nuclear particles, see page 630). A, B, and C are mesons. One meson has struck an atom at D and has produced an electron-positron pair, E and F. The opposite swervings of E and F reveal opposite charge. As the electron, F, passes through the carbon plate, HG, it hits an atom and produces a new shower of particles. One of them, I, is that of an atomic nucleus, as revealed by the dense track. Note how particle J traces a curve that becomes smaller as it approaches K after passing through the plate, indicating that it has slowed down. Observe the slow electron swerving in an almost complete circle, L.

Nuclear changes

Many transmutations can be produced in the laboratory by bombarding materials with various nuclear particles or gamma rays.

Rays from radium were used at first for this purpose. But a number of far more effective atom-smashing machines to hurl atomic "bullets" at nuclear targets have since been developed. One such machine, the Van de Graaff **electrostatic generator** (Fig. 26-13) builds up high-voltage electric fields which accelerate charged atoms (ions) and shoot them at a target. Another type of atom-smasher is the **cyclotron** (sī′klô·trŏn), invented by the American physicist Ernest Lawrence (Fig. 30-5).

6. The cyclotron

Figure 45-7 shows how the cyclotron works. A large, flat, hollow cylindrical metal box several feet across (the size varies) is cut in half. The two halves, A and B, are placed a short distance

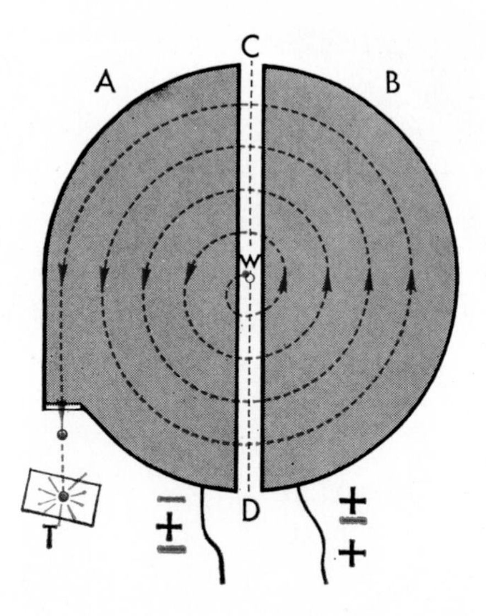

45-7 This merry-go-round type of atom-smashing machine, the cyclotron, gives atomic bullets high speed by a rapid succession of small pushes. The particle gains speed every time it crosses the gap CD until it is hurled through a metal window at the target, T.

apart, enclosed in a container from which the air has been removed, and mounted between the poles of a magnet (Fig. 30-5). A hot wire filament, W, in the space between the two halves of the box gives off electrons. These electrons bombard nearby atoms of gas (usually hydrogen or helium), ripping off one or more electrons from them so that the atoms of gas are turned into charged ions. The positively charged nuclei are the "bullets" of the cyclotron. To accelerate them, the two halves of the box are given opposite charges by an outside electric circuit. As a result, the negative half of the box attracts the positive ions, causing them to move.

The magnetic field through the box causes the positive ions to move in a circular path. Because of this curving motion the stream of particles returns to the center line, CD, between the two halves. At the moment the particles reach the center, the charge on the box halves is reversed so that the particles are pulled across the intervening space to the other half of the box.

The accelerating particles make one complete revolution around the inside of the box in about $\frac{1}{1,000,000}$ second. This means that the alternating voltage on the two halves must have a frequency of about 1,000,000 cycles per second.

With each crossing of the gap between the halves the stream of particles picks up speed and energy. As its speed increases, it moves in ever-widening circles. After many such trips between the two halves of the box, the spiral path of the stream of particles grows until the stream finally shoots out of a thin metal window at the edge of the box to hit a material serving as a target. By this time the speed of the particles may be tens of thousands of miles per second, so that the particles are very effective atom-smashers.

A great many transmutations of atoms have been produced in this way. High-speed positively charged particles from devices like the cyclotron can smash through the repulsive force of the positively charged nucleus to its center and change it to the nucleus of another element. Often the new atoms produced by this smashing process are "unstable" and begin to break down at a regular rate just like the radioactive atoms of radium. Thus man can artificially create atoms which are radioactive.

A number of new kinds of atom-smashing machines are now in existence (Fig. 45-8), some of which can produce particles with energies equivalent to several billion electron-volts (page 592).

An atom-smasher must hurl these charged particles or "atomic bullets" with enough kinetic energy to overcome the electrostatic repulsion of the protons or electrons of the atom in order to penetrate it. That is why big machines like the cosmotron are so useful.

Brookhaven National Laboratory

45-8 This atom-smasher at Brookhaven, Long Island, accelerates protons to enormous energy. When the protons then collide with atoms inserted as targets, the resultant fragments provide clues to the structure of the nuclei of the target atoms.

7. "Atomic billiards"

By using a cloud chamber or some other device to observe the paths resulting from collisions of high-speed atomic particles with atoms of a target material, scientists can figure out just what kinds of atoms have been formed by these artificial transmutations. The bombardment of atoms by high-speed particles might be thought of as a kind of "atomic billiards."

Figure 45-9 represents one such collision—and the resulting first man-made atomic transmutation, accomplished by the English scientist, Ernest Rutherford, in 1919. He bombarded nitrogen with the alpha particles ($_2He^4$) from radium. The "bookkeeping" for nuclear particles in this collision is shown at the foot of this page.

You notice here that the total mass of the nuclei of the different atoms is the same before and after transmutation. (There is a slight difference that we shall note later). Likewise the total positive charges before and after are equal.

The bombardment of the light metal, beryllium, with alpha particles (helium nuclei) led to the discovery of the **neutron** by the English scientist Chadwick

	$_2He^4$ helium	+	$_7N^{14}$ nitrogen	→	$_8O^{17}$ oxygen	+	$_1H^1$ hydrogen (proton)
protons (positive charge)	2		7		8		1
neutrons	2		7		9		0
mass number	4	+	14	=	17	+	1
atomic number (positive charge)	2	+	7	=	8	+	1

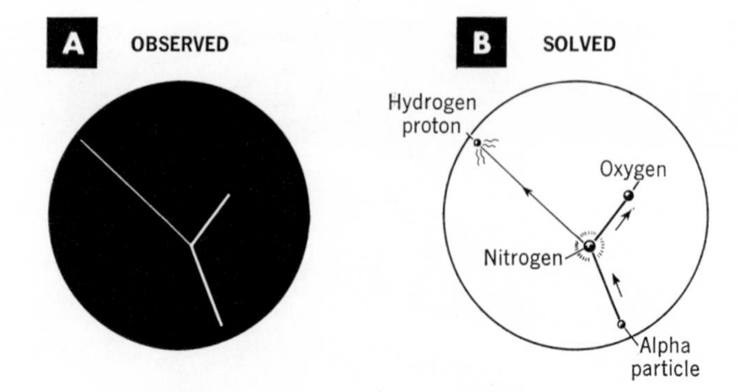

45-9 Atomic billiards! Calculations based on angles and speeds of particles in the observed track (A) led Rutherford to deduce that nitrogen and helium atoms were being transmuted into oxygen and hydrogen atoms.

in 1932 (see Chapter 27). Neutrons are ideal as "atomic bullets" in transmutation experiments. Because they have no electrical charge, they travel between the outer electrons of an atom right to the nucleus without being deflected. A positively charged atomic particle (such as an alpha particle or a proton) will be repelled by the positive nucleus of an atom as it approaches, and will therefore have difficulty hitting it. Similarly, negative particles (such as electrons) will be repelled by the negative charge of outer electrons of the atom and will rarely get in to strike the nucleus. Neutrons, having no charge, are not affected by the charges on either nucleus or electrons. When a nucleus is penetrated by a neutron, the neutron may be captured, and if the nucleus thus formed is unstable and therefore radioactive, it disintegrates and is transmuted.

Figure 45-10 shows how the collision of a neutron with an atom of platinum produces radioactive platinum that emits a beta ray to become gold.

$$Pt^{196} + n^{1} \rightarrow Pt^{197}$$
$$Pt^{197} \rightarrow Au^{197} + e^{-}$$

Similarly, many other elements are turned into isotopes of a different element by neutron bombardment. Isotopes, you will remember, are forms of the same element which differ only in mass because of differences in the number of neutrons. They are the same element because they have the same number of protons, that is, the same atomic number. Many (but not all) of these isotopes are radioactive, like radium, and explode to form still other atoms. **Radioactive isotopes,** used as **tracers,** are of great importance in biology and chemistry. These self-exploding atoms inside a plant or an animal can be detected in very tiny amounts from the outside by a Geiger counter (see page 627) so that one can find out exactly where the atoms have gone. In this way we can trace the wandering of atoms of a given element through a plant or an animal, find out where they go, how long they stay there, and when and how they leave. Radioactive isotopes may soon become a scientific tool as important as the microscope.

8. Measuring radioactivity

One of the earliest tools used to measure radioactivity, especially gamma radiation, was a charged **gold-leaf electroscope** (page 333). This instrument is still used in modified form. It is usually charged by momentarily connecting a high-voltage battery between the metal rod supporting the leaves and the metal case. If the leaves are charged negatively, any positive ions of air in the case will be attracted to them. Each ion that comes in contact with them will remove one or more electrons. Slowly the leaves

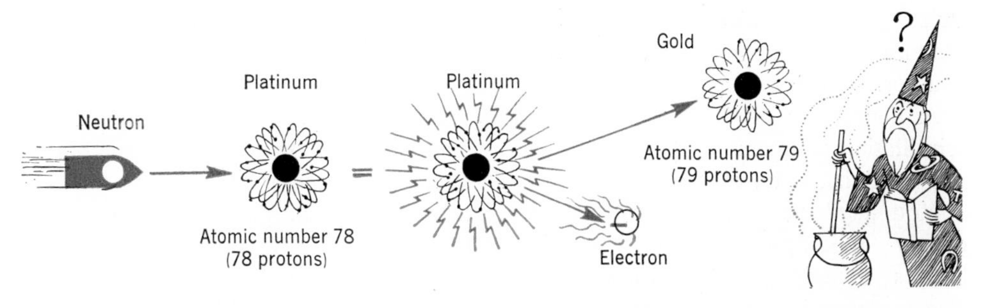

45-10 The ancient alchemist looks on with envy as platinum is transmuted into gold after it is struck by a neutron. Of course, there is no economic value to such a transmutation, but scientists were led to the discovery of practical atomic energy by means of such experiments.

will lose their charge. They will no longer repel each other as strongly, and gravity will cause them to collapse. Their rate of collapse is used as a measure of the strength of the radioactivity that caused the ionization. A scale may be mounted behind the leaves to measure their movement. Since the instrument depends on the ionization of air for its operation, it is called an **ionization chamber.**

A compact form of ionization chamber, similar to a fountain pen in size and shape, is carried by laboratory workers to determine their exposure to radiation.

Another important instrument used to detect and measure radioactivity is the **Geiger counter.** This device consists of a cylindrical metal tube containing a gas, with a metal wire (W in Fig. 45-11) down its center. The walls of the tube and the wire in the center serve as oppositely charged terminals of a high-voltage source. The distance between the outer metal cylinder and the wire is great enough to prevent the charge from sparking across the nonconducting gas between them. But when the tube is brought near some radioactive atoms, the high-speed particles given off by the atoms shoot through the tube, creating ions which make it possible for the gas to conduct a charge between the terminals of the tubes. This charge is amplified electrically to cause an earphone or a loud-speaker to click, a neon bulb to flash, a meter to register, or a counter to count.

Another early tool used to measure radioactivity was a device much like a luminous watch dial, which produced visible flashes or **scintillations** when struck by alpha particles. It consisted of a small glass screen coated with zinc sulfide (a phosphor), which momentarily fluoresced wherever a particle struck it. The experimenter observed the flashes with a magnifying glass and counted them. In the modern **scintillation counter** the flashes are detected by a very sensitive form of photoelectric cell which detects the feeblest flash.

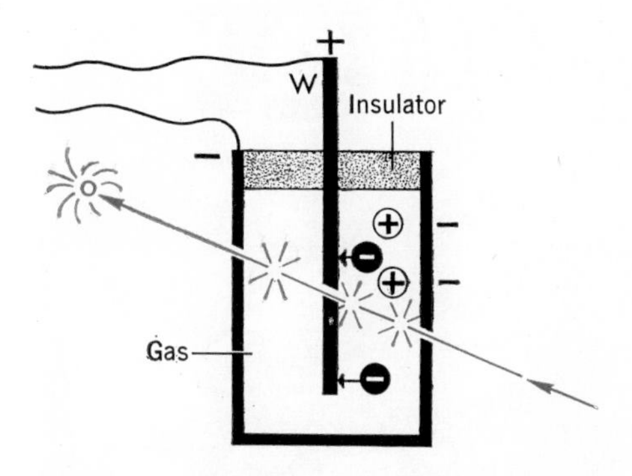

45-11 A simplified diagram of a Geiger counter. As atomic particles pass through, they ionize the gas molecules in the tube, allowing a current to flow through the chamber. The pulse of current is amplified and causes a click in a loudspeaker.

Pulses generated in the photoelectric cell are amplified and counted as in a Geiger counter. By using different phosphors or even fluorescent crystals for the screen, scintillation counters can be made sensitive to alpha, beta, or gamma radiation.

You may have seen pictures of atomic-energy laboratory workers wearing badges containing photographic film. The more radiation they have been exposed to, the more the film will appear to be blackened when it has been developed. This method of detecting radiation is used mainly for the protection of personnel in these laboratories.

9. Using tracers

Suppose that a physician wants to observe a patient suffering from a disease of the thyroid gland, located in the neck (Fig. 45-13). He knows that the element iodine is absorbed from the blood by the thyroid gland and is essential to its operation. Therefore, the patient is given a small, harmless quantity of radioactive iodine in his food. When a Geiger or scintillation counter is placed near his neck, the counts begin to increase in a short while, showing that the thyroid gland is removing iodine from the bloodstream. The doctor can tell how long the thyroid gland keeps the iodine, when it loses it, and how rapidly this occurs. From this information he can diagnose certain kinds of thyroid disease.

The radioactive iodine acts in this instance as a tracer. Similar tracer work is today of great importance in scientific research in the fields of biology and chemistry.

American Cancer Society

45-12 A sample of radioactive carbon is exposed to a Geiger counter to measure the degree of its radioactivity.

10. Mass into energy

In 1905, Einstein suggested that mass and energy are equivalent to each other. According to Einstein the amount of energy in ergs, *E,* equivalent to a given mass of matter, *m,* is given by the formula:

$$E = mc^2$$

where *m* represents the mass in grams and *c* stands for the enormous velocity of light (3×10^{10} cm/sec). As you observe from the formula, the velocity of light is multiplied by itself and then by the mass to give the equivalent amount of energy in ergs. Thus a very small amount of mass equals an enormous amount of energy. Translating this formula into familiar units we find that:

$$E \text{ (in kwh)} = 11{,}400{,}000{,}000\ m \text{ (in lb)}$$

This means that *if* we could destroy a pound of matter and convert all of it into energy we would get about 11.4 billion kilowatt-hours of energy! If we could destroy a mere 20 pounds of matter and convert it into energy, we would get enough energy to supply all of the electrical energy used in the world for one year! Or, if all the matter in one breath could be converted into energy it could

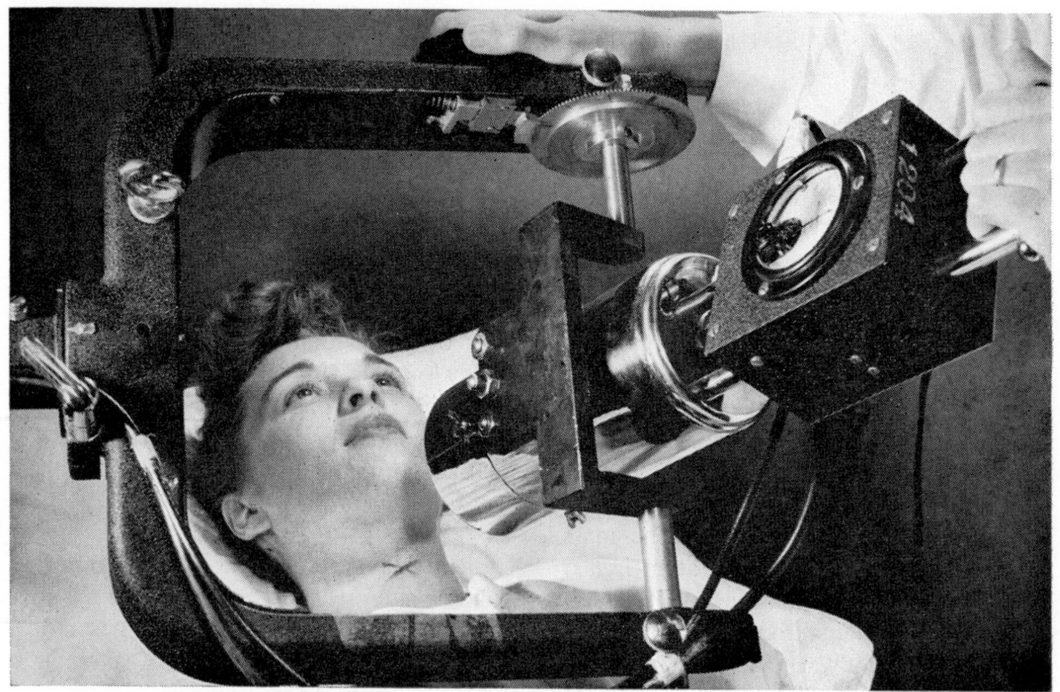

Brookhaven National Laboratory

45-13 X marks the location of the thyroid gland in the throat of this patient. The clicking of the Geiger counter reveals the way in which her thyroid is picking up the radioactive iodine that she has taken by mouth. Such experiments are important in studying and treating diseases of the thyroid gland.

drive an ocean liner for several years (Fig. 45-14).

Although Einstein did not predict that mass could be converted into energy in a practical way, atomic scientists have actually been able to do so, thus providing evidence for his theory and for the Law of Conservation of Mass-Energy (page 10). Today, atomic power plants are actually converting mass into energy on a large scale.

11. Where does the energy come from?

When the cyclotron was used to bombard the light metal lithium with protons, the results confirmed Einstein's theory about the equivalence of mass and energy. Two helium atoms were produced as a result of this collision. The bookkeeping for this transmutation, using extremely accurate values for the atomic masses involved, is shown in the table on page 630.

As you see, the atomic masses do not balance exactly. About 0.0185 units of mass have disappeared. The amount of energy this mass represents was calculated from Einstein's formula $E = mc^2$ and compared to the observed amount of

45-14 One breath of air doesn't weigh very much. But if all the mass in the breath of air could be converted into energy, it could drive an ocean liner for several years. Note that we said "if."

$$\text{Lithium} + \text{Hydrogen} \rightarrow \text{Helium} + \text{Energy}$$
$$_{3}Li^{7} + {}_{1}H^{1} \rightarrow 2\ {}_{2}He^{4} + \text{Energy}$$

$$\text{Atomic masses } 7.0182 + 1.0081 = 2 \times 4.0039 + \text{Energy}$$
$$8.0263 = 8.0078 + \text{Energy}$$
$$\text{Energy} = 0.0185 \text{ units of mass}$$

energy possessed by the fast-moving helium particles resulting from the collision. The two amounts of energy were almost identical. Here was dramatic evidence for the fact that mass had been converted into energy, confirming Einstein's formula.

There is evidence, too, that energy can be converted into mass. The mass of the entire nucleus of an atom is *less* than the sum of the masses of the individual protons and neutrons that compose it. This means that when a nucleus is broken up into its individual parts, some extra mass must be supplied from somewhere to account for the greater total mass of the individual particles. This extra mass is equivalent to the energy that holds the protons and neutrons of the nucleus together. This energy is called the **binding energy** of the nucleus.

In nearly every case the energy released in a nuclear reaction comes from the difference in binding energy between the original and the newly formed atoms and *not* from the destruction of any of the nuclear particles. That is why in a nuclear equation the left- and right-hand sides have the same total atomic number and the same total mass number.

We learned in Chapter 1 that radiation from the sun produces all the ordinary forms of energy we have on earth. To supply this energy, as well as the much greater amount that the earth does not capture, 4 million tons of the sun's mass have to be consumed in every second. It is believed that this conversion of mass into energy occurs inside the sun when atoms of hydrogen are changed (transmuted) into helium atoms at temperatures of many millions of degrees. The energy of a hydrogen bomb (H-bomb) is obtained by methods similar to the energy-producing changes that occur inside the sun (page 640).

12. Other atomic particles

In our discussion of the particles that make up atoms of the materials that exist on the earth we have concentrated upon three: the electron, the proton, and the neutron. As far as we know now, these three are the most important. But there are other particles which have been detected.

The **positron** (E in Fig. 45-6) is a particle of the same weight as an electron, but its charge is positive instead of negative. Positrons are released from some atoms that have been made radioactive by bombardment. They can also result from collisions between atoms and cosmic ray particles. Sometimes a gamma ray produces an electron-positron pair. Here the energy of gamma rays is being turned into matter.

A positron leads a very fleeting existence. When it collides with an oppositely charged electron, the positron and the electron annihilate each other. Both disappear, and in their place there appears an equivalent amount of energy in the form of a gamma ray. The energy released is c^2 times the mass destroyed, giving further confirmation of the Law of Conservation of Mass-Energy (page 10).

Cosmic-ray collisions can produce another kind of particle called the **meson** (mĕs′ŏn). The mass of the meson is between that of the electron and the proton—roughly 200 to 300 times the mass of an electron. It may have a positive charge, a negative charge, or no charge. There seem to be a number of varieties of meson. The life of some varieties is only about $\frac{1}{1,000,000}$ second. Immediately after the collision which produces them, they decay like radioactive atoms into electrons or positrons.

With better atom-smashers we can obtain particles whose energies approach that of cosmic rays. We can therefore produce mesons in large quantities for study in the laboratory. Mesons are responsible for the strong forces which hold neutrons and protons together in the nucleus; in other words, they furnish the "cement" of the nucleus. Once we can produce mesons at will we may find we have access to much greater supplies of atomic energy than we now get from the nuclei of atoms. What mankind does with such new sources of energy is an urgent question which we shall discuss in our next chapter.

Neutrinos, like neutrons, have no charge, but unlike neutrons they have practically no mass at all. Their existence was predicted to explain the variation in energy of beta rays emitted from disintegrating nuclei, but because they have no charge and little mass, they do not interact easily with other forms of matter and were not detected by instruments until recently.

The **antiproton** was also first predicted to explain a nuclear reaction and also has just recently been detected. It is just like a proton but has a negative charge. When an antiproton collides with a proton, the result is similar to what happens when a positron collides with an electron; both are annihilated and their mass is converted into energy—about 2,000 times as much as that produced by a positron-electron collision, since their masses are 2,000 times greater. Physicists are now speculating on the possibility of the existence of atoms of "antimatter," made up of antiprotons, antineutrons, and positrons! It is even conceivable that some of the galaxies we observe outside our own Milky Way system may consist entirely of stars made of antimatter.

Going Ahead

THE VOCABULARY OF PHYSICS

Define each of the following:

radioactive
half-life
ionization chamber
cloud chamber
transmutations
cyclotron
meson
mass number
antiproton
alpha rays
scintillation counter
beta rays
neutron
radioactive isotopes
tracers
Geiger counter
positron
electrostatic generator
atomic number
neutrino
binding energy
gamma rays
scintillations

THE PRINCIPLES WE USE

1. What property of radium atoms makes them different from most other atoms found in nature?

2. What happens when a radium atom breaks down?

3. What is meant by the statement that the half-life of radium is 1,600 years?

4. What causes the scintillation of a luminous watch dial?

5. How does the heat energy supplied from radium compare with that of an equal weight of TNT?

6. What is the final product of radium breakdown?

7. How do scientists try to duplicate the breakdown of atoms in the laboratory?

8. What was the first artificial transmutation performed in the laboratory? Write the equation which shows the products before and after the experiment.
9. What was Einstein's prediction as to the relationship between mass and energy?
10. According to Einstein's theory, how much energy could be obtained by converting 1 pound of mass into energy?
11. How was evidence of his theory obtained by experiments with atomic transmutations?
12. What is the source of the energy of the sun?
13. How can atoms of gold be produced from atoms of platinum?
14. Name five basic atomic particles, and mention at least two facts about each.
15. How are positrons produced?
16. How are mesons produced?
17. Under what circumstances is energy converted into mass?
18. Describe the properties of neutrinos.
19. Describe the properties of antiprotons.

Explain why

20. Droplets of water form around the path of a high-speed atomic particle in a cloud chamber at the moment the pressure is released.
21. Powerful magnets are sometimes placed above and below a cloud chamber while it is operating.
22. Physicists are building larger and larger atom-smashing machines.
23. It is impossible for an atom smasher to accelerate neutrons.

APPLYING THE PRINCIPLES

1. Draw a diagram of a simplified Wilson cloud chamber and explain the operation of the chamber.
2. Draw a diagram of a cyclotron and give a brief explanation of its operation.
3. What are the uses of atom-smashing machines like the cyclotron?
4. Draw a diagram of a Geiger counter and explain its operation.
5. Describe two uses for a Geiger counter.
6. Explain the construction and operation of an ionization chamber.
7. Explain the construction and operation of a scintillation counter.
8. Explain why atom-smashing machines are being built to produce higher and higher energy.

SOLVING PROBLEMS ABOUT ATOMIC ENERGY

1. If it takes 1,600 years for half of a given sample of radium to break down into other atoms, how much of 1 gram of radium will be left after 16,000 years?
2. The age of the earth is calculated from the radioactivity of uranium. It takes about $4\frac{1}{2}$ billion years for half of a given amount of uranium to break down into lead. If a certain sample of uranium on earth is found together with lead in ores in the proportion of 80% uranium to 20% lead, what is the age of the rock? If you can, make an exact calculation using logarithm tables.
3. The amount of energy radiated by the sun is sufficient to provide about 1 horsepower for every square yard of the earth's surface when the sun is overhead. What is the loss of mass of the sun each second? (Use the following facts: 1 hp equals about $\frac{3}{4}$ kw. The conversion of 1 lb of mass will produce about 11.4 billion kwh of energy. The distance to the sun is 93 million mi. The formula for the area of a sphere of radius r is $A = 4\pi r^2$.)

---

4. Plot a graph of the disintegration of 1 gm of polonium 218 (Po^{218}) whose half life is 3.0 min. Plot the mass, m, remaining after a time t. The time axis should include 0 to 12 min.
5. Repeat Problem 4, but plot the *logarithm* of the remaining mass against the time. Interpret the shape of your graph.

CHAPTER

46

Atomic Energy

When atoms were first bombarded with neutrons it was noticed that the new atoms which resulted often had an atomic number one higher than the original atom. Neutron bombardment, for example, changes platinum, number 78, into gold, number 79 (Fig. 45-10); and iron, number 26, becomes cobalt, number 27.

In 1934 the Italian scientist Enrico Fermi began to experiment with the bombardment of uranium. Of all the elements known at that time uranium had the highest atomic number, 92. Fermi wondered whether uranium, when bombarded with neutrons, might not change into a new element with an atomic number of 93.

As Fermi expected, the uranium sample was altered by the bombardment. Now the job was to identify the new atoms that had been produced. Fermi and other scientists all over the world worked at the problem. They wrongly assumed that whatever the new material was, it would be a heavy atom like uranium with approximately the same atomic number. It was not until the latter part of 1939 that two German scientists working on the problem announced the discovery that revolutionized atomic science and changed the course of history.

Nuclear fission

These two scientists, Otto Hahn and F. Strassman, found that the atoms of uranium were split, or divided, approximately in half by the bombardment of neutrons (Fig. 46-1). The new atoms produced were of much lower mass and atomic number than the original atoms of uranium. This is the process which today we call **nuclear fission.** Moreover, after each fission some mass disappeared, having been converted into energy.

1. Splitting the uranium atom

Soon other discoveries were made about the fission (splitting) of uranium atoms. It was found that the energy released by this breakdown was much greater than that which occurred with naturally radioactive materials—about 20 times as much per atom. Some isotopes of the element thorium (atomic number 90) could also be split in the same way, releasing about the same amount of energy per atom as uranium.

Then came the most important discovery of all. When the uranium atom was split, it produced additional neutrons! A single neutron thus caused an explosion which produced several other neutrons. It now appeared possible that a nuclear **chain reaction** could split many atoms in a given piece of uranium (Fig. 46-2). In such a chain reaction the exploding atoms would in turn explode others which would then explode others, and so on. In this way a self-sustaining chain of explosions would be set up in which a large part of the energy in the atoms could be liberated.

2. Starting a chain reaction

But there were difficulties to be solved before such a chain reaction could take place. Natural uranium is known to consist of two principal isotopes, uranium

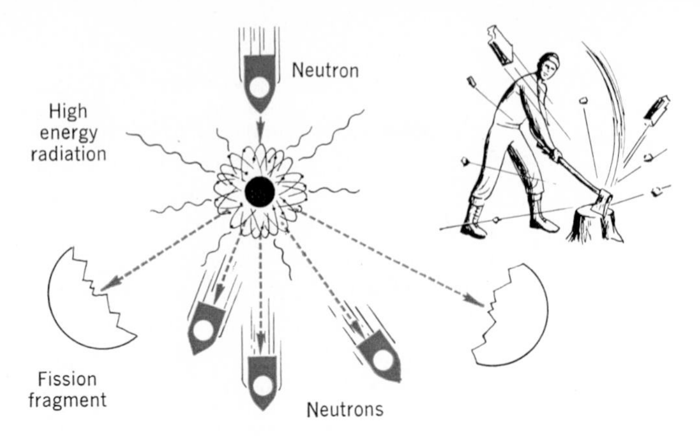

46-1 Most radioactive materials chip off small bits when they explode. But when uranium 235 is struck by a neutron, it splits into two large fragments and several neutrons. This splitting process (fission) gives off a great amount of energy in the form of heat and gamma radiation. And the newly formed neutrons make it possible to obtain a chain reaction.

238 and uranium 235: any given piece of uranium contains these two isotopes in the proportion of 140 atoms of U^{238} to 1 atom of U^{235} (Fig. 46-3). But it is only the "minority" atom, U^{235}, which can be split by neutron bombardment. Atoms of U^{238} capture and absorb the bombarding neutrons, but do not split.

A free neutron moving about in a piece of uranium strikes atoms at random. There is thus a 140-to-1 chance that it will strike an atom of U^{238} and be captured, instead of hitting a U^{235} atom and causing it to split and release other neutrons.

To keep a chain reaction going, every neutron which explodes a U^{235} atom must produce at least one other neutron which will explode still another atom of U^{235}. Otherwise the reaction will rapidly die out.

3. The nuclear reactor

A way was found to prevent neutrons from being captured by the nonsplitting atoms of U^{238}. It turned out that if the

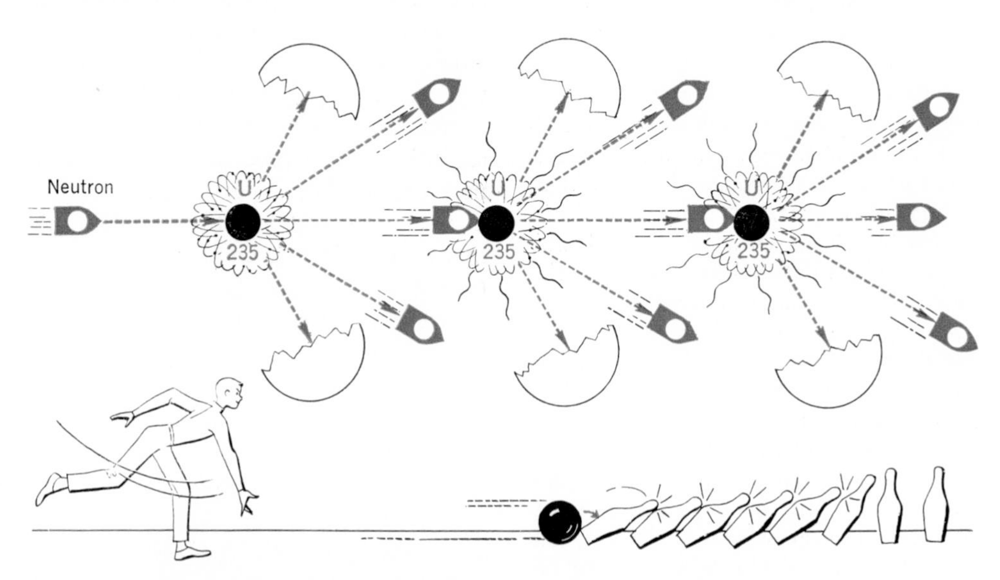

46-2 If one of the newly formed neutrons resulting from fission of a U^{235} atom is successful in splitting another atom, and this process continues, then a chain reaction will result. Like the bowler who makes a strike and knocks down all the pins, the original neutron hits one atom and starts the successive explosions of atom after atom.

neutrons were slowed down below a certain speed, very few were captured by the U^{238} atoms. Instead, the U^{235} atoms absorbed these slow-moving neutrons and then split, with the liberation of energy and the production of more neutrons.

Certain light materials such as carbon or "heavy water" (see Chapter 27) do not capture neutrons but instead let them bounce off. When the neutron hits an atom of such a material it gives up some of its energy to it and is slowed down. Thus if a neutron is set free in carbon, it bounces around from one atom of carbon to another until its speed is reduced below the point where it would be captured by U^{238}. A material used in this way to slow down neutrons is called a **moderator** (Fig. 46-4).

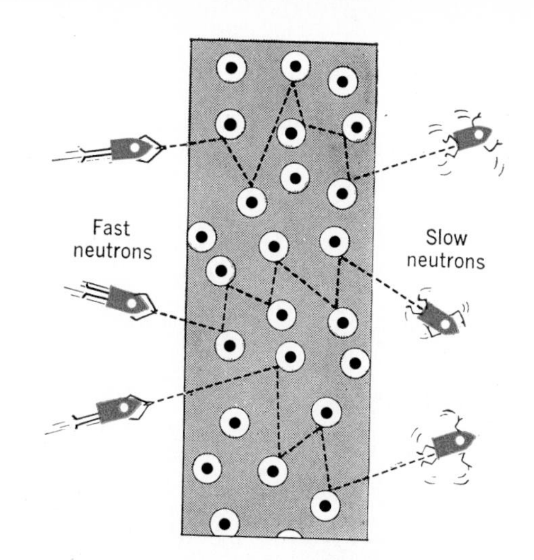

46-4 Moderator materials slow down neutrons by letting them bounce off the atoms without capture. After a number of bounces, the neutrons lose most of their energy and speed and are capable of splitting U^{235} atoms.

Figure 46-5 shows a **nuclear reactor**, which was developed by physicists from all over the world working in the United States during the Second World War. The black dots represent rods of uranium imbedded in blocks of carbon (actually graphite, a form of carbon). Since the air around us is constantly being bombarded by cosmic rays, it always contains a few stray neutrons. One such neutron, entering the blocks of carbon in the pile, may bounce about until it is slowed down and eventually wanders into the container of uranium. Once there, it is in little danger of capture by the atoms of U^{238}, because of its slow motion. Instead, after a number of collisions with nuclei of U^{238} atoms, it finally collides with an atom of U^{235} and produces an atomic fission (splits the atom). This releases energy and a new crop of neutrons. One or more of these may shoot into the carbon, slow down after a number of collisions, and strike more uranium to create another atomic fission. If this process continues we have a chain reaction. Atom after atom will be split and provide an enormous amount of energy in the form of heat (molecular motion) and radiation (gamma rays and high-speed particles).

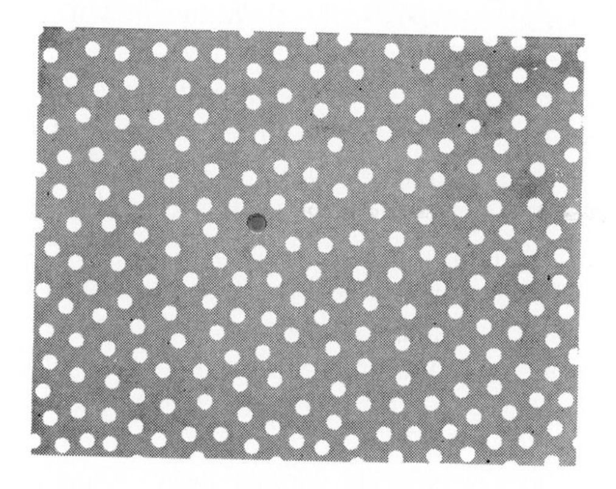

46-3 U^{235} atoms are rather rare compared to U^{238} atoms in natural uranium (1 to 140). But scientists found a way of getting the neutrons to be selective and bounce off the U^{238} harmlessly until one finally hit a U^{235} atom. This was done by slowing the neutrons down to a point where they had no effect on U^{238} atoms.

4. Controlling the reactor

But the nuclear reactor must be of a certain size. If it is too small, too many neutrons will escape from it, and the

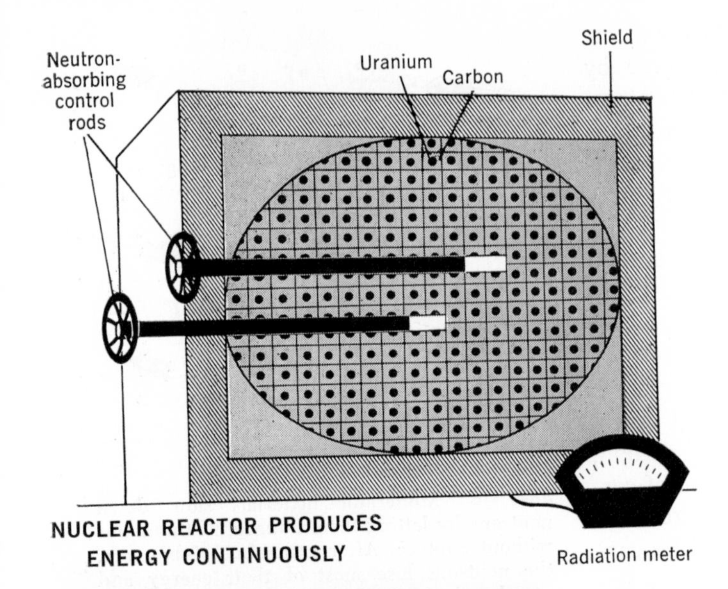

46-5 The nuclear reactor produces a slow, controlled chain reaction. The carbon (moderator material) surrounds the uranium and slows down the neutrons to a speed where they are captured only by U^{235} atoms. Such reactors can produce heat and power.

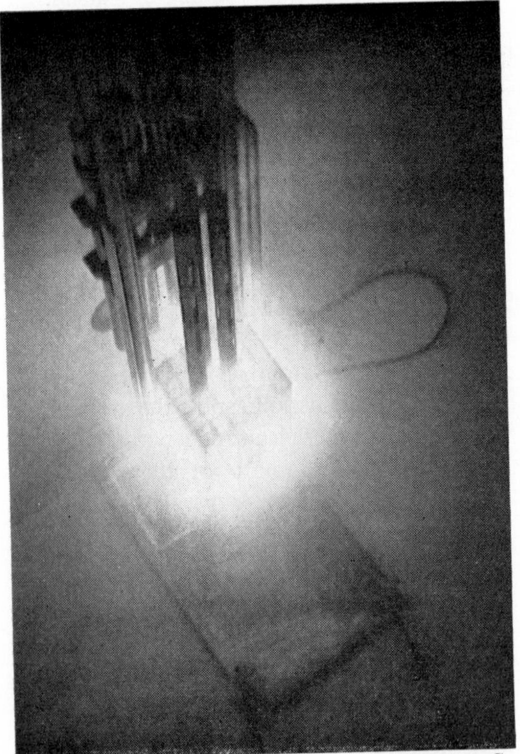

Union Carbide Chemicals Co.

46-6 If water is used as a moderator it is easy to see the core of a reactor that is functioning. Here the fuel rods at the bottom of a deep tank of water are surrounded by an intense blue glow created in the water by the fission of the uranium.

chain reaction will stop. On the other hand, if it is too large, the chain reaction may increase to the point where the reactor will melt, releasing dangerous amounts of radiation. Hence there is a certain "critical size" which is needed for the reactor to operate.

The danger of melting is controlled by rods made of certain materials, such as the metal cadmium or the element boron. Materials like cadmium or boron are excellent "neutron thieves." When a rod of cadmium is inserted into the reactor, it captures so many neutrons that the chain reaction is slowed down and stops. When the rod is pulled out, the chain reaction can increase again. These cadmium **control rods** are used to keep the chain reaction at a balance in which 100 atom-splitting neutrons are replaced by 100 others which split atoms in turn.

The first such self-sustaining reactor, which produced heat and radiation, was put into operation under the stands of Stagg Field at the University of Chicago on December 2, 1942 (Fig. 46-7). We may take this date as the opening of the Atomic Age.

5. Plutonium: a man-made element

At the beginning of this chapter we told how scientists were led to the discovery of atomic fission when they sought for a new element created by the bombardment of uranium with neutrons. They expected this element to have an atomic number higher than that of uranium. They finally turned out to be right in this expectation, but the new element so formed was not discovered until after nuclear reactors were put into operation. Some neutrons, before they are slowed down, are captured by atoms of U^{238}. As a result, the atom of U^{238} (atomic number 92) becomes U^{239}, which is radioactive and shoots out a beta ray. A beta ray is an electron created in the nucleus by the conversion of a neutron into a proton. Since an electron cannot exist in the nucleus it is ejected at once as a beta ray. And since the conversion of a neutron into a proton changes the atomic number, a new element results. In the case we are describing, an atom of U^{239}, whose atomic number is 92, changes into an atom of **neptunium,** Np^{239}, whose atomic number is 93. Thus

$$U^{238} + n^1 \rightarrow U^{239} \rightarrow Np^{239} + e^-$$

The atom of neptunium is also unstable and in a short time it too shoots out an electron, changing into another element whose atomic number is 94. This element has been named **plutonium** (Fig. 46-8).

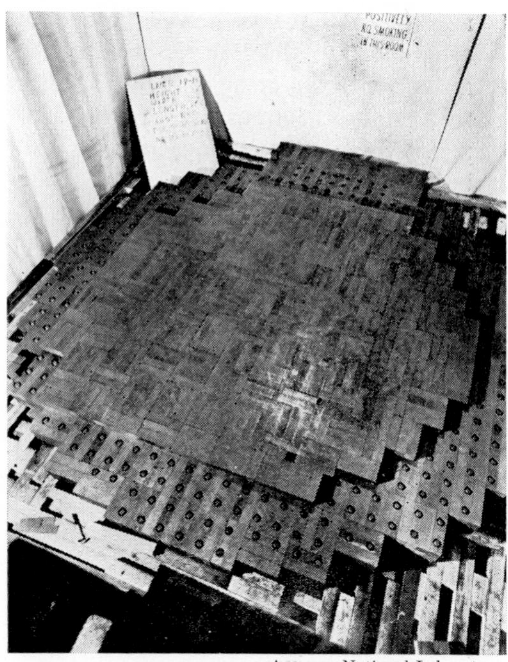

Argonne National Laboratory

46-7 Only photograph of the first history-making nuclear reactor (made in 1942). Note the uranium embedded in carbon blocks and covered with more carbon blocks. We see here the eighteenth layer of uranium. This piling-up process was continued for 57 layers.

$$Np^{239} \rightarrow Pu^{239} + e^-$$

It is interesting to note in passing that the names uranium, neptunium, and plutonium were taken from the planets Uranus, Neptune, and Pluto.

Plutonium acts very much like U^{235} in that it can be split by slow-moving neutrons. Hence it is atomic-energy mate-

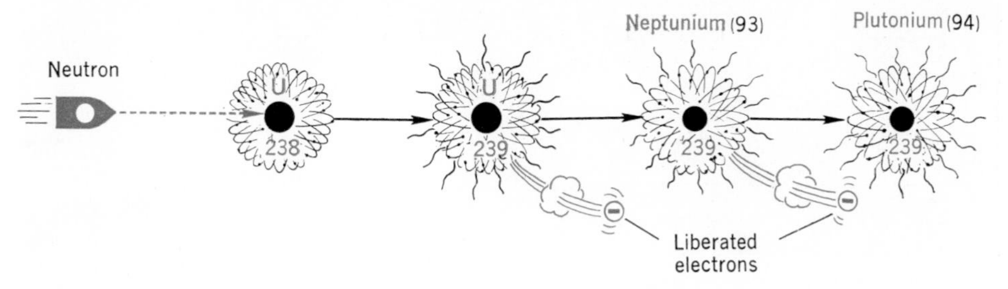

46-8 Plutonium is produced in a nuclear reactor from the capture of fast neutrons by U^{238} atoms. The capture changes U^{238} to U^{239}, which is radioactive and emits an electron to form neptunium 239 (element 93). Another loss of an electron by neptunium results in plutonium 239 (element 94).

rial. Furthermore, it can be produced artificially in a nuclear reactor. As the reactor operates, many of the U^{238} atoms that capture neutrons are changed into plutonium. By 1945, huge reactors had been constructed in Hanford, Washington, to produce considerable amounts of this new element.

Separation of the plutonium from the uranium from which it is formed is purely a *chemical* process, because we are dealing with two different chemical elements. This first large-scale artificial production of an element is one of the crowning achievements of modern science. One of the many problems that first had to be solved is shown in Fig. 46-9.

U^{235} can also be used to cause a chain reaction. Pure U^{235} is made with great difficulty by separating it from the U^{238} in natural uranium.

6. U^{235}, another fissionable material

U^{235} and U^{238} have the same chemical properties, because they are the same element; therefore they cannot be separated from each other by *chemical* means as plutonium and uranium can. The methods used to separate them must therefore be *physical* (mechanical) in nature. One method of separating the U^{235} atoms from the U^{238} is to convert the uranium into a gaseous compound and then to pass this gas through the tiny holes of a porous barrier. Because the molecules of the gas containing the lighter U^{235} atoms move more rapidly than those containing the heavier U^{238}, they pass through the holes in the barrier at a slightly greater rate. By repeating this process many thousands of times it is possible to separate usable quantities of concentrated U^{235} (Fig. 46-10).

The same thing can be accomplished by means of electromagnets. Again the uranium is put into gaseous form. The gas is ionized (charged) by bombardment with electrons, then the ions are speeded up by electrical attraction and shot through a small hole into a space between the poles of a powerful electromagnet. The magnet makes the ions

General Electric Co.

46-9 You can't go near atomic materials that are radioactive. This robot riding on a track alongside the Hanford nuclear reactor is used to turn wrenches and push materials into and out of the reactor. Twenty-four electric motors and numerous gears and other machines enable this robot to do its job.

U.S. Atomic Energy Commission

46-10 The largest building in the world, at Oak Ridge, Tennessee, houses the facilities for separating U^{235} from U^{238} by diffusion. The building is four stories high and is half a mile in length. Its roof is equal in area to about twenty large city blocks. If you look closely, you can see some tiny trucks and cars, dwarfed by the giant building.

swerve (see page 624), but because of their greater inertia, the heavier U^{238} ions do not swerve as much as the U^{235}. Hence two separate beams result. This method of separating the two kinds of uranium atoms from each other is similar to the separation of "positive" rays (Chapter 27) which led to the discovery of the proton.

The atomic bomb

The nuclear reactor gives up its energy relatively slowly; it may be months or even years before the reactor requires a new supply of uranium as "fuel." But in the atomic bomb a sudden explosion is desired—the chain reaction must take place in a fraction of a millionth of a second. The large amount of energy released in this short time produces temperatures of millions of degrees.

To get the greatest explosion possible, the fuel of the bomb should contain mainly atoms that can be split. This fuel can be of two kinds: plutonium or U^{235}.

7. Making the atomic bomb work

One way to make the atomic bomb is to use two pieces of either plutonium or U^{235}. Both pieces must be a bit more than half the **critical size.** By itself each piece is too small to keep a chain reaction going; neutrons are lost to the outside faster than they can be produced by fission. Hence nothing happens as long as the two pieces are kept apart.

But, as shown in Fig. 46-11, as soon as the pieces are rapidly brought together by some method, perhaps in a kind of "cannon," they make a lump that is larger than the critical size. This means that neutrons are now produced by fission faster than they can escape from the mass, and hence the chain reaction can continue to grow. The neutrons produced shoot out of the splitting atoms at speeds of thousands of miles per second,

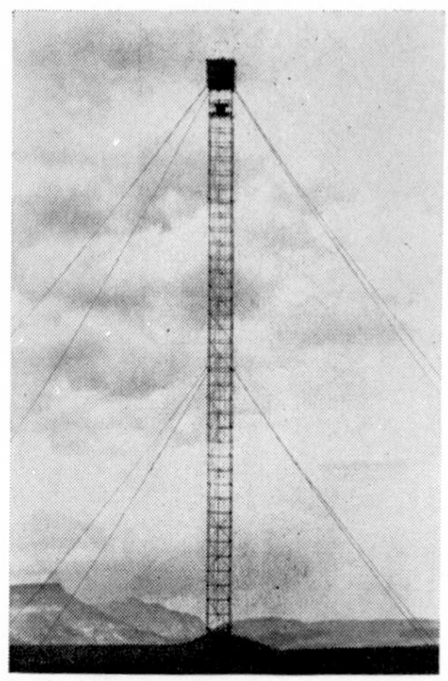

Lookout Mountain Laboratory, U.S.A.F.

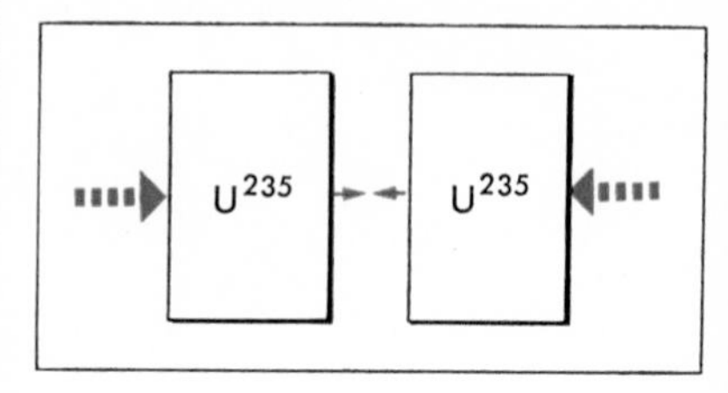

U.S. Atomic Energy Commission

46-11 **Bringing two separate pieces of U^{235} (or plutonium) together to form a single piece larger than the critical size causes the U^{235} to explode. This is the basic principle of the atomic bomb. *Left.* Testing tower on top of which a test bomb is exploded. *Right.* An atomic bomb is exploded.**

splitting one atom after another (Fig. 46-12). Most of the energy of this reaction is released in less than a millionth of a second!

8. Effects of the bomb

When an atomic bomb explodes high in the air, it gives off so much heat and light that at a distance of 1 mile it feels hotter than 50 suns. Wood chars at a distance of 2 miles. For miles around combustible materials catch fire. Fire, in fact, is one of the most destructive of the bomb's effects.

The enormous heat of the explosion produces temperatures of millions of degrees, which expand the air and thus create a terrific explosive blast that will blow over a wall 9 inches thick at a distance of $\frac{3}{4}$ mile. Billboards are knocked down at a distance of $2\frac{1}{2}$ miles. By the combined action of blast and fire, most houses within half a mile of the explosion are damaged beyond repair, and most of the people in the area are killed. In the average large city the blast would destroy an area of close to 200 blocks and kill thousands of people.

At the same time, the exploding bomb produces an enormous number of destructive gamma rays. A person standing in the open about $\frac{3}{4}$ mile from the explosion has only a 50-per-cent chance of surviving the "internal sunburn" caused by this radiation.

If the bomb is exploded near the ground or under sea water, the materials in and above the affected area are permeated with radioactive fission products, which may be carried many miles by a strong wind. This may mean slow death for many of the people caught in the area of the **fallout** of these radioactive particles after the explosion. And it may make a large area unsafe for some time.

9. Nuclear fusion

It would be difficult to make atomic bombs very much more powerful than those we have today out of U^{235} or plutonium. The reason is simple. As we have seen, the bomb is made of pieces of U^{235} or plutonium, each smaller than the critical size. When they are rapidly brought together and combined into one piece larger than the critical size, the chain reaction and resulting explosion take place. If any of the pieces were much larger, the bomb would explode by itself. Thus we probably cannot make a much bigger bomb by using larger pieces.

Yet it is possible to make a bomb with almost limitless explosive power. To do this we imitate the process by which the sun itself achieves its energy. As we have seen in Chapter 45, the sun gets its energy from atomic changes by building up atoms rather than splitting them apart. Inside the sun, atoms of hydrogen are combined to form helium atoms. This process is called **fusion,** in contrast to the fission, or atom-splitting, that occurs with uranium or plutonium. It requires a very rapid motion of the atoms, at temperatures of millions of degrees, for such a process to occur. For this reason fusion is sometimes referred to as a **thermonuclear reaction.** At such temperatures atoms attain such enormous velocities that their nuclei crash into each other despite the repulsion of the like positive charges in the nuclei. The fusion of hydrogen atoms to form helium atoms releases more than enough energy inside the sun to maintain the temperature required for the process itself. The process involves the conversion of four hydrogen nuclei (protons) into one helium nucleus and two positrons. Although this takes place in several steps, some of which involve carbon, the process can be summarized by the equation

$$4\ H^1 \rightarrow He^4 + \text{energy}$$

It has been estimated that 564 million tons of hydrogen become 560 million tons of helium in the sun every *second,* and that the 4 million tons of mass that disappear are converted into energy. However, the sun is so huge that it will take $1\frac{1}{2}$ billion years for it to lose only 0.01 per cent of its mass at this rate!

If this reaction or a similar one can be made to work on earth at a steady and controllable rate, and if we can devise ways to handle the enormous energy released, then man can make use of the water of the oceans for his fuel. The world need never again face a serious shortage of power. An immense amount of research is being devoted to the solution of this problem because of its enormous practical benefits, but the difficulties are formidable. In what kind of a vessel do you contain a reaction that is occurring at a million degrees? How do you control it? How do you withdraw the power produced?

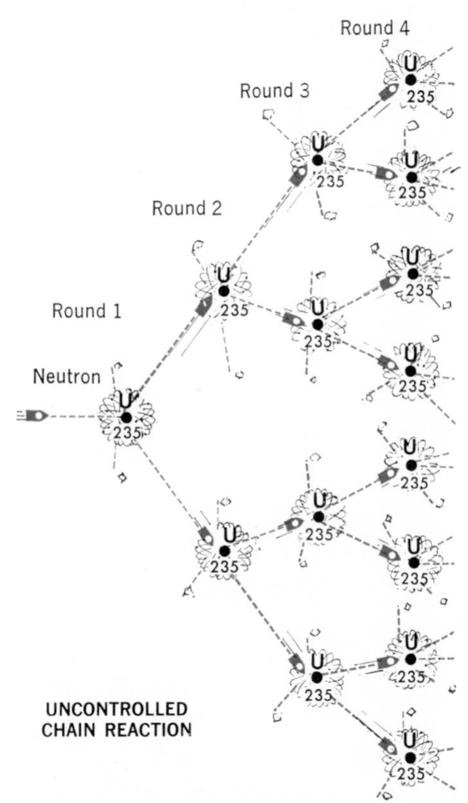

46-12 Unlike the slow, controlled explosion of atoms in the nuclear reactor, the chain reaction in the atomic bomb is incredibly rapid. This rapidity results when one group of neutrons produces a larger group on the next round of explosions and this process is continued.

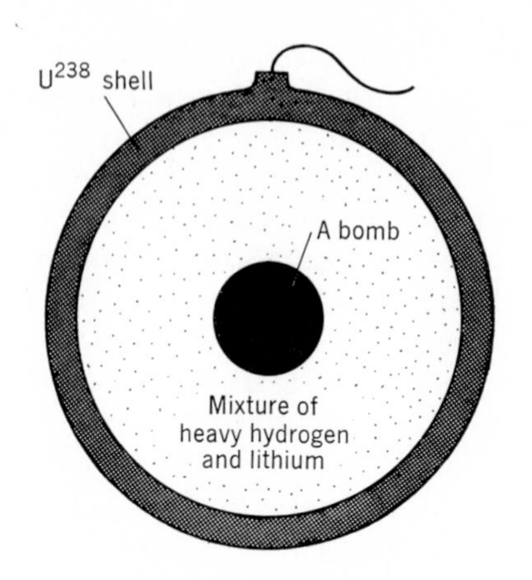

46-13 The A-bomb would be needed as a fuse to set off a much larger amount of explosive material in the fission-fusion-fission bomb. There is no limit to the explosive power of a bomb built in this way.

10. The hydrogen bomb

We could produce a bomb resembling a miniature sun if we used an ordinary atomic bomb to raise the temperature of some suitable material several millions of degrees. At this enormous temperature fusion of atoms becomes possible. Such a bomb is not limited by any critical size. We can use as much material as we like: the greater the amount of material the more terrible the explosion will be.

What materials are suitable for this agent of destruction? There are several possibilities. While the materials actually used are a military secret, a possible fuel may be lithium atoms of mass 6 mixed with hydrogen atoms of mass 2. Hydrogen of mass 2 (**deuterium**) can be manufactured by electrolysis of ordinary water (Chapter 27). Fusion of an atom of each kind produces two helium atoms and releases energy.

The hydrogen bomb is made to explode in much the same manner as a bomb of TNT. Nothing happens to TNT if the surrounding temperature is below the "kindling point." But when the molecules of TNT move fast enough, they rearrange themselves. This rearrangement of molecules creates further heat sufficient to explode the neighboring molecules of TNT. The exploding of the TNT spreads very rapidly from one molecule to the next.

In the same way, the ordinary atomic bomb is used as a kind of "match" to raise the temperature to the point where the atoms of lithium and deuterium undergo fusion (Fig. 46-13). As we have seen, such bombs can be of enormous size and destructive power. Bombs thousands of times more powerful than any made of U^{235} are not only possible—they exist and have been tested.

11. Fallout

The most powerful bombs yet built are probably of the so-called "fission-fusion-fission" variety, which consists of a plutonium fission bomb as a core which sets off an outer layer of lithium and hydrogen. The fusion of the lithium and hydrogen produces such high temperatures that the outermost layer, made of ordinary uranium, fissions, producing still more energy and large quantities of radioactive fission products (Fig. 46-13).

Ordinary uranium is, of course, mostly U^{238}, which does not ordinarily fission. But at the very high temperatures that result from the fusion reaction, U^{238} will fission like the rarer U^{235}. Since ordinary uranium will not fission at low temperature, it has no critical mass and hence there is no upper limit to the size and destructive power of such a bomb.

When a large fission-fusion-fission bomb explodes, releasing as much energy as several millions of tons of TNT, it forms a ball of fire three or four miles across, sufficient to destroy a large city. Where the fireball touches the surface of the earth it scoops up millions of tons of earth, and carries it skyward, contaminating it thoroughl[illegible]active atoms formed by the [illegible]ocess. Most of the radioactiv[illegible] soon settles back to the grou[illegible] the from the dust. This radioa[illegible] is one of the most serious d[illegible]d by the explosion of a "hyd[illegible]. For hours, and even days, af[illegible] sion it can threaten all life in [illegible] size of Connecticut.

Peacef[illegible] of atomic [illegible]

These are terrible possibilities [illegible] have been discussing. But we must [illegible] forget that atomic energy can [illegible] wonders of good as well as of evil.

12. Atomic energy for power

Today our major source of energy [illegible] coal. It is quite possible that this ma[illegible] be replaced by atomic energy in the future, giving us far more energy for peacetime use than we now possess, and at less cost. The nuclear reactor would probably be the basic "furnace" in producing this energy (Fig. 46-14). For this atomic furnace we need mainly purified natural uranium as the fuel, and pure carbon in the form of graphite (other materials, such as "heavy water," can substitute for the carbon) as a moderator to slow the neutrons.

Uranium is not a rare element. It is about as plentiful in the earth as copper, and more plentiful than zinc or lead. However, the ores of uranium are less concentrated than those of copper or zinc and are more difficult to refine.

Every continent and practically every [illegible]rge country possess some supply of [illegible]anium ore. It is very expensive to sepa[illegible]te uranium metal from its ore and then [illegible] process it still further in order to sep[illegible]te the U^{235} from the U^{238}. But only [illegible]ound of purified U^{235} is needed to [illegible]uce about the same amount of heat [illegible]gy as 1,500 *tons* of coal. Hence when [illegible] costs as little as about $10,000 a [illegible]d, it can begin to compete with coal [illegible] source of energy.

[illegible]eady (1959) Great Britain, a na[illegible]ith depleted coal supplies, has sev[illegible]tomic-energy power plants in op[illegible] to supplement the energy it now

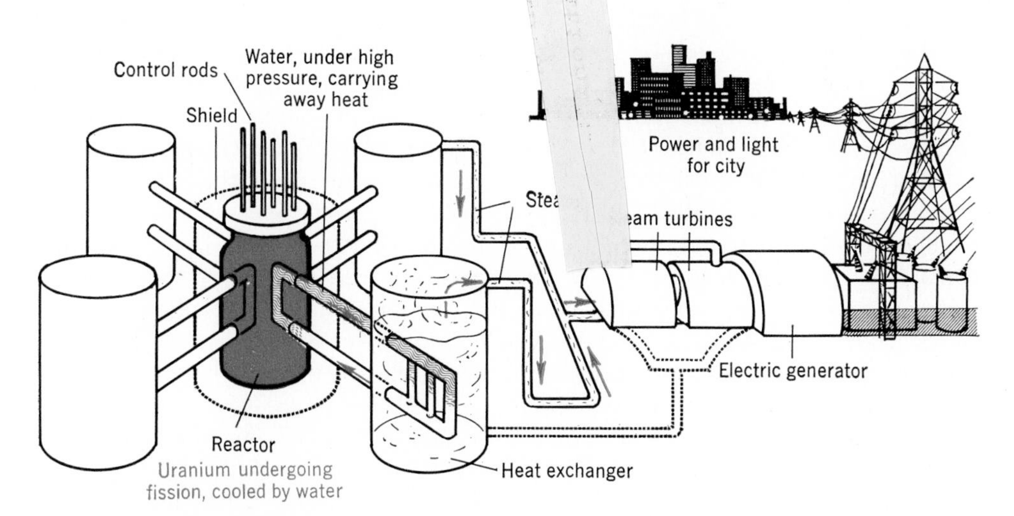

46-14 This is a diagram of the reactor used to generate electric power in the Shippingport, Pa., plant. Reactors can be used to drive ships or to provide valuable isotopes for use in research and in industry.

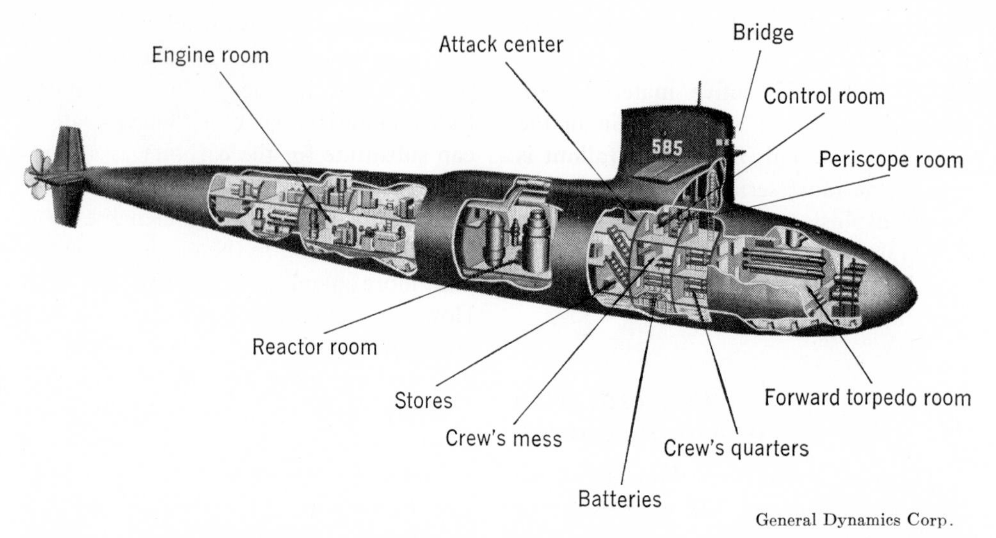

General Dynamics Corp.

46-15 **Atomic power already drives a number of submarines. Notice the location of the reactor. Do men need to spend much of their working time in nearby compartments? Why aren't reactors used to drive automobiles?**

obtains from burning coal, and other nations with limited supplies, such as India, are planning to build similar plants. In the United States an atomic energy power plant in Shippingport, Pennsylvania, went into operation in 1957. At the same time several atomic-powered submarines were in operation, and an atomic-powered merchant ship was being built.

When a reactor is used to produce power, it produces it in the form of heat. Air, water, or even molten metals circulating through the reactor carry this heat away. In a device called a **heat exchanger,** the heat is transferred to pipes containing water, and the resulting steam drives a conventional turbine and electric generator (Fig. 46-15).

The radiations produced by a nuclear reactor can injure people nearby. To guard against this danger, thick walls of materials such as concrete and lead must be built around the "furnace." These walls bring the weight of even a small reactor up to many tons. Thus the reactor at present is not suitable for automobiles or airplanes, although it is being used in powerhouses and ships and submarines (Fig. 46-15). But further research may improve this situation to make the nuclear reactor available for many uses.

A promising type of reactor still under development contains U^{235} mixed with U^{238}. Since only U^{235} fissions, the U^{235} is the fuel of the reactor. However, if enough fast neutrons are emitted during fission to convert the U^{238} into plutonium, which is also fissionable and is therefore a fuel, then the reactor is said to "breed" fuel. Such a reactor, which breeds new fuel while it produces energy, is called a **breeder reactor,** and is a very promising source of power.

13. The production of radioactive isotopes

In Chapter 27 we discussed "isotopes," or atoms of the same element that differ in mass. You will recall that the chemical properties of isotopes of the same element are exactly the same. But, because of the difference in the structure of the nuclei, isotopes often have very different nuclear properties. Thus many of them are radioactive.

Medical science has found an important use for the nuclear reactor in producing radioactive isotopes. These are manufactured mainly by neutron bombardment of materials inserted in the nuclear reactor—a bombardment which is much more intense than any obtained from other devices, such as the cyclotron. The atoms of most materials placed in the reactor are changed into new atoms by the bombardment of neutrons, and many of these new atoms are radioactive. In Chapter 45 we saw how medicine uses a wide variety of radioactive isotopes as tracers.

14. Radioactive isotopes in research

Radioactive-tracer methods are not limited to human beings. They are useful in industry, and in research with animals and plants. One of their important uses today is in attacking the problem of photosynthesis, that mysterious process by which plants produce all the world's food supply and coal from carbon dioxide, water, and sunlight. Here one of the radioactive isotopes of carbon, carbon 14, gives promise of enabling scientists to crack one of nature's most important secrets. By feeding plants carbon dioxide made with radioactive carbon 14, scientists are trying to trace the chemical changes that occur in the leaf and thereby unlock the secrets of photosynthesis.

Another of the interesting uses of radioactive carbon 14 is in measuring the age of ancient objects. Suppose the radioactivity of the cloth of an ancient mummy is measured and compared with the radioactivity of recently manufactured cloth. The modern cloth would show greater radioactivity. But why?

There is always a tiny amount of radioactive carbon 14 in the air, which results from cosmic-ray bombardment of atoms in the air. Some of this radioactive carbon is taken into plants as they absorb carbon dioxide. As a result, all plant (and animal) structures have a very feeble but measurable radioactivity due to carbon 14. As time goes on, the carbon 14 atoms explode, and in about 5,000 years half of them are gone. Thus the carbon 14 radioactivity becomes less and less with time, until finally it is no longer measurable. You can see how it would be possible for scientists to measure the age of the mummy by how much radioactive carbon 14 remains in it. Its use has helped immeasurably in finding the age of a number of relics of ancient civilizations.

Research with radioactive isotopes has its problems, the danger of injury to those who use them being one of the greatest, despite the precautions taken (Fig. 45-4). On the other hand, radioactive isotopes can be used to cure disease and save lives.

Which way?

Modern science began only a few centuries ago. It has continually increased its pace of achievement, and today is leaping ahead faster than ever before. But the very magnitude of the achievement is frightening if it is channeled into destructive purposes. Considering the horrors of atomic warfare, the faint of heart are apt to think that science is a Frankenstein monster and to wish that science had not been born at all.

Science has indeed brought us to a crossroads. The Federation of Atomic Scientists has put the problem clearly: "If the terror of the bomb is great, and properly great, the hope for man in the release of nuclear energy is even greater.

The fruit of that science which follows in the proud tradition of Galileo, the outcome of that complex organization of society which has made possible the city of New York and the Hanford plutonium plant, is the large-scale release of nuclear energy. We cannot see more than the faint shadow of what such a new force can mean for man. But it is our faith as scientists, and our experience as citizens of the twentieth century, that it will mean much. It will grow and develop. It will lead a life of its own. No influence that we have seen in our time can prevent this.

"Such a growth will bring death to the society that produced it if we do not adapt ourselves to it. This is the dilemma that the release of nuclear energy has brought to a world torn already by a horrible war. The nations can have atomic energy, and much more, but they cannot have it in a world where war may come."

In the last analysis it is really a question of man's ability to grow. Man must rise in his social behavior to the greatness of his own achievements in scientific discovery.

During the International Geophysical Year, modern science began to reach out into space with the launching of earth satellites. Arthur C. Clarke of the British Interplanetary Society states the problem for mankind in this way:

"Now, at the beginning of 1958, the Age of Space reaches out to touch every home, every school, every village everywhere. We shall not *survive* if we are lazy, soft, or complacent . . . The choice is clear: we choose between greatness—and oblivion. 'Where there is no vision, the people perish.' These words have an ominous, familiar ring. But reverse them and they become bright with hope and challenge. For history has demonstrated again and again that 'Where there is vision, the people triumph.' "

Going Ahead

THE VOCABULARY OF PHYSICS

Define each of the following:

nuclear (atomic) fission	fallout
chain reaction	plutonium
moderator	critical size
nuclear reactor	fusion
control rods	breeder reactor
neptunium	deuterium
thermonuclear reaction	heat exchanger

THE PRINCIPLES WE USE

1. What was the nature of the discovery made by Hahn and Strassman?
2. How does the energy released from uranium fission compare with the energy released by most naturally radioactive atoms?
3. What fact about the fission of uranium made it possible to produce a chain reaction?
4. What happens to a chain reaction if, starting with 100 splitting atoms, only 99 are split by the next crop of neutrons? What happens if 101 are split on the next round and this rate is maintained? What happens if 100 splitting atoms cause the splitting of another 100 in each round?
5. In natural uranium, what is the chance of a neutron hitting a U^{238} atom compared with its chance of hitting a U^{235}?
6. How was it found possible in the nuclear reactor to prevent the capture of most neutrons by U^{238} atoms?
7. How is the chain reaction in the reactor kept under control?
8. What happens to U^{238} atoms that capture neutrons?
9. What is the basic device in which atomic energy can be harnessed for peaceful use?
10. What materials are needed to make a nuclear reactor?
11. How much energy can be obtained from a pound of ordinary uranium by fission? From a pound of U^{235}?
12. How could the heat from a reactor be harnessed for heating homes?
13. How could it be harnessed for power?

14. How can radioactive carbon 14 be used to measure the age of very ancient objects?
15. Explain how a breeder reactor works.

Explain why

16. Fermi tried to bombard uranium with neutrons.
17. Neutrons pass easily through solid materials.
18. Nuclear reactors and atomic bombs have a certain critical size.
19. The control of atomic energy is an important question for humanity.

CONSIDERING ATOMIC ENERGY PROBLEMS

1. Explain why the use of atomic energy for automobiles is impractical under present conditions, while the use of atomic energy for large ships is more practical.
2. Explain why it is probable that for some years to come coal furnaces will be used in preference to nuclear reactors for the generation of electricity in large cities.
3. How might atomic energy be used to take X-ray-like pictures of materials from inside the materials?

FURTHER READING YOU WILL ENJOY

Explaining the Atom, Selig Hecht. Viking Press, New York, 1957. 237 pp.

Mr. Tompkins Explores the Atom, George Gamow. Macmillan Co., New York, 1944. 97 pp.

Laboratory Experiments with Radioactive Isotopes. Twenty experiments for high school demonstrations. Superintendent of Documents, United States Government Printing Office, Washington, D.C. 30¢.

The Hydrogen Bomb: The Men, the Menace, the Mechanism, J. R. Shepley and Clay Blair. David McKay Co., New York, 1954. 244 pp.

Dawn Over Zero: The Story of the Atomic Bomb, William L. Laurence. Alfred Knopf, New York, 1946. 274 pp.

FOR REFERENCE AND FURTHER STUDY

MATHEMATICAL STATEMENT OF MAJOR CONCEPTS

The science of physics is the study of mass and energy and their relationships. Some of the major facts of physics together with their mathematical statement are summarized in the pages that follow.

MACHINES

1. WORK = force applied × distance through which the force acts:

$$W = FD$$

2. THE PRINCIPLE OF WORK states that work input = work output, neglecting friction and other losses.

3. MECHANICAL ADVANTAGE expresses the relationship between an effort force and a resistance or load.

a. FOR THE PULLEY:

$$\text{M.A.} = \frac{\text{Resistance, } R}{\text{Effort, } E} = \frac{\text{Distance effort moves, } D_e}{\text{Distance resistance moves, } D_r}$$

b. FOR THE LEVER:

$$\text{M.A.} = \frac{\text{Length of effort arm, } A_e}{\text{Length of resistance arm, } A_r}$$

c. FOR THE WHEEL AND AXLE:

$$\text{M.A.} = \frac{\text{Radius of effort wheel, } A_e}{\text{Radius of resistance wheel, } A_r}$$

d. FOR THE INCLINED PLANE:

$$\text{M.A.} = \frac{\text{Length of plane, } L}{\text{Height of plane, } h}$$

e. FOR THE SCREW:

$$\text{M.A.} = \frac{D_e}{D_r} = \frac{\text{Circumference of effort circle, } 2\pi r}{\text{Pitch of screw, } p}$$

4. M.A. AND THE PRINCIPLE OF WORK. The work input in a pulley system, for example, is the effort force, E, $\times$ the distance the effort moves, D_e, and the work output is the resistance, R, $\times$ the distance the resistance moves, D_r:

$$RD_r = ED_e \text{ (neglecting friction) or } \frac{R}{E} = \frac{D_e}{D_r}$$

Similar relationships can be derived mathematically for all the simple machines. Full explanations together with worked-out sample problems are in the text. Refer to the index for individual machines.

5. THE PRINCIPLE OF MOMENTS: The moment, M, of a force around a fulcrum can be expressed as force $\times$ perpendicular distance, L, from force to fulcrum: $M = FL$. According to the Principle of Moments, when an object is in equilibrium the sum of the clockwise moments, CM, is always balanced by the sum of the counterclockwise moments, CCM:

$$CM = CCM$$

6. FRICTION: In many machines a force, F, is applied to overcome friction between a surface and a weight, W, resting against it.

   a. COEFFICIENT OF FRICTION (between horizontal surfaces): The coefficient of friction between two horizontal surfaces is found by:

   $$k = \frac{F}{W}$$

   See table on page 678.

   b. FRICTION ON AN INCLINED PLANE: The coefficient of friction, k, between an object and the surface it rests on can be determined by tipping the surface until the object on it, once started, continues to slide down the surface at constant speed. Then by trigonometry:

   $$k = \frac{\text{Height of plane, h}}{\text{Base of plane, b}} = \tan\theta$$

   where θ is the angle between the plane and the horizontal.

   c. FLUID FRICTION. The coefficient of fluid friction, K, relates the friction of a fluid, F_f, and the velocity, v, of the object moving through it. Fluid friction is found by:

   $$F_f = kv^2$$

7. EFFICIENCY: The efficiency of any energy converter is always the ratio (expressed as a per cent) between useful work output and work input:

$$\text{Efficiency} = \frac{\text{Useful work output}}{\text{Work input}} \times 100$$

The efficiency of simple machines may also be expressed by:

$$\frac{\text{Ideal effort}}{\text{Actual effort}} \times 100 = \frac{\text{Actual M.A.}}{\text{Ideal M.A.}} \times 100$$

VELOCITY AND ACCELERATION

The derivation of most of the formulas for velocity and acceleration will be found in Chapter 9.

8. VELOCITY. The formula relating the distance, s, an object moves at constant speed and the time, t, during which it moves is:

$$v = \frac{s}{t}$$

9. ACCELERATION. The definition of acceleration, a, is:

$$a = \frac{\text{Change in velocity } (v_f - v_0)}{\text{Change in time, } t} = \frac{\Delta v}{t}$$

a. FOR OBJECTS STARTING FROM REST AND MOVING WITH A CONSTANT ACCELERATION: Velocity is calculated from:

Final velocity: $v_f = at$

or $v_f^2 = 2as$

Average velocity during acceleration: $v_{av} = \frac{at}{2}$

Distance traveled during acceleration: $s = \frac{at^2}{2}$

b. ACCELERATION OF FALLING OBJECTS: In finding the acceleration of falling bodies, a in the above formulas has the value for the acceleration due to gravity, g. Thus:

$$v_f = gt; \ s = \frac{gt^2}{2}; \ v_f^2 = 2gs$$

10. THE PENDULUM: If we know the length, L, of a pendulum, we can calculate its period, T, from:

$$T = 2\pi\sqrt{\frac{L}{g}}$$

FORCE AND ACCELERATION

11. If a force is applied to give a mass an acceleration, the force applied = mass × acceleration produced, or:

$F = ma$ *(Newton's Second Law)*

Full explanation of Formulas 12 and 13 is given in Chapter 10.

12. CENTRIPETAL FORCE: If an object of mass, m, has a velocity, v, around a curve with a radius, r, the centripetal force, in poundals or dynes, is:

$$F = \frac{mv^2}{r}$$

13. THE LAW OF GRAVITATION. Every object in the universe is attracted to every other object in the universe with a force, *F*, which is proportional to the product of the two masses (m_1m_2) and inversely proportional to the square of the distance, *r*, between them. Thus (in dynes):

$$F = \frac{Gm_1m_2}{r^2}$$

where the constant, $G = 6.67 \times 10^{-8}$ dyne cm^2/gm^2 (if *m* is in grams and *r* is in centimeters). This gravitational constant, *G*, is not to be confused with the acceleration, *g*, due to gravity.

a. ON THE SURFACE OF THE EARTH: Since the force exerted by gravity on an object on the surface of the earth is called its "weight," *W*, and since this force gives the object an acceleration *g* when it is allowed to fall freely, the formula F = ma becomes:

$$W = mg$$

b. VELOCITY OF EARTH SATELLITES: By equating the expression for centripetal force (**11**) to the expression for the force of gravity (**12**) (which is responsible for the centripetal force), we can calculate the velocity an earth satellite must have in order to circle the earth at any distance, *r*.

$$v^2 = \frac{Gm_e}{r}$$

Note that the velocity depends on the mass of the earth, m_e, and is independent of the mass of the satellite.

ENERGY AND MOMENTUM

A mass in motion has momentum. It also has the ability to do work by reason of its motion; that is, it has kinetic energy.

14. KINETIC ENERGY: The following formula, which is derived in Section 1 of Chapter 12, gives kinetic energy in foot-pounds or gram-centimeters:

$$E = \frac{mv^2}{2g}$$

15. MOMENTUM: A force, *F*, exerted through a period of time, *t*, will give a mass, *m*, a velocity, *v*. We find that *mv*, the momentum, increases during the time the force is applied as given by:

$$Ft = \frac{\Delta mv}{g}$$

This formula is derived from the formula for acceleration, $a = \Delta v/t$, and from F = ma.

PRESSURE AND BUOYANCY

16. Pressure is the force which acts on a unit of area; that is:

$$P = \frac{F}{A}$$

17. LIQUID PRESSURE: At a depth, *h*, in a liquid whose density is *D*, the pressure is:

$$P = hD$$

18. SPECIFIC GRAVITY: This is the ratio between the mass of an object and the mass of an equal volume of water. The numerical values for density and specific gravity are the same when density is expressed in gm/cm^3 (see table on page 677), but are different when English units are used for density. Density has units, but specific gravity has none. Thus:

$$\text{Specific gravity} = \frac{\text{Mass of object}}{\text{Mass of equal volume of water}} = \frac{\text{Weight of object}}{\text{Loss of weight in water}}$$

19. THE HYDRAULIC PRESS. This is a machine in which a pressure applied to a small piston is transmitted through a liquid to a larger piston. The forces on the two pistons are in proportion to their areas. As a machine, the M.A. of the hydraulic press is:

$$\text{M.A.} = \frac{\text{Area of large piston}}{\text{Area of small piston}} = \frac{\text{Resistance, R}}{\text{Effort, E}}$$

TEMPERATURE

20. TEMPERATURE CONVERSION FORMULA: The relationship between Fahrenheit and Centigrade scales is expressed by:

$$C = \tfrac{5}{9}(F - 32)$$

21. BEHAVIOR OF GASES

a. BOYLE'S LAW: If the temperature of a gas remains constant, but the pressure and the volumes change, the relationship between the original and the final pressures and volumes is:

$$P_1V_1 = P_2V_2$$

b. CHARLES' LAW: If the pressure remains constant, but the temperature and volume change, the relationship between the original and the final temperatures and volumes is:

$$\frac{V_1}{V_2} = \frac{T_1}{T_2}$$

c. GENERAL GAS LAW: If pressure, volume, and temperature all change, the resulting relationship is:

$$\frac{P_1V_1}{T_1} = \frac{P_2V_2}{T_2}$$

22. LINEAR EXPANSION: If the temperature of an object is changed by an amount, ΔT, its *length*, L, changes by an amount, ΔL, which is given by:

$$\Delta L = k_e L \Delta T$$

where k_e, one coefficient of linear expansion, is a constant for the material used (see table on page 678).

23. CUBICAL EXPANSION: If the temperature of an object is changed by an amount, ΔT, its volume, V, changes by an amount, ΔV, which is given by:

$$\Delta V = k_c V \Delta T$$

where k_c, the coefficient of cubical expansion, is a constant for the material used (see table on page 678).

24. RADIATION. Any mass radiates heat at a rate proportional to the fourth power of its absolute temperature multiplied by a constant depending on the type of surface and its area. This relationship is known as *Stefan's Law:*

$$R = kT^4$$

25. HEAT EXCHANGE: The total heat gained or lost by a mass, m, is given by:

$$\Delta H = mS\Delta T$$

where the specific heat, S, is the number of calories of heat gained or lost in order to change the temperature of one gram (or one pound) of the object by one degree Centigrade (or Fahrenheit), and ΔT is the number of degrees in change of temperature.

26. HEAT OF FUSION: If a substance melts—ice melting to water, for example—a number of calories (or Btu), L, must be supplied to each gram (or pound) in order to melt it. L represents the latent heat of fusion. The total heat, H, required to melt an object of mass, m grams (or pounds), already at its melting point is therefore:

$$\Delta H = mL$$

27. HEAT OF VAPORIZATION: If a substance vaporizes—water turning to steam, for example—a number of calories (or Btu), L (called the heat of fusion), must be supplied to each gram (or pound) at its boiling point in order to turn it from a liquid to a vapor. The heat required to vaporize an object of mass m grams (or pounds) already at its melting point is therefore

$$\Delta H = mL$$

28. RELATIVE HUMIDITY: The ratio between the amount of water vapor in the air (absolute humidity) to the maximum amount of water vapor the air can hold is the relative humidity, which is calculated by:

$$\text{Relative humidity} = \frac{\text{Actual absolute humidity of air}}{\text{Absolute humidity of saturated air}} \times 100$$

Relative humidity is a per cent. See table on page 680.

29. POWER: The rate at which work is done by an engine, i.e., the power of the engine, is:

$$\text{Power, P} = \frac{\text{Work, W}}{\text{Time, t}}$$

a. HORSEPOWER: Engines are often rated by horsepower, *HP*:

$$\text{HP} = \frac{\text{Work done by engine (in ft-lbs)}}{550 \times \text{time (in seconds)}}$$

WAVE MOTION

30. WAVE LENGTH: The length of a wave, L, depends on how far it moves (at velocity, v) and how many times it vibrates (frequency, f) in a given period of time. Thus:

$$L = \frac{v}{f}$$

ELECTRICAL ENERGY

31. ELECTROCHEMICAL EQUIVALENT: In electrolysis, the mass, m, in grams, of an element deposited is proportional to the electric charge, Q, in coulombs, that has flowed through the solution. Hence:

$$m = ZQ$$

where Z is a constant (grams per coulomb). (See table on page 679.)

ELECTRICAL CIRCUITS

LETTER SYMBOLS: Symbols used in electricity are important to know in order to read formulas easily.

E = Electromotive Force (measured in volts)
I = Current (measured in amperes)
R = Resistance (measured in ohms)
P = Power (measured in watts)

SCHEMATIC SYMBOLS: The symbols in the table at the top of the next page are the ones used in the diagrams in this text. Other symbols are also to be found in other scientific writing.

SCHEMATIC SYMBOLS

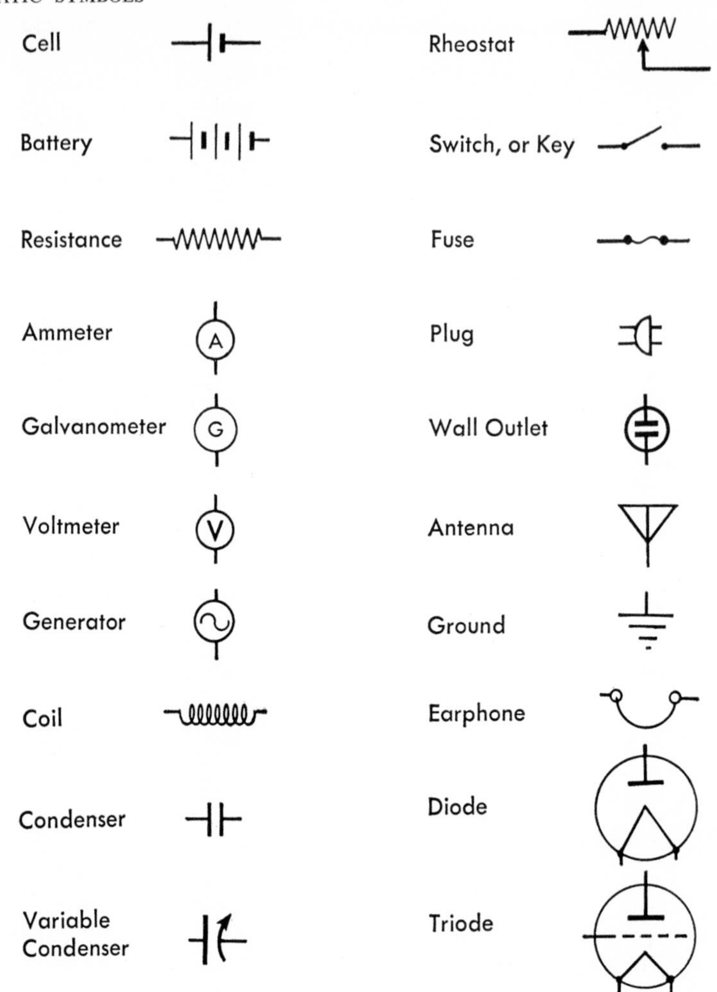

32. OHM'S LAW: Ohm's Law relates electromotive force to current and resistance. The current, *I*, flowing through a resistance, *R*, is proportional to the electromotive force, *E*, across the resistance and inversely proportional to the resistance. The formula is:

$$I = \frac{E}{R} \text{ or } E = IR$$

33. ELECTRICAL POWER: Power is the rate of doing work (see Formula 29). In electrical terms power (in watts) is given by:

$$P = EI$$

From Ohm's Law, power in watts can also be expressed as:

$$P = \frac{E^2}{R} \text{ or } P = I^2R$$

1 horsepower = 746 watts

34. EFFICIENCY: Efficiency is the ratio between useful work output and work input (see Formula 7). It is also:

$$\text{Efficiency} = \frac{\text{Power output, } P_o}{\text{Power input, } P_i}$$

35. SERIES CIRCUITS. If several devices are in series, the total resistance, R, is

$$R = r_1 + r_2 + r_3 + \cdots$$
$$I = i_1 = i_2 = i_3 = \cdots$$

If several voltages are in series, their combined voltage, E, is

$$E = e_1 + e_2 + e_3 + \cdots$$

36. PARALLEL CIRCUITS. If several resistances are in parallel, the total resistance, R, is

$$\frac{1}{R} = \frac{1}{r_1} + \frac{1}{r_2} + \frac{1}{r_3} + \cdots$$

and the current, I, is

$$I = i_1 + i_2 + i_3 + \cdots$$

If several voltages are in parallel, the combined voltage, E, is

$$E = e_1 = e_2 = e_3 = \cdots$$

37. TERMINAL VOLTAGE AND VOLTAGE DROP. All cells have an internal resistance, r or r_{int}, that causes an internal voltage drop, e, when current is flowing through an external resistance. Therefore inside the cell whose electromotive force is E the voltage drop, e, is Ir; and hence at the terminals appears a

$$\textit{Terminal voltage} = E - e, \text{ or } E - Ir$$

38. RESISTIVITY OF WIRES: The resistance, R, of a conducting material is:

$$R = \frac{\text{Length, } L \times \text{resistivity}}{\text{Cross-sectional area, } A}$$

where resistivity is the resistance of a unit length of a piece of the material of unit cross-section (see table on page 679).

39. HEAT ENERGY FROM ELECTRICITY: Heat in calories produced by P watts during a time t seconds is

$$H = 0.238\ Pt$$

and since $P = EI$ and $E = IR$ we also have:

$$H = 0.238\ EIt \text{ and } H = 0.238\ I^2Rt$$

ALTERNATING CURRENT

Symbols used in alternating-current formulas are:

Z = Impedance (corresponds to Resistance in DC circuits), measured in ohms

Q = Amount of charge transferred from one condenser plate to another, measured in coulombs

L = Inductance (of a coil), measured in henries

$\mathbf{X_L}$ = Inductive reactance of a coil (a form of impedance), expressed in ohms

C = Capacity (of condenser), measured in farads

$\mathbf{X_C}$ = Capacitive reactance of condenser (also a form of impedance), expressed in ohms

$\boldsymbol{\theta}$ = Phase angle, the difference in phase between voltage and current, expressed in degrees

f = frequency, in cycles per second

For schematic symbols, see pages 657–58.

40. EFFECTIVE VALUES: Alternating current and alternating voltage may be represented on a graph by a sine curve. The effective value of voltage and current is 0.707 × the peak value or:

$$E_{eff} = 0.707\ E_p$$
$$I_{eff} = 0.707\ I_p$$

41. OHM'S LAW FOR AN AC CIRCUIT: Compare with Formula 32.

$$I = \frac{E}{Z}$$

42. INDUCTIVE REACTANCE: X_L, the inductive reactance of a coil whose inductance is L, is given by:

$$X_L = 2\pi fL$$

where f is the frequency of the AC voltage and L is expressed in henries.

43. CONDENSERS: The amount of charge, Q, that can be transferred from one condenser plate to another by an electromotive force, E, is:

$$Q = CE$$

where C is the capacity of the condenser.

CAPACITIVE REACTANCE: The capacitive reactance, X_C, of a condenser whose capacity is C is given by:

$$X_C = \frac{1}{2\pi fC}$$

44. INDUCTIVE REACTANCE, CAPACITANCE, AND RESISTANCE IN AN AC CIRCUIT: In a circuit containing inductive and capacitive reactance and resistance the impedance, Z, is given by:

$$Z = \sqrt{R^2 + (X_L - X_C)^2}$$

and the phase angle, θ, by

$$\tan \theta = \frac{X_L - X_C}{R}$$

Refer to Chapter 35 for other formulas and sample solutions to problems.

45. RESONANCE: In a circuit containing both L and C there must be a frequency, f, at which $X_L = X_C$ and hence by the formula just above, $Z = R$. This frequency is known as the resonant frequency of the circuit, and is given by:

$$f = \frac{1}{2\pi\sqrt{LC}}$$

Since $X_L - X_C = 0$, the phase angle θ will equal 0 degrees at the resonant frequency.

46. POWER FACTOR: To find the power consumed in an AC circuit in which the phase angle is θ we must multiply EI by $\cos\theta$, the power factor, as follows:

$$P = EI \cos\theta$$

LIGHT AND LENSES

Light is a form of energy that acts sometimes as if it consisted of waves and at other times as if it consisted of particles (see Chapter 44). Statements of a few of the major concepts follow.

47. INDEX OF REFRACTION: The ratio of the speed of light in a vacuum, s_v, to the speed of light in a transparent substance, s_s, is a constant, k, for each transparent material, known as its index of refraction:

$$k = \frac{s_v}{s_s}$$

The index of refraction is also defined by Snell's Law as follows:

$$k = \frac{\sin i}{\sin r}$$

where i is the angle of incidence and r is the angle of refraction.

48. THE LENS FORMULA: Light rays from a source can be focused by a lens. The distance of the object from the lens, D_o, the distance from the lens to the image formed by the lens, D_i, and the focal length of the lens, F, are related by the expression:

$$\frac{1}{D_o} + \frac{1}{D_i} = \frac{1}{F}$$

49. MAGNIFICATION: Magnification, M, is defined as the ratio between the height of the image and the height of the object:

$$M = \frac{h_i}{h_o}$$

By applying a theorem about similar triangles to the ray diagram it can also be shown that:

$$M = \frac{D_i}{D_o}$$

50. ILLUMINATION: The amount of illumination on a surface decreases with the square of its distance from the source and is measured in foot-candles.

$$\text{Illumination, I} = \frac{\text{Candlepower of source, C}}{\text{Distance, D}^2}$$

51. EFFICIENCY OF LAMPS: The efficiency of the production of light energy by an electric lamp is:

$$\text{Efficiency} = \frac{\text{Light power output}}{\text{Electrical power input}}$$

52. DIFFRACTION INTERFERENCE PATTERNS: When light of wave length L passes through a narrow slit of width a, it is diffracted and produces interference effects. The direction along which the first node (cancellation of light) is seen is given by the angle r measured from a line perpendicular to the slit and:

$$\sin r = \frac{L}{a}$$

Two light sources of wave length L are just resolved by a lens of diameter a (can just be detected as separate sources) when the angle between them when viewed without the lens is:

$$\sin r = 1.22 \frac{L}{a}$$

MASS–ENERGY RELATIONSHIP

53. ENERGY CONVERSION: Einstein showed that matter and energy are equivalent to each other and that the amount of energy, E (in ergs or dyne-centimeters), equivalent to a given mass of matter, m, is given by:

$$E = mc^2$$

where c is the velocity of light (3×10^{10} centimeters per second) and m is the mass in grams.

MATHEMATICS FOR PHYSICS

To be fluent in a language you must know the vocabulary and syntax thoroughly. In the same way, to understand the principles of physics described in this book you must know some elementary mathematics, and you must know it thoroughly. For even in elementary work in physics, mathematics is a part of the language and syntax of the subject and *one of your most important tools.*

Very little mathematics is needed. Elementary algebra and the simplest principles of geometry are sufficient, though elementary knowledge of sines, cosines, and tangents is helpful for a few of the topics and needed for some of the more challenging problems. It is not important that you know a great deal of mathematics for your present work in physics; *it is only important that you know thoroughly what you need.* When you have mastered this section, you should be able to handle the mathematics in any of the problems in this book and get on with your central task of learning the principles of physics.

1. Symbols

In algebra you commonly used the letter x to indicate the unknown quantity you wished to solve for. Physicists try to avoid x for unknowns, preferring to use letters that give some hint of the nature of the unknown quantity; thus, v for velocity, t for time, P for pressure, F for force, etc. While this practice is a useful one for you to follow, you should realize that any letter at all will do. Indeed, Greek letters are often used to stand for angles; angle θ (theta), for example.

You will find it useful to use *subscripts* in order to distinguish between several different quantities of the same sort. Thus, if we wish to distinguish between the weight of your father and the weight of your brother we would use the letter W for weight for both, but would indicate the two different weights by W_f and W_b respectively.

Different forces might be indicated as F_1, F_2, F_3, and F_4. You will find subscripts in common use throughout this book.

2. Elementary operations

Multiplication is indicated in the following ways. Numbers to be multiplied are separated by an $\times$. Thus 20 times 4 is written 20×4. If a sum of numbers is to be multiplied, such as $4 + 7$ times 20, then parentheses are used, and we write $20(4 + 7)$, without the use of the times sign ($\times$).

Of course it makes no difference which number is written first. Thus 20×4 is the same as 4×20, and $20(4 + 7)$ is the same as $(4 + 7)20$.

When letters are used to represent quantities the rules are similar. When pressure, P, is to be multiplied by volume, V, this is written PV without the multiplication sign. Similarly we write abc for a times b times c.

If one of the quantities being multiplied (called a **factor**) is a sum such as

$P + 15$, the product of, say, $P + 15$ and V is written $(P + 15)V$. This means that *each* quantity inside the parentheses is to be multiplied by the quantity outside. $(P + 15)V$ can therefore be written $PV + 15V$, and $ab(c + d)$ can be written $abc + abd$.

Division is commonly indicated by a horizontal line between **numerator** (upper number) and **denominator** (lower number).

2 divided by 3 is $\frac{2}{3}$ (or, to save space, 2/3)

V divided by T is $\frac{V}{T}$ or V/T

$3(a + b)$ divided by $c + d$ is $\frac{3(a + b)}{c + d}$,

or $3(a + b)/(c + d)$.

Fractions written in this way can be simplified if both numerator and denominator have the same factor, for such a common factor may be canceled. Thus *8* is a common factor in

$$\frac{24a}{16c} = \frac{8 \times 3a}{8 \times 2c}$$

Dividing numerator and denominator by *8*,

$$\frac{\not{8} \times 3a}{\not{8} \times 2c} = \frac{3a}{2c}$$

Similarly

$$\frac{9ab(c + d)}{3b(c + e)} \text{ simplifies to } \frac{3a(c + d)}{c + e}$$

since *3b* is a factor common to both the numerator and the denominator. Although c appears in both numerator and denominator, note that it is *not a factor* and hence cannot be canceled out.

If either numerator or denominator (or both) is a fraction we can usually simplify it by inverting the denominator and multiplying. In this way

$\frac{7}{8}$ divided by $\frac{3}{4}$ is written $\frac{\frac{7}{8}}{\frac{3}{4}}$

and simplified to $\frac{7}{8} \times \frac{4}{3}$,

which can be further simplified as described in the preceding paragraph by dividing numerator and denominator by *4* (i.e. by canceling *4*'s).

$\frac{7}{2 \times \not{4}} \times \frac{\not{4}}{3}$ thus becomes $\frac{7}{6}$

$\frac{9}{11}$ divided by 3 can be written

$$\frac{\frac{9}{11}}{\frac{3}{1}} \text{ or } \frac{9}{11} \times \frac{1}{3}, \text{ which is } \frac{3}{11}$$

Similarly

$\frac{6ab}{5d}$ divided by $\frac{3ac}{20(c+d)}$ is $\frac{6ab}{5d} \times \frac{20(c+d)}{3ac}$,

and canceling reduces this to $\frac{8b(c + d)}{cd}$.

Practice Problems

Simplify the following:

1. $6(8 + P)$ [ANS. $48 + 6P$]
2. $3(5 + 2V)$
3. $3c(5d + 2e)$
4. $8ab(cd + 3ce)$
5. $\frac{3}{2}(5 + 2V)$ [ANS. $\frac{15}{2} + 3V$]
6. $\frac{3}{2} \times \frac{8}{15}$
7. $\frac{4}{5} \times \frac{3}{8}(2a + b)$
8. $\frac{4 \times 3 \times 5 \times 7}{5 \times 8 \times 2 \times 9}$
9. $\frac{4}{5} \times \frac{3}{8} \times \frac{5}{2} \times \frac{7}{9}$
10. $\frac{a}{b} \times \frac{c}{d} \times \frac{b}{e}$ [ANS. $\frac{ac}{de}$]
11. $\frac{x}{3} \times \frac{5}{y} \times \frac{12}{x}$
12. $\frac{9ab}{5} \div \frac{3ab}{5}$ [ANS. 3]
13. $\frac{4(l + m)}{3l} \div \frac{5m}{6l}$
14. $P_1P_2V \div P_1V$
15. $\frac{W_1W_2}{W_3} \times \frac{W_3}{W_1}$
16. 3% of 450
17. 1.4% of 83.2

3. Exponents

When a number, p, is to be multiplied by itself the process is written as pp or p^2, called *p squared*. Similarly, if it is to be multiplied by itself again, ppp is written

p^3, called *p cubed;* and *pppp* is written p^4 and is called *p to the fourth*, and so on. The number p is called the **base,** and the number which tells how many times it is multiplied by itself is the **exponent.**

Thus if one side of a cube has a length of 4 inches, the area of one side is

$$4^2 = 16 \text{ square inches (in}^2\text{)}$$

and the volume is

$$4^3 = 64 \text{ cubic inches (in}^3\text{)}$$

Of course p^1 is just p; likewise $4^1 = 4$.

Negative exponents are very convenient in physics. They are a shorthand way of indicating division by a number whose exponent is positive. Thus 1/10 can be written 10^{-1}; $1/10^2$ is also written 10^{-2}, and $1/5^8$ can be written 5^{-8}, etc. If a volume is to be divided into cubes whose sides are p long and whose volumes are therefore p^3, the number of cubes that would fit in the volume V might be written V/p^3. Such an expression is often written Vp^{-3}. In the same way

$\frac{a^2b}{cd^3}$ can be written as $a^2bc^{-1}d^{-3}$, and

$\frac{a+b}{c+d}$ can be written as $(a+b)(c+d)^{-1}$.

Multiplication. When numbers with the same base are multiplied together their exponents are added. We saw above that *ppp* is written p^3 and *pp* is written p^2. If p^3 is now to be multiplied by p^2, we are multiplying p by itself five times. Thus

$$p^3 \times p^2 = p^{3+2} = p^5$$

If $p = 3$, then $p^5 = 3^5$, which is 243.

$$y \times y^4 = y^{1+4} \text{ or } y^5$$

Division of exponential numbers with the same base is done by subtracting the exponents. When a^5 is divided by a^3 we can write

$$\frac{a^5}{a^3} = a^{5-3} = a^2$$

Again, $\frac{2^6}{2^4} = 2^{6-4} = 2^2$.

The use of negative exponents does not require any new rule: $p^3 \times p^{-2} = p^{3-2}$ or p. You can see that this will be so if you write out the quotient completely and then cancel

$$\frac{p \times \not{p} \times \not{p}}{\not{p} \times \not{p}} = p$$

$$3^3 \times 3^{-2} = 3^{3-2} \text{ or } 3^1 \text{ or } 3$$

which you can check by writing it as

$$\frac{3^3}{3^2} = \frac{27}{9} = 3$$

Consider the example p^n/p^n when n is any number (any exponent). Since we are dividing one thing, p^n, by itself, we know the answer is 1. But using our rule for exponents, we have

$$\frac{p^n}{p^n} = p^{n-n} = p^0$$

Thus we conclude that any quantity, p, to the zero power is equal to 1.

Similar letters with different subscripts are, of course, different numbers. Hence you cannot simplify a_1a_2 any further, and

$$a_1^2\, a_2 \times a_1a_2^3 = a_1^3a_2^4$$

If an exponential number, a^5, is to be cubed, the operation is written $(a^5)^3$. This means $a^5 \times a^5 \times a^5$, which is a multiplied by itself 15 times, or a^{15}. Apparently we could have found this exponent by multiplying *5* by *3*

$$(a^5)^3 = a^{5\times3} = a^{15}$$

Again, another example,

$$(a^3)^x = a^{3x}$$

(CAREFUL: Do not confuse this operation with a^3a^x, which equals a^{3+x}.)

As a final example,

$$(a^{-2})^3 = a^{-6}$$

You can only add and subtract exponents if the base of each is the same. For example, you cannot simplify $2^4/3^5$ by any operation we have described. Nor can you do anything with a^4b^3/c^2. But you can simplify a^4b^3/c^2 divided by ab^2/c by the rules which you learned earlier, as follows. You will first invert the denominator and then multiply

$$\frac{a^4b^3}{c^2}\bigg/\frac{ab^2}{c} = \frac{a^4b^3}{c^2} \times \frac{c}{ab^2}$$

Now divide the quantities that have the same base, which leads to the simplified form, a^3b/c. That is, both the numerator and the denominator are divided by ab^2c.

Again,

$$\frac{(a+b)^4(a-b)^2}{ab(a+b)^3} \text{ can only be } \frac{(a+b)(a-b)^2}{ab},$$

since the only base in *both* numerator and denominator is $a + b$.

Some simplification is possible if different bases each have the same exponent. Thus $(2)^2 \times (3)^2$ can be written as $(2 \times 3)^2$. The result will be 36 (which you can check by noticing that $2^2 \times 3^2 = 4 \times 9 = 36$).

Again,

$$a^4b^4c^4 \text{ is } (abc)^4, \text{ and } \frac{p^3q^3}{r^3} \text{ is } \left(\frac{pq}{r}\right)^3$$

Fractional exponents are often used to indicate roots; $\sqrt{p}$ can be written as $p^{\frac{1}{2}}$. This is reasonable enough when you remember that $\sqrt{p} \times \sqrt{p} = p^1$, which is also obtained by addition of exponents. The cube root of p can similarly be written as $p^{\frac{1}{3}}$. The cube of this expression, which is certainly just p, is found from $(p^{\frac{1}{3}})^3$ by the usual procedure of multiplying exponents.

If a cube has a volume, V, one side of the cube has a length $\sqrt[3]{V}$, which can also be written $V^{\frac{1}{3}}$. Likewise the area of one face of the cube will be $(V^{\frac{1}{3}})^2$, which can be written simply as $V^{\frac{2}{3}}$.

Practice Problems

Simplify the following:

1. $a^2 \times a^5$
2. $a^2b \times ab^3$ [ANS. a^3b^4]
3. $3^2 \times 3^6$
4. $3^2 \times 3^{-6}$ [ANS. 3^{-4} or $1/3^4$]
5. $5^3 \times 5^{-2}$
6. $8^5 \times 8^{-2} \times 8$
7. $y^{-3} \times y^5 \times y$
8. $(y^3)^2$
9. $(x^2y^3)^3$ [ANS. x^6y^9]
10. $(xy^4)^3$
11. $(xy^4)^0$
12. $(xy^{-4})^3$
13. $(xy^{-4})^p$
14. $x^2y^4 \div xy$
15. $\frac{a^2b^3c}{d} \div \frac{abc}{d}$ [ANS. ab^2]
16. $pqr^5 \div \frac{pq^2r^3}{s^2}$
17. $pq^{-3}r^5 \times p^{-1}q^4r$
18. $(pq)^3 \times (pq^2)^3$
19. $p_1p_2^{-3} \times (p_1p_2)^{-3}$
20. $(a + b)^4 \div (x - y)^4$
21. $\frac{(a + b)^2(a - b)^3}{9} \div \frac{(a - b)^3}{3}$
22. $(5p^2)^2$
23. $10^{-3} \times 10$
24. $8 \times 10^5 \times 2 \times 10^{-3}$
25. $8 \times 10^5 \div 2 \times 10^{-3}$
26. $(2 \times 10^6)^3$
27. $(p_1p_2)^2 \times \frac{{p_1}^2}{p_2}$
28. $a_1b_1c_1 \times a_2b_2c_2$
29. $\frac{W_A{W_B}^2}{40} \div \frac{{W_B}^2W_C}{15}$

4. Powers-of-ten notation

Very large numbers and very small numbers are much easier to handle if they are written as powers of ten. *Any* number may be written as a power of 10 multiplied by a second number between 1 and 10. Thus, 150, which is 1.5×100, may be written as 1.5×10^2. And 4,530,000 may be written as 453×10^4 or as 45.3×10^5 or, best of all, as 4.53×10^6. Notice how moving the decimal point one digit to the left is balanced out by increasing the exponent by one. It is helpful in writing numbers in powers-of-ten notation to remember that 1000 is 10^3 and 1,000,000 is 10^6.

Numbers far smaller than 1 can be written in this way using negative powers of 10. A decimal fraction like 0.00371 is the same as 371/100,000 or $371/10^5$. This may be written as 371×10^{-5} or better as 3.71×10^{-3}. Notice that 1/100 is 10^{-2}, 1/1000 is 10^{-3}, and 1/1,000,000 is 10^{-6}.

The powers-of-ten notation is particularly useful in physics because enormously large and exceedingly small quantities are dealt with constantly. Thus in a glassful of air there are about 7×10^{21} molecules, each about 10^{-8} cm in diameter, traveling at about 5×10^4 cm/sec, and weighing about 5×10^{-23} gm. Numbers like these are easy to multiply and divide by one another and easy to compare with each other if you use the powers-of-ten notation.

Multiplying 4,530,000 by 0.00371 may seem formidable. But it becomes easier if, as before, we write these numbers as $4.53 \times 10^6 \times 3.71 \times 10^{-3}$. Notice that $4.53 \times 3.71 = 16.8$ (to three significant figures—see next section) and $10^6 \times 10^{-3} = 10^{6-3} = 10^3$. Hence $4.53 \times 10^6 \times 3.71 \times 10^{-3} = 16.8 \times 10^3$.

To divide 4,530,000 by 0.00371 write

$$\frac{4.53 \times 10^6}{3.71 \times 10^{-3}} = 1.22 \times 10^{6-(-3)} = 1.22 \times 10^9$$

Here is a slightly harder example.

$$\frac{8 \times 5280 \times 0.870}{453 \times 2{,}080{,}000 \times 0.0140}$$

is first written as

$$\frac{8 \times 10^0 \times 5.28 \times 10^3 \times 8.7 \times 10^{-1}}{4.53 \times 10^2 \times 2.08 \times 10^6 \times 1.40 \times 10^{-2}}$$

Now we combine the powers of ten by applying the rules of exponents, and we combine the "ordinary" numbers by the usual laws of multiplication and division

$$\frac{8 \times 5.28 \times 8.70}{4.53 \times 2.08 \times 1.40} \times \frac{10^0 \times 10^3 \times 10^{-1}}{10^2 \times 10^6 \times 10^{-2}}$$

$$= \frac{368}{13.2} \times \frac{10^2}{10^6}$$

$$= 27.8 \times 10^{-4}$$

$$= 2.78 \times 10^{-3} \text{ (or 0.00278)}$$

It is sometimes helpful to rewrite the last step in order to simplify the final division. Thus, in another problem, if we had

$$\frac{1.38 \times 10^8}{2.30 \times 10^{-2} \times 6.92 \times 10^5}$$

we would get

$$\frac{1.38 \times 10^8}{15.9 \times 10^3}$$

If division is to give us a number between 1 and 10 times a power of ten, we must rewrite the denominator. Then

$$\frac{1.38 \times 10^8}{0.16 \times 10^1} = 8.67 \times 10^7$$

Similarly

$$\frac{1.28 \times 10^{-4}}{8.60 \times 10^3} = \frac{12.8 \times 10^{-5}}{8.60 \times 10^3} = 1.49 \times 10^{-8}$$

or, equally correct,

$$\frac{1.28 \times 10^{-4}}{8.60 \times 10^3} = \frac{1.28 \times 10^{-4}}{0.860 \times 10^4} = 1.49 \times 10^{-8}$$

If you need to take the square root of a number written in powers-of-ten notation, you must divide the exponent of 10 by 2. For this to be possible without producing a fractional exponent, the power of ten must be an even number. This is easily managed as shown by the following examples.

$$\sqrt{6.4 \times 10^9} = \sqrt{64 \times 10^8} = 8 \times 10^4$$

$$\sqrt{0.0016} = \sqrt{16 \times 10^{-4}} = 4 \times 10^{-2}$$

Likewise when extracting cube roots the exponent must first be made divisible by 3. Hence

$$\sqrt[3]{27{,}000} = \sqrt[3]{27 \times 10^3} = 30$$

and

$$\sqrt[3]{0.000064} = \sqrt[3]{64 \times 10^{-6}} = 4 \times 10^{-2}$$

Practice Problems

A. Write the following numbers in powers-of-ten notation.

1. 100

2. 840 [ANS. 8.4×10^2]

3. 0.6

4. 98,300

5. 273

6. 17

7. 4

8. 0.0036 [ANS. 3.6×10^{-3}]

9. 0.15

10. 0.0763

11. 0.00002

B. Write in ordinary notation.

12. 10^3

13. 10^{-3}

14. 8.03×10^5

15. 10^0

16. 6.21×10^1

17. 4.2378×10^3

18. 3.14×10^9

19. 3.14×10^{-4}

20. 7.00×10^{-8}

C. Perform the following operations, writing your numbers and answers in powers-of-ten notation.

21. 648×984

22. $428{,}000 \times 0.610$ [ANS. 2.61×10^5]

23. $\frac{0.327}{4960}$ [ANS. 6.59×10^{-5}]

24. $\frac{0.000296}{0.109}$

25. $\frac{6830}{247}$

26. 0.00296×0.109

27. $32 \times 43{,}860 \times 0.01$

28. $\frac{4380 \times 60}{0.038}$

29. $\frac{72 \times 5270}{0.0021}$

30. $\frac{3.00 \times 10^{10} \times 1695}{0.0008 \times 600 \times 4.03}$

31. $\frac{5 \times 4300 \times 0.000061}{386{,}000 \times 0.10 \times 93}$

32. $\sqrt{0.0081}$

33. $\sqrt{0.81}$

34. $\sqrt{640{,}000}$

35. $\sqrt{250 \times 10^5}$

36. $\sqrt{1.21 \times 10^{-8}}$

37. $\sqrt[3]{0.008}$

38. $\sqrt[3]{0.000000064}$

39. $\sqrt[3]{2.7 \times 10^{10}}$

40. $\frac{\sqrt{2.7 \times 10^{10}} \times 0.008}{43{,}860}$

5. Significant figures

A large part of a physicist's time is spent making measurements. You too will spend much of your laboratory time making measurements and recording them for future use. It is important that your numbers tell you *two* things: (1) the size of the quantity that you measured, and (2) the *accuracy to which you measured it.* It is easy to forget this second point and as a result waste a lot of time unnecessarily.

The accuracy of your measurements is indicated by the number of digits (i.e. 0, 1, 2, 3, 4 . . .) you record, not counting any zeroes that are needed to locate a decimal point. Thus a length written as 47 cm is a length you have measured to the nearest centimeter. But 47.1 cm is a measurement to the nearest tenth of a centimeter and is therefore ten times as accurate. 47.13 cm implies a measurement 100 times as precise as 47 cm.

If your measurement of 47 cm was accurate to the nearest tenth of a centimeter, *you should say so* by writing 47.0 cm. Zeroes at the end of a number should never be omitted simply because the number "comes out even." They are as important as any other digits in indicating the precision of a measurement.

The digits that express your measurement are called **significant figures.** The more significant figures in your measurement the more accurate it presumably is.

Zeroes used *before* a number merely to locate a decimal point are not significant figures and give no indication of the accuracy of your measurement. Such zeroes are merely determined by the size of the unit you use in your measurement. Your measurement of 47 cm (accurate to *two* significant figures) is not a bit more accurate when written as 0.47 m or as 0.00047 km. Indeed, 0.00047 km is only one tenth as accurate a measurement as 47.1 cm. Equally precise measurements are expressed by 47 cm, 0.030 gm, and 0.0014 sec, all being accurate to two significant figures.

On page 35 you can see how to deal with significant figures when making computations. It is worth repeating here two important points:

1. No calculation can produce an answer more accurate than the data or calculations that produce it. Hence you can never express your answers to more significant figures than the least accurate figure in the calculation. This rule applies even to a calculation in which two nearly equal numbers are subtracted, such as $67.34 - 67.32 = 0.02$. The result of this calculation is known to only *one* significant figure, since that is the least accurately known figure in the calculation. It *cannot* be written as 0.02000.

2. For simplicity we assume in this book that all data is accurate to 3 significant figures unless otherwise stated, and that answers should be worked out also to 3 significant figures wherever possible. Thus we may assume that a length of 3 cm can be considered as 3.00 cm for purposes of computation and that $\pi = 3.14159$ has the value 3.14.

6. The slide rule

You can greatly simplify the arithmetic of physics problems by using a slide rule. Indeed, slide rules are widely used by engineers, business men, navigators, and chemists as well as physicists. Whenever numbers are to be multiplied or divided (or squares, square roots, sines, cosines, tangents, or logarithms found) a slide rule does the job quickly. A slide rule 10 inches long (which need cost no more than $3) is accurate to three significant figures, which, as we have seen, is sufficient for all work in this book. Avoid rules less than 10 inches long as their accuracy is restricted to only two significant figures.

You will see how useful a slide rule is if you work out by direct multiplication and long division the solution to the illustrative problem in section 4: $4{,}530{,}000 \times 0.00371$. Five or six minutes is reasonable time to do this arithmetic. Ask a friend who uses a slide rule to check your answer; reasonable time for the same problem: one minute!

Since the essential purpose in studying physics is to learn *physics*, not arithmetic, any time you can save from purely routine calculations is a clear gain. You can then spend more of your time learning the principles of physics and learning how to organize the solutions to problems.

Slide rules commonly come with directions for their use. Also, anyone who is familiar with them can easily show you how to use one. With care and a little practice you will be able to use a slide rule to solve most of the problems in this book.

7. Rules of equations

The equations of elementary physics are no different from the equations of elementary algebra, and the rules for solving them are precisely the same. Some of the rules that you use over and over again are summarized below.

1. *Any quantity can be added to or subtracted from* both *sides of an equation.*

$$P + 15 = 45$$

Subtracting *15* from *both* sides gives

$$P = 30$$

Notice that this is the same thing as moving the 15 across to the right-hand side and changing its sign, giving

$$P = 45 - 15$$

whence again $P = 30$.

2. *Any quantity may be used to multiply or to divide* both *sides of an equation.*

$$7L = 49$$

Dividing both sides by 7 gives

$$L = 7$$

Or again, $3ab = 20a$

Dividing both sides by *3a* gives

$$b = 6.67$$

3. *Equations in the form of a proportion can be simplified by cross-multiplying.* Thus, if

$$\frac{r}{8} = \frac{9}{48}$$

we can write $r \times 48 = 8 \times 9$

whence $48r = 72$

$$r = 1.5$$

Notice that this rule is essentially Rule 2. Thus, we could have begun by multiplying both sides by *8*, giving

$$\frac{r \times 8}{8} = \frac{9 \times 8}{48}$$

which simplifies to $r = 1.5$.

4. Both *sides of an equation can be raised to the same power or have the same root taken.* Thus

$$\sqrt{m} = 23$$

is solved by squaring both sides to give

$$m = 529$$

Or again, if $n^3 = 64$ we can take the cube root of each side to find that $n = 4$.

Many of the problems you solve in physics will require the use of these steps several times in succession. Consider the following example which concerns the total current, I, flowing through two resistances, r_1 and r_2, across a voltage, E.

$$I = \frac{E}{r_1} + \frac{E}{r_2}$$

When known numerical values (say $I = 11$, $E = 28$, and $r_1 = 7$) are substituted, the equation becomes

$$11 = \frac{28}{7} + \frac{28}{r_2}$$

and the problem is to find r_2. Subtract 28/7 from both sides (Rule 1 to isolate the term containing the unknown, r_2)

$$11 - \frac{28}{7} = \frac{28}{r_2}$$

Simplify the left-hand side

$$7 = \frac{28}{r_2}$$

Multiply both sides by r_2 (Rule 2 to get r_2 out of the denominator)

$$7r_2 = 28$$

Divide both sides by 7 (Rule 2)

$$r_2 = 4$$

As another example consider the formula describing the flight of a projectile thrown from a platform moving with a velocity, v. Upon substituting the known quantities into the formula, the equation might assume the form

$$\left(\frac{v-5}{2}\right)^2 = (15)^2 + 32 \times 10$$

$$\left(\frac{v-5}{2}\right)^2 = 545$$

Take the square root of both sides (Rule 4)

$$\frac{v-5}{2} = 23.3$$

Multiply both sides by *2* (Rule 2)

$$v - 5 = 46.6$$

Add *5* to both sides

$$v = 51.6$$

Practice Problems

Solve these equations.

1. $a + 17 = 50$
2. $3a + 17 = 29$
3. $6a + 17 = 31 - a$
4. $\frac{a}{5} = \frac{6}{11}$ [ANS. 2.73]
5. $\frac{10P + 4}{5} = 3$
6. $8x - 2 = \frac{2x}{5}$
7. $6\left(3y - \frac{1}{2}\right) = \frac{3}{4}$
8. $\frac{T}{5} = \frac{20}{T}$
9. $T^2 - 6 = 75$
10. $\sqrt{T - 4} = 8$
11. $\sqrt{2m^2 - 4} = m$
12. $4y^2 + 3 = 3y^2 + 84$

8. Simultaneous equations

If two quantities are known in a physics problem, one equation alone is not sufficient to determine both quantities. In order to find two unknown quantities you need two independent equations.

If three quantities are unknown, three independent equations are required in order to find them. Two or more independent equations describing the same situation and containing the same unknown quantities are called **simultaneous equations.**

In this book you will occasionally meet a pair of simultaneous equations. They can always be solved easily by the following procedure. Rearrange one equation so that one of the unknown quantities is alone on one side of the equation. This expression for the unknown quantity is now substituted into the second equation, which can then be solved for the second unknown quantity. Once the second unknown quantity is found it may be substituted into either equation and the first unknown quantity solved for.

When you have followed the example worked out below, return and read the preceding paragraph once more, thinking how it applies to the worked-out example.

$$(1) \quad r + 7s = 29$$

$$(2) \quad 3r - s = 21$$

Get r alone on the left side of the first equation (by subtracting $7s$ from both sides)

$$r = 29 - 7s$$

Put this value of r into the second equation

$$3(29 - 7s) - s = 21$$

Notice that we now have an equation containing only s. Simplifying by first multiplying the parentheses through by *3* and then collecting terms

$$87 - 21s - s = 21$$

$$22s = 66$$

$$s = 3$$

Now substitute this value of s back into

either of the original equations (say, the first one)

$$r + 7 \times 3 = 29$$

$$r = 8$$

You can check your work by putting these two values into the second equation.

$$3 \times 8 - 3 = 21$$

$$21 = 21$$

Since this is a true statement you can be sure your answers are correct.

Now go back and re-read the preliminary paragraph.

Practice Problems

Solve the following pairs of equations:

1. $8r = s + 16$
   $s + 4 = 4r$
2. $23m - 8 = n$
   $m + 36 = n$
   [ANS. $m = 2$, $n = 38$]
3. $5w = 5y - 28$
   $w + y = 8$
4. $\frac{2a}{5} - b = 0.4$
   $3a + 1.5 = 30b$
5. $2(g - h) = 20$
   $g + 5(h - 1) = 41$
   [ANS. $g = 16$, $h = 6$]

9. Graphs

Throughout your study of physics you will be interested in how physical quantities depend upon one another. How does current, I, vary with voltage, E? How does the length of a spring, L, depend on the load, W?

The relationship between pairs of quantities is seen most clearly on a graph. Drawing and interpreting graphs is an important part of your work in physics.

If you have not had any practice drawing graphs, you should refer to the discussion of them on page 22.

10. Some useful geometry

Here, without proof, are listed the facts of elementary geometry you will find most useful for your work in this book.

1. *The sum of the angles in any triangle is 180°, which is two right angles.* In the figure,

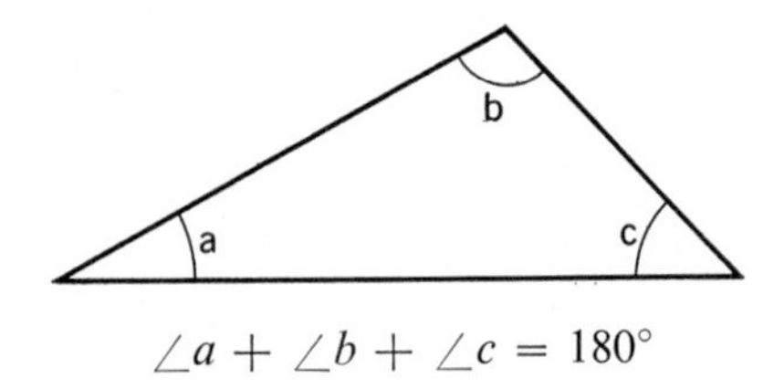

$$\angle a + \angle b + \angle c = 180°$$

2. *If the angles of one triangle are each equal to an angle of another triangle then the two triangles are said to be similar.* For a physicist the most useful fact about such similar triangles is that the ratio between a pair of corresponding sides is the same for all three pairs of sides. Thus if

$$\angle a_1 = \angle a_2, \quad \angle b_1 = \angle b_2, \text{ and } \angle c_1 = \angle c_2,$$

then

$$\frac{A_1}{A_2} = \frac{B_1}{B_2} = \frac{C_1}{C_2}$$

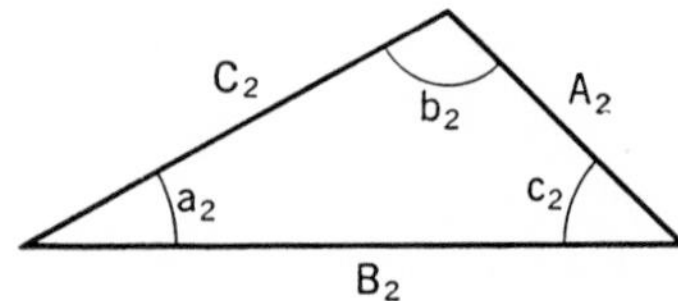

3. *Two angles are equal if (a) they are vertical angles* ($\angle a_1 = \angle a_2$; $\angle s_1 = \angle s_2$; *(b) their corresponding sides are parallel and in the same relative positions* ($\angle t_1 = \angle t_2$); *(c) their corresponding sides are perpendicular and in the same relative posi-*

tions ($\angle p_1 = \angle p_2$); or (d) they are corresponding angles of similar triangles.

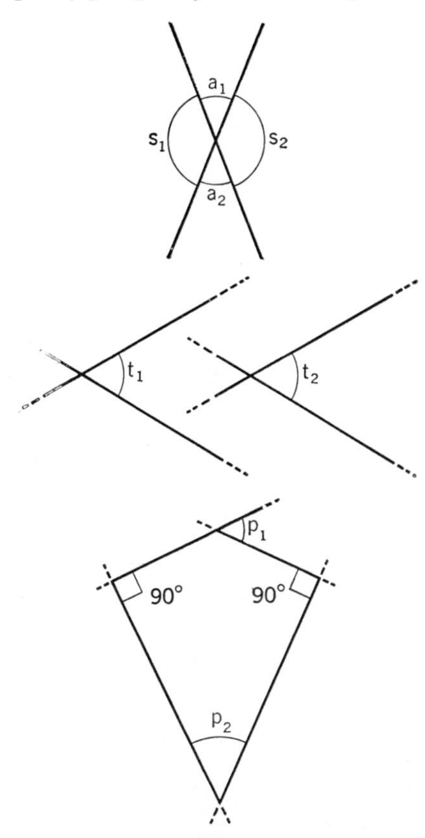

4. The relationship between the sides of right triangles is often important.

(a) *In* ANY *right triangle the lengths of sides A, B, and C are related as follows:* $A^2 + B^2 = C^2$. This very useful fact is called the **Pythagorean Theorem.** The long side, *C*, opposite the right angle is called the **hypotenuse.**

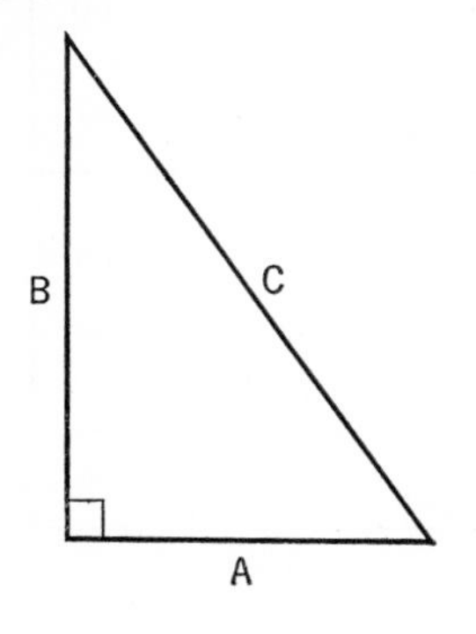

(b) *In a right triangle whose acute angles are* 30° *and* 60° *the sides have the relative lengths shown in the figure below.*

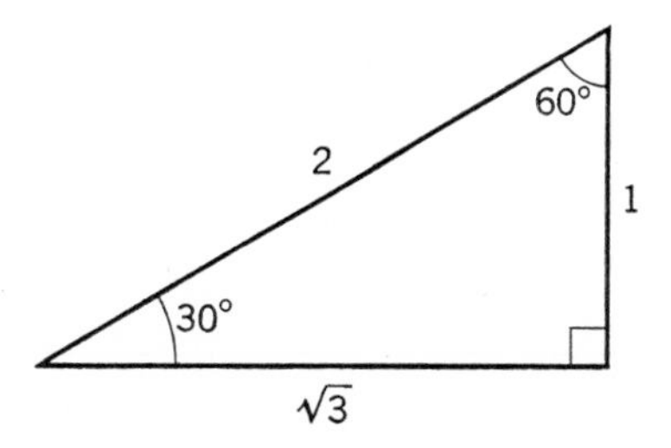

(c) *In a right triangle whose acute angles are both* 45° *the sides have the relative lengths shown in the figure below.*

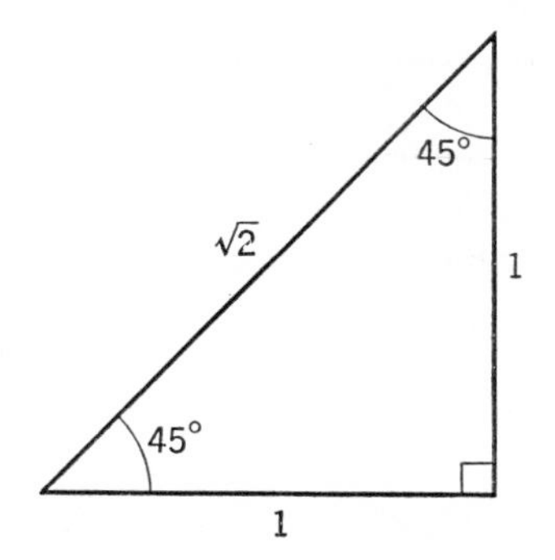

11. Elementary trigonometry

Many calculations in physics, particularly in optics and mechanics, involve the sides or angles of triangles, one of which is an unknown quantity.

If the triangle is a 30° 60° or a 45° 45° right triangle, the rules of the preceding section are usually of sufficient help in solving for an unknown side or angle. And in any right triangle, the Pythagorean Theorem is the key if two sides are known and the third side is sought.

But if angles are given or required in other right triangles than these, then you must use the sine, cosine, or tangent of an angle.

The sine of an angle of a right triangle is the ratio of the side opposite the angle to the hypotenuse (side opposite the right angle). Thus in the triangle shown here the sine of a (written: $\sin a$) $= A/C$, and $\sin b = B/C$.

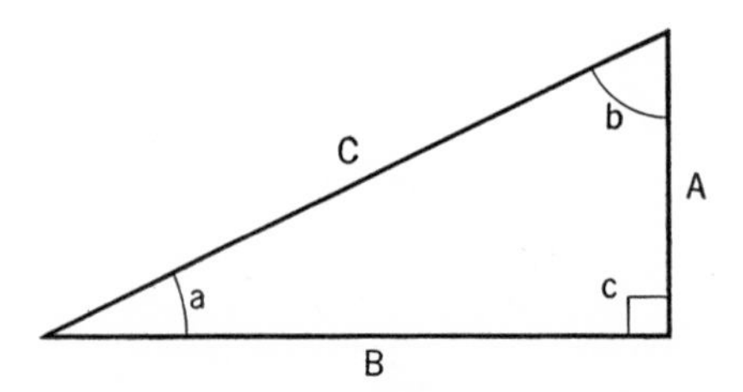

The cosine of an angle is the ratio of the side adjacent to the angle and the hypotenuse. The cosine of angle a (written: $\cos a$) $= B/C$, and $\cos b = A/C$.

Occasionally you know two sides, A and B, of a triangle and the angle between them ($\angle c$) and you need to find the third side, C. If the triangle is not a right angle, the Pythagorean Theorem is of no use, but the following formula will give you the unknown side, C, directly.

$$C^2 = A^2 + B^2 - 2AB \cos c$$

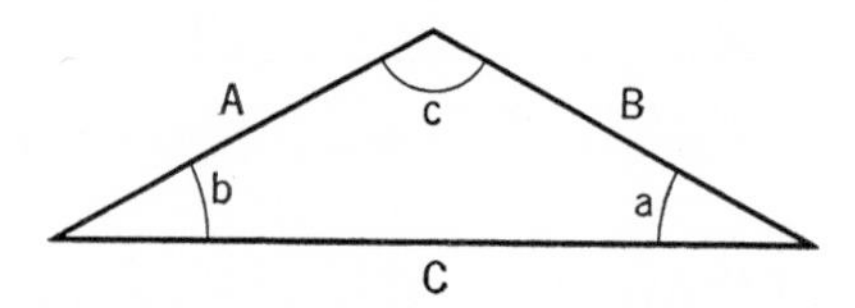

Finally, *the tangent of an angle is the ratio of the side opposite the angle and the side adjacent to the angle.* Hence $\tan a = A/B$, and $\tan b = B/A$.

The sines, cosines, and tangents of all angles from 0° to 90° are given in the table on p. 682.

You should notice now how sines, cosines, and tangents can be put to use solving physics problems.

SAMPLE PROBLEM

1. If angle a in the last figure is 13° and $A = 8$ cm, find the length of C. Notice that none of the facts about right triangles in Section 10 are of any use here. But remembering that

$$\sin a = \frac{A}{C}$$

we can find C by substitution

$$\sin 13 = \frac{8}{C}$$

Referring to the table on p. 682 (to 3 significant figures)

$$0.225 = \frac{8}{C}$$

$$C = 35.5 \text{ cm}$$

2. Find angle b in the same figure if $B = 12$ ft and $A = 3$ ft. Here we use the fact that

$$\tan b = \frac{B}{A}$$

$$= \frac{12}{3}$$

$$= 4.0$$

In the table on p. 682 we find that (to 2 significant figures) the angle whose tangent is 4.0 is 76°.

Practice Problems

Find the lettered quantities in the following triangles:

1.

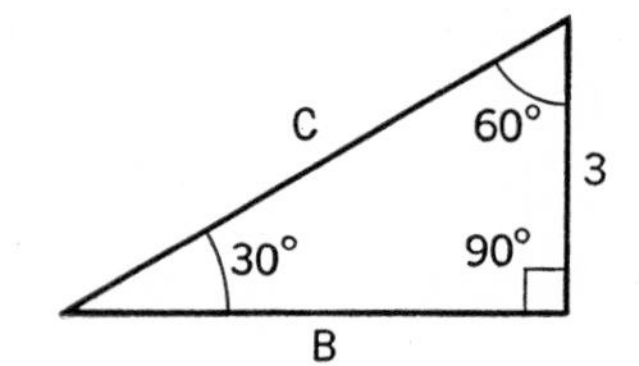

2.

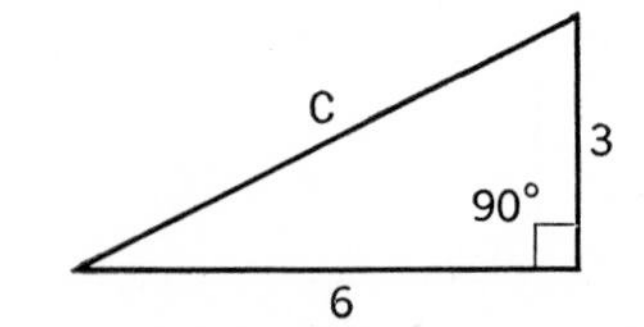

3.

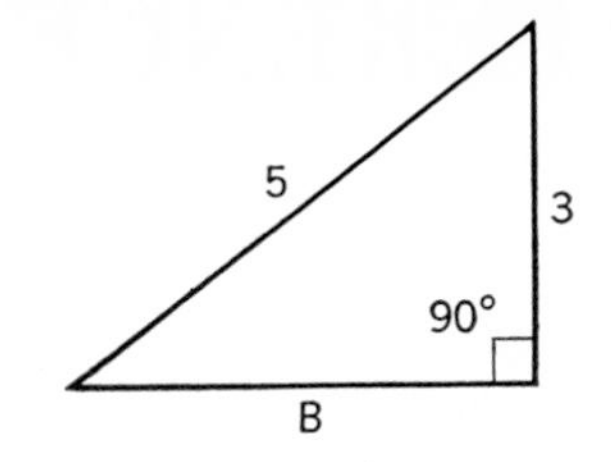

4.

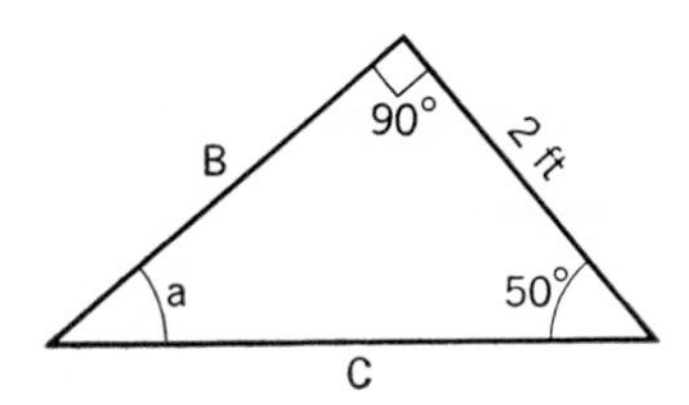

5.

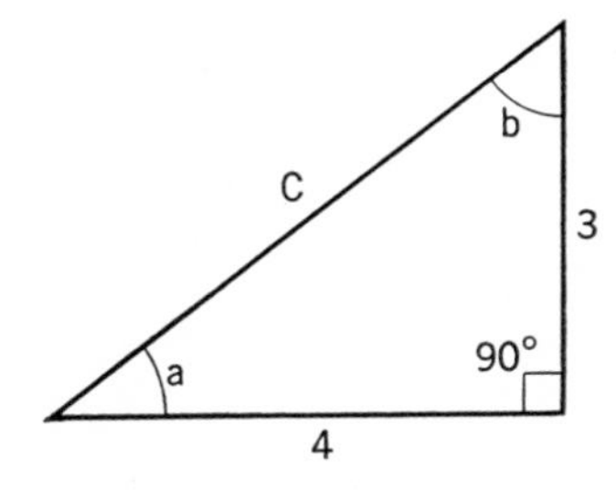

6.

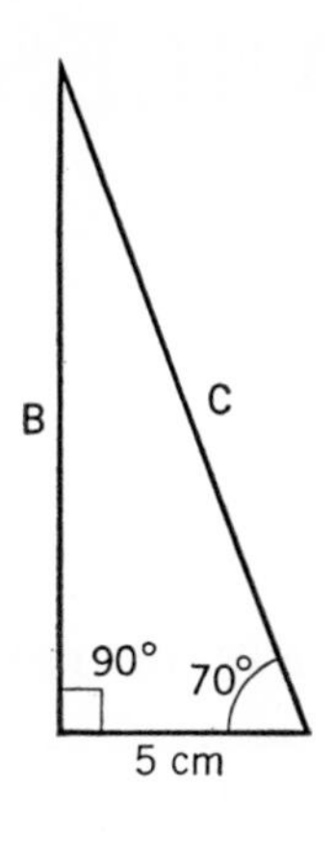

7.

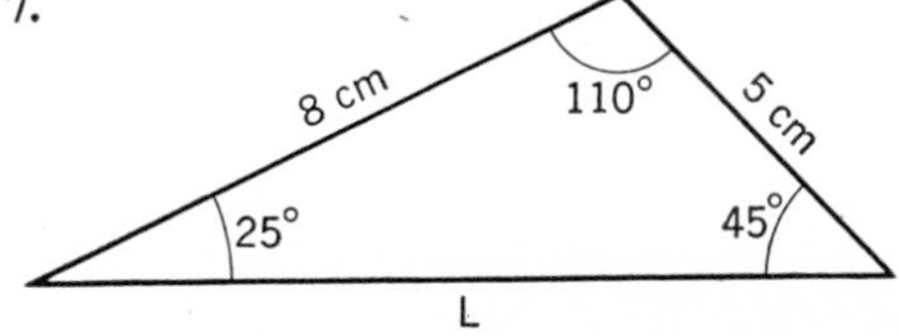

HINT. Draw a vertical line from the 110° angle down perpendicular to L, creating two triangles.

12. Summary of arithmetical formulas

Area of a square, sides l:

$$A = l^2$$

Area of a rectangle, sides l and w:

$$A = lw$$

Area of a triangle, base l, height h:

$$A = lh/2$$

Area of a circle, diameter d, radius r:

$$A = \pi d^2/4 = \pi r^2$$

Circumference of a circle, diameter d:

$$C = \pi d$$

Surface area of a sphere, radius r:

$$A = 4\pi r^2$$

Volume of a sphere, radius r:

$$V = \frac{4}{3}\pi r^3$$

Volume of a rectangular box, sides l, w, and h:

$$V = lwh$$

Volume of a cylinder, radius r, length l:

$$V = \pi r^2 l$$

TABLES FOR REFERENCE

MEASUREMENTS OF LENGTH, AREA, VOLUME, AND MASS

ENGLISH SYSTEM	METRIC SYSTEM	EQUIVALENTS
A. *Length* (*Linear Measurement*)		
12 in = 1 ft	10 mm = 1 cm	1 in = 2.54 cm
3 ft = 1 yd	100 cm = 1 m	1 ft = 30.48 cm
5,280 ft = 1 mi	1,000 mm = 1 m	1 m = 39.37 in
	1,000 m = 1 km	1 m = 1.09 yd
		1 km = 0.62 mi
B. *Area* (*Square Measurement*)		
144 in^2 = 1 ft^2	100 mm^2 = 1 cm^2	1 in^2 = 6.45 cm^2
		1 ft^2 = 929 cm^2
C. *Volume* (*Cubic Measurement*)		
1,728 in^3 = 1 ft^3	1,000 mm^3 = 1 cm^3	1 liter = 1.06 qt
231 in^3 = 1 gal	1 ml = 1 cm^3	1 gallon = 3.78 liters
	1 liter = 1,000 ml	
D. *Mass*		
16 oz = 1 lb	1,000 mg = 1 gm	1 kg = 2.2 lb
2,000 lb = 1 ton	1,000 gm = 1 kg	1 lb = 454 gm

COMMON CONVERSION FACTORS AND EQUIVALENTS

(a force of) 1 poundal	= 1.38×10^4 dynes	1 joule	= 0.238 cal
1 atmosphere	= 14.7 lb/in^2		= 10,000,000 ergs
	= 76 cm of mercury		= 0.739 ft-lb
	= 34 ft of water	1 hp	= 550 ft-lb/sec
60 mi/hr	= 88 ft/sec		= 746 watts
1 Btu	= 252 cal	1 Faraday	= 96,500 coulombs
	= 778 ft-lb		

IMPORTANT PHYSICAL CONSTANTS

1 ft^3 of water weighs 62.4 lb
1 liter of water weighs 1.00 kg
Density of air at 0°C and 76 cm = 0.0807 lb/ft^3
= 1.29 gm/liter
Charge of electron = 1.60×10^{-19} coulomb
Mass of electron = 9.11×10^{-28} gm
Mass of hydrogen atom = 1.6734×10^{-24} gm
Velocity of light = 3.00×10^{10} cm/sec
Gravitational constant, G = 6.67×10^{-8} dyne cm^2/gm^2
Acceleration of gravity (approximate) = 32 ft/sec^2
= 980 cm/sec^2
Solar constant = 1.94 $cal/cm^2/min$
Number of atoms per gram of hydrogen (Avogadro's Number) = 6.02×10^{23}

DENSITY AND SPECIFIC GRAVITY

The *density* of a material is the mass per unit volume, or:

$$D = \frac{m}{V}$$

The *specific gravity* * of a material is the same as the figure for density in column 2 with the units omitted.

Material	*In lb/ft³*	*In gm/cm³*	*Material*	*In lb/ft³*	*In gm/cm³*
Compressed gases in centers of dense stars	6×10^6	10^5	Glass	about 150	about 2.40
Gold	1,200	19.3	Sulfuric acid (conc.)	115	1.83
Mercury	849	13.6	Water	62.4	1.00
Compressed iron in core of earth	about 750	about 12	Alcohol (grain)	49.4	0.791
Lead	708	11.3	Gasoline	42.0	0.673
Silver	656	10.5	Wood (maple)	34.1	0.55
Copper	555	8.89	Air at 0°C at sea level	0.0807	0.00129
Iron	490	7.85	Air 15 miles up	5.6×10^{-3}	9.0×10^{-5}
Granite	170	2.72	Gases in inter-planetary space	about 10^{-19}	about 10^{-21}
Aluminum	168	2.70			

* Specific gravity may be defined as the ratio of the weight of a substance to the weight of an equal volume of water.

COEFFICIENTS OF FRICTION

The force of friction, *F*, is proportional to the force, *W*, pressing the two sliding surfaces together. The constant fraction is called the coefficient of friction, *k*, and is defined by the relation:

$$k = \frac{F}{W}$$

Wood on wood, dry	0.25 to 0.50
Wood on wood, soapy	0.20
Metal on metal, dry	0.15 to 0.20
Metal on metal, greased	0.03 to 0.05
Metal on metal, rolling	about 0.002
Rubber on concrete, dry	0.60 to 0.70

COEFFICIENTS OF LINEAR EXPANSION

These coefficients of linear expansion express the fractional increase in the length of a material for a temperature rise of 1 degree centigrade. Thus, the increase in length is expressed by the formula: change in length = coefficient of expansion × original length × change in temperature or:

$$\Delta L = k_e L \Delta T \, *$$

Aluminum	0.000025
Brass	0.000019
Copper	0.000017
Glass (ordinary)	0.000009
Glass (Pyrex)	0.000004
Invar	0.0000009
Iron (and steel)	0.000011
Lead	0.000028
Platinum	0.000009
Quartz	0.000005
Silver	0.000019
Tungsten	0.000043
Zinc	0.000026

* Δ means "change in."

COEFFICIENTS OF CUBICAL EXPANSION

The coefficient of cubical expansion is a number that expresses the fractional increase in volume of a substance per degree centigrade. Thus, the change in volume is expressed by the formula: change in volume = coefficient of cubical expansion × original volume × change in temperature or:

$$\Delta V = k_c V \Delta T \, *$$

Alcohol	0.00113
Aluminum	0.000075
Glass (ordinary)	0.000027
Iron (and steel)	0.000034
Mercury	0.000182
Oil (petroleum)	0.00899

* Δ means "change in."

HEAT CONDUCTIVITY

Air is taken as the standard, with a conductivity of 1. Thus silver conducts 19,300 times as many calories per second as an equal thickness of air of the same cross-sectional area. These values are approximate.

Good Conductors	
silver	19,300
copper	17,400
aluminum	8,800
iron	2,900
Poor Conductors	
glass (window)	44.0
concrete	35.0
water	25.0
wood	7.0
asbestos	4.8
linen	3.8
glass wool	2.6
cotton	2.4
cork board	1.9
rock wool	1.8
air	1.0

HEAT RELEASED BY BURNING OF FUEL These figures are approximate)

A calorie is the amount of heat which will raise 1 gram of water 1 centigrade degree; a Btu is the amount of heat which will raise 1 pound of water 1 Fahrenheit degree. The conversion formula is:

$$1 \text{ Btu} = 454 \times \frac{5}{9} = 252 \text{ cal}$$

	cal/gm	*Btu/lb*
Hydrogen	34,000	61,200
Gasoline	11,500	20,750
Butter	9,200	16,500
Coke	8,000	14,400
Coal, hard	7,400	13,300
Ethyl alcohol	6,500	11,620
Coal, soft	6,000	10,800
Oak wood	4,000	7,200
Sugar	4,000	7,200
Bread, white	2,600	4,800
Potatoes	1,000	1,800
Milk	720	1,300
Apples	630	1,130
Spinach	575	1,030
String beans	430	775

ELECTROCHEMICAL EQUIVALENTS

The electrochemical equivalent, Z, is a constant for the element used and expresses the mass of this material deposited on the negative electrode per unit of electrical charge. The mass, m, is measured in grams and the charge, Q, in coulombs. The total mass (in grams) deposited is expressed by:

$$m = ZQ$$

Hydrogen	0.00001045
Oxygen	0.0000829
Nickel	0.0003041
Copper	0.0003294
Zinc	0.0003387
Silver	0.001118

RESISTIVITY OF MATERIALS

The resistivity (specific resistance) is given in ohms per centimeter of length for a cross-sectional area of one square centimeter, at a temperature of 20°C. The formula for calculating the resistance of wires of any length is:

$$R = \frac{L \text{ (in cm)} \times \text{resistivity}}{A \text{ (in cm}^2\text{)}}$$

Silver	1.6×10^{-6}
Copper	1.7×10^{-6}
Aluminum	2.8×10^{-6}
Tungsten	5.5×10^{-6}
Iron	10×10^{-6}
German silver	33×10^{-6}
Nichrome	100×10^{-6}
Glass	2×10^{13}

INDEX OF REFRACTION

The index of refraction, k, of a substance is the speed of light in a vacuum, s_v, divided by the speed of light in the substance, s_s, or:

$$k = \frac{s_v}{s_s}$$

It may also be found by Snell's Law from the formula:

$$k = \frac{\sin i}{\sin r}$$

Vacuum	1.0000
Air	1.0003
Ice	1.31
Water	1.33
Alcohol	1.36
Glass (light crown)	1.51
Carbon disulfide	1.63
Glass (dense flint)	1.71
Diamond	2.42

The absolute humidity of saturated air is the maximum amount of water vapor the air can hold at a given temperature. Absolute humidity is expressed in number of grams of water vapor per cubic meter of air.

Temperature (°F)	*Absolute Humidity (gm/m^3)*	*Temperature (°F)*	*Absolute Humidity (gm/m^3)*	*Temperature (°F)*	*Absolute Humidity (gm/m^3)*	*Temperature (°F)*	*Absolute Humidity (gm/m^3)*
26	3.7	44	7.5	62	14.0	80	25.0
28	4.0	46	8.1	64	15.0	82	26.5
30	4.4	48	8.7	66	16.0	84	28.2
32	4.8	50	9.3	68	17.1	86	30.0
34	5.2	52	10.3	70	18.2	88	32.0
36	5.6	54	10.7	72	19.4	90	34.0
38	6.0	56	11.4	74	20.6	92	36.0
40	6.4	58	12.2	76	22.0		
42	6.9	60	13.0	78	23.5		

RELATIVE HUMIDITY
(Figures in intersection of columns give relative humidity in per cents)

Reading of Dry-Bulb Thermometer (°F)	*Difference between Readings of Wet- and Dry-Bulb Thermometer (°F)*															
	1°	2°	3°	4°	5°	6°	7°	8°	9°	10°	11°	12°	13°	14°	15°	16°
60°	94	89	83	78	73	68	63	58	53	48	43	39	34	30	26	21
61°	94	89	84	78	73	68	63	58	54	49	44	40	35	31	27	22
62°	94	89	84	79	74	69	64	59	54	50	45	41	36	32	28	24
63°	95	89	84	79	74	69	64	60	55	50	46	42	37	33	29	25
64°	95	90	84	79	74	70	65	60	56	51	47	43	38	34	30	26
65°	95	90	85	80	75	70	66	61	56	52	48	44	39	35	31	27
66°	95	90	85	80	75	71	66	61	57	53	48	44	40	36	32	29
67°	95	90	85	80	75	71	66	62	58	53	49	45	41	37	33	30
68°	95	90	85	80	76	72	67	62	58	54	50	46	42	38	34	31
69°	95	90	85	81	77	72	67	63	59	55	51	47	43	39	35	32
70°	95	90	86	81	77	72	68	64	59	55	51	48	44	40	36	33
71°	95	90	86	81	77	72	68	64	59	55	51	48	44	40	36	33
72°	95	90	86	82	78	73	69	65	61	57	53	49	45	42	38	34
73°	95	91	86	82	78	74	69	65	61	57	53	50	46	42	39	35
74°	95	91	86	82	78	74	69	65	61	58	54	50	47	43	39	36
75°	95	91	86	82	78	74	70	66	62	58	54	51	47	44	40	37
76°	96	91	87	82	79	74	70	66	62	59	55	51	48	44	41	38
77°	96	91	87	83	79	75	71	67	63	59	56	52	48	45	42	39
78°	96	91	87	83	79	75	71	67	63	60	56	53	49	46	43	39
79°	96	91	87	83	79	75	71	68	64	60	57	53	50	46	43	40
80°	96	91	87	83	79	75	72	68	64	61	57	54	50	47	44	41
82°	96	92	88	84	80	76	72	69	65	61	58	55	51	48	45	42
84°	96	92	88	84	80	76	73	69	66	62	59	56	52	49	46	43
86°	96	92	88	84	81	77	73	70	66	63	60	57	53	50	47	44
88°	96	92	88	85	81	77	74	70	67	64	61	57	54	51	48	46
90°	96	92	89	85	81	78	74	71	68	65	61	58	55	52	49	47

Element	Symbol	Atomic Number	Atomic Weight	Element	Symbol	Atomic Number	Atomic Weight
Hydrogen	**H**	1	1.008	Tellurium	**Te**	52	127.61
Helium	**He**	2	4.003	Iodine	**I**	53	126.91
Lithium	**Li**	3	6.940	Xenon	**Xe**	54	131.30
Beryllium	**Be**	4	9.013	Cesium	**Cs**	55	132.91
Boron	**B**	5	10.82	Barium	**Ba**	56	137.36
Carbon	**C**	6	12.011	Lanthanum	**La**	57	138.92
Nitrogen	**N**	7	14.008	Cerium	**Ce**	58	140.13
Oxygen	**O**	8	16.000	Praseodymium	**Pr**	59	140.92
Fluorine	**F**	9	19.00	Neodymium	**Nd**	60	144.27
Neon	**Ne**	10	20.183	Prometheum	**Pm**	61	[145]†
Sodium	**Na**	11	22.991	Samarium	**Sm**	62	150.35
Magnesium	**Mg**	12	24.32	Europium	**Eu**	63	152.0
Aluminum	**Al**	13	26.98	Gadolinium	**Gd**	64	157.26
Silicon	**Si**	14	28.09	Terbium	**Tb**	65	158.93
Phosphorus	**P**	15	30.975	Dysprosium	**Dy**	66	162.51
Sulfur	**S**	16	32.066*	Holmium	**Ho**	67	164.94
Chlorine	**Cl**	17	35.457	Erbium	**Er**	68	167.27
Argon	**A**	18	39.944	Thulium	**Tm**	69	168.94
Potassium	**K**	19	39.100	Ytterbium	**Yb**	70	173.04
Calcium	**Ca**	20	40.08	Lutetium	**Lu**	71	174.99
Scandium	**Sc**	21	44.96	Hafnium	**Hf**	72	178.58
Titanium	**Ti**	22	47.90	Tantalum	**Ta**	73	180.95
Vanadium	**V**	23	50.95	Tungsten	**W**	74	183.86
Chromium	**Cr**	24	52.01	Rhenium	**Re**	75	186.22
Manganese	**Mn**	25	54.94	Osmium	**Os**	76	190.2
Iron	**Fe**	26	55.85	Iridium	**Ir**	77	192.2
Cobalt	**Co**	27	58.94	Platinum	**Pt**	78	195.09
Nickel	**Ni**	28	58.71	Gold	**Au**	79	197.0
Copper	**Cu**	29	63.54	Mercury	**Hg**	80	200.61
Zinc	**Zn**	30	65.38	Thallium	**Tl**	81	204.39
Gallium	**Ga**	31	69.72	Lead	**Pb**	82	207.21
Germanium	**Ge**	32	72.60	Bismuth	**Bi**	83	209.00
Arsenic	**As**	33	74.91	Polonium	**Po**	84	210
Selenium	**Se**	34	78.96	Astatine	**At**	85	[211]†
Bromine	**Br**	35	79.916	Radon	**Rn**	86	222
Krypton	**Kr**	36	83.8	Francium	**Fr**	87	[223]†
Rubidium	**Rb**	37	85.48	Radium	**Ra**	88	226.05
Strontium	**Sr**	38	87.63	Actinium	**Ac**	89	227
Yttrium	**Y**	39	88.92	Thorium	**Th**	90	232.05
Zirconium	**Zr**	40	91.22	Protactinium	**Pa**	91	231
Niobium	**Nb**	41	92.91	Uranium	**U**	92	238.07
Molybdenum	**Mo**	42	95.95	Neptunium	**Np**	93	[237]†
Technetium	**Tc**	43	[99]†	Plutonium	**Pu**	94	[242]†
Ruthenium	**Ru**	44	101.1	Americium	**Am**	95	[243]†
Rhodium	**Rh**	45	102.91	Curium	**Cm**	96	[245]†
Palladium	**Pd**	46	106.7	Berkelium	**Bk**	97	[249]†
Silver	**Ag**	47	107.880	Californium	**Cf**	98	[249]†
Cadmium	**Cd**	48	112.41	Einsteinium	**E**	99	[254]†
Indium	**In**	49	114.82	Fermium	**Fm**	100	[252]†
Tin	**Sn**	50	118.70	Mendelevium	**Mv**	101	[256]†
Antimony	**Sb**	51	121.76	Nobelium	**No**	102	‡

* Because of the abundance of isotopes, the atomic weight varies between 32.063 and 32.069.
† Brackets indicate the mass number of the most stable known isotope of the element.
‡ The Atomic Weight of Nobelium had not been determined at the date of publication.

NATURAL SINES, COSINES, AND TANGENTS

(For sines and tangents, read angle at left; for cosines, read angle at right)

Angle	Tangent	Sine	Angle	Angle	Tangent	Sine	Angle
0	0.0000	0.0000	**90**	**46**	1.036	0.7193	**44**
1	0.0175	0.0175	**89**	**47**	1.072	0.7314	**43**
2	0.0349	0.0349	**88**	**48**	1.111	0.7431	**42**
3	0.0524	0.0523	**87**	**49**	1.150	0.7547	**41**
4	0.0699	0.0698	**86**	**50**	1.192	0.7660	**40**
5	0.0875	0.0872	**85**	**51**	1.235	0.7772	**39**
6	0.1051	0.1045	**84**	**52**	1.280	0.7780	**38**
7	0.1228	0.1219	**83**	**53**	1.327	0.7986	**37**
8	0.1405	0.1392	**82**	**54**	1.376	0.8090	**36**
9	0.1584	0.1564	**81**	**55**	1.428	0.8192	**35**
10	0.1763	0.1737	**80**	**56**	1.483	0.8290	**34**
11	0.1944	0.1908	**79**	**57**	1.540	0.8387	**33**
12	0.2126	0.2079	**78**	**58**	1.600	0.8480	**32**
13	0.2309	0.2250	**77**	**59**	1.664	0.8572	**31**
14	0.2493	0.2419	**76**	**60**	1.732	0.8660	**30**
15	0.2680	0.2588	**75**	**61**	1.804	0.8746	**29**
16	0.2868	0.2756	**74**	**62**	1.881	0.8830	**28**
17	0.2924	0.2924	**73**	**63**	1.963	0.8910	**27**
18	0.3249	0.3090	**72**	**64**	2.050	0.8988	**26**
19	0.3443	0.3256	**71**	**65**	2.144	0.9063	**25**
20	0.3640	0.3420	**70**	**66**	2.246	0.9136	**24**
21	0.3839	0.3584	**69**	**67**	2.356	0.9205	**23**
22	0.4040	0.3746	**68**	**68**	2.475	0.9272	**22**
23	0.4245	0.3907	**67**	**69**	2.605	0.9336	**21**
24	0.4452	0.4067	**66**	**70**	2.748	0.9397	**20**
25	0.4663	0.4226	**65**	**71**	2.904	0.9455	**19**
26	0.4877	0.4384	**64**	**72**	3.078	0.9511	**18**
27	0.5095	0.4540	**63**	**73**	3.271	0.9563	**17**
28	0.5317	0.4695	**62**	**74**	3.487	0.9613	**16**
29	0.5543	0.4848	**61**	**75**	3.732	0.9659	**15**
30	0.5774	0.5000	**60**	**76**	4.011	0.9703	**14**
31	0.6009	0.5150	**59**	**77**	4.332	0.9744	**13**
32	0.6249	0.5299	**58**	**78**	4.705	0.9782	**12**
33	0.6494	0.5446	**57**	**79**	5.145	0.9816	**11**
34	0.6745	0.5592	**56**	**80**	5.671	0.9848	**10**
35	0.7002	0.5736	**55**	**81**	6.314	0.9877	**9**
36	0.7265	0.5878	**54**	**82**	7.115	0.9903	**8**
37	0.7536	0.6018	**53**	**83**	8.144	0.9926	**7**
38	0.7813	0.6157	**52**	**84**	9.514	0.9945	**6**
39	0.8098	0.6293	**51**	**85**	11.43	0.9962	**5**
40	0.8391	0.6428	**50**	**86**	14.30	0.9976	**4**
41	0.8693	0.6561	**49**	**87**	19.08	0.9986	**3**
42	0.9004	0.6691	**48**	**88**	28.64	0.9994	**2**
43	0.9325	0.6820	**47**	**89**	57.29	0.9998	**1**
44	0.9657	0.6947	**46**	**90**	—	1.0000	**0**
45	1.0000	0.7071	**45**				

Sine The ratio of the side opposite the angle to the hypotenuse.

Cosine: The ratio of the side adjacent to the angle to the hypotenuse.

Tangent. The ratio of the side opposite the angle to the side adjacent to the angle.

SQUARE ROOTS (to 4 significant figures)

n	$\sqrt{n}$	$\sqrt{10n}$	n	$\sqrt{n}$	$\sqrt{10n}$
1	1.000	3.162	**51**	7.141	22.58
2	1.414	4.472	**52**	7.211	22.80
3	1.732	5.477	**53**	7.280	23.02
4	2.000	6.324	**54**	7.348	23.24
5	2.236	7.071	**55**	7.416	23.45
6	2.449	7.745	**56**	7.483	23.66
7	2.646	8.366	**57**	7.550	23.87
8	2.828	8.944	**58**	7.616	24.08
9	3.000	9.486	**59**	7.681	24.29
10	3.162	10.00	**60**	7.746	24.49
11	3.317	10.49	**61**	7.810	24.70
12	3.464	10.95	**62**	7.874	24.90
13	3.605	11.40	**63**	7.937	25.10
14	3.742	11.83	**64**	8.000	25.30
15	3.873	12.25	**65**	8.062	25.50
16	4.000	12.65	**66**	8.124	25.69
17	4.123	13.04	**67**	8.185	25.88
18	4.243	13.42	**68**	8.246	26.08
19	4.359	13.78	**69**	8.307	26.27
20	4.472	14.14	**70**	8.367	26.46
21	4.583	14.49	**71**	8.426	26.65
22	4.690	14.83	**72**	8.485	26.83
23	4.796	15.17	**73**	8.544	27.02
24	4.899	15.49	**74**	8.602	27.20
25	5.000	15.81	**75**	8.660	27.39
26	5.099	16.12	**76**	8.718	27.57
27	5.196	16.43	**77**	8.775	27.75
28	5.292	16.73	**78**	8.832	27.93
29	5.385	17.03	**79**	8.888	28.11
30	5.477	17.32	**80**	8.944	28.28
31	5.567	17.61	**81**	9.000	28.46
32	5.657	17.89	**82**	9.055	28.64
33	5.745	18.17	**83**	9.110	28.81
34	5.831	18.44	**84**	9.165	28.98
35	5.916	18.71	**85**	9.220	29.15
36	6.000	18.97	**86**	9.274	29.32
37	6.083	19.24	**87**	9.327	29.50
38	6.164	19.49	**88**	9.381	29.66
39	6.245	19.75	**89**	9.434	29.83
40	6.325	20.00	**90**	9.487	30.00
41	6.403	20.25	**91**	9.539	30.17
42	6.481	20.49	**92**	9.592	30.33
43	6.557	20.74	**93**	9.644	30.50
44	6.633	20.98	**94**	9.695	30.66
45	6.708	21.21	**95**	9.747	30.82
46	6.782	21.45	**96**	9.798	30.98
47	6.856	21.68	**97**	9.849	31.14
48	6.928	21.91	**98**	9.899	31.30
49	7.000	22.14	**99**	9.950	31.46
50	7.071	22.36	**100**	10.00	31.62

Notes:

1. To find the square root of any number n, use the first column ($\sqrt{n}$); to find the square root of 10n (any number between 100 and 1000), use the second column ($\sqrt{10n}$).
2. The square root of a number between 1000 and 10,000 is $10\sqrt{n}$; the square root of a number between 10,000 and 100,000 is $10\sqrt{10n}$.

To find the square of a number greater than 10, read the tens digit in the left column plus the units digit in the top row. Thus, to find 23^2, read 20 horizontally and 3 vertically; the square is 529.

		1	**2**	**3**	**4**	**5**	**6**	**7**	**8**	**9**
0		1	4	9	16	25	36	49	64	81
10	100	121	144	169	196	225	256	289	324	361
20	400	441	484	529	576	625	676	729	784	841
30	900	961	1024	1089	1156	1225	1296	1369	1444	1521
40	1600	1681	1764	1849	1936	2025	2116	2209	2304	2401
50	2500	2601	2704	2809	2916	3025	3136	3249	3364	3481
60	3600	3721	3844	3969	4096	4225	4356	4489	4624	4761
70	4900	5041	5184	5329	5476	5625	5776	5929	6084	6241
80	6400	6561	6724	6889	7056	7225	7396	7569	7744	7921
90	8100	8281	8464	8649	8836	9025	9216	9409	9604	9801

Notes:

1. If you are using powers-of-ten notation, $30^2 = (3 \times 10)^2 = 3^2 \times 10^2$. The first square can be read directly under the units in the top row and multiplied by 10^2.
2. To find squares of numbers greater than 100 (10^2), convert to powers-of-ten notation and use the table. Thus $960^2 = (96 \times 10)^2 = 96^2 \times 10^2$, which is 9216×10^2 (from 90 on the horizontal line plus 6 on the vertical line).
3. To find the square of a decimal such as 5.6 write it as 56×10^{-1}, find the square for 56 in the table, and write the square of 10^{-1} as 10^{-2}; thus $5.6^2 = (56 \times 10^{-1})^2 = 3136 \times 10^{-2} = 31.36$.

THE ADVANCED PLACEMENT EXAMINATION IN PHYSICS

Are you planning to go to college?

Are you planning to take College Entrance Examination Board exams for entrance into college?

Would you like to qualify for sophomore courses your very first year in college—or receive credit for required freshman courses on entrance?

Does your school offer courses equivalent to college freshman level in any subject in which you are particularly interested?

If your answer is "yes" to *all four* of these questions, then you will be interested in the Advanced Placement Program of the College Board.

Since 1954 it has been possible for able college-bound students from many schools to go directly into sophomore courses in various college subjects. They qualify for such courses by taking a special "Advanced Placement Examination" in each subject in May of their senior year. The results of the examination, now taken by thousands of students in the United States, are sent to the college to which a student has just been admitted. The college then decides whether or not to grant college credit for its freshman-level course or to allow the student to pursue studies in a sophomore course.

For further details you or your principal should write to The Director, Advanced Placement Program, College Entrance Examination Board, 425 West 117th Street, New York 27, N.Y.

The Advanced Placement Examination in Physics assumes that you have had a second course in general physics, or an accelerated college preparatory course, or the equivalent. Some students who have done high honors work in physics or who have participated in talent award contests may also be able to qualify for advanced placement by taking the examination. If you are thinking seriously of taking the Advanced Placement Examination in Physics, it is important that you obtain a copy of the syllabus from the director of the program and study further the details of what kind of preparation is expected. The following examination questions will also help you to see what is involved.

These questions, reprinted here by special permission, have been taken from the first half (Section I) of the 1958 examination and the second half (Section II) of the 1956 examination. As you read the collection through you should keep in mind two important points:

1. This collection is a composite intended to show you the structure of the examination and to give you some idea of its breadth of coverage and its length. Examinations since 1957 contain an *optional* section involving the use of calculus.

2. The collection contains questions—clearly beyond the scope of *Exploring Physics:* New Edition—which illustrate the level of difficulty to be expected (even apart from the optional questions requiring calculus) and make it quite clear that study of *Exploring Physics* alone is NOT adequate preparation for this examination.

If you bear these two points carefully in mind, the collection of test questions that follows will give you a reasonable idea of the difficulty and length of the examination and will give able students of *Exploring Physics* a chance to try some unusually challenging problems.

SAMPLE TEST QUESTIONS

SECTION I *—Time: 1 hour

Directions: Each of the questions or incomplete statements below is followed by five suggested answers or completions. Select the one which is best in each case.

1. Of the following "natural laws," the one which LEAST precisely represents the actual behavior of matter is
(A) Faraday's law of electrolysis
(B) Boyle's law for gases
(C) Coulomb's law of force between point charges
(D) Newton's law of gravitation
(E) Snell's law of refraction

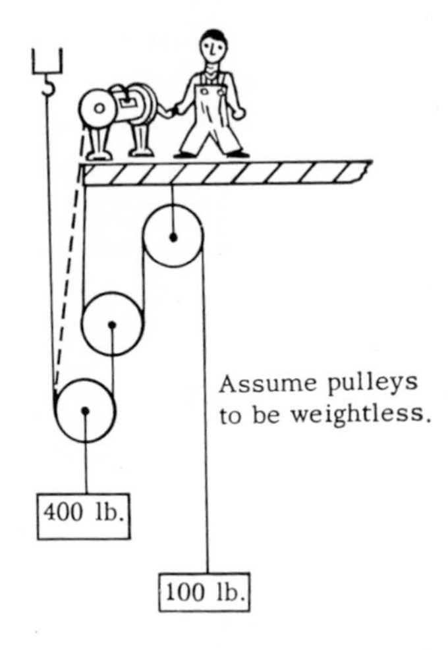

2. The system shown above is in equilibrium. Then the left-hand rope is attached to the drum and the man turns the drum. As he does so, if there is no friction, the 100-pound weight will
(A) remain motionless
(B) rise more slowly than the 400-pound weight
(C) rise as fast as the 400-pound weight
(D) rise faster than the 400-pound weight
(E) descend

3. Which is the LEAST precise of the following experimental determinations, each given with an accompanying limit of error?
(A) 22,800 $\pm$ 200 tons
(B) 20.4 liters $\pm$ 2%
(C) 0.042 $\pm$ 0.002 inch
(D) 630 lb. $\pm$ 1 part in 80
(E) $3.00(10)^8 \pm 1.5(10)^6$ meters/second

4. Dark lines appear in the solar spectrum because of
(A) dispersion
(B) inherent difficulties in instrumentation
(C) the absence of certain elements on the sun
(D) absorption of sunlight by the earth's atmosphere
(E) absorption of certain radiations by elements present in the outer regions of the sun

5. Two capacitors of different capacitances are charged separately with equal charges and are then connected together, plus-to-plus and minus-to-minus. All of the following are true EXCEPT:
(A) The final potentials across each are equal.
(B) The final energies stored in each are equal.
(C) The total net final charge equals the total net initial charge.
(D) The final charge on each is proportional to its initial capacitance.

* The questions here given comprise the first of two sections (except for two questions) of the May, 1958, examination. Reprinted with the permission of Educational Testing Service, Princeton, N.J.

(E) Initially the potential across the smaller capacitor was higher than the potential across the larger.

6. All of the following statements are true EXCEPT:

(A) Increase in temperature of a solid body causes a shift of the peak of the energy distribution toward longer wave lengths.
(B) The radiation which affects the eye as light has wave lengths extending roughly from 0.00004 to 0.00008 centimeter.
(C) Modern theory, based on the particle nature of radiation-energy, now affords a satisfactory and consistent explanation of the photoelectric effect and the radiation of energy.
(D) The Stefan-Boltzmann law states that the energy radiated by a "black body" is proportional to the fourth power of the absolute temperature.
(E) Increase in temperature of a body causes an increase in the total energy radiated.

7. A sample of gas is to be warmed up through a specified temperature range, either (I) with the pressure allowed to increase while the volume is kept constant or (II) with the volume allowed to increase while the pressure is kept constant. Experiment shows that less heat is required in case (I) than in case (II). This is evidence that

(A) molecules become smaller under increased pressure
(B) gas expanding under constant pressure requires molecules of higher average kinetic energy than gas expanding at constant volume
(C) a fundamental error in experimentation exists
(D) gas in expanding performs external work
(E) adiabatic phenomena are involved

8. A sinusoidal alternating current has a maximum value of 1.000 ampere and thereby a root-mean-square value of 0.707 ampere. The current will be measured as 0.707 ampere

(A) when sent through a D.C. ammeter (d'Arsonval moving-coil instrument)
(B) by any experimental method
(C) when indicated by a hot-wire ammeter
(D) when sent through a rectifier in series with a D.C. ammeter
(E) when calculated from the electrolytic deposition of silver

9. When a proton and an alpha particle both fall from rest through the same potential difference, the ratio of their final velocities, v_p/v_a, is

(A) $\sqrt{2}$
(B) 2
(C) $2\sqrt{2}$
(D) 4
(E) $4\sqrt{2}$

Questions 10–12

Assume that you have available appropriate wires, switches, uncalibrated rheostats, sources of E.M.F., and so forth, but only the following six calibrated instruments:

I. Dial resistance box, 0–100 ohms in steps of 1 ohm
II. Ten-ohm slide wire mounted on a graduated meter stick, maximum current 0.5 ampere
III. Galvanometer, 0–0.5 milliampere
IV. D.C. ammeter, 0–10 amperes
V. D.C. voltmeter, 0–12 volts
VI. Standard cell, E.M.F. 1.018 volts

What is the *smallest* combination of the above six instruments which would permit you to measure each of the following?

10. Resistance of an automobile headlight when at full brilliance
(A) I
(B) V
(C) IV, V
(D) III, VI
(E) I, II, III

11. Electromotive force of a 1.5-volt dry cell
(A) V
(B) I, VI
(C) I, II, III
(D) I, II, VI
(E) II, III, VI

12. Resistance of a 100-watt lamp with the filament cold
(A) I
(B) II, VI
(C) III, IV
(D) I, II, III
(E) II, III, VI

13. If two solid, circular discs having the same weight and thickness are made from metals having different densities, the disc having the larger moment of inertia about its center of mass is
(A) the disc made of the denser metal
(B) the disc made of the metal of lower density
(C) neither; the two discs will have the same moment of inertia
(D) the disc which is rotating faster
(E) not determinable from the data given

14. A 3000-pound car running 20 miles per hour due east hits a 4000-pound truck running 20 miles per hour due north, on an icy crossroad. The two lock together and skid with negligible friction. The speed of the combined wreck is approximately
(A) 28.2 miles/hour
(B) 20 miles/hour
(C) 14.3 miles/hour
(D) 10 miles/hour
(E) 5.7 miles/hour

15. Polarization is a phenomenon which
(A) distinguishes reflection from refraction
(B) occurs for light waves only
(C) occurs when interference takes place
(D) may be expected for every zone of the electromagnetic spectrum
(E) distinguishes sonic from ultrasonic waves

16. If our standard of mass were not 1 gram but some other unit 10 times larger, then
(A) the value of the gravitational constant G would be only $\frac{1}{100}$ times its present value
(B) the specific gravity of benzene, now 0.89, would be 8.9
(C) the value of the acceleration due to gravity g would be changed
(D) the units of volume remaining unchanged, the numerical value of all densities would become ten times as great
(E) none of the above would be true

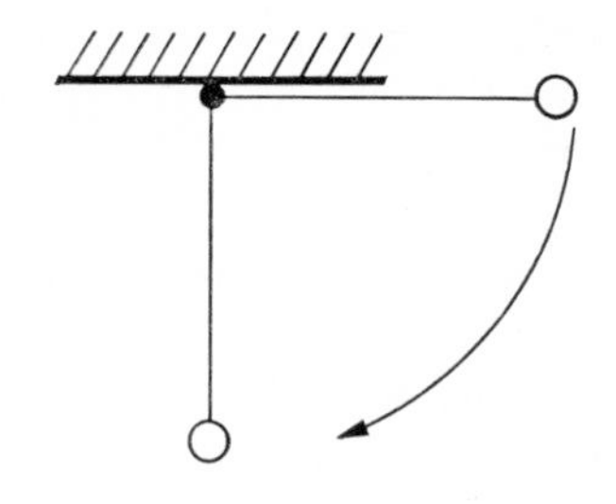

17. A simple pendulum consists of a bob of m grams of lead hung on a fine thread L centimeters long, in a region where the acceleration of gravity is g. The pendulum is drawn aside until it is horizontal and released. The thread breaks at the lowest point each time the experiment is tried. On the basis of this, the load M grams that would just break a sample of the same thread if hung on it at rest is given by

(A) M = m
(B) M = 2m
(C) M = 3m
(D) M = m + 2m/g
(E) M = (m + 2M)/g

Directions: For each of the incomplete sentences below, ONE or MORE of the completions given are correct. Decide which completion or completions are correct and indicate the correct answer by

A if only *1, 2, and 3* are correct,
B if only *1 and 3* are correct,
C if only *2 and 4* are correct,
D if only *4* is correct,
E if some other completion (or combination of completions) of those given is correct.

18. A large wood ball and a large lead ball of equal radii are dropped from rest from the top of a tower. They start together and are observed to reach the ground almost together. One may reasonably infer *from this observation* that

(1) air resistance is almost negligible for these balls
(2) the weights of the balls are proportional to their masses
(3) air buoyancy is almost negligible for these balls
(4) these balls fall according to the relationship $s = \frac{1}{2}gt^2$

19. A list of units of energy would include which of the following?

(1) 1 coulomb-volt
(2) 1 kilowatt-hour
(3) 1 electron-volt
(4) 1 calorie/second

20. If standing waves of definite frequency are set up in a medium between two points A and B, A being a source of energy,

(1) all particles of the medium between A and B will have the same amplitude of vibration
(2) all particles of the medium between A and B will vibrate with the same frequency
(3) the motion of all particles of the medium between A and B will be in phase
(4) there may be a reflector at B

21. In order that the principle of conservation of momentum be applicable to an interaction involving two or more bodies,

(1) the process must be a perfectly elastic collision
(2) no kinetic energy may be dissipated as heat
(3) the relative speed of separation must be the same as the relative speed of approach
(4) there must be no net external impulse

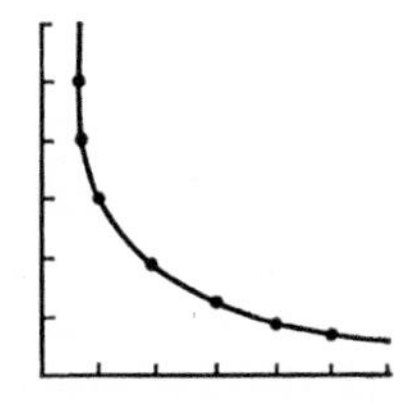

22. Each of the following sets of experimental measurements is plotted on a graph. Which of the graphs would have the form shown in the above sketch?

(1) Pressure *vs.* volume for a gas at constant temperature
(2) Pressure *vs.* density for a gas at constant temperature
(3) Acceleration *vs.* mass for an object pulled along a frictionless table with a constant force
(4) Power *vs.* current for a metal resistor

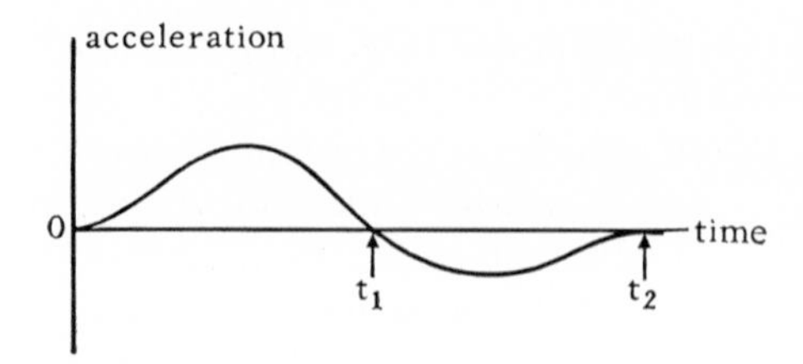

23. A body starts from rest in the positive direction on a straight-line path at time t = 0. The graph above shows the acceleration of this body as a function of time. One can accurately state that at the time t_1 the
(1) velocity of the body is constant
(2) velocity of the body is higher than at any other time before t_2
(3) forces on the body are balanced
(4) body is at rest

24. Suppose there are two closed containers of equal volume containing equal numbers of molecules of two gases A and B at the same temperature. Suppose also that the molecules of B have the same mass but twice the diameter of the molecules of A. One can accurately say that the
(1) density of gas B is half the density of gas A
(2) mean free path of the B molecules is half the mean free path of the A molecules
(3) average speed of the B molecules is lower than the average speed of the A molecules
(4) pressure is the same in each container

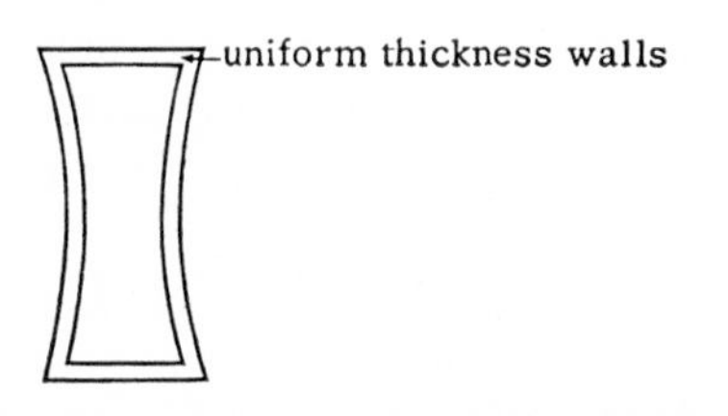

25. Two thin watch glasses are cemented together to form a hollow lens in the manner shown in the figure above. It is true that the lens will diverge a beam of parallel rays when filled with
(1) air and placed in air
(2) air and placed under water (index of refraction 1.33)
(3) water and immersed in carbon disulfide (index of refraction 1.62)
(4) carbon disulfide and immersed in water

SECTION II—* Time: 2 hours

Answer two and only two questions from each of the two Parts, A and B. Answer both given questions in Part C. Be sure you show clearly the steps by which you arrive at your answers. You will get part credit for the steps even if the answer is incorrect. If, after solving a numerical problem, you feel that the result is incorrect or unreasonable, and if you do not have sufficient time to search for the error, state briefly your opinion that it is incorrect and the reasons which made you feel so.

PART A

Answer 2 out of 3 questions.

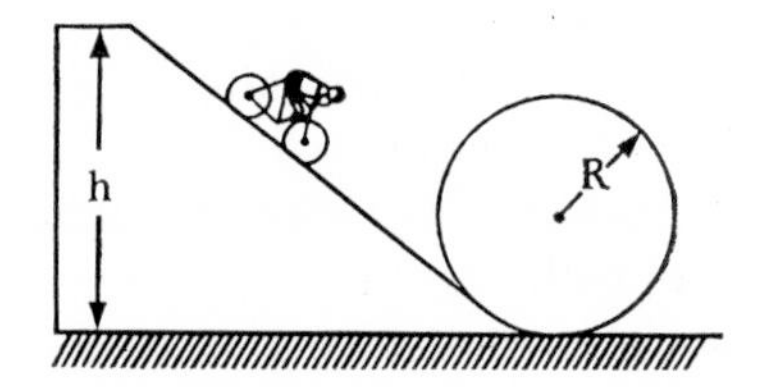

A-1. A circus stunt rider is going to coast down a steep track on a bicycle and loop-the-loop at the bottom as shown

* The questions here comprise the entire second (and final) section of the May, 1956, examination. Since 1957, Section II of the examination has contained an *optional* part which involves use of the calculus and might be chosen *instead* of one of the non-calculus parts. Reprinted with the permission of Educational Testing Service, Princeton, N.J.

above. The radius of the loop is R. In terms of R, find an expression for the least possible height h of the starting platform sufficient for him to travel around the loop without falling. Assume that there is no friction and that he does not pedal.

A-2. An 80-kilogram man, running with a speed of 6.0 meters per second, jumps from a level floor onto the rear of a 240-kilogram cart which was initially at rest. Both bodies then move together with a speed V meters per second.

(a) Find the initial kinetic energy of the man.

(b) Find the speed V.

(c) Find the final kinetic energy of the man.

(d) Find the kinetic energy which he imparted to the cart.

(e) The man now jumps off the back of the cart in such a way that he has no horizontal speed with respect to the floor. What do the speed and kinetic energy of the cart now become?

(f) Compare the results of (d) and (e), and comment on them.

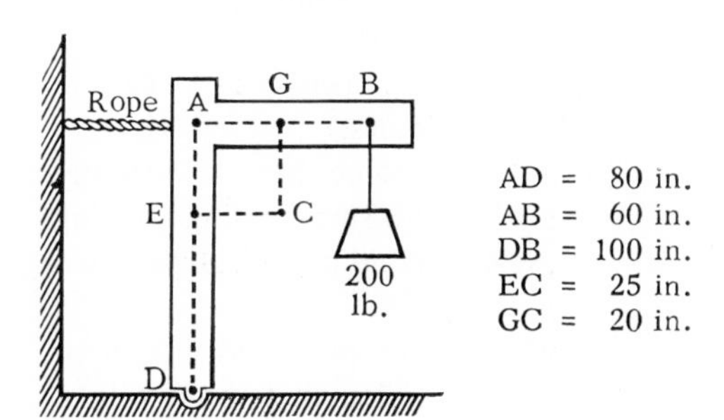

A-3. The fixed-arm derrick in the diagram above weighs 80 pounds. Its center of gravity is at C. A 200-pound object is suspended at point B. The lower end of the derrick rests in a lubricated cup at D. The derrick is supported by a horizontal rope at A.

(a) Determine the magnitude and direction of the force acting *on* the derrick at A.

(b) Determine the magnitude and direction of the force acting on it at D. (Express the direction in terms of the tangent of the angle with the horizontal.)

(c) Indicate on the diagram the approximate direction of the force at D.

PART B

Answer 2 out of 3 questions.

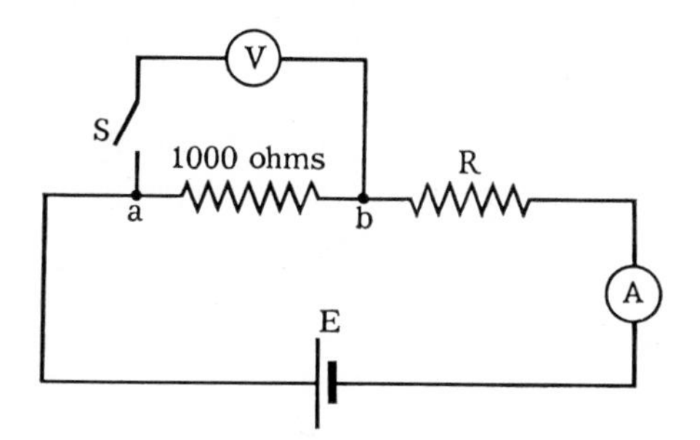

B-1. In the circuit shown above, the milliammeter A, of negligible resistance, reads 30 milliamperes with the switch S open. When switch S is closed, the voltmeter V reads 22.5 volts and the milliammeter reads 45 milliamperes. The E.M.F. E of the battery is constant and its internal resistance is zero.

(a) Find the E.M.F. E.

(b) Find the resistance R.

(c) Find the resistance of the voltmeter V.

B-2. A deuteron consists of a proton and neutron. The rest-masses of the particles (in atomic mass units) are

$$\text{deuteron} = 2.0147$$
$$\text{proton} = 1.0081$$
$$\text{neutron} = 1.0090$$

(a) Calculate the binding energy of the deuteron in Mev.

(b) What is the frequency of a gamma ray photon of this energy?

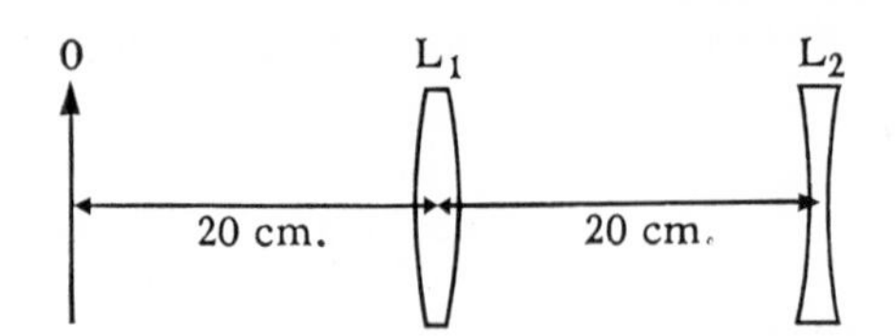

B-3. An object O, converging lens L_1 (focal length 10 cm), and diverging lens L_2 (focal length 8 cm) are arranged as shown in the diagram.

(a) Locate the image of O formed by L_1 alone. Give the distance of this image from L_1 and indicate on which side of L_1 the image appears.

(b) In the same way locate the image of O formed by the combination of lenses L_1 and L_2.

PART C

Answer both questions.

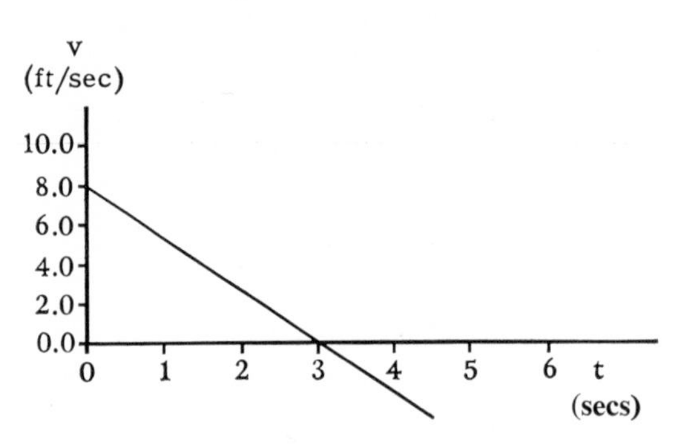

C-1. The above diagram represents a portion of the history of the straight-line motion of a body, with v denoting velocity and t time.

(a) Extract from this diagram and present in words all the information you can about the motion of the body, its acceleration, and the progressive changes in its position along its line of motion.

(b) Sketch a curve on an s *vs.* t diagram which is consistent with the above history.

C-2. (a) A small 12-ohm resistor can be connected across the terminals of a battery maintaining a constant potential difference of 60 volts. The resistor is placed in a can containing 300 grams of water at 20°C. The total heat capacity of the can itself is 25 calories per °C. A switch is closed and current allowed to flow for 45 seconds. Predict the final temperature of the water, giving a complete algebraic solution before using any numbers. Explain your work, indicating how definitions and laws of nature are applied in the reasoning. Describe the phenomenon in terms of energy transformation. State what idealizations about the physical situation are introduced. (Given: 1 calorie = 4.19 joules.)

(b) Suppose some heat is actually lost to the surroundings during the 45-second period. *Argue from the algebra of your solution* whether the predicted final temperature calculated above will be higher or lower than the final value which would actually be observed.

THE VOCABULARY OF PHYSICS

A battery. The battery that furnishes current to heat the filament or heater of a vacuum tube.

absolute humidity. The weight of the water vapor found in a unit volume of air. In the metric system, it is the number of grams of water vapor in a cubic meter of air.

absolute pressure. The total pressure exerted by a gas or a liquid. The absolute pressure of a vacuum is zero pounds per square inch, and of the atmosphere about 14.7 pounds per square inch.

absolute temperature scale. The scale of temperature whose zero is absolute zero and whose degree is the same size as a centigrade degree. On this scale, also called the *Kelvin scale,* water freezes at 273° and boils at 373°.

absolute zero. The lowest possible temperature. At this temperature, −273°C, the zigzag Brownian motion of atoms and molecules ceases.

absorption spectrum. A continuous spectrum crossed by dark lines caused by the absorption of some wave lengths by relatively cool gas or transparent liquid through which the radiation passes before entering the spectroscope. The sun's spectrum is of this type.

acceleration. The rate at which velocity is changing with time. Acceleration = the change in velocity divided by the time during which the change took place.

acceleration of gravity. The acceleration of a freely falling object in the absence of air resistance or other source of friction.

accommodation, power of. The ability of the lens of the eye to change its shape in order to focus upon the retina objects at various distances.

acoustics. The science of sound.

actual M.A. The mechanical advantage a machine is found to have when friction and other losses are taken into account. Actual M.A. is the ratio of the resistance force to the actual effort force.

adhesion. The force of attraction between molecules of two different kinds of substances. It is the force that holds water to skin, paint to wood, etc.

alloy. A fusion or mixture of two or more metals in which the metals cannot be distinguished from each other or separated by ordinary means.

alpha particle. A helium nucleus, consisting of two protons and two neutrons. It has a charge of +2 and a mass number of 4. Alpha particles are shot out by many, but not all, radioactive elements.

alternating current (AC). An electric current that travels first in one direction, then in the opposite direction through a circuit.

altimeter. Any device used to measure height, usually above sea level. Some altimeters using electronics measure height above the earth's surface. The simplest altimeter is a properly calibrated barometer.

ammeter. The most common device for measuring the flow of electric current.

ampere. An electric current of one coulomb per second. It is the current which flows when an electromotive force of one volt is applied across a resistance of one ohm.

amplifier. An electronic device whose purpose is to create strong electrical signals that are exact replicas of the weak signals fed into it.

amplitude. The amplitude of a sound or water wave is the maximum distance through which the vibrating particles are displaced from their rest positions. Amplitude is a measure of the energy carried by the wave.

amplitude modulation (AM). Modulation of a radio carrier wave by varying its amplitude. This, rather than frequency modulation, is the conventional method in most radio broadcasting.

anastigmat lens. A combination of several simple lenses commonly used in cameras and other optical instruments to reduce chromatic and spherical aberration.

aneroid barometer. A barometer whose pressure-sensitive element is an evacuated metal box with springy walls that are responsive to small changes in air pressure. Their motion, magnified by levers, moves a pointer across a calibrated scale.

angle of attack. The angle which the bottom of an airplane wing makes with the approaching air stream.

angle of incidence. The angle between the

normal to a surface and a ray falling upon the surface.

angle of reflection. The angle between the normal to a surface and a ray being reflected from the surface.

angle of refraction. The angle between the normal to a boundary between two transparent materials and a ray in the optically denser material just before or just after it has been refracted.

anode. The positive (+) terminal of an electroplating apparatus, an electrolysis apparatus, or a vacuum tube.

antifriction metals. Special metals, including alloys, which have very little friction with other more common metals such as steel. They are therefore very useful as bearings. Babbitt metal is a commonly used antifriction metal.

antiproton. A particle that has the same mass as the proton but has a negative charge. It exists for brief moments during various high-energy nuclear reactions.

armature. Commonly the rotating coils of wire in a motor or a generator.

astigmatism. A defect in vision in which the lens of the eye is curved differently in one direction than in another (like the side of a football), so that it is impossible to see all of an object in focus at once.

atom. The smallest piece of an element that can exist and still be that element. There are over 100 elements, each with its own kind of atom, but all made of the same electrical parts—electrons, protons, and neutrons—in various combinations.

atomic number. The number of protons in the nucleus of an atom. It is also equal to the number of electrons revolving around the nucleus. Each of the approximately 100 elements has, of course, a different atomic number.

Atomic Theory. The idea that matter cannot be divided into smaller and smaller pieces indefinitely—that in fact there is a limit beyond which matter cannot be divided and still be matter.

atomic weight. The weight of any atom compared with the weight of an oxygen atom (which is taken as exactly 16). The atomic weight is approximately equal to the number of protons and neutrons in an atom's nucleus (the mass number).

audio frequency (AF). Frequencies of alternating current in the range from about 10 cycles per second to 16,000 cycles per second. These are also the frequencies of audible sound waves.

automation. The replacement of human guidance by automatically operated machinery.

average speed. Distance covered divided by the time required to cover it. By this definition it does not matter whether the speed was constant throughout the trip.

Babbitt metal. An alloy of tin, copper, and antimony frequently used as a bearing metal to reduce friction, and hence wear, against a steel shaft.

back E.M.F. A voltage induced in the armature of a motor in the opposite direction to the voltage being applied to it. Back E.M.F. is generated as a result of the fact that the motor armature, rotating as it does in a magnetic field, is also acting as a generator.

ballast coil. A small iron-core coil placed in series with fluorescent lamps and other electrical devices to limit the flow of alternating current through them. Also called a *choke coil.*

banked road. A curving road whose edge on the outside of the curve is higher than the edge on the inside of the curve. The slope provides force inward (centripetal force) on a car going around the curve—a force that is a partial substitute for the force of friction.

barograph. An aneroid barometer that operates a pen held against a moving chart, thus making a permanent record of day-to-day atmospheric pressure.

barometer. The most common device used to measure the pressure exerted by the earth's atmosphere. There are several different kinds; among them, the mercury barometer and the aneroid barometer.

battery. A group of voltaic cells (wet or dry) connected in series to form a single source of voltage. The total voltage of the battery equals the sum of the voltages of the cells.

B battery. The battery that furnishes voltage to drive electrons around the plate circuit of a vacuum tube (or transistor) circuit.

beats. The throbbing effect caused by alternate reinforcement and interference between two sound waves of nearly the same frequency. Electromagnetic waves are also capable of producing beats.

beta particle. An electron shot out by a radioactive atom.

binding energy. The energy that holds together the protons, neutrons, and other particles in the nucleus of an atom.

binocular effect. The effect of depth created by two different views (one from each eye) fusing together in the brain. It gives an impression of how far away an object is.

blocking condenser (bypass condenser). An electrical condenser whose function is to allow the passage of alternating voltages but to block the passage of direct voltages.

boiling point. The temperature at which the vapor pressure of a liquid becomes equal to the pressure exerted on the liquid by the surrounding atmosphere. At this temperature bubbles of vapor can sustain themselves inside the liquid and grow.

breeder reactor. A reactor that creates (breeds) new fissionable material even while it is producing energy.

bright-line spectrum. A spectrum consisting of one or more discrete wave lengths of light (which appear as bright lines) separated by regions of darkness where there is no radiation. A bright-line spectrum is emitted by incandescent gases.

Brownian motion. The random zigzag motion of microscopic particles of visible matter as they are struck by single molecules of the gas or liquid in which they are suspended. Brownian motion is one of our few direct pieces of evidence of the thermal motion of atoms and molecules.

brush. The fixed contact that presses against the commutator rings on the rotating armature shaft of an electric motor or generator. The brushes and commutator permit the current to enter or leave the armature.

Btu (British thermal unit). The amount of heat energy needed to raise one pound of water one degree Fahrenheit.

buoyancy. The upward force exerted on an object in a fluid. The buoyant force, according to Archimedes' Principle, is equal to the weight of the fluid displaced by the object.

calibration. The process of marking off the correct divisions on a measuring instrument.

calorie. The amount of heat energy necessary to raise one gram of water one degree centigrade. The "large calorie," spelled *C*alorie, is the amount of heat energy necessary to raise one *kilogram* of water one degree centigrade.

calorimeter. An insulated container used to make heat measurements.

camera obscura. A dark room with a small hole in one wall through which light enters to form an image of the outdoor surroundings upside down upon the opposite wall. It is a pinhole camera.

candlepower. The unit by which the intensity of light sources is measured. An ordinary candle has about one candlepower, but more precisely we define it as one sixtieth of the light intensity from a 1 cm^2 hole in the top of a box of molten platinum at 1,773°C.

capacitance. See **capacity.**

capacitor. See **condenser.**

capacity (of an electrical condenser). The ability of a condenser to store electrical energy. The capacity of a condenser is measured in farads. A farad is the capacity of a condenser such that one volt of E.M.F. can transfer one coulomb of charge from one terminal to the other.

capillary action. The rising of liquids in thin tubes which they wet and the depression of liquids in tubes which they do not wet.

carrier current. In a radio transmitter or receiver, the high-frequency alternating current whose fluctuations in amplitude (or in frequency) correspond to the radio waves traveling from transmitter to receiver.

cathode. The negative (−) terminal of an electroplating apparatus or of a vacuum tube.

cathode rays. A stream of electrons given off at the cathode (negative terminal) of a vacuum tube.

cell. See **voltaic cell.**

center of gravity. The only point at which an object can be supported so that it will balance, no matter in what position it may be turned. It is the point where all the mass of an object *seems* to be concentrated.

centigrade scale. On this scale of temperature water boils at 100° at sea level and freezes at 0°; hence a centigrade degree is $\frac{1}{100}$ of the temperature difference between boiling and freezing water.

centrifugal force. The force equal and opposite to centripetal force. It is the force, therefore, that opposes efforts to pull a moving mass into a curved path.

centrifuge. A device for separating a mixture of liquids or liquids and solids by spinning them rapidly. The resulting centripetal force is greater upon the denser material and causes it to travel to the outside of the curved path while the less dense material travels towards the inside.

centripetal force. The force that must be exerted on an object at right angles to its direction of motion in order to make it move in a curved path. To force an object moving with a velocity v into a curved path whose radius is r requires a centripetal force $F = mv^2/r$.

chain reaction. The process in which the neutrons produced by fission of U^{235} or plutonium are able to cause fission of still more U^{235} or plutonium atoms, resulting in the production of still more neutrons, and so on.

charge. An accumulation of electricity of

only one kind (as on a cloud, a hard rubber or Bakelite rod).

choke coil. See **ballast coil.**

chromatic aberration. A defect of a lens in which various colors of light are brought to a focus at different distances from the lens.

cloud chamber. An instrument for revealing in the form of streaks of fog droplets the paths of ionizing particles produced by ionization, transmutation, and other nuclear processes.

clutch. A device in an automobile that allows the engine to be disconnected from the wheels of the car so the driver can shift gears. Two discs normally held together by springs form part of the drive shaft. When they are forced apart by the driver pressing on the clutch pedal, the connection between engine and rear wheels is broken.

coated lens. A lens on whose outer surface a hard transparent film has been formed of precisely the best thickness to eliminate by interference rays of light that would otherwise be reflected by the lens surface and lost instead of passing through it.

cochlea. The snail-shell-shaped organ in the inner ear responsible for converting sound waves into nerve impulses to the brain.

coefficient of cubical expansion. The change in volume of a unit volume of any material when its temperature changes by one degree. The coefficient of cubical expansion is three times the coefficient of linear expansion, using the same temperature scale.

coefficient of friction. The ratio between the force, *F*, needed to overcome friction and the force, *W*, holding the sliding surfaces together. The coefficient, $k = F/W$.

coefficient of linear expansion. The change in length of a unit length of a solid material when it is heated or cooled one degree of temperature. The coefficient for the Fahrenheit scale is $\frac{5}{9}$ the coefficient for the centigrade scale.

cohesion. The force of attraction between molecules of the same material. It is the force that pulls water into drops, keeps a stretched wire from breaking, and makes solid objects keep their shape.

collecting ring. The positively charged ring that collects electrons repelled from the dark areas on the screen of a television camera tube.

color addition. The process of combining two colors to produce a third color different from either.

color blindness. The inability to distinguish between certain colors because of a defect in the nerve endings in the retina of the eye.

color subtraction. The process of passing light through filters which absorb (subtract) one or more colors from the light.

commutator. A ring consisting of many insulated copper segments located on the rotating shaft of a motor or a generator. Current enters and leaves the rotating member (or armature) through the commutator and a motionless brush, against which the commutator slides.

component. The part of a vector quantity that is effective in a direction other than the original direction.

compound. A material made of two or more kinds of atoms held together in chemical combination; for example, carbon dioxide, CO_2.

compound bar. A bimetallic strip whose two metals have different coefficients of expansion. Changes in temperature cause the bar to bend. Its bending may be used to close or open electrical circuits, as in the thermostat.

compound microscope. A microscope consisting of at least two lenses, an objective and an eyepiece, in contrast to the single magnifying lens, which is sometimes called a *simple microscope*.

compression. As applied to sound, the half of a sound wave in which the air is compressed to greater than its normal density. Also called *condensation*.

Compton effect. When an electron is struck by X radiation or other very short wave-length radiation, it acquires some momentum, and the wave length of the radiation is increased. This is called the Compton effect.

concave lens (diverging lens). A lens that is thinner in the center than at the edges.

condenser. In electricity a device consisting of metal plates separated by a dielectric (insulation). One use for this device is the storage of electric energy (also called capacitor). In chemistry, a device for cooling a gas or a vapor in order to convert it into a liquid.

condensing lens. A lens whose function is to concentrate into a parallel beam the rays of light diverging from the light source in a slide or motion picture projector.

conduction (heat). The flow of heat energy from a warm place to a cool place by the process of successive collisions between adjoining molecules.

conductor. Any material through which electricity or heat flows easily. This property is the opposite to that possessed by insulators.

conservation of energy. The concept that energy is neither created nor destroyed, even when converted from one form to another.

conservation of mass. The concept that mass is never created nor destroyed, even while it is being changed from one form to another.

conservation of mass-energy. The concept that has replaced the Law of Conservation of Mass and the Law of Conservation of Energy in modern physics. It grants that mass may be destroyed, but when it is it is always converted into an equivalent amount of energy, and vice versa.

continuous spectrum. A spectrum containing *all* wave lengths within its range. It is emitted by all solids heated to incandescence.

convection. Motion of a gas or a liquid caused by uneven heating. The moving convection current carries heat from the warmer to the cooler regions of the gas or liquid.

converging lens. A convex lens. Converging lenses converge or bring together to a focus parallel rays of light.

convex lens (converging lens). A lens that is thicker at the center than at the edges.

cornea. The tough, transparent protective coating on the outer surface of the eye.

cosmic rays. The most penetrating rays with the shortest wave length known. They shower the earth from unknown sources in outer space.

coulomb. A unit of electrical charge. It is the charge on 6.28×10^{18} electrons, which is roughly the amount of charge that flows through a 100-watt light bulb in one second.

critical angle. The angle of incidence that has an angle of refraction of 90 degrees.

critical size. The least possible size of a reactor or atomic bomb which will permit the chain reaction to sustain itself.

crystal detector. A small crystal (usually of germanium or silicon) placed in a radio circuit through which the modulated carrier current flows. The crystal detects or demodulates the current, allowing rectified radio frequency current to pass through it. Only rectified radio frequency current can operate phones or a loud speaker.

current (electric). A flow of electric charge. It is usually measured in units called *amperes.*

cycle. One complete vibration of an alternating disturbance.

cyclotron. A machine for accelerating charged particles to high energy by repeatedly accelerating them through an electric field in which they are constrained to remain by the presence of a strong magnetic field.

deceleration. The negative acceleration of an object that is slowing down.

declination. The angle between true north and the direction in which the compass points.

demodulation. The process of rectifying or filtering out the radio frequencies in a radio signal, leaving the audio frequencies undisturbed. Also called *detection.*

density. The weight of a unit volume of a substance. Since 1 cubic centimeter of mercury weighs 13.6 gm, its density is 13.6 gm/cc. Similarly, the density of water is 62.4 pounds per cubic foot.

detection. See **demodulation.**

deuterium. Hydrogen whose atoms contain one neutron as well as one proton in their nuclei. Hence deuterium atoms weigh twice as much as ordinary hydrogen atoms and are sometimes called "heavy hydrogen."

dew point. The temperature to which air containing water vapor must be cooled in order to begin the formation of dew. At the dew point the relative humidity is 100%.

diaphragm. In optics, the adjustable opening behind the lens of the eye or a camera to control the amount of light entering. Also called *iris diaphragm.* In sound, the part of a microphone which is forced to vibrate in and out by an arriving sound wave. The vibrating disc of a telephone receiver.

diathermy machine. An electronic device which produces very short radio waves capable of heating the interior of solid objects. It is also used by doctors to produce local artificial fever.

dielectric. A synonym for *electrical insulator.*

diesel engine. An internal combustion engine in which only the air is compressed. Heated by the compression, the air ignites the fuel which is sprayed into it, thus making spark plugs unnecessary. A conventional automobile, by contrast, compresses a *mixture* of air and fuel and ignites the mixture by means of spark plugs.

differential. The complex set of gears at the rear of a car that enables the motion of the engine to reach the rear wheels and also enables each rear wheel to rotate at a different speed.

diffraction grating. A series of slits, often more than 15,000 per inch, ruled usually on glass or a film of transparent plastic.

Its function is to disperse a beam of radiation into a spectrum.

diffuse reflection. The scattered reflection of light by an irregular surface. This page is a diffuse reflector rather than a regular reflector. So are most objects (except smooth shining objects such as mirrors) in the room around you.

diode. A vacuum tube containing only two electrodes, a filament or heater and a plate. The prefix "di-" means "two." See also **triode.**

direct current (DC). An electric current, either pulsating or steady, that flows in only one direction around a circuit instead of reversing direction repeatedly as an alternating current does.

distributor. The mechanical rotating switch in an automobile engine that connects the high voltage of the induction coil to one spark plug after another in the proper sequence.

domain. A microscopic region in a magnetic substance such as a piece of iron, within which the atoms are lined up with their north poles pointing in the same direction. In magnetized iron the domains are all approximately aligned with one another; in unmagnetized iron their directions are random.

Doppler effect. The change in pitch of a sound wave or the change in frequency (hence color) of a light source as the source moves rapidly toward or away from an observer.

drag. The force of air friction resisting the horizontal motion of an airplane.

dry cell. A form of voltaic cell in which the electrolyte is in the form of a paste sufficiently thick to allow the cell to be carried about conveniently. Flashlight cells are dry cells, but a storage battery whose electrolyte is liquid, by contrast, is a wet cell.

dynamometer AC meter. A form of AC voltmeter or ammeter in which the permanent field magnet is replaced by a coil. The current to be measured passes through the field coil as well as through the movable coil in series with it.

dynamotor. A motor and a generator combined in a single unit having a low-voltage field winding in common.

dyne. The force that will give a mass of one gram an acceleration of 1 centimeter per second per second. The dyne is the fundamental unit of force in the metric absolute system. 980 dynes are equivalent to the weight of one gram.

eardrum. The small flexible membrane that separates the middle ear from the outer ear.

eclipse. The cutting off from view of one object by another. In an eclipse of the sun (solar eclipse), the moon comes between the earth and the sun, cutting off the sun from view. In an eclipse of the moon (lunar eclipse) the earth cuts off the light of the sun shining on the moon so that the moon cannot be viewed.

eddy currents. Unwanted electric currents induced in the core of transformers, motors, generators, and other devices by the changing magnetic fields passing through them.

effective value. The value of an alternating voltage (or current) that measures its ability to produce heat energy. A direct current having the same value would produce the same amount of heat. The effective value of AC current or voltage is 0.707 times the peak value.

efficiency. In a machine, the per cent of the work input that is converted into useful work output. In a lamp, the number of candlepower (or watts of light energy) given out for each watt of electric power consumed.

effort force. The force put into a machine; for example, the force of your finger on a typewriter key.

elastic limit. A point beyond which an object cannot be strained without being permanently deformed. When a spring is stretched past its elastic limit, it does not return to its original shape.

electric current. A flow of charged particles. For example, it may consist of a flow of electrons along a wire or through a vacuum or a flow of ions through a gas, a vacuum, or a liquid.

electrocardiogram. A record of the changing voltages that accompany the beating of the heart.

electrochemical equivalent. The mass of material deposited from an electroplating solution by a unit of electrical charge. Its units are commonly grams per coulomb. It is a constant for any one element but differs from element to element.

electrochemical series. A list of elements arranged in order of their chemical (and hence electrical) activity.

electrode. A positively or negatively charged terminal in an electrolysis tank, vacuum tube, or other electrical device.

electrolysis. The separation of positive and negative ions in a conducting solution by means of the attraction of positively and negatively charged terminals (electrodes) immersed in the solution.

electrolyte. Any compound that ionizes in water to form a conducting solution.

electromagnet. An electrically operated magnet made of a magnetic core (such as iron) around which is wrapped a coil of insulated wire. It becomes magnetized when a current flows through the wire.

electromagnetic waves. Transverse waves consisting of simultaneously alternating magnetic fields and alternating electric fields. They travel at 3.00×10^{10} cm/sec (186,000 mi/sec) through a vacuum. Light, radio waves, X rays, and ultraviolet rays are a few examples.

electromotive force (E.M.F.). For simplicity it may be considered to be the force that drives electrons and hence gives rise to an electric current. Its most common unit is the volt, an electromotive force that can drive a current of one ampere through a resistance of one ohm.

electron. The tiniest possible particle of negative electricity. Its mass is approximately $\frac{1}{1,840}$ that of the proton. Electrons form the outer parts of all atoms. An electric current is the flow of electrons along a wire. Beta rays resulting from radioactivity are a stream of electrons.

electron microscope. An enormously powerful microscope that uses a beam of electrons instead of light rays to view small objects. It is far more powerful than an ordinary light microscope.

electron-volt. The energy attained by an electron as it moves through a potential difference of 1 volt. It is equivalent to 1.60×10^{-12} ergs.

electroscope. A simple device for detecting electric charge. Its action depends on the fact that two similarly charged objects may repel one another strongly enough to be driven apart.

electrostatic generator. Any of several kinds of machines for accumulating a high-voltage electric charge by mechanically transporting a large number of small charges onto a single conductor. A Van de Graaff generator is an example.

electrostatic machine. A type of electrostatic generator for the continuous production of large quantities of static electricity.

element. Material made of one and only one kind of atom. Over 100 different elements are known. The most common element on earth is oxygen.

energy. The ability to do work. It may be either *potential* (stored-up energy) or *kinetic* (energy of moving objects). The unit of energy is the same as that for work—the foot-pound, gram-centimeter, erg, etc.

energy converter. A device that converts energy from one form to another. For example, a steam engine converts heat energy into mechanical energy, and a battery converts chemical energy into electrical energy.

enlarger. A photographer's darkroom device for making prints of various sizes from a single negative.

equilibrant. A balancing force on an object that keeps it in equilibrium.

equilibrium. The condition that exists when all the forces and moments acting on an object are in balance. When an object is in equilibrium, it is either motionless or moving in a straight line at constant speed—in any case not accelerating.

erg. The work done by a force of one dyne when it moves through a distance of one centimeter. One joule of work (a watt-second) is ten million ergs.

exposure meter. A photographer's instrument for measuring the amount of light falling on his subject. It indicates the values of exposure time and f-number (size of diaphragm opening) at which his camera should be set.

external-combustion engine. An engine, such as the steam engine, in which the fuel is burned outside the cylinder of the engine.

eyepiece. The lens in a microscope, telescope, or other optical instrument which forms a virtual enlarged image for inspection by the eye (or camera).

fallout. Radioactive particles formed by an atomic bomb explosion, which settle to earth over a period of time, often far from the scene of the explosion.

farad. The unit of electrical capacity. A condenser or other device has a capacity of 1 farad if 1 volt can succeed in transferring 1 coulomb of charge (6.28×10^{18} electrons) from one plate to the other.

farsightedness. A defect of the eye in which usually the eyeball is too short, so that nearby objects cannot be brought clearly into focus on the retina, though distant objects can be seen clearly.

feedback. See **regeneration.**

field coil. One of the coils of a field magnet.

field magnet. Commonly, the stationary coils of wire in a motor or a generator.

filament. In a lamp bulb or vacuum tube, the wire heated by electricity until it gives out light and, in some cases, electrons. In the X-ray tube, the negatively charged tungsten wire which is heated sufficiently (by an electric current) to give off a steady stream of electrons.

filter (electrical). An electrical circuit the purpose of which is to prevent the flow of some frequencies of alternating current but to permit the passage of others.

filter (optical). A piece of colored glass

placed in front of a beam of light in order to prevent certain colors of light from reaching the film or screen.

fission. The splitting of the nucleus of an atom, such as plutonium or U^{235}, with the release of several neutrons and a great deal of energy.

fluid. Any material that can flow, which means in general gases or liquids.

fluid friction. The force of friction exerted on an object by a fluid (such as air or water) through which it is moving.

fluorescence. The process by which phosphors and other materials give off radiation (usually visible) when they are struck by electrons or by visible or ultraviolet light. A television screen is fluorescent.

f-number. The ratio between the focal length of a lens and its diameter. The higher the f-number, the "slower" the lens and the dimmer the image.

focal length. The distance of the principal focus from the center of a lens or a curved mirror.

foot-candle. The illumination falling on a surface one foot away from a source of one candlepower.

foot-pound. A unit of work. It is the work done by a force of one pound moving through a distance of one foot.

four-stroke engine. An engine in which the piston makes four round-trip strokes to complete one cycle of operation—intake, compression, power, and exhaust.

Fraunhofer lines. Dark lines crossing the otherwise continuous spectrum emitted by the surface of the sun. They result from the absorption of some wave lengths by the outer layers of the sun's atmosphere.

Freon. A common type of refrigerant in household refrigerators.

frequency. The number of waves leaving or arriving at a given point each second.

frequency modulation (FM). Modulation of a radio carrier wave by varying its frequency rather than its amplitude. The resulting signal is far less affected by static than is the more conventional amplitude modulated (AM) signal.

friction. The force that resists the sliding or rolling of one object on another.

front. The boundary surface between two large air masses of different temperatures. A front is often hundreds of miles long.

fuel injector. A high-pressure pump that sprays fuel into the hot compressed air in a diesel engine cylinder.

fulcrum. The motionless point around which a lever turns.

fundamental tone. The tone of lowest pitch (usually also the loudest) produced by any freely vibrating source of sound.

fuse. A piece of low-melting-point wire in series in an electric circuit, designed to melt and open the circuit if the current through it becomes dangerously high.

fusion. The process in which nuclei of light atoms fuse together to form a heavier atom with release of energy. The energy of the sun is released by a fusion process.

g. The symbol for the acceleration of an object falling freely under the influence of gravity. Its value is approximately 32 ft/sec^2 or 980 cm/sec^2, varying slightly with one's location.

galvanometer. A device for detecting and measuring small electric currents by means of a current-carrying coil suspended in a magnetic field.

gamma rays. Waves similar to those of light but of much shorter wave length and much greater penetrating power. Gamma rays are given off by many, but not all, radioactive elements and by some artificial atom-smashing processes, including fission.

gauge pressure. Gas pressure read by a gauge adjusted to read zero pounds per square inch at atmospheric pressure. A deflated tire or an empty tank have a gauge pressure of zero pounds per square inch but an absolute pressure of 14.7 pounds per square inch.

Geiger counter. A radiation-measuring device whose action depends upon a single-ionizing particle of radiation to make the Geiger tube conducting, so that a charge can flow through it and cause a counting device to register.

generator. In electricity, a machine for creating electrical energy from mechanical energy.

germicidal lamp. A lamp that produces useful amounts of germ-killing ultraviolet light.

gram-centimeter. The work done by a force of one gram when it is moved through a distance of one centimeter.

graph. A chart showing the numerical relationship between two variable quantities.

gravimeter. A device for measuring tiny variations in the force of the earth's gravity.

gravitation. The force that every object in the universe exerts on every other object by reason of its mass.

grid. An electrode placed between the anode and cathode of an electron tube, the charge on it being intended to control the flow of current through the tube.

grounding. Using the earth or the metal

case containing electrical equipment as a conductor, partly in order to save wire in large or complex electrical circuits.

half-life. The length of time required for half the radioactive atoms of an element in a sample of it to decay. During the next half-life, half of those remaining will decay, and so on.

half-wave rectification. Conversion of alternating current into pulsating direct current by cutting off the current flowing in one direction but allowing current in the opposite direction to flow freely.

heat. For simplicity, it may be considered to be a measure of the *total* energy of motion of *all* the molecules and atoms in an object. The total amount of heat energy in an object is measured in calories or British thermal units (Btu).

heat of fusion. The heat released by a unit mass of a material as it changes from a liquid to a solid (freezes) without a change in temperature.

heat of vaporization. The heat needed by a unit mass of a liquid in order to convert it into a vapor at its boiling point without a change in temperature.

heavy water. Water whose molecules contain one or two atoms of deuterium (H^2) instead of ordinary hydrogen atoms.

henry. The unit of inductance. A coil has an inductance of 1 henry if an E.M.F. of 1 volt is induced in it when the current through the coil is changing at the rate of 1 ampere per second.

Hooke's Law. A description of the behavior of many (but not all) elastic objects. It states that the distance an object is stretched is proportional to the force exerted on it, provided it is not stretched beyond the elastic limit.

horsepower. A unit of power. It is the rate of doing work equal to 550 foot-pounds per second (or 33,000 foot-pounds per minute).

hydrometer. A device which floats upright in liquids at a depth which depends upon the density (or specific gravity) of the liquid. A scale on the hydrometer allows one to read the specific gravity of the liquid directly.

hygrometer. A device for measuring relative humidity.

ideal effort. The effort force that would have to be put into a machine if the machine were ideal; that is, weightless and frictionless, in order to produce a given output force.

ideal M.A. The mechanical advantage a machine would have in the absence of friction and other losses. It is equal to the effort distance divided by the resistance distance.

illumination. The amount of light energy falling on a unit area per second. A common unit is the foot-candle.

impedance. The effective resistance of an AC circuit.

impulse. The unbalanced force exerted on an object times the time during which the force acts. The impulse applied to an object is equal to the change of momentum that results.

index of refraction. A number that indicates the ability of a transparent material to bend a ray of light entering it from a vacuum (or, for approximate values, the air). It is defined as the speed of light in a vacuum divided by the speed of light in the substance.

induced current. Current produced in a conductor as a result of lines of a magnetic field moving across the conductor (or the conductor moving across a magnetic field).

inductance. A measure of the ability of a coil to produce voltage of self-induction as a changing current flows through it. The unit of inductance is called the henry.

induction (magnetic). The creation of magnetism in a magnetic material like iron by placing it in a magnetic field. It is not necessary that the material touch the magnet for it to be magnetized by induction.

induction coil. A form of transformer capable of creating very high voltages as a result of its high turns ratio and the rapidity with which the primary voltage is changed by a mechanical circuit breaker.

inertia. The tendency of an object, as a result of its mass, to continue in whatever motion it possesses. Thus, as a result of its inertia, an object at rest tends to remain at rest, and an object in motion tends to remain in motion at constant speed in a straight line.

infrared rays. Invisible radiation whose wave lengths are just longer than those of visible red light. Sometimes called heat rays because of their ability to heat up objects they strike.

inner ear. The part of the ear containing the cochlea, the organ ultimately responsible for converting sound waves into nerve impulses to the brain.

instantaneous speed. The speed a speedometer would show at any particular moment.

insulator. A very poor conductor (of heat or electricity).

interference. The cancellation of the effects of two waves as they collide or overtake each other under certain special conditions.

internal combustion engine. An engine whose power comes from the explosion of a fuel-and-air mixture inside the engine's cylinders. The explosion drives a piston down the cylinder, and the piston can then perform mechanical work.

ion. An electrically charged atom or molecule (by reason of having gained or lost one or more electrons) or an elementary electrical particle such as a proton or an electron.

ionization. The formation of ions (electrically charged particles) from atoms or molecules.

ionization chamber. A radiation-measuring device whose action depends upon the measurement of ionization created by the passage of radiation.

ionosphere. The layer of the atmosphere beginning about 50 miles up which contains electrically charged atoms (ions) able to reflect radio waves.

iris diaphragm. The device behind the lens of the eye or a camera, in which there is an adjustable opening whose size controls the amount of light entering.

isotopes. Forms of the same element which differ only in the number of neutrons in their nuclei and hence in their atomic weights. For example, hydrogen has three isotopes with none, one, and two neutrons in their nuclei.

jet propulsion. Forward propulsion of an airplane or rocket by the force equal and opposite to the force driving hot expanding gases (the jet) toward the rear.

joule. The work done in moving 1 coulomb through a potential difference of 1 volt. Hence joules = coulombs × volts. A joule is equal to 10^7 ergs, or about 0.7 ft-lb.

keeper. A piece of soft iron placed across the poles of a horseshoe magnet, whose effect is to help keep the domains of the magnet aligned.

Kelvin scale. Another term for *absolute temperature scale*. Named for Lord Kelvin, a British physicist.

kilocycle. One thousand cycles (or vibrations) per second.

kilogram. The standard unit of mass in the metric system. It is by definition the mass of a block of platinum preserved at the International Bureau of Weights and Measures near Paris.

kilowatt. A unit of power. One thousand watts is the power of 1,000 joules per second.

kilowatt-hour. A unit of work or energy. It is the work done in one hour at the rate of one kilowatt (one thousand watts). Hence a 1,000-watt bulb burning for one hour consumes one kilowatt-hour of energy.

kinetic energy. The energy an object has by reason of its motion. If it is moving with a velocity *v*, its kinetic energy in ft-lb or gm-cm, $E = mv^2/2g$.

laminations. The thin layers of soft iron or silicon steel of which the cores of transformers, motors, and generators are formed. Cores are laminated instead of being built in a solid piece in order to reduce eddy currents and thus increase efficiency.

large calorie (or Calorie). The amount of heat energy needed to raise the temperature of a kilogram of water one degree centigrade.

larynx (Adam's apple). The part of the air passage in the throat across which the vocal cords are stretched.

latent heat (of fusion). The heat needed by a unit mass of a solid to convert it directly into a liquid without a change in temperature.

latent heat (of vaporization). The heat needed by a unit mass of a liquid to convert it directly into a vapor without a change in temperature.

law (in science). A statement of fact that has always been observed to hold true in every tested circumstance and which presumably expresses a general truth.

Lenz's Law. The law that states that whenever a current is induced in a wire, the field produced by the current is always in such a direction as to oppose the motion causing it.

Leyden jar. A simple device for storing an electric charge. It consists of a glass jar having two metal foil plates, one on the inside and one on the outside.

lift. The upward force of an airplane wing caused by a combination of upward push of air on the bottom of the wing and the upward push resulting from low pressure on top of the wing.

light year. The distance light travels in one year—about 6 trillion miles. The light year is a unit of distance used in astronomy.

lines of force (magnetic). Imaginary lines whose direction and concentration are a convenient description of a magnetic field.

liter. The volume of a cubical box 10 centimeters on each edge. A liter is 1,000 cubic centimeters, or 1.06 quarts.

longitudinal wave. A form of wave motion in which the disturbance is *in the same* *direction* as the direction of travel of the waves. Sound waves are longitudinal waves. They are also called *compressional waves.*

lubricant. A material, commonly oil, used to reduce the friction, and hence wear, between moving parts of a machine.

magnetic. Capable of being magnetized or being attracted by a magnet.

magnetic field. A region of space in which a force can be exerted at every point upon a magnetic, but not upon a nonmagnetic, object.

magnetic induction. See **induction (magnetic).**

magnetism. The force by which a magnet attracts or repels magnetic materials such as iron or steel or another magnet.

magneto. A generator whose field magnet is a permanent magnet.

magnification (of a lens or a lens system). The ratio, *M,* between the image size, h_i, and the size of the object being magnified, h_o: $M = h_i/h_o$.

mass. The property of an object by which it resists changes in its speed or direction of motion. Inertia and the pull of gravity on an object (that is, its weight) are both results of an object's mass.

mass number. The number of protons plus the number of neutrons in the nucleus of an atom. It is approximately equal to the atomic weight.

matter. Anything that has mass and occupies space.

mechanical advantage (M.A.). The amount by which any machine multiplies a force. The ideal M.A. of any simple machine is the ratio between effort distance and resistance distance. The actual M.A. of any simple machine is the ratio between the resistance and the actual effort.

mechanical equivalent of heat. The amount of mechanical energy equivalent to a unit of heat energy. In particular, the mechanical equivalent has the value 778 ft-lb = 1 Btu or 42,700 gm-cm = 1 cal.

meson. A charged or neutral particle having about 200 times the mass of an electron. Mesons are released from the nuclei of atoms when they are struck by very high energy radiation.

microfarad. Unit of capacity of a condenser. One millionth of a farad.

microphone. A device for converting sound energy into electrical energy.

microwaves. Very short radio waves, a fraction of an inch to a few feet in wave length.

millibar. A unit of pressure commonly used by meteorologists. It is a pressure of 1,000 dynes/cm^2. Standard atmospheric pressure at sea level is 1,013.2 millibars. This is equivalent to 760 millimeters of mercury.

mirage. The virtual image seen when light coming to the eyes from a distant object is bent by passing through the boundary between layers of warm and cool air.

moderator. A material (such as graphite or heavy water) that slows down neutrons but does not absorb them. Moderators are essential in the operation of nuclear reactors.

modulation. In radio it is the process of using an audio-frequency voice current (or other useful signal) to change the amplitude or frequency of a radio-frequency carrier current for purposes of radio broadcasting.

modulation current. See **voice current.**

molecule. The smallest part of a compound that can exist and still have the chemical properties of the compound. The parts of which a molecule is made are atoms.

moment. A turning effect created by a force, *F,* acting at a perpendicular distance, *s,* from the center of rotation. The magnitude of the moment is *Fs.* Also called *torque.*

momentum. The product of the mass of an object and its velocity. The total momentum of all the objects before they collide equals the total momentum after the collision. Momentum is a vector quantity.

nearsightedness. A defect of the eye in which usually the eyeball is too long, so that distant objects must be brought nearer to the eye in order to be seen clearly.

neutrino. Particles with no charge and almost no mass, released from the nucleus of atoms along with the emission of beta rays.

neutron. A neutral (uncharged) particle found in the nucleus of the atom having a mass equal to that of the proton.

nonmagnetic. Incapable of being attracted by a magnet, or therefore of being magnetized.

normal. The perpendicular to a surface.

nucleus. The central core of an atom, composed largely of protons and neutrons. Although very tiny compared to the size of the entire atom, the nucleus contains most of the atom's weight.

objective (or **objective lens).** The lens through which light enters a telescope or a microscope.

ohm. The unit used to measure electrical resistance. It is the resistance that will allow just one ampere to flow if an electromotive force of one volt is applied across it. It is also the resistance of a "wire" of mercury 106.3 cm long and 1 mm^2 in cross section at 0°C.

Ohm's Law. The law that states that the current, *I,* that flows in a circuit is directly proportional to the applied voltage, *E,* and inversely proportional to the resistance, *R;* hence, $I = E/R$.

opaque. Refers to an object through which light (or some other kind of ray) will not pass.

oscilloscope. A cathode-ray tube connected to an amplifier. The cathode-ray tube has at one end of it a fluorescent screen on which a moving electron beam falls, creating visible light patterns of the electrical signal fed into the amplifier.

overtones. Vibrations of higher frequency than the fundamental produced by any freely vibrating source of sound.

parabola. The path followed by a freely falling body that is traveling near the surface of the earth with a horizontal component of velocity.

parabolic. Refers to the shape of a surface curved in such a way that it reflects to a single point all the parallel rays in a beam of light that strikes it parallel to its axis.

parallel circuit. An electrical circuit in which two or more electrical devices are connected so that the current must divide, some going through one device, some through another.

peak voltage. The maximum value of an alternating voltage.

penumbra. An area of partial shadow. Some, but not all, of a light source is visible from a penumbra.

period (of a pendulum). The time for a pendulum to make one complete swing back and forth. If the pendulum has a length, L, the period $T = 2\pi\sqrt{\frac{L}{g}}$, where *g* is the acceleration of gravity.

permeability. The ability of iron and other magnetic materials to concentrate lines of force.

perpetual motion machine. A machine in which friction and other energy losses are zero, so that actual work output equals work input. Hence the work output could be applied as input and such a machine would be able to drive itself forever with no external source of energy.

persistence of vision. The ability of the retina of the human eye to retain an image for a fraction of a second (about $\frac{1}{16}$ sec) after the object has vanished.

phase. The relationship between two regular alternating disturbances such as sound waves or AC voltages. Phase is commonly expressed in terms of degrees. A 90-degree phase difference means the two series of waves are one quarter ($\frac{90}{360}$) of a cycle out of step (out of phase). One cycle = 360°.

phosphor. A chemical compound that gives off visible light when it is struck by electrons, ultraviolet light, or some other kind of invisible radiation.

photoelectric effect. The release of electrons from the surface of some metals when light shines upon them. It is essentially the creation of electrical energy from radiant energy.

photometer. An instrument for measuring the candlepower of a light source.

photomicrograph. A photograph taken through a microscope.

photosynthesis. The process by which green plants capture the energy of sunlight and use it to create sugar and oxygen out of water and carbon dioxide.

physics. The science that studies energy and the nature of matter as matter.

piston. The plunger that slides smoothly up and down in the hollow cylinder of a pump or an engine.

pitch. The "highness" or "lowness" of a sound. A high pitch is created by sound waves of high frequency; a low pitch is created by waves of lower frequency. Also called *tone.*

pitch (of a screw). The distance between threads of a screw. It is therefore also the distance a screw advances when it is rotated once.

plate. The anode (positively charged terminal) in a vacuum tube.

plutonium. A fissionable man-made element, atomic number 94, which is useful as a source of atomic energy. It is made from U^{238} in a reactor.

polarization (electrical). The formation of free hydrogen gas on the positive terminal (anode) of a cell. The film of gas bubbles on the anode increases the internal resistance of the cell, thus decreasing any flow of current in the circuit.

polarized. Refers to light waves whose vibrations are all in planes parallel to one another.

pole (magnetic). The end of a magnet where the magnetism seems to be concentrated.

positron. A particle of the same weight as the electron but of positive charge. It is released from the nuclei of some atoms made radioactive by bombardment.

potential difference. The work done in moving a unit of electric charge between two points is defined as the potential difference between the two points. It is measured in volts.

potential energy. Stored-up ability to do work.

poundal. A force that will give a mass of one pound an acceleration of one foot per second per second. The fundamental unit of force in the English absolute system. 32 poundals are equivalent to approximately one pound of force.

power. The rate at which work is being done. It is commonly measured in foot-pounds per minute or kilogram-meters per second. Power = work/time.

power factor. The factor by which power in an AC circuit is reduced by reason of the fact that voltage and current are out of phase. In the relationship $P = EI \cos \theta$, $\cos \theta$ is the power factor. The angle θ is the phase angle between E and I.

pressure. The force exerted on a unit of area. It is usually measured in lb/in^2 and gm/cm^2.

primary coil. The input coil of a transformer or induction coil.

principal axis. The line perpendicular to a lens or a curved mirror through its center.

principal focus. The point on the axis of a lens or curved mirror at which light rays striking the lens or mirror parallel to its principal axis are brought to a focus. Its distance from the center of the lens or mirror is called the focal length.

Principle of Work. The work put into any machine equals the work gotten out if work against friction and other losses are included in the work output.

proton. A positively charged particle of electricity found in the nucleus of every atom. Together with neutrons, protons make up most of the weight of an atom.

psychrometer. A form of hygrometer consisting of a wet and a dry bulb thermometer both mounted so they can be spun around through the air by means of a handle.

pupil. The round hole in the iris diaphragm through which light enters the eye on its way to the retina.

quantum theory. A branch of physics concerned with the nature of radiation and the processes by which atoms emit it and absorb it.

radar. An electronic device for detecting distant objects at night or through fog or clouds by sending out very short radio waves and timing (by means of an oscilloscope) the return of the echo. The word comes from *RA*dio *D*etecting *A*nd *R*anging.

radiation. Energy in the form of electromagnetic waves. It travels (in a vacuum) at about 186,000 miles a second. For example, radio waves, infrared rays, and visible light are forms of radiation.

radioactivity. The spontaneous breaking down of the nucleus of an atom during which alpha particles, beta particles, and gamma rays are given off and a new kind of atom is formed as a result.

radiosonde. A device sent into the upper atmosphere by means of a balloon; it detects and radios to the ground such weather information as temperature, pressure, and humidity.

radio telescope. A large radar-like antenna connected to a sensitive radio-frequency receiver, whose function is to detect and measure the radio waves reaching the earth from outer space.

rarefaction. As applied to sound, the half of a sound wave in which the density of the air is less than normal.

reactance. The opposition to the flow of alternating current produced by the inductance and by the capacitance of a circuit. Reactance depends not only on inductance and capacitance but also on the frequency of the alternating current.

reactor. An assembly of fissionable atoms and material to slow down neutrons (the moderator), whose function is to release atomic energy at a controllable rate.

real image. An image formed by the *actual* intersection of light rays. Such an image can be made visible by placing a screen at the intersection.

reciprocating engine. An engine whose action produces a "back-and-forth" (reciprocating) motion (as in the case of the pistons of an automobile engine) rather than a rotary motion (as in the case of a turbine).

rectifier. A device that allows electric current to flow in only one direction, thus converting AC to DC.

reflecting telescope. A telescope whose objective is a parabolic mirror (instead of a lens).

refracting telescope. A telescope whose objective is a lens (rather than a parabolic mirror).

refraction. The bending of a beam of light as it goes from one transparent material into another having a different index of refraction.

regelation. The process of melting under pressure and refreezing when the pressure is released.

regeneration. The process in electronics of

feeding a small amount of the amplified signal in the plate circuit back into the grid circuit so that it can be still further amplified and fed back, and so on, thus creating either a greatly amplified signal or oscillation. Also called *feedback*.

regular reflection. Reflection from a surface so smooth that the directions of the reflected rays are not scattered. A mirror is a regular reflector.

relative humidity. The ratio between the amount of water vapor actually in the air and the amount the air could hold at the same temperature. The ratio is usually written as a per cent.

relay. A switch operated by means of an electromagnet. A small current can cause the switch to open or close, thus controlling the flow of a much larger current.

resistance (electrical). A measure of the opposition that a conductor offers to the flow of electric current. Its most common unit is the ohm.

resistivity. Commonly the resistance in ohms of a block of material 1 cm long and 1 cm^2 in cross section, usually at 20°C. Given the resistivity and the dimensions of any wire, its resistance can easily be found.

resolving power. The ability of a telescope or a microscope to show two very close objects as separate objects, thus revealing fine detail. Poor resolving power is the result of diffraction.

resonance. If the natural frequency of vibration of two objects or the natural frequency of oscillation of two electrical circuits is the same, the two are then said to be in resonance. At the resonant frequency the maximum transfer of energy can occur.

resultant. The vector quantity arrived at by combining two or more vectors (called the components).

retina. The layer of light-sensitive nerve endings lining the back of the eye.

reverberation. Multiple echoes which cause speech and music to sound distorted in a confined space.

rheostat. A variable electrical resistance. The volume control of a radio is a familiar example.

root-mean-square current (or voltage). Another term for *effective value*.

saturated (magnetically). Refers to a piece of magnetic material in which all the domains are aligned and hence no further magnetization is possible.

saturated solution. Refers to a solvent that contains all the solute it can dissolve at a given temperature. It can apply, for example, to water containing dissolved salt or air containing water vapor.

saturated vapor pressure. The saturated vapor pressure of a liquid is the pressure of the vapor when vapor and liquid are in equilibrium.

scalar quantity. A physical quantity that does not have a direction associated with it. Temperature and candlepower are scalar quantities. So is speed. Velocity, however, is a vector quantity.

scanning. The back-and-forth motion of an electron beam sweeping over the entire surface of a television or cathode-ray tube screen in a small fraction of a second.

scintillation counter. A radiation-measuring device in which scintillations produced in a special crystal by the radiation are detected by a sensitive form of photoelectric cell, amplified, and counted.

secondary coil. The output coil of a transformer or induction coil.

selective reflection. A colored object appears colored because it absorbs some colors from the incident light and reflects others. The reflected colors are the color of the object. The process is called selective reflection.

series circuit. An electrical circuit in which two or more electrical devices are connected so that the same current flows through each one in turn. Contrast with a *parallel circuit*.

shock wave. A longitudinal wave of very large amplitude (and hence of large energy), usually created by the motion of an object through the air at a speed faster than sound.

short circuit. An undesirable, even dangerous, low-resistance connection between two points of differing potential in an electrical circuit. If there is little or no resistance around the path so formed, then very large currents will flow.

shunt. A low-resistance bypass connected in parallel with a second resistance such as a galvanometer coil. A shunt resistance converts a galvanometer into an ammeter.

shunt motor. A motor whose armature and field magnet coils are connected in parallel. Its particular advantage is that it tends to run at nearly constant speed for a wide variety of loads.

shutter. In a camera, the cover which is briefly removed from behind the lens to allow light to enter.

significant figure. One of the digits in a measurement, not including any zeroes required merely to locate a decimal point. In $0.0\underline{3017}$ the underlined digits are significant.

sine curve. The form of the graph of $y = \sin x$. A graph of the voltage generated by an AC generator has the form of a sine curve.

sinusoidal voltage (or current). Alternating voltage (or current) whose magnitude varies with time in the same way that y varies with $\sin x$.

solar battery. A large photosensitive surface that generates useful amounts of electric energy when sunlight falls upon it.

solar constant. A measure of the radiant energy reaching the earth's surface from the sun. Its value is 1.94 cal/cm^2/min.

solenoid. A cylindrical coil of wire.

sonar. A system for finding the distance of nearby objects by measuring the time it takes for a high-frequency (ultrasonic) sound wave to be reflected from them. Similar to radar (which uses electromagnetic waves), the greater the distance, the longer the sound wave takes to return.

sound barrier. The increasing resistance to further speed as an airplane approaches the speed of sound (about 750 miles an hour at sea level).

sound wave. One compression and one rarefaction of the air following one another at such a rate that they can be heard by the human ear.

specific gravity. The ratio between the density of a liquid and the density of water. Liquids denser than water therefore have a specific gravity greater than 1.00; those less dense have a specific gravity less than 1.00.

specific heat. The amount of heat needed to raise a unit mass of a material through one degree of temperature. Its units are calories/gm/°C or Btu/lb/°F.

specific resistance. See **resistivity.**

spectroscope. A device for sorting out into a spectrum the various wave lengths composing a beam of light or other radiation. A prism or a diffraction grating is used to produce the spectrum.

spectrum. When a beam of rays is sorted out into the waves that compose it, the array of waves arranged in order by wave lengths is called a spectrum.

spherical aberration. A defect of a lens in which light passing through the center is brought to a focus at a distance different from that of light passing through the edges.

stalling angle. The angle of attack at which the smoothly flowing air breaks away from the upper surface of an airplane wing and becomes turbulent. At this angle the wing loses its lift and the airplane stalls.

starting current. The large current that flows in an electric motor as it is started, before it has picked up enough speed to create a back E.M.F. It can produce enough heat to destroy the motor unless it is reduced by a temporary starting resistance.

static. Noise from a radio, created by unwanted sources of radio energy such as nearby automobile ignition systems, electric switches, or lightning.

static electricity. An isolated charge of electricity which may either remain motionless on an object such as a cloud or a condenser or may flow in a brief time from one object to another. Unlike current electricity, it is not capable of continuous flow.

streamlining. Shaping a high-speed object such as a bullet, boat, or plane so that there is the least possible air or water resistance to its motion.

sublimation. The change in state in which a solid turns directly into a gas without any intermediate liquid state, or, vice versa, the conversion of a gas directly into a solid. An example is the sublimation of moth balls.

supercooled. A term applied to air that has been cooled below the dew point but whose water vapor has failed to begin to condense due to the lack of solid nuclei on which water droplets can form.

surface tension. The greater cohesive force between molecules at the surface of a liquid which causes the surface of the liquid to act like an elastic skin.

switch. A device for easily and safely completing and breaking an electrical circuit.

sympathetic vibration. The forced vibration of an object whose natural frequency corresponds to the frequency of waves striking it from a source of sound. Also called *resonant vibration.*

synchronous motor. An AC motor whose rotation is synchronized with the alternations of the applied voltage, and which therefore maintains a constant speed. Clock motors and phonograph turntable motors are synchronous motors.

temperature. A measure of the *average* energy of motion of the molecules of an object. It is commonly measured in degrees centigrade, Fahrenheit, or Kelvin.

terminal velocity. The final constant velocity reached by a falling object. At terminal velocity the downward force of gravity on an object is precisely balanced by the upward force exerted on it by the gas or liquid through which it is falling.

terminal voltage. The potential difference between the terminals of a cell or battery. The terminal voltage is equal to the E.M.F. minus the Ir drop inside the cell or battery. Hence, when no current, *I*, is flowing, the terminal voltage also equals the E.M.F.

terrestrial telescope. A telescope similar to an astronomical telescope, but containing an extra lens arranged so that an *erect* image of a distant object is formed.

thermionic emission. The emission of electrons by hot solids, such as the filament of a vacuum tube or a lamp bulb.

thermocouple. A junction between two wires of different materials which develops an E.M.F. when heat is applied.

thermocouple meter. A form of AC voltmeter or ammeter whose action depends on the thermoelectric effect. AC flowing through the meter heats a thermocouple junction whose voltage is then indicated by a DC voltmeter.

thermometer. Any device for measuring temperature whose operation depends upon the expansion and contraction of some temperature-sensitive material.

thermonuclear reaction. A nuclear reaction which takes place as a result of enormous thermal energies possessed by the reacting nuclei. Hydrogen fusion is a thermonuclear reaction started (in a hydrogen bomb) by the high temperature of a fission reaction.

thermopile. A series of thermocouple junctions that produces electrical energy from small quantities of heat energy.

thermostat. A temperature-operated switch used to control mechanical or electrical equipment automatically. Examples are the thermostat of an automobile cooling system, and the refrigerator type of thermostat.

thrust. The forward force exerted upon an airplane by the propulsion system.

tone. See **pitch.**

torque. A turning effect; also called a *moment.*

tracers. Radioactive isotopes added to stable isotopes of atoms taking part in biological, chemical, or even physical processes which make it possible to follow (trace) the process by Geiger counter or other radiation-measuring device.

transformer. An electrical device for changing the voltage of an alternating current.

translucent. A term describing any material that transmits light but scatters it at the same time. Frosted glass and thin paper are both translucent, but not transparent.

transmission (in an automobile). The complex set of gears controlled by the gearshift lever of a car by which the relative speeds of engine and rear wheels can be changed.

transmutation. The conversion of an atom of an element into an atom of a different element either by radioactive decay or as a result of bombardment by high-energy radiation.

transparent. Refers to an object that allows light (or some other kind of ray) to pass through it freely and without scattering.

transverse wave. A form of wave motion in which the disturbance is at right angles (transverse) to the direction of travel of the waves. Electromagnetic waves are transverse waves.

triode. A vacuum tube containing three electrodes, usually a filament (or a cathode), a grid, and an anode (or plate).

tritium. Hydrogen whose atoms contain two neutrons in their nuclei in addition to the usual single proton and therefore weigh three times as much as ordinary hydrogen atoms.

tungsten. A metallic element whose melting point is so high that it can be heated enough to give off light without melting. It is used extensively for filaments of lamps and vacuum tubes. It is also called *wolfram.*

tuning. The process of selecting one radio station's signals out of the many that are "on the air" by adjusting a resonant circuit in a radio to the carrier frequency of the desired station.

turbine. A form of engine in which a bladed wheel (or rotor) is forced to rotate by means of jets of water or of hot gas directed against the blades. Because it eliminates the back-and-forth motion of a piston, it is simpler and smoother-running than a reciprocating engine.

turbojet. A common form of jet engine in which air is drawn into the engine and is compressed by means of a fan or turbine rotated by power drawn from the jet exhaust gases.

ultrasonics. The study of sound waves whose frequency is higher than the human ear can hear—from about 20,000 vps to several hundred thousand vps.

ultraviolet waves. Invisible radiation whose wave lengths are just shorter than those of visible violet light. Sometimes called **black light.**

umbra. The completely dark part of a shadow. Compare with **penumbra.**

unbalanced force. A force on an object that is not balanced by an equal and opposite force. Only an unbalanced force can cause an object to accelerate.

uniform velocity. Velocity that is constant in both direction and magnitude.

universal motor. A motor (usually small) designed to run with reasonable efficiency on either alternating or direct current.

unstable. Tending to move with ever-increasing velocity as a result of the action of an unbalanced force. Not in equilibrium.

vacuum. A space containing little or no air or molecules of any other kind.

vacuum tube. An evacuated glass or metal tube containing two or more electrodes, and through which an electric current is passed for any one of a wide variety of purposes.

vacuum-tube voltmeter (VTVM). A type of meter in which AC is amplified by vacuum tubes, rectified, and then applied to a DC meter.

vapor pressure. The pressure exerted on their surroundings by molecules evaporating from the surface of a liquid.

vector. An arrow whose length and direction describe a physical quantity (such as force or velocity or electric field strength) which possesses both magnitude and direction. Temperature is not a quantity that can be described by a vector.

velocity. The time rate at which distance is being traversed in a particular direction. Velocity is a vector quantity, and hence it involves not only speed but also the direction of motion.

vibration. Back-and-forth motion of a body producing sound waves. One vibration = one cycle.

virtual image. An image which the eye sees by following diverging rays back to their apparent source. Since the rays do not actually start where the image is seen, it is called a virtual image.

vocal cords. The tightly stretched membranes in the throat whose vibrations produce the sound waves of the human voice.

voice coil. The solenoid in a loud-speaker through which the fluctuating audio current is passed. Its varying magnetic field pushes on a field magnet and causes it and the diaphragm to which it is attached to vibrate, thus creating sound waves in the air.

voice current. In a radio transmitter or receiver, the varying electric current whose fluctuations correspond to the sound waves being transmitted. Also called *modulation current.*

volatile. Easily evaporated. A term applied to liquids such as gasoline or perfume, which have a high vapor pressure.

volt. The unit used to measure electromotive force. It is the work required to drive one ampere through a resistance of one ohm.

voltage of self-induction. The voltage induced in a coil by its own moving lines of force such as appear when an alternating current flows through the coil.

voltaic cell (or **cell).** A combination of an electrolyte and two electrodes of different materials chosen in such a way that a voltage appears across the electrodes when they are immersed in the electrolyte. Dry cells and storage batteries are voltaic cells.

watt. A power of 1 joule per second; 746 watts = 1 H.P. or 550 ft-lb/sec.

wave length. The distance from any point on a wave to the corresponding point on the next wave.

wave mechanics. A branch of modern physics in which atoms are treated interchangeably as either waves or particles.

weight. The force that gravity exerts on a mass.

Wheatstone bridge. A circuit containing resistances, a galvanometer, and a cell used for making precise measurements of electrical resistance.

wolfram. Another name for the metal *tungsten.* See **tungsten.**

work. That which is accomplished by a force moving through a distance in the direction of the force. The units of work are always the product of force and distance, e.g., foot-pounds, dyne-centimeters (ergs), etc.

work input. The product of effort force and the distance through which it moves. The work input into a machine always equals the work output if work against friction is included with the latter.

work output. The product of the force exerted and the distance through which it moves. Taken together with work done against friction it is always equal to the work input.

X rays. Very penetrating invisible rays given off by the metal target of an X-ray tube when it is struck by a stream of electrons (cathode rays).

INDEX

A page reference in bold type indicates a drawing or picture.